Paleoclimates

Paleoclimates

Understanding Climate Change Past and Present

Thomas M. Cronin

GEORGETOWN UNIVERSITY

COLUMBIA UNIVERSITY PRESS NEW YORK

Columbia University Press
Publishers Since 1893
New York   Chichester, West Sussex
Copyright © 2010 Columbia University Press
All rights reserved

Library of Congress Cataloging-in-Publication Data

Cronin, Thomas M.
    Paleoclimates : understanding climate change past and present / Thomas M. Cronin.
        p.   cm.
    Includes bibliographical references and index.
    ISBN 978-0-231-14494-0 (cloth : alk. paper)—ISBN 978-0-231-51636-5 (ebook)
    1. Paleoclimatology.   I. Title.
  QC884.C739   2009
  551.609′01—dc22   2009019740

Columbia University Press books are printed on permanent and durable acid-free paper.

This book is printed on paper with recycled content.

Printed in the United States of America

c 10 9 8 7 6 5 4 3 2

References to Internet Web sites (URLs) were accurate at the time of writing. Neither the author nor Columbia University Press is responsible for URLs that may have expired or changed since the manuscript was prepared.

15182475

mh

Contents

Tables

Preface

Climate change is a concern for everyone. Such is the case for the public at large, policy makers, and research scientists in climatology as well as in fields as diverse as meteorology, oceanography, astronomy, chemistry, geochemistry, computer modeling, and biology, among many other disciplines. Climate science is truly an integrated field in which research efforts are made to understand the modern and future climate system and the impact of climate on a variety of social systems.

As a "soft-rock" geologist by training, I have, not surprisingly, taken a geological view of climate change in writing this book. Geological evidence about climate—that is, evidence deduced from any natural archive, including glacial ice, tree rings, speleothems, and other "nonrock" systems—rests on the foundations build by generations of field geologists, stratigraphers, geochemists, micropaleontologists, and geophysicists who, just like climatologists, physical oceanographers, and astronomers, operate according to principles that guide their respective fields.

Paleoclimatology is inherently a retrospective and interpretive field built on an extensive empirical database of measurements obtained from physical, geochemical, paleontological, and other sources called proxies that are compiled into paleoclimate reconstructions. Proxies are the "bread-and-butter" empirical sources of data for paleoclimatology, similar to the satellite observations or buoys that provide the observational framework for atmospheric scientists and oceanographers.

In addition to relying on empirical observations, like any other natural science, paleoclimatology is hypothesis driven. An abundance of climate hypotheses, some sufficiently mature to be called theories, are available to paleoclimatologists, who go to great pains to test them. Deep-sea coring expeditions and multiyear ice-sheet coring programs are not carried out randomly. Each project is planned, often years in advance, to test one or several major theories about the earth's climate. Much of the content of this book is derived from these large-scale programs as well as the efforts of individual scientists.

One theme in this book is the pluralism of causal mechanisms in climate change. Take the example of efforts today to distill modern atmospheric and ocean warming into those portions of temperature change that have been caused by greenhouse gas, volcanic, solar, and aerosol forcing—that is, to detect and attribute causality to modern temperature patterns. Such a view, admirable and fruitful as it has been, seems simplified given the complexity of the system, the unknowns about natural variability and feedback processes, and the scale of the greenhouse gas perturbation now with us. This book takes a much wider view of climate change than that reflected in recent temperature trends.

Geological archives of past climate incorporate not only the effects of forcing from solar variability and greenhouse gases, but also multiple feedback mechanisms that often produce spectacularly unusual climate states, cyclic environmental variations, and massive impacts on ecosystems. As valuable as it is to understand modern processes, like many of my contemporaries, I threw uniformitarianism aside long ago and adhere to the view that the past tells us as much as or more about the present and future as the other way around. In this sense, modern climate trends in temperature, rainfall ice sheets, and sea ice do not seem so unusual to many of us, for we have the perspective of time unavailable to others.

This book is a hybrid, neither a standard textbook nor a comprehensive review of the literature, but rather an effort to summarize the state of the art of paleoclimatology specifically as it pertains to contemporary issues in climate science in general. Introductory material on the climate system and paleoclimate methods and principles given in the first two chapters is of necessity brief, meant to introduce the reader to terms and concepts discussed in later chapters, but not to be taken in lieu of formal, quantitative study of these fields. As such, sections may be too elementary for experts in specific fields, and some readers may wish to proceed directly to chapters on climate changes over a particular timescale, or changes caused by specific climate-forcing mechanisms. Although many common themes run throughout paleoclimatology, each chapter can be read independent of the others and in any order.

Several colleagues have told me I might be a little crazy to undertake such an ambitious effort, so when I started a few years ago I was well aware that a book covering 4 billion years of earth-climate history was bound to contain imbalances, shortcomings, and gaps. It is worth emphasizing, then, that this book is merely an entrée into an immense, exponentially growing field that is certain to experience major advances in the next few decades. With that said, I hope at least to instill an appreciation for the contributions of paleoclimatologists to climate science and for one of the major environmental issues of our time.

Acknowledgments

This book grew out of my "oceans" course at Georgetown University. Sincere thanks go to Tim Beach and Chuck Weiss of Georgetown University's Edmund A. Walsh School of Foreign Service (SFS) for their liberal views on what future diplomats and professionals in the Science, Technology and International Affairs program ought to be learning and for giving me a chance to teach them. Georgetown's SFS has come a long way since my father attended the program after returning from the U.S. Marines after World War II, and hopefully climate-ocean topics will continue to attract the best nonscience students, as well as those in the natural sciences.

My sincere thanks go to countless colleagues for discussion, data, and especially the use of graphics: Jinho Ahn, Pat Bartlein, Michael J. Behrenfeld, Karen Bice, David Black, Ron Blakey, Henk Brinkhuis, Phil Brohan, Anders Carlson, Mark Chandler, Dennis Chesters, Inge Clark, Peter Clark, Ed Cook, Rosanne D'Arrigo, Rob DeConto, Peter deMenocal, Harry Dowsett, Gary Dwyer, Art Dyke, Ed Erwin, Jan Esper, Frédérique Eynaud, Jane Ferrigno, J. Foley, Dave Franzi, Alexandre Ganachaud, Felix Gradstein, H. Grissino-Mayer, Galen Halverson, Sidney Hemming, Tim Herbert, Fritz Hilgen, Paul Hoffman, Jim Hurrell, Martin Jakobsson, Phil Jones, C. Kucharik, Curt Larsen, Carrie Lear, Mark Leckie, Dave Leverington, Zheng-Xiang Li, Valerie Masson-Delmotte, Mike McPhaden, Jerry McManus, Ken Miller, Alan Mix, Isabel Montañez, Raimund Muscheler, Robert T. O'Malley, Mark Pagani, Didier Paillard, Heiko Pälike, Uwe Pflaumann, Michael Pidwirny, Paolo Pirazolli, Dave Pollard, John Rayburn, Stefan Rahmsdorf, Jackie J. Richter-Menge, Eelco Rohling, Chris Scotese, Dan Sigman, Linda Sohl, Appy Sluijs, Lowell Stott, John-Inge Svendsen, Lynne Talley, Jim Teller, David W. J. Thompson, Harvey Thorleifson, Kerstin Treydte, Jan Veizer, Pinxian Wang, Debra Willard, Richard S. Williams Jr., Rob Wilson, Jim Wright, Carl Wunsch, Moriaki Yasuhara, and Jim Zachos. You all made the good parts of this book possible; I'll take the blame for the missteps and offenses.

Another group worthy of thanks from me and the entire community includes all those paleoclimatologists who freely contribute their data to the National Oceanic and Atmospheric Administration's (NOAA) World DataCenter (WDC) for Paleoclimatology, and to Dave Anderson and his NOAA staff for helping WDC grow and flourish. Credit also goes to PANGEA (Publishing Network for Geoscientific & Environmental Data), and other online databases for free and open access to published data. Many institutions granted graphics permission, including the Intergovernmental Panel on Climate Change (IPCC), England's Met Office Hadley Centre for Climate Change, the American Association for the Advancement of Science, the Geological Society of America, Elsevier Publishing, NOAA, the U.S. Geological Survey, the National Aeronautics and Space Administration, the Geological Society of America, and others cited in the text and captions. Some maps were made courtesy of Martin Weinelt's online map creation program (http://www.aquarius.geomar.de/) and the Generic Mapping Tool (GMC), developed by Paul Wessel (University of Hawaii at Manoa) and Walter H. F. Smith (NOAA) (http://gmt.soest.hawaii.edu/). Paleogeographical maps were kindly provided by Ron Blakely of Northern Arizona University. The American Geophysical Union (AGU) deserves much credit for their open-use policy regarding AGU publications and graphics.

I am grateful to Pat Fitzgerald of Columbia University Press for editorial guidance and encouragement throughout this endeavor. Colgate University's wisdom has conferred upon us both a penchant for thinking big. Columbia's Marina Petrova provided invaluable support with the manuscript text. I also cannot emphasize enough the contributions of three anonymous reviewers whose insight and wisdom added greatly to the final version of this book.

Finally, to Margarita, Jason, Nicolas, and Anthony, I owe much gratitude for pushing me to write this book and giving me the chance to finish it. May future climate hold no surprises for them and their children.

Abbreviations

Abbreviation	Term	Major Usage in Chapters
AABW	Antarctic Bottom Water	Ch. 4–7
ACC	Antarctic Circumpolar Current	Ch. 4
ACEX	Arctic Coring Expedition	Ch. 4
ACR	Antarctic Cold Reversal, just before YD	Ch. 7
AIM	Antarctic isotopic maxima	Ch. 6
AMO	Atlantic Multidecadal Oscillation	Ch. 10
AO	Arctic Oscillation/North Annular	Ch. 10
AOGCM	Atmosphere-ocean general-circulation model	Ch. 9–12
B/A	Bølling-Allerød	Ch. 7
BIT	Branched and isoprenoid TetraEther index	Ch. 7
B/M boundary	Brunhes-Matuyama paleomagnetic boundary (670 ka)	Ch. 5
BWT	Bottom-water temperature	Ch. 7
CAI	Central American Isthmus	Ch. 4
CCD	Carbonate compensation depth	Ch. 4, 5
CFR	Climate field reconstruction, mainly tree rings	Ch. 11
CIE	Carbon isotope excursion (e.g., PETM)	Ch. 4
CLIMAP	Climate: Long-Range Investigation, Mapping, and Prediction	Ch. 5
COHMAP	Cooperative Holocene Mapping Project	Ch. 8
CRE	Catastrophic Reef Event (coral reef drowning)	Ch. 7
DMS	Dimethylsulphide	Ch. 12
DO events	Dansgaard-Oeschger events	Ch. 6
DSDP	Deep Sea Drilling Program	Ch. 3, 4
DVI	Dust veil index	Ch. 11
EAIS	East Antarctic Ice Sheet	Ch. 4–7
EAM	East Asian Monsoon	Ch. 6
EDC	EPICA Dome Concordia Antarctic ice core site	Ch. 5–8
EDML	EPICA Dronning Maud Land Antarctic ice core site	Ch. 5–8
EEP	Eastern equatorial Pacific	Ch. 4–10
EIS	Eurasian Ice Sheet	Ch. 5–7
ENSO	El Niño–Southern Oscillation (internal climate variability)	Ch. 10
E-O	Eocene-Oligocene boundary, climate shift 34 Ma	Ch. 4
EPICA	European Ice Core Project	Ch. 5
EPILOG	Environmental Processes of the Last Ice Age, Land, Oceans, and Glaciers	Ch. 7
EPR	Exxon Production and Research	Ch. 3, 4
GCM	General-circulation model (climate model)	Ch. 9
GIA	Glacio-isostatic adjustment	Ch. 7
GIS	Greenland interstadial	Ch. 6
GISP	Greenland Ice Sheet Project (also GISP1 and GISP2)	Ch. 6–8
GLAMAP	Glacial Atlantic Ocean Mapping	Ch. 7
GOM	Gulf of Mexico	Ch. 9
GRIP	Greenland Ice Core Project	Ch. 6–8
GS	Greenland stadial	Ch. 6
Gt	Gigatons	Ch. 6–12
H-event	Heinrich event	Ch. 6
HNLC	High-nutrient-low-chlorophyll (hypothesis for glacial CO_2)	Ch. 5
ICDP	International Continental Scientific Drilling Program	Ch. 4
IDAG	International Ad Hoc Detection and Attribution Group	Ch. 11

Abbreviation	Term	Major Usage in Chapters
IDP	Interplanetary dust particles	Ch. 5
IODP	Integrated Ocean Drilling Program	Ch. 4
IPCC	Intergovernmental Panel on Climate Change	Ch. 11–12
IPO	Interdecadal Pacific Oscillation	Ch. 10
IRD	Ice-rafted debris	Ch. 6–9
ITCZ	Intertropical Convergence Zone	Ch. 6–8
LGM	Last Glacial Maximum (22–20 ka)	Ch. 6, 7
LIA	Little Ice Age	Ch. 8–12
LIPs	Large igneous provinces (influence Mesozoic OAE)	Ch. 3
LIS	Laurentide Ice Sheet	Ch. 5–7
LPIA	Late Paleozoic Ice Age	Ch. 3
MBE	Mid-Brunhes Event (~450 ka)	Ch. 5
MIS	Marine Isotope Stage	Ch. 5, 6, 7
MLD	Maximum latewood density (tree ring)	Ch. 11
MMCO	Mid-Miocene Climatic Optimum (18–14 Ma)	Ch. 4
MOC	Meridional overturning circulation	Ch. 4–9
MPT	Mid-Pleistocene Transition (~1.0 Ma–700 ka)	Ch. 5
MPTO	Mid-Pliocene Thermal Optimum (3–4 Ma)	Ch. 4
MSA	Methanesulphonic acid	Ch. 12
MWP	Medieval Warm Period	Ch. 8–12
MWP	Meltwater Pulse (rapid deglacial sea-level rise)	Ch. 5–7
NADW	North Atlantic Deep Water	Ch. 4–7
NAIS	North American Ice Sheet	Ch. 7
NAO	North Atlantic Oscillation	Ch. 10
NGRIP	North Greenland Ice Core Project	Ch. 6–8
OAE	Oceanic Anoxic Event	Ch. 4
ODP	Ocean Drilling Program	Ch. 3, 4
OMZ	Oxygen minimum zone (ocean margins)	Ch. 3
PAGES	Past Global Changes	Ch. 8
PBO	Preboreal Oscillation	Ch. 7
PDI	Power dissipation index	Ch. 12
PDO	Pacific Decadal Oscillation	Ch. 10
PEP	Pole-Equator-Pole	Ch. 8
PETM	Paleocene-Eocene Thermal Maximum	Ch. 4
PMIP	Paleoclimate Model Intercomparison Project	Ch. 8
PNA	Pacific North American	Ch. 10
PRISM	Pliocene Research, Interpretation and Synoptic Mapping	Ch. 4
PW	Petawatts	Ch. 6
QUEEN	Quaternary Environments of the Eurasian North	Ch. 7
RCO_2	Atmospheric CO_2 concentration, multiple of preindustrial level	Ch. 3
RCS	Regional Curve Standarization (tree rings)	Ch. 11
RG	Reef Generation (during sea-level rise)	Ch. 7
RSL	Relative sea level	Ch. 7
SAM	South American Monsoon	Ch. 6
SAP	Surface air pressure	Ch. 10
SAT	Surface air temperature	Ch. 11
SI	Stomatal index (for fossil leaves, CO_2 estimation)	Ch. 3, 4, 7, 8
SLR	Sea-level rise	Ch. 12
SOI	Southern Oscillation Index	Ch. 10
SPECMAP	Spectral Climate Mapping Project	Ch. 5
SSS	Sea-surface salinity	All
SST	Sea-surface temperature	All
TEX_{86}	TetraEther Index	Ch. 4–7
THC	Thermohaline circulation	Ch. 4–9
VEI	Volcanic explosivity index	Ch. 11
VSMOW	Vienna Standard Mean Ocean Water	Ch. 2
WAIS	West Antarctic Ice Sheet	Ch. 4–7
WEP	Western equatorial Pacific	Ch. 4–10
YD	Younger Dryas	Ch. 7

Paleoclimates

1

Paleoclimatology and Modern Challenges

Introduction

The earth's climate is changing in ways that raise fundamental questions about how the climate system functions, why these trends are occurring, and what future changes will occur. Atmospheric and ocean warming, Arctic Ocean sea-ice decline, Antarctic ice-shelf collapse, the retreat of alpine glaciers, the surging of ice-sheet margins in Greenland and Antarctica, extreme rainfall patterns, and ocean acidification are a few highly publicized trends. Are these trends typical of our planet? Can we say they are within the natural variability of our relatively mild interglacial climate state? Have similar events occurred in the past and, if so, why? Will they continue or even accelerate in the future?

This book describes patterns and causes of past climate change—i.e., the topic of paleoclimatology. Our main objective is to provide a context for understanding today's changing climate and future changes predicted by climate models under elevated greenhouse gas concentrations. The word *paleoclimatology* comes from the prefix "paleo," meaning old, and "klimat," from the Greek word for inclination or latitude. The foundations of paleoclimatology—the principles and methods on which it rests—include *archives, chronology, proxies,* and *paleoclimate modeling.*

Archives are natural features of the earth—tree rings, tropical and deep-sea corals, speleothems (cave stalagtites, precipitated calcite), ice from Greenland and Antarctica and smaller ice caps, deep-sea manganese nodules, lake sediments, wetlands, ocean sediments, geological and geomorphological features, geophysical records, and a variety of groups of organisms—that preserve clues about past climate and environmental change. Chronology involves the application of geological time, using long-term records, to date, correlate, and understand climate changes over various timescales.

Proxies are geochemical, physical, geological, and biological measurements obtained from archives. With few exceptions, like the Roda Nilometer, a record of Nile River flow near Cairo from A.D. 645 to 1920 (Hassan 1981), most observational records extend back a few decades to a century or so ago, a miniscule fraction of earth's 4.6-billion-year history. Observations are also spatially limited, and only since the advent of satellites and remote sensing methods has the earth been monitored at a global scale. Proxies contain information about atmospheric and oceanic temperature and circulation, sea-level positions, ice-sheet and sea-ice distribution, precipitation, atmospheric chemistry, biological productivity, and other climate parameters over all timescales. Climate proxies extracted from these archives are surrogates for measurements from ground-based instruments, the satellites and tools that observe the modern system. Paleoclimate modeling involves using the synergy between proxy reconstructions and computer or mathematical models to gain insights into causes and mechanisms and to improve climate model forecasting.

Paleoclimatology has emerged over the past few decades as a major integrated field of scientific investigation devoted to the understanding of the fundamental patterns and processes of the earth's climate over all temporal and spatial scales. This chapter introduces basic processes of the earth's climate system, the causes of climate change, feedbacks that amplify or dampen the climate signal, and major challenges surrounding current and future climate changes needed to understand climate changes of the past. Chapter 2 in this volume describes the methods of paleoclimatology, and the rest of the chapters sequentially examine patterns and causes of climate change, moving from those occurring over hundreds of millions of years to recent changes of the past millennium.

The Earth's Climate System

Climate is often defined as average weather conditions, the mean temperature and precipitation, atmosphere pressure,

wind, etc., measured over several decades for a particular region. It represents the large-scale and long-term framework within which weather changes occur. Whereas meteorologists can predict weather from an initial condition for a few days, chaotic behavior in the atmosphere renders long-term forecasts impossible. In contrast, changes in mean climate state and variability around this mean, due to factors discussed below, are in theory predictable, at least in a probabilistic sense. A given change in solar energy, for example, will produce a range of predictable outcomes in terms of mean global surface air temperature (SAT). This fundamental premise surrounds modern climate modeling research and forecasting.

Climate changes on timescales considered in paleoclimatology, however, are not limited to changes in mean atmospheric conditions. The earth's integrated climate system also includes the oceans, ice, lithosphere, and biosphere, which constantly interact with each other over all timescales. The major components of the climate summarized schematically in Figure 1.1 should be referred to throughout the following discussion.

Solar Radiation

We begin with a discussion of the radiation budget of the atmosphere. Except for small amounts of energy generated from the earth's interior, solar energy drives climatic systems in the form of radiation. Solar energy reaches the atmosphere and the earth's surface as photons in the form of shortwave radiation (~300–1000 nm; peak wavelength is 500 nm in the visible part of the spectrum). Energy from radiation is expressed as

$$E = hc/\lambda = h\mathrm{v}$$

where h is Planck's constant (6.626×10^{-34} joules sec^{-1}), c is the velocity of light, λ is the wavelength of radiation, and v is the frequency of the radiation.

Figure 1.2 shows the average energy balance between incoming short-wavelength solar radiation (insolation) and outgoing long-wavelength radiation measured in Watts (W) per square meter. Daytime solar radiation, called the solar "constant," averages 1368 W m^{-2} sec^{-1} measured in a cross-sectional area at the top of the earth's atmosphere. Importantly, this value actually varies daily, over 11-year sunspot cycles, and over longer timescales, with still poorly known implications for climate (see Chapter 11 in this volume). When the incident angle of radiation and the earth's rotation are taken into account, the average radiation is calculated as one quarter the solar constant, or 342 W m^{-2}. Of this total, about 107 W m^{-2} is reflected back to space—77 W m^{-2} by clouds, atmospheric gases, and aerosols (particulate material) and 30 W m^{-2} by the earth's surface. Aerosols pro-

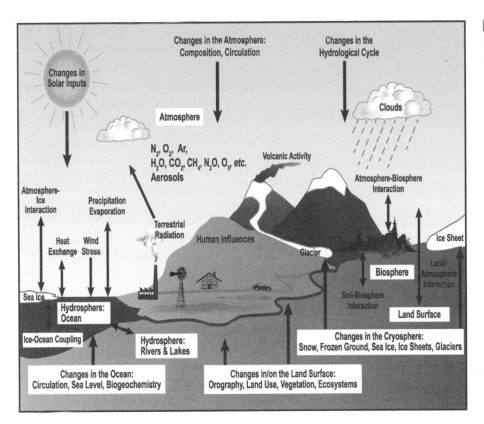

FIGURE 1.1 Schematic view of the components of the climate system, their processes, and their interactions. Courtesy of the Intergovernmental Panel on Climate Change (IPCC). Figure and caption from FAQ 1.2, figure 1, IPCC chapter 1, "Historical Overview of Climate Change Science" (Le Treut et al. 2007), Cambridge University Press. Courtesy of the Intergovernmental Panel on Climate Change.

duced by human pollution have a short-term impact on reflected radiation, but explosive volcanic events eject aerosols above cloud layers and affect earth's radiation and temperature for one to three years, depending on the event. Therefore, the contribution of volcanic activity as a climatic forcing mechanism over the last millennium must be considered to assess the contribution from anthropogenic greenhouse gas emissions.

The reflectivity by the earth's surface is called its albedo. Albedo can be expressed as a unitless number between 0 (darker surface) and 1 (lighter) or as a percentage of insolation. Globally averaged albedo today is about 0.3, which means 30% of the total insolation is reflected back to space. Lighter regions of the earth's surface (ice sheets, sea ice, snow cover, deserts) reflect relatively more radiation than dark regions (ocean surface, vegetated continents). As a consequence, changes in the earth's surface conditions during large-scale climatic cycles, such as ice ages, and anthropogenic land-use changes influence regional and perhaps larger-scale climate change through their effect on albedo (Pielke et al. 2002).

On average, the remaining unreflected 235–240 $W\,m^{-2}$ radiation is absorbed or scattered by various parts of the climate system, including gases in the atmosphere (mainly molecular nitrogen, N_2, oxygen, O_2, and ozone, O_3). Each gas has its own unique radiative properties; oxygen and ozone, for example, absorb relatively more radiation at ultraviolet

(shorter) wavelengths. To achieve radiative balance, the earth radiates energy back to space in the form of infrared (long) wavelength radiation because cooler bodies radiate longer wavelengths. If all absorbed energy were reflected back to space, the average surface air temperature would be about −16°C. However, some re-emitted long-wavelength radiation is trapped by atmospheric trace gases and clouds, resulting in a mean annual surface temperature of about +14°C. Water vapor (H_2O) and carbon dioxide (CO_2) are especially strong absorbers of infrared wavelengths, and this is why they are important greenhouse gases.

The Hydrosphere

Many aspects of past climate changes involve the earth's hydrosphere. The hydrosphere includes water contained in the world's oceans (97.3%); ice sheets and glaciers (2.1%); groundwater in aquifers (0.6%); and surface water, soil moisture, water vapor in the atmosphere, and water locked up in the biosphere (<0.1%). Water in the earth's crust is also circulated through volcanoes and hydrothermal vents via plate tectonic processes. The water cycle forms the nucleus of the global climate system, and the flux of water is controlled by processes central to paleoclimatology: evaporation (largely in tropical oceans), condensation of water vapor in the atmosphere into clouds, precipitation, snow, sea-ice and glacial ice formation, and cycling in the lithosphere.

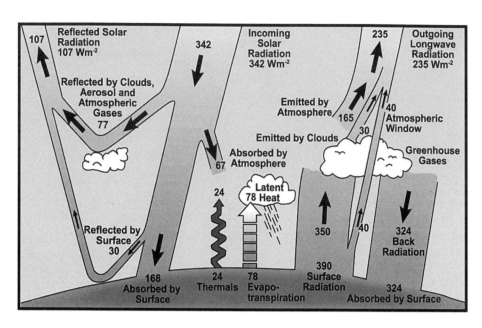

FIGURE 1.2 Estimate of the earth's annual and global mean energy balance. Over the long term, the amount of incoming solar radiation absorbed by the earth and atmosphere is balanced by the earth and atmosphere releasing the same amount of outgoing long-wavelength radiation. About half of the incoming solar radiation is absorbed by the earth's surface. This energy is transferred to the atmosphere by air warming in contact with the surface (thermals), by evapotranspiration, and by long-wavelength radiation that is absorbed by clouds and greenhouse gases. The atmosphere in turn radiates long-wavelength energy back to the earth as well as out to space. Source: Kiehl and Trenberth (1997). Figure and caption from FAQ 1.1, figure 1, IPCC chapter 1, "Historical Overview of Climate Change Science" (Le Treut et al. 2007), Cambridge University Press. Courtesy of the Intergovernmental Panel on Climate Change.

Water is the medium through which major and minor chemical species are circulated and transported from one part of the climate system to the next in complex biogeochemical cycles. Over millions of years, for example, chemical weathering on continents plays a critical role in global fluxes of key elements like carbon, silicon, and strontium. Over shorter timescales, regional or global hydrological balance is reflected in lake-level oscillations that record large-scale changes in precipitation. It is essential to keep the hydrosphere in mind when viewing other parts of the climate system.

The Atmosphere

General Concepts The lower two divisions of the atmosphere, the troposphere and stratosphere, are most important for discussing climate change. The troposphere extends from the surface to a height of 10–15 km and experiences daily, seasonal, and long-term variability in temperature, cloud cover, circulation, and other features that dominate mean regional weather conditions. Stratosphere-troposphere interactions are especially important in understanding the impact of short-term changes in solar energy input.

Atmospheric circulation, like that of the oceans, involves fluid motion on a rotating planet heated by solar energy and driven by a combination of factors: pressure gradients caused by differential heating from solar radiation in low and high latitudes, and Coriolis force, caused by the earth's rotation. Energy absorbed by oceans and continents near the equator are three times greater than that in polar regions, causing a strong equator-to-pole thermal gradient and a constant transfer of energy by atmosphere and oceans from low to high latitudes. Pressure gradients created by differential heating lead to the major features of atmospheric circulation described below.

In addition to geographical gradients in solar energy distribution, the earth's rotation affects atmospheric circulation. Coriolis force is the deflection of objects (fluids) that are not bound by friction, resulting from the earth's eastward rotation and decreasing velocity going from equator to poles. It is a function of mass, speed, angular velocity, and latitude, calculated as

$$\text{Coriolis} = m \times 2\Omega \sin \varphi \times u$$

where m is mass, u is speed, Ω is the angular velocity of the earth, and φ is latitude. The term 2Ω is also called f, or the Coriolis parameter. Coriolis "force" is not a force sensu stricto, but the result of the eastward rotation of the earth. In simplistic terms, Coriolis force deflects atmospheric systems and ocean currents as they move poleward. In the southern hemisphere, Coriolis force deflects fluids toward the left, or westward, and in the northern hemisphere toward the right, or eastward.

Major Atmospheric Features The major planetary features of the atmosphere are shown in Figure 1.3. Atmospheric circulation in a north-south direction is called meridional flow, whereas east-west flow is referred to as zonal flow. Shifts from predominantly meridional to zonal flow

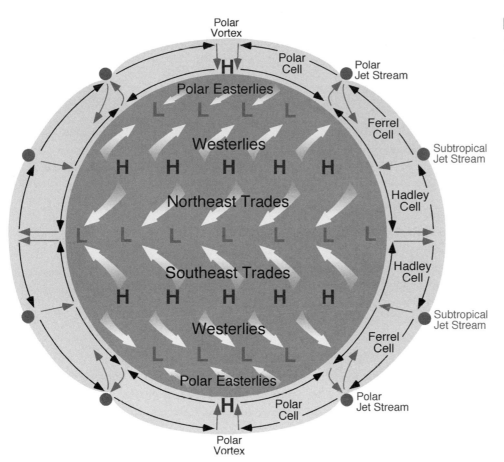

FIGURE 1.3 Planetary atmospheric circulation features. See text. Courtesy of Michael Pidwirny (http://www.physicalgeography .net/fundamentals/8q_1.html).

have occurred during rapid climatic reorganizations of the late Quaternary deglaciations.

The northeast and southeast trade winds, the westerlies, and the polar easterlies are dominant circulation patterns that separate high- and low-pressure regions. The westerlies are the prevailing midlatitude winds characterized by strong cyclonic (low-pressure) cells. Large-scale undulations in the westerlies cause shifts in atmospheric pressure, storm tracts, and temperature, and form a dominant decadal and multidecadal mode of variability called the North Atlantic Oscillation.

Meridional circulation is divided into Hadley, Ferrel, and Polar cells that form alternating high- and low-pressure systems that influence geographical variability in mean precipitation. For example, strong energy flux near the equator causes air to rise, forming the rising limb of Hadley cells; this results in heavy precipitation in tropical regions. The low-latitude belt of atmospheric convective clouds is called the Intertropical Convergence Zone (ITCZ), a narrow zone of maximum rainfall in oceanic equatorial regions that forms at the convergence of northern and southern trade winds and results from warm, moist air rising and releasing moisture into the atmosphere. The position of the ITCZ varies seasonally; today its mean position is in

the northern hemisphere, but it shifts to south of the equator seasonally. The ITCZ experienced major oscillations and net southward migration during the Holocene epoch (see Chapter 8 in this volume). The sinking limb of the Hadley cell near 30°N and 30°S latitude is characterized by weak winds and high pressure in subtropical regions. The world's modern deserts are located in regions of sinking dry air and low precipitation. The subtropical and polar jet streams are located near the Hadley-Ferrel cell and Ferrel-Polar cell boundaries because of temperature and pressure gradients at low elevations.

Zonal circulation in the equatorial Pacific is called Walker circulation, or the Walker cell. In simplistic terms, Walker circulation forms a large low-latitude loop across the Pacific Ocean, with a westward-flowing lower limb and eastward-flowing upper limb. However, Walker circulation is complicated by interaction with the Hadley cell and deflection by Coriolis force. Walker circulation is also dynamically linked to the upper layers of the equatorial Pacific Ocean through the behavior of wind-stress-driven planetary-scale Kelvin and Rossby waves. Low-amplitude eastward-propagating Kelvin waves tend to cause downwelling and westward-propagating Rossby waves tend to cause upwelling in upper ocean layers. We will return to discussions of Walker circulation when we discuss Pliocene global warmth (Chapter 4), the Holocene development of the Asian monsoon (Chapter 8), and dynamics of the El Niño–Southern Oscillation mode of variability (Chapter 10).

Atmospheric Chemistry The earth's atmospheric composition is molecular nitrogen (N_2, 78.1%), oxygen (O_2, 20.9%), argon (Ar, 0.93%), carbon dioxide (CO_2, 0.038%), and trace amounts of various inert gases and reactive gases such as methane (CH_4) and nitrous oxide (N_2O). These can also be expressed volumetrically; in the case of CO_2 the concentration in the year 2008 was about 385 parts per million volume (ppmv). Analysis of temporal changes in the concentration of three trace greenhouse gases, CO_2, CH_4, and N_2O, is a critical sector of ice-core paleoclimatology because their radiative properties influence planetary energy balance and climate. For example, ice-core paleoclimatology shows that the preindustrial and glacial-age concentrations of CO_2 were 280 ppmv and 200 ppmv, respectively. Many other chemical species also are studied in ice cores. Some are chemically active species and include nitrates, sulfates, chlorine, and several aerosols grouped under the term *glaciochemical species.* Trace gases and glaciochemical records will be emphasized in chapters 5–8 in this volume.

The Oceans

General Concepts The oceans transport mass, heat generated by solar radiation, salt, nutrients, and other dissolved chemical constituents. Ocean mass transport is measured in Sverdrups ($1 Sv = 10^6 m^3 sec^{-1}$), named after pioneer oceanographer Harald Sverdrup. Sverdrups measure the volume of water flowing past a theoretical point per second. For example, the World Ocean Circulation Experiment (WOCE), a global analysis of ocean mass and heat transport, estimated net deep water production rates of 15 ± 12 Sv in the northern hemisphere and 21 ± 6 Sv in the southern hemisphere (Ganachaud and Wunsch 2000).

Like the atmosphere, oceanic heat transport (measured in Petawatts, PW, $1 PW = 10^{15} W$) redistributes heat from low to high latitudes in near-surface layers. The WOCE estimates of global, zonally integrated, meridional heat transport by the oceans are almost 2 PW for the northern hemisphere and 0.7–0.8 PW for the southern hemisphere. The oceans also have the capability of interhemispheric and global heat transport via deep-ocean circulation, and paleoceanographic records show significant changes in the strength of deep-ocean circulation and heat transport over millennial timescales.

Surface Circulation Surface ocean circulation is driven by wind stress, heating and cooling, evaporation, and precipitation; is ultimately controlled by the earth's rotation and solar radiation; and is influenced by the position of land masses. Wind stress also influences ocean-surface circulation and is expressed as

$$\tau = c W^2$$

where τ (tau) is wind stress, W is wind speed, and c is a constant that depends on atmospheric conditions. In general, c increases with increasing wind speed; greater wind speed increases turbulence and surface roughness, and a typical value for c is 2×10^{-3}.

Coriolis force deflects major ocean currents, such as the northward flowing Gulf Stream in the North Atlantic and the Kuroshio Current in the North Pacific, toward the east. In the southern hemisphere, Coriolis force deflects currents toward the left, or westward, and in the northern hemisphere toward the right, or eastward.

Surface currents in the upper ocean are affected by friction (transfer of momentum) and turbulence. The term *eddy viscosity* refers to the internal friction for any parcel of fluid. The Ekman theory for wind-driven currents holds that currents are a combination of Coriolis force and friction from wind stress acting on surface layers of ocean. Ekman layer flow is thus perpendicular to the main wind direction. These factors decrease exponentially with depth, causing what is called the Ekman spiral.

Two major types of Ekman flow are upwelling (Ekman pumping) and downwelling. Upwelling is important to climate reconstruction in coastal and equatorial regions where colder nutrient-rich waters rise into the surface layers to replace the wind-driven surface waters. For example, in the Pacific Ocean near the equator, strong Ekman flow northward and southward creates a zone of equatorial divergence, or upwelling, that is linked with climate changes occurring over various timescales. Oceanic divergence and strong upwelling zones along coastal margins, such as those off Peru and Ecuador, result in extremely high biological productivity. These areas are among the world's most biologically productive regions where detailed paleoceanographic records can be recovered.

Current speed is controlled by wind stress, the density of water, eddy viscosity, and Coriolis force. Geostrophic flow is another characteristic feature of ocean circulation. If the ocean were infinitely wide, Ekman flow would dominate. In reality, ocean boundaries along coastal regions impede flow and lead to slopes in the sea-surface levels. As water piles up in one region, it causes horizontal pressure gradients to form. Like wind, water flows along high- to low-pressure gradients, and this influences ocean-surface currents. Oceanographers calculate the hydrostatic pressure acting on a unit area because of different loads as

$$p = -\rho g z$$

where p is pressure, ρ (rho) is the density of seawater, g is acceleration due to gravity, and z is depth. The horizontal pressure gradient force per unit area is calculated from this relationship for two locations a given distance apart, and with a given angle representing the slope. The product of pressure gradients and Coriolis force is called geostrophic flow, in which ocean circulation proceeds along isobars, or lines of equal pressure. Geostrophic flow predominates in ocean interior regions away from the equator and leads to the primary surface ocean currents described in the following section.

Surface Ocean Currents   We illustrate the location of some of the major surface ocean currents in Figure 1.4. The subtropical gyres are located between about 20–40°N and 20–40°S latitude. Surface water in subtropical gyres circulates clockwise in the northern hemisphere and counterclockwise in the southern hemisphere. These gyres are bounded by equatorial currents and western and eastern boundary currents. The boundary between the North Atlantic subtropical gyre and subpolar North Atlantic water is called the Polar Front (or Subarctic Front), a critical zone during glacial-interglacial and millennial climate changes.

Gyre circulation driven by anticyclonic wind patterns includes relatively fast, deep, and narrow surface western boundary currents and slower, more diffuse eastern boundary currents. Western boundary currents include the Pacific's Kuroshio and the Atlantic's Gulf Stream (also called the North Atlantic Current in high latitudes) and in the southern hemisphere the Brazil and East Australian Currents. These are major exporters of warm water from low to high latitudes. Poleward heat transport via these currents is often cited as one reason that northeastern Atlantic and Pacific coastal regions in northern Europe and the Gulf of Alaska currently have relatively warm climates despite their location in high latitudes, although the atmosphere plays an important role as well. Changing strength of surface western boundary currents can have significant impacts on the climate history of high-latitude regions. Eastern boundary currents include the Canary and California Currents in the northern hemisphere off Europe and Africa and North American and the Benguela and Peru (Humboldt) Currents in the southern hemisphere off Africa and South America.

The subtropical gyres in the southern hemisphere are influenced by the Antarctic Circumpolar Current (ACC). The ACC is a 24,000-km westerly wind-driven current that represents the largest wind-to-geostrophic flow exchange of energy in the world's ocean (Wunsch 1998; Ganachaud and Wunsch 2000). ACC mass transport estimates range from 120 to 157 Sv. The ACC actually consists of several complex water masses: the Subantarctic Front, the Polar Front, and the Continental Water Boundary near Antarctica. Low-salinity Antarctic Intermediate Water lies north of the Subantarctic Front. A mixed-layer water mass called Subantarctic Mode Water is the source of several intermediate water

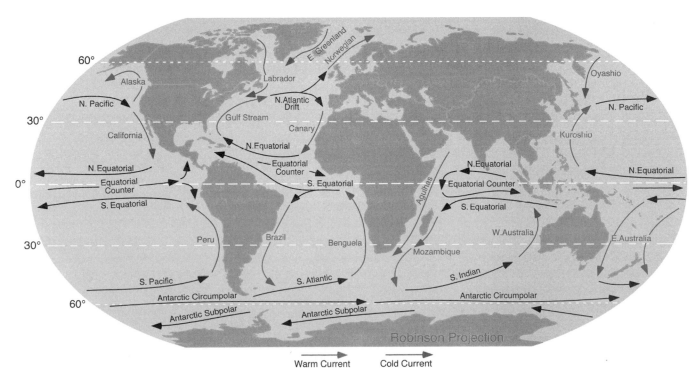

FIGURE 1.4 Major features of surface ocean circulation. See text. Courtesy of Michael Pidwirny (http://www.physicalgeography.net/fundamentals/8q_1.html).

masses throughout the Pacific and Indian Oceans influencing the oxygen minimum zone and carbon cycling.

In the high-latitude North Atlantic region, the northward-flowing Norwegian Current is an extension of the Gulf Stream and brings warm, salty water into the Norwegian-Greenland Seas (Nordic Seas), and eventually into the Arctic Ocean through the Barents and Greenland Seas. The southward-flowing East Greenland Current is an extension of the Transpolar Drift in the central Arctic Ocean and exports cold water from the Arctic into the North Atlantic. The Nordic Seas are today the critical regions of deep-water convection, although the location of convective cells shifted significantly during glacial periods.

Deep-Ocean Water Masses Meridional overturning circulation (MOC) refers to the characteristic feature of global circulation where surface currents such as the Gulf Stream flow from low to high latitudes, then cool, sink, and circulate around the globe. The Atlantic component (AMOC) is a wind- and buoyancy-driven two-layer flow consisting of the upper limb in the thermocline and surface layers and the deeper limb formed by North Atlantic Deep Water (NADW), also called northern-source deep water (Figure 1.5). The upper limb of AMOC originates in the South Atlantic in the thermocline and shifts position to the surface mixed layer near the equator before flowing northward to high latitudes. The lower limb of NADW is formed through convection in the Labrador Sea and the Norwegian-Greenland Seas (Nordic Seas), with contributions from the Arctic Ocean. NADW formation is mainly driven by density changes caused by winter cooling of surface source water of high salinity.

The precursor of NADW from the Nordic Seas is the combination of Iceland Scotland Overflow Water and Denmark Strait Overflow Water that spills over sills in the Denmark Strait between Greenland and Iceland and the Iceland-Faroes Rise between Iceland and the Faroe Islands (Mauritzen 1996). The Labrador Sea component flows southeastward and becomes entrained with that originating in the Nordic Seas south of Greenland to form NADW. NADW is relatively warm (2–4°C) when compared with underlying Antarctic Bottom Water (AABW) (0–2°C), although changes in the relative strength, location, and temperature of NADW formation is one of the most intensely studied topics in paleoceanography. Antarctic Intermediate Water occupies mid-depths below the warmer surface layers. AMOC varies in strength over seasonal and interannual (Cunningham et al. 2007), decadal (Bryden et al. 2005), and longer (see Chapter 6 in this volume) timescales.

AABW forms around the Antarctic continent mainly through brine formation and density increase from cooling at the ocean-atmosphere boundary. Brine formation occurs on the Antarctic continental shelf during sea-ice formation when salt is expelled, increasing surface ocean salinity. This dense, salty water sinks near the shelf-slope break and, because AABW is colder and denser than NADW, it sinks beneath it and flows northward, filling the deepest basins in the Atlantic. Antarctic Intermediate Water forms at mid-depths and flows northward below warmer surface currents (Figure 1.5).

Deep-water masses do not form in the North Pacific Ocean because evaporation rates are low, due in part to cooler sea-surface temperatures (SSTs). Cool North Pacific SSTs result from upwelled water supplied from deeper-water masses and reduce the ability of the overlying atmosphere to hold moisture, lowering evaporation rates, and, even in the case of relatively high salinities, preventing deep convection.

The ocean also stores and transports salt such that changes in the flux of freshwater from rivers, precipitation, or melting ice can alter the salinity of critical regions of the world's oceans, potentially altering circulation patterns. Together, heat and salt influence the density of surface waters; water that is colder, more saline, or both tends to sink more rapidly than warmer water or freshwater. This buoyancy effect from heat and salt is widely referred to as thermohaline circulation (THC). In contrast to MOC, which refers to the circulation of mass, THC describes ocean circulation in the context of the physical properties of temperature and salinity as described in classic papers by Stommel (1961) and Rooth (1982). Wunsch (2002) points out that the distribution of heat and salt are themselves not the drivers of ocean circulation, but rather "passive" consequences of the mechanical wind-and-tide-driven flux of fluid mass. One commonly sees the term *THC* in the paleoceanographic literature because some proxy methods are indicators of ocean temperature and salinity and not strictly flow.

This brief description might suggest a somewhat "sluggish" global ocean circulation with large-scale density-driven fluid flow over centuries to millennia, a topic commonly encountered in the paleoclimate literature on glacial climate events. However, observations and modeling in physical oceanography indicate that buoyancy-driven ocean circulation originating in the high-latitude convective diffusion cells is not sufficient to drive the complete cycle of ocean circulation, including upwelling of deep water, back to the surface. A more dynamic, turbulent ocean circulation, varying over shorter temporal and fine spatial scales, governed by winds and tides (Munk 1997; Munk and Wunsch 1998) and bottom topography (Polzin et al. 1997; Ledwell et al. 2000), is needed to more fully explain global patterns of circulation. For example, Ledwell et al. (2000) measured diffusivity (Kp) at various locations in the Brazilian Basin of the Atlantic and found relatively low values (one-tenth those assumed for the mean global background level of $Kp = 10^{-4}\,m^2\,sec^{-1}$) for the generally flat abyssal plain, but much higher values over the

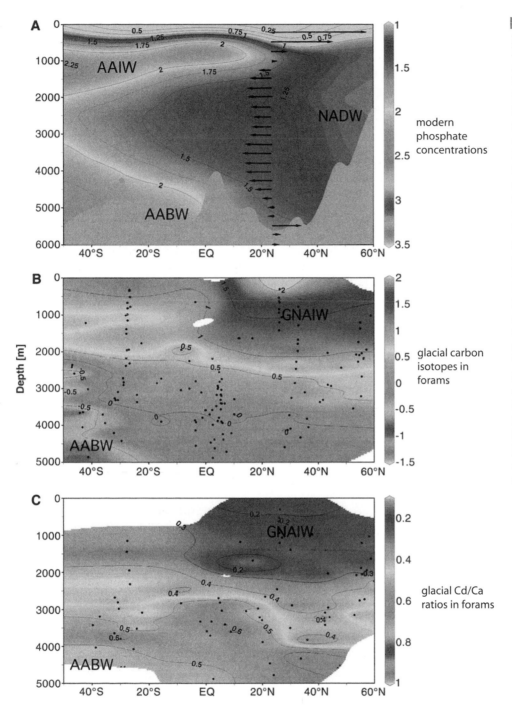

FIGURE 1.5 (A) Modern dissolved phosphate concentrations (μmol liter⁻¹) in the western Atlantic. Arrows show southward flow of North Atlantic Deep Water (NADW), which is compensated by the northward flow of warm near-surface water underlain by Antarctic Bottom Water (AABW). (B) Distribution of the carbon isotopic composition ($\delta^{13}C$ VPDB) of benthic foraminifera in the western and central Atlantic during the Last Glacial Maximum (LGM) projected onto a single meridional section. (C) LGM cadmium concentrations (nmol kg⁻¹) based on cadmium/calcium (Cd/Ca) in the shells of benthic foraminifera. Today, the isotopic composition of dissolved inorganic carbon and the concentration of dissolved Cd in seawater both show "nutrient"-type distributions similar to that of PO_4. AAIW = Antarctic Intermediate Water; GNAIW = Glacial North Atlantic Intermediate Water. From Lynch-Stieglitz et al. (2007). Courtesy of J. Lynch-Stieglitz and American Association for the Advancement of Science.

mid-Atlantic Ridge. They postulated that this turbulence was driven by tide-generated waves and was sufficient to bring deep water up from within the basin. Garabato et al. (2004) also found diffusivity values in parts of the southern ocean 10 to 1000 times greater than the background level, demonstrating a key role for topography in mixing processes in the ACC.

In addition to tides, wind, and bottom topography, "salt fingers"—those areas of the ocean (mainly in temperate regions) where warm, salty water sinks into cooler, fresher

layers—redistribute heat and, to a lesser degree, salt vertically in upper layers of the ocean, thereby influencing diffusivities (Schmitt et al. 2005). Salt fingers provide an additional mechanism driving circulation. For our discussion, these dynamic processes partially explain the upwelling of deeper waters to the surface at depths of less than 1000 m, influence the large-scale ocean circulation system, and have implications for climate-ocean modeling and paleoceanography (see Wunsch and Ferrari 2004). In fact, in light of these new developments in physical oceanography and ocean circulation, Wunsch

(2006) questioned whether the bipolar seesaw hypothesis of Broecker was a realistic explanation of observed millennial-scale patterns (see Chapter 6 in this volume). Although such a view is not universal (Alley 2007), we will see that new proxy methods and paleoceanographic reconstructions are beginning to provide more direct and spatially robust measures of dynamic aspects of ocean circulation during millennial oscillations.

Thermocline and Upwelling The oceanic features discussed above are located in the uppermost 50–150 m, called the surface mixed layer. Below the mixed layer, water temperature decreases in a zone called the thermocline. The seasonal thermocline at the bottom of the mixed layer forms when seasonal warming and cooling cause oceanic overturning. The permanent oceanic thermocline is the steep temperature gradient from surface to deep abyssal regions.

Upwelling of cooler, nutrient-rich waters occurs in low-latitude regions where trade winds drive oceanic divergence. Equatorial upwelling is especially relevant to climatic cycles caused by changes due to the earth's precession cycles. Enhanced biological productivity and carbon sequestration in equatorial upwelling zones was at one time considered a factor in decreased glacial-age atmospheric CO_2 levels, although there is now consensus that the Southern Ocean around Antarctica was the site of greater air-ocean carbon transfer during glacial periods (see Chapter 5 in this volume).

The Cryosphere

General Concepts The earth's cryosphere includes all frozen water and is the realm of specialists in glaciology, oceanography, permafrost, and other periglacial processes. Like other parts of the climate system, the cryosphere is constantly in a state of flux, interacting with the oceans and the atmosphere in still poorly understood ways. Here we introduce several important glaciological concepts and calculations that illustrate controlling factors on ice growth and decay and then outline the major features of the cryosphere.

Mass Balance The mass balance (MB) budget of ice sheets is the most important cryospheric parameter in terms of past and future climate changes that are due to a number of complex factors. A basic expression to compute ice-sheet flow for a grounded ice sheet is what is called the depth-integrated mass continuity equation,

$$\partial H / \partial t = -\nabla \times HU + M$$

where H is ice thickness, U is velocity (also expressed as q, or discharge), ∇ is called the divergence operator, and M is the net accumulation and ablation rate. Flow is controlled according to Glen's flow law, which incorporates strain rate, stress, gravity, ice-density-specific heat capacity, temperature, geothermal heat flux, and thermal conductivity. Thermodynamic equations are used to compute temperature distribution within the ice sheet itself.

Computing MB, however, is a difficult chore because mass change in the modern Greenland and Antarctic Ice Sheets varies greatly in time and space and gradients must be estimated from scattered observations. Today, three approaches are used to estimate MB for modern ice sheets: balance melting and ice discharge loss with snow accumulation; measure elevation change through time from satellite altimetry; and measure the weight of the ice sheet through time using gravity measurements from satellites (Rignot and Thomas 2002; Chen et al. 2006; Rignot et al. 2008; see also Chapter 12 in this volume).

The surface mass balance (SMB) of ice sheets is an equally complex parameter to measure. SMB is measured using satellites, meteorological data, ground-penetrating radar, borehole strain rate measurements, and modeling. In addition to surface temperature and precipitation, wind-driven sublimation, surface scouring, and internal climate variability (El Niño–Southern Oscillation) also influence surface conditions of ice sheets.

Ice-dynamic processes in the interior, and especially along the margins of ice sheets, play a major role in ice-sheet behavior. These processes include basal lubrication by surface meltwater percolated to the ice-bedrock interface, melt-induced velocity changes, ice-margin calving, ice-shelf collapse, and glacier surging (DeAngelis and Skvarca 2003; Thomas et al. 2004; Howat et al. 2007; Truffer and Fahnestock 2007; van de Wal et al. 2008). Ice dynamics has emerged as a central topic in efforts to quantify ice-sheet response and the resulting sea-level rise (SLR) due to climatic warming. This is because alterations to SMB due to radiative changes and precipitation at the ice-sheet surface are negligible compared to dynamic changes at the margins of the ice sheet in glaciers terminating below sea level. For example, along margins of the Greenland Ice Sheet (GIS), outflow velocity and discharge rate are controlled by basal melting and bedrock configuration, which constrain MB (e.g., topography and cross-sectional area). Similarly, in addition to dynamic processes affecting Greenland ice, marine-based sectors of the West Antarctic Ice Sheet (WAIS), those regions where ice is grounded below sea level, are more sensitive to rapid disintegration (Rignot et al. 2008). Some marine-based glaciers of the WAIS are melting faster than other parts of the Antarctic Ice Sheet (AIS) because of accelerated flow as grounded portions are dislodged by the thinning or collapsing ice shelves that buttressed them and because of warming ocean and atmospheric temperatures.

In the Antarctica Peninsula where large ice shelves have disintegrated over the past ten years, land-based glaciers that feed ice shelves are no longer buttressed by the ice shelf and have higher velocities than those still blocked by it. These processes may have contributed to rapid rates of SLR as northern-hemisphere ice sheets decayed during deglacial warming (Zweck and Huybrechts 2003; see also Chapter 7 in this volume).

Glacier Budgets   The budget of any particular glacier, computed from accumulation and ablation rates of ice and snow, is used to estimate its MB. Changes in MB are estimated from measurements on meltwater discharge, surface elevation, and glacial density using a variety of field and remote sensing methods. Changes in a glacier's MB can be influenced by many climatic (atmospheric temperature, precipitation) and nonclimatic (bedrock configuration, internal glacial dynamics) factors. The Equilibrium Line Altitude is that point on a glacier where mean accumulation and ablation are about equal.

Many glaciers have their terminus in or near the land-ocean margin. The rate of glacier retreat and thinning (M) for marine-terminating glaciers with their beds below sea level is controlled by three factors expressed in the equation

$$M = M_b + M_h + M_L$$

where M_b is climatically controlled (temperature, precipitation) surface MB, M_h is thinning or thickening from local divergence of glaciers, and M_L is expansion or retreat of glaciers at their terminus. M_L is considered the most important factor in terms of future ice loss for particular marine-terminating glaciers because it represents nonlinear dynamic behavior in contrast to the near steady-state behavior of SMB. M_L is controlled by the rate of calving, the velocity of ice flow, and the width, thickness, and ice density at the glacier's terminus.

Major Features of the Cryosphere   The study of the cryosphere has advanced rapidly in recent decades with improvements in remote sensing from imagery, altimetry, and gravity methods using satellites. Some useful resources are the Satellite Image Atlas of Glaciers of the World (Williams and Ferrigno 1988), the National Snow and Ice Data Center, the inventory of global glacier and small ice caps (Dyurgerov and Meier 2005; Meier et al. 2007), and the Intergovernmental Panel on Climate Change (Lemke et al. 2007).

Ice Sheets   Figure 1.6 shows the GIS in relation to several smaller ice caps on Canadian islands and the sea-ice-covered Arctic Ocean. The roughly 80% of Greenland that is covered by ice constitutes an area of about $1.7 \times 10^6 \, km^2$ and a volume of $2.6 \times 10^6 \, km^3$, equivalent to a SLR of about 6 m should it melt (Weidick 1995). Figure 1.7 shows an overview map made from satellite images of the AIS, by far the earth's largest repository of freshwater. The figure also shows surrounding ice shelves and adjacent oceans. The AIS, including ice shelves and ice, covers $13.9 \times 10^6 \, km^2$, about 85% of which is inland ice, and has a volume of $30.1 \times 10^6 \, km^3$ (Drewry 1983). The AIS is divided into the larger, continental East Antarctic Ice Sheet (EAIS, inland ice volume $26.9 \times 10^6 \, km^3$) and the smaller, marine-based WAIS (inland ice volume $3.2 \times 10^6 \, km^3$). The AIS includes a number of ice shelves around its periphery, parts of which are grounded below sea level. Total AIS ice volume would raise sea level about 70 m should it melt (Shepherd and Wingham 2007). The Antarctic continent has experienced large spatial variability in temperature, snow accumulation, MB, and other climatological trends over the past 50 years, in part influenced by large-scale atmospheric dynamics caused by internal modes of variability (see Chapter 10 in this volume) and ocean-ice dynamics along parts of the WAIS. In contrast to the WAIS, which is losing mass at a rate of $0.13–0.41 \, mm \, yr^{-1}$ sea-level equivalent, the EAIS does not seem to have contributed to SLR during the past 50 years, although meteorological stations are sparse and longtime series are not available (Monaghan and Bromwich 2008).

Glaciers and Small Ice Caps   There are roughly 160,000 alpine glaciers and small ice caps in the world, including those in Iceland and the Tibetan Plateau but excluding the Greenland and Antarctic Ice Sheets. They cover a total area of $785 \pm 100 \times 10^3 \, km^2$ and a volume of $260 \pm 65 \times 10^3 \, km^3$ (Cogley 2005; Dyurgerov and Meier 2005). If they melted completely, they would cause sea level to rise $0.65 \pm 0.16 \, m$. Meier et al. (2007) estimated that alpine glaciers and small ice caps would dominate the contribution of melting parts of the cryosphere to 21st-century SLR, with a $0.1–0.25 \, m$ sea-level equivalent (see Kaser et al. 2006). Evidence from alpine glaciers with the longest monitoring records shows negative MB trends during the late 19th and 20th centuries, a pattern consistent with warming atmospheric temperatures and observed SLR (Oerlemans 2005).

Sea Ice   Arctic Ocean sea-ice extent, defined as areas with greater than 15% surface ocean ice cover, averaged a seasonal minimum (September) $7 \times 10^6 \, km^2$ to a maximum (March) $16 \times 10^6 \, km^2$ from 1979 to 2005. September ice cover has been decreasing rapidly at a rate of $100,000 \, km^2 \, yr^{-1}$ against the mean (Serreze et al. 2007), and an even greater sea-ice decline occurred in the summer of 2007 (Figure 1.8). Arctic sea ice is controlled by several complex factors

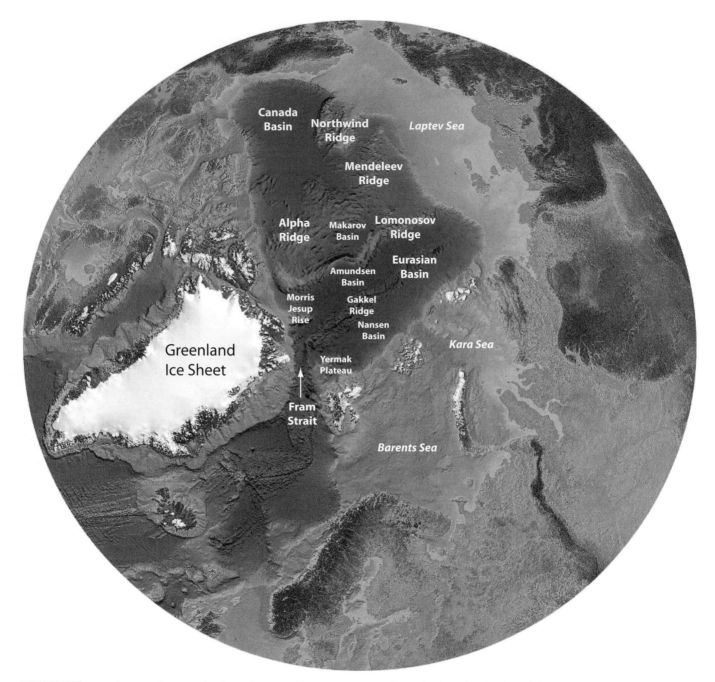

FIGURE 1.6 Map showing the Greenland Ice Sheet, small ice caps on Canadian islands and Iceland, and the Arctic Ocean Bathymetry. From Jakobsson et al. 2008. Courtesy of M. Jakobsson and International Bathymetric Chart of the Arctic Ocean.

divided into thermal and dynamic processes. Thermal factors include radiative flux, surface air temperature, and ocean temperatures; dynamic processes include winds, ocean circulation, and currents (e.g., Rothrock and Zhang 2005). Interannual variability in sea ice is balanced between production, largely in continental shelf regions for multiyear ice, and export largely through the Fram Strait between Svalbard and Greenland. This balance reflects many factors—e.g., ice and ocean circulation, such as the strength of the Transpolar Drift Current, and spring air temperature. The Arctic Oscillation, a distinct mode of climate variability operating over decadal timescales (see Chapter 10 in this volume), exerts atmospheric control on interannual variability in sea-ice extent, as does the volume of runoff from large Siberian rivers. Thinning of ice from below, where the inflowing Atlantic Water layer about 100–500 m below the surface is near 0°C even in the central Arctic Ocean, is also an important process influencing sea-ice cover.

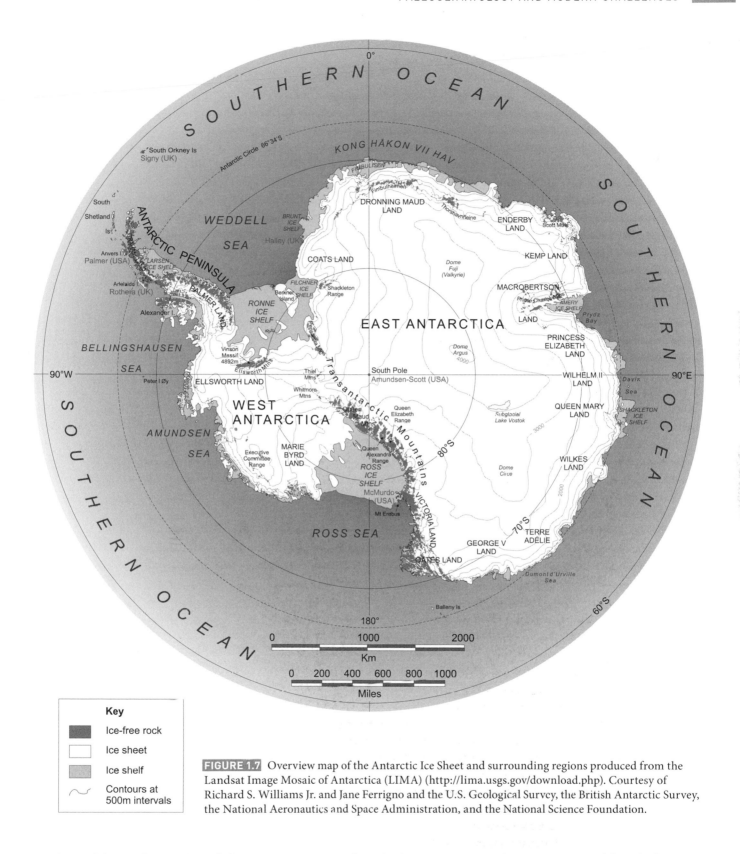

FIGURE 1.7 Overview map of the Antarctic Ice Sheet and surrounding regions produced from the Landsat Image Mosaic of Antarctica (LIMA) (http://lima.usgs.gov/download.php). Courtesy of Richard S. Williams Jr. and Jane Ferrigno and the U.S. Geological Survey, the British Antarctic Survey, the National Aeronautics and Space Administration, and the National Science Foundation.

Around Antarctica, sea-ice exhibits enormous seasonal variability. For the period from 1979 to 2007, Antarctic sea ice averaged $4.3 \times 10^6\,\text{km}^2$ in March to $18.7 \times 10^6\,\text{km}^2$ in September. Unlike the trends in Arctic sea ice, Antarctic sea ice in 2007 was near its mean extent, although there was substantial spatial variability due to regional atmospheric and ocean conditions. Sea-ice-ocean-biogeochemical interactions are a central theme in efforts to understand global

FIGURE 1.8 Arctic Ocean sea-ice cover during March and September 2006 and 2007. Sea-ice extent during March 2006 (seasonal maximum) and September 2007 (seasonal minimum) were the lowest in the instrumental record since 1979. From Fetterer et al. (2002). Courtesy of J. Richter-Menge and the National Snow and Ice Data Center (http://www.nsidc.org/data/seaice_index).

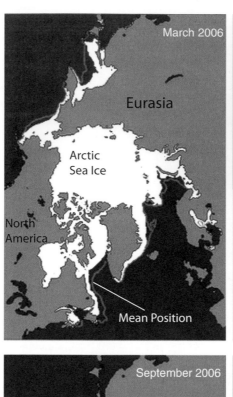

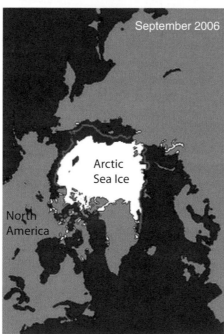

carbon cycling during periods of rapid, large-scale climate change.

Summary The earth's cryosphere has special significance for many aspects of paleoclimatology covered throughout this book. The elevation and volume of Pleistocene ice sheets influenced hemispheric atmospheric circulation, sea level, albedo, and other climate factors. Evidence suggests that the AIS has experienced dynamic fluctuations in volume on million-year timescales. The influence of glacial meltwater on ocean-surface salinity and temperature regimes very likely changed in North Atlantic MOC and caused abrupt climate change. Sea ice around the Antarctic and in the Arctic Ocean, Nordic Seas, and northern North Atlantic has been extremely dynamic over short and long timescales. Even answers to the most fundamental question about ice sheets—when and why do they grow and decay—remain enigmatic. These challenging questions are addressed further in chapters 3–9 in this volume.

The Lithosphere

The earth consists of an outer crust, a mostly solid mantle, and a core, with inner solid and outer liquid layers distinguished mainly on the basis of geophysical and mineralogical properties. Continental crust is lighter, older, and thicker than oceanic crust. The upper layers of the earth are also distinguished on the basis of their rheology or ability to flow, and this distinction lies at the heart of the theory of plate tectonics and sea-floor spreading (also called continental drift). The crust and upper mantle form the outmost 50–100-km-thick rigid lithosphere, which is composed of eight major plates and about 26 smaller plates (Figure 1.9).

Lithosphere plates "ride" on a more ductile, viscoelastic dynamic asthenosphere. Lithospheric plates are constantly shifting position, mainly as a result of unequally distributed subcrustal heat and convective processes in the mantle. There are three types of plate boundaries. Convergent boundaries occur where plates collide and exhibit active earthquake and volcanic activity. Here we find subduction zones, deep-sea trenches, and island arcs typical of the circum-Pacific region. At divergent boundaries, plates move apart from mantle upwelling, basalt formation, and sea-floor spreading. Plates at transform boundaries slide past each other laterally, like the San Andreas, California fault system. Relatively stable intraplate regions are usually characterized by slow uplift and subsidence due to processes such as isostasy, the force of gravity acting upon crustal materials of different densities.

Long-term climate history is linked to plate tectonic processes and the changes in ocean-land distribution and topography they cause, for several reasons. Changes in topography, for instance, are hypothesized to be one possible factor in long-term climate cooling during the Cenozoic (Ruddiman and Kutzbach 1989). One effect of mountain building on climate is that extremely high rates of uplift can alter global weathering patterns and geochemical fluxes from continents to oceans (Ruddiman and Raymo 1988; Raymo et al. 1988), ultimately changing atmospheric levels of CO_2 and contributing to global cooling (Raymo 1991; Berner 1994). A completely opposing view holds that climate changes influence the uplift of mountains and plateaus (Molnar and England 1990; Molnar 2004; see also Chapter 4 in this volume).

The location within a plate or near a plate boundary can significantly influence the interpretation of paleoclimate records. For example, in some tectonically active zones, emerged tracts of fossil coral reefs formed during past high stands of sea level are uplifted, providing an excellent record of sea-level history once the rate of uplift is subtracted from the reef's modern elevation. In contrast to tectonically active areas, sea-level history determined from relatively stable regions provides a partial solution, because the local sea-level record is unencumbered by major tectonic movements over the past few hundred thousand years. The deformation of the asthenosphere is a critical component of the study of glacio-isostatic vertical movements caused by mass redistribution during continental ice sheet growth and decay. Changing continental positions also impose constraints on

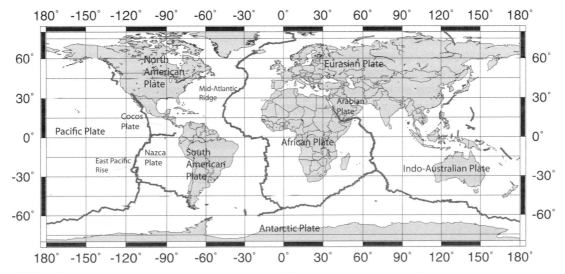

FIGURE 1.9 The earth's major lithospheric plates and mid-ocean ridges. There are about 26 minor plates not shown. Produced with Online Map Creation, Martin Weinelt (http://www.aquarius.ifm-geomar.de/omc_intro .html) and The Generic Mapping Tool (GMT) (http://gmt.soest.hawaii.edu/) software developed by Paul Wessel (University of Hawaii at Manoa) and Walter H. F. Smith (National Oceanic and Atmospheric Administration).

past patterns of oceanic circulation and the redistribution of solar energy by ocean currents and deep-water circulation. Many geophysical processes operate in earth's lithosphere, playing various roles in the reconstruction and interpretation of climate history.

The Biosphere

The earth's biosphere is a key component of the climate system because it is central to global biogeochemical cycling of carbon, nutrients, and other chemicals that influence climate either directly or indirectly. Biogeochemical cycling, both in the terrestrial and marine biosphere, involves interactions of biomes and ecosystems with the atmosphere and oceans at various timescales (figures 1.10 and 1.11). The biosphere, however, is governed by unique processes quite distinct from those for the rest of the climate system because it consists of living organisms and the communities and ecosystems they form. Take the most basic stochastic process of a hypothetical genetic mutation that might confer a competitive advantage of nutrient-uptake capability for a species of ocean phytoplankton, or the rapid evolutionary expansion of vascular land plants in the Middle Devonian period 385 million years ago, with large ramifications for

the global carbon cycle. It is clear that organism-organism and organism-environment interactions studied in the fields of evolutionary biology and ecology become relevant for studies of long-term climate change. We will see many examples in later chapters; an in-depth discussion of biological aspects of climate change can be found in Cronin (1999, ch. 3).

Emergent Climate Features

One final concept about the earth's climate deserves mention. Emergent features of climate are large-scale phenomena that one might not predict from purely physical laws and equations (Schmidt 2007). Unlike many aspects of the climate system (heat, atmospheric pressure, momentum, mass, water masses, etc.), emergent features cannot easily be measured directly, nor can they be calculated using equations from physical laws. They develop from complex, still partially understood processes in the climate system and represent important phenomena for climate modeling and prediction and paleoclimate reconstruction. Continental ice sheets, major ocean current systems, the ITCZ, and tropical monsoon systems are among the more obvious. Emergent climate phenomena are transitory over longer timescales and they

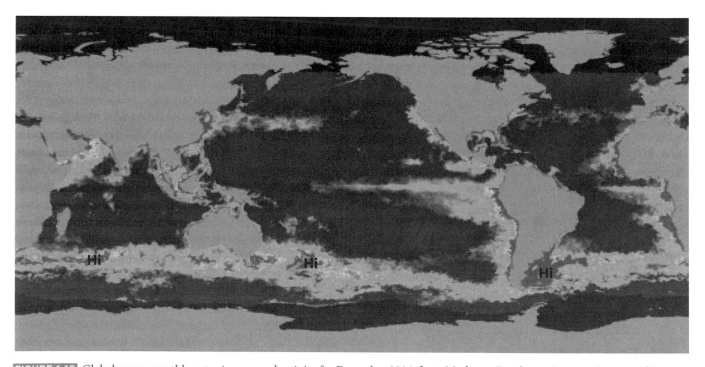

FIGURE 1.10 Global ocean monthly net primary productivity for December 2004, from Moderate Resolution Imaging Spectroradiometer (MODIS) satellite chlorophyll, and sea-surface temperature, made from the Vertically Generalized Production Model (VGPM), from Behrenfeld et al. (2005, 2006). Higher production is along coastal regions and in the Southern Ocean circumventing Antarctica. Central ocean regions have low productivity. Dark regions labeled "Hi" are highest in productivity, reaching $1000 \, mg \, C \, m^{-2} \, day^{-1}$. Courtesy of M. Behrenfeld and R. O'Malley and Oregon State University (http://www.science.oregonstate.edu/ocean.productivity/).

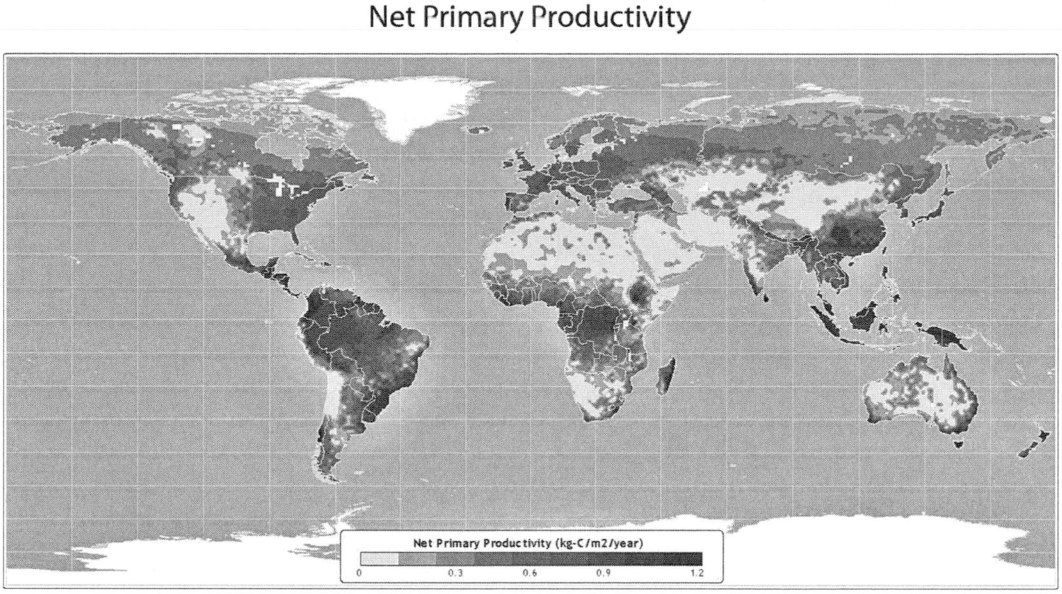

Net Primary Productivity

FIGURE 1.11 The total net primary productivity in each grid cell over one year for the terrestrial biosphere, in kg carbon m^{-2} yr^{-1}. Data taken from 1815 Simulation (Foley et al. 1996; Kucharik et al. 2000). Courtesy of J. Foley and C. Kucharik, the Atlas of the Biosphere, Center for Sustainability and the Global Environment (SAGE), University of Wisconsin–Madison.

will resurface many times in later chapters as we discuss large-scale, long-term climate evolution.

Causes of Climate Change

This section introduces factors called external forcing mechanisms and internal processes that cause climate to change. At the outset, we stress that the climate and paleoclimate communities, while sometimes having common research goals, define the terms *external* and *internal* in opposing and sometimes confusing ways. It is useful, therefore, to briefly distinguish their respective definitions and viewpoints (Table 1.1).

External Forcing

Climatological View Climatologists divide the factors that cause climate to change into two categories—external factors and internal processes. With reference to modern climate change, external forcings include those that affect the atmosphere's radiative properties and global energy balance, as illustrated in Figure 1.2: greenhouse gas concentrations, solar radiation changes, volcanic emissions, and sulfate aerosols. These are the four primary forcing mechanisms used to evaluate anthropogenic climate change against natural background variability (see chapters 11 and 12 in this volume).

Paleoclimatological View Most paleoclimatologists consider external forcing mechanisms more broadly and in-

clude those stemming from extraterrestrial processes such as the solar system's gravitational effects on the earth's orbit, extraterrestrial dust, and meteor impacts. For example, changes in the earth's orbital geometry because of gravitational forces in the solar system influence the seasonal and geographical distribution of insolation. Orbital climate changes are clearly an external factor driving glacial-interglacial-scale climate oscillations (see chapters 3 and 4 in this volume), although not directly related to changes in total radiative forcing, such as caused by solar irradiance changes or elevated greenhouse gas concentrations. Nonetheless, orbital changes are indeed relevant to debates about human-induced changes in greenhouse gases during the Holocene, prior to the industrial revolution (Ruddiman 2007; see also chapters 8–12 in this volume).

Internal Processes

Climatological View Internal climate processes influence a broad range of climatic variability that does not involve changes in radiative balance. Climatologists define these "modes" of variability as "unforced" changes caused by dynamic interactions between the atmosphere and oceans typically over annual to multidecadal timescales. Interannual variability caused by the El Niño–Southern Oscillation phenomenon is the most famous, but there are many other modes of internal variability considered critical for understanding human-induced climate change (see Chapter 10 in this volume).

TABLE 1.1 Causes of Climate Change*

Climatology Emphasis	Paleoclimatology Emphasis
External	**External**
Solar radiation	Eccentricity
Sunspot variation and irradiance changes	Obliquity (tilt)
Solar diameter	Precession of equinoxes
Solar ultraviolet wavelength variability	Axial precession, asteroid impacts, internal plate tectonics
Carbon dioxide (CO_2), methane (CH_4), nitrous oxide (N_2O), halocarbons, tropospheric nitrogen oxides, carbon monoxide (CO), sulfate aerosols	Galactic dust
Aerosols	Solar irradiance
Explosive volcanism	Meteor impacts
Internal	**Internal**
Internal modes of variability, "unforced" variability; see Chapter 10	Orogenesis, mountain building, weathering, biogeochemical changes
	Ocean-volume changes, sea level
	Ice-sheet and glacial lake meltwater influx
	Explosive volcanism
	Dimethylsulfide (cloud condensation nuclei)
	Mid-ocean ridge outgassing

*See text for feedback mechanisms.

Internal unforced changes may also occur over longer timescales and involve the earth's cryosphere as well as its ocean and atmosphere. For example, Heinrich events, the growth and collapse of portions of glacial-age ice sheets over 5–10 kiloannum (ka), might be self-sustaining oscillations caused by ice-sheet dynamics with no external forcing but with significant global climate repercussions (see Chapter 7 in this volume).

Paleoclimatological View In addition to these internal modes of variability, other processes inherent to the earth as a dynamic planet also affect climate. These include orogenesis (mountain building), plate tectonics (sea-floor spreading, mid-ocean ridge basalt formation and outgassing), volcanism, and long-term biological changes (i.e., evolution of land plants). Such processes have major impacts on the earth's global-scale weathering and erosion, ocean and atmospheric circulation, sea level, and biogeochemical cycling of carbon, nutrients, and other key elements throughout parts of the climate system. Volcanic emissions of CO_2 and other gases can influence atmospheric greenhouse gas concentrations.

One can see that there are different viewpoints on what is an external or internal "cause" of climate. Paleoclimatology considers large volcanic eruptions as an internal planetary process, but climatology views them as an external

forcing mechanism because of their cooling influence on atmospheric temperature. A similar situation exists for mid-ocean ridge outgassing of CO_2 over millions of years, clearly a geological process, but one that affects radiative properties of ancient atmospheres by changing greenhouse gas concentrations.

Feedbacks
General Concepts

Feedback mechanisms (or feedbacks) are processes that respond to external forcing by amplifying (positive) or dampening (negative) the initial strength of the forcing. Most reconstructed climate changes from paleoclimate records are the cumulative product of one or several feedback mechanisms, as well as of external forcing and internal variability. Without feedbacks, for instance, direct forcing from geographical and seasonal changes in insolation caused by cycles in the earth's orbital geometry would not produce the scale and sawtooth shape of late Pleistocene ice age cycles (see Chapter 5 in this volume). Feedbacks often operate until a new equilibrium climate state is reached and are often viewed as self-regulating processes. In the case of temperature, several types of feedbacks act like thermostats preventing climate from runaway cold or hyperthermal conditions. Extreme climate states, such as the Neoproterozoic "Snowball Earth" almost a billion years ago, when the earth's surface may have completely frozen, or nearly so, may signify times where negative feedbacks do not counterbalance cooling, resulting at least temporarily (millions of years) in a runaway climate effect (see Chapter 3 in this volume).

In some cases, feedbacks operate until a climatic threshold is crossed when the system switches, often abruptly, to a very distinct, quasi-stable climate state. This "hysteresis" effect applies to several types of climate behavior, like the millennial-scale changes in the ocean's MOC and the growth and decay of large Cenozoic ice sheets. Feedbacks operate over all timescales and are usually distinguished by the physical, chemical, biological, and geological processes controlling the speed at which they act on the climate system. We group them into three informal categories—fast, slow, and slower feedbacks.

Fast Feedbacks

Clouds and Water Vapor   Clouds and water vapor are sometimes called Charney fast feedback mechanisms (after Jule Charney, who reviewed the influence of rising atmospheric CO_2 concentrations on global temperature; see Charney et al. 1979). The influence of clouds, which some climate models indicate have a positive feedback on temperature as CO_2 concentrations increase, is one of the most intensely studied aspects of human-induced climate change and climate prediction (Soden et al. 2005; see Weart 2003 for a historical review). The role of water vapor in climate has also been a major point of contention (Lindzen 1997; Held and Soden 2000). The timescales for cloud and water vapor feedbacks are generally too short for paleoclimate analysis. However, Greenland, Antarctic, and high-elevation low-latitude ice-core proxy records yield clues about changes in surface ocean biological processes, wind and dust concentrations, and other atmospheric conditions related to cloud formation during the last 800,000 years. On longer timescales, Kump and Pollard (2008) hypothesized that during the Cretaceous period, reduced ocean production of dimethylsulfide, a precursor to atmospheric condensation nuclei, amplified the effects of elevated CO_2 concentrations through positive albedo feedbacks, causing extreme global warmth.

Sea Ice   Sea ice can influence climate in many ways. Sea ice has a higher albedo than ocean water, and greater sea-ice extent creates a positive feedback during periods of cooling. Sea ice can also prevent deep-water convection in high-latitude ocean regions and thus can weaken deep-water mass formation, MOC, and equator-to-pole heat transport. Sea ice also insulates the underlying ocean from heat lost to the atmosphere and can control the ocean moisture source for growing land-based ice sheets. Finally, it prohibits photosynthetic activity in regions of the ocean where carbon sequestration is important in regulating atmospheric concentrations. Because seasonal sea ice grows and melts rapidly, it can be viewed as one type of fast-feedback mechanism influencing climate over timescales shorter than ~1–10 ka.

Slow Feedbacks

Land-Surface Vegetation   Vegetation and land-surface feedbacks are important over several timescales. Leaf phenology (i.e., the time of year leaves emerge) influences climate at seasonal-annual timescales; terrestrial ecosystem dynamics operate over annual to decadal timescales; and continental-scale biome changes occur over glacial-interglacial cycles. Change in land-surface cover (grasslands, forests, ice, etc.) can have direct effects on albedo and hydrology, and indirect effects on carbon cycling, methane production and sequestration, and greenhouse gas concentrations. Observational records and modeling reveal vegetation-climate linkages at global and continental scales (Notaro et al. 2005, 2006, 2007). Liu et al. (2006) found that between the years 1982 and 2000, atmospheric temperature change in middle to high latitudes produced a strong positive vegetation feedback of up to 10–25% of total monthly variance in northern middle to high latitudes. In contrast, tropical vegetation was influenced more by precipitation but the feedback response to warming was much smaller than in higher latitudes. Over centennial to millennial timescales, vegetation-precipitation feedbacks play an even larger role in climate dynamics.

Sea-Surface Temperature and Ocean Circulation   Colder ocean temperatures and greater sea-ice extent during glacial periods can lead to stronger ABW formation and weaker NADW, diminishing heat flux to high northern hemisphere latitudes. This can produce a positive feedback situation where reduced equator-to-pole heat transport can lead to colder temperatures at higher latitudes, greater land and sea ice, reduced albedo, and so on (see chapters 5–7 in this volume).

Carbon Cycle   Changes in the global carbon cycle affect the concentrations of greenhouse gases in the atmosphere and thus climate. For example, the atmospheric residence time of CO_2 varies constantly because of changing sources and sinks but generally is on the order of 50–200 years. Rising CO_2 concentrations due to outgassing from oceans to atmosphere amplifies climatic warming triggered by orbital insolation during glacial-interglacial transitions. During the last four deglacial intervals, mean annual temperature rose about 5°C over 10,000 years, in part because of atmospheric

CO_2 concentrations rising from 200 to 280 ppmv. The ocean-atmosphere transfer of carbon during these climatic transitions is an important feedback in the global carbon cycle.

Carbon-cycle-climate feedbacks over long timescales may be even more important than previously believed. For example, Schmittner et al. (2008) modeled future climate changes over multicentennial and millennial timescales under what is called the "business as usual" scenario—i.e., a situation in which human carbon emissions continue to increase without intervention or ultimate stabilization near 650 ppmv (see "Climate Sensitivity" in the next section). Their integrated climate-carbon-cycle-ecosystem model simulations suggest that ocean carbonate chemistry (strengthened alkalinity gradient) has greater positive feedback on climate, showing that the oceans will progressively absorb less CO_2 from the atmosphere than they did over the last century. Additional changes in ocean warming and circulation lead to greater atmospheric nitrous oxide concentrations, which also exert a positive feedback on climate. In essence, the state of the climate centuries from now ultimately will be determined by atmospheric CO_2 concentrations and may be unlike any seen in historical times. Obviously studies such as those of Schmittner et al. justify consideration of past climate response to elevated CO_2 concentrations and the role of carbon cycle feedbacks over various timescales.

Ice-Sheet Dynamics Ice sheets influence climate in a number of ways. Pleistocene ice sheets reached several kilometers in thickness and directly affected global atmospheric circulation patterns. They also created strong albedo feedbacks; snow and ice albedo values range from 0.7 to 0.9, whereas those for ocean surfaces are only 0.1 or less. Radiative forcing during Pleistocene glacial maxima may have been reduced $3.5 \pm 1 \, W \, m^{-2}$ in part because of the greater albedo of ice sheets. As a result, the earth absorbed an estimated 1.5% less solar irradiance. By storing freshwater on continents, ice sheets also lower the global sea level by up to 160 m during the largest Pleistocene glacial cycle, exposing continental shelves to erosion, changing total land-surface albedo, and allowing the release of methane trapped in marine sediments.

Other Feedbacks Several other feedbacks deserve mention. Stratosphere-troposphere interactions may amplify small changes in solar irradiance and produce centennial climate anomalies such as the Little Ice Age (A.D. 1400–1900; Shindell et al. 1999, 2001). There is substantial evidence from windblown material in marine sediments and continental loess deposits that the atmosphere during glacial periods was dustier, although not necessarily windier, than during interglacial periods (Rea 1994; Hesse and McTainsh 1999). Dust aerosols can transport nutrients like iron to marine ecosystems, enhancing productivity and carbon uptake, and are an especially important positive feedback over glacial-interglacial climate cycles (Winckler et al. 2008).

Slower Feedbacks

Geological processes cause feedbacks that affect long-term climate change. Over multimillion-year timescales, CO_2 uptake by the lithosphere from the atmosphere through continental silicate weathering is counterbalanced by CO_2 production from carbonate rocks and organic-rich sediments during volcanic activity (ocean ridge CO_2 outgassing). This silicate weathering-CO_2 negative feedback mechanism acts like a self-regulating global thermostat. If continental uplift and weathering rates increase, this feedback is less effective; CO_2 is consumed at a faster rate than it is produced, atmospheric CO_2 concentrations decline, and climate cools. This mechanism has been invoked to explain the Cenozoic greenhouse-to-icehouse transition; however, many factors are involved in global biogeochemical cycles over geological timescales: volcanic CO_2 outgassing, seafloor spreading rates, carbonate formation and distribution, temperature, the proportion of silicate-based rocks exposed, organic carbon burial, relief and exposed area, biological evolution of land plants and carbonate-secreting organisms, vegetation cover, precipitation, and others. These uncertainties reflect the importance of the geological processes addressed in Chapter 4.

Modern Challenges and Paleoclimatology

Most research in paleoclimatology is carried out because understanding climate variability over all timescales sheds light on interactions among parts of the climate system and the causes of climate change as already described here. In this section, we introduce six specific topics that pose formidable challenges to the broad climate-research community and, consequently, they form the rationale behind a large segment of paleoclimate research. These topics are not mutually exclusive of one another and encompass interrelated themes described in later chapters.

Climate Sensitivity

Globally averaged CO_2 concentrations were 280 ppmv during preindustrial time, are near 385 ppmv today, and will reach or surpass 500–600 ppmv sometime in the 21st century. Consequently, it is of utmost importance to estimate the global temperature response to greenhouse gas forcing. This response—called the earth's *climate sensitivity* to radiative forcing caused by CO_2 and other greenhouse gases (e.g., methane, CH_4, nitrous oxide, N_2O)—is an elusive measure, and quantifying sensitivity has been a major obstacle to predicting future climate for the past few decades. Charney et al. (1979) pointed out the wide discrepancy among climate model sensitivity of mean annual temperature to a doubling of CO_2 concentrations. An even greater discrepancy exists in model-based estimates of low- and high-latitude temperature sensitivity (Rind 2008).

Sensitivity is expressed as the change in global temperature (T) due to doubling of atmospheric CO_2 concentrations, or

$$\Delta T_{2 \times CO_2}.$$

There are several ways to define climate sensitivity (Gregory et al. 2002; Annan and Hargreaves 2006). One widely used definition is the equilibrium response in global mean surface temperature (ΔT) that would result from radiative forcing at the top of earth's atmosphere because of a doubling of atmospheric CO_2 concentrations. Forcing is often expressed as

$$-\lambda \Delta T \ (t).$$

The term lambda, λ, or the climate feedback parameter, is critical in the estimation of sensitivity. Lambda incorporates feedback due to changes in albedo, atmosphere (clouds, water vapor), hydrological budget (including the cryosphere), equatorial temperature gradients (polar amplification), sea-level change, and other mechanisms to reach an equilibrium climate state. A slightly different view of climate sensitivity considers forcing at the troposphere-stratosphere boundary (tropopause).

Climate sensitivity is evaluated using coupled atmosphere-ocean general circulation model simulations of the climate response to radiative forcing by solar and volcanic activity as well as by greenhouse gases. Estimates of $\Delta T_{2 \times CO_2}$ are usually expressed as a probability distribution with a range of 1.5–4.5°C and a central value of 2–3°C (Forest et al. 2002; Boney et al. 2006; Randall et al. 2007; Roe and Baker 2007). The wide range reflects uncertainty in feedbacks from clouds, heat uptake and circulation by oceans, and continental land-atmosphere interactions. Many researchers believe that a certain level of future warming from greenhouse gas emissions produced since the year 1750 is already "in the pipeline" (Hansen et al. 2005). Given that a doubling of the pre-industrial CO_2 levels from continued emissions is nearly inevitable in this century, constraining climate sensitivity is obviously a high priority for climate science and for society in general. Paleoclimatology provides a natural baseline for understanding the response of global temperature, precipitation, and other parameters to the ways in which past changes in CO_2 and other greenhouse gases have helped constrain climate sensitivity.

Polar Climate

The extent of sea ice in the Arctic Ocean reached its lowest recorded level in 2007, raising serious concerns about human influence on Arctic temperatures and polar climate. In the southern hemisphere on the Antarctic Peninsula, where the rise in temperatures since 1950 have been among the largest on the planet, the catastrophic disintegration of the Larsen Ice Shelf along its eastern side between 1995 and 2002 is considered by some glaciologists to be another manifestation of the sensitivity of polar ice to climate change. The Wilkins Ice Shelf on the peninsula's western side is likewise considered vulnerable to rising temperatures.

The polar-amplification hypothesis holds that high latitudes respond to radiative forcing from greenhouse gases earlier and with greater sensitivity than do the tropics. During periods of warming, reduced snow and sea-ice cover lowers the earth's albedo, creating a positive feedback, which when combined with other factors (sea-ice thickness, clouds) tends to amplify polar-climate response to externally forced climate change (Holland and Bitz 2003; Masson-Delmotte et al. 2006). Declining Arctic sea ice, disintegration of ice shelves, and surging Greenland glaciers raise the distinct possibility that these patterns support the polar-amplification hypothesis and are signs of anthropogenic forcing.

Polar paleoclimatology has a rich tradition and has made great strides the past few years. Geological and ice-core records show that high latitudes become proportionally colder during glacial periods and warmer during global warmth. During the Last Glacial Maximum ~21,000 years ago, polar regions cooled much more (7–20°C) than the tropics (2–4°C). During past periods of global warmth, polar temperatures rose more than those in the tropics. The CAPE Project, for instance, confirmed polar amplification during the peak warmth of the last interglacial interval about 125,000–130,000 years ago (CAPE–Last Ingterglacial Project Members 2006). Even larger pole-to-equator temperature gradients occurred during the Cretaceous period (100 million years ago; Bice et al. 2003) and the mid-Pliocene epoch (3 million years ago; Dowsett 2007).

Paleoclimatologists have examined polar amplification in several large, international projects. Ice-core projects in Antarctica and Greenland have produced proxy records of atmospheric properties covering the last 800,000 and 100,000 years, respectively. Deep-sea sediment-coring projects have recovered the more than 40-million-year history of the AIS in sediments from the surrounding Southern Ocean. The ANDRILL (ANtarctic geological DRILLing) Project is now focused on the direct reconstructions of AIS history from the coasts of Antarctica (Naish et al. 2007). Deep-sea coring in the Greenland and Norwegian Seas and, for the first time, in the central Arctic Ocean itself (Moran et al. 2006; Backman et al. 2006) has extended the Arctic's climate record back almost 60 million years. High-latitude paleoclimatology yields direct information about how fast continental ice sheets grow and decay, how fast sea level can rise, how glacial meltwater influences global ocean circulation and climate, and how much the poles and the tropics warm and cool during large-scale climate changes, shedding light on pole-to-equator thermal gradients.

The Sun's Role in Climate Variability

Solar energy varied from 1366 to 1367 W m^{-2} during the last two 11-year sunspot cycles, a fluctuation of only 0.1% of total irradiance. During the Maunder sunspot minimum between A.D. 1645 and 1715, corresponding with the Little Ice Age, solar irradiance fell only 0.2–0.35% (Lean 2000). The contribution of solar forcing to global mean radiative forcing of climate since the Maunder Minimum is relatively small compared to that from greenhouse gases (Forster, Ramaswamy et al. 2007b). There is nonetheless great interest in how such small changes in irradiance can translate into detectable climate change through amplifying feedback mechanisms from stratospheric processes, tropical ocean temperature, land- and sea-ice albedo and thermal insulation, clouds, water vapor, or other factors.

Solar variability is the focus in many areas of paleoclimatology. Cosmogenic radionuclides (e.g., beryllium-10, carbon-14) produced in the atmosphere by high-energy galactic cosmic rays can be measured in sediment and ice-core samples, providing our most direct, albeit imperfect, evidence for past solar variability. In addition to variations in total solar output, cosmic-ray flux and radionuclide production rates are modulated by solar wind and the strength of the earth's geomagnetic field. Consequently, establishing direct correlation between cosmogenic isotopes and climate proxy measurements remains a high priority, although a challenging task.

Another solar-climate link comes from paleoclimate reconstructions for the past 100,000 years. Speleothems, ice cores, and lake and ocean sediment records sometimes exhibit quasi-cyclic climate patterns most commonly near 88-, 200-, and 1500-year frequencies. These frequencies resemble hypothesized solar cycles, and many researchers have argued that solar forcing influences climate at decadal to millennial timescales.

Over intermediate timescales (10^5–10^6 years), geographical and seasonal changes in solar insolation are the principal forcing agent in the orbital, astronomical, or Milankovitch theory of climate. Orbital geometry varies in its obliquity (tilt), precession of axis (wobble) orbit (precession of the equinoxes), and shape (eccentricity). These combine to produce cyclic change in insolation at various latitudes and during different seasons. Orbital theory is one of the most successful paleoclimate applications to climate research.

Over the 4.5-billion-year history of the earth, solar output has become progressively stronger by 30%, but prokaryotic organisms nonetheless evolved during the Archean Eon more than 3.5 billion years ago, in spite of reduced solar energy and extreme environments at that time. One hypothesis to explain this "Faint Sun" paradox holds that extremely high atmospheric greenhouse gas concentrations may have negated a 30% lower solar luminosity billions of years ago. These and other hypothesized solar-climate links involve complex, partially understood processes and certain assumptions.

Carbon Cycle and Climate

As mentioned above, carbon cycling is a critical aspect of climate change. The element carbon is stored in and transferred between land, ocean, atmosphere, and biological reservoirs over various timescales. Understanding the global carbon cycle is critical because human activity has put a total of ~300 gigatons of carbon (Gt C, 10^9 metric tons, = one petagram [Pg C] or 10^{15} g) into the atmosphere, almost all since the 19th century (Houghton 2007). About 2.1 Gt C equals 1 ppmv. Barring unforeseen events, it is expected that humans will add 6–7 Gt C annually this century (Hansen and Sato 2004).

What is the fate of this carbon? A full treatment of the global carbon cycle requires volumes, but many excellent sources are available (Schimel et al. 2000; Sabine et al. 2004; Forster, Ramaswamy et al. 2007b; Houghton 2007). Carbon-cycling mechanisms vary depending on the interval of time under consideration. Between 2000 and 2005, for instance, when excellent records are available, about 50–60% of carbon from human sources remained in the atmosphere, where it influenced radiative forcing and climate. About 30–31% entered the ocean carbon "sinks," affecting carbonate chemistry, productivity, and ecosystem functioning. Of the total carbon emitted by humans over the past few centuries, Sabine et al. (2004) estimated that ~48% has been taken up by the oceans. Land areas disturbed by humans are a net source of carbon, whereas natural (undisturbed) terrestrial ecosystems take up atmospheric carbon through photosynthesis and "CO_2 fertilization." Estimates for terrestrial sources and sinks vary greatly depending on methodology. For example,

Houghton (2003) calculated a release of 156 Gt C between 1850 and 2000 because of human-induced land-use changes. This amount is larger than the 38 Gt C estimated from carbon budget analyses that reflect uncertainty in undisturbed terrestrial ecosystem uptake, altered land-use sources, and ocean uptake rates. Rates of carbon sequestration, in both oceans and land areas, are governed by biogeochemical processes not fully understood, and rates are expected to change with future emissions, land-use changes, forest regrowth, terrestrial and marine ecosystem response, and ocean carbonate chemistry changes.

Paleoclimate records provide insights to global carbon cycling and climate. For example, climate variability over the past 500 million years is linked to atmospheric CO_2 concentrations on the basis of solid evidence from a variety of proxy records of CO_2 and temperature. Several periods of extreme warmth are directly associated with massive injections of carbon into the global oceans and atmosphere over several thousand years (see chapters 3 and 4 in this volume). Direct information about the global carbon cycle during ice-age (glacial-interglacial) cycles comes from measurements of atmospheric CO_2 and CH_4 from Antarctic ice cores for the last 800,000 years. This interval spans the last eight orbital climatic cycles. Atmospheric CO_2 concentrations fluctuate between a minimum of 177–200 ppmv during glacial intervals and 280–300 ppmv during the last four interglacials, including the Holocene preindustrial period. Prior to about 430,000 years ago, interglacial CO_2 concentrations were 30–40 ppmv lower than those in the last four cycles.

The discovery of low glacial atmospheric CO_2 concentrations measured in ice cores nearly 30 years ago was a landmark event that led to a large body of research on the role of ocean circulation, productivity, and nutrient cycling in the carbon cycle. Based on ice-core greenhouse gas records, we know that human greenhouse gas emissions have been injected into the atmosphere at rates at least 10 times faster than any known natural flux during even the most rapid climate changes. Methane concentrations vary as well over glacial cycles, although with a different pattern due to the atmospheric residence times, sinks, and sources for each gas.

Abrupt Climate Change

Humans have affected the earth's climate through greenhouse gas emissions and land use, but the exact nature of the future response is still unknown. As a consequence, societal and ecosystem adaptation to an uncertain changing climate has grown into an immense field at the crossroads of basic research and social policy. At the heart of adaptation is this question: How fast can mean climate state change? Years, decades, centuries, or millennia?

Paleoclimatology provides indisputable evidence for abrupt changes in at least regional climate over mere decades. It has been known for several years that atmospheric temperatures in high-latitude regions rise and fall by more than 10°C within decades. Recent analyses of high-resolution Greenland ice-core records indicate that shifts in sources of moisture from precipitation during the most rapid deglacial warming intervals occurred in only a few years and that these initiated a 50-year-long shift in atmospheric temperature (Steffensen et al. 2008). These abrupt transitions are associated with major reorganization of global atmospheric circulation, including migrations of the ITCZ.

Some abrupt climate changes were caused by the discharge of massive volumes of glacial lake water into the world's ocean in catastrophic outbursts. During glacial lake outbursts, the strength of ocean MOC was reduced in less than a century, changing equator-to-pole heat transport and mid- to high-latitude mean climate. Will future changes in MOC due to an enhanced hydrological cycle affect European and North American climate?

The Asian monsoon system also experienced several rapid perturbations altering seasonal precipitation during the last 10,000 years. Midlatitude "mega-droughts" lasting decades punctuated our current interglacial climate, signifying major climate regime shifts. In high latitudes, sea ice froze and melted over decades to centuries, affecting carbon cycling, ocean circulation, and climate. Glaciers and portions of ice sheets melted over timescales that glaciologists still struggle to understand in terms of ice-sheet dynamics and glaciological processes. During the global climate changes of the Mesozoic and Cenozoic eras, huge amounts of carbon were injected into the climate system in periods of global warmth, raising CO_2 concentrations to at least 10 times preindustrial levels. What do these mega-droughts, sea- and land-ice dynamics, and carbon-cycle anomalies tell us about future climate? Understanding past abrupt climate changes is of utmost importance for evaluating adaptation options to rising temperature, droughts, SLR, and other potential consequences of changing climate.

Detection of a Human Fingerprint

The detection and attribution (D&A) of climate change caused by human activity has grown into a large field because it is essential to separate out the various human and natural causes behind today's rapidly changing climate. Paleoclimatology provides a baseline against which to compare recent observations and climate model simulations and tease out anthropogenic influence. The most direct and widely publicized application of paleoclimatology to D&A is the northern-hemisphere SAT proxy reconstructions covering the last 1000 years (see Chapter 11 in this volume). Based mainly on tree-ring, ice-core, and a few sediment records spliced together with instrumental data, SAT reconstructions used in tandem with volcanic and solar forcing estimates and climate models give the strongest evidence to date that recent warming is due in part to greenhouse gases. D&A is no simple task, given large uncertainties stemming from internal climate variability and the nonlinear behavior of the climate system. Challenges also lie ahead for the detection and attribution of climate variables other than SAT, such as hurricane frequency, ocean acidification, ice-sheet and sea-ice decline, and many others (see Chapter 12 in this volume).

Perspective

Decision makers in government and other organizations use modern climatic observations and forecasts of future climate change generated by climate models to make major societal and economic choices. The stakes are very high. This book describes past climatic changes reconstructed from natural archives that provide, at the very least, a context for current and predicted future climate trends.

Perhaps, past climate changes also provide some lessons and guidelines regarding future climate change. One might therefore be tempted to ask whether reconstructed climate changes inferred from paleoclimate records are relevant for modern or future climate. Let us play devil's advocate and look at two situations where, at first glance, it might seem that they may not be so important. The first revolves around efforts to detect and attribute historical and current climate changes to human-induced factors already mentioned. D&A studies combine the use of observations and estimates of external forcing from greenhouse gases, solar irradiance, and volcanic activity to evaluate past climate patterns. Some D&A studies also use "pseudo-proxy" time series of internal climate variability as input in lieu of long-term observations or proxy records (e.g., Barnett et al. 2008). *Pseudo-proxy* is a relatively new term used to designate synthetic, statistically, or computer-simulated records of annual to multidecadal temperature or precipitation variability (Mann and Rutherford 2002; von Storch et al. 2004). Mann and Rutherford used pseudo-proxy data to test the performance of tree-ring-based proxy records; others have used them in coupled general-circulation modeling analyses of recent climate patterns (International Ad Hoc Detection and Attribution Group 2005; Barnett et al. 2008). A primary rationale for using pseudo-proxies instead of empirical proxy reconstructions is stated by von Storch et al. (2004): "A number of modeling studies of the evolution of the climate in the past centuries pose some questions about the reliability of empirical reconstructions based on regression methods.... This apparent discrepancy poses a question as to whether model simulations overestimate secular climate variability or regression-based reconstructions underestimate it."

Climate modelers know that model-simulated pseudo-proxy time series may not completely reflect internal variability, so various statistical methods are used to increase confidence in them. Nonetheless, the main point is that many proxy records of temperature or precipitation, or reconstructed indices of internal variability, are usually not used in D&A efforts, either because they are too "noisy," sparsely distributed, stochastic, or have limited periods of record. This means that the detection of human-induced climate change is carried to a large degree with minimal empirically derived longtime series of internal climate patterns.

As we see in chapters 10–12, there are a growing number of archive records of climate variability over the past millennia at annual to centennial timescales, overlapping and extending the instrumental record. Several paleorecords suggest that modes of climate variability known from instruments do not exhibit persistent behavior during preinstrumental periods; this adds further uncertainty to future forecasts. More generally, it is still not certain how external factors influence internal variability under different mean climate states.

Another example of skepticism toward proxy records of climate variability concerns Dansgaard-Oeschger (DO) events (see Chapter 6 in this volume). DO events are large (>10°C) and abrupt (over decades) changes in atmospheric temperature over Greenland that occurred repeatedly, every 500–2000 years, during the last glacial period (~100–20 ka). DO temperature variability has been reconstructed from several Greenland ice-core isotopic records, and other proxy records around the world correlate—with varying degrees of statistical certainty—with Greenland DO events. Most paleoclimatologists agree that DO events are a fundamental mode of millennial climate variability typical of glacial periods, albeit not fully understood.

Wunsch (2003a) suggests, mainly on the basis of statistical analyses and modern atmospheric (wind) processes, that millennial-scale DO isotopic fluctuations may signify "unstable water vapor trajectories" or "wind trajectory shifts" and not climate change. With reference to low-frequency (10,000 years) isotopic variability in Greenland and Antarctic ice cores, he suggests it may signify real climate changes but that two high-latitude data points do not provide sufficient spatial resolution to infer bipolar asynchronous behavior of the climate system.

Pseudo-proxies and DO variability provide food for thought. There is clearly healthy skepticism about both the limited spatial coverage and climatic significance of some proxy records. These examples reflect a growing synergy between climate modelers, oceanographers and atmospheric scientists, and paleoclimatologists that generates both intense scrutiny of and productive debate about the patterns and causes of climate change. It seems fair to say they raise the most basic questions: "What is climate change? How do we measure it?"

There are no simple answers to these questions. It may be that most paleoclimatologists are too busy worrying about calibrating their mass spectrometers, getting a valid age-depth model for a sediment core, or avoiding contamination when cleaning foraminiferal shells to be concerned with such issues. Still, one wonders whether the Akkadians, Maya, Anasazi, or any other ancient culture, whose demise has been linked to climate change, considered persistent Holocene droughts mere weather extremes or something more serious. Or did early Holocene coastal communities view a rising sea level that inundated continental shelves as just tidal surges, or as something larger (deglaciation)? Let the reader decide if and how proxy records from geological archives offer fundamental and relevant insight into how the integrated climate system works.

2

Methods in Paleoclimatology

Introduction

Paleoclimatology uses a diverse group of methods and strategies to reconstruct and interpret climate changes of the past. In this chapter we review major *archives* of climate data, *geochronology* and *age dating* methods, *proxies* taken from within archives that serve as surrogates of climate measurements, and *paleoclimate modeling* research. Paleoclimatology is a rapidly expanding field, drawing on advances in many widely disparate disciplines and technologies. This growth is evident in the recently published four-volume *Encyclopedia of Quaternary Paleoenvironments* (Elias 2006) and the *Encyclopedia of Paleoclimatology and Ancient Environments* (Gornitz 2009). Not surprisingly, there is also a great deal of specialization in paleoclimatology, with research communities focused on various archives of climate information—ice cores, tree rings (dendroclimatology), ocean sediments (paleoceanography), lake sediments (paleolimnology), corals (scleroclimatology), and speleothems, among others. Each specialty could merit a full text. Our emphasis is on general concepts, principles, and examples that dovetail with challenges outlined in Chapter 1 and on anticipating topics covered in later chapters.

As with any scientific endeavor, a full appreciation of methods and principles requires hands-on experience in the field and laboratory; there is no substitute for working with experienced researchers in a program designed to improve a proxy method or to recover a paleoclimate record from the field. In fact, the paleoclimate community recognizes this need. New programs have been developed to expand on traditional instruction in marine, atmospheric, and earth sciences. These include the Urbino (Italy) Summer School for Paleoclimatology (USSP, http://www.uniurb.it/ussp/), the Integrated Ocean Drilling Program-sponsored "School of Rock" geared toward hands-on deep-sea coring experience (http://www.iodp-usio.org/Education/SOR.html), the European Science Foundation EuroCLIMATE Spring School on Quaternary Timescales aimed at chronology (http://www .esf.org/activities/eurocores/programmes/euroclimate/ events.htm), and even programs on climate modeling (Chandler et al. 2005, http://edgcm.columbia.edu/). These types of resources are ideal opportunities for those seeking intensive instruction in paleoclimate methods.

Archives of Past Climate Changes

Evidence for changes in climate are preserved in what we call archives, natural features of the earth that can be sampled and analyzed using an enormous number of physical, chemical, and biological methods. The ultimate goal of analyzing an archive is to reconstruct temporal and spatial patterns of environmental change linked to regional or global climate. Paleoclimate archives are found virtually everywhere, in oceans, trees, caves, lakes, soils, etc. Together, they constitute the natural history books that record the earth's climate changes.

Sediment Records

Sediment accumulates in oceans, lakes, bogs, marshes, on land, and even in caves. Sediment sequences recovered through coring or exposed in natural outcrops are the most commonly studied archives of climate. The primary means of sediment transport include water, wind, ice, and occasionally biological agents. Water transports sediment in fluvial systems, oceans, estuaries and bays, and in lakes. Wind transports dust (aerosols) from continents, volcanic eruptions, and ocean surfaces (sea salt). Land ice (glaciers and ice sheets), sea ice, and icebergs (calved from land-based ice) are all mechanisms that transport sediment. Sedimentology is a major discipline in the geosciences, and a firm background in sediment processes is a prerequisite in paleoclimatology.

Continuous Sedimentation For our discussion here, we note that an idealized situation in which to use sediment records in paleoclimate reconstruction would involve sediment accumulated continuously at a near-constant rate over the selected period of interest. Continuous sedimentation allows researchers to design a sampling strategy of climate proxies preserved in sediments that focuses on the nature of climate variability attributable to a particular hypothesized mechanism. One example would be the analysis of slowly accumulating deep-sea sediments in the Caribbean to understand the climatic and oceanographic impact of the 10-million-year emergence of the Central American Isthmus. Sediment cores from old, large lakes—such as Lakes Titicaca in Bolivia, Baikal in Siberia, and Malawi in Africa—provide records of the last several million years of continental climate. Continuously deposited sediment sequences, with few or no temporal breaks (called depositional hiatuses) due to cessation of sedimentation, erosion, or both, provide the most complete record of environmental change during a particular geological interval. Although such ideal conditions are rarely met completely, continuous sedimentation is one guiding principle for site selection of key sediment records described in this book.

Sedimentation Rate The rate at which sediment accumulates is another important factor because this determines the temporal resolution of a proxy record derived from a sediment sequence. For example, in many deep-sea regions, hemipelagic sedimentation averages about 2–4 cm per 1000 years, allowing researchers to reconstruct paleoceanographic history over timescales of tens of thousands to millions of years. Except in the deepest parts of the ocean's abyssal plains (e.g., the deep Pacific Ocean), where calcite is dissolved and red clay dominates sediment, most hemipelagic sediment covering the seafloor away from continental margins is biogenic in origin, consisting of shells of planktic foraminifera, coccolithophores, siliceous microfossils, organic-walled organisms (e.g., dinoflagellates), and, to a lesser degree, benthic foraminifera, ostracodes, and pteropods (planktic mollusks). Smaller contributions to deep-sea sediments come from windblown dust, silt transported from continental margins, and ice-rafted debris (IRD), which is especially prominent in polar and subpolar seas and during glacial intervals. Slowly accumulating hemipelagic sediment provides a firm foundation for Mesozoic and Cenozoic paleoclimatology.

In the past 20 years there have been major efforts to improve the temporal resolution of deep-sea paleoceanographic records. Sediment "drifts"—deposits in regions of the oceans where sediment has accumulated at rates of ~10–15 cm per 1000 years, at least for certain intervals—have become the focus of paleoclimatologists seeking to understand millennial-scale variability (McCave et al. 1995; Hunter et al. 2007). Wold (1994) outlines the Cenozoic history of the major North Atlantic drifts—the Eirik, Bjorn, Feni, Gardar, Hatton, Snorri, and Gloria Drifts—that originated at various times between the Eocene-Oligocene boundary (34 million years [Ma]) and the Pliocene (5–1.8 Ma).

One complication when interpreting sediment drift records is that, in contrast to typical low-sedimentation-rate hemipelagic sequences, material introduced into high-rate drift areas from other regions via bottom currents can obscure the regional paleoceanographic signal from locally derived plankton shells and other proxies (McCave et al. 2002). On the other hand, silt-sized material can yield important clues about bottom-current direction and velocity and thus add another proxy about deep-sea circulation (Ohkouchi et al. 2002). Complications notwithstanding, sediment drifts have added a new dimension to paleoceanography at millennial timescales.

Near ocean margins, sedimentation rates are often even higher because of the influx of sediment from adjacent land areas. Some ocean-margin sediment records allow researchers to investigate ocean history at centennial-scale resolution. In addition, because of their proximity to continents and preservation of terrestrial proxies such as pollen, continental-margin sediment records also permit the direct integration of ocean and terrestrial paleoclimate records for a particular region. Downslope sediment transport, sea-level changes, and other ocean processes can cause unconformities, changes in sedimentation rates, and other problems, requiring caution in core-site selection and analysis.

Bioturbation
Another factor influencing the temporal resolution of any sediment record is the activity of organisms that live on or in sediment and bioturbate or that burrow after sediment accumulates on the seafloor or lake bottom. Bioturbation tends to mix sediments, blurring the temporal resolution of paleorecords. Other factors being equal, the slower the sediment rate, the deeper organisms bioturbate, or both, the more substantial the time averaging of sediment deposited over a particular interval.

Rhythmites
Rhythmite is a general term used to refer to regularly deposited sediment sequences consisting of two or more alternating sediment types. In some semi-enclosed marine basins, the effects of bioturbation on sediment are mitigated by depletion of dissolved oxygen levels in bottom water. Hypoxic (60–120 μmol kg^{-1}), suboxic (<10 μmol kg^{-1}), or anoxic (no oxygen) bottom water prevents all or most bottom-dwelling, burrowing organisms from inhabiting

the seafloor, depending on the tolerances of individual species. Hemipelagic sediment deposited in low-oxygen environments is often laminated into distinct layers, often in regular, quasi-cyclic sequences. Notable examples of semi-isolated basins containing laminated sediments include the Cariaco Basin off Venezuela, the Santa Barbara Basin off California, and the Guaymas Basin in the Gulf of California. A distinct class of rhythmite is laminated sediment called varves—annual couplets of light and dark, fine and coarse-grained sediment formed in proglacial lakes located near ice-sheet margins. Both laminated hemipelagic sediments and varves are also used for high-resolution paleoreconstructions.

Evaporites
An entire class of sediments called evaporites consisting of minerals such as sulfates (gypsum, barite), halides (halite, fluorite), borates, and carbonates (trona) formed from dissolved salts. Evaporites form in a variety of situations, such as semi-enclosed marine or continental basins, usually in tropical or subtropical regions and often in arid climates, where evaporation exceeds precipitation. Evaporites are used widely in pre-Quaternary paleoclimatology of continental regions (Parrish 1998) and in tropical regions to study Holocene paleoprecipitation records (Hodell et al. 2001).

Paleolimnology and Terrestrial Records
In addition to ocean records, sediment records from a variety of terrestrial environments are used in paleoclimatology. Lake sediments have become an important archive of the continental record of atmospheric temperature and precipitation (Cohen 2003; Fritz 2008). Windblown loess is an important archive of atmospheric variability during orbital climate cycles (see Chapter 5 in this volume) and monsoon evolution (see Chapter 8). Wetland environments—bogs, fen, marshes, and peatlands—are also actively studied for specific elements of the climate system, such as paleohydrology, methane sequestration and production, and the global carbon cycle.

Paleosols
Another important terrestrial paleoclimate archive is paleosol. Paleosols are not, strictly speaking, sediment deposits; they are ancient soils that capture several important components of the climate system through their genesis and secondary geochemical overprinting after soils are formed and buried. For example, pedogenic carbonates found in many paleosols are used to reconstruct atmospheric carbon dioxide (CO_2) concentrations from the Paleozoic, Mesozoic, and Cenozoic eras (Retallack 1990; Ekart et al. 1999; Royer et al. 2001b). Pedogenic soil CO_2 reconstructions are essential because they cover periods of the geological record older than those covered by Quaternary ice-core CO_2 records (chapters 3 and 4). Paleosols are also used to estimate

changes in the amount and seasonality of precipitation (Retallack 2005).

Ice Cores

The study of paleoatmospheres and biogeochemical cycling comes to a large degree from the study of ice cores from the Greenland and Antarctic Ice Sheets (GIS and AIS) and smaller ice caps and alpine glaciers. Greenland and Antarctic ice-core records presently cover the last 110,000 and 800,000 years, respectively. Some smaller glaciers and small ice caps yield records back to the Last Glacial Maximum (LGM) ~21,000 years ago, but most contain records of the last 10,000 years.

Ice-core records fill many important needs in paleoclimatology and in climate science in general. They have provided a wealth of evidence about topics as diverse as global aridity and elevated atmospheric dust during glacial periods, the role of tropical atmospheric water vapor during rapid climate change, solar variability, and biomass burning, to mention a few. The discovery of natural oscillations in greenhouse gases—(CO_2), methane (CH_4), and nitrous oxide (N_2O)—from fossil air trapped in polar ice ranks as one of the most important recent advances in the earth sciences (see chapters 5–7 in this volume). Ice-core records provide a robust, unambiguous, and quantitative chronicle of natural variability in atmospheric trace gases over the last 800,000 years, against which recent changes due to human activity can be compared. In particular, ice-core greenhouse gas records provide clues about land-atmosphere methane flux and ocean-atmosphere CO_2 flux during past climate changes (see chapters 6 and 7 herein).

In addition to direct measurements of CO_2, CH_4, and N_2O concentrations, dozens of chemical species and physical properties have been measured in ice cores. These include stable isotopic composition of both the ice matrix itself and trapped molecular oxygen within the ice, cosmogenic isotopes (^{10}Be, beryllium, ^{36}Cl, chlorine), insoluble particulate matter (dust), ice and air geochemistry (soluble and insoluble anions and cations), and electrical conductivity (ECM), among others. From these, inferences can be made about past winds and atmospheric circulation changes, sea-ice dynamics, rapid atmospheric temperature change, bipolar climate change, solar activity, global biogeochemical cycles, terrestrial vegetation and marine phytoplankton activity, volcanic activity, biomass burning, wetland evolution, and many other factors. Ice-core paleoclimatology intersects almost every aspect of climate history described in the later chapters.

Ice Coring Programs Since the 1960s there have been a number a large international ice-coring programs, including the Greenland Ice Sheet Project (GISP 1, GISP 2), Greenland Ice Core Project (GRIP), North GRIP (North Greenland Ice Core Project Members 2004), and, in Antarctica, the Vostok ice-core project (Petit et al. 1999), European Project for Ice Coring in Antarctic (EPICA Community Members 2004, 2006; Spahni et al. 2005; Jouzel et al. 2007; Loulergue et al. 2008), and the Dome Fuji project (Kawamura et al. 2007; Lüthi et al. 2008). The International Partnerships in Ice Core Sciences (IPICS), a group of ice-core scientists, engineers and coring specialists, plans to recover a 1.5-million-year climate record as one of their highest priorities (Brook and Wolff 2006).

Several low-latitude high-elevation glaciers have been intensely studied by teams led by Lonnie Thompson and Ellen Mosley-Thompson of Byrd Polar Research Center, Ohio State University: the Quelccaya (Thompson et al. 1984, 1985) and Huascarán ice caps (Thompson et al. 1995b) in Peru, the Dunde ice cap, Qinghai, China (Thompson et al. 1989); the Guliya and Puruiganjri ice caps, Tibet (Thompson et al. 1995a, 1997, 2006a), and Mount Kilimanjaro, Africa (Thompson et al. 2002). The Penny, Mount Logan, Agassiz, and Devon ice caps in the Canadian Arctic have also been studied by David A. Fisher and colleagues at the Geological Survey of Canada (Fisher et al. 1995, 1998). These are discussed further in appropriate chapters in the context of climate over various timescales.

Tree Rings—Dendroclimatology

Dendroclimatology has a rich tradition in North American paleoclimatology and archaeology, beginning with Andrew E. Douglass's classic studies in the desert southwest, published in the first issue of *Ecology* in 1920. Dendroclimatology rests on key principles about tree physiology and ecology and how trees grow in response to changing climatic conditions (Fritts 1991; Scweingruber 1996). Henri Grissino-Mayer (http://web.utk.edu/~grissino/principles.htm) summarizes these principles as follows: *uniformitarianism* (in geoscience, the concept that present processes operated in the past and that they are key to interpreting past events), *limiting factors* (either rainfall or temperature is the primary factor controlling ring growth), *aggregate tree growth* (various factors influencing tree-ring growth such as age, climate, ecological disturbance, and noise, which can be "decomposed" from one another), *ecological amplitude* (essentially, the ecological niches of tree species), *site selection* (based on paleoclimate objectives), and *cross-dating and replication* (the use of multiple trees to confirm the calendar year of growth for each ring and to maximize the environmental signal, respectively). In many ways, these principles apply to most paleoclimatic research.

Tree-ring climate reconstructions have been central to recent debates about the contribution of anthropogenic greenhouse gas forcing to rising surface air temperatures (SATs) in the northern hemisphere during the past century (see Chapter 11 in this volume). They have also been instrumental in the reconstruction of natural climate variability in middle and high latitudes and high elevations, particularly in North America and Europe, to study paleoprecipitation and droughts (in chapters 10 and 12 herein).

Tree-Ring Databases   The field has exploded in the past decade, largely because of the need to extend instrumental temperature and precipitation records back in time as part of a worldwide effort to distinguish natural climate variability from anthropogenic changes. Today, an extensive repository of tree-ring paleoclimatology data is available at the National Oceanographic Atmospheric Administration's International Tree-Ring Data Bank (ITRDB) (http://www.ncdc .noaa.gov/paleo/treering.html). The ITRDB includes contributions from numerous individual researchers around the world, including almost 500 sites in the WSL-Birmensdorf Tree Ring Data base from Fritz Schweingruber (Swiss Federal Institute for Forest, Snow, and Landscape Research), the spatially gridded North American Drought record presented in the format of the summer Palmer Drought Severity Index for 286 sites from Ed Cook (Lamont-Doherty Earth Observatory, Columbia University), and many others. The related field of dendrochronology is the science of using tree rings as a means to date physical and archaeological records.

Speleothems

Speleothems are mineral (mainly calcite) deposits precipitated from groundwater in caves that include stalagtites, stalagmites, vein calcite, and flowstones. Speleothem research may be the most rapidly growing field in paleoclimatology because speleothems meet several important criteria for reconstructing climate: they are high resolution, often continuously deposited, stable isotopic, trace elements, noble gas proxies, and they have excellent chronology (from uranium-thorium dating). Unlike dendroclimatology and ice-core paleoclimatology, which began in the early and mid-20th century, respectively, speleothem paleoclimatology is a relatively new field. With a few exceptions, the majority of speleothem contributions to paleoclimatology have come in the past 10–15 years.

The most important discoveries in speleothems revolve around evidence they provide for centennial and millennial variability in tropical-subtropical monsoon systems and migration of the Intertropical Convergence Zone (ITCZ) during the Holocene epoch (see Chapter 8 in this volume) as well as absolute dating of Heinrich-event (H-event) millennial climate variability between 18,000 and 60,000 years (see Chapter 6). Some speleothem records suggest solar forcing

of centennial-millennial climate variability and are central to debates about the causes of rapid climate changes emanating in the tropics. As such, they have added a major dimension to climate dynamics during these types of rapid climate change.

In addition to speleothems, vein calcite precipitated on the walls of some subterranean cave systems contain important climate information. The most notable vein calcite paleoclimate record is the Devils Hole, Nevada record analyzed by Isaac Winograd and colleagues at the U.S. Geological Survey (Winograd et al. 1988, 1992). The Devils Hole record has played an important role in understanding orbital climate changes over the past 500,000 years.

Coral Skeletons and Mollusks

The need to understand climate variability at annual and decadal timescales has fostered research into physical and chemical records preserved in organisms that grow annual layers called growth lines or banding. Coral skeletal banding provides an archive of tropical ocean-climate dynamics for several reasons. First, many corals are colonial and sedentary, so a colony records a particular region's temperature or hydrographical history. Second, coral skeletons accumulate calcitic skeletons in annual layers that provide both chronology and a source of proxies. Coral skeletal chemistry is affected by changes in the oceanic environment caused by climate changes; these changes become incorporated into the permanent skeletal record. The main drawback of modern coral colonies is that most extend back only 50–300 years and thus they are mainly pertinent to late Holocene internal modes of climate variability (see Chapter 8 in this volume) and the assessment of human-induced warming over the last century (see chapters 11 and 12). Some fossil corals dated by uranium-thorium age dating are also used to study annually resolved records during brief intervals in the Quaternary system.

Another growing field is the study of mollusk shells and their geochemistry. One of the most useful is *Astarte islandica*, a species that lives for up to 300–400 years and contains excellent geochemical proxy records of ocean temperature and circulation (Weidman et al. 1994; Wanamaker et al. 2008).

Geochronology

In the field of paleoclimatology, it is often necessary to think in terms of timescales typically reserved for geologists and astronomers. Geological time is not, however, limited to slow processes or multimillion-year timescales. Decadal, annual, and even seasonal timescales are as important as millions of years in the field of geochronology as applied to climate changes of the past. Here we introduce three aspects of geochronology: *the geological timescale, dating and correlation methods*, and chronological *conventions* used in paleoclimatology. Just as one needs a background in sedimentology to study sediment records, hydrology and geochemistry to study speleothems, ecology and physiology to study tree rings, and chemistry and physics to study ice cores, geochronology requires a background in stratigraphy and geological dating methods.

Geological Timescale

The standard framework for any discussion of climate changes of the past is the geological timescale. The International Commission on Stratigraphy, a subgroup of the International Union of Geological Sciences, provides the most recent standard chronostratigraphic scheme, shown in Figure 2.1, adopted from Gradstein, Ogg, and Smith's recent timescale revision (Gradstein et al. 2004).

Chronostratigraphy is the branch of stratigraphy devoted to aspects of geology dealing with time and the temporal relationships of sediments, ice cores, and other geological features. The units in Figure 2.1 represent rocks, sediments, and other features formed during a specific period in geological time. Many chronostratigraphic terms were first introduced during geological investigations in the 19th and early 20th centuries and are still in use today, only with much better age control.

The timescale shows standard names for geological periods, epochs, and ages/stages, the paleomagnetic polarity chrons delineating flip-flopping of the earth's magnetic field, and stage boundary ages based on radiometric dating. The term *chronozone* refers to a chronostratigraphic unit without rank in the formal hierarchy (sediments, ice, speleothems, etc.) that formed during a certain interval of time. The Younger Dryas (YD) chronozone between ~12.9 and 11.5 kiloannum (ka) is an example (see Chapter 7 in this volume). In addition to chronostratigraphic terms, paleoclimatologists use cyclostratigraphic and climatostratigraphic terms defined on the basis of geological-climatic features dominated by their apparent cyclicity and the predominant climatic control of their genesis, respectively.

We have arranged later chapters sequentially from older to younger periods of geological time: Chapter 3 covers the Precambrian through Cretaceous Periods, Chapter 4 the Cenozoic, chapters 5–10 various intervals during the Quaternary system (the last 1.81 Ma), and chapters 11 and 12 the last 2000 years, the later part of the Holocene interglacial. The last 2000 years include the Anthropocene, that recent period of the earth's history when human activity has influenced the global environment and climate. The definition of the Anthropocene is under debate; it covers the last two cen-

turies or the last few millennia, depending on which hypothesis about the duration of human influence on climate one accepts (see chapters 10–12 herein).

The status of the Quaternary system and the Holocene epoch, two of the most important periods for paleoclimatology, is in flux (Clague 2005). Working Groups of the International Commission on Stratigraphy will soon decide on how to define their stratigraphic boundaries and ages. At issue is whether to define the Quaternary as a "climatically bounded" interval or a chronostratigraphic unit using more traditional criteria—a physical point in a stratigraphic section (a boundary stratotype) and clear criteria (lithological, paleontological, geochemical, etc.)—to identify it. The first option would define its base coincident with the European Gelasian Stage at 2.59 Ma (Figure 2.1). The conventional definition for the base of the Quaternary is 1.81 Ma, coincident with the base of the Pleistocene as defined by its boundary stratotype in Vrica, Italy. Pending decisions by these groups, we adopt 1.81 Ma for the base of the Quaternary. We use 11,500 calibrated years for the base of the Holocene epoch coincident with the end of the YD climate event and the beginning of the Preboreal interval, which are widely recognized in Greenland ice cores and other marine and terrestrial records from the North Atlantic region (Björck et al. 1998; see also chapters 7 and 8 in this volume).

Dating and Correlation Methods

Geochronology provides a framework to analyze sediment, coral, ice-core, tree-ring, and other archive records in the *time domain*. Knowing the age of a particular proxy record and how to correlate it to those from other regions is essential for understanding causes of climate change. Several major categories of age dating and correlation methods are described here, moving from longer to shorter timescales (Walker 2005).

Radiometric Dating   Radiometric dating involves measurements of naturally occurring products of radioactive decay in sediments, basalts and other geological archives. Atoms of many elements, especially heavier ones with atomic numbers higher than 83, contain several forms called isotopes that have different atomic weights because of additional neutrons in the nucleus. In one type of radioactive decay of unstable isotopes, an alpha particle (α, 2 protons, 2 neutrons) is emitted, forming a new element and reducing both the atomic number and mass. Beta (β) emission involves the conversion of a neutron to a proton, thus increasing the atomic number of the daughter product by one.

The rate of radioactive decay for any unstable isotope is a characteristic of the atom's physical properties, not the

Geologic Timescale

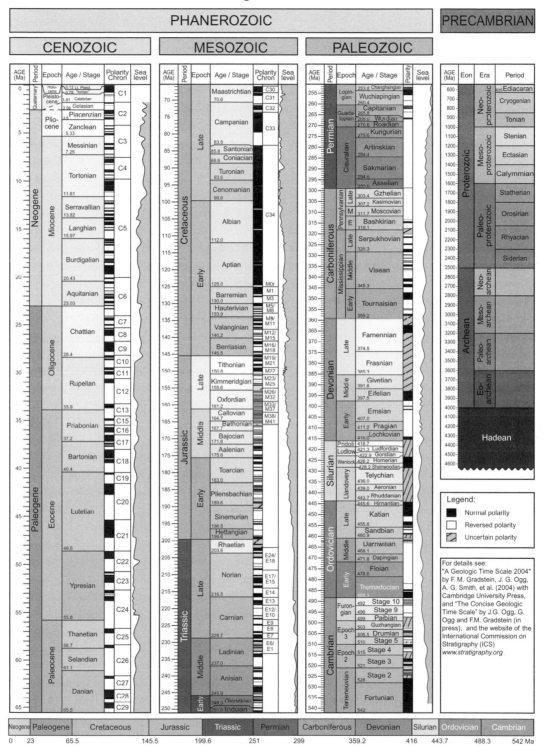

FIGURE 2.1 Geological timescale from the International Commission on Stratigraphy. The status of the Quaternary is not yet decided. Its base may be assigned as the base of the Gelasian and extend the bases of the Pleistocene to 2.6 Ma. The "Tertiary" comprises the Paleogene and Neogene and has no official rank. Courtesy of F. Gradstein, copyright © 2008 International Commission on Stratigraphy.from Gradstein et al. (2004).

environmental conditions, and is expressed as a probability of decay as

$$dn/dt = -\lambda\, n$$

where n is the number of atoms at any time, lambda (λ) is the decay constant, and t is time. The decay rate determines the age range over which a particular isotope is most useful in geological age dating. The half-life of an isotope is the amount of time required for half the original number of atoms in the isotope to decay into the daughter product. Half-life ($t_{1/2}$) can be expressed as

$$t_{1/2} = \ln 2/\lambda.$$

One of the most important radioactive decay series is the uranium-thorium-lead decay series in which the parent isotope ^{238}U (atomic number 92) decays through a series of steps including ^{230}Th (atomic number 90), ultimately to ^{206}Pb (atomic number 82). This is often called the U-series dating method. The half-life of the decay series from ^{238}U to ^{206}Pb is about 4.5×10^9 years.

Over long timescales, uranium-lead analyses of zircons from volcanic rocks are one preferred way to date Neoproterozoic "Snowball Earth" deposits and Paleozoic and Mesozoic records, as described in Chapter 3 of this volume (Hoffman et al. 2004; Gradstein et al. 2004). If there are no volcanic rocks in a particular region, some Neoproterozoic and Paleozoic sediments can also be dated by ^{187}Re (rhenium)–^{187}Os (osmium) radioisotopes in black shale and other sediments (see, e.g., Kendall et al. 2004; Selby and Creaser 2005). Increasing numbers of radiometric dates have contributed to, but not completely resolved, debates about the number, geographical extent, synchroneity, and causes of Neoproterozoic glaciations (see Chapter 3 in this volume).

A common application of the uranium-series dating uses ^{234}U, ^{230}Th, and $^{231}Protactinium$ isotopes to date Quaternary material. For example, uranium-thorium (U/Th) concentrations are used to date mid-to-late Pleistocene and Holocene coral reefs, speleothems, and vein calcite. One important criterion in U/Th dating of corals and other material is that the material dated has remained a closed chemical system since formation; that is, it has not been diagenetically altered and has remained closed to episodic or continuous uranium uptake or loss since formation. Geochemical modeling can, however, help constrain U-series ages on samples suspected of open-system chemical behavior (Cheng et al. 1998). U/Th dating has been applied to orbital-scale climate variability of the past 500,000 years (see Chapter 5 in this volume) and late Pleistocene and Holocene millennial-scale variability (see chapters 6 and 7).

Potassium-argon dating ($^{40}K/^{40}Ar$, K/Ar dating) can be used to date various minerals in mica, clays, volcanic deposits (tephra), and evaporitic sediment (McDougall and Har-

rison 1999). The half-life of ^{40}K is 1.248×10^9 years, so it is a useful method back to the Precambrian era. The argon-argon ($^{40}Ar/^{39}Ar$) dating method is a newer technique developed to improve the accuracy of K/Ar dating, although it only provides a relative age that must be calibrated using K/Ar dating. Ar/Ar dating is applied to timescales over tens of millions of years and is especially important in dating Mesozoic- and Cenozoic-age igneous rocks related to periods of global warmth (see Chapter 3 in this volume).

Absolute ages from $^{40}Ar/^{39}Ar$ dating and other methods, combined with standard paleomagnetic polarity reversal patterns and biostratigraphic events, form part of the integrated Cenozoic timescale already discussed. For example, combination of $^{40}Ar/^{39}Ar$ dating with astronomical tuning has reduced the absolute uncertainty from >1% to 0.25% for key Neogene sections in Morocco and pinpoints the age of the Cretaceous-Tertiary Boundary at 65.95 Ma (Kuiper et al. 2008).

Radiocarbon Dating Natural, unstable radiocarbon (^{14}C) is produced in the atmosphere from cosmogenic ray bombardment of molecular nitrogen (N_2). Once formed in the atmosphere, carbon is eventually taken up in carbon monoxide (CO) and then CO_2 before decaying as

$$^{14}CO_2 \rightarrow {}^{14}NO_2 + \beta.$$

Radiocarbon has a half-life of 5568 ± 30 years (the original half-life estimate of ^{14}C discoverer W. Libby; the more recent "Cambridge" half-life is 5730 ± 40 years). This means that radiocarbon can be used to date material back to about 50,000 years, typically using the accelerator mass spectrometry method.

Radiocarbon dating has complications not relevant to other radiometric dating techniques because the element carbon, including both the lighter, stable (^{12}C) and the unstable (^{14}C) isotopes, has complex biogeochemical pathways in the atmosphere. For example, radiocarbon is stored in $^{14}CO_2$ (carbon dioxide gas), $Ca^{14}CO_3$ (carbonates), and $^{14}C_6H_{12}O_6$ (organic molecules, photosynthetic products). As a consequence, ^{14}C-dating can be applied to a variety of carbon-containing material (wood, bone, charcoal, foraminifera, mollusks, bulk sediment, etc.).

A radiocarbon "date" therefore refers to the age of the carbon in the dated material and not necessarily the age of the material itself. This is due to the time it takes for carbon to circulate through various parts of the carbon cycle. Radiocarbon ages are thus routinely converted to an absolute age, or what is called a calibrated or "calendar" year timescale.

Calibration of radiocarbon dates is not a routine matter, for three primary reasons—changes in *production rate* in the atmosphere, global and regional *reservoir effects* due to the residence time of radiocarbon in various parts of the climate system, and *changing reservoir effects* due to climate and ocean circulation. The production rate of atmospheric ^{14}C varies according to changes in solar activity and the earth's magnetic field (Stuiver and Quay 1980), and these changes must be computed and factored into calibrations. Efforts to quantify changing production rates utilize other cosmogenic isotopes from the same Greenland ice cores and paleomagnetic intensity records as independent proxies of solar irradiance flux. For example, Muscheler et al. (2004) concluded that, once production-rate changes estimated from ^{10}Be and paleointensity records (Laj et al. 2002 are factored in, measured changes in ^{14}C activity in the Summit Greenland ice core between 25 and 10 ka can be attributed to carbon-cycle changes associated with millennial and centennial climate variability (see chapters 6–8 in this volume).

Reservoir effects add complications because ^{14}C is taken up into various components of the global carbon cycle, all of which have complex residence times subject to change during climate events. Take the example of marine organisms that secrete a calcium carbonate ($CaCO_3$) shell from dissolved ions in seawater. After CO_2 gas dissolves in seawater, it combines with water (H_2O) to form carbonic acid (H_2CO_3), which releases hydrogen ions into the water to form bicarbonate (HCO_3^-) and carbonate ions. Mollusks and other calcite-secreting organisms fix bicarbonate with calcium (Ca^{+2}) to produce ($CaCO_3$):

$$Ca^{+2}(aq) + 2HCO_3^-(aq) \leftrightarrow CaCO_3 + CO_2 + H_2O.$$

The carbon captured in the process of shell secretion enters the ocean mainly in high latitudes at sites of deep-water formation. Following atmosphere-ocean transfer, carbon circulates in ocean basins over roughly 1000–1500 years until it contacts the atmosphere in upwelling regions. The mean age of dissolved ocean ^{14}C is about 440 years (the global marine reservoir effect), but carbon reservoir effects vary spatially and temporally because of local and regional atmosphere-ocean processes. Carbon in some deeper parts of the ocean is more than 1000 years old because of the slow circulation of deep water. The true age of dissolved carbon in the modern ocean, therefore, varies greatly geographically and bathymetrically (Broecker et al. 1985). Stuiver and Braziunas (1993) used the term ΔR to refer to regional differences in reservoir effects.

The third complication is that the mean ocean reservoir effect can decrease during periods of rapid climate change when rates of deep-water formation diminish and the rate of ocean uptake of ^{14}C slows. Bondevik et al. (2006) provide compelling evidence from paired shell and wood ages that the ocean's radiocarbon reservoir age increased early in the YD climate reversal from 400–600 years, remained stabilized for 900 years during the rest of the YD, and then rap-

idly (within a century) fell again by 300 years at the YD-Holocene transition ~11,500 years ago.

Radiocarbon Calibration Programs Despite these complications, radiocarbon calibration has made great strides in the past decade. Marine and terrestrial calibration curves go back about 50,000 years, with greatest accuracy for the last 11,000–12,000 years. In the case of dating terrestrial material such as wood or carbon from lacustrine sediments, an atmospheric calibration curve is used to convert radiocarbon to calendar-year ages. Atmospheric ^{14}C is taken up directly into organic matter by terrestrial vegetation, so records of ^{14}C from tree-ring cellulose produce a useful calibration of ^{14}C ages from terrestrial material to a calendar-year chronology (Stuiver et al. 1998).

Calibration of dates from marine material (e.g., foraminifera, mollusks) incorporates the global marine reservoir correction and ΔR values available for some regions. Several efforts have quantified the global and local reservoir effects for marine radiocarbon dates. These include comparison of paired U/Th and radiocarbon dates from the same fossil coral specimens (Bard et al. 1990b; Fairbanks et al. 2005), radiocarbon dating of paired benthic and planktonic foraminifers from deep-sea sediments (Broecker et al. 1990a; Keigwin 2004), and comparison of annual tree-ring records (Becker 1993; Kromer et al. 1998) and lake sediment (Kitagawa and van der Plicht 1998) records with high-resolution marine records (Hughen et al. 2004a, b), dating paired terrestrial and marine material from marine cores (Bondevik et al. 2006; Richard and Occhietti 2005; Cronin et al. 2008).

There are now several calibration programs used to convert radiocarbon ages to calendar-year timescales: the IntCal program from Minze Stuiver and Paula Reimer at the University of Washington and Queens University (Stuiver and Reimer 1993) and the University of Oxford (http://c14 .arch.ox.ac.uk/), and the Lamont-Doherty Earth Observatory (LDEO) program developed by Richard Fairbanks (http://www.radiocarbon.ldeo.columbia.edu/research/ radiocarbon.htm). A local reservoir correction database is also available from Queens College (http://calib.qub.ac.uk/ marine/).

IntCal 98 was initially a tree-ring calibrated curve extending back to 11.8 ka, with lower resolution to 24 ka. The newer IntCal 04 calibration curve for terrestrial samples greatly improves the resolution back to 26 ka (Reimer et al. 2004). IntCal 04 for marine samples extends the calibrated ^{14}C timescale back to 26 ka at a higher resolution, using a combination of tree rings and carbon modeling (to 10.5 ka), foraminiferal ages from laminated sediments from the Cariaco Basin off Venezuela, and uranium-thorium dated corals (to 26 ka) (Hughen et al. 2004a, b). The LDEO program mainly uses coral uranium-series to extend the IntCal calibration from 12.4 ka back to 50 ka (Fairbanks et al. 2005; Chiu et al. 2007). In addition, the LDEO curve incorporates a global carbon model designed to determine the true age of the world's carbon reservoirs in the preindustrial and glacial ocean (Butzin et al. 2005). Hughen et al. (2006) provide a comparison between the Cariaco Basin ^{14}C records and U/Th-dated coral (Bard et al. 2004; Cutler et al. 2004; Fairbanks et al. 2005) and speleothem records from Hulu Cave, China (Wang et al. 2001) for the interval 10–60 ka.

Cross-correlation radiocarbon chronologies using U/Th-dated speleothems and corals and cosmogenic isotope records, combined with improved understanding of the carbon cycle through process and modeling studies, has led to an understanding of carbon cycling and millennial climate variability not anticipated even a few years ago. Current international coordination efforts to incorporate the rapidly growing radiocarbon database, standardize radiocarbon chronologies, and improve calendar-year age models beyond 26 ka (Reimer et al. 2004; Reimer and Hughen 2008) will significantly improve chronologies for both terrestrial and marine paleoclimatology over the next few years.

Lead-210 and Cesium-137   Two radioisotopic methods are commonly used to date records covering the last century. These are lead-210 (^{210}Pb) and cesium-137 (^{137}Cs). Lead-210 is a short-lived isotope (half-life 22.8 years) used in ecosystem studies and to calibrate paleoclimate records to instrumental records. Cesium-137 (half-life 30.3 years) relies on the spike in cesium produced by nuclear testing that peaked in 1963 and its declining concentrations since, and is also used to date recent climate records.

Annual Layer Counting   Annual layer counting is a means to date paleoclimate archives covering relatively young paleoclimate records. Annual banding in tree-rings (dendrochronology) cover the past few decades back to ~11,000 years in European tree-ring records. Dense networks of annual tree-ring records of temperature and precipitation have become available for North America and Europe covering the last 400–500 years, with decreasing numbers of older trees back to 1000 years (see chapters 11 and 12 in this volume).

Two complications accompany some tree-ring records. These are called the "divergence" problem and the "segment length curse." Divergence is the apparent reduced sensitivity of tree growth to temperatures. It occurs in some tree species mainly in high latitudes of the northern hemisphere in rings grown over recent decades. The segment length curse refers to the influence of environmental conditions and tree physiology and biology during growth on annual growth rings and their interpretation. A regional tree-ring chronology is developed from several trees, and age-related ring variability

can potentially complicate real environmental signals in the rings. Divergence and the segment length curse are critical issues for interpreting causes of recent climate warming (see Chapter 11 in this volume).

Living coral reefs are mainly used to study tropical climate variability for the past few decades to several centuries, with emphasis on El Niño–Southern Oscillation (ENSO) climate dynamics. Sclerochronology uses the annual growth lines in corals for chronology. "Floating" annual resolution timescales are also available in fossil corals and mollusks (Scourse et al. 2006) when independent ages can be obtained by U/Th dating, volcanic markers, and other methods.

Layer counting in ice cores also provides an annual resolution in cases where snow accumulated rapidly and seasonal changes in physical or chemical characteristics of the ice can be identified. Annual layers are identified by visual stratigraphy, ECM (a measure of acidity), stable isotopic (δ^{18}O, δD), and glaciochemical (i.e., soluble ions, NH_4^+, Ca^{2+}) measurements. Using these methods requires an understanding of the processes that govern atmospheric processes over ice sheets in source regions of moisture, snow accumulation on ice sheets, the snow-firn-ice transition, and postdepositional processes such as glacier flow. In the case of Greenland ice cores, layer counting involves sophisticated procedures in which many physical and chemical proxies are measured to understand the deposition and postdepositional history of the ice. A relatively consistent annual chronology is currently available for several Greenland ice cores going back 41,000 years (see chapters 6 and 7 in this volume). The error bar on the chronology is 2–3%, depending on the interval under consideration (i.e., glacial versus deglacial).

Excellent sediment varve chronologies from the Baltic Sea region and parts of the northeastern United States also provide an annual resolution for the climate changes during the last deglacial interval (see Chapter 7 in this volume).

Volcanic Events   Volcanic ashes (tephras) constitute nearly instantaneous time markers useful for dating and correlating sediment and ice-core records. Ashes are most useful for the last 1000 years because many individual ash events from Icelandic and other volcanoes are historically dated and preserved in marine sediments in the North Atlantic region (Eiríksson et al. 2006) and Greenland ice cores (Hammer 1984; Zielinski et al. 1994, 1997). Prehistorical ashes, including some that resulted from extremely large eruptions, can also be dated using radiometric techniques and used for correlation. The Vedde ash (~12 ka; see Chapter 7 herein) and the Toba ash (~74 ka) are two of the most famous.

Astronomical Tuning   Tuning is a method used to date and correlate paleoclimatic events using computations of the changes in geographical and seasonal distribution of insolation caused by the influence of gravitation in the solar system on the earth's orbital eccentricity (shape), the precession of the equinoxes, axial obliquity (tilt), and axial precession. The orbital or astronomical theory of climate change, described fully in Chapter 5, holds that long-term changes in seasonal and geographical distribution of solar radiation strongly control, or at least modulate, global climate during much of the earth's history. Astronomical tuning is mainly used for the Cenozoic era (the last 65 Ma).

The reasoning behind astronomical tuning is as follows. Geological dating of climate events through radiometric methods is not always as accurate as astronomers' calculations of the history of the earth's orbital variations. For example, when tuning first appeared as a quantitative dating method for Quaternary orbital cycles, Andre Berger (1978; Berger and Loutre 1991) calculated that the accuracy of the astronomical calculations of their frequencies over the past 5 Ma ranges from 1% for the precession cycle, <0.01% for the tilt cycle, and about 3% for the eccentricity cycle of the earth's orbit. These errors were smaller than those for most radiometric dates. In addition to age errors, material for radiometric dating of sediment sequences is sparsely distributed in time and space.

The tuning of the geological record to the astronomical timescale was developed to test the orbital theory of climate by John Imbrie and colleagues (Imbrie et al. 1984, 1992) using mainly deep-sea oxygen isotope stratigraphy. The continuous $\delta^{18}O$ curve, anchored to key tiepoints such as the ^{14}C-dated LGM, ~21 ka, the U/Th-dated last interglacial maximum ~127 ka, and the Brunhes-Matuyama paleomagnetic boundary, was "tuned" using the frequencies of the two orbital parameters, precession and tilt (22 and 41 ka, respectively). When they observed a phase difference in the isotope curve, an adjustment was made so the curve "fit" the astronomical timescale. Shackleton et al. (1995b) summarized tuning's role in paleoclimate research: tuning (1) allows the study of leads and lags among various climate proxies; (2) yields an estimate of the amount of time represented within a sedimentary section; and (3) provides a more accurate geological timescale than that provided by traditional radiometric dating methods.

Tuning involves certain assumptions about the initial forcing by insolation changes and the climate system response due to feedbacks introduced in Chapter 1. In the case of a linear climate response, such as obliquity cycles during the late Pleistocene epoch, one assumes that the phase relationship between the climate forcing (i.e., obliquity) and the system response (i.e., the proxy record) is constant and known. Tuning is also complicated when sedimentation rate varies through the studied interval. The assumption that observed climatic cycles are the products of orbitally induced changes in insolation, though widely accepted as valid given the of-

ten spectacular physical and chemical cycles observed in the geological record, still introduces some circularity into the tuning procedure, which might eliminate some of the real climate signals not forced by orbital cycles. For example, tuning can pose problems when radiometric ages are obtained for specific climate records that conflict with astronomically tuned ages.

Nonetheless, the astronomically tuned timescale has progressively expanded back to ~500 ka (Martinson et al. 1987), 3.0 Ma (Ruddiman et al. 1989; Raymo et al. 1989; Shackleton et al. 1990), 5.0 Ma (Shackleton et al. 1995a; Hilgen and Langereis 1989; Hilgen 1991a, b; Lourens et al. 1996; Tiedemann et al. 1994). Since the publication of the insolation curves by Laskar et al. (2004) going back more than 50 Ma, many Cenozoic deep-sea records have been tuned to this timescale (e.g., Lourens et al. 2004; see also Chapter 4 in this volume).

Tuning is used mainly in marine sediment, ice-core records, and some long lake sediment records. In ice cores, direct dating of ice older than about 40,000 years requires correlation of the marine oxygen isotopic stratigraphy by means of the ratio ^{18}O to ^{16}O ratios in O_2 molecules trapped in bubbles in ice. Ice core $^{18}O/^{16}O$ records are then correlated to the marine oxygen isotope stratigraphy dated indirectly by U/Th ages on emergent coral reefs and tuning. An additional factor in ice-core chronology is the fact that ice sheets are extremely dynamic, with deeply buried sections of the ice sheet flowing large distances from the original site of snowfall and ice formation. Consequently, in addition to the oxygen isotope stratigraphy, glaciological modeling of deep ice-flow patterns is required to establish chronology of ice-core records.

Age Models Most paleoclimate studies develop an age model for a particular time series of climate proxies using dating methods appropriate for the archive and timescale in question. Creating an age model usually entails converting linear depth (distance) in a core or a physical archive (speleothem) into time (years before present) using linear or polynomial equations constructed from down core age tie points. This approach is useful, but hiatuses or changes in the rate of deposition (sediment), accumulation (ice), or precipitation (speleothems) can create age offsets in certain intervals. More recently, some researchers have applied Bayesian statistical methods to building age models from radiocarbon-dated records (e.g., Ramsey 1998; Ramsey et al. 2001).

Correlation Methods The correlation of paleoclimate records from various regions is accomplished using a variety of time markers (biostratigraphic events, paleomagnetic reversals, volcanic tephras, short-term, widespread isotopic excursions, etc.) or proxy time series (e.g., oxygen isotope stratigraphy). Correlation can be accomplished using a variety of statistical methods that take into account the error bars on age models and sampling density. A less sophisticated but nonetheless popular method is simply to visually match two proxy records that are each independently dated. This curve matching (also called wiggle matching) can be justified when there is a firm understanding of the underlying climate processes that caused the observed variability in a climate parameter.

Spectral Analyses In addition to climate patterns being analyzed in the time domain, climate variability is also analyzed in the *frequency domain* to search for periodic variability that might be linked to a specific forcing mechanism. Spectral analyses became routine during the 1960s in studies of orbital-scale climate dynamics and are now applied to most timescales encountered in paleoclimatology. It may not be surprising that a wide spectrum of frequencies has been reported in the paleoclimate literature. In many or even most cases, the physical mechanisms that might produce a particular periodicity are not clear, and in some cases a proxy's age control does not merit statistical evaluation. Many experienced researchers realize these and other limitations in the analyses of frequency and adopt a conservative view when interpreting what are apparent "cyclic" or "quasi-cyclic" patterns. We will see many examples in later chapters.

Conventions and Abbreviations

Conventions used here to designate time are yr = years, ka = kiloannum, Ma = millions of years, and Ga = billions of years (Rose 2007). These abbreviations are used to designate both "years before present" and a span of years. That is, "100 ka" means 100,000 years ago and the hyphenated term "100-ka" means a period of time 100,000 years in duration. The same convention is used for Ma and Ga, millions and billions of years, respectively. Many other terms are used to designate time in the published literature. One sees, for example, that some literature uses "a" for annum, "kyr" for thousands of years, and "yr BP" or "kyr BP" to refer to years and thousands of years before present.

Defining the term "present," as in "years-before-present" (or years ago), also raises problems pertaining to Holocene climate change over the last 11.5 ka. Radiocarbon dates are usually given either in years before the year 1950, when nuclear-bomb carbon perturbed the ratio of naturally produced radiocarbon in various systems, or in calendar years (B.C., A.D., or C.E. [common era]). Some ice-core and tree-ring studies, especially those spliced with instrumental records, count back from the year the proxy records were recovered, or from the most recent year of observation data. This

can lead to confusion when correlating recent climate events over short timescales. In chapters 11 and 12 in this volume, paleoclimate records covering the last 2000 years use age models from the original publications without standardizing to a specific date. However, in the future, the community should settle on a common approach, such as "counting" back from the year 2000 designated as "b2k" (Wolff 2007). This convention has already appeared in recent papers.

Paleoclimatology's lexicon includes an enormous number of terms so engrained in the literature as to be easily recognizable from their acronym: the Last Glacial Maximum (LGM), Younger Dryas (YD), Dansgaard-Oeschger (DO) events, and the Paleocene-Eocene Thermal Maximum (PETM) are just a few of these. Other terms have been borrowed from modern climatology and oceanography, like ENSO, sea-surface temperature (SST), and North Atlantic Deep Water (NADW), or signify a major paleoclimate program like Climate: Long-Range Investigation, Mapping, and Prediction (CLIMAP) or Pliocene Research, Interpretation and Synoptic Mapping (PRISM). These and other abbreviations are introduced as needed and a full listing is given in the frontmatter.

We make one final point on chronology before proceeding. Throughout the text and in the figures, we present age models of the original authors who applied the best methods available at the time. These and other age models are subject to change with new information and technological advances.

Proxies in Paleoclimatology

Proxies are biological, physical, or chemical measurements made in sediments, fossil shells, ice cores, trapped air, speleothems, corals, and other archives that serve as surrogates for climate parameters studied by modern climatologists with observations, instruments, satellites, and computer models. Proxies constitute the primary tools for reconstruction of climate-related parameters. We introduce basic concepts and important groups of proxy methods in this section. A more complete list of seminal papers and review articles on each method is given in the Appendix. Papers chosen for the Appendix provide the reader with an entrée into the literature where a much fuller critical discussion of the strengths and weaknesses for each method can be found. General discussions of proxy methods can be found in the following sources: for pre-Quaternary records, Frakes et al. (1992) and Parrish (1998); for Quaternary records, Bradley (1999) and Cronin (1999); for geochemical methods used in paleoceanography, Fischer and Wefer (1999), Henderson (2002), Frank (2002), and Rohling and Cooke (1999); for isotopes in continental systems, Swart et al. (1993); for speleothems, McDermott (2004) and McDermott et al. (2006); and for tree rings, Fritts (1991) and Scweingruber (1996), in addition to later chapters as appropriate.

General Concepts

The most important prerequisite for using a particular geochemical, biological, or physical proxy method is an understanding of first principles governing processes that link it to some aspect of the climate system. For example, sometimes proxies are relatively direct measurements of a climate parameter. Greenhouse gas concentrations in trapped air in ice cores are one example, although these too require consideration of complex postformation physical and chemical processes within the trapped air bubbles. Many other proxies are less direct, such as carbon isotopes and cadmium/calcium ratios in deep-sea foraminifera used as proxies for ocean nutrient concentrations and large-scale changes in ocean circulation. Every proxy method goes through years of rigorous development, and even the most mature methods, such as oxygen isotopes in foraminifers, still pose challenges. Thus, it should be kept in mind that paleoclimatologists are constantly improving the methods listed in the Appendix and that new ones are sure to be developed.

Calibration Calibration of proxies to a climate parameter is one of the most complex areas of paleoclimatology. Calibration is usually carried out through field or laboratory investigations, or both, often using statistical methods and modeling of key processes controlling the preservation of the signal in an archive. A typical example of proxy calibration is shown in Figure 2.2 for long-chained, algae-produced molecules called alkenones used in the $U^{k'}_{37}$ paleothermometry method (Herbert 2003; Conte et al. 2006. It shows independent temperature calibrations derived from water-column (Figure 2.2A) and ocean core-top samples (Figure 2.2B). A third calibration comes from laboratory culturing of alkenone-producing algae.

Calibration can also be accomplished by statistically comparing a measured proxy time series to overlapping instrumental records, called the calibration period. In some cases, calibration statistics are accompanied by *verification* procedures. The verification period refers to an interval predating the calibration period. For example, Mann et al. (1999) utilized a calibration period for surface temperature between 1902 and 1980 for tree-ring proxy records and a verification period from 1854 to 1901 (Wahl et al. 2006).

Climate Indices In addition to reconstructing specific parameters like temperature, some proxies are designed to reproduce a particular index used in modern climatology. Climate indices are themselves derivatives of atmospheric or oceanic measurements that capture certain modes of climate variability. For example, the North Atlantic Oscillation (NAO) refers to the oscillation of atmospheric pressure between the Iceland low-pressure and Azores high-pressure systems in the North Atlantic region (see Chapter 10 in this volume). NAO-related variability is closely tied to processes in the Arctic region and involves large-scale decadal changes in precipitation, SST, and other parts of the integrated climate system. The instrumental record of the NAO extends back to the mid-19th century, and paleo-indices of NAO variability have been developed from tree-ring records that match this instrumental record (Cook 2003). Some ice-core records also provide measures of NAO variability indirectly tied to the instrumental record (Fischer and Miedling 2005)

Multiple Proxies Using multiple proxies to reconstruct a more robust picture of climate variability is an increasingly popular approach. Multiple proxies are especially valuable when they are obtained from the same archive record. Hughen et al. (2004c), for example, using geochemical proxies preserved in Cariaco Basin sediments, showed that terrestrial vegetation response to abrupt climate changes took only a few decades during the last deglaciation. Methane synchronization in Greenland and Antarctic ice cores is another example. Synchronization refers to the interpolar correlation of high-latitude climate records for isotopic and other ice proxies taken from the same samples as ice-core methane measurements. It is based on the fact that methane has a short (~10-year) residence time in the atmosphere. By measuring methane concentrations and temperature proxies from the same ice-core samples, we now have convincing evidence for asynchronous climate changes in the two hemispheres during millennial-scale climate events. Multiple proxy measurements in ice cores also yield clues into the glacial-interglacial temperature-CO_2 relationship during deglaciation (see Chapter 7 in this volume).

In paleoceanography, multiple proxies allow a sophisticated record of ocean structure and circulation. For example, species of planktic foraminifera have distinct ecological requirements such that different species live at different water depths in the upper ocean. This has allowed the reconstructions of upper-ocean structure from the same samples in deep-sea cores (Spero et al. 2003). Similarly, by combining isotopic and trace-element records from same-sample benthic and planktic foraminiferal shells, time leads and lags between surface and deep-ocean temperature and circulation patterns can be evaluated (Stott et al. 2007).

Atmospheres

Stable Isotopes: General Concepts Stable-isotope ratios of oxygen ($^{18}O/^{16}O$), carbon ($^{13}C/^{12}C$), hydrogen (D, deu-

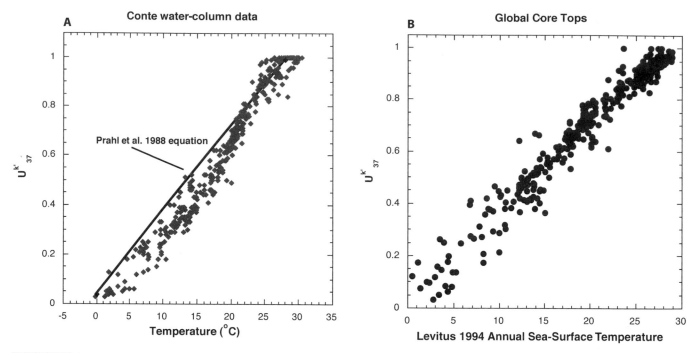

FIGURE 2.2 (A) Calibration of alkenone proxy to temperature, using sediment column samples (Conte et al. 2004). (B) Calibration to temperature, using sediment core-top samples (Herbert 2003). Courtesty of T. Herbert and American Geophysical Union.

terium, $^2H/^1H$), nitrogen ($^{15}N/^{14}N$), and other elements constitute the single most important family of paleoclimate tools used to reconstruct past climate parameters in many types of archives. A complete treatment of stable-isotope paleoclimatology is beyond our scope here; the following brief section on the terminology gives the basic tenets of the stable-isotope method and a foundation for its use in specific applications mentioned below and in later chapters. An excellent discussion of oxygen isotopes, the most widely applied method, is given in Rohling and Cooke (1999).

Stable isotopes of elements occur naturally in various proportions. Oxygen ^{16}O is the most common (99.76%), followed by ^{18}O (0.20%) and ^{17}O (0.04%). Heavier isotopes have additional neutrons giving them a larger atomic mass than lighter ones. The use of stable isotopes as proxies is based on the fact that the ratio of heavy to light isotopes of an element in any material is a function of many variables—among them, climate-related variables such as ocean temperature, hydrography, ice volume, atmospheric temperature, and moisture source.

Figure 2.3 illustrates this principle in the context of the global hydrological cycle whereby the ratios of ^{16}O and ^{18}O vary as they are transferred from one part of the climate system to another. Fractionation refers to the differential uptake of lighter or heavier isotopes during the formation and breakdown of chemical compounds because of slight differences in mass during this transfer. Two factors influence fractionation—isotopic chemical reactions, mainly re-

lated to temperature, and kinetic effects, which involve non-equilibrium processes. By identifying the dominant factors that control isotopic fractionation in a particular material, paleoclimatologists obtain a quantitative measure of secular and geographical variability in key parameters.

Several conventions are used in stable-isotope geochemistry. The ratio oxygen isotopes is expressed as

$$\delta^{18}O\ (‰) = \frac{(^{18}O/^{16}O)_{sample} - (^{18}O/^{16}O)_{standard}}{(^{18}O/^{16}O)_{standard}} \times 1000.$$

The value δ ("per mil," ‰) is the standard convention used in paleoclimatology for other elements. The original standard in the pioneering study of Urey et al. (1951) was a fossil Belemnite from the Cretaceous Pee Dee Formation of South Carolina. This sample has been replaced by what is called the Vienna Pee Dee Belemnite. Another common standard used in hydrology (the $\delta^{18}O$ in water, ice, and water vapor) and climatology is the $\delta^{18}O$ of Vienna Standard Mean Ocean Water (VSMOW).

Surface Air Temperature Air temperature at the surface of the earth is reconstructed from several proxies in ice cores, tree rings, speleothems, ocean and lake sediments. In ice cores, for example, several methods are used to estimate atmospheric temperature (Jouzel 1999). Oxygen and deute-

Processes involved with oxygen isotope fractionation

Raleigh distillation

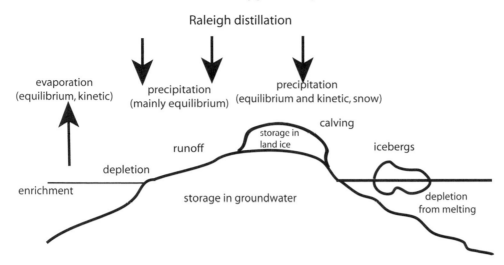

FIGURE 2.3 Schematic diagram of the global hydrological cycle as it pertains to the distribution of oxygen isotopes in precipitation, ice, seawater, and groundwater. Fractionation occurs when oxygen moves from one state to another and is controlled by equilibrium effects (mainly temperature) and kinetic effects. The fixing of the lighter isotope ^{16}O occurs in groundwater and continental ice, influencing various applications of $\delta^{18}O$ as a paleoclimate proxy. See text. Modified from Rohling (2007). Courtesy of E. Rohling.

rium isotopic values of glacial ice are sources of information on atmospheric temperature. Note that we refer here to the $\delta^{18}O$ of ice (designated $\delta^{18}O_{ice}$) as opposed to the $\delta^{18}O$ of molecular oxygen, O_2, trapped within air bubbles (designated $\delta^{18}O_{air}$), used to correlate ice-core and other climate records (see previous paragraphs). Deuterium and $\delta^{18}O_{ice}$ have a relationship with temperature (Dansgaard et al. 1993; Grootes et al. 1993; Cuffey et al. 1995) that depends on location, temperature, and atmospheric circulation. For oxygen, this relationship can range from a 0.3–0.6 ‰ $\delta^{18}O$ shift per 1°C of temperature. Classic studies of $\delta^{18}O_{ice}$ from multiple Greenland and Antarctic cores indicate robust patterns of isotopic change during late Quaternary climate changes, leading Grootes (1995:551) to conclude that "The isotopic composition of precipitation reflects primarily the temperature difference between the ocean surface at which the water vapor formed as well as the place of precipitation (air temperature at condensation level some distance above the surface). Thus the isotopes do not provide a simple, direct local temperature record, although the precipitation temperature often dominates."

However, isotopically derived temperature estimates reflect both the local atmospheric temperatures above the atmospheric inversion layer, where precipitation is formed, and temperature at the source of precipitation. This is because other factors affect oxygen isotope ratios—sea-surface conditions, cloud temperatures, the season when the precipitation fell, and changing source area and storm tracks of the moisture. In effect, the kinetic effects (evaporation and condensation) mentioned above override the isotopic equilibrium (temperature) effect more so for $\delta^{18}O$ than for deuterium. Such is the case both for Antarctic (Stenni et al. 2001) and Greenland (Masson-Delmotte et al. 2005) ice-core records.

To evaluate both site and source temperature trends, researchers calculate a value called "deuterium excess," $d = \delta D - 8 * \delta^{18}O_{ice}$ (Johnsen et al. 1989; Jouzel et al. 2007) to estimate atmospheric temperature. The value d is the difference between the two isotopes and is sensitive to SST in the source area of the moisture, relative humidity, and wind speed. Deuterium and oxygen isotopes together can track both local temperatures over the ice sheet and ocean-surface conditions near the source area, important properties that vary during periods of rapid climate change.

Three other geochemical proxies of atmospheric temperature are the $^{15}N/^{14}N$ ($\delta^{15}N_{air}$) and $^{40}Ar/^{36}Ar$ ($\delta^{40}Ar_{air}$) (Severinghaus and Brook 1999) and $^{29}N_2/^{28}N_2$ ($\delta^{29}N_{air}$) ratios from air trapped in ice (Lang et al. 1999). These methods are especially useful to document large (>8–10°C), rapid (centuries or less) temperature shifts over Greenland and Antarctica and their relationship with greenhouse gas variability by overcoming problems associated with the ice-air age difference during the firn-to-ice transition.

Surface air temperatures (SATs) in high-latitude land areas reconstructed from dense networks of tree-ring records cover the last 500–1000 years. As with paleoprecipitation reconstruction from tree rings, several prerequisites are needed for quantitative paleotemperature reconstruction from tree rings: a firm understanding of tree physiology, replication and cross-dating with multiple trees to minimize noise and confirm chronology (i.e., the calendar year of ring formation), standardization procedures to account for age and growth of trees, and instrumental records for calibration and verification.

Precipitation   Direct and indirect records of precipitation are produced from a growing number of proxy methods, including tree rings (ring width and density, cellulose isotopes), lake levels and lake chemistry (diatoms, ostracodes), speleothems (stable isotopes), sediments (pollen, plant assemblages, deuterium isotopes), paleosols, and ice cores. Paleoprecipitation reconstruction is often region specific because of the high spatial variability of global precipitation patterns over all timescales. Taking the case of dendroclimatology as an example, tree-ring widths are commonly used to reconstruct rainfall in arid and semiarid regions where drought stress can produce distinct growth patterns.

One rapidly growing group of proxy methods variously referred to as organic biomarkers includes compound-specific isotopic measurements on long-chained organic molecules. Organic biomarkers are resistant to postdepositional degradation and have great potential for long-term paleoclimate reconstruction in lake and marine sediments. An example is deuterium measurements (δD) of n-alkanes organic molecules formed in terrestrial plant leaf waxes transported to and preserved in lake sediments (Sachse et al. 2005; Smith and Freeman 2006; Schimmelmann et al. 2006). N-alkanes from leaf waxes accumulate in lakes as organic matter and monitor regional hydrological variations driven by climate. Complicating factors include plant physiology and photosynthesis: C_3 plants (95% of terrestrial plants) fractionate hydrogen more than most C_4 plants (grasses, other arid-region plant types). However, if the C_3/C_4 ratio of plant types is known, then δD is a valuable paleoclimate proxy of terrestrial climate processes. Paleoprecipitation and fluvial discharge to oceans can also be measured using high-molecular weight n-alkane and other organic biomarker molecules preserved in marine sediment as a proxy for the influx of terrestrial organic material (Pancost and Boot 2004).

Atmospheric Circulation and Wind   Changes in atmospheric circulation are inferred from a number of proxies. Some chemical species derived from aerosols include anions and cations in the soluble fraction of snow in ice cores. Aerosol particles come in various sizes: windblown dust (from continents and volcanic activity) ranges from 1–20 μm, sea salt from 8–12 μm, and pollen grains from 10–100 μm. Aerosol species affect the acidity of the snow and ice, and generally each chemical species has a different source. Some important anion species and their sources are nitrate (NO_3^-) from nitric acid (HNO_3), sulfate (SO_4^{2-}) from hydrogen sulfate (H_2SO_4), and chloride (Cl^-) from sea salt.

Changes in precipitation can affect snow accumulation such that the rate of deposition of the ice itself can be a sensitive paleoclimate indicator. The mass accumulation rate is not exactly the same as the precipitation falling on the ice because of three complicating factors (Mosley-Thompson et al. 1993): deflation, redeposition, and sublimation, all of which can result in the removal of mass (precipitation) from the ice

surface. To use accumulation and ice-layer counting as a chronological or climatological tool, Mosley-Thompson and colleagues devised a quantitative relationship to estimate the original layer thickness $L(t)$ at time (t) and to express the relationship between ice layers and accumulation rate. The value $L(t)$ is a function of the current ice layer thickness, the current accumulation rate, the thickness of the ice sheet at the core site, and the age of the particular ice layer. When variable ice accumulation rates do not correspond to decadal changes in climate, the discrepancy is likely due to differences in regional precipitation (Mosley-Thompson et al. 1993). In other cases, a positive correlation between net accumulation and $\delta^{18}O$ in several Greenland ice-core records shows that accumulation can be a valuable indicator of climate changes (Meese et al. 1994).

Particulate aerosols (dust) play a major role in global climate through their influence on atmospheric radiative properties and indirectly through their effects on nutrient flux to and carbon sequestration by the oceans. Dust is preserved in ice cores (e.g., O'Brien et al. 1995) and marine sediments (Rea 1994; Murray et al. 1995; Winckler et al. 2008) and is used to evaluate changing wind patterns, dust flux during glacial-interglacial cycles, and other processes.

Atmospheric Chemistry   A growing number of chemical measurements are used in ice cores to identify changes in atmospheric chemistry related to climate. Sulfate is a good example to illustrate the many factors that can affect the deposition of glaciochemical species in snow. There are three main sources of sulfate—sea salt, dimethylsulfide (DMS) produced by marine organisms, and volcanic activity. Each varies, independently leading to multiple sources of natural variability in atmospheric sulfate concentrations. Seasonal factors also complicate sulfate deposition in snow. In Antarctica, sulfate maxima are reached during spring and summer, minima in winter, partly because of atmospheric circulation and chemical processes. All these factors influence sulfate concentrations found in Antarctic snow and those measured in paleoclimate studies of ice. Still, processes affecting many glaciochemical species are understood well enough to apply them to climate reconstruction.

Ammonium (NH_4^+) is a soluble gas that is a byproduct of biomass burning; it can be a useful indicator of fire history on continental areas near source areas of air masses passing over Greenland (Mayewski et al. 1993; Taylor et al. 1997).

Explosive volcanic eruptions also produce nitrates and sulfates found in glacial ice. Fine sulfurous ash can remain in the stratosphere for 6–18 months; coarser particles settle within a few months (Hammer 1977). By subtracting the sea-salt background component from the sulfur in ice cores, one can calculate excess sulfur and identify historical volcanic events, which can be used for cross-checking age models in cores (Mosley-Thompson et al. 1993). Many studies have discovered volcanic products in Greenland (Hammer 1984) and Antarctic (Delmas et al. 1985) ice cores. Classic studies by Langway et al. (1985, 1995) demonstrated interhemispheric correlation of major volcanic events, and Clausen and Hammer (1988) showed that even a single event, such as the massive Tambora eruption in 1815, has impacts on the sulfate record throughout various parts of the Greenland Ice Sheet (GIS). Sulfate aerosols are also excellent proxies for evaluating anthropogenic influences on atmospheric chemistry compared to baseline variability over the last millennium (Duan et al. 2007).

Greenhouse (Trace) Gas Concentrations CO_2 is a water-soluble gas present in today's atmosphere in a concentration of 0.038%, a very small proportion relative to gases such as oxygen and nitrogen. CO_2 has an atmospheric residence time between 50 and 200 years, depending on terrestrial and oceanic sources and sinks that exchange with the atmosphere. This long residence time makes measurements of "fossil" CO_2 concentrations from ice cores a proxy for mean global atmospheric concentration.

Atmospheric CO_2 reveals clues about the earth's carbon budget, including changes in sinks and sources of carbon during climatically induced reorganizations of the terrestrial and marine biosphere. One famous example is the confirmation from ice cores that atmospheric concentrations of CO_2 gases were lower during glacial periods than during interglacials (190–200 vs 280–300 parts per million by volume, or ppmv). This discovery forced the scientific community to explain the causes of reduced glacial CO_2 levels and fostered intense research efforts on the role of the oceans in sequestering carbon through ecosystem dynamics (productivity), chemical fluxes (nutrient), ocean-circulation changes, or all of these (Broecker 1982; Knox and McElroy 1984; Siegenthaler and Wenk 1984; Berger and Kier 1985; Boyle 1988b; Broecker and Peng 1989). Research on glacial-interglacial carbon flux continues unabated today (see chapters 5–8 in this volume).

There are several proxies to reconstruct past CO_2 concentrations. Direct measurements of paleo-CO_2 concentrations come from air trapped in Antarctic ice cores for the last 800 ka. This proxy method involves analysis of the ice close-off age (the transition time from snow to firn to ice formation) and postdepositional processes such as diffusion within the air bubbles. Ice-core paleo-CO_2 records and their relationship to global climate changes are discussed in chapters 5–9 herein. CO_2 trapped in Greenland ice cores is not considered a reliable record of paleo-CO_2 because of the effects of dust and geochemical alteration of the original gas concentration in the air bubbles.

Four proxy methods are used to reconstruct paleo-CO_2 for intervals prior to 800 ka. These are $\delta^{13}C$ values from pedogenic (soil) carbonates, the density and size of fossil leaf stomatal pores, $\delta^{13}C$ values from long-chained alkenones (algal-produced organic molecules), and $\delta^{11}B$ isotopic values from foraminifera (related to seawater pH and alkalinity). Chapter 3 presents additional explanation of these methods and the paleo-CO_2 curves produced from them.

Methane (CH_4) is the second major atmospheric trace gas directly linked to biological activity and climate change. The primary naturally occurring source of atmospheric CH_4 are anaerobic bacteria in tropical and mid- to high-latitude wetlands (Senum and Gaffney 1985). Other sources of CH_4 include biotic (e.g., coal, lignite, natural gas) and abiotic (e.g., volcanic) processes, termites, animals, marine gas hydrates (clathrates) as well as permafrost (see Schaefer and Whiticar 2008). Anthropogenic methane contributions come mainly from agriculture.

The primary methane sink is oxidation by the hydroxyl radical (OH) in the troposphere, ultimately producing formaldehyde. The OH radical itself forms from solar radiation and has the effect of cleansing the atmosphere of many chemical species. The concentration of methane has increased from 715 parts per billion by volume (ppbv) in the year 1750 to 1774 ppbv in 2005.

Methane concentrations are reconstructed from trapped air in Antarctic and Greenland ice cores. Changes in CH_4 concentrations in ice-core records are a proxy indicator of tropical and high-latitude wetland activity, depending on the time interval (Chappellaz et al. 1990; Brook et al. 1996). Temperature and precipitation influence the formation of CH_4 such that, other factors being equal, the warmer and wetter the regional climate, the greater the production of CH_4. Ice-core records show that methane concentrations oscillated between 320 and 790 ppbv during glacial-interglacial cycles of the last 650,000 years (see Chapter 5 in this volume). Methane flux from wetland activity is also closely tied to short-term climate changes during glacial, deglacial, and Holocene periods (see chapters 6–8).

Nitrous oxide (N_2O) is a long-lived atmospheric gas produced by the earth's oceans and soils. In the atmosphere, nitrous oxide is removed by photodissociation from sunlight. Nitrous oxide varies in gases obtained from ice cores, and these records, combined with monitoring data, show its concentration has risen from preindustrial concentrations of 265–270 ppbv to 319 ppb in 2005. A further discussion of major and minor greenhouse gases and their role in climate can be found in Forster et al. (2007b) and in later chapters in this volume.

Oceans

Paleoceanography is the study of the history of ocean temperature, circulation, salinity, density, productivity, chemistry and its relation to all aspects of past climate changes. Paleoceanography uses a large group of proxy methods, including a growing number of geochemical tools. Henderson (2002), for example, grouped marine geochemical proxies into four categories: (1) long-chained organic molecules (i.e., alkenones), also called biomarkers, (2) stable isotopes, (3) radiogenic isotopes (tracers), and (4) trace metals. In addition to geochemical methods, there are many micropaleontological tools available to track oceanic changes from sediment core records. Here we outline ocean proxies most useful for analysis of sea-surface and bottom temperature, surface salinity, ocean circulation, productivity, and acidity.

Sea-Surface Temperature   The primary methods to extract SST records from deep-sea sediment cores are $\delta^{18}O$ and Mg/Ca in planktic foraminifers, alkenone biomarkers, assemblages of foraminifers, dinoflagellates, and diatoms, and, in shallow water, Sr/Ca and other geochemical methods in tropical corals, and in mid-depth ocean regions, mollusks.

Oxygen isotope ratios in planktonic foraminiferal calcite shells were originally used to reconstruct Quaternary SSTs because ocean temperature (Urey 1947; Craig 1965) affects the oxygen isotopic composition of precipitated calcite (Emiliani 1955). If a species secretes its shell in thermodynamic equilibrium with the seawater in which it lives (some species do, others do not), other things being equal, the greater the temperature of the seawater, the less enriched the foraminiferal $\delta^{18}O$.

It was soon realized that foraminiferal $\delta^{18}O$ values are also influenced by other factors, such as global ice volume in continental ice sheets. Continental ice volume affects foraminiferal $\delta^{18}O$ because, during glacial periods, freshwater with negative $\delta^{18}O$ values ($< 30‰$) is stored on the continents so that the isotopic composition of seawater changes (Figure 2.3). During interglacial periods there is less continental ice and proportionally more of the light ^{16}O in seawater. Thus, interglacial seawater $\delta^{18}O$ is isotopically more negative (lighter); glacial seawater is preferentially enriched (heavier) and, other factors being equal, glacial foraminiferal calcite $\delta^{18}O$ will be more positive.

This "glacial effect" (Olausson 1965; Shackleton 1967) dominates foraminiferal $\delta^{18}O$ values, especially in the tropical Pacific Ocean and deep-sea environments where glacial-interglacial temperature variations are smaller than those in high latitudes. It is accepted today that on average about two-thirds of the total observed $1.8‰$ glacial-interglacial shift in $\delta^{18}O$ of foraminifera is due to ice-volume changes. Seawater $\delta^{18}O$ also varies temporally and spatially for a variety of reasons, complicating the interpretation of foraminiferal records (Rohling and Cooke 1999; LeGrande and Schmidt 2006). Vital effects in individual species also affect the $\delta^{18}O$-seawater relationship. In particular, carbonate ion concentration (Spero et al. 1997) and disequilibrium effects influence many foraminiferal species, although this relationship is predictable in some species (Bemis et al. 1998).

In addition to isotopic studies of foraminifera, the interpretation of oxygen isotopes from other carbonates has seen major revision in Neoproterozoic and Paleozoic paleoclimatology—most notably, rethinking of the $\delta^{18}O$-temperature relationship in favor of a hypothesis that $\delta^{18}O$ fluctuations signify changes in the $\delta^{18}O$ of ancient seawater (Kasting et al. 2006; see also Chapter 3 in this volume).

Magnesium substitution for calcium in the $CaCO_3$ shell secretion is a function of temperature and other physical-chemical and biological (secretion rate) factors. Magnesium/calcium ratios in several planktic species are routinely used as a paleotemperature proxy (Elderfield and Ganssen 2000; Anand et al. 2003; Lea et al. 2003). The most important factors to consider in Mg/Ca paleothermometry are species-specific vital effects, shell-cleaning procedures, postmortem and postdepositional differential dissolution effects, seasonality in shell secretion, and still partially understood kinetic effects. Complexity notwithstanding, progress in the application of Mg/Ca paleothermometry has been rapid in the last decade, and the method has been particularly important in quantifying 2–4°C tropical cooling in the glacial ocean and reconstruction of hyperthermal ocean temperatures during the mid-Cretaceous period.

Alkenone and alkenoate biomarkers produced by haptophytic algae are an accurate method used to estimate SST between 0 and almost 30°C. Also called the $U^{k'}_{37}$ paleothermometer method for the 37 carbon atoms in organic molecules, its use in paleo-SST reconstruction reflects a broader ability of geochemists and paleoclimatologists to extract complex, organic molecules from marine and lacustrine sediments. In the case of the $U^{k'}_{37}$ method, these biomarkers are mainly temperature-dependent proxies.

In contrast to shell geochemical methods (oxygen isotopes, Mg/Ca ratios), which rely on an understanding of both biotic and physical-chemical thermodynamic relationships of calcite secretion, the $U^{k'}_{37}$ method assumes a more direct relationship between the physiological activity of the algae that produce the molecules and the water temperature in which they grow (see Herbert 2003). Another advantage is that alkenone-producing organisms require sunlight for photosynthesis and growth. Thus, like many dinoflagellates but unlike some foraminiferal species, haptophyte algae live in the upper water column and record near-surface SSTs often closely correlated to surface air temperature (SAT). The most common algae producing these alkenones are *Emiliana huxleyi* and *Gephyrocapsa oceanica*, two species of coccolithophores (the microfossil group called calcareous nannoplankton) whose calcareous plates (coccoliths) blanket the midlatitude and low-latitude ocean floor.

Prahl and Wakeman's alkenone paleotemperature equation (Prahl and Wakeman 1987; Prahl et al. 1988) is expressed as

$$C_{37:2}/(C_{37:2}+C_{37:3})$$

where $C_{37:2}$ and $C_{37:3}$ are the di- and tri-unsaturated forms of the molecule. Calibration of this ratio to water temperature comes from laboratory culturing experiments as well as sediment trap and core-top sediment analyses, and the method has produced some of the most impressive SST records in the last decade.

Another method to estimate SST is called the TEX_{86} method (Schouten et al. 2002, 2003). The method is based on a TetraEther index for organic molecules with 86 carbon atoms and is a useful proxy that is independent of salinity variations that might influence $\delta^{18}O$ values. Its application to Mesozoic and early Cenozoic deposits has led to major revision in reconstructions of SST in equatorial and polar regions (see chapters 3 and 4 in this volume).

Bottom-Water Temperature   The quantification of deep-ocean cooling during Quaternary glacial periods and long-term Cenozoic greenhouse-to-icehouse climate evolution has played a major role in understanding deep-sea circulation and decoupling the ice volume and temperature signal from benthic foraminiferal oxygen isotope records. The most commonly used deep-ocean temperature proxies are Mg/Ca ratios in benthic foraminifera (Elderfield et al. 2006) and ostracodes (Dwyer et al. 2002). Developed during the past 15 years, these trace-element proxies have used culturing and core-top calibrations and deep-sea sediment core records to confirm a 1–3°C deep-sea cooling during late Pleistocene glacial periods (see chapters 3–6 in this volume).

Salinity   Sea-surface salinity is a difficult parameter to measure directly. Oxygen isotopes in foraminifera record hydrological change due to regional evaporation, precipitation, and mixing of freshwater from melting glacial ice and river discharge, and these can be associated with changes in salinity. Algal biomarkers and deuterium excess are more direct measures of salinity variability, and many estuarine species of benthic foraminifera, ostracodes, and dinoflagellates are physiologically limited to certain salinity regimes and provide semiquantitative paleosalinity tools, especially for maximum and minimum salinity values.

Ocean Circulation   Six methods are used to reconstruct large-scale features of global circulation during Quaternary climate changes. The first two are $\delta^{13}C$ values (Oppo and Lehman 1993, 1995) and cadmium/calcium (Cd/Ca) ratios of benthic foraminifers (Lehman and Keigwin 1992; Boyle and Keigwin 1987; Rickaby et al. 2000). These were considered deep-water proxies because their variability reflects, in part, changes in bottom-water chemistry related to the source of deep-water formation. In general, there is preferential uptake of the lighter isotope ^{12}C in surface waters and this is taken up by organisms and released into seawater when decomposition occurs. As a consequence, all other factors being equal, older bottom water has a lower $\delta^{13}C$ value, and this characteristic can be observed in carbon isotope values in benthic foraminifers. Cadmium concentration in seawater is related to the nutrient phosphorus. Surface ocean biological activity takes up cadmium and phosphorous, sinks, and remineralizes, resulting in higher Cd/Ca ratios in deep water than at the surface. The concentration of Cd varies in modern Antarctic Bottom Water (AABW) and AABW water masses, suggesting these proxies might be used to reconstruct deep-water paleoceanography. Consequently, carbon isotopes and Cd/Ca ratios were first used to reconstruct Atlantic Ocean deep water oscillations over glacial-interglacial cycles under the premise that they were proxies for the relative strength in isotopically depleted, nutrient-rich AABW and isotopically enriched, nutrient-poor NADW (Boyle and Keigwin 1982; Boyle 1990, 1992; Elderfield and Rickaby 2000).

Carbon isotopes and Cd/Ca ratios are complicated by biological, ecological, and geochemical processes. In the case of $\delta^{13}C$ values in foraminifers, microhabitat environments in porewater, respiration, and photosynthesis by symbionts all influence the isotopic signal (Mackensen et al. 1993; Spero et al. 1997; McCorckle et al. 2008). Four additional proxy methods are considered more direct "dynamic" tracers of the strength of Quaternary ocean circulation. These are (1) the ratio of protactinium and thorium isotopes ($^{231}Pa/^{230}Th$) in marine sediment (Marchal et al. 2000), (2) radiocarbon activity ($\Delta^{14}C$) in marine sediment, foraminifers, and corals, (3) neodymium isotopes ($^{143}Nd/^{144}Nd$), and (4) sortable silt (SS), the fraction of deep-sea sediment between 10 and 63 microns in size. The ultimate source of SS is continents, but once deposited in oceans, its transport and deposition is governed by the strength of deep-sea currents. These proxies are most useful in evaluating the response of meridional overturning circulation (MOC) to abrupt climate changes (see chapters 7 and 9 in this volume). Pre-Quaternary paleoceanography uses a suite of radiogenic isotopes to document long-term changes in ocean circulation related to the evolution of ocean basins, continental orogenesis, mid-ocean ridge formation, and global biogeochemical cycling. Radiogenic tracers are typically analyzed in ferromanganese crusts and foraminiferal shells. Each isotopic proxy is governed by factors other than deep circulation, such as continental weathering and ocean ridge basalt (Frank 2002; see also Chapter 4 in this volume).

Nutrients and Productivity   Dissolved nutrient concentration is an especially important segment of marine chemistry relevant to the issue of carbon cycling and human-induced climate change. Proxy methods that address nutrient flux to the oceans and utilized by ocean phytoplankton have been the focus of intense research and modeling efforts since reduced glacial-age atmospheric CO_2 concentrations were first discovered in Antarctic ice cores nearly 30 years ago. This interest reflects the ocean's role in sequestering carbon from the atmosphere and the influence of glacial-age atmospheric circulation, aridity, wind patterns and dust flux, and complex changes in surface, mid-depth and deep-ocean circulation on nutrients fueling phytoplankton growth and carbon storage. Geochemical nutrient proxies include $\delta^{15}N$ of nitrate, cadmium/phosphate ratios, protactinium/thorium ratios, uranium concentration, several radionuclides, and faunal and floral (phytoplankton) assemblages.

Acidity-pH-Alkalinity   Rising CO_2 concentrations from anthropogenic activity are changing the oceans' carbonate chemistry and carbon cycling on a global scale (Royal Society of London Working Group 2005). These trends raise concern about their impact on biotic systems as well as global climate (Kleypas et al. 2006). Two major proxy methods address these concerns by studying changes in the ocean's hydrogen ion concentration (pH) and alkalinity, the ability of the ocean to take up carbon ions under constant pH, controlled by bicarbonate ion concentration. The first method focuses on boron isotopes ($\delta^{11}B$) to document changes in pH over recent centuries that are due to rising CO_2 concentrations, increased uptake of anthropogenic CO_2 by the world's oceans, and resulting ocean acidification (Pelejero et al. 2005). The second application is $\delta^{11}B$ in planktic foraminiferal shells used as an indirect estimate of atmospheric CO_2 concentrations over millions of years (Pearson and Palmer 2000; see also Chapter 4 in this volume). These relatively new geochemical proxy methods will no doubt receive increasing attention in future years.

Cryosphere

Continental Ice Sheets   Continental ice sheets play a central role in paleoclimatology. There are three distinct approaches to reconstructing the volume and spatial extent of continental ice sheets. The first is glacial geological mapping and dating on formerly glaciated parts of continents involving a large number of geological and geomorphological features formed by ice sheets. In regions adjacent to ice sheets, an entire field of "geocryology" studies a variety of periglacial features, including permafrost. Reconstruction of ice-sheet volume is accomplished through analyses of isostatically uplifted (postglacial rebound) features such as lake and marine shorelines and glacio-isostatic modeling of the earth's viscoelastic response to the weight of ice sheets (Peltier 2004). Glacial geology is discussed further in Chapter 9 in our treatment of abrupt climate change due to catastrophic glacial lake drainage.

A second approach is the use of oxygen isotopes in marine foraminifera as a proxy for continental ice volume once the effects of ocean temperature are subtracted, as previously discussed herein.

The third method uses the elevation of radiometrically dated paleoshorelines formed by coral reefs or tidal marsh peat. The current elevation of a paleoshoreline, once corrected for vertical uplift or subsidence from isostatic and tectonic processes, can provide a sea-level datum. The difference between a corrected paleo-sea-level elevation and modern sea level is then converted to sea-level and ice-volume equivalents.

The integration of glacial geology, oxygen isotopes, and paleoshorelines to estimate ice-sheet volume during the LGM has been one of the finest achievements in the understanding of Quaternary ice-age cycles (see Chapter 5 in this volume). Still, there are many sources of uncertainty: error bars on shoreline indicators and radiometric dates, regional effects on glacio-isostatic adjustment, differences in glacio-isostatic adjustment models due to mantle viscosity, and uncertainty in temperature effects on oxygen isotopes, among others.

Glaciers   Glacial moraines formed in alpine areas (e.g., European Alps, Andes, Rocky Mountains) dated by radiocarbon of cosmogenic isotopes provide important high-elevation records of glacier advance and retreat, which is often controlled by temperature and climate. In fact, alpine glaciers and snowline positions are considered among the most sensitive indicators of climatic warming in high elevations (Porter 2000a, 2001a). Alpine glacial history can also be reconstructed from sediment records in lakes adjoining large glaciers (Seltzer et al. 2002).

Sea Ice   In contrast to continental glaciers, sea ice forms and melts in polar seas over short timescales from seasons to centuries. Reconstructing sea-ice history from proxy records is an important field in paleoclimatology given the sea-ice feedbacks discussed in Chapter 1. It is also one of the most challenging because of the complexity of sea-ice dynamics and the difficulty of distinguishing between ice calved from land-based glaciers and ice sheets from ice formed by frozen seawater.

A growing number of innovative methods are used to infer changes in sea ice. Curran et al. (2003) used dimethylesulfide produced in large quantities in seawater during periods of melting Antarctic sea ice to study sea-ice extent around Antarctica. They measured methanesulphonic acid (MSA) concentrations from the Law Dome ice core to document a 20% sea-ice extent since 1950 superimposed on large-scale decadal variability. Because MSA is a derivative product of phytoplankton-produced DMS, it is a biochemical proxy directly linked to sulfur cycling in the surface ocean as opposed to sulfates produced by other atmospheric sources.

A micropaleontological method for reconstructing sea ice comes from fossil dinoflagellate assemblages (de Vernal et al. 2001). De Vernal's transfer function method is based on hundreds of modern and core-top dinoflagellate samples and allows reconstruction of the number of months per year of sea-ice cover from sediment core records.

Indirect methods to infer past sea-ice conditions include calcareous nannoplankton (Gard 1993), diatoms (McMinn et al. 2001), subpolar-polar planktic foraminifera (*Neogloboquadrina pachyderma, Globigerina bulloides*) (Dieckman et al. 1991; Hillaire-Marcel and de Vernal 2008), inorganic calcium carbonate (Jennings et al. 2002), sea-ice dwelling pelagic ostracodes (Cronin et al. 1995), sediment mineralogy (Andrews and Eberl 2007), iron oxide sediment grains (Darby 2003), and sedimentary organic matter (Gibson et al. 2003). Many studies use multiple geochemical and micropaleontological methods to relate sea-ice dynamics to oceanographic and climatic patterns.

Icebergs   Icebergs calved from glaciers and the margins of large ice sheets deposit sediment layers in oceans, forming one type of IRD. Research on IRD exploded when Bond et al. (1992) proposed that catastrophic discharge of large numbers of icebergs was a characteristic feature of abrupt millennial climate events during the last glacial period ~60–18 ka. These H-events deposited IRD layers up to a meter thick across large parts of the North Atlantic (see Chapter 6 in this volume). Researchers have since applied a suite of innovative proxy methods to Heinrich layers to determine their age, distribution, provenance (locus of origin), and relationship to large-scale climate variability (Hemming 2004).

Lithosphere

The earth's lithosphere interacts with other parts of the integrated climate system over all timescales, and geological records from the lithosphere are especially critical for deciphering long-term climate records. The most important contribution from geological evidence is the reconstruction of several critical boundary conditions necessary for interpreting and modeling pre-Quaternary climate: paleogeography, ocean-land distribution, location and bathymetry of key ocean gateways, paleo-elevations of major plateaus and mountain ranges, global biogeochemical cycles, and ocean and atmospheric chemistry. These pertain most to climate extremes over millions of years and longer (see chapters 3 and 4 in this volume).

An important component of the earth's lithosphere is the surface of continents, because land-surface conditions influence climate through various vegetation-biogeochemical feedbacks and albedo effects. Land-surface conditions are reconstructed primarily through paleovegetation analyses of pollen assemblages obtained from lake sediments and packrat middens. Paleovegetation reconstructions reveal major changes in terrestrial biomes during climate changes, notably during the LGM, the last deglacial interval, and the Holocene epoch. These reconstructions are often integrated with general-circulation climate models to evaluate patterns and causes of climate changes over millennial timescales and key feedback processes.

Paleoclimate Modeling

General Concepts

Paleoclimate modeling combines proxy reconstructions with climate model simulations and has grown into a fundamental approach in paleoclimatology for the understanding of climate dynamics (Hargreaves et al. 2006). In its most generic form, paleoclimate modeling integrates empirically derived climate proxy records with various types of climate models to simulate climate and parse out causes and feedback mechanisms. This section first distinguishes paleoclimate modeling from climate forecasting, then describes types of models, strategies for data-model analysis, and applications to challenges. Although coupled atmosphere-ocean general-circulation models (AOGCMs) represent the major class of climate models used to forecast future climate, we cast a wider net here and include brief discussions of carbon-cycle, ocean-circulation, sea-ice, glaciological, and biogeochemical models. Excellent discussions of the development of paleoclimate modeling can be found in Crowley and North (1991), Barron and Moore (1994), Parrish (1998), Saltzman (2002), Rind (2008), Cane et al. (2006), a special volume of *Climates of the Past* (Rousseau et al. 2008).

Paleoclimate Modeling, Weather, and Climate Forecasting

We begin by distinguishing the large field of climate modeling from weather prediction. Weather prediction takes an initial atmospheric state and computes future conditions, which with present capabilities cannot be predicted more than a week or so because of the chaotic nature of atmospheric circulation. Recent efforts like the three-dimensional modeling in the Estimating Ocean Circulation and Climate Program (ECCO) (Wunsch and Heimbach 2006, 2007) seek to integrate the ocean system into predictions longer than a week. Programs like ECCO shift priorities from weather prediction to understanding climate over decadal timescales and include topics such as sea-level variability (Wunsch et al. 2007).

In contrast to weather, the earth's climate state at any time is a mean condition forced by external factors and modified by many feedbacks. Climate incorporates the major boundary conditions of atmospheric and ocean circulation, land distribution, ice sheets, vegetation, and other large-scale features of the earth, each shaped over different timescales by distinct physical processes. Forecasting climate hinges on the assumption that understanding the response of these systems to radiative forcing such as greenhouse gases and all attendant feedbacks is possible, even though, as we see in Chapter 7, chaotic behavior may also explain millennial-scale climate patterns.

Next we distinguish climate modeling used to forecast future climate changes from paleoclimate modeling (Trenberth 1993; Randall et al. 2007). Gavin Schmidt of NASA categorized the physics of climate modeling into three types of computations: (1) fundamental physical laws of conservation of momentum, mass and energy, (2) processes known in theory but computed with varying degrees of accuracy (equations of fluid flow), and (3) relationships known empirically through measurements and statistics (evaporation, humidity). Schmidt points out that modelers address the latter two through what is called parameterization—the attempt to compute the behavior of small-scale phenomena though approximation or averaging, often on larger spatial scales.

Many climate models in forecasting have been used to simulate the response of the earth's climate to changing radiative balance from human greenhouse gas emissions and natural radiative changes from volcanic events and solar output. In the 2007 Intergovernmental Panel on Climate Change (IPCC) report, the results of 23 different modeling research groups were presented (Randall et al. 2007). These include simulations from the National Aeronautics and Space Administration's (NASA) Goddard Institute of Space Studies (GISS) (Schmidt et al. 2006; Hansen et al. 2007a), the National Oceanographic Atmospheric Administration's (NOAA) Geophysical Fluid Dynamics Laboratory (GFDL) (Delworth et al. 2006) model, and National Center for Atmospheric Research (NCAR) Community Climate System Models (CCSM) (Blackmon et al. 2001).

These models solve equations for physical laws, fluid motion of the atmosphere and ocean, and biogeochemical processes between adjacent Cartesian grid boxes in three-dimensional models having varying degrees of vertical and horizontal spatial refinement (Figure 2.4). In the case of the GISS GCM, simulations can be performed at several horizontal and vertical resolutions of latitude and longitude and vertical layers, the finest being 2°×2.5° latitude×longitude and more than 30 vertical layers of the atmosphere and ocean. Differences among model parameterizations of processes that operate at sub-grid scales, such as cloud formation and physics, are one reason that simulations from different modeling groups often produce different results.

Paleoclimate data-modeling research involves somewhat different strategies and an array of model types. Usually paleoclimate modeling is simpler from a computational standpoint and often targets a specific large-scale process rather than solely radiative forcing. More than 20 years ago, Eric Barron commented on the use of models in paleoceanography as follows (Barron 1986): "At the very least such models can be applied to explore the limitations or appropriateness of various assumptions in paleoceanography. At the very best, paleoceanographic data may be used as a verification of simulations of past oceans."

Barron anticipated an explosion in the modeling of past climate patterns. Today, paleoclimate modeling examines large-scale changes in boundary conditions, like the distribution of continents and oceans, elevated greenhouse gas concentrations, and the presence of large ice sheets, to examine the most basic aspects of climate dynamics over various timescales. The rest of this chapter introduces the types of paleoclimate models, strategies, and selected applications.

Types of Climate Models

Energy-Balance Models Energy-balance models (EBMs) are considered a relatively simple type of model because they are concerned with the earth's temperature and radiative balance between incoming solar radiation (mainly ultraviolet, short-wavelength) and outgoing reflected radiation (mainly infrared, long-wavelength) and its interplay with greenhouse gases and particulates in the atmosphere and their radiative properties. EBMs have various degrees of complexity, ranging from nondimensional (viewing the planetary-scale mean annual temperature), one-dimensional (in which latitudinal gradients are incorporated using zonally averaged temperature—i.e., mean temperature for latitudinal zones), and two-dimensional models (which introduce geographical features—distribution of land, ocean, and elevation—and seasonal changes in temperature.

Fully Coupled Atmosphere-Ocean General-Circulation Models AOGCMs are used mainly in the evaluation of multiple external forcing factors (greenhouse gases, aerosols, volcanic emissions, and solar irradiance) and various feedbacks described earlier. As such, one powerful application of AOGCMs is analysis of the relative contribution of human

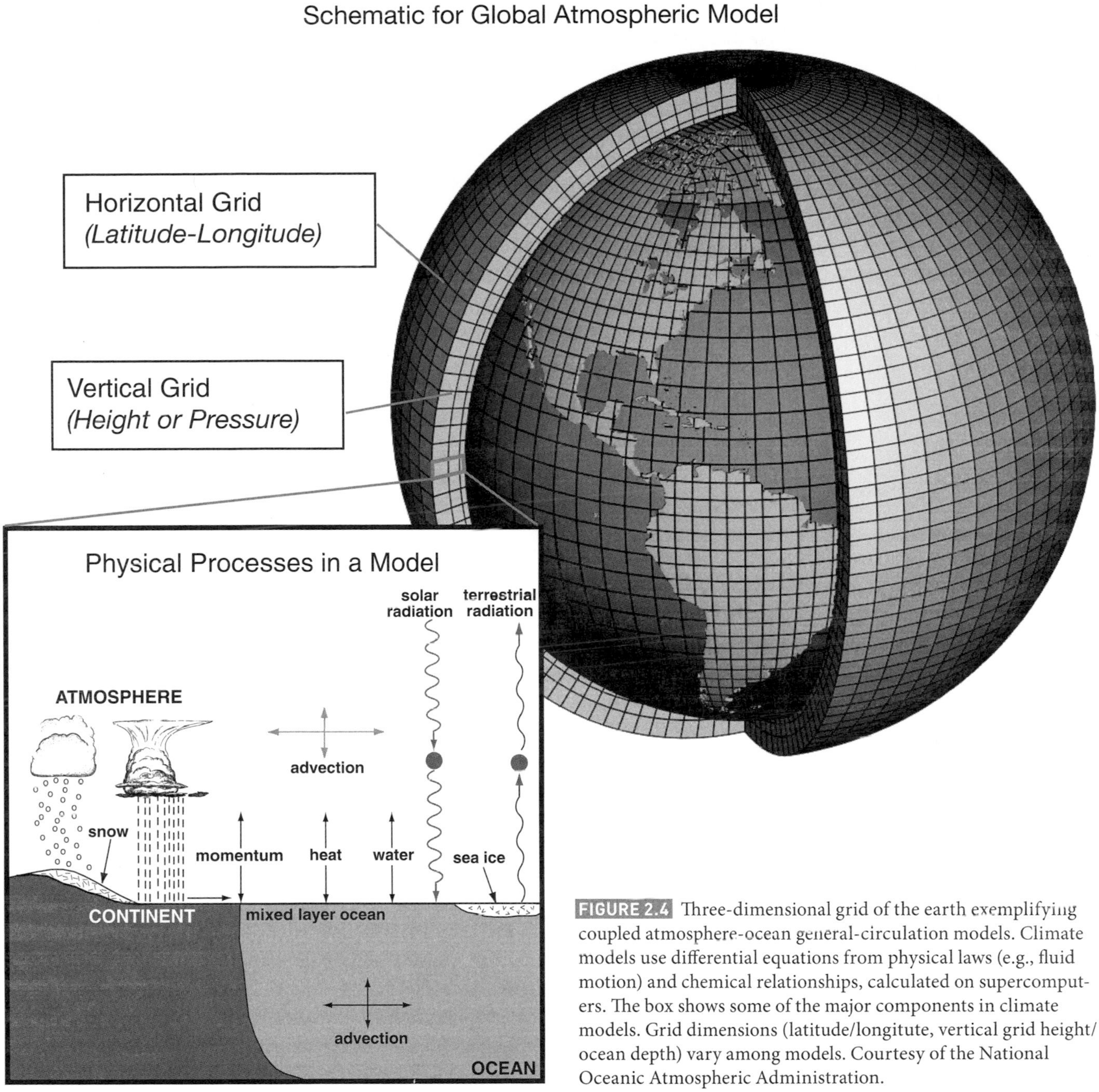

Schematic for Global Atmospheric Model

Horizontal Grid
(Latitude-Longitude)

Vertical Grid
(Height or Pressure)

Physical Processes in a Model

solar radiation terrestrial radiation

ATMOSPHERE

snow

advection

momentum   heat   water   sea ice

CONTINENT   mixed layer ocean

advection

OCEAN

FIGURE 2.4 Three-dimensional grid of the earth exemplifying coupled atmosphere-ocean general-circulation models. Climate models use differential equations from physical laws (e.g., fluid motion) and chemical relationships, calculated on supercomputers. The box shows some of the major components in climate models. Grid dimensions (latitude/longitute, vertical grid height/ocean depth) vary among models. Courtesy of the National Oceanic Atmospheric Administration.

activity to climate variability in the context of natural baseline volcanic and solar forcing. Modeling the last 1000–2000 years of proxy-based and observational temperature records from volcanic, solar, and greenhouse gas forcing is the focus of chapters 11 and 12 in this volume.

Another major application of AOGCMs is the investigation of fresh-water forcing of MOC from catastrophic discharge of glacial lake water into the North Atlantic Ocean. Abrupt freshwater discharge events occurring over only 1–10 years are an excellent way to test model perfor-

mance and climate response to future changes in global hydrology and precipitation. Consequently, model simulations provide a useful source of the short-term transient response of MOC over several centuries following a discharge event (Stouffer and Manabe 2003). Freshwater forcing in the Southern Ocean around Antarctica has also been investigated using AOGCMs (Weaver et al. 2003). Chapter 9 herein describes this paleoclimate-model synergy in depth for deglacial freshwater discharge events between about 15 and 7 ka and the general topic of abrupt climate change.

Intermediate-Complexity Models The growth of this generic class of models reflects the need to simulate large-scale and long-term climate experiments that are too expensive and time consuming to run on finely gridded, fully coupled AOGCMs in multidecadal transient simulations. Called Earth System Models of Intermediate Complexity (EMICs) by the IPCC (Randall et al. 2007), such models are increasingly used for a variety of paleoclimate applications. EMICs vary greatly in design and capability. They usually integrate separate models of different parts of the climate system (e.g., sea ice, terrestrial biomes) into a coupled simulation, often with emphasis on resolving a particular climate response. The University of Victoria Earth System Model is a good example to illustrate the components of an EMIC and the difference from fully coupled AOGCMs (Weaver et al. 2001). The University of Victoria model includes a three-dimensional ocean-circulation model coupled to thermodynamic sea ice, energy and moisture balance, a land-surface model with a single layer, and a land-vegetation model. It does not include a fully operational gridded atmospheric circulation model component. The University of Victoria model has been used to evaluate the influence of mean climate state on MOC under rising greenhouse gas concentration (Weaver et al. 2003, 2007).

Another global coupled EMIC is the ECBilt model (Goosse and Fichefet 1999) used widely in paleoclimate-data studies. For example, ECBilt has been coupled to a sea-ice model (ECBilt-CLIO) and a land-vegetation (grassland versus forest) model (VECODE) to investigate MOC and climate response to freshwater forcing (Renssen et al. 2001) and solar forcing (Renssen et al. 2007).

Biogeochemical Models Sometimes called geochemical models, biogeochemical models attempt to simulate the large-scale interaction of various components of the earth's climate system over long timescales. For example, the GEO-CARB model of Robert Berner and colleagues has 10-million-year time steps and several subroutines that compute the flux of key elements from and to the earth's lithosphere (Berner 1991, 1994). Biogeochemical models have become increasingly sophisticated, and are aimed at a variety of applications such as the study of the global flux of methane (Bartdorff et al. 2008), sulfur (Berner 2006), and oxygen (Berner 2004) over the Phanerozoic eon.

Glaciological Ice-Sheet Models Ice-sheet modeling is an immense field pertinent to many timescales. In essence, ice-sheet modeling aims to understand why, how, and when ice sheets form and decay. Like atmospheric-ocean models, ice-sheet models incorporate fundamental physical laws of conservation of mass, momentum, and heat with various parameterizations of force balance, boundary conditions between the base of the ice sheet and bedrock, and the margins with the ocean and other details that influence ice flow and temperature (Huybrechts et al. 1996; Huybrechts 2006). As discussed in Chapter 1 in this volume, many factors govern the growth and decay of ice sheets, most notably surface air temperature (SAT), precipitation, basal melting and velocity, rates of sea-level rise, location of the ice-margin grounding line, ice-sheet elevation, and viscosity changes in the ice sheet itself, among others.

Testing ice-sheet models involves comparison to either modern-day GIS and AIS or ancient ice sheets reconstructed from glacial geology and other methods. A good example of paleo-ice sheet modeling is the simulation of late Quaternary orbital-scale ice sheets by Huybrechts (2002). Huybrechts's model includes forcing of Antarctic ice-sheet behavior derived from atmospheric temperature at the ice-sheet surface, precipitation above the ice sheet, basal melting, and sea level computed as follows:

$$\text{Temperature: } \Delta T_{VA}(t) = a\,\Delta T_{VS}(t) - \Delta T_{VC}(t)$$

$$\text{Precipitation: } T_I(t) = 0.67\,T_S(t) + 88.9$$

$$\text{Basal melting: } dS/dt = 1/ts\,(S - Sr)$$

where ΔT_{VS} is temperature change at the Vostok ice-core site based on deuterium isotopes; ΔT_{VC} is a correction for ice-sheet-elevation change; T_I and T_S are mean annual temperatures above the atmospheric inversion layer and surface temperature, respectively; S is the basal melting rate in m yr^{-1}; and Sr is the equilibrium melting rate, a function of temperature (ΔT_{VA}) and several measures of sensitivity to melting. This particular model study found glacial-age Antarctic ice-sheet sensitivity to changes in the grounding line in the West Antarctic Ice Sheet and first-order model correspondence to orbital-scale ice volume and climatic oscillations over the last 400 ka.

Another type of modeling effort relevant to ice sheets and paleoclimate is Richard Peltier's ICE-5G glacio-isostatic model, which simulates the earth's glacio-isostatic response to the loading and unloading of ice sheets and meltwater during the last glacial-deglacial cycle since the LGM ~22,000 years ago (Peltier 2004). The ICE-5G model is basically an integrated geophysical model, based on geophysical equations calibrated to paleo-sea-level records, that produces global paleotopographic reconstructions with 1000-year time steps. ICE-5G allows researchers to distinguish between vertical isostatic movements caused by ice-sheet growth and decay from those changes due to global sea level. The size and thickness of LGM ice sheets, as well as estimates for mantle viscosity, continental shelf effects, and other factors determine isostatic response regions near and far from the Laurentide and Fennoscandian Ice Sheets.

The modeling of ice-sheet behavior is often integrated with the modeling of ice shelves and ice streams (MacAyeal et al. 1996; Hulbe and MacAyeal 1999; Hulbe 2001; Hulbe and Fahnestock 2007; Schoof 2007) to determine the relationship between these components of ice sheets and the larger ocean-atmosphere system. One specific activity involves efforts to couple ice-sheet modeling with general-circulation climate models to constrain the contribution from melting ice sheets to future sea-level changes, a factor poorly constrained in the IPCC 2007 report. Ice-sheet modeling projects include the European Ice Sheet Modeling Initiative (EISMINT) (Huybrechts et al. 1996) and the more recent Ice Sheet Model Intercomparison Project (ISMIP) (http://homepages.vub.ac.be/~phuybrec/ismip.html).

Ice-sheet models are most relevant to later discussions on the inception and instability of the AIS during the Cenozoic era (Chapter 4) (DeConto and Pollard 2003a, b; DeConto et al. 2008), orbital-scale glacial cycles (Chapter 5) (Huybrechts 2002; Saltzman 2002), H-events and catastrophic ice-sheet decay (Chapter 6) (MacAyeal 1993a, b), deglaciation (Chapter 7) (Zweck and Huybrechts 2003), abrupt climate changes (Chapter 9) (Rahmstorf et al. 2005), and late Holocene ice-sheet behavior (Chapter 12).

Strategies in Paleoclimate Modeling

Time-Slice Reconstructions "Time-slice" or synoptic proxy-based reconstructions are designed to run model simulations forced with a particular set of boundary conditions, such as SST, atmospheric CO_2 levels, or global terrestrial land cover, for a specific period. In time-slice studies it is assumed that the reconstructed climate signifies an equilibrium climate state including all feedback responses. Initial boundary conditions derived from proxy data are gridded, and unknown conditions, such as land-ice distribution or continental elevation, may be "prescribed" on the basis of present-day conditions.

Take the well-studied case of the LGM ~21 ka (see Chapter 7 in this volume). Large ice sheets covered parts of North America, Eurasia, Patagonia, and greater areas of Antarctica than today, global sea level was 125 m lower, sea ice reached much farther south (north) in the North Atlantic and Southern Oceans, atmospheric greenhouse gas concentrations were 30% lower than interglacial levels, terrestrial ecosystems and surface- and deep-ocean circulation were modified, and insolation reaching high northern latitudes was lower because of changes in the earth's axial tilt and precession (wobble). Proxy reconstructions integrated into a globally gridded map of LGM ocean temperature, land-surface conditions (ice sheets, terrestrial biomes), atmospheric chemistry, etc., are analyzed through climate model simulations. Many simulations of LGM climate have been performed since the CLIMAP

Project first published maps of the ice-age world (CLIMAP Project Members 1981). In the case of fully coupled AOGCMs, simulated output using a synoptic proxy database can be used to evaluate the entire climate system, including parts not directly reconstructed from proxies.

Another well-studied interval for paleoclimate modeling is the Holocene epoch. Diffenbaugh et al. (2006) combined a three-dimensional regional (RegCM3) model of the United States with a recent version of the NCAR Community Climate System Model (CCSM3) to investigate the summer precipitation response to mid-Holocene changes in insolation and ocean SSTs. Earlier modeling studies had documented the large-scale precipitation and response to insolation forcing including complex vegetation and land-surface feedbacks. The application of the latest generation of finely gridded models reveals more detail about continental summer-climate dynamics and the models are generally a good match to paleoclimate data on Holocene precipitation. Chapter 8 in this volume discusses additional paleoclimate modeling of the Holocene epoch carried out under the Paleoclimate model intercomparison projects called the PMIP and PMIP2.

Transient-Climate Simulations In contrast to global synoptic paleo-datasets, most climate proxy reconstructions constitute time series representing periods of climate change that is transient, quasi-cyclic, or both for a particular region. Until recently, modeling transient changes over longer timescales was difficult because of computational constraints and the limited spatial distribution of proxy records, sometimes limited to relatively few sites. These problems are being overcome using more sophisticated modeling approaches, including intermediate-complexity models and networks of proxy time series that are denser, more widely distributed, or both (Liu et al. 2007).

Ensemble-Model Reconstructions The use of ensemble-model simulations involves running multiple simulations to determine the most likely response in a probabilistic sense. One example is the study of the progressive Holocene intensification of ENSO by Brown et al. (2008), who examined ENSO response to mid-Holocene insolation using the Hadley Met Office (HadCM3) coupled atmosphere-ocean model. This model has 19 vertical atmospheric levels at a 3.75° by 2.5° spatial resolution and 20 levels at 1.25° for the ocean grid.

Applications of Paleoclimate Modeling

We now outline the major applications of paleoclimate modeling as they pertain to the challenges and causes of climate change already described. No one category is mutually exclusive of the others. In later chapters, we will see how each fits into the broader framework of proxy-based reconstructions and their interpretation.

Climate Sensitivity Paleoreconstructions of at least five periods of time have been used to constrain climate sensitivity to greenhouse gas forcing: the LGM, the last millennium, global warmth of the Cretaceous period, the Phanerozoic eon, and the Little Ice Age (LIA). Studies of the LGM applied to climate sensitivity go back to early studies by Hansen et al. (1984) and Hoffert and Covey (1992). Today, with improved temperature reconstructions, we know that LGM cooling was roughly 3–9°C, with the tropics cooling by ~2.5–3°C, Antarctica by 9°C, and high-latitude northern hemisphere by 11–20°C. The radiative forcing was reduced by about 3.5–4 W m^{-2} compared to today's, due to insolation, CO_2 concentrations, and feedbacks associated with lower sea level, albedo of large ice sheets, and atmospheric dust. Recently, Roche et al. (2007) expressed the paleoclimate-sensitivity link with reference to the LGM as follows: "The Last Glacial Maximum climate is one of the classical benchmarks used both to test the ability of coupled models to simulate climates different from that of the present-day and to better understand the possible range of mechanisms that could be involved in future climate change."

The LGM is ideal as a benchmark because its age is constrained by radiocarbon dating and layer counting; LGM atmospheric CO_2 and CH_4 concentrations are known from ice-core records; and continental ice sheets and vegetation, sea level, and surface-ocean temperature have been reconstructed on global and regional scales (see Chapter 7 in this volume). Crucifix (2006) ran LGM simulations using four different climate models that were also used for forecasts of future climate. His main conclusion is that global LGM temperature is not sufficient to constrain climate as a simple scaling exercise sensitivity, because of imprecise LGM forcing and LGM-CO_2 feedbacks (low-latitude clouds, midlatitude to high-latitude clouds and albedo) not being adequately resolved in climate models. In other words, we cannot assume a simple ratio between forcing and temperature over the range of boundary conditions provided by LGM paleoreconstructions.

Paleoclimate modeling studies of the last millennium contribute directly to climate sensitivity and include SAT (Hegerl et al. 2006; Crowley et al. 2008). Emphasis in these studies is placed on climate response to forcing from a combination of volcanic events and solar variability reconstructed from several proxy methods that cover the important period of cooling ~ A.D. 1400–1900 known as the Little Ice Age, attributed to the Maunder Minimum in solar activity (see Chapter 11 in this volume, and Crowley 2000; Rind et al. 2004).

Global warmth of the Cretaceous period and the Eocene epoch shed light on the following question: are estimates of climate sensitivity that are derived from short-term observations and modeling of the last century applicable to long timescales and large-scale changes similar to those some expect in the future? Early applications of paleoclimate data to climate sensitivity analysis were studies of global warmth during the Cretaceous and Eocene (Barron et al. 1981, 1995; Sloan and Barron 1992; Covey et al. 1996). Bice et al. (2006) recently focused on multi-proxy based temperature and CO_2 estimates during oceanic anoxic events (see Chapter 4 in this volume) in the peak of Cretaceous warmth to test the GENESIS climate model's ability to simulate hyperthermal conditions. They found that the GENESIS model underpredicted tropical temperatures when forced by estimated paleo-CO_2 concentrations. One possible explanation for the data-model temperature discrepancy is the influence of atmospheric methane concentrations 30 times those of today (50 ppmv versus 1.65 ppmv) during the Cretaceous period. But even with elevated methane concentrations, atmospheric CO_2 concentrations of 3500 ppmv (~10 times 2008 values) are still required to produce extreme warmth. Such integrated studies emphasize the need for improved temperature and CO_2 proxies and highlight the possibility that the sensitivity of climate models to elevated greenhouse gas forcing is too low.

There is growing evidence that atmospheric CO_2 concentrations oscillated during the Phanerozoic eon, often in parallel with global temperature. Royer et al. (2007) used the GEOCARB and GEOCARBSulf models of paleo-CO_2 (Berner and Kothavala 2001; Berner 2004, 2006) for the last 420 Ma and found that that climate sensitivity $\Delta T > 1.5$°C has been a robust feature of the earth's atmosphere. Importantly, sensitivity of $\Delta T > 6$°C is unlikely given our understanding of the long-term silicate weathering–CO_2 relationship. Royer's approach incorporates a wide range of paleo-CO_2 concentrations, and it is likely that periods warmer than today represent equilibrium climate integrated over several million years.

Finally, Annan and Hargreaves (2006) used multiple sources of observation (instrumental records, volcanic events) and paleo-data for the LGM and LIA, and Bayes Theorem, a probabilistic approach that adds new information to a previous probabilistic sensitivity estimate. They concluded that a climate sensitivity to doubling of CO_2 concentrations above 6°C and 4.5°C was unrealistic and highly unlikely, respectively. Their evaluation of the LGM indicates a sensitivity of 2.7°C with a standard deviation of 1.7°C.

In sum, as pointed out by Rind (2008), in low and high latitudes there are significant consequences regarding uncertainty about future climate sensitivity to greenhouse gas forcing that require improved paleo-data sets.

Carbon Cycling   Carbon-cycle modeling is obviously of significance to future climate patterns, given elevated CO_2 and CH_4 concentrations and the scale and impact of past oscillations shown in ice-core records. Paleoclimate modeling of the carbon cycle includes a diverse array of methods and topics all pertaining to various aspects to the global carbon fluxes over tens of millions of years to millennial-scale changes.

The most sophisticated carbon-cycle model over geological, multimillion-year timescales is the GEOCARB developed by Robert Berner (Berner 1991, 1994, 2004). GEOCARB incorporates several slow feedbacks such as silicate weathering on continents, volcanic outgassing of CO_2, continental uplift to produce a ~500-million-year reconstruction of atmospheric CO_2 at a temporal resolution of 10–30 Ma (Berner and Kothavala 2001; Edmond and Huh 2003; Bickle 2005). The GEOCARB model predicts multimillion-year fluctuations in atmospheric CO_2 in line with CO_2-proxy based reconstructions from leaf stomata, pedogenic carbon, and marine phytoplankton isotopic chemistry, although the error bars on proxy-based paleo-CO_2 reconstructions can be substantial (see chapters 3 and 4 in this volume).

The importance and complexity of global carbon fluxes during deglacial transitions cannot be overstated (Sigman and Boyle 2000; Archer et al. 2000). Köhler et al. (2005), for example, used a box model and proxy records to rank the contribution of various carbon-cycle processes to account for the 80 ppmv rise in CO_2 during the last deglaciation (18– 11.5 ka):

- Greater vertical ocean mixing rates in the Southern Ocean: >30 ppmv
- Decreases in ocean alkalinity and carbon inventories: >30 ppmv
- Reduced strength of the ocean's biological pump: about <20 ppmv
- Rising ocean temperature: about 15–20 ppmv
- Stronger meridional ocean circulation: about 15–20 ppmv
- Growth of coral reefs: about 5 ppmv

Note the total contribution from these factors exceeds the 80 ppmv CO_2 rise measured in ice cores because other factors were at work: the expansion of terrestrial vegetation into formerly glaciated regions as ice sheets retreated, the rise in global sea level as ice sheets melted, and greater oceanic uptake of CO_2 due to reduced sea-ice cover. Kohler et al. (2005) estimated that these led to a combined lowering of CO_2 by <30 ppmv during the last termination.

At odds with the hypothesis of ocean sequestration of carbon during glacial maxima is the view of Zeng (2003, 2007). Zeng used a "semi-empirical" earth-system model that combines a number of different model approaches, such as interpolated output from the fully coupled general-circulation

model, climate (temperature, precipitation) and ice sheet (Peltier 2004) models, and terrestrial carbon, land-surface, and other models to evaluate the storage of carbon during glacial periods and release during deglaciation. The release of carbon from land, stored underneath large northern-hemisphere ice sheets during glacials, occurs through basal melting, lubrication, and basal ice-sheet flow and represents a key element in the simulations.

In an intriguing application of paleomodeling to the global carbon cycle, Archer and Ganopolski (2005) addressed the question, given past and expected future anthropogenic greenhouse gas emissions, of when we can expect to experience the next glaciation, in light of glacial-interglacial cycles that have occurred over the last 500,000 years caused by orbital variability, atmospheric CO_2 and other feedbacks. Based on output from one version of the intermediate-complexity CLIMBER-2 model designed to simulate climate-ice-sheet behavior, their startling conclusion was that release of 5000 Gt C might prevent another glaciation for the next 500,000 years. In other words, rising greenhouse gas concentrations, whether due to combustion of fossil fuels or methane release from ocean clathrates (see the paragraphs that follow), will prevent the onset of another glacial age through ice-sheet nucleation processes into the foreseeable future, taking into account all available proxy records and the relatively mature understanding of glacial-interglacial cycles observed in orbital-ice-sheet models of Imbrie and Imbrie (1980) and Paillard (1998), among others. This result means that future climate might be similar to climate regimes not seen in the last 2.6 Ma when the high-amplitude glacial-interglacial cycles of the late Cenozoic era began.

Another important application of paleoclimatology involves the impacts of carbon-cycle feedbacks in forecasting future climate. Carbon-cycle-climate models show that rising CO_2 concentrations will affect future climate but that climate change in turn affects the storage of carbon, especially in the world's oceans, leading to greater amounts of anthropogenic carbon remaining in the atmosphere (Cox and Jones 2008). The problem is there is no agreement from model simulations on how large this carbon-cycle feedback effect will be, with estimates ranging from a relatively small increase in atmospheric CO_2 concentration of 30 ppmv to a very large increase of 250 ppmv (IPCC WG I 2007). Preanthropogenic climate and CO_2 variability during the LIA (A.D. 1500–1900) suggests that CO_2 concentrations fell by about 7 ppmv during LIA cooling, lagging climate by about 50 years (Scheffer et al. 2006). Similar abrupt changes in CO_2 concentration and temperature accompany DO climate variability (Ahn and Brook 2008). Many questions about these short-term climate-carbon feedbacks remain unanswered: what was LIA temperature variability particularly in the oceans, what was the contribution of human activity such as land-use changes on LIA carbon fluxes, and what is the climate sensitivity to small changes in CO_2 concentrations? These issues are discussed in chapters 11 and 12 in this volume.

In addition to CO_2, methane (CH_4) is another major greenhouse gas involved in long- and short-term climate changes. The global inventory of methane stored in gas clathrates frozen in ocean sediments is 3000–15000 Gt C, roughly equivalent to the amount of carbon stored in fossil fuels. Paleoclimate evidence indicates that major changes in methane flux occurred during climate changes during the PETM ~55 Ma (Dickens et al. 1994; see also Chapter 4 in this volume), progressive cooling since the Miocene epoch (~14 Ma), and glacial-interglacial cycles of the last few million years (chapters 6 and 7 in this volume). Warming oceanic temperatures of ~1.5°C or more over millennia are considered a major forcing for the release of oceanic clathrate-bound methane. Buffett and Archer (2004) and Archer and Buffett (2005) simulated changes in the global clathrate reservoir during steady-state and transient climate changes over geological timescales using combined output from ocean carbon, sediment diagenesis, and methane clathrate geophysical models. They discovered significant clathrate release during past periods of climatic warming consistent with independent proxy-based data and, more important, that anthropogenic perturbation of the carbon cycle can potentially lead to the release of thousands of Gt C.

Abrupt Climate Change and MOC Modeling the impact of abrupt glacial lake drainage on MOC is one of the most extensive applications of paleoclimate modeling because the freshwater volume, timing, and outlet routes of drainage events are known quantitatively from decades of geological field mapping in North America and Europe (Teller and Leverington 2004; see also Chapter 9 in this volume). Some of the more informative studies have shown a North Atlantic system extremely sensitive to freshwater forcing at the onset of the YD cold reversal and the 8.2 ka event, and during glacial-age DO events. Although the strength and duration of MOC weakening and the broader climate response vary greatly as a function of the location of freshwater discharge and the particular model simulations, there is nonetheless firm evidence that lake discharges influenced hemispheric and probably global climate (Rahmstorf 1996; Manabe and Stouffer 1997; Stocker and Marchal 2000; Ganopolski and Rahmsdorf 2002; Rahmstorf et al. 2005; LeGrande et al. 2006; Stouffer et al. 2006; LeGrande and Schmidt 2008).

Climate Feedbacks Many types of feedback mechanisms are investigated in paleoclimatology. One is the vegetation-biophysical-climate response to seasonal changes in orbital insolation during the Holocene epoch, which is now well established from paleoreconstructions and model simula-

tions (Kutzbach et al. 1996; Prentice and Webb 1998, Prentice et al. 2000). Harrison et al. (2003) showed that the mid-Holocene (6 ka) dynamic response to orbital changes in seasonality involved a wetter southwestern United States and a drier midcontinent because of positive ocean feedbacks related to La Niña-like conditions in the tropical Pacific Ocean. These North American mid-Holocene changes were part of a global shift in the ITCZ and global monsoon systems in both northern and southern hemispheres (see Chapter 8; Braconnot et al. 2007).

The sea-ice-climate feedback is another important factor critical to climate changes over several timescales. Gildor and Tziperman (2000, 2001), Tziperman and Gildor (2003), and Sayag et al. (2004) proposed that a sea-ice switch is important in millennial and orbital climatic cycles. Some of their model simulations indicate a sea-ice switch can turn "on" in only 50 years after initial sea-ice formation through positive feedback of sea-ice albedo. During glacial periods when sea and land ice are at their maximum extent, sea ice slows moisture transport by limiting evaporation. Once a threshold is crossed, glaciers are starved of precipitation and begin to retreat from ablation. The process reverses itself, albedo decreases, and deglaciation begins. In the case of abrupt millennial climate reversals such as DO events (see Chapter 6 in this volume) and the YD (see Chapter 7), sea ice can melt within decades and the sea-ice switch is thrown off. Hansen et al. (2007) also described the importance of the sea-ice-water "albedo flip" feedback mechanisms for abrupt Pleistocene glacial terminations. Here, snow and ice become wet during spring melt season, leading to an abrupt shift in albedo and triggering catastrophic ice-sheet disintegration in as little as 100 years under the right forcing conditions.

Take the example of the Holocene epoch. Renssen et al. (2006) ran 9000-year climate model simulations forced by greenhouse gases, total solar irradiance (TSI), and orbital insolation. Their results demonstrated the importance of sea-ice and water-column stratification, especially in the Arctic, in slowing MOC response to TSI external forcing. MOC slowdown creates a positive feedback, more cooling and sea ice, weaker MOC, and changes in the location of deep-water convection.

Intermodel Comparisons

The skill of climate models to simulate natural systems is often judged on the basis of intermodel comparisons. Intermodel comparisons are used in the assessment of 23 climate models by the IPCC (Randall et al. 2007), and so too paleoclimatologists carry out comparisons. The PMIP projects (Joussaume et al. 1999; Braconnot et al. 2007) are a coordinated effort carried out under the auspices of the Past Global Change

Project. PMIP1 employed an atmospheric circulation model, whereas more recent advances allow PMIP2 to use fully coupled AOGCMs (Otto-Bliesner et al. 2007). PMIP efforts will include pre-Quaternary climate reconstructions such as the Pliocene PRISM global data set (Chandler et al. 2008).

Perspective

Before proceeding, a word about future modeling strategies is in store. The World Climate Research Programme is striving toward what is called "seamless" prediction. Such an idealized, integrated model system would link all timescales and processes—short-term radiative and temperature response to greenhouse gases, decadal-internal climate variability, ocean-atmosphere-land interactions, ice sheets, and biogeochemical processes (vegetation, carbon cycling, etc.) (Palmer et al. 2008). Although years from fruition, such an ambitious effort explicitly recognizes the importance of both short- and long-term climate prediction and the dynamics of climate variability over all timescales, including those reconstructed by paleoclimatologists, the subject of the remainder of this book.

LANDMARK PAPERS Paleoclimate Data-Modeling Research

Barron, Eric J., Starley L. Thompson, and Stephen H. Schneider. 1981. An ice-free Cretaceous? Results from climate model simulations. Science 212:501–508.

Barron, Eric J. and W. M. Washington. 1985. Warm Cretaceous climates—High atmospheric CO_2 as a plausible mechanism. In E. T. Sundquist and W. S. Broecker, eds., *The Carbon Cycle and Atmospheric CO_2: Natural Variations, Archean to Present*, pp. 546–553. Washington, DC: American Geophysical Union.

Global warmth is on many people's minds today, as it was in 1981 when Eric Barron, Starley Thompson, and Steve Schneider conducted their pioneering paleoclimate modeling study of hyperthermal climates of the mid-Cretaceous about 100 million years ago. Their dual purposes were to use computer models of climate to investigate the effects of long-term changing geography on global climate and paleoclimate data from fossil plants to evaluate the performance of the models in simulating empirical geological evidence for warmth. Both climate models and proxy-based reconstructions were in the early stages of development. By adopting a strategy that used models and data in tandem to quantify extreme climate change and to understand underlying mechanisms, they took advantage of obvious evidence for extreme warmth (fossil reptiles and warmth-loving plants in high-latitude sediments) and advances in computers.

They used a zonally averaged energy-balance model at the National Center for Atmospheric Research in Boulder, Colorado, where Schneider was a researcher and deputy director. Incorporating various values for land-surface vegetation and land-sea distribution, which both influence albedo and radiative balance, they asked whether these differences alone could account for global warmth and an ice-free planet. The model simulations conflicted with the paleoclimate data when forced with Creta-

ceous paleogeography. So, either the model structure or the calculations of physical processes (parameterization of clouds, other feedbacks) were incorrect, the paleoclimate proxy data were misinterpreted, or a factor other than paleogeography was responsible for global warmth. Their in-depth assessment of what factors caused an estimated radiative forcing +12 W m² for Cretaceous warmth led them to conclude that "A change in heating of +12 W/m2 is approximately equivalent to a 5 percent increase in solar constant or a 700 percent increase in atmospheric CO_2, based on typical model sensitivity."

In the wake of the discovery of reduced glacial-age carbon dioxide (CO_2) concentrations from ice cores, Barron and the National Center for Atmospheric Research's climate modeler Warren Washington did a follow-up study using an early general-circulation model to evaluate whether Cretaceous warmth could be explained by elevated CO_2 concentrations. They found that CO_2 concentrations four times preindustrial levels were a plausible scenario to explain Cretaceous warmth but that paleogeography also played a major role.

The impacts of the Cretaceous climate modeling by Barron and colleagues are still felt today. Recent improvements in paleothermometry now show that tropical and polar temperatures during the Cretaceous and Paleogene periods were even warmer than Barron's estimates of 30°C and 5–19°C, respectively. The Cretaceous tropics were probably 35°C or higher, and the North Pole region was 20–24°C during the hyperthermal Paleocene-Eocene Thermal Maximum. We will see in Chapter 3 that much remains to be done—even state-of-the-art climate models cannot simulate ancient pole-to-equator thermal gradients, even with CO_2 concentrations 10 times preindustrial levels. More generally, paleoclimate modeling at all timescales epitomizes one of the principles of modern paleoclimatology outlined in this chapter and throughout our text.

3

Deep Time: Climate from 3.8 Billion to 65 Million Years Ago

Introduction

The earth's 4.8-billion-year (Ga) history involved complex geological, biogeochemical, and climatic changes during its evolution from gaseous cloud to present geography and climate. Pre-Quaternary climatic (Hambrey and Harland 1981; Crowley and North 1991; Frakes et al. 1992) ocean and atmospheric evolution (Holland 1984, 2003; Grotzinger 1990; Kasting and Howard 2006) has long attracted a variety of specialists, but it is mostly the domain of geologists, geochemists, paleobiologists, and geophysicists. More recently, it has gained the attention of paleoclimate modeling groups because of the importance of understanding the dynamics of extreme climates and the role of atmospheric carbon dioxide (CO_2) and methane (CH_4).

The details of the earth's geological and climatic history is far beyond the scope of any single text, but certain aspects of climate between 3.8 Ga and 65 million years ago (Ma)—today commonly referred to as Deep Time—are the focus of research efforts to understand extreme climatic states and the evolving relationship between atmospheric CO_2 and global climate. This chapter covers those aspects of the long-term geological record of ancient climate that are most germane to greenhouse gas forcing and extreme climate states. Geological processes are critical to the evolution of climate. Paleogeography is the field that reconstructs continental positions and ocean basins long destroyed or severely modified by plate-tectonic processes, using paleomagnetic, geophysical, paleontological, and other lines of evidence. Table 3.1 lists some of the major continents and oceans used in deep-time paleoclimatology preserved in fragmentary form on today's continents. Several paleogeographical maps are provided for key climatic intervals, and a full discussion of global paleogeography can be found in Frakes et al. (1992), Veevers (2004), Li et al. (2008), and the PALEOMAP reconstructions of Chris Scotese (http://www.PALEOMAP.com).

Reconstructing climate history from the incomplete geological record is a challenging task that requires integrated geochronology, geochemistry, and lithological, paleomagnetic, and paleobiological analyses. Nonetheless, the single dominant theme in the earth's geological and climatic history is captured in the prefix "super": supercontinents (e.g., Gondwanaland, Pangea), supergreenhouse (the end of "Snowball Earth" climate, much of the Mesozoic), supercycles (the Wilson tectonic cycle), supersequences (stratigraphic and geophysical packages of sea-level history), supergroups (packages of geological formations formed over millions of years), and superplumes (core-mantle igneous activities). A similar portrayal of climatic changes over the last few billion years would not be hyperbole.

Early Earth, Faint Sun Paradox, and Atmospheric Carbon Dioxide

Background

The most important issue pertaining to the earliest evolution of the earth's climate is that energy emitted by the sun has progressively increased over 4.8 Ga as it evolved from the fusion of hydrogen to helium and helium into heavier elements. Carl Sagan (Sagan and Chyba 1997) proposed that, compared to today, solar radiation was 30% lower ~4 Ga during the Archean eon (3.8–2.5 Ga) in the middle part of Precambrian time (4.5 Ga–544 Ma). Solar energy is calculated as follows:

$$T_e = \sqrt[4]{S/4\,(1^{-a})}$$

TABLE 3.1 Phanerozoic Paleogeography*

Name	Feature	Era	Approx. Age (Ma)[†]	Comment
Rodinia	Landmass	Neoproterozoic		
Gondwanaland	Landmass	Neoproterozoic-Carboniferous	650–350	Southern hemisphere, merged into Pangea
Pangea	Landmass	Permian-Jurassic	350–185	Breakup, 185 Ma
Laurussia	Landmass	Paleozoic	650–330	Merged with Gondwanaland
Laurentia	Landmass	Early Paleozoic	540–450	Merged with Gondwanaland
Laurasia	Landmass	late Paleozoic-Mesozoic	350–185	Northern part of Pangea
Atlantic	Ocean	Jurassic-Cenozoic	185–0	Opening in Jurassic
Panthalassia	Ocean	Paleozoic-Jurassic	540–160	Predecessor of paleo Pacific
Iapetus	Ocean	Late Neoproterozoic–Permian	560–320	
Tethys	Ocean	Mesozoic-Cenozoic	250–3	Various stages, Proto-, Paleo, neo-Tethys

Note: For major landmasses and oceans, see Veevers (2004); for smaller landmasses, see Ford and Golonka (2003) and their references.

*Other minor oceans and land masses are not listed.

[†]All ages are approximate, depending on author and definition.

where a is albedo (about 30% today) and S is solar flux (today about $1370 \, W \, m^{-2}$). Lower T_e during the Precambrian should have led to a glaciated planet unless albedo was near zero or other factors influenced planetary mean temperature. Sedimentological, geochemical, and paleobiological evidence, however, indicates that the earth supported liquid water where the organisms evolved beginning about 3.5 Ga (Knoll and Carroll 1999; Anbar and Knoll 2002; Knoll 2003; Catlin and Buick 2006). Oxygen isotopes suggest that the earth may have experienced hyperthermal environments, with temperatures for the Archean eon as high as 55–85°C (e.g., Knauth 2005). This is referred to as the Faint Sun Paradox, and the geological record of the Archean eon sheds light on climate during a dimmer sun.

Archean Paleotemperature

Oxygen isotopes of ancient carbonates and cherts have been used to reconstruct Archean temperatures. The $\delta^{18}O$ method is a valuable paleoclimate proxy applied to a variety of materials ranging from carbonate rocks to speleothems to ice cores throughout the geological record. Temperature is a dominant control over the ratio of heavy and light oxygen isotopes, but other factors such as seawater pH and hydrology (isotopic composition of source water), continental ice volume, and postdepositional diagenetic processes complicate the simple interpretation of $\delta^{18}O$ as a temperature proxy.

Many Archean carbonates and cherts have $\delta^{18}O$ values that are 10–20‰ lower than that of standard mean ocean water (30‰) (Burdett et al. 1990). Knauth (2005) and Knauth and Lowe (2003) interpreted low Archean $\delta^{18}O$ values from chert as evidence that, at least at times, Archean temperatures exceeded 55°C. Such extremes exceed any estimated for the Phanerozoic eon covering the last 543 Ma.

Kasting et al. (2006) and Jaffrés et al. (2007) reinterpreted the Precambrian isotope record, suggesting that oxygen isotopic values reflect lower $\delta^{18}O$ values of ancient seawater and not temperature. The argument is based on the idea that higher heat flow from the earth's interior caused shallower Archean oceans, especially along mid-ocean ridges, and contributed to seawater $\delta^{18}O$ values as much as 10‰ lower than those typical of more recent geological periods. If correct, this idea reconciles evidence for isotopic composition of cherts, Neoproterozoic glaciations (see the subsection "Snowball Earth—Neoproterozoic Climate Changes"), and paleobiological evidence for early evolutionary biotas. It means that temperatures of 50°C to >80°C may be overestimates and that the Archean was not necessarily a "hyperthermal" environment.

Archean Greenhouse Gases

Whatever the exact temperature history of the Archean eon, one hypothesis regarding keeping the planet warm enough to support water and life with a fainter sun holds that atmospheric CO_2 and methane (CH_4) levels were elevated over those of the last few hundred million years (Walker et al. 1981; Kasting and Ackerman 1986). According to Energy Balance Models, today the earth's temperature would be about 0°C if it were not for natural greenhouse forcing, mainly by atmospheric CO_2, resulting in the actual ~15°C mean annual surface temperature. Assuming constant albedo and water vapor near saturation during the Archean eon, elevated CO_2 concentrations are a favored explanation to counterbalance the weaker solar output.

Some estimates of Precambrian CO_2 concentrations are as high as 100–1000 times modern levels (Kasting 1987). Using the current atmospheric CO_2 concentration of about 385 parts per million by volume (ppmv) for comparison, this would mean Precambrian CO_2 concentrations were somewhere between 38,500 and 385,000 ppmv, or 3.8–38% of the atmosphere's gas volume. There is no consensus on Precambrian CO_2 concentrations, in part because proxies for paleo-CO_2 concentrations are indirect and have large error bars (Kasting 1993). For example, soil geochemistry suggests concentrations toward the lower end of the range (Rye et al. 1995). Sheldon (2006) estimated that CO_2 concentrations were closer to 23 times modern values from about 2.5–1.8 Ga, then fell significantly between 1.8 and 1.1 Ga.

The sources of Archean CO_2 are also uncertain. One source might have been degassing from the lithosphere through volcanic activity. In addition, smaller Archean continents might mean there was a smaller terrestrial sink for carbon, and lower weathering-CO_2 feedbacks may have influenced CO_2 levels during the Archean-Proterozoic time frame. Methane is another greenhouse gas that may have contributed to keeping the earth's climate warm during its early evolution under a reduced solar influx until just after 1 Ga. Methane concentrations from about 3.8 Ga until the rise of atmospheric oxygen ~2.3 Ga were probably as high as 1000 ppmv and may have been caused by the serpentinization of rocks, methanogenic bacteria, or other factors (Kasting 2005).

In sum, despite the difficulty in obtaining quantitative paleo-atmospheric chemistry and temperature reconstructions for Archean climate, we see from this brief discussion that greenhouse gases were a likely factor in preventing the early earth from attaining a glacial state and allowing aquatic organisms to evolve.

Snowball Earth: Neoproterozoic Climate Cycles

Background

The Neoproterozoic era from 1 Ga to 544 Ma is the youngest part of the Proterozoic Eon and represents a time when the earth experienced extraordinary climatic extremes embodied in the Snowball Earth hypothesis. Snowball Earth at its simplest holds that glaciation enveloped all or most of the planet in solid ice, including oceans and equatorial regions, at least twice and possibly three or four times, over the course of a 170-Ma period between about 750 and 580 Ma. The paleogeography of this period was characterized by the breakup of the supercontinent Rodinia into a number of smaller land masses by about 600 Ma (Li et al. 2008) (Figure 3.1). A more subdued glacial model called "Slushball Earth," in which some tropical ocean regions remained ice free in an otherwise glaciated planet, has also been proposed. Solar energy

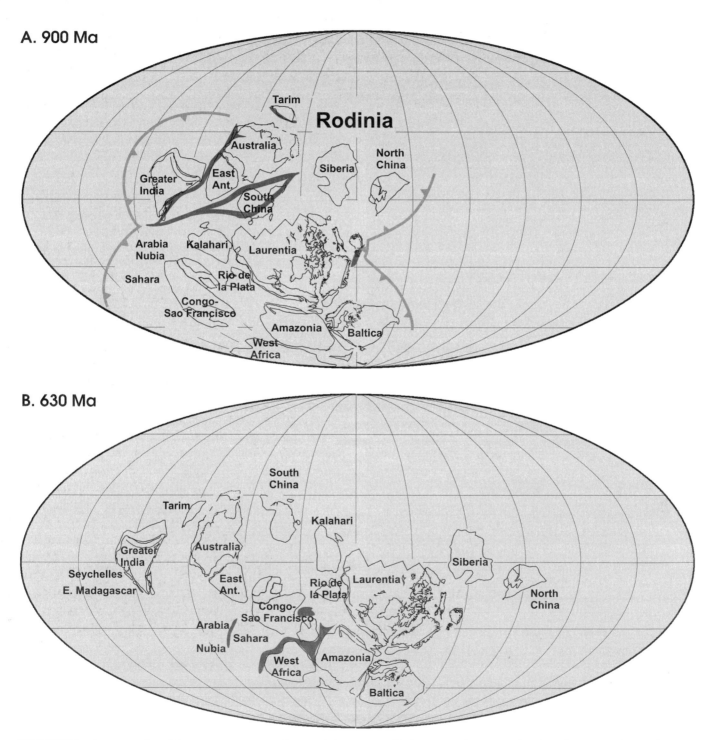

A. 900 Ma

B. 630 Ma

FIGURE 3.1 Paleogeography of the supercontinent Rodinia at (A) 900 Ma and (B) 630 Ma, showing its breakup, from Li et al. (2008). Paleo-landmasses (Laurentia, Amazonia) and modern geography are indicated. Courtesy of Z. X. Li and Elsevier.

FIGURE 3.2 Outcrop of Ghaub Formation (635 Ma) showing proglacial marine debris flow and ice-rafted debris sediments with large glacial dropstones overlaid by cap carbonate dolostones of Cryogenian age, from the Otavi Group, Namibia. Courtesy of P. Hoffman and Snowballearth.org.

during the Neoproterozoic era was still about 6% less than that today.

Kirschvink (1992) is credited with the initial Snowball Earth hypothesis, although earlier workers (e.g., Harland 1964) had postulated extensive glaciations on the basis of geological evidence from Precambrian deposits (Figure 3.2) (see Fairchild and Kennedy 2007). In fact, the geological era between ~850 and 630 Ma has been given the formal name the Cryogenian period because of its predominantly global cold climate state (Plumb 1991).

In addition to evidence for a glacial world, a second major feature of Neoproterozoic climate is geological evidence for massive and, according to some, rapid global warming and deglaciation following glacial episodes. Pierrehumbert (2002) notes from our understanding of atmospheric processes, energy balance, and the radiative effects of clouds that getting the earth's climate out of a global Neoproterozoic glacial state is climatologically a more difficult task than getting into one.

The literature on Neoproterozoic climate has exploded in the past decade. Papers focused on the geological and geochemical evidence can be found in Hoffman et al. (1998), Hoffman and Schrag (2002), Halverson et al. (2005), Hoffman (2005), McCay et al. (2006), Fairchild and Kennedy (2007), and the volume edited by Jenkins et al. (Jenkins 2004a, b) is devoted to this topic. The details of the Rodinia breakup are incompletely known and subject to various paleogeographical interpretations (Karlstrom et al. 2000; Torsvik 2003; Goddéris et al. 2007; Li et al. 2008;). In this section, we briefly outline the basic geological evidence for glaciation and deglaciation in the Neoproterozoic era and examine several theories to explain the evidence.

Stratigraphy and Lithology

Evidence for Neoproterozoic glaciations has long been known from a variety of glacial deposits on all modern continents except Antarctica. The stratigraphy and carbon isotope records of Neoproterozoic glacial and deglacial sediments are illustrated in Figure 3.3 from Halverson et al. (2005). The two most studied Cryogenian glaciations are called the Sturtian (~720–700 Ma) (Fanning and Link 2004) and the Marinoan (640–635 Ma), and there may have been several distinct glacial intervals within each. The third glacial age is called the Gaskiers (also Squantum-Gaskiers, ~585–580 Ma), within the Ediacaran era. This glaciation preceded the evolution of the famous metazoan Ediacaran fauna discovered first in Australia and now known from other continents. Some Neoproterozoic sequences such as the British-Irish Caledonides contain evidence for three distinct glacial intervals (McCay et al. 2006).

Although Neoproterozoic sequences vary greatly in detail, the basic lithostratigraphy consists of glaciogenic deposits overlaid conformably by mainly dolomitic sediments called cap carbonates, which represent postglacial climatic warming. The glacial deposits include diamictites (sometimes called tillites), glacio-marine sediments, ice-rafted dropstones, and other glacially derived sediments and glacial features such as abraded rocks, striations, and polished rock surfaces. Hoffman (2005) also describes a ridge in the Ghaub Formation, Namibia, interpreted as a transverse medial moraine located near paleo-ice streams similar to those seen today along margins of the Greenland and Antarctic Ice Sheets. Glacial deposits and features are known from the Congo, Sahara, and Oman in Africa, Australia, China, and many sites in North

FIGURE 3.3 Carbon isotope records of Neoproterozoic glaciations, from Halverson et al. (2005). (A) and (B) represent two possible age calibrations for carbon isotopic records. Courtesey of G. Halverson and the Geological Society of America.

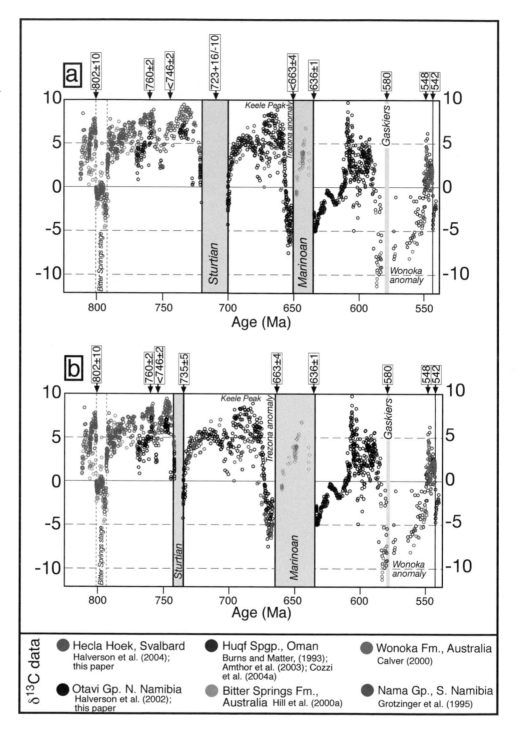

FIGURE 3.3 Carbon isotope records of Neoproterozoic glaciations, from Halverson et al. (2005). (A) and (B) represent two possible age calibrations for carbon isotopic records. Courtesey of G. Halverson and the Geological Society of America.

America, Ireland, Scotland, and Norway (Harland 1964; Hambrey and Harland 1981, 1985; Eyles and Januszczak 2004; Hoffman 2005; Etienne et al. 2007). Ice-rafted dropstones like those illustrated in Figure 3.2 are among the most dramatic examples from glacio-marine environments. Although the evidence for major Neoproterozoic glacial activity is indisputable, the genesis of some deposits is debated, in particular whether suitable analogs can be found in modern glacial environments.

Cap carbonate sediments (cap dolostones and limestones) overlie most Neoproterozoic glacial sequences. Typically, a dolostone conformably overlies the glacial sediments and the dolostone is often overlain by a limestone. Most cap carbonates are several meters to tens of meters thick (Hoffman et al. 2007). They vary in structure, mineralogy, grain size, geochemistry, and postdepositional diagenetic modifications. Some carbonates are intermixed with siliciclastic facies.

The formation of carbonates at a time when there were no major marine organisms poses complex geochemical and geological issues relevant to their paleoclimatic interpretation. Most believe that cap carbonates represent a major shift in depositional environment and that they signify changes in the global hydrological cycle, increased weathering, a change in the global carbon cycle, and a eustatic sea-level rise (Hoffman et al. 1998; Kennedy et al. 1998; Ridgwell et al. 2003; Condon et al. 2005), although some may represent more localized marine transgressions in semi-isolated sedimentary basins.

Geochemical Records

Carbon isotope stratigraphy is available for many Neoproterozoic sections, and $\delta^{13}C$ excursions may signify global-temperature, sea-level, carbon-cycle, or other environmental changes (Knoll et al. 1986; Hoffman et al. 1998; Karlstrom et al. 2000). The carbon isotopic curve in Figure 3.3 shows isotopic depletions of up to 10–15‰ during Sturtian, Marinoan, and Gaskiers glacial-deglacial cycles. Carbon isotope stratigraphy has been used for interregional correlation of Neoproterozoic deposits in the same way oxygen isotope stratigraphy is used to correlate Cenozoic ice volume fluctuations. However, this implies synchroneity of isotopic variations. Hoffman et al. (2007) suggested that sedimentary and isotopic evidence from cap carbonates in the Keilberg (Ediacaran) sequence in Namibia indicates diachronous deposition and signifies glacio-eustatic transgression across regions of different elevation. Depending on the thickness of ice during the previous glacial episode, the sea-level rise may have been 600 m (uncorrected for isostatic adjustment) over only 2–10 ka, although the timescale is uncertain and depends on the interpretation of the isotopes (Kennedy et al. 2001a, b; Higgens and Schrag 2003). Some bulk carbon isotopic measurements from cap carbonates may also be influenced by diagenesis, making interregional correlation difficult. The development of additional geochemical proxies in the next few years should resolve the issue of synchroneity.

Age and Duration of Glacial Intervals

In contrast to the variety of methods used to age-date and correlate Phanerozoic records, the Precambrian era relies mainly on fairly sparse radiometric dating for absolute ages. The absolute dating of Neoproterozoic deposits has improved greatly with the advent of uranium-lead isotopic radiometric dating of zircons from volcanic ashes, providing a precision approaching 1 Ma. Relatively few U/Pb dates are available, and the stratigraphic position of dated material in relationship to glacial sediments and cap carbonates introduces ad-

ditional uncertainty regarding the age and duration of most glacial deposits and overlying cap carbonates. As a consequence, the estimated durations of glacial periods range from several Ma to 35–40 Ma. In an attempt to narrow this range, Bodiselitsch et al. (2005) reasoned that extraterrestrial iridium, the element found in sediments at the Cretaceous-Tertiary boundary proving a meteor (bolide) impact at 65 Ma, would accumulate in ice on a glaciated planet. By measuring iridium anomalies in Marinoan and Sturtian deposits, they estimated these glacial episodes lasted between 3 and 12 Ma.

An additional complication stems from the fact that it is difficult to distinguish multiple short-term glacial episodes within a major glacial period on the basis of so few dates. For example, geochemical evidence for active hydrological cycling and weathering implies multiple glacials within the Neoproterozoic era of Oman (Rieu et al. 2007). Chronostratigraphic uncertainty should be reduced in future studies to achieve better understanding of the timing of glaciations preserved in sediments on various continents.

Models and Mechanisms

The most extreme Snowball Earth state would envision a planet, including low latitudes and ocean regions, completely covered by rigid ice (Hoffman et al. 1998; Hoffman and Schrag 2002). Kirschvink (1992) suggested that, in its original form, global glaciation might occur under certain continental configurations such as the supercontinent Rodinia. If extensive land areas were located in middle and low latitudes, ice sheets could grow through the gradual accumulation of snow. Tropical land areas reflect more solar insolation than open ocean surfaces and in theory could enhance cooling through albedo feedback mechanisms. Sea-ice formation limits evaporation and winds, perhaps leading to large-scale ocean stagnation and oxygen depletion in the oceans, explaining the abundant iron deposition in the Neoproterozoic sediment.

A less extreme version is the Slushball Earth model. In this model, low-latitude oceans may have been cool but not completely ice covered, and tropical coastal regions may have had glaciers but not thick, rigid continental ice sheets (e.g., see Chandler and Sohl 2000; Hyde et al. 2000; Leather et al. 2002; Sohl and Chandler 2007). Fairchild and Kennedy (2007) recount that an early hypothesis offered by Brian Harland (1964) held a slushball-type view in which equatorial oceans remained ice free. They also note that the slushball concept comes mainly from the climate modeling perspective rather than from field evidence. From the geological viewpoint, there is clear lithological evidence for complex glacial dynamics during the Neoproterozoic era, including glacio-marine sedimentation and ice streams as well as other processes.

A third viewpoint is that the Neoproterozoic earth was not completely glaciated and that some sediments interpreted as glaciogenic are actually tectonically derived. Eyles and Januszczak (2004) disputed the Snowball Earth model, basing their argument largely on the sedimentological analyses and tectonic setting of the Otavi Group, Namibia. They instead suggested that glaciogenic sediments were deposited in regional sedimentary basins largely controlled by local tectonics and topography and that cap carbonates may not be conformable with underlying coarse-grained sediments. This Zipper-Rift hypothesis implies that Neoproterozoic glaciations may have been local, diachronous, and absent in low latitudes. Most workers hold that the Zipper-Rift hypothesis is not supported by the bulk of available field evidence, but this conflicting view nonetheless exemplifies difficulties in correlating and dating Neoproterozoic sequences and interpreting some lithologies.

Several causal factors are critical to understanding Neoproterozoic glacial and deglacial climate (Williams 2000; Poulsen et al. 2002; Pavlov et al. 2003; Donnadieu et al. 2004; Jenkins 2004a, b; Goddéris et al. 2007). The most important factors are reduced solar influx, atmospheric CO_2 and methane forcing, paleogeography (Rodinia breakup), tectonic activity (mountain building, relief, volcanic degassing), and orbital obliquity (the earth's axial tilt). Feedback mechanisms include silicate weathering, ice albedo, CO_2 drawdown and carbon burial, and ocean circulation.

A large climate modeling literature attempts to parse out these factors (Chandler and Sohl 2000; Hyde et al. 2000; Schrag et al. 2002; Pollard and Kasting 2005; Goodman 2006; Lewis et al. 2006, 2007; Goddéris et al. 2007; Sohl and Chandler 2007). The skill of models to simulate snowball or slushball-like conditions hinges on their sensitivity to geographical boundary conditions, reduced solar flux, continental relief, and atmospheric CO_2 and CH_4 concentrations, as well as the strength of feedback mechanisms. The ability of ice in high latitudes or on high-elevation tropical continents to flow toward sea level and eventually cover oceans, or both, is an especially important issue. This is particularly significant given the revolution in modern glaciology prompted by historically unprecedented behavior of Greenland outlet glaciers, Arctic Ocean sea ice, Antarctic ice shelves, and other dynamic features (see Chapter 12 in this volume).

While the existence of a near-global glacial state poses challenges to modelers and paleoclimatologists alike, the cap carbonates are more difficult to explain (Crowley et al. 2001; Pierrehumbert 2004; Pollard and Kasting 2005). Rapid deglaciation due to greenhouse gas forcing and ice-albedo feedbacks implies major instability in the global hydrological balance for an undetermined amount of time. An additional complication is what impact total ice cover would have on evolving organisms (Knoll 2003). Thin sea ice or "sea glaciers," in places transparent to sunlight, might allow biological processes to continue even in a glaciated state (McKay 2002,; Warren et al. 2002; Pollard and Kasting 2005).

In sum, the earth was severely glaciated several times between about 750 and 580 Ma, each time exiting a global or near-global glacial state into warmer climate. Uncertainty remains about several aspects of the severity of glaciation: the growth, geographical distribution, and thickness of land and sea-ice; the number of glaciations; temporal variability within each glacial era; the synchroneity of deglaciation; the significance of geochemical proxies; and the role and source of CO_2 and methane in postglacial warming. There is no universal agreement on the ultimate trigger or dominant feedbacks, and those that apply to one glacial cycle may not apply to others.

Phanerozoic Climate Change
Paleogeographical Framework

Nearly continuous geochemical proxy records have been compiled for all or most of the Phanerozoic Era, providing clues into the nature of long-term atmospheric, oceanic, and climate evolution. These records must be viewed against the backdrop of the paleogeography of continental and ocean basins compiled in projects like the Paleogeographic Atlas Project by Chris Scotese, Fred Ziegler, and Dave Rowley at the University of Chicago; the PLATES Project by Scotese, John Sclater, and Larry Lawver at the University of Texas Institute for Geophysics; and the PALEOMAP Project by Scotese at the University of Texas, Arlington. Descriptions of their paleogeographical reconstructions can be found in Zielger et al. 1983, 1996, Scotese and Sager 1988, Scotese 1997, and http://www.scotese.com/ for references, and a review of Phanerozoic paleogeography and tectonics in Veevers (2004). Ron Blakey of Northern Arizona University has also compiled global and map reconstructions for 41 geological time slices, showing past continent and ocean geography (http://jan.ucc.nau.edu/~rcb7/RCB.html).

In Figure 3.4, three reconstructions are shown depicting Gondwanaland, Pangea, and the post-Pangea continental breakup and formation of the Atlantic Ocean. Continent-ocean basin distributions provide critical boundary conditions for climate modelers, as well as geologists, to evaluate the impacts of plate tectonics on ocean heat transport, atmospheric circulation, carbon cycling, biotic evolution and extinction, and a host of other climate-related factors.

122222222222222222223222222222322222222

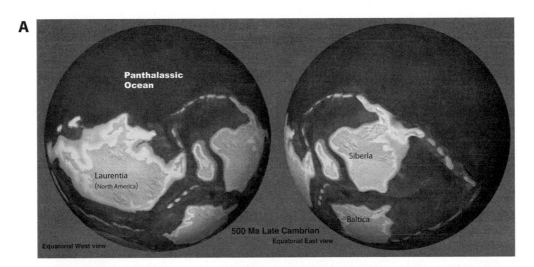
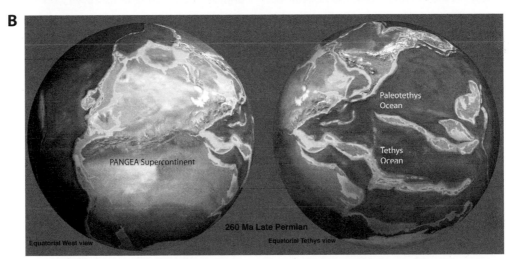
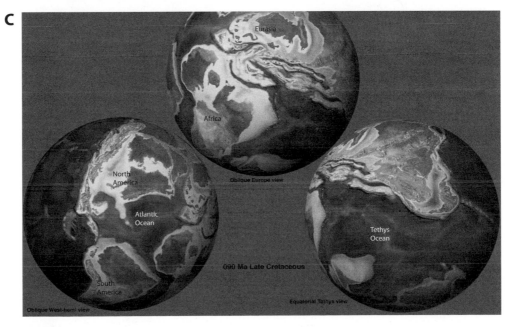

FIGURE 3.4 Paleogeographic maps showing (A) Late Cambrian, (B) Early Permian, and (C) Late Cretaceous periods and major continents and oceans. Plate tectonic maps courtesy of R. Blakey, Northern Arizona University.

Oxygen Isotopes and Temperature Curve

Veizer et al. (1999) measured $\delta^{18}O$ values from more than 1600 Cambrian to Cretaceous brachiopod, belemnite, and other fossils in Bochum, Germany and Ottawa, Canada (Bochum/Ottawa dataset) and compiled almost 4000 additional published measurements from numerous sources into a grand Phanerozoic oxygen isotope curve (Figure 3.5). One major feature of the curve is a steady increase from a mean $\delta^{18}O$ value of $-8‰$ in the Cambrian ($\sim$543–450 Ma) to 0‰ in the Quaternary periods. Superimposed on this trend are multimillion-year oscillations related to major glaciations during the late Ordovician ($\sim$445 Ma) and Carboniferous ($\sim$300 Ma) periods (Frakes et al. 1992) and the late Cenozoic era (see Chapter 4 in this volume). The late Ordovician glaciation was relatively short ($\sim$1–2 Ma), may have coincided with relatively high atmospheric CO_2 concentrations (eight to 12 times preindustrial levels; see below) (Gibbs et al. 1997), and possibly had orbital control on glaciations (Sutcliffe et al. 2000; Herrmann et al. 2003). The Carboniferous glaciation was the most extensive during the Phanerozoic.

Several factors control the oxygen isotopic composition of carbonates, so the Phanerozoic curve is subject to differing interpretations. Depleted values in the Paleozoic imply elevated ocean temperatures, different oxygen isotopic composition of seawater, oceanographic conditions favoring density stratification, or postdepositional diagenesis. Veizer et al. (2000) interpreted the oxygen isotopes in terms of tropical ocean temperatures and observed two intervals when temperature and atmospheric CO_2 levels were decoupled—the high CO_2 world of the Silurian, when temperatures were only a few degrees C above those today in tropical oceans, and the Carboniferous, when CO_2 concentrations were low and major glaciations occurred.

Royer et al. (2004) adjusted the oxygen isotope curve to account for changes in ocean chemistry and pH and found a better fit between $\delta^{18}O$-derived temperature and atmospheric CO_2. Came et al. (2007) reanalyzed the Paleozoic $\delta^{18}O$ curve using a new carbonate clumped isotope method (Ghosh et al. 2006). This method yields a temperature curve from carbonate (low-magnesium brachiopod shells and aragonitic mollusks) independent of the isotopic composition of seawater. Came concluded that Silurian sea-surface temperatures (SSTs) between 443 and 423 Ma were 5–11°C warmer than those of today, whereas Carboniferous SSTs were similar to today's temperatures. These new analyses and the reinterpretation of Veizer's $\delta^{18}O$ dataset provide a better fit between long-term temperature and atmospheric CO_2 levels.

Phanerozoic oxygen isotope curve

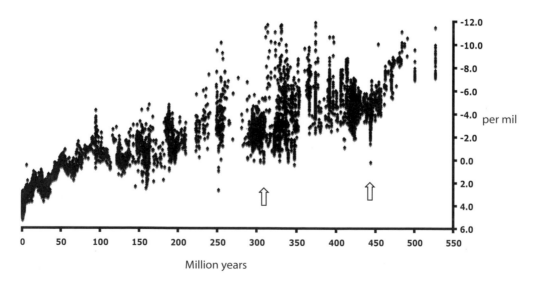

FIGURE 3.5 Phanerozoic oxygen isotope curve compiled from numerous sources by Veizer et al. (1999) (updated online in 2004; see Kasting et al. 2006). The plot shows mainly benthic foraminifers (many compiled by Zachos et al. 2001), belemnites, and brachiopods (planktonic foraminifer isotope data omitted) plotted against the age model of Gradstein et al. (2004). Major Ordovician and Carboniferous glaciations are shown by arrows, from Frakes et al. (1992); see Chapter 4 in this volume for Cenozoic glaciations. Data sources are given at http://www.science.uottawa.ca/~veizer/isotope_data/. Courtesy of J. Veizer.

Strontium Isotope Curve

The ratio of heavy (^{87}Sr) and light (^{86}Sr) strontium isotopes in seawater is controlled by input from continental sources through chemical weathering, transport by rivers, and input from chemical alteration of oceanic crust at mid-ocean ridges and tectonic zones (Veizer 1989; Capo and DePaolo 1990). Today, the mean ocean ^{87}Sr/^{86}Sr value is about 0.709 (Elderfield 1986). Riverine sources have relatively high (0.712) and oceanic sources lower (0.703) ratios, although changes in source-rock weathering, rates of sea-floor spreading, and ocean hydrothermal activity modulate strontium inputs over long timescales (Shields 2007). In general, the strontium isotopic curve is used extensively as a correlation tool (McArthur and Howarth 2004) as well as a paleoclimate proxy.

Veizer's Phanerozoic strontium curve incorporates more than 4000 measurements (Figure 3.6). Except for parts of the Mesozoic era, when mid-ocean ridge strontium sources may have dominated (Goddéris and François 1995; Jones and Jenkyns 2001), the strontium curve signifies changes in mean ocean ^{87}Sr/^{86}Sr caused by mean global weathering rates related to continental topography, atmospheric conditions, temperature, and factors such as ocean magmatism. In addition to the oxygen and strontium curves, Veizer compiled a Phanerozoic carbon isotopic curve. Together the three curves suggest that, to a first approximation, there is dominant tectonic control over global biogeochemical cycling and that higher-frequency variations in the strontium curve for specific geological intervals may involve additional climatic factors.

Atmospheric CO$_2$

One major issue in deep-time paleoclimatology is whether atmospheric CO$_2$ concentrations directly control or amplify large-scale climatic fluctuations over millions of years (Kump 2001, 2002; Crowley and Berner 2001). Complex as this topic may be, there are two fundamental approaches—(1) the empirical reconstruction of paleo-CO$_2$ concentrations and (2) the modeling of long-term global carbon cycle incorporating Phanerozoic biogeochemical and tectonic cycles. Four primary proxy methods are used to reconstruct CO$_2$ concentrations for geological intervals prior to the last 800 ka, the interval for which direct measurements of CO$_2$ are available from air trapped in Antarctic ice (see chapters 5–7 in this volume). This summary is adopted from Royer et al. (2001b), Beerling and Royer (2002), and Royer (2006).

Pedogenic Carbon Isotopes The δ^{13}C of pedogenic (soil) carbonates is one primary method for CO$_2$ reconstruction (Ekart et al. 1999). Carbon dioxide in paleosols (fossil soils) comes from both atmospheric and biologically produced sources. The carbonate ion in soil CaCO$_3$ is derived largely from CO$_2$ produced by biological respiration during decomposition of organic matter and respiration by roots (Cerling et al. 1989; Quade et al. 1989). Computing paleo-CO$_2$ from soil carbonates must take into account temperature, soil porosity, geochemical depth profiles in soils, diffusion, respiration rate, the δ^{13}C of respired CO$_2$, and other factors. If these processes can be constrained and modeled, then pedogenic soil δ^{13}C values should reflect large-scale changes in atmospheric CO$_2$.

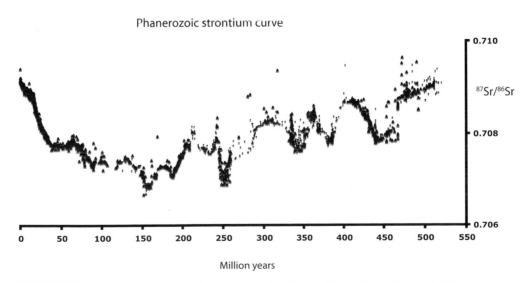

Phanerozoic strontium curve

^{87}Sr/^{86}Sr

Million years

FIGURE 3.6 Phanerozoic strontium isotope curve (^{87}Sr/^{86}Sr) published by and compiled from numerous sources by Veizer et al. (1999, updated 2004), plotted against the age model of Gradstein et al. (2004). Data sources are given at http://www.science.uottawa.ca/~veizer/isotope_data/. Courtesy of J. Veizer.

This method has been tested in Holocene soil profiles (Cerling et al. 1989) and is especially useful in pre-Cenozoic paleo-atmospheric reconstruction (Cerling 1991; Mora et al. 1996; Ekart et al. 1999).

Leaf Stomata Fossil leaf stomatal pores are used as a paleo-CO_2 method over all timescales (Mora et al. 1996; Royer 2001b; Beerling and Royer 2002). The method hinges on basic plant physiology whereby stomata help regulate and minimize water loss during photosynthesis. Higher atmospheric CO_2 concentrations in theory allow less stomatal pore area (size, density, or both) and minimize water loss. One stomatal index (SI) relies on the empirical relationship between the number of stomatal epidermal cells on leaf cuticles from plants collected over the past 150 years and CO_2 concentrations that have risen during this time because of anthropogenic activity (Beerling 2002). McElwain and colleagues (McElwain and Chaloner 1995, 1996; McElwain 1998; McElwain et al. 1999) called their index a stomatal ratio. Stomatal methods have complications arising from physiological variability among tree taxa and the nonlinear stomatal response at elevated CO_2 concentrations.

Algal Biomarkers $\delta^{13}C$ from di-unsaturated long-chained alkenones (organic biomarkers in sediment) are linked to specific phytoplankton algal groups and are used to reconstruct paleo-CO_2 back to the late Cretaceous period (Hayes et al. 1989; Jasper and Hayes 1990). This method relies on the relationship between atmospheric CO_2, CO_2 dissolved in seawater, and carbon isotopic fractionation in organic material during photosynthesis (Freeman and Hayes 1992). It is most useful in Cenozoic marine sediments where alkenones of Prymnesiophyceae (haptophytic) algae are preserved (Pagani et al. 1999, 2005) and, with less precision, in the Cretaceous period. Like the pedogenic carbonate method, alkenone paleo-CO_2 reconstructions involve extremely complex, partially understood, organic geochemical relationships. Complications arise, for example, from biological factors related to algal cell growth, nutrient conditions, independent paleotemperature estimates, and diagenesis. This method is discussed further in Chapter 4 in this volume.

Boron Isotopes The boron isotopic composition in foraminifers ($\delta^{11}B$) was introduced as a way to reconstruct atmospheric CO_2 because boron composition is related to ocean pH and alkalinity, which in turn are influenced by atmospheric CO_2 and its uptake by the ocean (Pearson and Palmer 2000). Boron isotopes have certain drawbacks related to changes in the mean ocean boron isotopic composition, seawater pH, temperature, and postdepositional diagenetic effects.

Phanerozoic CO_2 Curves Figure 3.7 shows a CO_2 reconstruction based on pedogenic carbonates and SIs as compiled by Royer et al. (2004) and Royer (2006). This plot

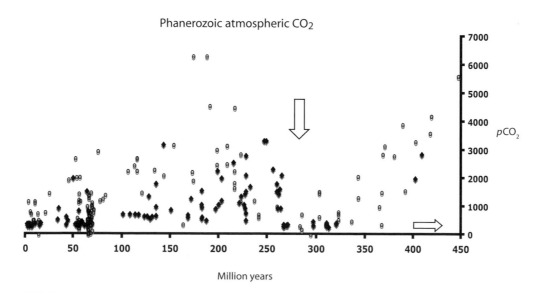

Phanerozoic atmospheric CO_2

pCO_2

Million years

FIGURE 3.7 Estimates of atmospheric CO_2 concentrations (pCO_2) based on stomatal index (solid diamonds) and soil carbonates (open circles), compiled by Royer 2006 (see also Royer et al. 2001b, 2004) from more than 40 publications. The timescale is from Gradstein et al. (2004). The CO_2 concentration in the year 2008 was about 385 ppmv (horizontal arrow). Error bars on many estimates are considerable, depending on the method used and the material analyzed (see text). The large vertical arrow shows late Carboniferous low CO_2 and cold climate. See Chapter 4 in this volume for the Cenozoic CO_2 reconstruction from phytoplankton geochemistry and chapters 5–7 for CO_2 reconstructions from ice cores.

does not show the error bars on paleo-CO_2 estimates, which can be substantial especially for pre-Cenozoic intervals. Some highlights of the curve are the low values during the late Paleozoic era, when extensive glaciations occurred, and the exceptionally high values during much of the Mesozoic era, when climate generally was warm although punctuated by several cooler episodes. As already mentioned, Veizer et al. (2000) suggested a decoupling of CO_2 and climate at the Ordovician-Silurian boundary and the late Jurassic-early Cretaceous periods. However, Beerling (2002) used arborescent lycopod leaf SI values to infer that Carboniferous 330–300 Ma and Permian 270–260 Ma values were 344 and 313 ppmv, respectively, roughly halfway between preindustrial and modern levels. These analyses supported a first-order link between Paleozoic climate and atmospheric CO_2. Royer (2006) also argued that, at least for geological intervals where data are sufficient, there is a first-order relationship between temperature and CO_2 since the Ordovician period (~420 Ma). Cooler Phanerozoic temperatures correspond to atmospheric CO_2 concentrations below 1000 ppmv, large-scale glaciations occur below a threshold level of 500 ppmv (perhaps higher when solar luminosity is lower), and cool but nonglacial episodes occur during Mesozoic global warmth when concentrations dip below 1000 ppmv. Royer's critical analysis leads to the conclusion that CO_2 forcing of climate occurs over all timescales though

detailed pre-Cenozoic CO_2-temperature relationships at timescales <10 Ma are sparse.

Rothman (2002) recently gave another interpretation for Veizer's $^{87}Sr/^{86}Sr$, $\delta^{18}O$, and $\delta^{13}C$ curves and the Phanerozoic CO_2-climate relationship. Rothman argued that when the effects of sediment recycling are removed from the strontium curve, an integrated $^{87}Sr/^{86}Sr$-$\delta^{13}C$ analysis yields a new paleo-CO_2 curve that suggests a decoupling of climate and CO_2. There is "no systematic correspondence with the geologic record of climatic variations at tectonic timescales." This analysis used the widely cited compilation of glacial periods from Frakes et al. (1992), an excellent source of glacial geology but not necessarily a continuous or quantitative record of global, tropical, or high-latitude surface temperature. Acknowledging these new results does not disprove a climate-CO_2 link. Rothman suggests additional work is needed.

GEOCARB Model

In a series of studies, Yale's Robert Berner developed a model of global carbon and other biogeochemical cycles to produce the Phanerozoic CO_2 curve in Figure 3.8, often compared to empirical CO_2 reconstructions for timescales of 10–30 Ma (Berner 1990, 1991, 1994). The GEOCARB model incorporates more than 11 biogeochemical equations for weathering rates (especially for weathering of continental silicate rocks); CO_2 flux from volcanic,

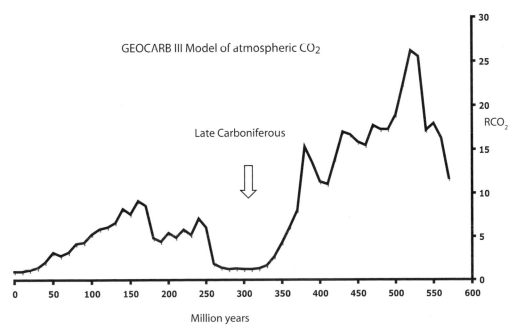

FIGURE 3.8 GEOCARB III model of Phanerozoic CO_2 concentrations. The y-axis plots CO_2 concentrations as a multiple of preindustrial levels of 280 ppmv. Thus an RCO_2 value of 5 means 5×280 ppmv. From Berner and Kothavala (2001). GEOCARB III is a low-resolution (~30 Ma) model of atmospheric CO_2; the error bar (not shown) varies through the Phanerozoic eon. The late Carboniferous (314–300 Ma) interval had low CO_2 and ocean temperatures like today's, shown by the arrow (see Came et al. 2007).

metamorphic, and carbonate diagenetic processes; carbon burial and uptake in carbonate rocks; mountain uplift; and global hydrological cycling, among other processes. GEO-CARB III improves on the quantitative relationships between weathering and uplift, and weathering and land plants, paleogeography, the hydrological cycle, and river runoff (Berner and Kothavala 2001).

The estimated atmospheric CO_2 concentrations produced by the GEOCARB III model are given as RCO_2 (multiples of preindustrial values). Exceptionally high CO_2 concentrations occurred during the early to mid-Paleozoic era (10–25 times preindustrial levels), falling dramatically to near preindustrial levels during the late Paleozoic, and in the Mesozoic era rising to 3–8 times preindustrial values. In general, there is good agreement between empirical and modeled patterns in figures 3.7 and 3.8.

Bartdorff et al. (2008) recently modeled Phanerozoic methane concentrations, showing a significant rise to 10 ppmv during the Carboniferous period (~290–300 Ma), when coal-forming swamps dominated much of the terrestrial landscape. Elevated methane concentrations may have accounted for radiative forcing of atmospheric temperature of 1°C. Moreover, methane may have played a role in moderating cooling that occurred during the Permo-Carboniferous glacial interval, when atmospheric CO_2 concentrations decreased.

Global Sea Level

In 1977 Peter Vail and colleagues at Exxon Production and Research (EPR) published a monographic study of relative sea-level changes during the Phanerozoic eon based on several dozen seismo-stratigraphic records from several dozen continental margins (Vail et al. 1977). Influenced by Lawrence Sloss and other predecessors' classic studies of North American stratigraphy of marine transgressive-regressive sedimentary sequences, EPR incorporated advances in seismic geophysical methods to produce what became known as the Vail or EPR sea-level curve. Their widely cited pre-Cenozoic curve published in *Science* (Vail et al. 1977) was a summary of detailed methodological descriptions of seismic interpretations and sea-level curves for the Paleozoic, Mesozoic, and Cretaceous-Cenozoic, published in 10 chapters of an American Association of Petroleum Geologists Memoir (Payton 1977).

The Vail curve was in essence a compilation of marine sedimentary sequences from relatively rapidly subsiding sedimentary basins from passive (relatively tectonically stable) continental margins. These are called onlap-offlap sequences, referring to alternating stratigraphic sequences formed during periods of onlapping (rising, or high) sea level and offlapping (falling, or low) sea level. Most basins

used in the construction of the Vail curve were located in North America, but there were 4 from West Africa, 3 from Europe, 3 from Indonesia, 3 from Australia and New Zealand, and 5 from Central and South America. Based on hypothesized synchroneity of global sea-level changes, the Vail curve included several "orders" of cycles: first order, 100 Ma or longer; second order, 10–100 Ma; third order, 1–10 Ma; and high-frequency events <1 Ma. They acknowledged that third- and higher-order cycles since the Oligocene epoch were probably influenced by glacio-eustatic sea-level fluctuations, which in Chapter 4 we see turned out to be largely correct. About the same time, Pitman (1978) published a sea-level curve for the late Cretaceous period through late Miocene epoch based on rates of sea-floor spreading and resulting changes in mid-ocean ridge volume. A revised EPR sea-level curve published by Bil Haq, Vail, and colleagues covering the Triassic and younger periods and integrating additional age control (Haq et al. 1987) became known as the Haq curve. Soon after their publication, the EPR and Haq curves received scrutiny and criticism (e.g., Miall 1997) revolving mainly around the synchroneity and scale of sea-level oscillations. An excellent review of these developments can be found in Hallam (1992).

In the past 20 years, the use of oxygen isotopic records mainly from deep-sea foraminifera has revolutionized paleoceanography in general, and refined the long-term sea-level record in particular. Combining isotope records from more than 40 deep-sea core sites, many along the New Jersey margin, Ken Miller and colleagues from Rutgers University produced a new Phanerozoic sea-level curve that attempts to reconcile the oxygen isotope record of ice volume with continental-margin stratigraphy and the Haq curve (Miller et al. 2005). The nearshore sediment record was corrected for postdepositional subsidence caused by compaction and thermal cooling, using what is called the backstripping method (Kominz et al. 1998; Van Sickel et al. 2004). The oxygen isotope record corrected for changes in water temperature and calibrated to the coastal stratigraphy at key chronological tiepoints provides a revised sea-level curve.

Figures 3.9 and 3.10 show the Miller curve since 170 Ma and the Haq curve since 240 Ma. One can see that the Haq curve shows a sustained high sea level from 100–200 m above those of the present day, with a notable Cretaceous maximum when a continuous seaway inundated western North America and other low-lying continental regions (Figure 3.4). Progressive but stepwise drops in sea level characterized the Cenozoic era. In contrast, the Miller curve shows significantly lower absolute values for sea level during the Mesozoic era, exceeding ~50 m above the present level only briefly during the Cretaceous period and Eocene epoch. Recently, several

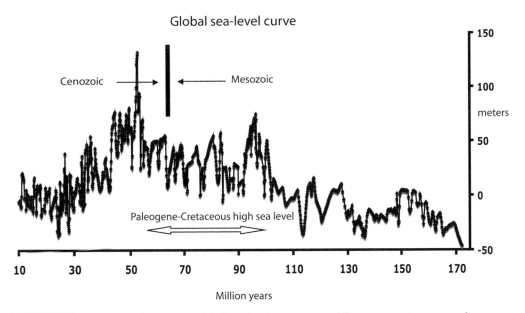

FIGURE 3.9 Mesozoic and Cenozoic global sea-level reconstructed from oxygen isotopes and backstripped continental margin stratigraphy and subsidence model, from Miller et al. (2005). See Chapter 4 in this volume for the detailed sea-level record during the last 65 Ma. Note the high sea level in the Cretaceous and Paleogene periods.

studies have refined the Mesozoic and Paleozoic portions of the global sea-level curve. Muller et al. (2008) reconstructed the Mesozoic sea level and proposed that the New Jersey margin has been subjected to more subsidence than accounted for by Miller, which led to an underestimated sea level for this interval. They proposed that Mesozoic global sea level was actually somewhere between the higher estimates in the Haq curve and the lower estimates of Miller and colleagues.

Haq and Schutter (2008) revised the Paleozoic estimates from 542–251 Ma using unconformity-bounded stratigraphic sequences from several continents. They discovered, among the highlights of this study, 172 distinct eustatic sea-level events ranging in amplitude from a few meters to 125 m, as well as that a global high sea level about 225 m above present sea level was reached in the late Ordovician period, followed by a rapid fall in the latest Ordovician.

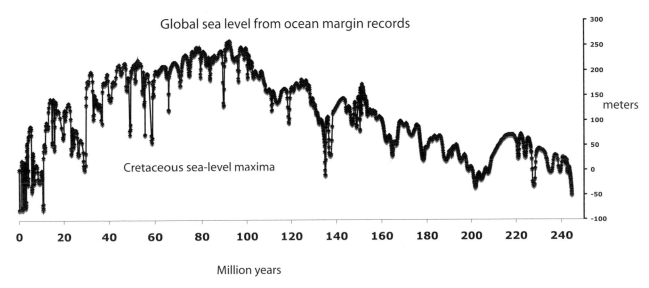

FIGURE 3.10 Sea-level curve from Haq et al. (1988), constructed from ocean and continental margin stratigraphic onlap-offlap record of marine transgression. Note high sea-level intervals in the Cretaceous period. Sea-level maxima in the Haq curve are more than 100 m higher than those in the Miller et al. (2005) New Jersey curve in Figure 3.9.

Permo-Carboniferous Climate: The Role of Carbon Dioxide

The end of the Paleozoic era from 300–270 Ma experienced one of the largest Phanerozoic transitions when global climate emerged from an extended glacial world, unparalleled until the late Cenozoic era, to a much warmer climate. This global change is marked by the shift from the Late Paleozoic Ice Age (LPIA) at the end of the Carboniferous and earliest Permian periods to middle-Permian warmth. Emerging evidence suggests the global change can be attributed to a 5–10-fold increase in atmospheric CO_2 concentrations, from near preindustrial levels (200–300 ppmv) to greater than 2000 ppmv.

Evidence for this greenhouse gas-driven climate change comes from stratigraphic, paleobotanical, paleoceanographic, sea-level, and geochemical records throughout the world. Glacial deposits are known from relict terranes of the supercontinent Gondwanaland preserved in southern Africa, Argentina, Tasmania, Australia, and Antarctica (Banks 1981; Frakes et al. 1992; Scotese et al. 1999; Isbell et al. 2001, 2003; Veevers 2004). Late Paleozoic glacial history inferred from glaciogenic sediments and glacial features has traditionally been linked to cyclic sedimentary sequences in low-latitude regions in the midcontinent of North America that was part of the Pangea supercontinent (Wanless and Shepard 1936; Crowell 1999; Scotese et al. 1999; Cecil and Edgar 2003). The "cyclothems" are believed by many to be caused by glacio-eustatic sea-level oscillations as large as 60–200 m in amplitude, possibly driven by orbital climatic cycles (see Chapter 5 in this volume).

Isbell et al. (2003) revised the traditional interpretation of Permo-Carboniferous Gondwanan glacial and eustatic sea-level history based on geological evidence their group obtained mainly from Antarctica. In this scenario, the Gondwanan ice sheet in Antarctica may have been at times smaller than previously believed. Though glaciers and ice caps existed in the late Paleozoic, they may have been thinner and covered less area and might not have exerted primary glacio-eustatic control over sea level.

Late Paleozoic equatorial climate has been reconstructed from sediments deposited in the Midland, Anadarko, Pedregosa, Paradox, and Grand Canyon Embayment basins of the western United States (Montañez et al. 2007). Paleogeographical reconstructions show that this region was part of a large equatorial belt on and around Pangea, which formed when Gondwanaland merged with Laurussia (Figure 3.4) (see, e.g., Frakes et al. 1992; Golonka and Ford 2000; Torsvik et al. 2004; Veevers 2004).

Proxy records for atmospheric CO_2 are derived from carbon isotopes from soils, plant fragments, coal and charcoal, (Ekart et al. 1999; Royer et al. 2001b; Beerling 2002;

Royer 2006; Montañez et al. 2007) and SSTs from $\delta^{18}O$ measurements mainly from tropical brachiopods (Mii et al. 1999, Veizer et al. 1999; Korte et al. 2005). Most of these and other isotope records are available in the databases of Royer (2006) and Veizer (1999, 2000). After accounting for salinity, pH, and ice volume contributions to the $\delta^{18}O$ signal, Montañez estimates that tropical late Paleozoic SSTs varied 4–7°C during glacial cycles. From trough (~300 Ma) to peak (~280 Ma), mean tropical SST temperature rose more than 15°C at the end of the LPIA, though there is a large uncertainty associated with Paleozoic $\delta^{18}O$-based paleothermometry. In general, the Montañez group concluded that there is a strong co-variance between temperature, atmospheric CO_2, and glaciogenic sedimentation during the Permo-Carboniferous transition, suggesting CO_2 forcing of climate and biotic diversity at least at >1 Ma timescales for this period.

For a century, research on Paleozoic climate was segmented into sedimentological, stratigraphic, glaciological, paleogeographical, paleobiological, and other disciplines. Integrated studies such as those highlighted here provide a more cohesive, though still incomplete, picture of the interplay between geography, climate, sea level, glaciation, and atmospheric CO_2. Pending more precise paleo-CO_2 proxies, the "CO_2 paradigm" proposed by Crowley and Baum (1992) on the basis of climate modeling of the late Paleozoic remains a leading hypothesis to explain the largest global warming in the last 500 Ma. Solar luminosity, paleogeography, continental topography, glacial dynamics, and orbital insolation also influenced climate during the late Paleozoic climate transition.

Jurassic and Cretaceous Oceanic Anoxic Events and Greenhouse Climate

Background

Jurassic and Cretaceous Oceanic Anoxic Events (OAEs) were extraordinary global geological and climatic events witnessing extreme climatic warmth, massive organic-carbon burial and carbon-cycle disruption, near-global anoxia in the world's oceans, and major biotic crises. The name OAE comes from Schlanger and Jenkyns's (1976) paper on Cretaceous (Aptian-Albian and Cenomanian-Turonian) deep-sea black shales. Black shale is a lithology typical of marine sediments deposited in shallow-water Cretaceous environments preserved and exposed on today's continents. The discovery of black shales in deep-sea deposits led to the proposal that they signified global ocean anoxia caused by expansion of the ocean's oxygen minimum zone (OMZ) to abyssal depths as a mechanism to explain these distinct and pervasive Cretaceous facies. Evidence came mainly from not only the distinct lithology, but also the geochemistry, micropaleontology, and global distribution of OAE sediments. Cretaceous OAE sediments are also known for their characteristic large positive carbon isotope excursion ($\delta^{13}C$) and major faunal and floral changes (Arthur et al. 1987; Schlanger et al. 1987). Another reason Jurassic and Cretaceous black shales are so well studied is they contain immense amounts of buried organic carbon that constitutes a large proportion of the world's known petroleum source rocks (Arthur and Schlanger 1979; Ulmishek and Klemme 1990).

Ironically, OAEs are now under intensive study because they signify an extreme paleoclimate state when large, rapid injections of CO_2 into the ocean-atmosphere system coincided with oceanographic change and global climate warming. The very carbon buried during past global warmth, recovered and burned as fossil fuel, is now returning to the world's atmosphere and ocean, with potentially serious oceanographic, biotic, and climatic consequences (see chapters 11 and 12 in this volume). A similar situation holds for the Paleocene-Eocene Thermal Maximum (PETM) at 55 Ma, another period of global warmth and rapid carbon-cycle disruption, considered an "incipient" OAE by Cohen et al. (2007) (see Chapter 4 herein). In this section we describe the stratigraphy, geochemistry, and causes of OAEs.

Ages of OAEs

Since Schlanger and Jenkyns's paper, additional Jurassic and Cretaceous OAEs have been identified and assigned numbers and names based on the region in which they were identified in local stratigraphic sequences (Table 3.2). The major OAEs considered to be global oceanic events occurred during the early Toarcian (Jurassic, 183 Ma), the Valanginian-Hauterivian (early Cretaceous, 132 Ma), the early Aptian (OAE 1a, 120 Ma), and near the Cenomanian-Turonian boundary (OAE 2, 93 Ma). Other regional OAEs occurred during the earliest Albian-Aptian (OAE 1b–1d, 109–113, 102, 99–100 Ma) and at the Coniacian-Santonian boundary (86 Ma). OAE 2 (93 Ma) and the subsequent mid-Turonian stage (91.5 Ma) experienced some of the warmest global temperatures and highest sea levels of the last 115 Ma (Huber et al. 2002) and constituted a major turning point in climate history.

Characteristics of OAEs

Like Neoproterozoic glacial-carbonate sequences, not all OAEs are alike but certain features are common to most events (Arthur et al. 1987). Each is recognized by a large positive carbon isotope excursion in carbonate and organic carbon. Figure 3.11 shows the carbon isotopic record from Erbacher et al. (1996) as compared to the $^{87}Sr/^{86}Sr$ curve of Bralower et al. (1997) as compiled by Leckie et al. (2002).

TABLE 3.2 Jurassic and Cretaceous Ocean Anoxic Events

System	OAE Number*	Stage	Age (Ma)	Event Name	Characteristics
Cretaceous	OAE 3	Coniacian-Santonian	86		
Cretaceous	OAE 2[†]	Cenomanian-Turonian Boundary	93	Bonarelli	
Cretaceous	OAE 1d	Late Albian	99	Breistroffer	
Cretaceous	OAE 1c	Late Albian	102	Toolebuc	
Cretaceous	OAE 1b	Late Aptian–early Albian	109–113	Leenhardt-Urbino, Paquier Monte Nerone, Jacob-113	Tripartate peaks, various names
Cretaceous	OAE 1a[†]	Early Aptian	120	Goguel, Selli	
Cretaceous	Weissert[†]	Valanginian-Hauterivian	132	Weissert	
Jurassic	OAE[†]	Toarcian	183	Posidonienschiefer	5 per mil ^{13}C shift, 6–7°C, 1000 ppmv CO_2 rise

Note: Several other possible Jurassic and Cretaceous regional OAEs have also been identified. See Schlanger and Jenkyns (1976); Jenkyns (1980), Arthur et al. (1987); Bralower et al. (1994); Leckie et al. (2002); and Takashima et al. (2006). OAE 2 dated through astronomical tuning (see Meyers and Sageman 2007). OAE = Oceanic Anoxic Event.

*Some authors use roman numerals, e.g., OAEII.

[†]Global events.

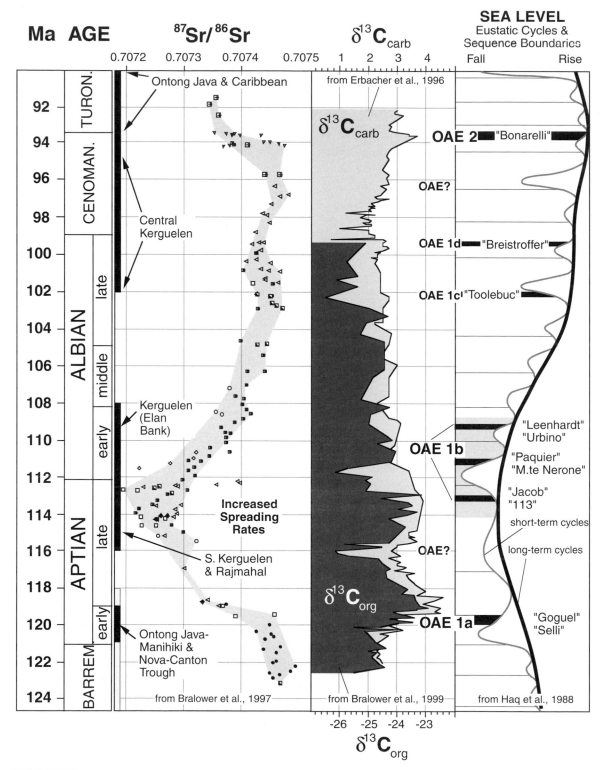

FIGURE 3.11 Paleoclimate record of Cretaceous Oceanic Anoxic Events (OAEs) 1a–d and 2, from Leckie et al. (2002). The strontium and carbon isotope curves are from Bralower et al. (1994, 1997). The "eustatic" sea-level curve is from Haq et al. (1988) (see also Figure 3.9). Courtesy of M. Leckie and American Geophysical Union.

Note that peak carbon isotopic excursions (CIEs) occurred during OAE 2 and OAE 1a. The $\delta^{13}C$ excursion is usually 2–3‰ but was as high as 5–7‰ during the Toarcian event and is separated by a short-term negative excursion (Jenkyns and Clayton 1997; Hesselbo et al. 2000). It is recognized in fossil wood as well as marine carbonates and organic carbon and thus signifies a global excursion (Hesselbo et al. 2007). Wagner et al. (2008) found a −1.5‰ change in $\delta^{13}C$ in leaf wax n-alkanes at the beginning of OAE 1b (early Albian) at Deep Sea Drilling Program (DSDP) Site 545 off northwest Africa. When viewed with isotopic data from other records, these results are consistent with the hypothesis of rapid release of light carbon into the atmosphere. Some OAE CIEs last about 200,000 years and exhibit short-term fluctuations around a positive mean (Kemp et al. 2005; Cohen et al. 2007).

Abrupt warming of 5–10°C and extreme temperatures occurred during OAEs. Maximum ocean temperatures reached as high as 42°C on the Demerara Rise in the South Pacific, and 31°C on the Falkland Plateau at a paleo-latitude of about 60°S (Huber et al. 2002; Bice et al. 2003). These temperatures are roughly 10–15°C and 25–30°C higher than modern tropical and subpolar temperatures at similar latitudes. These and other records show there was a reduced pole-to-equator thermal gradient during Cretaceous OAE events. Bathyal-depth ocean temperatures were also relatively warm, up to 20°C in the North Atlantic (Norris and Wilson 1998; Fassell and Bralower 1999; Huber et al. 1999, 2002). OAE 1b experienced an abrupt warming of 3–5°C (Wagner et al. 2008).

As seen here, SIs and pedogenic carbonate proxy records show that atmospheric CO_2 levels were elevated during the Mesozoic era. Bice and Norris 2002 and Bice et al. 2003, 2006 estimated a range of 600–2400 ppmv for at least some OAE intervals. Global increases in CO_2 are probably related to flux from enhanced igneous activity (tectonic forcing) (Bice and Norris 2002), although P. Wilson et al. (2002) suggest that subduction of carbonates at plate boundaries might have contributed to elevated CO_2 levels.

The high organic content, the disappearance or dissolution of microfossil phytoplankton and zooplankton groups, deformed foraminifera, and depauperate benthic faunas in stratigraphic intervals dominated by OAEs indicate widespread depletion of dissolved oxygen in the oceans. Changes in surface ocean microfaunas and floras, productivity, nutrient utilization, and ocean stratification are also evident, although there is spatial and bathymetric variability as well as distinct patterns among OAEs (Bralower 1988; Premoli Silva et al. 1999; Erba 2004; Eleson and Bralower 2005; Hardas and Mutterlose 2007). Arthur et al. (1987) noted "benthic-free zones" and depauperate faunas associated with the Cenomanian-Turonian OAE sediments, but the response of benthic faunas to oceanic change during OAEs varied greatly with paleodepth (Holbourn and Kuhnt 2001). In addition to micropaleontological data, molybdenum concentrations also indicate that there were precession-driven oscillations in dissolved oxygen at least for the Toarcian OAE (Pearce et al. 2008).

The strontium curve offers insight into OAE events. A decrease in $^{87}Sr/^{86}Sr$ ratios may mean enhanced hydrothermal strontium sources, perhaps related to the emplacement of the South Kerguelen Plateau or other igneous bodies. Younger OAEs might correspond with ocean-crust production in the central Kerguelen and Ontong-Java Plateaus, but views vary on which oceanic plateau or flood basalts are associated with each OAE. The strontium curve for the Toarcian OAE is linked to increased seafloor hydrothermal production rather than riverine strontium sources (Jones and Jenkyns 2001).

Biological changes in most major marine fossil groups, including extinction and speciation, accompanied most OAEs (Fischer and Arthur 1977; Bralower et al. 1994; Leckie et al. 2002; Erba 2004). Leckie et al. (2002) noted massive marine faunal turnover (speciation plus extinction) of planktonic microfossil groups at OAE events. Premoli Silva et al. (1999) found major biotic crises in the tropical Tethyan Ocean during Cretaceous OAEs. In contrast, Holbourn and Kuhnt (2001) found taxonomically similar benthic faunas stratigraphically above and below a barren black shale zone but no major extinction during OAE 1b. Although beyond our scope here, suffice it to say that anoxic events had major impacts on the evolution of some marine biotas, with repercussions for global biogeochemical cycling.

Causes of OAEs

Ocean Processes Two primary models have been proposed to explain ocean-wide anoxia during OAEs. The first, proposed by Schlanger and Jenkyns, involves a deepening of ocean margin OMZs reaching abyssal depths in the most extreme cases (Pedersen and Calvert 1990, Meyers 2006, Takashima et al. 2006). In this scenario, enhanced surface ocean productivity in surface water above the OMZ due to stronger upwelling decrease dissolved oxygen levels. The second idea involves the development of a stagnant ocean from density stratification similar to that in a partially mixed estuary where a strong pycnocline inhibits exchange between well-aerated surface and oxygen-depleted deep waters. Deep-bottom temperature warming and/or increased thermohaline circulation may have also been factors (Huber et al. 1999; Erbacher et al. 2001).

Oxygen depletion is a complex phenomenon in isolated marine (euxinic) basins such as the modern Black Sea and in partially mixed estuaries like Chesapeake Bay, as well as in the OMZ on ocean margins. Mechanisms that contribute to oxygen depletion include density stratification, terrestrial-to-ocean organic-carbon transport, surface ocean productivity, ocean-crust production, and CO_2 outgassing (Ryan and Cita 1977; Thierstein and Berger, 1978; Dean and Gardner 1982; Jones and Jenkyns 2001). At present, there is no consensus on a single preferred ocean-oxygen model and it is possible that different OAE events experienced different conditions.

Carbon Source The source of elevated atmospheric CO_2 during OAEs is an important unresolved issue. For the Toarcian event, the heating of massive coal deposits in India by major igneous activity might release methane and produce a carbon anomaly (McElwain et al. 2005; Beerling and Brentnall 2007). The disassociation of methane gas trapped in ocean sediments may also have played a role in Mesozoic OAEs and their abrupt climate changes. Hesselbo et al. (2000) and Kemp et al. (2005), among others, proposed that large quantities of the light isotope ^{12}C injected into the earth's oceans from biogenic methane frozen in continental shelf sediments caused negative CIEs. The release of methane hydrates would oxidize, reduce dissolved CO_2, and elevate atmospheric CO_2 concentrations, as proposed by Dickens et al. (1995, 1997) for the PETM. Beerling et al. (2002) modeled the mass balance of carbon exchange during seafloor methane release and estimated that during Toarcian and Aptian OAEs, there was a release of 5000 and 3000 Gt C, an atmospheric CO_2 rise of 900 and 600 ppmv, and warming of 2.5–3.0°C. Kemp's evidence also suggests Toarcian carbon anomalies were controlled over shorter timescales by astronomical forcing.

If the OAE isotopic excursions are not global in extent, they might represent local phenomena and not global carbon-cycle perturbations. Wignall et al. (2006) questioned the timing, magnitude, and geographical distribution of the Toarcian events, contradicting the interpretation that these were truly global events (see van de Schootbrugge et al. 2005; Kemp et al. 2006). Hesselbo et al. (2007) argued that, in addition to the shape of the strontium curve, the exact match between wood and marine CIEs in widely distributed sections supports a massive injection of methane during the Toarcian OAE, and they add that greater atmospheric water vapor and CO_2 may have contributed to this climate anomaly.

Kump et al. (2005) proposed an elaborate hypothesis for OAE warming and biotic extinctions in which hydrogen sulfide (H_2S) gas upwelled to the surface ocean and was transferred to the atmosphere, leading to mass extinctions. If the hydrogen sulfide influenced the ozone layer, it might have diminished its protective role in shielding the earth from ultraviolet radiation, further contributing to its biotic impact. Ultimately, CO_2-forced ocean warming was the cause of the H_2S gas release. Additional work sorting out the primary and secondary factors that contribute to carbon isotopic anomalies is needed, especially at high-resolution timescales finer than 1 Ma during various OAEs.

Large Igneous Provinces Tectonic forcing by large igneous provinces (LIPs) is one possible triggering mechanism for OAEs and associated ocean-climate changes. LIPs are intrusions that form in both continental flood basalts and large oceanic plateaus that were common in many regions during the Mesozoic era (Coffin and Eldholm 1994, 2005). Larson's (1991a, b) influential proposal that LIPs caused by "superplume" core-mantle processes account for Mesozoic OAEs prompted intense investigation into this forcing mechanism (Condie 2001; Condie et al. 2002). Evidence for superplumes came from the lack of magnetic anomalies in oceanic basalts, the existence of major Cretaceous igneous provinces (plateau basalts), and paleoceanographic and climatic features of OAEs (Bralower et al. 1997; Leckie et al. 2002).

There are differing viewpoints on how to define LIPs. Sheth (2007), for example, proposed a formal classification of LIPs emphasizing their size and the mechanisms causing them and including relatively small igneous deposits. Bryan et al. (2002) and Courtillot and Renne (2003) include only deposits with large areal extent and volume: LIPs are those igneous bodies having areas larger than 0.1 million km^2 and volumes greater than $0.1 \times 10^6 km^3$, although most LIPs exceed $1 \times 10^6 km^2$ and $1 \times 10^6 km^3$. Coffin and Eldholm (2005) summarized the global distribution and age for the earth's LIPs and showed that a large proportion formed during the Cretaceous period between 120 and 80 Ma (see Courtillot and Renne 2003).

LIPs linked to OAEs and Mesozoic climate extremes include the African Karoo series, the Kerguelen Plateau in the Indian Ocean, the Ontong-Java Plateau in the Pacific Ocean, the Falkland Plateau off southern Africa, and the Caribbean Plateau (during the Cretaceous period located in the eastern equatorial Pacific); the Rajmahal (Jurassic) and Deccan (late Cretaceous) Traps of India; and the Parana flood basalts of South America. In his initial LIP hypothesis, Larson explained marine geophysical, climatic (temperature), oceanographic (carbon isotopic), and sea-level (high eustatic sea level) data through the superplume hypothesis. Many workers accept mantle superplumes as a possible mechanism, but Sheth (1999, 2005) has argued that a plume mechanism is not appropriate at least for some continental basalt such as the Indian Deccan Traps.

Jones and Jenkyns (2001) proposed that strontium, CIEs, and the asynchroneity of anoxia and sea-level changes meant that the CO_2 degassing from hydrothermal activity caused by higher rates of ocean-crust production on mid-ocean ridges and plateaus was a critical factor for the development of OAEs. By modeling strontium isotopes during the Mesozoic era, either a reduction in riverine strontium by changes in sea level, climate, or continental source rock or a massive increase in hydrothermal strontium could explain observed patterns. They favored a mid-ocean ridge hydrothermal strontium source as opposed to oceanic plateaus.

One critical issue—the subject of a large literature in igneous geology and volcanology—is the duration of emplacement of igneous provinces. The well-studied late Cretaceous Deccan Traps may have formed in only about 1 Ma. The short duration, episodic nature, and concentration of Cretaceous LIPs supports the view that they are potential mechanisms influencing the Cretaceous ocean-atmosphere system at least over timescales of millions of years.

OAE Summary   Mesozoic global warmth and OAEs involved the almost unfathomable interplay between massive igneous activity, biogeochemical cycling, ocean productivity, dissolved oxygen, carbon exchange between sediment-ocean-and-atmospheric reservoirs, ocean and atmospheric temperature, and tectono-eustatic sea-level changes. Superimposed on ~>1 Ma-year processes is orbital control of abrupt cyclic climate-ocean oscillations possibly including small to mid-size continental ice sheets. The causal link between LIP events and Mesozoic global warmth involves testable hypotheses that require better constraints on the age and duration of LIPs, especially those located in modern ocean basins; better multi-proxy reconstructions for ocean, climate, and atmospheric carbon concentrations during OAEs; and sophisticated climate and carbon-cycle modeling.

Mesozoic Climate and Pole-to-Equator Thermal Gradients

Background

The earth's pole-to-equator thermal gradient represents a fundamental climatological property related to equator-to-pole heat transport efficiency, hydrological cycling, and atmospheric chemistry. The meridional thermal gradient, or ΔT, is a measure of the pole-to-equator thermal gradient that can be estimated for the most extreme climate states. Today, during a relatively warm period in an otherwise cold late Cenozoic era, ΔT is about 33°C. During a globally warm Mesozoic climate, ΔT was about 19–23°C, depending on which tropical SST estimate one accepts (e.g., Huber et al. 1995; P. Wilson et al. 2002; Bice et al. 2003). On the basis of new data from the central Arctic Ocean from Ocean Drilling Program (ODP) Leg 302, the nearly as warm PETM at 55 Ma had a ΔT of 15°C (Sluijs et al. 2006). Ocean heat transport has been a leading hypothesis to explain polar warmth and a reduced ΔT for the Cretaceous

period (Barron et al. 1995), but new discoveries suggest that greenhouse gases may have also played a role (see the subsection "Late Cretaceous Cooling" that follows). In stark contrast to the Cretaceous period, during the Last Glacial Maximum 21–22 ka, polar regions cooled relatively more than the tropics (the tropics did, however, cool) and ΔT was about 50°C. It is believed that oceanic heat transport was generally similar between Quaternary glacial and interglacial states, thus requiring decreased atmospheric CO_2 concentrations, confirmed from Antarctic ice cores (see Chapter 5 in this volume) to account for cooler glacial tropics (Lea et al. 2003).

Turonian Global Warmth

Several extraordinary temperature reconstructions based mainly on oxygen and strontium isotopes suggest a warmer tropical ocean during parts of the Cretaceous period. For example, estimates of tropical North Atlantic SST from the index of tetraethers with 86 carbon atoms (TEX_{86}) paleothermometry indicate temperatures of 32–36°C (Schouten et al. 2003). The Turonian Stage (~93–89 Ma) following the Cenomanian-Turonian boundary OAE 2 event is one of the most notable. Figure 3.12 shows low- and high-latitude Turonian temperatures compared to temperature gradients simulated by the GENESIS climate model forced by elevated CO_2 concentrations. Turonian temperatures for the equatorial Atlantic SST were as high as 33–42°C and 30–33°C for the high southern latitudes on the Falkland Plateau (60°S) (Huber et al. 1995; Bice et al. 2003), James Ross Island (65°S) (Ditchfield et al. 1994), and the equatorial Atlantic on the Demerara Rise (Norris et al. 2002; P. Wilson et al. 2002; Bice et al. 2006).

The proxy-model comparison indicates that extremely high CO_2 concentrations 6500 ppmv or higher—much greater levels than those of proxy reconstructions—are needed to attain the pole-to-equator temperature gradient. To address this conflict, Bice et al. (2006) reconstructed atmospheric CO_2 between 102 and 82 Ma from ODP Leg 207 organic layers using lipid biomarkers (phytane $\delta^{13}C$) corrected for temperature with Mg/Ca ratios and oxygen isotopes. Their results show a CO_2 curve in the 600–2400 ppmv range, with maximum concentrations in the late Turonian (89 Ma) and Coniacian (97 Ma) stages. These estimates are in general agreement with estimates from other proxy methods and Berner's GEOCARB model. The GENESIS model could not produce the reconstructed temperature pattern with these CO_2 concentrations.

The implication is that climate models are not yet capable of simulating extreme warmth and low-latitude thermal gradients of the Cretaceous period on the basis of CO_2 forcing. The models may require revision in terms of sensitivity to CO_2 forcing, methane may have contributed to warming (Bice et al. 2006), or additional factors were involved. One

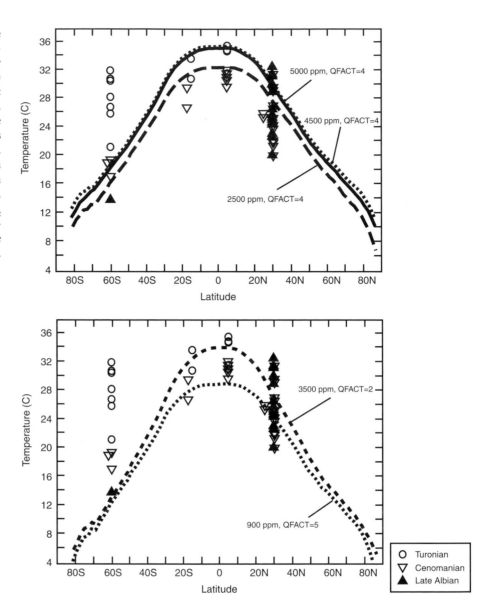

FIGURE 3.12 Latitudinal temperature gradient for the Turonian (late Cretaceous) stage based on various proxy records and model-simulated (QFACT) temperature under different atmospheric CO_2 concentrations, from Bice et al. (2003). Proxy temperature estimates are from foraminifera, except the James Ross Island oyster (Ditchfield et al. 1994); note that the high southern latitudes are much warmer than simulated temperatures at 4500 ppmv. Recent work suggests Cretaceous CO_2 concentrations may have been 600–2400 ppmv to achieve a low equator-pole gradient. Courtesy of K. Bice.

possibility is that the opening of a major tropical deep-water oceanic gateway between the North and South Atlantic Oceans at the Turonian-Cenomanian boundary may have influenced ocean circulation and temperature (Poulsen et al. 2003; Forster et al. 2007a).

Short-Term Cycles and Sea Level

Short-term climate variability also occurs during Cretaceous periods of warmth based on the TEX_{86} method. Turonian records from the tropical Demerara Rise suggest mean SST near 35–36°C but also period cooling of 1–3° C (Forster et al. 2007; Bornemann et al. 2008). Bornemann proposed that SSTs fell during a 200,000-year-long glacial episode at 91.2 Ma, consistent with stratigraphic (Sahagian et al. 1996)

and oxygen isotopic (Miller et al. 2003, 2004) evidence for a large (25–40 m), abrupt (<1 Ma) fall in sea level at this time. Tibert and Leckie (2004) also found evidence for mid-Turonian sea-level oscillations of 1–10 m from coastal deposits in the Western North American Interior Basin (see Dean and Arthur 1998). Bornemann envisioned an enhanced hydrological cycle caused by orbital forcing during this period of sustained global warmth.

Temperature variability within the Turonian and other mid-Cretaceous intervals is now well established, but direct evidence (i.e., ice-rafting deposits) for the existence of mid-Cretaceous ice on Antarctica remains to be discovered, and some researchers question isotopic evidence for ice-volume changes on the basis of preservation biases in foraminifera (Moriya et al. 2007).

Late Cretaceous Cooling

Late Cretaceous tropical cooling signified a change in the global pole-to-equator thermal gradients, raising the question of causality. Two favored explanations involve declining atmospheric CO_2 concentrations or changes in oceanic heat transport (D'Hondt and Arthur 1996; Pucéat et al. 2003, 2007). New discoveries complicate the picture. For example, Nordt et al. (2003) found evidence for two greenhouse climate events during the late Cretaceous period about 70 and 65.5 Ma based on analyses of continental sediments from the Big Bend area, Texas. Their reconstructions from paleosols yielded temperatures of 21–23°C and CO_2 concentrations 1000–1400 ppmv, supporting isotopic evidence for elevated oceanic temperatures during at least part of the late Cretaceous (Barrera and Johnson 1999).

Perspective

Deep-time geology provides a unique context for modern and future climate change. Pre-Cenozoic climatic extremes exceed any seen in the last 50 Ma and raise several themes for the remainder of this book. The first is the association of elevated atmospheric CO_2 and methane concentrations, global warmth, and high sea level over millions to tens of millions of years. The second is growing evidence for sub-million-year climate variability, including orbital-scale variations within periods of sustained global warmth of the Mesozoic and Late Paleozoic eras. The third theme is that the earth's climate sensitivity to thresholds in atmospheric CO_2 can be deduced from paleoclimatology. Royer et al. (2007), for example, used paleoclimate records to constrain minimum sensitivity to a doubling of preindustrial CO_2 concentrations at about 1.5°C for all or most of the Phanerozoic. The fourth theme is that trigger, amplification, and dampening feedback mechanisms for deep-time paleoclimatology are not well known for major climatic transitions, such as that during the Permo-Carboniferous. Multi-proxy climate reconstructions, more complete temporal and spatial control, and model simulations have proved to be a valuable means to achieve new perspectives on climate sensitivity and carbon-cycle-climate feedbacks.

LANDMARK PAPER Oceanic Anoxic Events and Global Warmth

Schlanger, Seymour O. and Hugh C. Jenkyns. 1976. Cretaceous oceanic anoxic events: Causes and consequences. Geologie en Mijnbouw 55:179–184.

What happens to the earth's ecosystems—its oceans, atmosphere, and biological systems—during periods of extreme climatic warmth? The answer to this question has obvious repercussions for future generations. It can, at least in part, be answered by the study of past hyperthemal periods in the geological record.

The discovery of Cretaceous black shales, which approach 30% organic carbon content, in sediment cores obtained in early Deep Sea Drilling Program expeditions has provided some answers. Seymour Schlanger and Hugh Jenkyns, two experts in Mesozoic stratigraphy and paleoceanography, made the important connection between climate, oceanic anoxia, and widespread black shales. The near-global distribution of age-correlative organic-rich sediments in sedimentary facies representing all paleobathymetric settings, in outcrops as well as in deep-sea cores, solidified the hypothesis that the earth's oceans experienced episodes of near-ocean-wide anoxia and organic-material accumulation during globally warm climates. Schlanger and Jenkyns proposed the term *Oceanic Anoxic Events* (OAEs) for these episodes, interpreting them as the interplay of geological and climatic factors, or "late Cretaceous transgression which increased the area and volume of shallow epicontinental and marginal seas and was accompanied by an increase in the production of organic carbon, and secondly the existence of an equable global climate which reduced the supply of cold oxygenated bottom water to the world ocean."

Jurassic and Cretaceous OAEs have since been actively researched for their fossil-fuel content—a large percentage of the earth's petroleum source rocks are OAE sediments. OAEs also yield insights into the global carbon cycle, the ocean's role in sequestering carbon, extreme climate states, and large-scale biological extinctions caused by climate change. Somewhat ironically, because massive amounts of carbon were injected into the earth's atmosphere during OAEs and similar early Cenozoic events, these fascinating geological phenomena have taken on new climatic significance because they shed light on potential environmental response to anthropogenic carbon emissions.

4

Cenozoic Climate

Introduction

The earth experienced global cooling during the Cenozoic era, the last 65 million years (Ma), culminating in extensive polar and midlatitude continental ice sheets during glacial periods of the last half-million years. Cenozoic climate evolved in steps, apparently crossing thresholds into colder climate regimes with greater pole-to-equator thermal gradients, rapid growth of the Antarctic and northern-hemisphere continental ice sheets, and changes in atmospheric carbon dioxide (CO_2) concentrations. Revolutionary tectonic, oceanographic, biogeochemical, and atmospheric changes contributed to this transformation, at times leading to extremes in global climate that had implications for future climate. Fischer (1981) called the Cenozoic transition from a warm, nearly ice-free earth to one with large, dynamic polar ice sheets the "greenhouse-icehouse" transition. In simplest terms, the major boundary conditions that influence today's climate—the distribution of the earth's oceans, continents, mountain chains, and dominant patterns of ocean and atmospheric circulation and chemistry—were attained during the Cenozoic era. This chapter outlines Cenozoic climate history, covering the following topics: (1) programs and proxies used in Cenozoic paleoclimatology, (2) temperature, sea-level ice volume, atmospheric CO_2, and deep-sea temperature history, (3) causal mechanisms (tectonic events, ocean gateways, atmospheric CO_2), and (4) extreme climate states and "hyperthermal" warmth.

Cenozoic Paleoclimate Programs and Proxies

Deep-Sea and Continental Coring Programs

Advances in Cenozoic climate history have come about largely through the international ocean drilling programs called the Deep Sea Drilling Project (DSDP, 1964–1983), Ocean Drilling Program (ODP, 1984–2003), and Integrated Ocean Drilling Programs (IODPs, since 2003). DSDP, ODP, and IODP carried out 96, 111, and 16 expeditions, or Legs, respectively, visiting almost 1300 sites using the research vessels *Glomar Challenger*, *JOIDES Resolution*, and several site-specific platforms. In general, these programs were designed to investigate many fundamental geophysical and tectonic aspects of the world's oceans. Twenty-three specific Legs that targeted key aspects of Cenozoic paleoceanographic and paleoclimate history discussed in this chapter are listed in Table 4.1. DSDP, ODP, and IODP cruises recovered continuous sedimentary records, many deposited at high accumulation rates, leading to major discoveries about Cenozoic climate.

In recent years, the International Continental Scientific Drilling Program (ICDP) has filled the important gap in continental paleoclimatology and fundamental geophysical and geological topics related to energy and mineral resources, biotic evolution and extinction, and earthquake and tectonic processes (Harms et al. 2007). ICDP cosponsors or coordinates more than 20 coring projects related to paleoclimate and global environmental change (Table 4.2). These programs investigate topics as diverse as an Eocene/Oligocene impact crater in Chesapeake Bay in the eastern United States to paleoclimate records from the earth's largest and oldest lakes, including Lakes Baikal and El'gygygytgyn in Siberia, Malawi in equatorial Africa, Petén-Itzá in Central America, and Titicaca in South America.

Though it is obviously impossible to review such a large literature, most of the major features of Cenozoic history described in this chapter are the outcome of these programs, and some of the most insightful reviews can be found in original Leg reports.

Foraminiferal Stable Isotopes

Our current understanding of Cenozoic climate evolution rests firmly on a foundation provided by early studies of deep-sea oxygen ($\delta^{18}O$) and carbon ($\delta^{13}C$) stable isotopes in the $CaCO_3$ shells of foraminifera. Pioneering Cenozoic reconstructions using foraminiferal stable isotopes include those of Douglas and Savin (1975), Shackleton and Kennett (1975), and Miller et al. (1987). As introduced in Chapter 2, isotope ratios of oxygen ($^{18}O/^{16}O$) and carbon ($^{13}C/^{12}C$) in climate archives constitute the single most important family of paleoclimate proxy methods. The $\delta^{18}O$ composition of oxygen in foraminiferal $CaCO_3$ is a function of several climate-related factors—the $\delta^{18}O$ of the seawater in which the organism secretes its shell (Gat 1996), water temperature during shell secretion (the greater the temperature, the less enriched the foraminiferal $\delta^{18}O$ [Urey 1947; Emiliani 1955; Craig 1965]), global ice volume (Shackleton 1967), and, to a lesser extent, ocean salinity (Craig and Gorden 1965). If a species secretes its shell in thermodynamic equilibrium with the seawater in which it lives (some species do, others do not), then the shell reflects seawater isotopic signature (Bemis et al. 1998).

The influence of continental ice volume on foraminiferal $\delta^{18}O$ (the "glacial effect") (Olausson 1965; Shackleton 1967), applied first to Quaternary glacial-interglacial cycles (see Chapter 5 in this volume), is perhaps the single most important metric of Cenozoic climate changes. Ice volume affects foraminiferal $\delta^{18}O$ because, during glacial periods, freshwater with negative $\delta^{18}O$ values ($<-30‰$) is removed from the oceans and stored on the continents (the equivalent to ~125 m of sea-level change), changing the isotopic composition of seawater. This change is reflected in foraminiferal shells. During interglacial periods, there is less continental ice, proportionally more ^{16}O in seawater, and its composition is more negative (i.e., lighter). Conversely, during glacial periods, continental ice preferentially stores the light isotope (^{16}O), seawater is enriched in $\delta^{18}O$, and foraminiferal calcite $\delta^{18}O$ will be less negative. The glacial effect is responsible for a large portion of the isotopic signal in foraminifera living in regions with relatively small temperature and salinity variability, such as the tropical Pacific Ocean (Shackleton and Opdyke 1973; Matthews and Poore 1980) and deep-sea benthic environments, although both tropical and deep-sea temperatures vary and this contributes to the $\delta^{18}O$ signal. Hydrological (evaporation, precipitation, seawater density), biological ("vital effects," nonequilibrium shell secretion, symbiont photosynthetic effects, metabolism, and growth rate), and postmortem dissolution also influence oxygen isotopes.

If tropical sea-surface temperatures (SSTs) remained relatively stable during glacial periods, and salinity effects were minor, then the glacial-interglacial $\delta^{18}O_{foram}$ for Pacific Ocean

TABLE 4.1 Ocean Drilling Program Legs and Cenozoic Paleoclimatology

Leg	Sites	Region	Objectives/Discoveries
113	689–697	Weddell Sea	Maud Rise: Transition from calcareous to siliceous microfossils at E-O boundary
			Major WAIS expansion in middle Miocene
119, 120	736–746	Kerguelen Plateau Prydz Bay	Late/middle-late Eocene glaciation in E. Antarctica ~42 Ma, large EAIS earliest Oligocene
150 (also 150 X onshore)	902–906	New Jersey offshore	Oligocene to Holocene stratigraphy and glacio-eustatic sea level from 180 records
174, 174A, AX, suppl. 174B	1071–1073		
151	907–913	Yermak Plateau, Fram Strait, Greenland Sea	Deep-water flow from Arctic and Nordic seas in late Miocene
152	914–919	East Greenland margin	Onset of ventilated North Atlantic inflow at 11–13 Ma, late Miocene glaciations 6–7 Ma
154	925–929	Ceara Rise, Atlantic	Orbital forcing back to 7 Ma, sediment flux due to climate and Andean uplift 8 Ma
162	907, 980–987	Atlantic Gateways	High-resolution Neogene paleoceanography of glacial inceptions, CAI closing
165	998–1002	Caribbean Sea	Miocene-Pliocene CAI closing
166	1003–1009	Bahamas transect	Neogene sea level and CAI closing
172	1054–1064	Blake Ridge Bermuda Rise	Late Pliocene-Quaternary paleoceanography, onset of large glacial cycles
175	1075–1082	Southeast Atlantic	Late Neogene Benguela current and upwelling history
177	1088–1094	South Atlantic Transect	Neogene paleoceanography
178	1095–1103	Antarctic margin	Neogene Antarctic Ice Sheet and sea-level history
181	1119–1125	Southwest Pacific gateways	Cenozoic Gateways, deep western boundary current, Antarctic Circumpolar Current
184	1143–1148	South China Sea	Cenozoic east Asian monsoon
188	1165–1167	Prydz Bay, east Antarctica	Cenozoic EAIS history
189	1168–1172	Tasmanian Gateway	Opening of Tasmanian Gateway and Antarctic Circumpolar Current
198	1207–1217	Shatsky Rise	Cretaceous-Paleogene paleoclimate
199	1215–1222	Paleogene equatorial Pacific	Paleogene paleoclimate
208	1262–1267	Walvis Ridge, S. Atlantic	PETM, Eocene-Oligocene boundary
302	M0001, M0002, M0003, and M0004	ACEX-Lomonosov Ridge, Central Arctic	Cenozoic Arctic ice and climate history
306	U1302–U1308	North Atlantic Drifts	High-resolution late Neogene climate

Source: http://www.odplegacy.org/science_results/leg_summaries.html.

Note: ACEX = Arctic Coring Expedition; CAI = Central American Isthmus; EAIS = East Antarctic Ice Sheet; E-O = Eocene-Oligocene boundary; PETM = Paleocene-Eocene Thermal Maximum; WAIS = West Antarctic Ice Sheet.

foraminifers of about 1.3‰ (glacial-interglacial $\delta^{18}O_{foram}$ varies regionally) would be due mainly to ice volume changes, i.e., the glacial effect. Fairbanks and Matthews (1978) calibrated the oxygen isotopic composition of foraminifers to sea level and estimated that a 0.1‰ change is equivalent to a ~10-m drop in sea level, or a little less than one-tenth the total glacial-interglacial change in ice volume. Because tropical surface and deep-sea bottom-water temperatures vary, they

TABLE 4.2 Completed and Ongoing Cenozoic Continental Paleoclimate Coring Projects

Lake Region	Region	Coordinates	Project Duration	Age Interval	Objective
Lake Baikal	Siberia	53° 29′ N, 108° 10′ E	1989–1999	0 to 20–25 Ma	Long-term Cenozoic climate
Lake Bosumtwi	Ghana	6° 30′ N, 1° 25′ W	2004	0–1.07 Ma	Monsoon history, dust-atmosphere, tropical SST
Lake El'gygytgyn	Siberia	67° 30′ N, 172° 5′ E	2008–ongoing	0–3.6 Ma	Arctic paleoclimate
Lake Qinghai	Tibetan Plateau	6° 48.67′ N, 100° 8.22′ E	2005	post-late Miocene	Late Miocene-Holocene, paleoclimate various timescales
Lake Petén Itzá	Guatemala	16° 59′ 58″ N, 89° 47′ 44″ W	2006	Quaternary	Quaternary paleoclimatology
Lake Malawi	Tanzania, Mozambique, East African Rift Valley	12° 00′ 09″ S, 34° 06′ 45″ E	2005	0–800 ka	Tropical climate
Lake Titicaca	Bolivia, Peru	5° 55′ S, 69° 48′ W	2001	late Quaternary	High-elevation tropical paleoclimate
Potrok Aike Maar Lake	Argentina, Patagonia	51° 58′ 58″ S, 70° 22′ 42″ W	2008	0–770 ka	Midlatitude southern-hemisphere climate

Source: http://www.icdp-online.org/contenido/icdp/front_content.php.

SST = sea-surface temperature.

are taken into account in constructing the isotopically derived Cenozoic sea-level curve discussed in the subsection "Sea Level and Global Ice Volume" (Pekar et al. 2002).

Carbon isotopes in Cenozoic foraminifera have also been used to infer changes in ocean circulation. However, $^{13}C/^{12}C$ ratios in seawater are governed by complex processes, including nutrient cycling, carbonate chemistry, seawater alkalinity, physiological processes during photosynthesis by marine algae, and others (Broecker and Peng 1982; Spero and Williams 1988; Spero 1992; Rohling and Cooke 1999). As a consequence, the climatic interpretation of $^{13}C/^{12}C$ ratios is a complex enterprise (Spero et al. 1997).

Trace Elements

Trace elements incorporated into the calcium carbonate ($CaCO_3$) shells of many marine and freshwater organisms are also used as proxies in paleoclimatology. Magnesium substitutes for calcium in inorganically and organically precipitated calcite. The magnesium-to-calcium-ion ratio (Mg/Ca) is influenced by temperature in a number of foraminiferal species (Nurnberg et al. 1996; Hastings et al. 1998; Dekens et al. 2002). Empirical temperature-Mg/Ca relationships have been applied to several deep-sea benthic (Martin et al. 2000,

2005) and planktic (Rosenthal et al. 2004; Barker et al. 2005) foraminiferal and ostracode (Dwyer et al. 1995, 2003) records on the basis of calibrations from field studies of living organisms, core-top specimens, and laboratory culturing of individuals from different water temperatures.

Applying Mg/Ca ratios to Cenozoic marine records, researchers have quantified surface and deep-sea ocean temperature history, separating the contribution to foraminiferal oxygen isotopes of temperature from that due to ice volume. Salinity, water chemistry, biological processes, and postmortem changes also influence Mg/Ca ratios, and cleaning procedures, analytical methods, and intra-shell chemical variability are also sources of uncertainty (Mekik et al. 2007). Several other elemental-calcium ratios used in paleoceanography include cadmium (Boyle 1992), lithium (Hall and Chan 2004a), barium (Hall and Chan 2004b), and strontium (Beck et al. 1992; Guilderson et al. 1994; Cohen et al. 2002).

Combining elemental ratios with stable isotopes is a common approach in Cenozoic paleoclimatology. Lear et al. (2003), for example, combined Sr/Ca ratios with $^{87}Sr/^{86}Sr$ isotopes to separate the weathering component of the Cenozoic strontium isotope curve from that due to hydrothermal activity. They identified a twofold increase in riverine strontium flux, starting at 35 Ma and increasing $^{87}Sr/^{86}Sr$ from 29–13 Ma, linked to Himalayan uplift. Hathorne and James (2006) used Li/Ca ratios and δ^7Li in planktic foraminifers to reconstruct the last 18 Ma of silicate weathering. Between 16 and 8 Ma, the flux of dissolved Li from continents decreased and δ^7Li values increased, indicating a lower silicate weathering rate and intensity associated with rising atmospheric CO_2. During the last 8 Ma, climatic cooling was linked to decreasing atmospheric CO_2 concentrations. Other paleoceanographic applications of isotopic ratios in "nontraditional" elements (e.g., iron, molybdenum, and copper) are given in Anbar and Rouxel (2007).

Paleo-CO_2 Concentrations

Atmospheric CO_2 is an important driver of Cenozoic climate history. As we saw in Chapter 3, several methods are used to estimate CO_2 concentrations for geological intervals predating the Antarctic ice core CO_2 record (see Chapter 5 in this volume). Analyzing the carbon isotopic composition of long-chained alkenones (37 or 39 carbon atoms in a chain) produced by marine haptophyte algae is a primary method. The alkenone isotope ($\delta^{13}C_{37:2}$) is expressed as epsilon, ε_p (also $\varepsilon_{37:2}$), and the alkenone-CO_2 relationship as

$$\varepsilon_p = \varepsilon_f - b/[CO_2 \text{ (aqueous)}]$$

where ε_f is carbon fractionation during carboxilation and b is an index related to algal cell growth rate, size, and shape. Culturing and field experiments (Pagani et al. 2002) on common coccolithophorid algae (called calcareous nannoplankton by micropaleontologists) indicate that ε_p varies as a function of cell growth rate and the concentration of aqueous CO_2 (Popp et al. 1998; Henderiks and Pagani 2007). The alkenone method is applicable under oligotrophic ocean conditions when phosphate concentrations are limited and nutrient concentrations and water temperature can be independently constrained (Pagani et al. 2005). Uncertainty comes from evolutionary changes in alkenone-producing algal groups, cell growth rates and geometry, and nutrient levels. Henderiks and Rickaby (2007) proposed another nannoplankton-based method of CO_2 reconstruction. They suggest that the "calcification tolerance" of calcareous nannoplankton—the ability of the two major extant groups, *Emiliana huxleyi* and species of *Gephyrocapsa* to grow their calcitic skeletons during photosynthesis—can also be used as a paleo-CO_2 proxy.

Carbon isotopes from pedogenic carbonates in fossil soils are also used to estimate Cenozoic CO_2 (Cerling 1991). This method relies on understanding the chemical processes governing the formation of $CaCO_3$ in soils from windblown dust carrying calcium ions and CO_2 derived from a combination of atmospheric sources and soil respiration. It is based on the principle that the isotopic composition of organic material accumulated in soils, which in many regions is influenced by carbon derived from grasses (called C_4 plants because of the distinct metabolic pathway for carbon uptake) since their evolution during the late Neogene, reflects the mixing of atmospheric and soil-produced CO_2. Empirical studies show that the $\delta^{13}C$ of modern soil carbonate is a CO_2 proxy if soil carbonates form in isotopic equilibrium with soil CO_2 (Quade et al. 1989; Cerling 1999). Pedogenic carbonates are useful in pre-Cretaceous paleoclimatology because continuous deep-sea sediment sequences are less common (Mora et al. 1996; Ekart et al. 1999; Royer 2006).

Vascular plant (conifers and broad-leaved angiosperms) leaf stomatal thickness and density are also atmospheric CO_2 proxies (Van der Burgh et al. 1993; Kürschner et al. 1996; Wagner et al. 1999; Royer et al. 2001a; Beerling and Royer 2002; Kouwenberg et al. 2003). Stomatal characteristics change historically, in response to anthropogenic increases in CO_2 concentration and in trees grown under experimental conditions. One disadvantage of stomatal methods is they are usually applied to discontinuous sediment sequences or short intervals of time, unlike the alkenone CO_2 curve obtained from continuous deep-sea sediment records.

The boron isotopic composition of planktonic foraminifera shells rests on the principle that as atmospheric CO_2 concentrations rise, more CO_2 is dissolved in ocean-surface layers, reducing ocean pH and increasing acidity (Spivak et al. 1993). Because the ratio of dissolved $B(OH)_3$ and $B(OH)_4^-$ ions are pH dependent, boron, $\delta^{11}B$ might serve as a paleo-pH proxy. Caldeira et al. (1999) challenged the assumptions about ocean pH, temperature and carbonate chemistry citing other factors such as species' vital effects, diagenesis, and biogeochemical processes in ocean-surface layers.

Each paleo-CO_2 proxy method has advantages and limitations and can produce contradictory results for the same geological interval. This field is experiencing rapid growth, and a number of excellent discussions describing methods and reconstructions are available (Royer et al. 2001b, 2004; Ehleringer et al. 2005).

Radiogenic Tracers

Stable isotopes and trace elements are influenced by biological and physicochemical processes complicating interpretations about past ocean circulation. During the past few years several unstable radiogenic tracers have been developed to evaluate ocean chemistry, circulation, and climate. Radiogenic tracers include the metals strontium, osmium, lead, neodymium, hafnium, and beryllium (Burton et al. 1997;

Pegram and Turekian 1999; Veizer et al. 1999; Peucker-Ehrenbrink and Ravizza 2000; Frank 2002; Burton 2006; Frank et al. 2006).

Each tracer has a different modern distribution, residence time, and input sources to the oceans. Strontium isotopes ($^{87}Sr/^{86}Sr$) and osmium isotopes ($^{187}Os/^{186}Os$), for example, have residence times longer than the mean ocean mixing of ~1000 years. Input sources are influenced by continental weathering and intensity, mid-ocean ridge hydrothermal activity, and, for osmium, by meteor impacts. Sr and Os isotopes can be measured in shells of foraminifera and other marine organisms and ferromanganese crusts that grow over millions of years on the ocean floor. These two elements have been useful in reconstructing long-term Cenozoic trends related to continental uplift, denudation, and resulting changes in continent-ocean flux that can influence biogeochemical cycling and climate. Neodymium ($^{143}Nd/^{144}Nd$, designated e_{Nd}) is not influenced by hydrothermal activity and is useful to distinguish changes in continental source rocks from those due to weathering flux. Neodymium, lead (^{206}Pb, ^{207}Pb, ^{208}Pb), hafnium ($^{176}Hf/^{177}Hf$), and beryllium ($^{10}Be/^9Be$) are used to evaluate the timing and ocean response to the opening and closing of ocean gateways and regional uplift and erosion (Burton et al. 1997; Frank and O'Nions 1998; Reynolds et al. 1999; Frank et al. 1999a; Frank 2002; Thomas et al. 2003; Scher and Martin 2006).

Major Features of Cenozoic Climate

Major features of Cenozoic climate history are illustrated in the global temperature, sea-level ice volume, atmospheric CO_2, and deep-sea temperature reconstructions in figures 4.1–4.4. These compilations are the cumulative products of decades of deep-sea sediment core research, improvements to the geological timescale, and refinement of proxy methods. We review the climatic framework provided by these proxies and examine details and mechanisms used to explain them.

Cenozoic Climate, Temperature, and Tectonic Events

Figure 4.1 is a global Cenozoic oxygen isotopic curve compiled from benthic foraminifera from deep-sea cores (Zachos et al. 2001). The total 5.4‰ increase in benthic foraminiferal $\delta^{18}O$ over the last 55 Ma represents the growth of continental ice sheets and ocean cooling. The $\delta^{18}O$ signal can be broken down into 3.1‰ due to deep-ocean cooling and 2.3‰ due to the long-term buildup of small and intermittent and, later, large and permanent continental ice sheets.

The Cenozoic oxygen isotopic stratigraphy developed by Miller and colleagues (Miller et al. 1987; Wright and

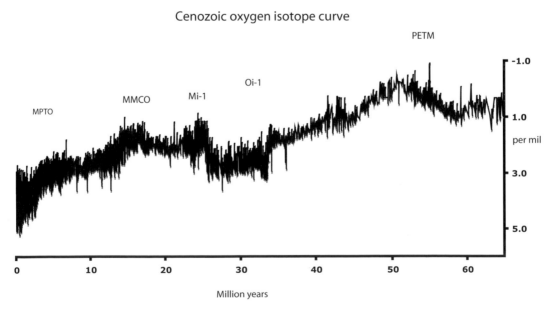

Cenozoic oxygen isotope curve

Million years

FIGURE 4.1 Cenozoic oxygen isotope curve based on benthic foraminifers from many Deep Sea Drilling Project and Ocean Drilling Program sites, compiled by Zachos et al. (2001) and updated by J. Veizer. Increasing values signify progressive planetary cooling during the last 65 Ma. Several key Cenozoic climate transitions—Paleocene-Eocene Thermal Maximum (PETM), oxygen isotope events Oi-1 and Mi-1, Mid-Miocene Climatic Optimum (MMCO), and Mid-Pliocene Thermal Optimum (MPTO)—discussed in this chapter are indicated. Courtesy of J. Zachos and J. Veizer.

Miller 1992 Miller et al. 1991, 1996) includes two prominent excursions shown in Figure 4.1—Oi 1 near the Eocene-Oligocene (E-O) boundary at 33.5–34 Ma and Mi-1 just after 23 Ma in the early Miocene epoch. These were major periods of cooling and glaciation, marked by abrupt positive shifts in foraminiferal $\delta^{18}O$. The Cenozoic isotopic curve is firmly integrated with the standard geomagnetic polarity timescale and continental-margin sea-level records using chronology "tuned" to changes in solar insolation caused by long-term changes of the earth's orbital geometry in the solar system.

Several other proxy methods capture long-term changes in various aspects of global Cenozoic climate and biogeochemical cycling, including climatic transitions during the Oi-1 and Mi-1 excursions and short-term climate events. Examples include calcium isotopes (De La Rocha and DePaolo 2000), lead isotopes (Christensen et al. 1997), deep-sea carbonate deposition and calcium carbonate compensation depth (CCD) (Tripati et al. 2005), osmium isotopes (Peucker-Ehrenbrink and Ravizza 2000), strontium isotopes (Veizer et al. 1999), and neodymium, lead, and osmium isotopes (Burton et al. 1997, Burton 2006). Radiogenic records cover various segments of the 65-Ma Cenozoic era (see the review in Frank 2002).

Sea Level and Global Ice Volume

Sea level oscillates because of plate-tectonic processes that affect seafloor spreading rates, mid-ocean ridge and ocean-basin volume (tectono-eustasy, 10^6–10^8 years), changes in continental ice-sheet and glacier volume (glacio-eustasy, 10^4–10^7 years), thermal expansion and contraction of ocean water (thermosteric effect, 10^1–10^3 years), internal climate dynamics (e.g., El Niño–Southern Oscillation [ENSO], 10^0–10^2 years), and the formation and destruction of shallow epicontinental seas and groundwater storage. Tectono-eustasy and glacio-eustasy are central to Cenozoic paleoclimatology.

The global Cenozoic sea-level curve in Figure 4.2 was constructed from continental-margin stratigraphy and deep-sea oxygen isotopic records of ice volume (Miller et al. 1987; Kominz and Pekar 2001; Miller et al. 2005). Cores from ODP Legs 150 and 174 from the New Jersey margin were instrumental in building the Miocene (Browning et al. 1996), Oligocene (Pekar and Miller 1996; Pekar et al. 2001, 2002), and Oligocene-Miocene (Miller et al. 1996) parts of this curve. Maximum sea level more than 50 m above that of the present day was reached during the early Eocene epoch, followed by a progressive fall during the Cenozoic era to minima of 120–125 m below the present level during late Pleistocene glacial intervals. These sea-level minima represent maximum Cenozoic ice volume when large ice sheets grew in Antarctica, North America, and Eurasia. There is independent support for the continental-margin sea-level curve from deep-sea benthic foraminiferal $\delta^{18}O$ values corrected for bottom-temperature water changes using Mg/Ca paleothermometry (e.g., Billups and Schrag 2003). Sea-level oscillations over 1- to 10-Ma timescales are superimposed on the general Cenozoic trend.

The sea-level curve must be reconciled with direct evidence for land-based ice. For years, it was thought that the growth of a large, permanent East Antarctic Ice Sheet (EAIS) began ~34 Ma at the E-O boundary and that contributions from the West Antarctica Ice Sheet (WAIS) and northern-hemisphere ice sheets were more important for Neogene sea-level fluctuations. There is some evidence for Eocene glacio-eustastic sea-level fluctuations. Pekar et al. (2005) drew an

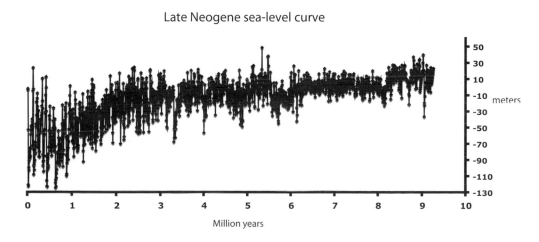

Late Neogene sea-level curve

Million years

FIGURE 4.2 Late Neogene isotope sea-level curve based on ice-volume estimates from deep-sea foraminiferal isotopic data. From Miller et al. (2005).

Eocene (~52–42 Ma) sea-level curve from ODP Leg 189 to the South Tasman Rise in which first-order changes were synchronous with those along the eastern United States and in Europe. Sea-level oscillations attributed to changes in EAIS volume were ~20 m during the early Eocene, 25–45 m during the Oligocene, and more than 50 m during the early Miocene (Pekar et al. 2005; Pekar and DeConto 2006). Tripati et al. (2005) also argued for a "dynamic middle Eocene cryosphere" on the basis of deep-sea isotopic and bulk carbonate records from equatorial Pacific ODP sites 1218 and 1219 from ODP Leg 199. Tripati concluded that ice-sheet accumulation at 42–38 Ma coincided with intensification in the ocean's CCD and that ice-sheet growth was not limited to Antarctica but required contributions from northern-hemisphere ice as well.

The conventional view that land ice did not develop in the northern hemisphere until the late Neogene has also been modified, though not conclusively. Evidence from the Arctic Ocean from the IODP Leg 302 Arctic Coring Expedition (ACEX) to the Lomonosov Ridge suggests there was northern-hemisphere ice before the Miocene epoch (Moran et al. 2006 Backman et al. 2008). Using the sedimentology of ACEX cores as a basis (see Frank et al. 2008; Darby 2008), St. John (2008) proposed that there was sea- and land-based ice in the Arctic as early as the Eocene epoch. Eldrett et al. (2007) also found evidence for Eocene (38–30 Ma) continental ice on East Greenland in cores from the Norwegian-Greenland Sea.

Although there is no continuous Cenozoic ice-volume reconstruction for either the WAIS or the EAIS, the geological records of Antarctica and the Southern Ocean provide evidence about the stability of Antarctic ice. One view based largely on geomorphological evidence and cosmogenic dating calls for EAIS stability for at least the past ~14 Ma (Denton et al. 1991, 1993; Sugden et al. 1993, 1995; Sugden 1996; Stroeven and Kleman 1999). Sugden and Denton (2004) reviewed the Cenozoic history of one well-studied region of Antarctica, the Convoy Range to Mackay Glacier area of the Transantarctic Mountains. They propose that there was complex tectonic activity and local glaciation between 55 and 14 Ma, followed by a hyperarid climate and tectonic stability during the last ~13.6 Ma. Others hypothesize a more dynamic Antarctic Neogene ice-sheet history, with partial warmer climate, wet-based glaciers, and partial deglaciation (Webb et al. 1984; Webb and Harwood 1991; G. Wilson et al. 2002; Hambrey et al. 2003; Naish et al. 2007); large reductions (25–70%) in the EAIS during the late Miocene epoch (Pekar and DeConto 2006); and orbital climatic cycles during the Miocene (Flower et al. 1997; Zachos et al. 1997) and the Oligo-Miocene transition at 24.1–23.7 Ma (Naish et al. 2001). Ehrmann et al. (1998) found lithological evidence from Mc-

Murdo Sound in the Ross Sea region of Antarctica for glacial activity during the late Eocene epoch. ODP Legs around Antarctica (Florindo et al. 2003; Barrett et al. 2006) and nearshore drilling programs such as the Cape Roberts Project (Barrett et al. 2000) and the ANtarctic geological DRILLing Project (ANDRILL) yield insight into the Cenozoic history of Antarctica. Ehrmann et al. (2005) showed stepwise evolution from early Oligocene humid climate to Pleistocene polar conditions. One ANDRILL project aimed at recovering the past 14 Ma of climatic history from the McMurdo Sound area has already recovered a 1300-m-long Miocene-Pleistocene record (Naish et al. 2007). This ANDRILL records shows an extremely dynamic WAIS margin during the Pliocene.

Atmospheric Carbon Dioxide

Pagani and colleagues produced a continuous reconstruction of Cenozoic atmospheric CO_2 from 45 to 5 Ma ago using alkenones preserved in deep-sea sediments (Pagani et al. 1999, 2005) (Figure 4.3). CO_2 values reached 1500 parts per million by volume (ppmv) during the Eocene epoch—more than five times higher than preindustrial concentrations. Eocene-to-early-Oligocene CO_2 oscillated sharply over million-year timescales, remaining within a general range of 1000–1500 ppmv. A sharp, permanent decrease from 1500 to below 500 ppmv occurred during the early to mid-Oligocene ~33–28 Ma. After a brief late Oligocene rise to 350 ppmv, CO_2 approached preindustrial levels at the Oligo-Miocene boundary ~23 Ma. Until this time, it appears that Cenozoic CO_2 and oxygen isotopic proxies were strongly coupled with each other. After about 20 Ma, however, Miocene CO_2 values remained in a narrow range of 190–260 ppmv, roughly similar to that for the last 700 ka. Maximum mid- to late Miocene CO_2 concentrations are comparable to preindustrial values of 275–280 ppmv. Pagani et al. did not find a large increase in atmospheric CO_2 during the mid-Miocene climatic optimum, nor a large decrease associated with the Neogene buildup of the Antarctic Ice Sheet (AIS). Assuming the alkenone-based curve is accurate, the ramifications are either that Miocene CO_2 and climate were partially decoupled at the million-year timescale, or the earth's climate had a greater sensitivity to relatively small changes in CO_2 concentrations.

The boron isotopic CO_2 curve of Pearson and Palmer (2000) suggests that the Paleocene and early Oligocene (~60–52 Ma) concentrations were as high as 2000 ppmv, and that decreases in CO_2 concentrations until 40 Ma were caused by reduced production of CO_2 from ocean ridges and volcanic activity and increased burial of carbon in sediments. The sharpest drop in their record was at ~52 Ma.

Cenozoic atmospheric CO_2

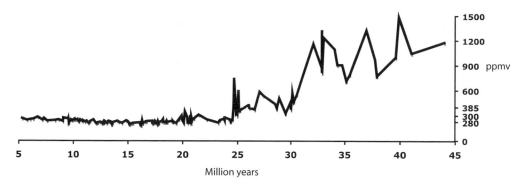

FIGURE 4.3 Cenozoic paleo-CO_2 concentrations in ppmv, based on phytoplankton alkenone geochemistry. Preindustrial (280 ppmv) and 2008 (385 ppmv) levels shown for comparison. From Pagani et al. (1999, 2005).

Since ~22 Ma, CO_2 concentrations oscillated around a narrow range of about 200–355 ppmv, generally in accord with the alkenone-based estimates. Demicco's et al. (2003) Cenozoic pCO_2 curve is based on major ion compositions in seawater (Ca and Mg) and ocean carbonate sediment mineralogy. Their computed pCO_2 values ranged from 1200–2500 ppmv from 60 to 40 Ma during parts of the Paleocene and Eocene to 100–300 ppmv during the Miocene to Recent. Lowenstein and Demicco (2006) used another method—sodium carbonate minerals (nahcolite $NaHCO_3$) precipitated in water in contact with the atmosphere chemistry—to estimate CO_2 values >1125 ppmv from the early Eocene Green River Formation in Wyoming (51.3–49.6 Ma). Other evaporates from Beypazari, Turkey (21.5 Ma) and Searles Lake, California (1 Ma) showed that CO_2 levels reached near modern-day concentrations during the early Miocene through Pleistocene epochs, supporting the alkenone-based estimates. To a first approximation, the Cenozoic paleo-CO_2 curve in Figure 4.3 matches that from Berner's GEO-CARB model introduced in the last chapter (Berner and Kothavala 2001; Berner 2004).

DeConto et al. (2008) analyzed the sensitivity of polar-region ice-sheet development to Cenozoic paleo-CO_2 levels by using a global climate ice-sheet model forced by orbital insolation and CO_2. Their results suggest that northern and southern hemispheres have very different CO_2 thresholds for ice-sheet growth. AIS buildup began during Oi-1 near a CO_2 threshold of about 750 ppmv; in contrast, northern-hemisphere ice-sheet development began about 25 Ma, when CO_2 concentrations reached near pre-industrial levels of 280 ppmv. AIS growth and subsequent oscillations are supported by evidence for Oligocene-Miocene eustatic sea-level fluctuations of up to 70 m (Pekar et al. 2002; Pekar and DeConto 2006).

Cenozoic paleo-CO_2 reconstructions represent major strides in our understanding the evolution of Cenozoic climate. There are, nonetheless, stratigraphic gaps in the paleo-CO_2 curve, large uncertainties for some reconstructions, and a need for additional proxies and high-resolution records (Royer et al. 2001a, 2004).

Deep-Sea Temperatures and Ocean Circulation

It had long been thought that the Cenozoic ocean went from a relatively weakly stratified global ocean with relatively warm surface and bottom temperatures to a strongly stratified system with warmer surface temperatures and cooler deep-sea temperatures that were separated by strong vertical thermal gradients. Oxygen isotopes and, more recently, magnesium-calcium paleothermometry are largely responsible for what is now widely accepted as significant deep-sea cooling during the Cenozoic era. Lear et al. (2000) and Lear (2007) analyzed Mg/Ca ratios of benthic foraminifera from four DSDP-ODP sites from modern water depths of 2080–4400 m (Figure 4.4). These sites were estimated to represent paleo-water depths between 1500 and 4300 m and thus cover a large vertical profile of the deep-ocean basin. This 50-Ma record features a progressive 12°C cooling, occurring in four distinct steps in the early to mid-Eocene, at the E-O boundary, during the late-mid Miocene, and the Plio-Pleistocene. These patterns generally match climate trends inferred from deep-sea foraminiferal isotopes and a number of other marine and continental records. Moreover, three major phases of ice-sheet growth are indicated by comparison of the Mg/Ca and oxygen isotope curves. The first Mg/Ca excursion in the late Oligocene corresponds to

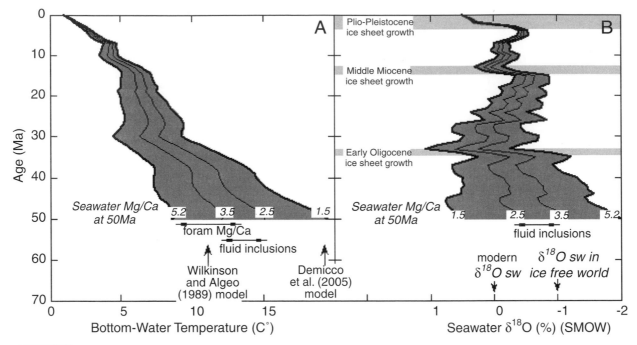

FIGURE 4.4 Cenozoic deep-sea paleoceanography: (A) Bottom-water temperature; (B) oxygen isotope composition of seawater, from Lear (2007). Envelope of temperature and seawater oxygen isotopes reflects mainly changes in the magnesium content of seawater, estimated by Wilkinson and Algeo (1989) and Demicco et al. (2003). Courtesy of C. Lear, The Geological Society (London), and The Micropaleontological Society.

event Oi-1; the second in mid-Miocene ~14 Ma (Zachos et al. 1996). Both events are associated with periods of Antarctic ice-sheet growth. The third Mg/Ca transition began during the late Pliocene epoch when large northern-hemisphere ice sheets became prominent.

Summary

These global climate, sea-level, CO_2 and deep-sea temperature curves form the primary pattern of first-order climate evolution over the past 65 Ma subscribed to by most of the paleoclimate community. Continual refinements are being made in the relationship between climate, sea level, and biogeochemical cycles. An illustrative example is the apparent late Oligocene warming seen in Figure 4.1 about 25–23 Ma. Pekar et al. (2002, 2006) estimated that the late Oligocene 26–23 Ma EAIS ice volume might have varied by 50–125% from modern volume. This implies that late Oligocene sea-level oscillations were on the order of ~60 m in amplitude. A related aspect to late Oligocene climate is orbital-scale glacial-interglacial variability when continental ice volume oscillated from 80% (glacials) to 50% (interglacials) of modern EAIS volume. DeConto and Pollard (2003a) estimated glacio-eustatic sea-level changes of ~20 m during these cycles, much smaller than the 120-m change during late Pleistocene climatic cycles, but nonetheless evidence for a dynamic Oligocene AIS.

Mechanisms of Cenozoic Climate Change
Orogenesis, Weathering, Atmospheric Circulation, and Biogeochemical Cycling

Orogenesis—mountain building and continental plateau uplift induced by lithospheric plate collisions and other tectonic processes—has characterized large regions of Asia, South and North America, and East Africa during the Cenozoic era. The largest continental physiographical feature on the earth, the Tibetan-Himalayan Plateau, is more than a million square kilometers, covering 10° of latitude by 20° of longitude above 5000 m elevation. Similarly, the East Africa Rift is a massive 6000-km-long mountain-plateau region between 1500 and 5100 m high. The Tibetan Plateau, East Africa Rift, and the Andes along western South America are relatively young products of plate-tectonic collisions. The Rocky Mountains and Colorado Plateau in western North America have been the center of more than 100 years of debate about their tectonic and elevation history. Some recent studies suggest that following the Laramide orogeny 70–35 Ma, mantle processes (doming or uplift), extensional tectonic activity causing subsidence and incision, a mosaic of regional uplift patterns, and climatic changes combined to induce elevation changes (McMillan et al. 2006). An equally complex Cenozoic history of subsidence characterizes the Great Basin and Sierra Nevada in the western United States (Horton et al. 2004).

Himalayan and, to a lesser degree, East African and Colorado Plateau uplift have been implicated as mechanisms for Cenozoic cooling through their influence on atmospheric circulation, biogeochemical cycling, and ocean-climate interactions (Raymo et al. 1988; Ruddiman and Raymo 1988). A full treatment of the uplift-climate connection can be found in the comprehensive volume of the *Journal of Geophysical Research* edited by Bill Ruddiman and Warren Prell (Ruddiman and Prell 1997) and in papers by Raymo (1994a) and Hay et al. (2002). The Himalayan uplift hypothesis has several components. Emerging mountains may have increased precipitation over continents, resulting in silicate weathering and erosion and blocking of zonal atmospheric flow over Asia (Ruddiman et al. 1989; Broccoli and Manabe 1997; Rind et al. 1997). It is known that the Asian monsoon experienced major changes during the Neogene period (An et al. 2001). Marine records from the Indian (planktonic foraminiferal upwelling proxies) and Pacific Oceans (windblown dust) and eolian sediments from continental Asia support the idea that the evolution of monsoon systems evolved in response to Himalayan uplift beginning ~10–20 Ma. The east Asian and Indian monsoons began roughly at 8–9 Ma, summer and winter monsoon systems intensified at 3.6–2.6 Ma, and during the last 2.6 Ma, the east Asian monsoon grew even stronger.

One critical uplift-climate mechanism involves the possible impact from greater intensity of chemical weathering and enhanced erosion of silicate bedrock on atmospheric CO_2 concentrations (Kutzbach et al. 1997). Partly on the basis of the Cenozoic radiogenic isotope curves, Raymo (1994a) and Raymo et al. (1992) theorized that continental silicate weathering led to oceanic carbonate formation and possible changes in nutrient flux (phosphorous). At its simplest, this model holds that slow uplift during early phases of collision between India and Asia occurred at 55–20 Ma followed by an increased rate of uplift beginning between 20 and 10 Ma coincident with global cooling. As introduced in Chapter 3, Cenozoic strontium and osmium isotope records from marine invertebrates (Veizer et al. 1999), planktic foraminifers (Hodell et al. 1989), and ferromanganese crusts (Burton 2006) might be a proxy for plateau and mountain weathering reflected in a greater flux of continentally derived isotopes. Pacific Ocean strontium and osmium records show inflexion points coincident with oxygen isotope events Oi-1 and Mi-1, and about 10 Ma (Burton 2006). They also show large changes in lead and neodymium isotopes near 23 Ma and 10–12 Ma, respectively.

Since it was formulated in the 1990s, the uplift-climate hypothesis has been actively debated. The tectonic and geo-logical history of the Himalayan Plateau is now better known (Harrison et al. 1992; Yin and Harrison 2000; An et al. 2001; Yin 2006). This massive reorganization of continental geometry involved an estimated 1400 km of north-south shortening in the Tibetan-Himalayan orogen since collision began ~70 Ma. Harrison et al. (1992) showed significant Miocene uplift in southern Tibet about 21–17 Ma, with some regions reaching their current elevation about 8 Ma, but little was known of its Paleogene uplift history. Chung et al. (1998) dated potassium-rich volcanism at 37–33 Ma, showing a much older age for Himalayan uplift than had previously been thought. Yin and Harrison (2000) found three periods of enhanced exhumation rates at 55–45 Ma (rate of 40 mm yr^{-1}), 12–8 Ma (6–7 mm yr^{-1}), and since 7 Ma (3–5 mm yr^{-1}). The emergence of the Himalayan Crystalline Complex—a weathering source, according to some authors—was exposed sometime after 11–5 Ma and thus coeval with proxy records for exhumation and denudation.

Quade et al. (1997) measured strontium concentrations in soils and fossil shells from fluvial deposits of the late Neogene Siwalik Group in low-lying regions of India and found a complex spatial and temporal variability in the strength and source of weathering since 14 Ma. There was an important shift in continental strontium flux at 7–8 Ma, leading to the conclusion that the marine strontium record was not a simple indicator of continental exhumation. Derry and France-Lanord (1996) analyzed clay mineralogy and $^{87}Sr/^{86}Sr$ ratios in sediments from the thick Bengal Fan on the eastern side of the Indian subcontinent taken during ODP Leg 116. They concluded there had been no major change in Himalayan sediment provenance since ~17 Ma but found a shift from a physical to chemical style of erosion about 7.4 Ma, suggesting that reduced erosion rates explained the strontium ocean budget after this time. O'Nions et al. (1998) and Frank and O'Nions (1998) reconstructed a 26-Ma radiogenic isotope record from ferromanganese crusts from the abyssal plain of the central Indian Ocean. A large increase in lead isotopes between 20 and 8 Ma was attributed to uplift and erosion of high Himalayan metamorphic and igneous rocks. All these studies added complications about the age and rate of uplift and erosion and the provenance of eroded material.

There is also uncertainty about the elevation history of the Himalayan Plateau (Hay et al. 2002; Molnar 2004; Sahagian 2005). At least four general methods are used to estimate elevation: (1) paleobotanical analysis of fossil plants, (2) vesicules preserved in ancient basalts (Sahagian and Maus 1994; Sahagian et al. 2002), (3) pedogenic carbonates (Quade et al. 1997), and (4) oxygen, carbon, and hydrogen isotopes in paleosol carbonates and clays (Ghosh et al. 2006; Quade et al. 2007; Garzione et al. 2008). Paleobotany relies on an understanding of the preferred habitat of various plant groups based either on ecological tolerance or the morphology of leaves or other preserved parts. The basalt vesicle method measures paleoatmospheric pressure and is not subject to environmental factors that influence plant physiology and morphology. Basalts and paleobotanical records, however, are generally stratigraphically discontinuous compared to records from deep-sea cores and ferromanganese crusts. Isotopic methods hold potential, but it is still difficult to make direct comparison of Himalayan elevation to long-term marine records, except for specific intervals.

The idea that the strontium curve reflected global biogeochemical cycling due to uplift resulting in atmospheric CO_2 drawdown required modification when Pagani et al. (1999) published a Miocene pCO_2 curve. If the strontium curve were directly tied to CO_2 drawdown, one would expect a strong correlation between the two proxies. The record of Pagani et al. showed a decrease from the Oligocene levels of ~350 ppmv starting at 25 Ma, followed by lower values (190–260 ppmv) during the early and middle Miocene epoch. In general, these results indicated at times that the strontium curve corresponded to CO_2 concentrations.

In addition, a completely opposing theory exists to explain the relationship between mountain building and climate. Molnar and England (1990) and England and Molnar (1990) proposed that Cenozoic climate changes themselves drove plateau uplift by enhancing physical and chemical weathering rates. The "apparent" uplift would have been caused by isostatic adjustment of continental regions in response to reduced load. Molnar (2004) cites evidence for a sharp increase in Pliocene sedimentation rates in many regions coincident with the onset of high-amplitude glacial-interglacial cycles. In fact, he stresses that most evidence cited for plateau or mountain "uplift" for a number of the world's high-elevation regions is not reliable at all (Molnar 2007). The avid structural geologist might also find interesting yet another view based on a review of exhumation of the Himalayan Plateau. Yin (2006) concluded that Himalayan exhumation is dominated by the mode and magnitude of structural deformation driven by tectonics and not by climate changes, at least over multimillion-year timescales. McMillan et al. (2006) came to a similar conclusion—that regional North American tectonics controlled elevation in the Rocky Mountains during the Cenozoic until regional subsidence occurred about 6–8 Ma in the central Rockies and 3–4 Ma in marginal regions.

The jury is still out on the hypothesis that Neogene climatic cooling was forced by Himalayan uplift, changes in the rate or intensity (or both) of continental erosion, and CO_2 drawdown, or whether climate changes drove plateau erosion. Some support for climate impact comes from the Miocene intensification of the Indian and Asian monsoon (Prell and Kutzbach 1992, 1997; An et al. 2001), the development of Chinese "red earth" and soil deposits that place the onset and intensification of east Asian aridification at 22 Ma and 8 Ma

(Guo et al. 2002, 2004), and foraminiferal lithium/calcium and lithium isotope records showing biogeochemical changes between 16 and 8 Ma (Hathorne and James 2006). New paleoelevation proxy methods suggest at least that parts of central Tibet were uplifted and remained at high elevations since the Eocene epoch. For example, Rowley and Currie (2006) presented data from oxygen and hydrogen isotopes in calcitic material from lake sediments and soils suggesting the Tibetan Plateau has remained more than 4 km high since 35 Ma. Molnar et al. (2006) suggested Rowley's data do not discount a mantle-thickening model for plateau development nor preclude appreciable uplift after 35 Ma. The argument revolves around conflicting models for the behavior of crustal and mantle portions of the lithosphere during continental collision. The regional tectonic histories of the western and eastern regions of the Tibetan Plateau are quite distinct, and paleoelevation evidence from one region may not be representative of the entire plateau (Royden et al. 2008). Suffice to say that additional work is needed to establish the Tibetan elevation history and the tectonic and climatic processes that caused it.

In other high-elevation regions, progress has also been made on elevation history and climatic processes. For example, Sepulchre et al. (2006) recently made a case linking post-late Miocene (8 Ma) east African uplift to regional aridification and changes in ecosystems, including the shift from C_3 to C_4 (grass) plant groups. Garzione et al. (2008) reviewed growing evidence that the Andes central plateau rose episodically during the Neogene, with episodes of rapid (1–4 Ma) uplift as much as 1.5 km.

In sum, the uplift-climate theory is much more complex than initially conceived, and improvements in paleo-elevation proxies will remain an important topic for Cenozoic paleoclimatology. In fact, this topic is especially germane to efforts to reduce fossil-fuel emissions and stabilize CO_2 concentrations near 450–500 ppmv, because those were the concentrations during the mid- to late Miocene, when rapid uplift and weathering has been proposed (Kutzbach and Behling 2004).

Ocean Gateways and Ocean Circulation

Another tectonic climate mechanism involves the opening and closing of ocean gateways and their influence on ocean circulation and heat transport. Today's continental positions, the product of motions by the earth's 12 major and many minor rigid lithospheric plates, influence global ocean circulation and heat transport. Ocean gateway-climate hypotheses emerged when early deep-sea drilling Legs provided better age constraints on ocean sediments, their relationship to critical gateway opening and closing, and continuous paleoceanographic proxy records (Berggren and Hollister 1977; see also Berggren 1982). We focus here on five ocean gateways shown in Figure 4.5.

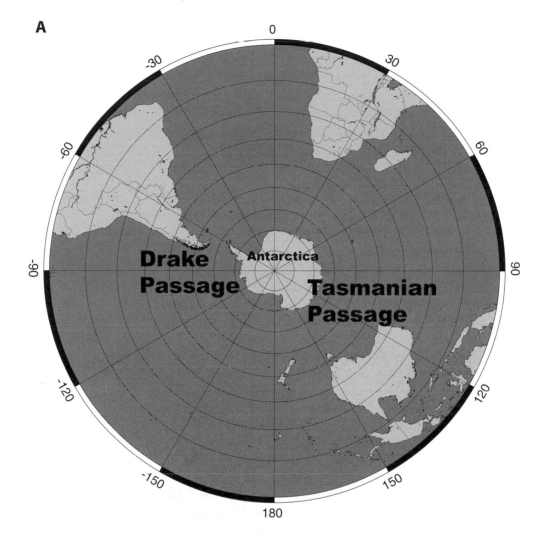

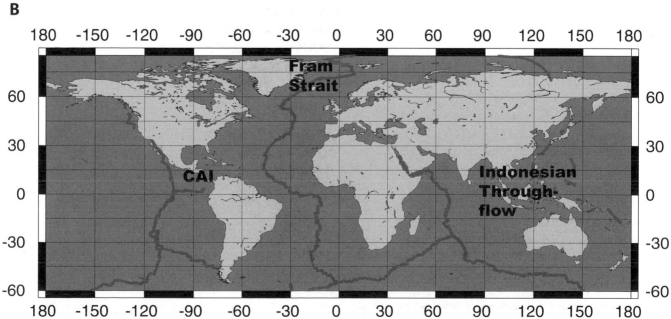

FIGURE 4.5 Map showing major Cenozoic ocean gateways: (A) The southern hemisphere (polar view) and the Tasmanian and Drake passages; (B) ocean gateways: The Central American Isthmus (CAI), Fram Strait, and Indonesian througflow region. Mid-ocean ridges are shown as gray lines. See text.

Tasmanian Gateway   Antarctica is isolated from the rest of world by the strong Antarctic Circumpolar Current (ACC), which transports more than 145 Sverdrups (Sv, 1 Sv = 10^6 m^3 per second) of water from west to east in a broad 550-km-wide zone near the polar front at 50–60°S latitude. The isolation of Antarctica and the development of strong zonal flow by the ACC are the result of the opening of two major oceanographic gateways—the Tasmanian Gateway between Australia and Antarctica and the Drake Passage between South America and Antarctica. It has been a fundamental tenet of theories about Cenozoic climate that these events had large impacts on global climate by preventing meridional oceanic heat flow from low to high southern-hemisphere latitudes fostering growth of the AIS. Seminal studies of deep-sea drilling cores led to the widely accepted hypotheses that the opening of these ocean gateways led to the thermal isolation of Antarctica and the early Cenozoic growth of a large, permanent AIS (Kennett and Shackleton 1975; Kennett 1977). The Tasmanian Gateway opening is linked to deep-sea oxygen isotope excursion Oi-1 marking the termination of sustained Cretaceous-Eocene greenhouse climate (Zachos et al. 1996).

The age of the opening of the Tasmanian Gateway is constrained by geophysical surveys, plate reconstructions, and several deep-sea coring Legs. The most direct evidence comes from magnetic anomalies and fracture zone lineations, allowing reconstruction of the complex plate motions. Lawver and Gahagan (2003) provided geophysical evidence that when the Tasmanian Gateway opened at ~32 Ma, there was a deep (2000 m) water connection. ODP Leg 189 targeted the Tasmanian Gateway to better constrain its tectonic and paleoceanographic history and role in global climate (Exon et al. 2001, 2002). Core sites were chosen on the Tasman Plateau, off the west side of modern Tasmania and along the South Tasman Rise, to capture the gradual disappearance of the Tasmanian Land Bridge, the former connection between the Tasmanian-Australian region and Antarctica and the beginning of oceanic flow through the gateway. Plate motions along this transform plate margin began to open in the Paleocene epoch, but by the Eocene epoch a shallow-water opening in the Tasmanian passage region had still not yet opened. Leg 189 confirmed that the Tasmanian Land Bridge was replaced by shallow marine sedimentation in the earliest Oligocene epoch, when a shallow ACC was established in this region. Most paleoceanographic evidence indicates that the deep-water passage was fully established during the Oligocene, possible just after 34 Ma (Stickley et al. 2004; Lyle et al. 2007). Kennett and Exon (2004) propose that the complex tectonic and oceanographic events in the Tasmanian region during the E-O transition led to a series of oceanographic and climate events including increased albedo, AIS elevation, enhanced Southern Ocean productivity, intensified deep-water formation and cooling, and atmosphere-ocean drawdown of atmospheric CO_2.

Drake Passage   The opening of the Drake Passage established full circumpolar oceanic flow around the entire Antarctic continent. Estimates of its age on the basis of geophysical data and plate-tectonic reconstructions range from as young as 17–22 Ma (Barker 2001) to 29–33 Ma (Lawver and Gahagen 2003) to 34–48 Ma (Livermore et al. 2005). Livermore et al. (2007) assessed the tectonic history of the Drake Passage, focusing on the Scotian Sea in the vicinity of South Georgia and the South Orkney Islands and surrounding oceanic regions. As with the Tasmanian passage, a landbridge between South America and Antarctica permitted terrestrial animals and plants to freely interchange at about 55–52 Ma. During the late Eocene ~50 Ma, the reconstructions of Livermore et al. (2007) show that there were shallow-water pathways across subsiding continental shelf areas and that by 34–30 Ma a continuous (1000–3000 m) deeper-water connection had formed.

A minimum age of 34–30 Ma for the Drake Passage is older than prior estimates (see Pfuhl and McCave 2005), but is supported by several geochemical and lithological studies from deep-sea cores. For example, terrigenous sediment influx and paleoproductivity studies from ODP Site 1090 suggest that the Drake deep passage opened by 32.8 Ma (Latimer and Filippelli 2002). Using neodymium isotope ratios from the Agulhas Ridge in the southern Atlantic Ocean, Scher and Martin (2004, 2006) argued that the Drake Passage opened even earlier, ~41 Ma, and then deepened over the next several million years. Tripati et al. (2005) reconstructed CCD in the tropical Pacific and the South Atlantic, showing a synchronous initial CCD deepening ~41 Ma and a permanent deepening at 34 Ma. The CCD pattern correlated with foraminiferal oxygen isotopic records, leading to the idea that continental ice sheets developed in both northern and southern hemispheres at this time, lowering global sea level by a total of ~100–125 m. All these events are at least indirectly linked to the Drake Passage opening.

The approximate synchroneity of major regional oceanographic and global biogeochemical transitions and the opening of the ACC gave support for the gateway-opening mechanism as a trigger for climate change. Nonetheless, Barker and Thomas (2004) warn that synchroneity does not imply causality, and there are strong arguments that can be made against the importance of ocean gateway opening in global climate. One line of evidence comes from modern oceanography in the Antarctic Polar Front, which suggests fine spatial scale variability that is not captured in paleorecords or simulated in climate models. They also point out that simulations by ocean-climate models indicate that ocean-heat transport was less important than atmospheric CO_2 concentrations as a cause for cooling (DeConto and Pollard 2003b).

Future studies of Southern Ocean isolation of Antarctica require integrated paleoceanographic reconstructions and ocean-climate model simulations using more sophisticated tectonic reconstructions. Brown et al. (2006), for example, produced eight digital circum-Antarctic paleobathymetric maps for Cenozoic time slices at 61, 52, 43, 32, 25 10, 5, and 0 Ma from geophysical surveys. The construction of digital paleobathymetric maps such as these will promote testing of the Southern Ocean gateway-climate hypothesis because they provide spatially well-resolved boundary conditions that allow ocean modelers to simulate the vertical and lateral plate motions during gateway openings as well as their influence on climate.

Indonesian Gateway (Eastern Tethys)   The Indonesian Archipelago is a jigsaw puzzle of islands and ocean seafloor that today forms a complete barrier to deep-ocean circulation and a partial barrier to surface ocean flow between the equatorial Indian and west Pacific Oceans. The closure of deep-water Indian-Pacific Ocean exchange about 18-14 Ma represents a major step in the closure of the Mesozoic Tethyan Ocean (Kennett et al. 1985, Linthout et al. 1997). Surface ocean exchange called Indonesian throughflow is today limited because of the partial closure of ocean straits from tectonic activity of the Pacific and Philippine Sea plates. The modern situation is the result of rapid late Neogene tectonics causing lateral geographical (northward shift of Australia and New Guinea and compression of the island of Halmahera) and vertical topographical changes (uplift of Halmahera, shoaling of the Molucca Sea).

Cane and Molnar (2001) proposed that between 5 and 3 Ma, south equatorial Pacific water flowed westward into the Indian Ocean at a more southerly position, such that Indonesian throughflow was stronger than it is today. Partly on the basis of geophysical evidence for the timing of the Indonesian ocean gateway, they hypothesized that tropical climate processes force larger-scale climate change. The resulting circulation pattern led to cooler tropical Indian Ocean SSTs (~3°C warmer) and less rainfall over east Africa. Citing modern ocean-atmosphere dynamics during ENSO oscillations and climate modeling, they proposed that major SST changes occurred when the land area in the western Indo-Pacific region was shifted 5° in latitude. A corollary to their hypothesis is that, before closure, the middle Pliocene climate ~3–5 Ma was in a permanent El Niño state (Molnar and Cane 2002).

The restriction of the Indonesian throughflow is linked to a number of well-known climatic patterns (e.g., see Kuhnt et al. 2004). One is the late Pliocene aridification over tropical Africa, when formerly humid topical climate was transformed toward much drier mean conditions (deMenocal 1995). Altered Indonesian throughflow also coincides with the inception of high-amplitude glacial-interglacial cycles and the growth of large northern-hemisphere ice sheets between 4 and 2.75 Ma. Cane and Molnar's idea is that decreased atmospheric heat transport due to events in Indonesia was an important factor in causing these global climate events, although we see later that the Central American Isthmus (CAI) closure is also linked to late Pliocene glaciation.

Support for a mid-Pliocene climate shift caused by cessation of ocean exchange in the Indonesian throughflow comes from a large shift in lead isotopes in hydrogenous ferromanganese crusts from the Indian Ocean (Frank et al. 2006) that suggest a change in source region about 3.5 Ma. Such a shift might reflect the uplift and formation of Indonesian islands including Papua New Guinea, consistent with Cane and Molnar's proposition.

Fram Strait   Because of its exchange of freshwater with the world's oceans and its extensive sea-ice cover, the Arctic Ocean and adjacent Nordic Seas (Norwegian and Greenland Seas) hold a critical but still poorly known role in global Cenozoic climate changes (Aagaard et al. 1991). The 400-km-wide Fram Strait, the only deep-water connection between the Arctic and the Nordic Seas, lies between northwestern Greenland and the island system of Svalbard. Warm North Atlantic water flows into the Arctic via the west Spitzbergen Current while cold, ice-laden Arctic water flows south in the East Greenland Current (Rudels et al. 2002). Deep-water exchange is governed by bathymetry and oceanography on either side of the strait.

Until recently, the Cenozoic tectonic, geological, and paleoceanographic history of the Arctic Ocean and the Fram Strait region was poorly constrained because of the logistical problems of deep-sea coring in regions covered by sea-ice and the lack of solid chronology from biostratigraphy and magnetostratigraphy. This data gap has had serious implications for understanding Cenozoic climate. It has meant that ocean-model simulations could not include realistic Arctic-wide climatic and sea-ice boundary conditions, hindered reconstruction of Neogene northern-hemisphere ice-sheet growth, and prevented assessment of bipolar glaciation hypotheses and forcing mechanisms of key Cenozoic climate transitions.

Recently paleoceanographers have developed well-dated ocean and tectonic reconstructions that provide insight into the geological history of the Fram Strait gateway. Several

ODP Legs to the Nordic Seas (Thiede and Myhre 1996), geophysical studies of the Lomonosov Ridge, the IODP Leg 302 ACEX cruise to the Lomonosov Ridge (Moran et al. 2006; Backman et al. 2008), and improved age control from dinoflagellate biostratigraphy and paleomagnetic chronology (Eldrett et al. 2007; Brinkhuis et al. 2006) have improved Arctic and Nordic Seas paleoceanography. The central Arctic Cenozoic environment evolved in several phases from a freshwater to brackish-water anoxic-hypoxic phase during parts of the Eocene (Brinkhuis et al. 2006) to a well-ventilated ocean with an open connection to the Nordic Seas and the Atlantic Ocean. Jakobsson et al. (2007a) integrated interbasinal seismo-stratigraphic records from the Lomonosov Ridge and the Nordic Seas, with paleoceanographic records from the ACEX cruise, and dated the opening of the deep-water connection in the Fram Strait as early Miocene 18.2–17.5 Ma and a full deep-water connection developed by about 13.6 Ma. Haley et al. (2008) also analyzed ACEX cores for radiogenic isotopic tracer neodymium isotopes to trace Arctic Ocean intermediate-depth history the past 15 Ma. High neodymium ratios in sediments deposited between 15 and 2 Ma and during large glacial periods since 2 Ma indicated that weathering from the Siberian basalts may have caused brine formation along the wide Eurasian continental shelf, the source of Arctic intermediate water. This situation contrasts with the modern and Pleistocene interglacial Arctic Ocean, which is influenced by inflowing mid-depth North Atlantic warm water. Tectonic forcing near the Fram Strait gateway may have been responsible for some of the observed trends.

Despite this progress, the significance of the deep and intermediate Fram Strait opening for global ocean circulation and climate requires additional analysis and modeling—in particular, regarding its relationship to major events such as the mid-Miocene climate optimum and late-Miocene-through-Pleistocene cooling.

Central American Isthmus The CAI consists of a tectonic mélange formed over 15 million years by complex motions of parts of the Pacific, Caribbean, and South American plates and several small microplates. Its history has been determined from onshore stratigraphic studies of thick turbidites and shallow marine sediments on the west and east sides of the isthmus, respectively (Duque-Caro 1990; Coates et al. 1992, 2003, 2004) and deep-sea cores from the eastern Pacific and Caribbean (Keigwin 1982; Haug and Tiedemann 1998; Haug et al. 2001a; Steph et al. 2006). The onshore geological history includes an initial fusion of the Central American island arc with the northern edge of South America about 12.8–9.5 Ma, the emergence of the CAI during the late Miocene, and the complete emergence of the CAI between 4.2 and 3 Ma (Duque-Caro 1990; Coates and Obando 1996).

Paleontological records led Jain and Collins (2007) to divide its history into an early shoaling phase (8.3–7.9 Ma), a preclosure interval (7.6–4.2 Ma), and a postclosure phase (4.2–2.5 Ma). The formation of the CAI is noteworthy in biological circles for creating a landbridge between North and South America, allowing what has been called the Great American Interchange of mammals and other vertebrates (Marshall et al. 1982), and at the same time for erecting an interocean land barrier leading to genetic isolation and speciation in marine organisms on either side (Collins et al. 1996) and calibration of molecular clocks (Marko 2002).

For more than three decades the CAI was also thought to be a factor in, if not the primary mechanism triggering, the intensification of late Pliocene and Pleistocene glacial cycles about 2.8 Ma (reviewed in Molnar 2008). Early deep-sea core records on the Caribbean and Pacific sides of the modern Isthmus confirmed the timing of CAI emergence and its impact on regional oceanography. Keigwin (1982) used foraminiferal oxygen and carbon isotopic records from DSDP sites 502 and 503 to infer a drop in surface salinity in the Caribbean beginning about 4 Ma and that by 3 Ma complete isolation of the Pacific and the Caribbean was complete. Keigwin's carbon isotopic records showed a two-step decrease in the $\delta^{13}C$ gradient between Caribbean and Pacific in which Pacific values fell more than those in the Caribbean, at 6 and 3 Ma, respectively. These showed greater salt and moisture transport to high latitudes, eventually leading to enhanced meridional overturning circulation and North Atlantic Deep Water (NADW) formation by 3 Ma.

Recent ODP Legs to the Caribbean, eastern Pacific, Bahama Banks, and subpolar North Atlantic Ocean have produced an integrated picture of the late Pliocene CAI closure and its basin-wide impacts. Frank et al. (1999b) and Billups (2002) provided neodymium and oxygen and carbon isotopic data for deep-water closure about 8 Ma. Oxygen isotope records from ODP Site 851 (Pacific) and Site 999 (Caribbean) suggest that the establishment of modern Pacific-Atlantic salinity gradient (~1‰) was complete by 4.7–4.2 Ma and relative global warmth until about 3 Ma was caused by greater poleward heat and salt transport (Haug and Tiedemann 1998; Haug et al. 2001a). The Pliocene salinity transition in the Pacific signified CAI closure to 100 m water depth or less. The development of the interocean carbon isotopic gradient in benthic foraminifers suggests nearly simultaneous, though not fully coupled, changes in deep-ocean circulation, including stronger NADW formation and eventual greater deep-water ventilation in the Caribbean following final CAI closure about 3–2.5 Ma. The series of events leading to CAI closure prompted Haug and Tiedemann (1998) to suggest that their impact was to "precondition" the high-latitude North Atlantic region for large ice-sheet buildup since ~2.8 Ma.

Driscoll and Haug (1998) agreed that the CAI forced greater equator-to-pole moisture and heat transport but hypothesized that enhanced moisture supply to northern regions in Eurasia led to greater freshwater riverine runoff into the Arctic, leading to greater sea ice and albedo on the basis of carbonate deposition in the Caribbean and extratropical late Pliocene climate records. Greater freshwater export from the Arctic would also influence the strength of NADW formation in the Greenland Norwegian Seas. These complex ocean-atmosphere-ice feedbacks would have produced the large northern-hemispheric ice-sheet buildups since 3 Ma, also attributed to the Indonesian throughflow blockage previously mentioned. Steph et al. (2006) examined surface and subsurface (400 m) salinity and SST patterns during the CAI closing. They confirmed the results of Haug et al. (2001a) in finding that an interoceanic salinity gradient developed as the isthmus closed, and they also identified a strong 23-ka orbital precession signal in salinity and SST variability within the Caribbean and in regions of Gulf Stream flow fed by water from the Caribbean. The characteristic salinity gradients and cycles began after 4.4 Ma, as closure of surface water began, and became fully developed by 4.1–3.7 Ma and may reflect a number of factors such as hydrodynamic Pacific-Caribbean flow through the emerging isthmus, Caribbean-Atlantic surface exchange, upper-ocean salinity variability associated with the Intertropical Convergence Zone, precipitation, and wind patterns, some related to uplift in the Andes Mountains.

Additional evidence about the impacts of CAI closure comes from the subpolar, subtropical, and deep North Atlantic. Bartoli et al. (2005) reconstructed submillennial-scale paleoceanography from the Bjorn Drift at ODP Site 984, south of Iceland. They discovered a 2–3°C warming at 2.95–2.82 Ma, just before the abrupt inception the first large northern-hemisphere glacial represented by marine oxygen isotope stage G6 at 2.74 Ma (Tiedemann et al. 1994).

Marlow et al. (2000) documented an increase in productivity and organic sediment accumulation at ODP Site 1084 off southwest Africa in the modern Benguela Current system during the late Pliocene epoch. Against a backdrop of long-term SST cooling of 10°C since 3.2 Ma, they showed that intensified trade winds induced upwelling, which was amplified by greater aridity, and windblown, nutrient-carrying dust characterized progressive cooling between 3.2 and 2.3 Ma (Hovan et al. 1991; Tiedemann et al. 1994). A sharp 2°C cooling at 2.8–2.5 Ma occurred at the same time that other Atlantic proxy records show intensified northern-hemisphere glacial activity. Marlow suggests that advection of surface water from Antarctic regions explained a diatom bloom at 3.2–2.3 Ma, ultimately connected to enhanced NADW formation. They also suggest a link between CAI-closure-induced ocean productivity changes during the late

Pliocene epoch and drawdown of atmospheric CO_2 further driving the global climate into stronger glacial periods. Prange and Schultz (2004) further proposed that the CAI closure influenced eastern Atlantic upwelling systems by causing a seesaw effect whereby the relative strength of upwelling off northwest Africa weakened while that off southwest Africa strengthened. In effect, as the isthmus closed completely, it caused large changes in cross-equatorial heat transport in the Atlantic. Burton et al. (1997) showed from radiogenic isotope tracers that the closing of the CAI dramatically altered NADW production ~3–4 Ma, essentially creating the circulation pattern seen during recent interglacial periods.

Schmittner et al. (2004) and Schneider and Schmittner (2006) used a global ocean model to simulate the impacts of CAI closing on Atlantic and Pacific productivity, ocean circulation, and nutrient cycling, using different simulated sill depths. Their basic conclusions confirmed that CAI closure intensified thermohaline circulation (THC), and in addition, as the CAI sill depth shoaled, current velocity increased, and North Atlantic and Pacific productivity increased as the Pacific-to-Atlantic nutrient flow was cut off. On a global scale, net primary productivity in the world's oceans increased with progressive closure, supporting a positive feedback mechanism for CO_2 drawdown. As with previous modeling studies (Mikolajewicz et al. 1993), they note that the CAI closure was a contributing factor, but not the sole cause of northern-hemisphere ice-sheet growth, as changes in high-latitude precipitation, orbitally driven insolation, uplift-induced continental weathering and flux to the oceans, and carbon-cycle feedbacks were also significant.

In sum, the CAI is the most intensely studied Cenozoic ocean gateway event and its closing appears to have had Atlantic-wide and perhaps global climatic repercussions. Nonetheless, Molnar (2008) points out that causality between CAI emergence and the development of northern-hemisphere ice sheets is not assured. Model simulations show a variety of ocean-circulation impacts, and the timescales for CAO emergence are more gradual and chronology still uncertain for final closure, as opposed to the timing of major paleoclimate shifts, the most notable of which is 2.7 Ma. Paleoceanographic reconstructions throughout the North Atlantic and Pacific Oceans and additional study of ocean and climatic response and potential feedback mechanisms during and following its closure are warranted.

Deep-Sea Fracture Zones In addition to landbridges and oceanic gateways, deep-ocean circulation is tightly linked to tectonic processes surrounding fracture zones at lithospheric plate boundaries. Frank et al. (2003) studied ferromanganese crusts from the central Atlantic Ocean near the Romanche and Vema Fractures zones, two conduits for circulation of deep-water masses of north and south origin. Figure 4.6 shows 31-Ma-long lead and neodymium curves from one crust, providing clues about deep-ocean circulation and terrigenous influx. The curves show relatively modest long-term changes in the two proxies, and Frank interpreted these, along with other manganese crust data from the Atlantic, as signifying terrigenous flux from the Orinoco/Amazon Rivers and from atmospheric dust from the Sahara Desert to the deep central Atlantic. Changes in the relative strength of AABW and NAWD, particularly Labrador Sea Water and AABW and their circulation through the Romanche and Vema Fracture zones, were also factors.

Orbital Forcing During the Cenozoic Era

Until now, we have focused mainly on Cenozoic climate changes occurring over millions of years and slow tectonic forcing mechanisms. In this section, we consider high-resolution rec-

ords, some with temporal resolution every 5–10 ka, for several Cenozoic climatic transitions. It would not be hyperbole to state that the last decade has witnessed a revolution in high-frequency Cenozoic paleoclimatology leading to paradigm shifts about the causes of climate changes. As important as tectonic processes are in altering atmospheric and ocean circulation and climate, it is now recognized that changes in the geometry of the earth's orbit and resulting changes in seasonal and geographical distribution of solar insolation due to gravitational forces in the solar system, combined with oceanic and biogeochemical feedbacks, are important drivers of Cenozoic climate variability (Herbert 1997).

The orbital theory of climate has been applied in one form or another to Pleistocene ice ages for more than 150 years. Five factors are responsible for greater acceptance of orbital forcing of pre-Pleistocene Cenozoic climate: (1) the recovery of high-sedimentation rate ODP cores, (2) better paleoceanographic proxy methods, (3) the availability of the Cenozoic CO_2 curves, (4) the publication of numerical solutions for

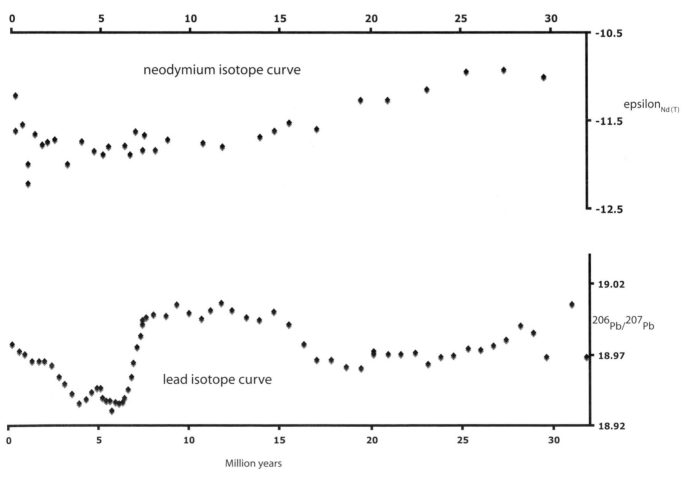

FIGURE 4.6 Neodymium (epsilon, $\varepsilon Nd(T)$) and lead (Pb) isotope curves from the 85-mm-thick ferromanganese nodule near the Romanche Fracture Zone (nodule ROM46) in the Atlantic Ocean, from Frank et al. (2003). When combined with other lead isotope data and radiogenic curves from other nodules, radiogenic isotopes indicate that Cenozoic water mass history cannot be explained solely by the mixing of northern and southern deep-water endmembers, but require inputs from African and South American continents. See text.

orbital insolation cycles extending back more than 50 Ma (Laskar et al. 2004), and (5) the tuning of the paleomagnetic polarity timescale to orbital cycles (Shackleton et al. 1990; Billups et al. 2004; Lourens et al. 2004). Primary orbital cycles are associated with the eccentricity of the earth's orbit (405-, 127-, 96-ka), and low (1.2-Ma) and high (41-ka) frequency cycles related to the earth's axial tilt (obliquity). Precession cycles of 19–26-ka are discussed in Chapter 5 of this volume.

Table 4.3 lists several deep-sea records spanning Cenozoic transitions for which orbital cycles have been recognized. They use a variety of proxy methods from different ocean regions, but all have several commonalities: detailed sampling of ODP cores from high-sedimentation-rate regions, allowing a high temporal resolution; orbital "tuning" of proxy records to Laskar's orbital timescale; generally the same conclusion that orbital forcing was important during major Cenozoic climate transitions, occurring abruptly, over one or two orbital cycles; "preconditioning" of global climate, possibly by tectonic events; and climatic amplification by atmospheric CO_2.

Holbourn et al. (2007) found that major cooling at the end of the Mid-Miocene Climatic Optimum coincided with a shift from a dominant control of insolation at 80°S latitude from obliquity- to eccentricity-dominated cycles. The transition from warm to cold climates was steplike, with cooling beginning at 14.05 Ma and culminating during two 41-ka obliquity cycles at 13.93 and 13.88 Ma. These coincided with a fall in atmospheric CO_2 concentrations, inferred from a decrease in foraminiferal $\delta^{13}C$ values at the end of the Monterey carbon isotopic excursion (CIE), and a simultaneous buildup of Antarctic ice. Holbourn hypothesized that obliquity-paced changes in moisture sources to Antarctica were responsible for these trends, invoking mechanisms similar to those used to explain late Quaternary orbital climate variability.

Flower et al. (1997) and Zachos et al. (1997, 2001) examined orbital forcing across the Oligocene-Miocene boundary and found there was a dual insolation minima in both eccentricity and obliquity sustained over a period of 100–200 ka near the Mi-1 event. These conditions, possibly combined with carbon-cycle feedbacks, led to an extended period of low seasonality, cool polar summers, and rapid ice-sheet buildup at 23.2–23.0 Ma. Climatic thresholds were crossed at 18.2, 23, and 27 Ma, when there was a coincidence of low eccentricity and obliquity.

TABLE 4.3 Pre-Quaternary Orbital-Scale Climate Variability*

Era	Age (Ma)	Source	Climate Event[†]	Primary Periodicity (ka)	Proxy	Reference
Mid-Miocene	12–17	ODP 1170A, 1171C	MMTO	100, 400	Mg/Ca C and O isotopes	Shevenell et al. 2004
Mid-Miocene	12.7–17.1	ODP 1146, 1237	MMTO	405-Carbon, 1200, 41 oxygen	Carbon, oxygen isotopes, iron	Holbourn et al. 2005a, b
Late Oligocene	26.5–30.5	ODP Site 1218	Oi2	405, 1200	Carbon, oxygen isotopes	Wade and Pälike 2004
Oligo-Miocene	17.86–26.5	ODP Leg 154 Ceara Rise, Site 926, 929	Mi1	All, 41, 96, 405	Carbon, oxygen isotopes, sediment	Pälike et al. 2006a
Eocene-Oligocene	31–35	ODP Leg 199	Oi1	405	Stable isotopes, mass accumulation	Coxall et al. 2005
Late Oligo–Miocene	16–25	ODP 1090 others	O12, Mi1, Mi2	405, 1200	Benthic isotopes	Billups 2002; Billups et al. 2004
Late Oligo–Miocene	20–26	ODP 926, 929	Mi1	41, 95, 406, 1300 plus	Oxygen, carbon isotopes	Zachos et al. 1997, 2001
Eocene	53–57	ODP Sites 1262, 1267	PETM, *Elmo*	100, 405, 2,250	Magnetic susceptibility, color	Lourens et al. 2005
Late Paleocene–Eocene	50–60	ODP 1051, 690, DSDP 550, 577		Mainly eccentricity	Carbon isotopes, cyclostraatigraphy	Cramer et al. 2003

Note: See prior studies: Flower and Kennett (1993a, b, 1995); Hodell and Woodruff (1994); Wright and Miller (1992); Woodruff and Savin (1989, 1991). DSDP = Deep Sea Drilling Program; MMTO = Mid-Miocene Thermal Optimum; ODP = Ocean Drilling Program; PETM = Paleocene-Eocene Thermal Maximum.

*Focus on papers since Laskar et al. (2004) solutions.

[†]See text. Orbital analyses lead to refinement of original isotope stratirgraphy.

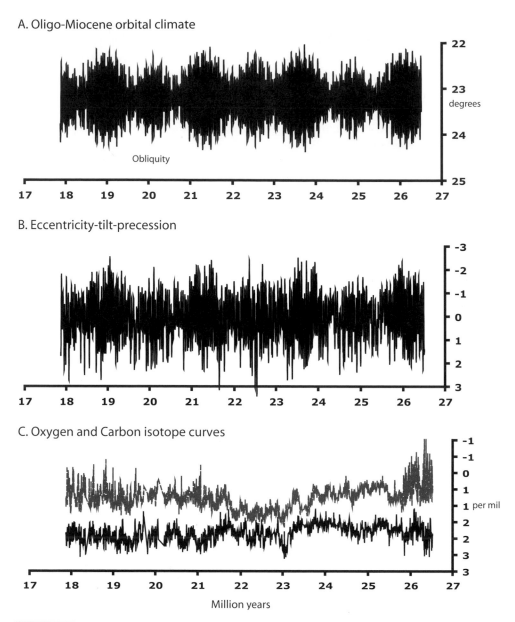

FIGURE 4.7 (A) Late Oligocene-Miocene obliquity, from Laskar et al. (2004) as presented by Pälike et al. (2006b). (B) Eccentricity-tilt-precession calculated by Pälike et al. (2006b). (C) Oxygen and carbon isotope curves from Ocean Drilling Program Leg 154 Site 926B, showing event Mi-1 and orbital cycles. Courtesy of H. Pälike.

Oligocene orbital cycles are the "heartbeat" of Oligocene climate, according to Pälike et al. (2006a, b). Figure 4.7 shows their 13-Ma-long (34–21 Ma) isotopic, carbonate, and mass accumulation rate records from ODP Site 1218 taken at 4800 m water depth in the equatorial Pacific. Eccentricity cycles and the 1.2-Ma obliquity cycle are superimposed on long-term climate changes. Four Oligocene climate phases were identified: (1) Phase I, ~34–31.5 Ma, was a recovery from the large Oi-1 excursion, with a 405-ka cycle in $\delta^{13}C$, (2) Phase II was a 2-Ma-long interval of decreasing $\delta^{13}C$ val-

ues, with simultaneous $\delta^{18}O$ variability, (3) Phase III experienced 405-ka cycles for another 2.5 Ma, and (4) Phase IV was characterized by the 405-ka $\delta^{13}C$ cycle and 1.2–2.4-Ma periods in $\delta^{18}O$. More extreme $\delta^{18}O$ excursions coincided with low-amplitude obliquity cycles and minimum eccentricity values and, to a first approximation, matched ice-volume–sea-level oscillations (Pekar et al. 2002). The basic pattern seen before, during, and after the Oi-1 event was described as an abrupt nonequilibrium climate "overshoot" separating orbitally driven quasi-equilibrium states of the Oligocene and Miocene epochs. Although the orbital influence on Oligocene climate seems unmistakable, continental silicate weathering and atmospheric CO_2 drawdown was probably also involved (Zachos and Kump 2005).

The greenhouse world of the late Paleocene and early Eocene epochs also experienced orbital forcing of climate. In fact, Cramer et al. (2003) concluded that between 50 and 56 Ma, the amplitude of eccentricity-driven cycles equaled that seen in the late Quaternary system. Their cyclostratigraphy, however, suggested that orbital variability was not directly responsible for the hyperthermal warmth of the Paleocene-Eocene Thermal Maximum (PETM) (see the subsection "Paleocene-Eocene Thermal Maximum" that follows). In contrast, Lourens et al. (2005) identified an orbital pattern during the early Eocene epoch in which the PETM and younger hyperthermal periods coincided with 405- and 100-ka eccentricity maxima. In this scenario, orbital forcing was considered important in its influence on Eocene CO_2 and methane cycling.

There is growing evidence for orbital forcing of climate during progressive planetary Cenozoic cooling. Eccentricity and obliquity changes occur concurrently with tectonic events and changes in the biogeochemical cycle, such that any theory to explain Cenozoic climate variability during critical events must consider multiple causes.

Extreme Climate States

The most important feature of the Cenozoic paleoclimate record is the existence of extreme climatic states, primarily periods of global warmth, low pole-to-equator thermal gradients, ice-free polar regions, and elevated CO_2 concentrations (Zachos et al. 2008). We describe four hyperthermal periods and their possible causes.

Paleocene-Eocene Thermal Maximum

The PETM was a brief period of global warmth about 55.5 Ma that involved an abrupt and extraordinarily large reorganization of the entire climate system. In addition to global warmth, the most telling signature of the PETM is a distinct clay layer marked by decreased $CaCO_3$ deposition (Figure 4.8) and a large negative CIE reaching 2–3‰ in marine and terrestrial sediments (Figure 4.9). The PETM CIE was identified in marine sediments in the South Atlantic Ocean at ODP Site 690 (Kennett and Stott 1991) and terrestrial sediment in the Bighorn Basin of Wyoming (Koch et al. 1992). Kennett and Stott estimated the age of the PETM, which is also called the Late Paleocene Thermal Maximum, at 57.3 Ma; it is now thought to be ~55.5 Ma based on $^{40}Ar/^{39}Ar$ dating of volcanic ashes and tuffs in the North

Atlantic (Storey et al. 2007) and tuning to orbital timescales (Lourens et al. 2005; Westerhold et al. 2007).

Past periods of warmth are intrinsically important from a geological perspective. In fact, the PETM event has been formally proposed as the boundary stratotype for the Paleocene-Eocene transition. But the PETM has attained a special status. As with Mesozoic Oceanic Anoxic Events, the stratigraphic coincidence of abrupt changes in carbon cycling and extreme warmth suggests a connection with, if not direct forcing of, global warming by greenhouse gases. Greenhouse gas forcing, however, is an overly simplistic interpretation of the PETM and associated climatic events (Higgins and Schrag 2006). Although a consensus exists that global warmth was caused, or at least amplified, by high atmospheric CO_2 concentrations, there are a number of hypotheses offered to explain various observations (Koch et al. 2003). First, we examine the vital statistics of the PETM before discussing possible causes.

Climate During the PETM The first aspect of the PETM is that warming occurred during an already warm period that began in the late Paleocene epoch ~59 Ma and continued through the early Eocene climatic maximum 52–50 Ma, before progressive global cooling began. On the basis of foraminiferal oxygen isotopic analyses, Kennett and Stott (1991) estimated that PETM surface ocean temperature rose as much as 6°C in only 6000 years. Additional proxy methods have since shown that the tropical and high-latitude surface ocean temperatures rose about 5–9°C (Zachos et al. 2003). Based on oxygen isotopes and TEX_{86} paleothermometry (Schouten et al. 2002), Zachos et al. (2006) estimated that coastal ocean temperatures in the Atlantic off New Jersey increased at least 8°C, to values near 33°C. IODP Leg 302 to the central Arctic Ocean recovered sediments from the PETM and showed a temperature rise from 18°C to a peak of at least 23°C, based on dinoflagellate cysts and TEX_{86} paleothermometry (Sluijs et al. 2006). These results show that the PETM had a small pole-to-equator thermal gradient as opposed to that of the late Cenozoic era.

The PETM was a fairly brief climatic excursion; its duration is reported as lasting 150,000 to 220,000 years. Its onset, however, as defined by rapid warming and the CIE event was abrupt, occurring in 20,000 years or less. This is only about twice the duration of a late Pleistocene deglaciation.

As with OAEs, there was major disruption of the global carbon cycle during the PETM. Many interpret the CIE as an influx of ^{12}C-enriched carbon into the global carbon cycle either from CO_2 or CH_4. Dickens et al. (1995) proposed that this influx was caused by 1000–2100 Gt C into the ocean-atmosphere system from methane trapped in ocean sediments. This is known as the clathrate or methane

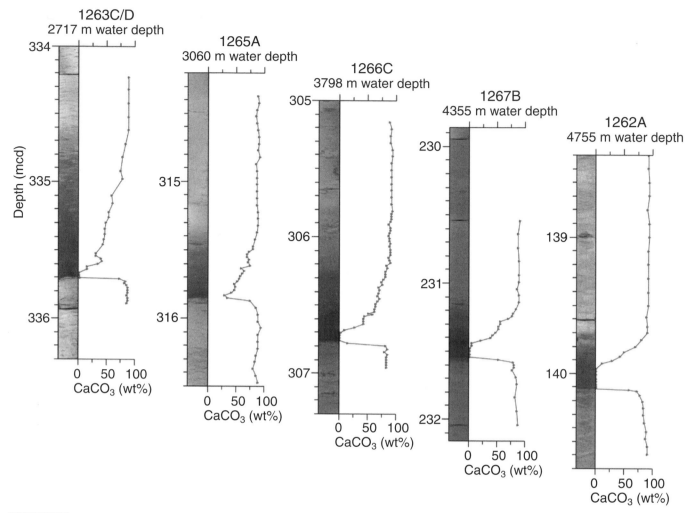

FIGURE 4.8 Photos of Ocean Drilling Program cores through Paleocene-Eocene Thermal Maximum, showing clay layer corresponding to decrease in percent carbonate. From Zachos et al. (2005). Courtesy of J. Zachos and American Association for the Advancement of Science.

disassociation hypothesis. Dickens proposed that seabed methane, which has carbon isotopic values of ~−60‰, was a possible source of PETM carbon. Seabed methane is a type of submarine clathrate in which gas molecules are trapped in upper layers of sediments in an ice-like matrix of water by hydrogen bonds under high pressure and low temperature. Clathrates constitute the single most common form of hydrocarbon on the earth. Dickens's theory holds that its release from the seafloor, possibly triggered by ocean temperature rise and oxidized to CO_2, would alter the global carbon cycle reflected in the carbon isotopic values.

The impact of carbon influx on atmospheric CO_2 concentrations was at first thought to be modest (Dickens 1997), but recent estimates suggest that at least 2000 Gt C and possibly as much as 4500 Gt C entered the atmosphere (Zachos et al. 2005). PETM CO_2 concentrations were between 600 and 2800 ppmv (Zachos et al. 2005; Pagani et al. 2006a). The scale of the PETM carbon event can be compared to the anthropogenic carbon footprint. Humans have put approximately 300 Gt C into the atmosphere since the 19th century; the modern oceans have taken up ~118 Gt C, or 40%, of the cumulative total of anthropogenic CO_2 produced between 1800 and 1994 (Sabine et al. 2004). Modern fossil-fuel burning continues at a rapid pace, producing ~7 Gt C per year (Hansen and Sato 2004). By the year 2400, the total carbon put into the atmosphere could reach 5000 Gt C if current rates continue and methods are not developed to draw down carbon (Caldeira and Wicket 2003).

Deep and intermediate ocean temperature, chemistry, and circulation changed significantly during the PETM. Tripati and Elderfield (2005) estimated a deep-ocean temperature increase of 4–5°C during the PETM, roughly in line with tropical surface ocean warming. They also documented intermediate-depth warming just prior to the PETM isotopic excursion, possibly due to greater downwelling in the

PETM carbon isotopic excursion (CIE)

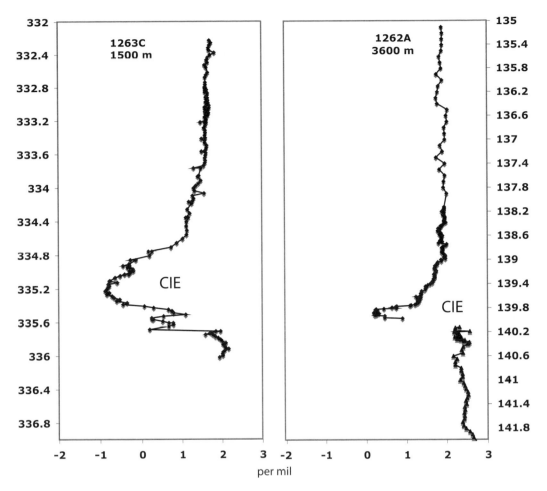

FIGURE 4.9 Paleocene-Eocene Thermal Maximum (PETM) record of carbon isotopic excursion (CIE) in Ocean Drilling Program core Sites 1263C (3600 m paleodepth) and 1262A (1500 m paleodepth), from Zachos et al. (2005).

North Pacific and reduced deep-water formation in high southern latitudes. Deep-ocean warming is one hypothesis to explain the proposed release of seabed methane as a carbon source for the PETM event.

Pagani et al. (2006b) analyzed PETM carbon (δ^{13}C) and hydrogen (δD) isotopes formed from long- and short-chain organic molecules called alkanes, which originate in terrestrial and aquatic plants and were deposited in organic-rich sediments in the central Arctic Ocean recovered during IODP Leg 302. The alkane record suggested that the CIE excursion was actually as large as −6‰, a value much larger than that estimated from low-resolution marine isotopic records. It also indicated a moist and warm north polar region, which supported climate model simulations of greater poleward moisture during global warmth. Enhanced precipitation during the PETM has also been documented from subtropical latitudes (e.g., Schmidtz and Pujalte 2007).

Zachos et al. (2005) examined the PETM across a transect of cores (2800–4700 m paleodepth) from the Walvis Ridge in the South Atlantic Ocean, taken on ODP Leg 208, to reconstruct the impact of changing carbon budget on ocean chemistry. One objective was to examine changes in the lysocline, the zone where carbonate begins to dissolve, during the PETM. The Leg 208 results showed that the PETM lysocline shoaled by as much as 2000 m, an immense perturbation of the global ocean carbonate chemistry. Ocean acidification through the uptake of CO_2 at the air-sea boundary would be expected if atmospheric CO_2 concentrations rose dramatically (see Chapter 12 in this volume).

Like many major climatic transitions, the PETM is associated with a biotic crisis in the deep sea as large as any in the past 90 Ma (Thomas 1998; Wing et al. 2005). Kennett and Stott (1991) estimated that benthic foraminifera and ostracode extinctions eliminated 35–50% of the total species in only a few thousand years. A sharp rise in deep-sea temperature, a fall in dissolved oxygen concentrations, or both were proposed as likely causes. Thomas (2003) and Thomas and Shackleton (1996) also documented benthic foraminiferal extinctions at the Paleocene-Eocene boundary, and Steineck and Thomas (1996) documented a major faunal turnover in the deep-sea ostracodes. The faunal patterns are clear, but the exact causes of the Paleocene-Eocene deep-sea extinction are not fully understood (Thomas 2007).

Causes of the PETM Mechanisms to explain the global PETM CIE, and specifically methane disassociation, are still widely debated (Thomas 2007). Greenhouse gas forcing during the PETM is still a leading hypothesis to explain abrupt and widespread PETM warmth. The gas hydrate hypothesis is one of several ways to produce elevated greenhouse gas concentrations, but clathrate release from deep-sea ocean warming could be due to THC changes, downwelling of surface water, underwater slope failure, and landslides among other things. Thomas et al. (2003) analyzed ODP neodymium records in fossil fish teeth during the PETM. Prior to and after the PETM, there was evidence for Southern Ocean deep-water formation; during the PETM, deep-water warming was relatively gradual. They did not find evidence for substantial changes in THC during the PETM and instead, they proposed that warming was the consequence of downwelling of warmer surface waters.

Bowen et al. (2004) supported a greenhouse gas forcing of PETM warmth, but suggested that the atmospheric CO_2 increase did not coincide with, nor was of sufficient magnitude to produce, the mean warming of 5–10°C. They proposed large midlatitude humidity increases and enhanced terrestrial carbon cycling to explain the differences between terrestrial and marine $\delta^{13}C$ records. Carbon release could in theory also be caused by comet impacts or extensive volcanism, including basalt formation at mid-ocean ridges.

The role of sea-level change during the PETM is unclear because continental ice volume was small in comparison to the later part of the Cenozoic. Sea-level change from oceanic thermal expansion, melting of continental ice, or both may have played a role. For example, a fall in sea level might expose a potential source of carbon-rich sediments along continental margins. On the basis of ostracodes from shallow-water sediments in New Jersey, Speijer and Morsi (2002) estimated there was a 20-m sea-level oscillation during the PETM. Gibbs et al. (2006) estimated that sea level began to rise ~20 ka prior to the CIE and continued rising for 10 ka after it. These oscillations are similar to high-frequency sea-level events already discussed, but many processes can change relative sea level along a coast, making it difficult to assign causality to a small oscillation. Precise identification of a past sea-level position is another problem due to large species' depth ranges, postdepositional processes (compaction), and other factors. Even using extant species as sea-level indicators, it is difficult to attain precision of less than a few meters in Pleistocene sea-level positions.

Pagani et al. (2006a) called the PETM a carbon-cycle "mystery" because the most important issue—the ultimate source of atmospheric CO_2—is unresolved. Higgens and Schrag (2006), among others, proposed that methane release could not account for observed isotopic and lysocline changes. Instead, they identified four possible causes for the oxidation of 5000 Gt C of organic carbon: (1) metamorphism associated with the intrusion of igneous rocks into organic-rich sediments, which forms thermogenic CH_4 and CO_2 (Svensen et al. 2004), (2) destruction of peatlands observed from pyritic sulfur ratio changes (Kurtz et al. 2003), (3) tectonically driven isolation of a large epicontinental sea, leading to desiccation, and (4) bacterial respiration of organic matter and CO_2 release.

In addition to identifying the sources of PETM carbon, it is significant that the PETM is only one of several Cenozoic hyperthermal events set in a framework of orbitally induced quasi-periodic climate changes (Lourens et al. 2005; Westerhold et al. 2007). The interplay between orbital forcing; global carbon cycling; atmospheric, ice-sheet, and oceanic feedbacks; and greenhouse gas forcing—the central theme of research on Pleistocene glacial-interglacial cycles for decades—will occupy the paleoclimate community in years to come. Whatever caused the PETM carbon isotopic anomaly and global warmth, and, more generally, early Cenozoic hyperthermal climate states, they pose a great challenge for climate modelers to simulate such extreme conditions.

Eocene-Oligocene Transition

The Oi-1 oxygen isotopic excursion near the E-O boundary is a major step from a global greenhouse into an icehouse world (Prothero et al. 2003). The opening of the Tasmanian Gateway ~33.5–33.7 Ma and near-simultaneous climatic cooling has been considered by many researchers the end of the Cenozoic greenhouse climate. Building on decades of research, Kennett and Exon (2004) argue that the Tasmanian Gateway event isolated the Antarctic continent, opened the Antarctic circumpolar current, increased albedo and Southern Ocean productivity, and intensified THC.

During the past few years, there have been major developments in several fields that complicate this relatively straightforward explanation of the Oi-1 event. First, there is debate about the level of extreme warmth of the Eocene epoch about 54–35 Ma just before Oi-1, revolving around paleotemperature proxies (Zachos et al. 2002, 2008; Huber 2008). Recent application of the TetraEther$_{86}$ (TEX$_{86}$) method and reanalyses of exceptionally well-preserved planktonic foraminifera suggest the tropics may have been as warm as 35°C or higher (Pearson et al. 2007; Huber 2008). Such extreme warmth may have impacted Eocene tropical terrestrial vegetation, but perhaps more important, may pose significant challenges for climate modelers in simulating what the possible causes may have been. Gradual cooling at the E-O transition was about 3–5°C, according to TEX$_{86}$ analyses (Schouten et al. 2008).

Second, until recently, the conventional view held that initial ice-sheet growth in the northern hemisphere began no earlier than about 15 Ma, accelerating during the late Miocene (Larsen et al. 1994; St. John and Krissek 2002) and mid-Pliocene (Fronval and Jansen 1996) epochs. As described above, growing evidence suggests earlier land and sea-ice-

buildup in Greenland (Eldrett et al. 2007) and around the Arctic Ocean (Moran et al. 2006; Backman et al. 2008). Third, Mg/Ca paleothermometry first indicated minimal change in deep-sea bottom temperature across the E-O boundary (Lear et al. 2000, 2008), but recent analyses of exceptionally well-preserved foraminifera indicate a cooling of near-surface and bottom water of about 2.5°C.

Other reconstructions of the Oi-1 transition indicate deepening of the ocean's CCD, marked by the deposition of CaCO$_3$ (Lyle et al. 2005; Coxall et al. 2005). Lear and Rosenthal (2006) also showed, on the basis of lithium/calcium ratios in benthic foraminifera, that the Oi-1 transition involved major changes in ocean saturation state and a deeper CCD (see also Rea and Lyle 2005; Lyle et al. 2007). Tripati et al. (2005) showed CCD deepening of 1000 m or more (Figure 4.10), signifying a change in ocean alkalinity linked with global carbon cycling and atmospheric CO$_2$ concentrations. Coxall

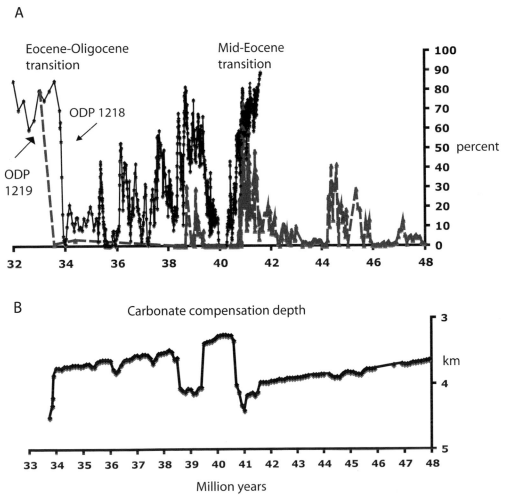

FIGURE 4.10 (A) Equatorial Pacific CaCO$_3$ weight percent from Ocean Drilling Program (ODP) core Sites 1218 (solid) and 1219 (dashed). (B) Equatorial carbonate compensation depth during Eocene and Eocene-Oligocene greenhouse-to-icehouse transition about 34 Ma, from Tripati et al. (2005).

et al. (2005) also showed that the CCD event was abrupt, occurred in two 40-ka-long steps, lasted a total of 300 ka, and coincided with orbitally driven insolation changes (Pälike et al. 2006a, b). Complicating matters, there is evidence that CCD deepening actually began as early as 41.6 Ma during what is called carbon anoxic event 3 just after the Mid-Eocene Climatic Optimum (Bohaty and Zachos 2003; Lyle et al. 2005). Large changes in ocean productivity also accompany the E-O boundary.

Fourth, Pekar et al. (2002) cross-calibrated continental-margin stratigraphy and deep-sea isotopic records, taking the Mg/Ca temperature curve into account, and hypothesized a 70-m sea-level drop at 34 Ma. This is 10 m greater than the total ice volume held by the modern EAIS. Ice-rafted debris in deep-sea cores from ocean regions off the Antarctic continent and glacial-marine deposits and lodgment tills discovered on Seymour Island off the Antarctic Peninsula (Ivany et al. 2006) support the idea of Antarctic ice growth near 34 Ma. So does the deep-sea temperature shift of Lear (2007). Tripati et al. (2005) proposed a substantial sea-level change at the E-O boundary of 100–125 m as support for the hypothesis of bipolar ice-sheet growth at the Oi-1 event and perhaps earlier.

The issue of ice-sheet growth before 34 Ma and its relation to isotopically derived global sea-level changes, THC, and atmospheric CO_2 is unresolved. Edgar et al. (2007) conducted modeling simulations and concluded there was no need to invoke bipolar ice between 41 and 34 Ma. Via and Thomas (2006) proposed that the initiation of bipolar deep-water formation occurred near 34–33 Ma, when a major shift in oceanic circulation occurred. The Cenozoic CO_2 curve indicates that the drop in atmospheric CO_2 concentrations near the E-O boundary occurred in steps (Pagani et al. 1999, 2005). Ocean-climate-ice modeling by DeConto and Pollard (2003b) demonstrated that, as CO_2 concentrations fell from four to five times to two to three times (~600–900 ppmv) preindustrial levels, orbitally driven insolation played a more important role in governing ice-sheet growth (Figure 4.11). Pollard and DeConto (2005) modeled the combined effects of orbital forcing and CO_2 to explain the 34-Ma cooling. They hypothesized that it represented a nonlinear threshold-induced climate response to changing CO_2 concentrations in which the EAIS exhibited hysteresis-like oscillatory behavior switching between two quasi-equilibrium stable states. Pollard and DeConto's theory of Antarctic ice-sheet hysteresis emphasizes atmospheric greenhouse gas thresholds and a

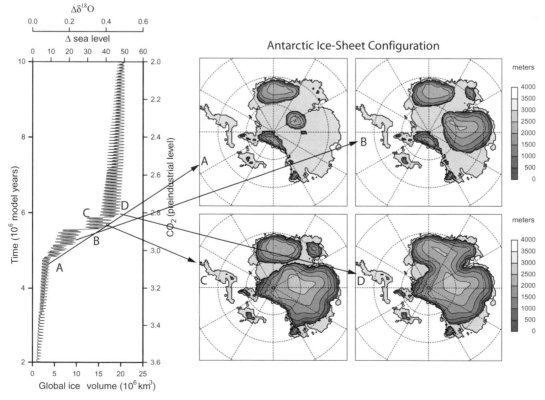

FIGURE 4.11 Antarctic ice-sheet evolution during greenhouse-to-icehouse transition, as reconstructed by a 10-million-year model simulation modified from DeConto and Pollard (2003a, b). The left panel shows oxygen isotopes, sea-level change, and modeled ice volume. The letters A–D designate four time intervals that correspond to the Antarctic maps at right, showing progressive growth in the area and the thickness of the ice sheet, influencing global sea level. Courtesy of R. DeConto and D. Pollard and Macmillan.

minor role for ocean-heat transport in controlling EAIS volume.

Barker and Thomas (2004) reviewed more than 30 years of research on the E-O climate transition and the isolation of Antarctica. Their main conclusion is that the "smoking gun" of synchroneity between the Oi-1 event, the Tasmanian Gateway opening, and the first EAIS ice-sheet growth does not justify an ocean-heat transport mechanism as a cause. One major reason is that the modern physical oceanography suggests that the ACC operates in narrow jets rather than large oceanic frontal zones. This means that the inception of biosiliceous sedimentation near 34 Ma is not necessarily indicative of a deep-water connection and circum-Antarctic throughflow. Instead, Antarctic ice sheets themselves may have been the cause for cooling of circumarctic surface water. They also point out two discoveries mentioned above— that the Oi-1 event was too rapid to be caused by a series of tectonic events, and that climate modeling simulating the impact of falling CO_2 concentrations provide a viable alternative explanation. New discoveries to test the hypothesized role of the ACC must come from paleovelocity reconstructions in key regions around the Antarctic, which can provide quantitative information on the strength and timing of deep-water flow.

In sum, there is a widespread reevaluation of the ACC-Oi-1 climate paradigm and several hypotheses exist, not necessarily exclusive of one another, to explain mid- to late Eocene and E-O boundary climatic changes. Reconstructions of oceans using dynamic current proxies, new tectonic reconstructions for the Tasmanian and Drake passages, and improved climate modeling will be the focus of future studies.

Mid-Miocene Climatic Optimum

The Monterey Formation is a striking sequence of organic-rich calcareous-siliceous sediments found along the central and southern California coastline deposited between 18 and 7 Ma during a critical period of climatic transition. The Monterey hypothesis proposed by Vincent and Berger (1985) holds that, following a period of global warmth known as the Mid-Miocene Climatic Optimum (MMCO) between 18 and 14 Ma, global cooling was the result of drawdown of CO_2 into organic material trapped in sediments of the Monterey Formation and other organic-rich formations along continental margins. Raymo (1994b) proposed that silicate weathering rather than organic carbon burial was the major factor in Miocene CO_2 drawdown and cooling. The evidence for climatic cooling comes largely from increases in oxygen and a decrease in carbon isotopic values in foraminifera from the Monterey Formation and a positive $\delta^{13}C$ excursion in marine carbonates from DSDP cores.

The preferential storage of light carbon (^{12}C) from organic material in Monterey-like sediments would enrich mean ocean ^{13}C and drawdown atmospheric carbon to the ocean. There is also evidence for changes in deep-ocean circulation and the growth of Antarctic ice at this time (Woodruff and Savin 1989).

Ben Flower and Jim Kennett analyzed foraminferal isotope records from the Monterey Formation and correlative deposits in deep-sea cores, providing support for the link between mid-Miocene cooling ~14–13.8 Ma, EAIS growth, and ocean circulation (Flower and Kennett 1993a, b, 1994, 1995). Their argument was that ocean-heat transport during the MMCO prevented large ice-sheet development on Antarctica, but beginning at 14 Ma deep-ocean-circulation patterns changed and the EAIS grew in volume. As shown by Woodruff and Savin (1991), the interval between 17 and 13.5 Ma is characterized by a series of inphase shifts in oxygen and carbon isotopes, suggesting a link between carbon cycling, cooling, and ice volume.

The Mid-Miocene optimum ended abruptly with a major cooling event about 14 Ma, at the end of the MMCO (Figure 4.12). High-resolution Mg/Ca ratios and isotopic measurements in planktic foraminifers from the South Pacific suggest a 6–7°C high-latitude cooling coincident with ice-sheet expansion on Antarctica (Shevenell et al. 2004). A significant finding was that during the 14-Ma cooling, ocean temperatures were antiphased with reconstructed atmospheric CO_2 concentrations, leading Shevenell to argue that the cooling was driven by the influence of orbital insolation on meridional heat and atmospheric moisture transport. In addition to oceanic cooling, Pagani et al. (1999) showed that CO_2 concentrations fell about 3 Ma before the MMCO and remained low thereafter. The 14-Ma cooling is also recognized by changes in deep circulation reconstructed from neodymium records (Frank et al. 1999a).

Global Warmth in the Pliocene Epoch

The early to middle Pliocene epoch, about 4.5–3.0 Ma, was the last time the earth's climate experienced what some paleoclimatologists consider a sustained-equilibrium state of global warmth for more than a half million years. In fact, the current Holocene interglacial might also last as long as 500,000 years because of anthropogenic greenhouse gas forcing and the disruption of natural orbital climatic cycles (Berger and Loutre 2003; Archer and Ganopolski 2005). Consequently, patterns and causes of sustained Pliocene warmth have been vigorously investigated the past 20 years.

Pliocene warmth goes by many informal names; here we will call it the Mid-Pliocene Thermal Optimum (MPTO). Low-amplitude climate variability during the MPTO, evident in marine records, stands in stark contrast to the high-

FIGURE 4.12 Mid-Miocene cooling: (A) Oxygen isotope curve for Miocene planktic (*G. bulloides*) and benthic (*C. mundulus*) foraminifers; (B) Mg/Ca-derived sea-surface temperature (SST) curve. Both curves are from Ocean Drilling Program Site 1171C, from Shevenell et al. (2004). The overall SST pattern is controlled by orbital forcing, and mid-Miocene SST cooling precedes Antarctic ice buildup by about 60 ka, shown by arrows.

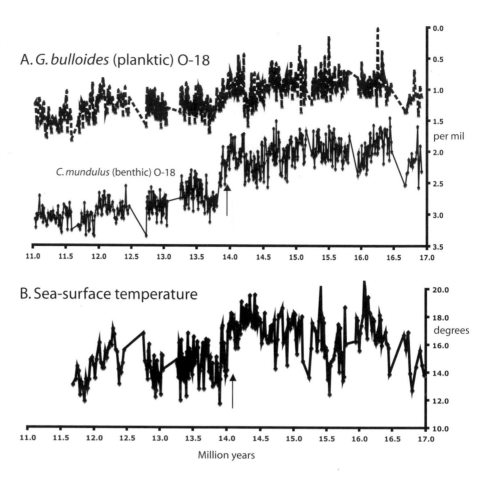

amplitude glacial-interglacial cycles of the last 2.8 Ma and especially the last 500 ka. In contrast to older periods in the Cenozoic, however, Pliocene continental topography, ocean-basin configuration, major atmospheric and ocean-circulation patterns, and atmospheric CO_2 concentrations were, to a first approximation, similar to modern boundary conditions, except for the CAI and Indonesian gateways already discussed. Thus, the MPTO represents a period similar to the present but with a warmer climate and reduced pole-to-equator gradients (Dowsett et al. 1992). This apparent contradiction has been called the "Pliocene Paradox" (Federov et al. 2006). Here we outline Pliocene climate reconstructions and hypotheses to explain MPTO warmth.

Paleoclimate Records of MPTO Pliocene climate records are available from a number of ODP sites and reveal several differences between the MPTO and modern climate: high-latitude ocean SSTs up to 8–10°C warmer (Cronin 1991; Dowsett et al. 1996), seasonally ice-free Arctic Ocean (Cronin et al. 1993; Knies et al. 2002), enhanced North Atlantic meridional ocean circulation (Raymo et al. 1992, 1996), less restricted Atlantic-Pacific ocean exchange through the Panamanian seaway (Haug et al. 2001a; Steph et al. 2006), reduced east-west symmetry in tropical Pacific Ocean SSTs, and a shallower thermocline (Wara et al. 2005; Ravelo et al. 2007). Studies of Pliocene shorelines show that global sea level was 15–25 m above present (Dowsett and Cronin 1990; Dwyer and Chandler 2008). Reduced Antarctic ice volume has also been proposed from studies of the Sirius Formation in Antarctica (Webb and Harwood 1991; Barrett et al. 1992), although Pliocene AIS stability is subject to debate (Marchant and Denton 1996), atmospheric pCO_2 levels are estimated to have been about 340–370 ppmv (Raymo et al. 1996; Kürschner et al. 1996), but these values are poorly constrained by limited data.

In addition to time series records of Pliocene climate, a synoptic paleoclimate reconstruction called PRISM (Pliocene Research, Interpretation and Synoptic Mapping) has shed light on global climate during a narrow time slice of the MPTO. The PRISM project began 20 years ago with the objective of characterizing global climate between 3.2 and 2.9 Ma, a period covering Marine Isotope Stages G19/18 through M2/1. PRISM used a variety of proxy-derived terrestrial (Thompson and Fleming 1996), open marine (Dowsett et al. 1996), and coastal (Cronin 1991) paleoclimate reconstructions and produced several iterations of global climate reconstruction that have been summarized in several volumes (Cronin and Dowsett 1991; Poore and Sloan

1996; Dowsett et al. 1999, 2005) and used in climate modeling studies by Chandler et al. (1994) and Sloan et al. (1996). PRISM reconstructions indicate a Pliocene climate state that lacks high-amplitude cycles (low-amplitude orbital cycles do occur) and exhibits sustained mean global warmth exceeding that reconstructed for other warm periods such as Marine Isotope Stage (MIS) 11, the last interglacial MIS 5e, the early Holocene, and the Medieval Warm Period. PRISM reconstructions have been used by several climate model groups in efforts to better understand patterns and causes of global warmth (Jiang et al. 2005; Haywood and Valdes 2004; Haywood et al. 2002, 2007a; Chandler et al. 2008).

Causes of Pliocene Warmth   Given concern about anthropogenic greenhouse gas forcing, the MPTO may be the closest "analog" to future climate, and it is worth examining its possible causes. Hypotheses to explain the MPTO can be broken down into four categories: (1) tectonic uplift, (2) ocean circulation and heat transport, (3) greenhouse gas forcing by elevated CO_2 concentrations, and (4) tropical forcing through sustained El Niño conditions. Crowley (1996) reviewed mid-Pliocene warmth and pointed out that two leading mechanisms, high pCO_2 and enhanced poleward ocean-heat transport, are not necessarily independent of one another. Similarly, the tropical forcing of the Pliocene climate is not independent of atmospheric chemistry and circulation or ocean processes.

The tectonic uplift theory (Ruddiman and Raymo 1988) is focused more on the climatic transition and inception and the increase in large northern-hemisphere ice sheets beginning at 3.2–2.8 Ma as a direct explanation for mid-Pliocene warmth. As already discussed here, midlatitude orogenesis, especially rapid uplift of the Himalayan and Colorado Plateaus the past few Ma, altered planetary atmospheric circulation, and during the Pliocene cold air masses circulated in high latitudes over North American and Europe. Beginning about 3.15 Ma these air masses migrated southward in two steps, at 2.8 and 0.9 Ma, because of changing orography, causing large ice sheets to grow under certain solar insolation conditions caused by changes in orbital geometry.

Enhanced North Atlantic Ocean circulation and a strengthened Gulf Stream might explain elevated temperatures in the polar and subpolar North Atlantic and Arctic Oceans (Dowsett et al. 1992) as well as more vigorous meridional overturning circulation (Raymo et al. 1996). Greater poleward heat transport during the MPTO would produce many observed high-latitude paleoceanographic patterns. Haug et al. (2001a) and Haug and Tiedemann (1998) proposed that the emergence of the CAI shoaled the straits connecting the Pacific and Caribbean to less than 100-m water depth and altered surface salinity on both sides of the Isthmus. CAI shoaling at 4.7 and 4.2 Ma would initially lead to mid-Pliocene warmth, especially in high northern-hemisphere regions, because of increased heat and perhaps salt transport. Increased poleward moisture transport is expected from the CAI closing (Driscoll and Haug 1998). Haug postulates that these tectonic events and the atmospheric and oceanographic response preconditioned the climate system prior to the eventual inception of large northern-hemisphere glacials. Steph et al. (2006) documented intra-Caribbean and Pacific-Caribbean salinity variability at orbital (precession) and longer timescales during the closure. The first-order signal supports the CAI-ocean-climate mechanism to explain long-term Pliocene trends. In addition, a strong precession signal was interpreted as evidence for low-latitude forcing of Pacific-Caribbean hydrography with ENSO dynamics as a possible trigger for orbital-scale changes.

Greenhouse gas forcing is also considered a factor in Pliocene warmth. Direct CO_2 reconstructions from European fossil leaf stomata records (van der Burgh et al. 1993; Kürschner et al. 1996) range from about 340–370 ppmv, similar to estimates from deep-sea carbon isotopes (Raymo et al. 1996). Pliocene CO_2 thus may have been about 10–30% higher than typical late Pleistocene interglacial concentrations (280–300 ppmv) and slightly higher than the 250–350 ppmv estimates for the late Miocene epoch (Pagani et al. 1999), but the stomatal estimates need confirmation from other methods. Methane concentrations for the MPTO are not available.

Haywood et al. (2005) combined tropical SST estimates with a coupled ocean-atmosphere climate model and found that the pattern of Pliocene climate could be explained by slightly elevated CO_2 concentrations rather than enhanced ocean circulation. They propose that a ~2 Wm^{-2} mean global radiative increase from a modest CO_2 increase would be enhanced by land and sea-ice albedo effects amplifying the warming effect. The scarcity of more precise estimates for mean Pliocene paleo-CO_2 concentrations and high-frequency CO_2 variability means that uncertainty about greenhouse gas forcing during the Pliocene remains a major gap in understanding global warmth at 4.5–3.0 Ma.

A new hypothesis gaining consideration is that the earth's climate during all or part of the MPTO represented a semipermanent El Niño climate state. ENSO is the primary mode of interannual climate variability caused by ocean-atmosphere interactions emanating in the tropical Pacific. Modern climate experiences a "warm" El Niño event every few years when Pacific tradewinds diminish, eastern equatorial ocean temperature increases, and the depth of the thermocline shoals in the western Pacific. Large El Niño events have global climatic impacts on rainfall, temperature, and atmospheric circulation; these impacts are called teleconnections. Molnar and Cane (2002) proposed that global

Pliocene marine and terrestrial climatic patterns resemble those during modern El Niño events, implying tropical forcing of global climate during at least part of the MPTO. Importantly, paleoclimate records cited by Molnar and Cane span a long time slice from 5–2.7 Ma and include many records with minimal age control. Nonetheless, the contrast between pre- and post-2.7-Ma Pliocene climate is stark and strong evidence exists for a qualitatively different mid-Pliocene mean climate state. Molnar and Cane (2007) updated their hypothesis, and note that not all regional Pliocene paleoclimate records support a canonical El Niño event but that there is good agreement between the Pliocene and 1997–1998 El Niño in terms of global temperature and, to a lesser degree, precipitation patterns.

On the basis of modern ENSO dynamics, tropical SST patterns, and atmospheric climate model simulations (Barreiro et al. 2006), Philander and Federov (2003) and Federov et al. (2006) also proposed that a semipermanent El Niño state is the most reasonable explanation of Pliocene warmth. Their main hypothesis is that the absence of cool SSTs in the eastern tropical Pacific can induce warming in extratropical regions through changes in tropical evaporation-precipitation and atmospheric feedbacks.

One key test of the Pliocene permanent El Niño hypothesis lies in a better understanding of tropical paleoceanography. Tropical and subtropical SST reconstructions based on faunal assemblages, alkenones, planktonic foraminiferal Mg/Ca, and oxygen isotope ratios are available from the western and eastern equatorial Pacific Ocean (WEP, EEP) (Chaisson and Ravelo 2000; Ravelo et al. 2004, 2006; Rickaby and Halloran 2005; Wara et al. 2005; Lawrence et al. 2006), the Peru and California margins (Dekens et al. 2007),

and the Atlantic Ocean (Marlow et al. 2000). These studies show that the EEP was up to 3–5°C warmer and the WEP was about the same or slightly cooler than their modern state. This means that the west-to-east SST gradient was only ~1.5°C during the Pliocene epoch, until a major climate shift occurred near 1.7 Ma, when the gradient increased to ~5°C. On the basis of planktic foraminiferal Mg/Ca ratios, Rickaby and Halloran (2005) proposed that the Pliocene thermocline was shallow in the EEP. However, Ravelo et al. (2006) reinterpreted the depth habitat of the foraminiferal species *Globorotalia tumida* and concluded that the bulk of the evidence indicates a deepened EEP thermocline.

Lawrence et al. (2006) found support for the El Niño hypothesis and a link between tropical and southern-hemisphere high-latitude regions (Figure 4.13). Orbital insolation cycles in eastern tropical Pacific SST, productivity, and thermocline depth at the 41-ka obliquity frequency seem to be tightly coupled during the mid-Pliocene epoch. They postulate a southern polar-oceanic link to the tropical thermocline and a link between nutrient cycling in the Southern Ocean and tropical SST's.

Ravelo et al. (2004) studied Pliocene-Pleistocene climatic transitions since 4 Ma and concluded that tropical and high-latitude climates exhibit asynchronous patterns during the progressive cooling, with major steps in the Pliocene at 4.5 and 4.2 Ma and the establishment of strong Walker Circulation and an east-west Pacific SST gradient, important for ENSO variability, at about 2.0–1.5 Ma. These transitions contrast with the age for the onset of northern-hemisphere ice sheets and high-amplitude glacial cycles at 2.75 Ma. The timing of these events suggests that high-latitude glacial ages began during a protracted El Niño–like state in the Pacific

Eastern equatorial Pacific sea-surface temperatures

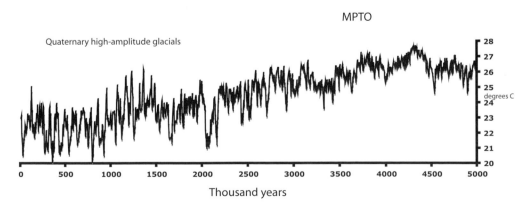

FIGURE 4.13 Evolution of eastern equatorial Pacific sea-surface temperature (SST) during the last 5 Ma, from the alkenone paleotemperature record of Ocean Drilling Program Site 846B, from Lawrence et al. (2006). Note the maximum SSTs several times during the early and middle Pliocene (Mid-Pliocene Thermal Optimum, or MPTO).

and a decoupling of high- and low-latitude climate, at least on these timescales.

Climate modeling provides evidence against a semipermanent El Niño–like state during the MPTO. Haywood et al. (2007b) used a coupled atmosphere-ocean general-circulation model and PRISM boundary conditions and concluded that El Niño conditions would lead to a mean annual temperature only 0.6°C higher than modern conditions. This value is lower than reconstructed temperatures from proxy data, implying slightly elevated atmospheric CO_2 concentrations, oceanic circulation, or some other factor caused Pliocene warmth. Additional modeling studies incorporating better reconstructions of CO_2 and deep-sea circulation and temperature are currently under way.

In sum, the potential causes of the MPTO include modestly higher atmospheric CO_2 concentrations, changes in high-latitude Southern Ocean circulation and nutrient cycling, different Indonesian and Central American paleogeography, or a combination of these. The primary message to be drawn from the Pliocene epoch, however, is that mean climate state is extremely sensitive to small changes in boundary conditions and has the potential for sustained global warmth.

Perspective

Cenozoic paleoclimatology has shifted its emphasis from the study of progressive global cooling to the understanding of patterns and mechanisms during climatic transitions and periods of global warmth. Cenozoic cooling involved a variety of possible causes: seafloor spreading, continental uplift, erosion, weathering, changes in ocean circulation and productivity, atmospheric circulation and biogeochemical cycling, and decreasing atmospheric CO_2 concentrations. There were very likely important thresholds crossed during Cenozoic climatic transitions and also during orbital modulation of climate, in ways we are only beginning to appreciate.

Periods of global warmth in the Paleocene, Eocene, Miocene, and Pliocene epochs pose a distinct set of problems. The PETM and Eocene hyperthermal periods show the potential for extreme warmth when CO_2 concentrations are elevated to 2–5 times preindustrial levels. Unabated, future CO_2 concentrations may approach what are now our best estimates for those during intervals in the Eocene.

In contrast, Miocene and Pliocene warmth tell us that, under the right conditions, more moderate but still warmer-than-modern global climate and reduced polar ice occurs even when greenhouse gas concentrations are near or only slightly above preindustrial levels. Is this because other factors caused the warmth or because the earth's climate sensitivity to greenhouse gases was greater in the past? To answer this question, paleoceanographic and tectonic reconstruction and improved skill in climate model simulation of extreme climate states is required.

LANDMARK PAPER Greenhouse and Icehouse Climate States: The Cenozoic Era

Fischer, A. F. 1981. Climatic oscillations in the biosphere. In M. H. Nitecki, ed., *Biotic Crises in Ecological and Evolutionary Time*, pp. 103–131. New York: Academic Press.

One irony about the global warming issue is that during the last 45 million years the earth progressively entered a relatively cold global climate condition called an icehouse state, characterized by significant amounts of polar ice and strong equator-to-pole temperature gradients. Accompanying cooling was enhanced by short-term variability in global temperature and continental ice volume. The Holocene interglacial we live in, though warm compared to recent glacial (ice-age) episodes, is nonetheless still chilly compared to the climate of the early Eocene epoch. What caused Cenozoic cooling and what were the impacts?

Alfred G. Fischer of Princeton University and the University of Southern California was, by training, a vertebrate paleontologist, but he published extensively on the topics of long-term global climate cycles, biotic crises, and extinction patterns. He was at the forefront of the resurgence in catastrophism in the earth sciences in the 1980s and was an expert in the sedimentological record of climatic cycles.

Among his most lasting legacies is his characterization of the Cenozoic transition from "greenhouse" to "icehouse state" in the *Biotic Crises* volume edited by Matthew Nitecki. Fischer's 1981 paper emphasized that the last 700 Ma experienced two icehouse-to-greenhouse mega-cycles and, with regard to the late Cenozoic era, the beginning of a third. He also drew on geochemical and climatological literature to support a hypothesis that mantle convection processes drove these mega-climatic cycles when atmospheric carbon dioxide concentrations fluctuated because of changes in volcanic emissions counterbalanced by carbon sequestration in the lithosphere through weathering and sedimentation.

The 1981 paper secured the greenhouse-to-icehouse climate concept as a fundamental characterization of global climate for the earth's entire history and epitomizes an early attempt to integrate disparate lines of evidence to reconstruct patterns, causes, and biological impacts of global climate changes. The early Cenozoic greenhouse climate is today one of the most intensely studied periods in paleoclimatology, for the obvious reason that we need to better understand its climate dynamics, the role of atmospheric carbon dioxide, and other forcings.

5

Orbital Climate Change

Introduction

The orbital theory of climate explains glacial-interglacial cycles and other environmental changes caused by changes in geographical and seasonal insolation from variations in the earth's orbital geometry over tens of thousands to hundreds of thousands of years. In its basic form, orbital theory holds that three aspects of the earth's orbit vary because of gravitational forces in the solar system and influence climate in a wide variety of ways. The first two—the tilt of the earth's axis, or obliquity, and the precession, broken down into axial precession (wobble) and precession of the equinoxes (the shifting of the earth's orbital revolution around the sun)—lead to changes in seasonal and geographical distribution of insolation reaching the earth's upper atmosphere. The third is eccentricity, the shape of the orbit around the sun, which changes from an ellipse to a circle, altering total insolation. Over the past century, orbital forcing of climate has been the most intensely studied extrinsic climate-forcing mechanism, invoked to explain many paleoclimate patterns.

Orbital theory is also called the astronomical theory, the theory of the ice ages, and the Milankovitch theory of climate, named after the early 20th-century Serbian Milutun Milankovitch, who popularized one widely accepted version to explain late Pleistocene ice ages (Milankovitch 1941). We use the more inclusive term *orbital* here because many paleoclimate patterns linked to changes in the orbit, such as pre-Quaternary cyclic environmental change during periods of minimal continental ice, are distinct from those addressed by Milankovitch in his original theory. Moreover, Milankovitch's emphasis on northern-hemisphere summer insolation as the key to the growth and decay of large ice sheets has been modified in many important respects such that there is no universal definition of the Milankovitch theory (Roe 2006).

Orbitally driven climate change goes by many descriptive terms—the Pacemaker, Heartbeat, Rhythm, Metronome, and Fundamental Driver of the ice ages—reflecting a consensus about the importance of orbital geometry for climate. Unlike radiative output from the sun, for which long-term variability remains incompletely known, orbitally driven insolation can be calculated—"retrodicted"—back in time using analytical (trigonometric, Berger and Loutre 1991) and numerical (Laskar et al. 1993, 2004) solutions based on solar system gravitational attractions among the planets and the earth-moon system. Orbital theory thus offers a distinctly deterministic view of climate variability, in contrast to aperiodic or stochastic modes described in other chapters. Still, it is useful to remember that the complex interplay among the planets and the moon renders changes in orbital variations quasi-periodic in nature, not perfectly cyclic, and subject to larger uncertainty the farther one goes back in time (Berger et al. 2005).

Consensus nonetheless exists that orbital forcing catalyzes the growth and decay of high- and midlatitude ice sheets, reorganizes terrestrial and oceanic ecosystems, and modifies global atmospheric, oceanic circulation and biogeochemical cycling, to mention a few perturbations to the climate system. These produce spectacular cyclic environmental records in many sedimentary sequences (Figure 5.1), a primary source of empirical observation needed to test orbital theory. Quasi-cyclic environmental changes combined with astronomical calculations raise complex questions for astronomy, geology, oceanography, glaciology, climate modeling, and atmospheric sciences. What processes link insolation changes at the top of the atmosphere to climate change at the surface? How are small changes in energy input and distribution amplified into massive perturbations of the global hydrological cycle, causing enhanced snow accumulation, reduced ablation during winter seasons, and the growth of large ice sheets extending to middle latitudes? What is the role of atmospheric greenhouse gas forcing in amplifying insolation? How do the ice sheets themselves affect the planetary radiation budget and atmo-

spheric circulation? What internal climatic mechanisms (ice-sheet and sea-ice albedo effects, ice-sheet bedrock interactions, precipitation and temperature feedbacks) amplify or dampen externally forced insolation changes? What caused changes in the frequency and amplitude of orbital climatic cycles over the past several million years?

To address such complexity, Berger and Loutre (1991) broke orbital theory down into four components: (1) the theoretical computation of long-term orbital and geometric changes of insolation, (2) the design of climate models to transform insolation changes into climate, (3) the collection of geological data to test models, and (4) modeling orbitally induced climate and paleoclimate-climate model data comparison studies. This chapter approaches orbital climate change as follows. We devote the first section to the astronomical parameters themselves and present insolation curves for the last million years. We then discuss the historical development of orbital climate during the 19th and early 20th centuries as well as "modern" orbital theory that emerged since the 1960s with the advent of deep-sea paleoceanography and better dating methods. Next we will outline important paleoclimate records of orbital-scale climate variability covering the last 2.5–5 Ma. The next section focuses on several key mysteries and paradoxes that surround the orbital theory of climate: the onset of high-amplitude glacial-interglacial cycles near 2.8 Ma, the Mid-Pleistocene Transition (MPT) from 1.2–0.7 Ma and the "100-ka" enigma, the Mid-Brunhes Event (MBE) and the Marine Isotope Stage 11 (MIS 11) paradox, the phasing between temperature and atmospheric greenhouse gas concentration during glacial terminations, and processes controlling atmospheric CO_2 during glacial periods. The last section discusses hypotheses and mechanisms to explain orbital climate patterns.

Astronomical Processes and Calculations

Understanding orbital variability over long timescales is a central theme in orbital theory. Figure 5.2 shows the geometry of the earth-sun relationship and the parameters used to compute insolation. The earth's orbital plane around the sun, called the ecliptic, serves as a frame of reference. The three components of orbital theory—precession ($\bar{\omega}$), tilt (ε), and eccentricity (e)–can be described from this frame of reference. The earth's axial tilt of 23.44° measures the angle between the ecliptic and the earth's equatorial plane and leads to the earth's seasons. Precession is often expressed as $e \sin \bar{\omega}$. The following discussion of the earth's orbital geometry as it relates to climate change is derived from Berger and Loutre (1991), Paillard (2001), and Berger et al. (2005).

A

FIGURE 5.1 Photos of continental sediments formed during orbital climate cycles: (A) The Orera section in Spain reveals the dominantly precession-controlled Miocene lacustrine limestone, floodplain/mudflat cycles. (B) The Ptolemais section in northern Greece shows Pliocene precession-related lignite-lacustrine marl cycles. Cyclic marine sediments are also common around the Mediterranean. Courtesy of F. Hilgen.

B

Precession

Precession cycles reflect secular changes in the perihelion—i.e., the time when the earth moves closest to the sun. The precession of the equinoxes was recognized by the Greek astronomer Hipparchus from minute changes in the position of the North Star. In the 16th century, Johannes Kepler explained the precession of the equinoxes in terms of the elliptical orbit of the earth. Precession is separated into two components—axial and elliptical precession. Axial precession results from torque of the sun and moon on the earth's

equatorial bulge. The torque causes the earth's axis of rotation to wobble, which in turn causes the North Pole to prescribe a circle over a 26-ka period. This is expressed as the orbital position γ in Figure 5.2 but the angle $\bar{\omega}$ is considered to be most important in terms of its climatic impacts.

Elliptical precession results from planetary effects on the earth's mass and causes its orbit to rotate independently about one focus. Today, summer in the northern hemisphere occurs when the earth is farthest away from the sun; the perihelion occurs during the northern-hemisphere winter, when 10% more winter radiation is received than will be 11 ka

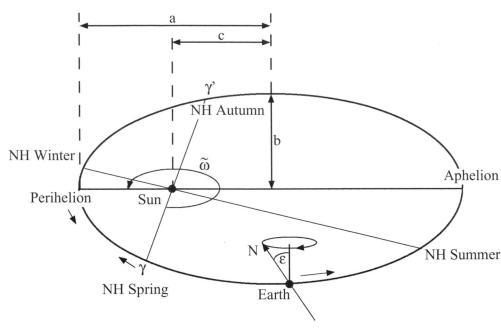

from now. Conversely, 12 ka ago, when northern-hemisphere ice sheets were melting, summer radiation was greater in the northern hemisphere than it is today. The net effect is that the precession of equinoxes (March 20 and September 22) and solstices (June 21 and December 21) shift slowly around the orbit with a period of 22 ka.

The climatic impact of precession changes is greatest in low latitudes and equals about ±10% insolation change for a season. At summer solstice, the change is 8% (40 W m⁻²) around the mean. This value is quite large compared with the climatic forcing of several W m⁻² (climate sensitivity) that would occur with a doubling of atmospheric CO_2 concentrations. However, atmospheric greenhouse gas forcing occurs year-round, whereas precession changes are seasonal, so the earth's climatic sensitivity to greenhouse gases is in theory greater than it is to the effects of precession (see Chapter 11 in this volume).

The climatic effect of precession changes can be imagined if the eccentricity and obliquity were set at zero. The seasonal cycle would disappear and the latitudinal gradient would be accentuated. In paleoclimatology, a precession signal is often identified by its impact on low-latitude terrestrial and oceanic processes and on the southern hemisphere. Ruddiman (2006b) identifies two important precession impacts: (1) a relatively small (compared to tilt) forcing of ice sheets, with a late response in the precession cycle, and (2) the widespread early-phase response of atmospheric greenhouse gases. Precession forcing influences tropical and boreal wetland activity, reflected in fluctuations of atmospheric methane (CH_4) concentrations in polar ice cores (Raynaud et al. 1988, 1993; Chappellaz et al. 1990, 1993), equatorial upwelling, and tropi-

cal monsoons, evident in paleoceanographic records (e.g., Prell and Kutzbach 1987), among other processes. Nonlinear climatic response to precession forcing may also explain suborbital-scale climatic variability observed at 10.3-ka and higher frequencies (Hagelberg et al. 1994; McIntyre and Molfino 1996).

Obliquity

Today the axis of rotation is tilted about 23.44° relative to the ecliptic plane and is gradually declining at a rate of about 0.5″ yr⁻¹. The inclination of the axis of Venus is, by contrast, tilted only 3°. When the earth is tilted toward the sun, there is summer in one hemisphere and winter in the other. Over long timescales, tilt varies between 22.05° and 24.5°, owing to planetary effects on the position of the ecliptic in space. One complete cycle of obliquity takes about 41,000 years.

The climatic effect of greater tilt is amplified seasonal cycles at high latitudes, because polar regions receive relatively more radiation than the tropics. For example, at an eccentricity of 4%, a change in tilt of 1° causes a 2.5% increase in summer insolation at 65°N, a 1.2% increase at 45°N, and an overall average increase of only 0.8% for the entire northern hemisphere. The net change in solar radiation reaching the top of the atmosphere in high latitudes during a 41-ka cycle is about 17 W m⁻².

Summer insolation at 65°N latitude, the approximate location of glacial-age northern-hemisphere ice sheets, is often considered to be a standard measure to evaluate the influence of obliquity on ice sheets and climate. Traditionally, obliquity forcing is thought to produce a linear and lagged response in high-latitude climate parameters, such as ice-sheet growth and decay and deep-water formation (Imbrie et al. 1992, 1993), but as we discuss below, more complex relationships are also known.

Eccentricity

Orbital shape called eccentricity, e, is measured by two elliptical parameters—the semimajor axis (a) and the focus of the ellipse (c) such that $e=c/a$. Eccentricity has changed from a circle where $e=0$ and an ellipse where $e=0.05$. Today the earth has a semi-elliptical orbit, with $e=0.0167$. In general, the more elliptical the orbit, the greater the climatic influence on seasons. The main periodicities for the changes in orbit are about 405 ka and 100 ka.

The influence of changes in eccentricity on insolation is quantitatively very small. A reduction in elliptical orbit from its present value of 0.017 to about 0.04 translates into a net change of only about 0.1% of total solar radiation reaching the upper atmosphere (~0.5 W m^{-2}). A change of this magnitude might alter mean annual global temperature by only a few tenths of a degree Celsius. This is an order of magnitude less than radiative forcing from glacial-interglacial changes in atmospheric CO_2 concentrations, which contribute several degrees to mean global warming during deglaciation. Thus, the small net insolation due to a 100-ka eccentricity cycle cannot directly cause the magnitude of glacial-interglacial climatic cycles occurring every ~100 ka for the last 500,000 years. This is called the "100-ka" enigma. Some accept that eccentricity modulates and amplifies the signal from precession and tilt through feedbacks internal to the climate system (Imbrie et al. 1993; see the section "Insolation Calculations").

Orbital Solutions

The computation of orbital configurations and insolation strength is a complex process, with elements of subjectivity and uncertainty often overlooked in paleoclimatology. Widely cited orbital records are those of Berger and Loutre (1991) (see also Berger 1978), who calculated what are called analytical (trigonometric) solutions for insolation from planetary masses and earth-moon gravitational pull. Berger's solutions recognize the 19- and 23-ka precession, 41-ka tilt, and 100- and 400-ka eccentricity cycles observed in some paleoclimate records. The analytical solutions are convenient for Quaternary paleoclimatologists because they allow comparison of the amplitude, frequency, and phasing of proxy and astronomical records. These solutions are most accurate back to ~1.5 Ma and are considered reliable back to 5 Ma in the time domain and 10 Ma in the frequency domain.

Laskar et al. (1993) and Laskar (1999) computed insolation values using a different method. They point out that an analytical method involves the assumption of a regular long-term solar system and quasi-periodic cycles of insolation, but in the computations does not take into account certain planetary elements. Their approach, called a "numerical" solution, involves integration of secular equations, some with as many as 150,000 polynomial terms, in a long-term analysis with a large integration time step of 200–500 years. Laskar's "La93" solution from the 1993 paper has been used in many paleoclimate comparisons.

Laksar also came to the conclusion that planetary motions exhibit long-term chaotic behavior, exponentially increasing by a factor of 10 every 10 million years. The implication might be that orbitally driven insolation forcing might be limited in paleoclimatology to about the past few million years. Laskar et al. (2004) recently extended numerical solutions (La04) for the last 65 Ma, using improved computer and integration methods of gravitational equations and advances in the earth-moon tidal system. The La04 solution has already been used to tune the Neogene timescale (Lourens et al. 2004) and holds potential for the entire 65-Ma-long Cenozoic era. Laskar et al. (2004) also provide solutions for the largest eccentricity cycle of 405 ka spanning the Mesozoic 65–250 Ma, though there are cautions due to chaotic behavior.

Though subject to uncertainties, Berger et al. (2005) point out that first-order insolation computations of the primary orbital elements produce similar results from both Lasker and Berger's own methods. For additional detail and quantitative computations, refer to discussions in Berger and Loutre (1991), Laskar et al. (1993, 2004), and Saltzman (2002).

Insolation Calculations

Figure 5.3 presents insolation curves for eccentricity, obliquity, and precession over the past 1 million years based on calculations by Berger and Loutre (1991). Insolation, expressed as the energy reaching the earth from solar irradiance, is calculated as follows. Solar radiation at the top of the atmosphere is calculated using the inverse square law:

$$I = E \ (4\pi \times R^2)/(4\pi \times r^2)$$

where I=insolation at the outer sphere of the atmosphere, E=radiation at the surface of the sun, $4\pi \times R^2$ is the surface area of the object, and $4\pi \times r^2$ is the surface area of outer sphere. The solar constant is calculated as

$$So = E \ (sun) \times (R \ (sun)/r)^2$$

where So is the solar constant, E is solar irradiance at the surface, R is the radius of the sun, and r is the distance between the earth and the sun. Insolation is usually expressed as watts (W) per square meter. Solar output actually varies

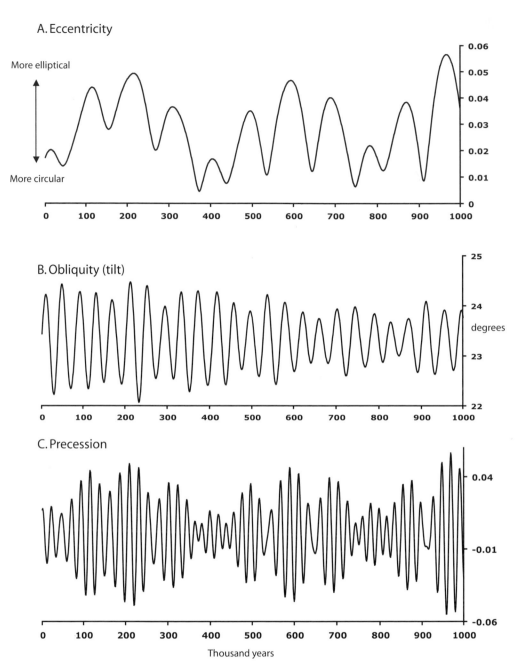

A. Eccentricity

B. Obliquity (tilt)

C. Precession

Thousand years

FIGURE 5.3 Orbital parameters during the last 1 Ma: (A) eccentricity, (B) obliquity, and (C) precession at 65°N latitude, from Berger and Loutre (1991).

over annual, decadal (11-year sunspot cycles), and longer timescales and is extremely relevant to climate changes over the last 100 ka, and especially the last millennium (see Chapter 11 in this volume). The solar constant is assumed to be invariant in orbital calculations.

What is immediately apparent is the small amplitude of eccentricity-induced energy variability ranging between ~342 and 343 W m^{-2}. For comparison, total solar irradiance during a modern 11-year sunspot cycle varies between ~365.5 and 366.5 W m^{-2}. In contrast to eccentricity, obliquity causes

large changes in the energy reaching polar regions, from 165 to 180 W m^{-2}, and much less variability at the equator (~415–418 W m^{-2}). Obliquity-forced insolation changes are symmetric but antiphased in the northern and southern hemispheres. Precession during summer solstice at the poles induces cyclic insolation changes of 500–600 W m^{-2} and smaller-scale changes at the equator (360–440 W m^{-2}). Precession forcing operates in a hemispherically asymmetric pattern.

The insolation patterns shown in Figure 5.3 represent only a few of many solutions of orbitally induced energy changes at the top of the earth's atmosphere at any particular latitude or day of the year. The choice of which orbital parameter—precession, obliquity, eccentricity, or all of them—is used in the modeling and interpretation of a particular geological sequence depends on the location and nature of the record. It has long been known that the choice of latitude can influence whether one finds concordance or mismatching between astronomical cycles and paleoclimatic records. In one early study of the Barbados coral-reef sea-level record, Broecker et al. (1968) showed the importance of insolation curves. By plotting the insolation curve for 45°N rather than 65°N, they found a more realistic fit of insolation to the observed record of high interglacial sea levels. At a 4% eccentricity, the insolation during the perihelion is 4.8% higher than when the perihelion is reached on the March 21 or September 21 equinox. Because a 1° increase in the earth's tilt increases the insolation more at high latitudes than at the equator, the choice of an insolation curve for 45°N instead of 65°N led to better agreement between the Barbados sea-level record of the last 150 ka and astronomical cycles.

Northern-hemispheric summer (June 21) insolation is often used as an index to evaluate late Pleistocene glacial-interglacial cycles. Precession influences low-latitude climate and hydrological processes such as monsoon systems and is usually emphasized in these studies. Insolation values for eccentricity, tilt, and precession are sometimes integrated into a single orbital "ETP" curve as a convenient measure of summed insolation signal to compare with paleoclimate records (Imbrie et al. 1992, 1993).

In addition to the use of seasonal orbital insolation values for a single day, an alternative proposed by Huybers (2006) is to integrate insolation over several summer months, rather than for a single day, because this measure represents a more realistic measure of insolation as it pertains to early Pleistocene climate changes dominated by the 41-ka period. He showed that by taking both insolation intensity and seasonal duration, a more realistic modeling of ice-sheet behavior and early Pleistocene glacial cycles could be achieved. It bears repeating that because of chaotic behavior in the solar system, insolation time series—especially for precession and tilt—are subject to increasing uncertainty the farther one goes back in time (Laskar et al. 2004). These are a few reasons why a comprehensive theory to account for the direct and indirect responses of various parts of the climate system to insolation changes is a daunting challenge.

Astronomical Tuning and Long-Term Orbital Solutions

Testing orbital theory with geological records requires continuous paleoclimate reconstructions and absolute ages from radiometric dating as tie points. Continuous paleoclimate records are widely available from deep-sea sediment cores, ice cores, and continental deposits and uranium-series ages of coral reefs and speleothems provide absolute ages on the last few glacial-interglacial cycles. However, direct absolute dating of the most continuous sequences from deep-sea oxygen isotopes and Antarctic ice cores is not available before about 40 ka (the limits of radiocarbon dating of sediments and layer counting in Greenland ice). In part because of this limitation, the procedure called astronomical "tuning" introduced in Chapter 4 was developed, whereby paleoclimate records are dated by tuning them to astronomical parameters computed from celestial mechanics and anchored by radiometric or dated paleomagnetic tie points.

One application of tuning has achieved near-legendary status in paleoclimatology. The paleomagnetic reversal at the Brunhes/Matuyama (B/M) boundary was long thought to be 730 ka in age, but on the basis of tuning deep-sea oxygen isotope and paleomagnetic records to Berger's astronomical solutions for insolation, Nicholas Shackleton and colleagues proposed that the B/M boundary was actually older (Shackleton et al. 1990). Soon thereafter, new radiometric $^{40}Ar/^{39}Ar$ dates confirmed the B/M boundary age at ~780–790 ka (Baksi et al. 1992; Spell and McDougall 1992), affirmed Shackleton's prediction, and provided strong support for both orbital forcing of global ice volume and the tuning method as a stratigraphic tool supported by later isotopic records.

Cyclicity is prominent throughout the pre-Quaternary geological column and interpreted by many as a manifestation of orbital forcing. The study of "cyclostratigraphy" goes back to the 19th century (Gilbert 1895) and remains an active research field today (Fischer and Bottjer 1991; Hinnov and Ogg 2007; Meyers and Sageman 2007). In fact, orbital tuning has become an integral part in the development of the Global Chronostratigraphic Scale used to designate and date unit stratotypes for global stages. A protocol using astronomically defined chronozones as formal chronostratigraphic units has been proposed for the Neogene period (Miocene, Pliocene, Pleistocene) covering the last 23 Ma (Lourens et al. 2004) and holds promise for older sedimentary sequences (Hilgen et al. 2006).

In a similar vein, Paleozoic (De Boer and Smith 1994), Mesozoic (Olsen 1984; Olsen and Kent 1996), and Cenozoic (Shackleton et al. 1999; Pälike et al. 2006b) geological records have been used to test the new astronomical calculations of Laskar et al. (2004). These efforts represent a kind of "reverse tuning" procedure where radiometrically dated sediment and isotopic cycles are used to cross-check insolation calculations. The cross-fertilization of cyclostratigraphy and celestial mechanics in the field of chronostratigraphy improves both the accuracy of stratigraphic correlation of sedimentary rocks and our understanding of the astronomical-climate link throughout the earth's history.

Historical Development of Orbital Theory

Orbital theory is one of the oldest facets of paleoclimatology, going back at least 160 years and, given that many concepts proposed by its pioneers are still widely accepted today, it is useful to describe the theory's history in some detail. Additional historical aspects of orbital theory can be found in Imbrie and Imbrie (1979), Hinnov (2000), and Paillard (2001).

Nineteenth and Early 20th Century

The parallel development of astronomy and geology includes important breakthroughs now widely applied in paleoclimatology. During the 19th century, discoveries in the geological sciences and celestial mechanics set the stage for revolutionary theories of an astronomical-climate link. In 1837, Swiss geologist Louis Agassiz developed his landmark theory of the ice ages described in his book *Etudes sur les Glaciers* (Agassiz 1840). Agassiz discovered first in Switzerland (and later in Scotland, North America, and elsewhere) widespread evidence that continental ice sheets covered large midlatitude regions of the northern hemisphere. This discovery immediately led to the question of what caused the ice ages.

Astronomer Jean le Rond d'Alembert calculated in the 18th century that the gravitational pull of the sun and the moon on the earth's equatorial bulge caused the 26-ka cycle of the precession of the equinoxes. French mathematician J. Adhémar proposed in 1842 that changes in the earth's orbital parameters may have caused Agassiz' ice age. Adhémar theorized that the two components causing the precession of the equinoxes—the 26-ka wobble and the 19-ka rotation of the elliptical orbit itself—shifted the positions of the equinoxes along the earth's orbit with a 22-ka period. These changes might in theory cause an "ice age" in whichever hemisphere experienced winter. Ice ages under Adhémar's theory would occur every half cycle— that is, every 11 ka, and would occur in the northern hemisphere while ice decay was occurring in the southern hemisphere and vice versa. Under this theory, the number of hours of daylight and darkness in each hemisphere was considered a key factor in explaining ice ages.

Scotland's James Croll (1864, 1875) studied the orbital influence on climate after Agassiz's theory had been accepted by most geologists. French astronomer Urbain Le Verrier determined in 1843 that the earth's eccentricity changed about every 100 ka and Croll postulated that high eccentricity led to ice ages because decreased winter sunlight created a positive

feedback loop. As more radiation was reflected, the climate became cooler. Croll also reasoned that the precession of the equinoxes was a factor. When the earth's northern-hemisphere winter is close to the sun, as it is currently, we experience an interglacial period. Conversely, 11,000 years ago, when the northern hemisphere was farther from the sun in boreal winter season, the potential existed for a glacial period because the eccentricity of the orbit was also elongate. Thus, Croll postulated the simultaneous interplay of two orbital parameters influencing climate: the elongate orbit and a winter solstice far from the sun. Croll's work pre-dated modern radiometric dating methods, but he nonetheless graphically displayed a hypothetical paleoclimate history showing the last glacial period lasting from 250 to 80 ka.

In the early 20th century, Pilgrim (1904) and Milankovitch (1941) studied the orbital climate link. Milankovitch's lifelong work focused on the major theoretical underpinnings explaining the influence of orbital changes on solar radiation in various seasons and latitudes. One of his contributions was a mathematical description of insolation changes and their influence on climate and ice sheets. His efforts to quantify the contributions of changes in the earth's eccentricity, tilt, and precession represent the advent of modern orbital theory. Milankovitch argued that orbital variations affecting insolation in high latitudes during northern-hemisphere summer were critical for the formation of continental ice sheets. He proposed a positive feedback mechanism whereby as summer insolation fell, snow from the previous winter would be preserved, in turn increasing the earth's albedo, reflecting more radiation back to space, causing further cooling (Pisias and Imbrie 1986). The interplay between solar radiation and the ice sheet itself became emblematic of Milankovitch's version of orbital theory, and Milankovitch himself remains an icon in paleoclimatology.

Deep-Sea Isotope Records and Paleoceanography

Testing orbital theory using geological records required three major advances: new proxies of climate change, more continuous records than those provided by continental glacial geology, and absolute dating methods to crosscheck the timing of reconstructed climate changes with astronomical solutions. The first two needs were filled by paleoceanography, a field that grew rapidly with the development of deep-sea sediment coring and foraminiferal assemblages as well as stable isotopic proxies of sea-surface temperature (SST) and global ice volume.

Modern paleoceanography began with early studies by Schott (1935), Arrhenius (1952), Ericson and Wollin (1968), and others (e.g., Phleger 1976). Cesare Emiliani's classic deep-sea oxygen isotope ($\delta^{18}O$) measurements on foraminifers (1955), which he believed to signify SST oscillations, explicitly supported orbital theory. Applying the temperature-isotope relationships of Epstein et al. (1953), Emiliani postulated that light isotope ratios signaled warm and heavy ratios and cooler SSTs and that isotope values oscillated at orbital frequencies. In another seminal study, Shackleton (1967) suggested instead that oxygen isotope ratios recorded mainly changes in continental ice volume. Shackleton reasoned that during ice ages, continental ice sheets preferentially uptake the light isotope ^{16}O, leaving seawater and shells secreted by foraminifers enriched in the heavier isotope ^{18}O. Glacial foraminiferal oxygen isotope ratios thus are heavier (more positive); interglacial foraminiferal δ^{18}O values are lighter (more negative).

The typical marine isotope curve has several notable features (Figure 5.4). First, during the last few hundred thousand years, there is an apparent 100-ka cycle. Second, the ages of interglacial periods, defined by the more negative oxygen isotopic ratios, generally match those for periods of high insolation. Interglacial ages also coincided with the uranium-series ages of emerged coral-reef tracts found in tropical regions. Third, the isotope curve exhibits a sawtooth pattern characterized by rapid periods of ice decay (glacial terminations) but more progressive, stepwise decline from interglacial to glacial climates (Broecker and Van Donk 1970).

Oxygen isotope stratigraphy became the standard tool for correlating deep-sea core records and deciphering the oceanic record of orbital cycles during the Quaternary system (Berger 1979; Mix 1992; Shackleton 2000). Following Emiliani and Shackleton, Marine Isotope Stages (MISs) have been assigned numbers, even numbers for glacial periods and odd numbers for interglacials, counting from MIS 1 (the Holocene) and MIS 2 (the Last Glacial Maximum, or LGM) back to late Pliocene MIS 104 about 2.62–2.52 Ma. Isotope stages older than MIS 104 are given a preface designating the magnetic chron in which they are located. Thus, glacial

MIS G6 is in the Gauss Chron, is dated at 2.74 Ma, and signifies a "climate crash" during one of the first high-amplitude northern-hemisphere glacial ages. The numbering of MISs and their relationship to climatic cycles can be found in many papers, including Shackleton and Opdyke (1973, 1976), Shackleton et al. (1990), Pisias et al. (1984), Pisias and Imbrie (1986), Ruddiman et al. (1989), Raymo et al. (1989), Tiedemann et al. (1994, 2006), and Lisiecki and Raymo (2005).

The paper by John Imbrie and Nilva Kipp (1971) ushered in planktonic foraminiferal assemblage analysis for SST analysis and testing of orbital theory (Ruddiman and McIntyre 1981; Imbrie et al. 1984, 1992). The Imbrie-Kipp transfer function method became a standard tool in paleoclimatology applied to diatoms (Koç Karpuz and Jansen 1992), radiolarians, ostracodes (Cronin and Dowsett 1990), dinoflagellates (de Vernal et al. 2001), and other groups. Modified versions of the foraminiferal transfer function method (Dowsett et al. 2005) and related modern analog techniques (e.g., Pflaumann et al. 2003) are still used today, usually in conjunction with geochemical temperature proxies and climate modeling (Haywood et al. 2005).

CLIMAP and SPECMAP

By the 1970s, the SST and ice-volume records from deep-sea foraminifera and other proxy records covering the past few hundred thousand years led to a "grand synthesis" in support of the orbital theory. Much of this research was carried out under the auspices of the CLIMAP and SPECMAP Projects. CLIMAP (Climate: Long-Range Investigation, Mapping, and Prediction) represented the first large-scale multidisciplinary effort to understand the causes of Quaternary ice ages that incorporated both empirical paleoreconstructions and climate modeling. By mapping the climate of the earth during both glacial and interglacial periods, CLIMAP culminated

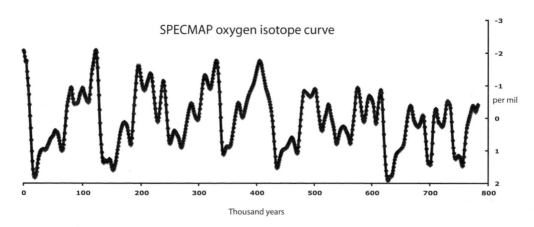

SPECMAP oxygen isotope curve

Thousand years

FIGURE 5.4 Marine oxygen isotope Spectral Climate Mapping Project (SPECMAP) stack, from Imbrie et al. (1984), showing predominant 100-ka orbital cycles during the last 800 ka. Other versions of marine isotope curves are available (see text and Figure 5.5).

in major publications that included the global reconstruction of ice age (CLIMAP Project Members 1981) and interglacial (CLIMAP Project Members 1984) climate. These studies also resulted in the influential paper "Variations in the earth's orbit: Pacemaker of the ice ages" (Hays et al. 1976), the CLIMAP volume (Cline and Hays 1976), the reconstruction of global ice sheets during the LGM (Denton and Hughes 1981), and the volume *Milankovitch and Climate: Understanding the Response to Astronomical Forcing* (Berger et al. 1984), as well as many other publications (Pisias and Shackleton 1984).

SPECMAP (Spectral Climate Mapping Project) advanced orbital theory by investigating climate variability in the frequency domain, that is, the periodicity of major Quaternary climatic cycles for the past 300 ka. SPECMAP accepted orbital variations as the driver of Quaternary ice-volume changes and, in lieu of direct geochronological age control, the deep-sea isotope curve and other proxy records were "tuned" to the astronomical timescale (Imbrie and Imbrie 1980; Imbrie et al. 1984).

Pisias et al. (1984; see also Pisias and Moore 1981) produced the first stacked oxygen isotopic curve from seven deep-sea cores. This standard isotope time series was tuned to the astronomical timescale using four approaches and became known as the SPECMAP "stack" (Martinson et al. 1987). A stack refers simply to a composite benthic or planktonic foraminiferal $\delta^{18}O$ time series averaged from multiple deep-sea sites, often tuned to orbital cycles, and used to study orbital climate dynamics, to correlate deep-sea records, and provide a chronology for nonmarine records (Imbrie et al. 1984; see also Pisias and Moore 1981 and Martinson et al. 1987). The SPECMAP stack became the standard against which many Quaternary paleoclimate records were dated and the foundation for Imbrie's model of orbital climate described in the section "Imbrie's Process Model."

Orbital tuning methods pioneered by SPECMAP are applied through the entire Quaternary and much of the Cenozoic for marine and some continental sedimentary sequences. Some notable studies are those of the Pliocene (Ruddiman et al. 1989; Raymo et al. 1990; Bassinot et al. 1994; Tiedemann et al. 1994; see also Lisiecki and Raymo 2005), Miocene (Hilgen et al. 1995; Shackleton et al. 1995a), Oligocene-Miocene (Shackleton et al. 1999), Oligocene (Pälike et al. 2006b), and Paleocene-Eocene (Cramer et al. 2003; Lourens et al. 2005) epochs.

Coral Reefs, Sea Level, and Absolute Dating

Absolute dating of paleoclimate events is critical for testing orbital theory. Age estimates for glacial features have alternatively been invoked to support and to refute orbital theory.

One example involves the relationship between glacio-eustatic sea-level history and orbital climate change. Glacio-eustasy, a concept traced back to Maclaren (1842), holds that sea level fluctuates between glacial periods (low sea level) and interglacial periods because ice sheets store huge volumes of water on continents. Whittlesey (1868) hypothesized that the storage of water in continental ice sheets caused global sea level to fall by more than 100 m, not far from the modern estimate of ~125 m. However, radiometric methods were not available to date glacial deposits or periods of high sea level, making it impossible to accurately test orbital forcing of the ice ages.

Coral-Reef Records of Sea Level

Coral-reef terraces along tropical coastlines provide a direct test of orbitally induced glacio-eustasy because they represent periods of high sea level, minimal continental ice, and climatic warmth. Corals can be dated because their skeletons take up radioactive elements from seawater, including radioactive elements used in uranium-series dating (^{234}U, ^{230}Th, and ^{231}Pa, with half lives of 244.5 ka, 74.4 ka, and 32.7 ka, respectively). The alpha counting method of ^{230}Th age dating used in early studies of Quaternary sea-level history has a large analytical age uncertainty of several thousand years, and the development of thermal ionization mass spectrometry (TIMS) (Edwards et al. 1987; Bard et al. 1990a, b) vastly improved geochemists' ability to date corals. TIMS analytical age uncertainty is <50 years for corals from the last postglacial sea-level rise (SLR) (<15 ka) and <±1000 years for the last interglacial corals (~117–134 ka). Some coral skeletons are only partially closed chemical systems, and postmortem exchange of uranium or its daughter products after the skeleton forms means that a correction factor is necessary (Thompson and Goldstein 2005).

Huon Peninsula and Barbados

Reefs located on the Huon Peninsula, New Guinea (Veeh and Chappell 1970; Bloom et al. 1974; Gallup et al. 1994) and on Barbados in the Caribbean (Broecker et al. 1968; Mesolella et al. 1968; Fairbanks and Matthews 1978; Bender et al. 1979) were important for testing orbital theory. Because of the Huon Peninsula's high uplift rate, at least 20 separate reef tracts are exposed along it. Reef tract VII formed during the last interglacial, when reduced continental ice volume caused a global sea level 6 m higher then that of the present, lies approximately 260 m above sea level (ASL). Most Huon reefs record relatively high eustatic sea levels, although some formed when sea level was slightly lower than modern sea level, and are now exposed because of uplift. Early U-series dating of New Guinea reefs indicated that a high sea level occurred about every 20 ka, suggesting an apparent precession periodicity.

On Barbados, the Rendezvous Hill, Ventnor, and Worthing terraces at 125, 103, and 82 ka represent marine oxygen isotope substages 5e, 5c, and 5a of the last interglacial. A younger terrace dated at 60 ka lies just above the present surf zone. In general, the ages of coral-reef tracts in Barbados support orbital theory.

In contrast to uplifted coasts, emerged fossil reefs along stable regions record only high stands of sea level higher than modern sea level and thus less continental ice volume than today. Early dating of the Miami Oolite and Key Largo Limestone in Florida and correlative features in the Bahamas (Broecker and Thurber 1965) suggested that maximum sea level during the last interglacial was about 4–6 m ASL, a value still accepted today. These results have been confirmed by dating reefs, solitary corals, mollusks, windblown sediments, and terraces in other stable regions such as the east coast of the United States (Cronin et al. 1981; Szabo 1985; Wehmiller et al. 2004), Hawaii and Bermuda (Muhs and Szabo 1994; Szabo et al. 1994; Muhs 2002; Muhs et al. 2002), the Bahamas (Neumann and Hearty 1996), and the Balearic Islands in the Mediterranean (Hillaire-Marcel et al. 1996).

Bloom et al. (1974) correlated the New Guinea and Barbados sea-level records and provided a strong case for glacio-eustatic control of reef development since 150 ka. By pinning the 125-ka-old reefs to a +6-m eustatic sea-level position for reefs from stable areas, they constructed a sea-level curve showing a stepwise fall from a high sea level ~125 ka to a sea-level minimum during the LGM ~21 ka.

Conflicting Sea-Level Records   Despite general support for orbital theory from coral reefs, the ages and durations of the interglacial high stands have been in dispute, posing issues for orbital forcing. Some evidence exists that MIS 5e sea level rose before 130–135 ka during Termination II, that is, several thousand years before the insolation inflexion point, and remained at or near the present level until about 118–117 ka (Szabo et al. 1994; Neumann and Hearty 1996). On the basis of the ice-volume record in the tuned SPECMAP chronology, Winograd et al. (1992, 1997,) argued that the MIS 5e interglacial high sea level lasted about 5–10 ka longer than expected. Hillaire-Marcel et al. (1996) also found evidence for an extended interglacial high sea level during the Tyrrhenian stage in the Balearic Islands of the Mediterranean. Mollusks are generally unreliable for U-series dating because they typically take up uranium after death, allowing possible migration of ^{230}Th. Hillaire-Marcel et al. (1996) demonstrated that uranium uptake occurred for only a few thousand years before diagenetic processes closed the chemical system, and under these conditions, the Tyrrhenian U-series ages produced reliable ages and evidence for an extended 17-ka period of high sea level. On the basis of reefs from Hawaii and Bermuda, Muhs et al. (2002) also pointed out the long duration of MIS 5e. Blanchon and Eisenhauer's (2000) restudy of Barbados reefs was inconclusive regarding the duration of high sea-level stands, and sea-level fluctuations associated with MIS 5e coral reefs and other proxy records remain actively researched (Hearty et al. 2007; Rohling et al. 2008).

Still, many conclude that tropical-reef sea-level record supports orbital theory (Stirling et al. 1995). Gallup et al. (1994) restudied Huon reefs, concluding that "for the last three interglacial and two intervening interstadial periods, sea level peaked at or after peaks in summer insolation in the northern hemisphere. This overall pattern supports the idea that glacial-interglacial cycles are caused by changes in orbital geometry." Thompson and Goldstein (2005, 2006) corrected coral ages for possible open-geochemical behavior (Thompson et al. 2003) and evaluated the last 240 ka of coral-reef ages in the context of orbital theory and tuning of deep-sea isotope records to astronomical insolation. They recognized the suborbital variability of sea level and noted a lack of one-to-one correlation between any particular season or latitude and sea level. Nonetheless, they concluded that the reef sea-level record supports first-order northern-hemisphere insolation forcing and provides the most reliable absolute age tie points for tuned SPECMAP isotope records for the last 240 ka.

It is also noteworthy that rapid changes in sea level occurred within the last interglacial period. By comparing isotopically derived sea-level history for MIS 5 from Red Sea foraminifera to U/Th-dated coral-reef records, Rohling et al. (2008) estimated maximum rates of SL rise of 25, 20, and 11 mm yr^{-1}, although minimum estimates ranged from 6–13 mm yr^{-1}. Nonetheless, MIS 5 had less ice volume than the Holocene interglacial, about 4–10 m sea-level equivalent, and thus the Red Sea coral records indicated potentially rapid rates of SLR—several times the current rate of SLR—even during periods of warmth and reduced ice volume (see chapters 10 and 12 in this volume).

Uranium-Series Dating of Sediments   Uranium-series dating of sediments has been applied to the testing of orbital theory (Slowey et al. 1996). Henderson and Slowey (2000) found evidence against northern-hemisphere forcing of sea level during MIS 7, and Henderson et al. (2006) dated MIS 7 and 9 sediments from ODP Site 1008A off the Bahamas. By comparing the U-series-dated sediment to tuned deep-sea isotopic and other (speleothem) records, they concluded that the tuned δ^{18}O timescale of Martinson et al. (1987) and Lisiecki and Raymo (2005) was inaccurate by up to 10 ka, at least for these two interglacials. The Bahaman record included a late MIS 7 high stand inconsistent with midlatitude orbital forcing.

Paleomagnetic Timescale   In addition to coral sea-level records, a second key chronological tie point for dating

Pleistocene deep-sea isotopic records is the Brunhes-Matuyama magnetic reversal. Shackleton and Opdyke (1973) first used this horizon to identify the late Pleistocene 100-ka cycles in foraminiferal isotope records from the Ontong-Java Plateau in the equatorial Pacific. As already discussed, the B/M boundary is about 780–790 ka, consistent with the results obtained by Shackleton et al. (1990) by orbital tuning of the marine isotope record.

Imbrie's Process Model

John Imbrie and colleagues (1992, 1993) described a process model for orbital climate variability based largely on marine records in two papers titled "On the structure and origin of major glaciation cycles: Parts 1 and 2." Part 1 addressed the origin of the 23- and 41-ka cycles, and part 2 the 100-ka cycle. Many recent hypotheses about orbital forcing at least for the last several glacial cycles are derivations of this model. Imbrie focused mainly on the causes of mid- to high-latitude northern-hemisphere glaciation, the polar and subpolar response to seasonal and geographical radiative forcing, and less so on low-latitude climate changes at the 23-ka precession frequency, now known from paleorecords of monsoons, methane, tropical dust, and other parameters (Ruddiman 2006b). Imbrie recognized that most parts to the climate system are affected by orbital forcing but that the response is neither simultaneous nor equal in magnitude. The model relied on the phase relationships between 16 paleoclimate indicators to depict the sequence of events within an idealized cycle (Imbrie et al. 1989). The indicators include southern-hemisphere deep-sea circulation, SSTs, terrestrial aridity (dust), ocean dissolution, and nutrients, among others, all tied together by the marine oxygen-isotope curve as a proxy for ice volume. Thirteen proxy-record indicators extend to 300 ka or earlier. Several were limited to ^{14}C-dated records of the last deglacial termination in the Nordic Seas and the Arctic Ocean in lieu of a long time series. All proxies are pinned to the isotope/ice-volume curve with various phasing patterns.

The cornerstone of Imbrie's model is the amplification of small insolation changes into major ice-age cycles through the action of the ice sheets. Imbrie concluded that the earth's climate responded to 23- and 41-ka insolation cycles in a linear, continuous way. In contrast, large midlatitude ice sheets amplify the initial insolation signal through feedback in the nonlinear response of the 100-ka cycle. The concept of "early" and "late" responders to insolation forcing within a particular cycle is central to Imbrie's interpretation. For example, an early response to obliquity insolation forcing would occur in high-latitude atmospheric temperatures and SSTs. Orbital forcing would influence greenhouse gas concentrations in different ways. Methane's (CH_4) response

would be rapid because methane's atmospheric residence time is ~10 years and its flux is influenced by atmospheric temperature and hydrological changes. Atmospheric carbon dioxide (CO_2), on the other hand, responds more slowly because the long-term global carbon cycle involves ocean sequestration during glacial periods and degassing during terminations. The ocean's role in air-to-sea CO_2 flux through physical (carbonate production), chemical (dissolution), and biological (photosynthesis) processes results in a slower response to orbital forcing than that of methane. Northern-hemisphere ice-sheet growth is relatively slow but its decay is more rapid. Opinions vary, but the ice-response time to orbital forcing may be as short as 1–4 ka or as long as 15 ka and is variable over the past 5 million years, depending on the extent, thickness, and location of the ice sheets.

Each ice-age cycle was characterized as a sequence of four end-member climate states: glacial, deglacial, interglacial, and preglacial. The trigger for each complete cycle is northern-hemisphere (65°N) June insolation. The theory recognizes the role of oceanic circulation for translating insolation variability into local responses (Rind and Chandler 1991). High-latitude ocean regions in the Nordic Seas and the Labrador Sea, where deep water forms North Atlantic Deep Water (NADW), are referred to as "heat pumps" because low-latitude heat is transferred to higher latitudes by ocean circulation. Each state in an idealized orbital cycle is characterized by a different mode of deep-ocean circulation. The glacial period is characterized by a shutdown or weakening of the North Atlantic deep-water formation; both the polar Nordic Sea and boreal Labrador Sea heat pumps are thought to be inoperative. During deglaciation, the Nordic heat pump is turned on. Under full interglacial conditions, the Labrador Sea boreal heat pump is also operative. Finally, during the preglacial state, insolation decreases in high latitudes; fields of ice and snow expand from their minimal conditions; and the Nordic pump is shut down or diminished.

Many modifications to Imbrie's model stem from longer and more highly resolved paleorecords, improved proxies of tropical climate and deep-ocean circulation, and more sophisticated climate modeling, outlined in the rest of this chapter.

Paleoclimate Records of Orbital Variability

Paleoclimate records exhibiting orbital variability come from many marine, continental, and Antarctic ice-core records obtained in the past 20 years covering the last 750 ka to more than 5 Ma. We present here a few that exemplify key aspects of orbital variability and form the framework to discuss still controversial issues surrounding orbital theory.

Marine Oxygen Isotope Records

Figure 5.5 shows the marine $\delta^{18}O$ isotope curve for the past 5 million years based on benthic foraminiferal $\delta^{18}O$ records from 57 deep-sea cores from the earth's oceans, compiled by Lorraine Lisiecki and Maureen Raymo (2005). Following convention, interglacial and glacial periods are labeled with odd and even numbers, respectively, with those before 2.6 Ma having the prefix identifying the paleomagnetic subchron. This curve is called the LR04 stack and clearly shows a progressive increase in amplitude and a decrease in frequency of isotopic cycles over the past few million years. As discussed elsewhere, benthic $\delta^{18}O$ isotope values in the LR04 stack are influenced by the $\delta^{18}O$ of seawater, a function of the amount of water stored in continental ice sheets (ice volume), and by bottom-water temperature (BWT), which itself is spatially variable (Skinner and Shackleton 2005). Important climatic regime shifts signifying different "modes" of orbital forcing are centered near 4.2 (the closure of the Central American Isthmus, Indonesian throughflow), 2.8 (the inception of large northern-hemispheric ice sheets), 0.9–1.2 (the MPT), and 0.45 Ma (the Mid-Brunhes Transition).

Ice-Core Records

Ice-core records of orbital-scale atmospheric temperature, composition, and circulation have led to momentous discoveries about the earth's climate. Figure 5.6 shows the locations of some of ice cores discussed here and in later chapters. The three most important Antarctic ice-core records of orbital climate variability are the 430-ka Vostok (Jouzel et al.

1987; Sowers et al. 1993; Petit et al. 1997, 1999), the 750-ka EPICA Dome C (EDC) (EPICA Community Members 2004, 2006; Spahni et al. 2005; Jouzel et al. 2007), and 360-ka Dome Fuji records (Kawamura et al. 2007).

The temperature proxy δD from the EDC and Vostok ice cores is shown in Figure 5.7, in which it is compared to the marine $\delta^{18}O$ record, the EDC dust record, and insolation curves for the northern and southern hemispheres (Masson-Delmotte et al. 2004). The highlights of the EDC record are the close match between EDC and Vostok deuterium records for the last 420 ka, the increase in the amplitude of cycles after MIS 12 and Termination V (the MBE), and the inverse relationship between dust peaks and δD (notable during MIS 16.2, an exceptionally cold glacial period). Before the MBE, glacials (except MIS 16.2) were not as cold but interglacials were much cooler and somewhat longer than their counterpart climate extremes after ~430 ka. This pattern agrees with that seen in the marine record.

Amplification of orbitally forced insolation changes by greenhouse gases is a leading explanation for the large amplitude of late Pleistocene glacial-interglacial cycles. Figure 5.8 shows the CO_2 record by Siegenthaler et al. (2005), compiled from EDC, Vostok, and Taylor and compared to the deuterium record from EDC and Vostok. It clearly shows high-amplitude 100-ka cycles evident in the marine and EDC ice-core isotopic records, the shift in amplitude at Termination V about 430 ka, and the reduced (30%) glacial-interglacial amplitude of atmospheric CO_2 variations before Termination V. This long-anticipated and unique record of atmospheric CO_2 concentrations spanning the MBE amplitude shift in orbital-scale climate variability reflects, to a first approximation,

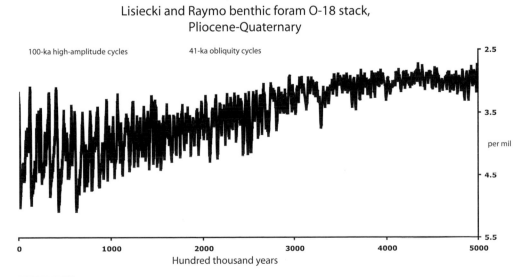

Lisiecki and Raymo benthic foram O-18 stack, Pliocene-Quaternary

FIGURE 5.5 Marine benthic oxygen isotope stack LR04, from Lisiecki and Raymo (2005). Isotope curve "stack" is a composite from 57 different deep-sea core benthic isotope records. Note the shift from late Pliocene–early Pleistocene 41 ka obliquity to late Pleistocene 100-ka "eccentricity" cycles.

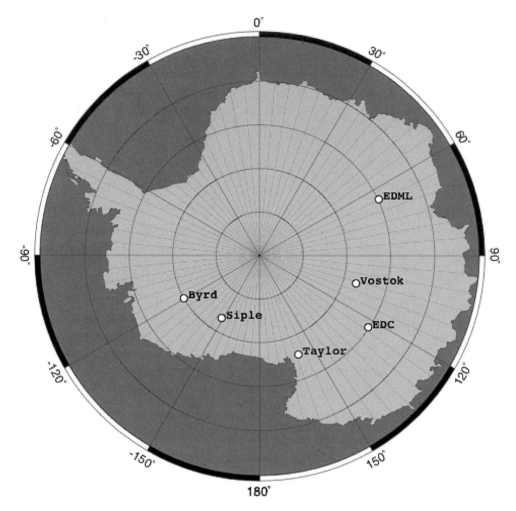

FIGURE 5.6 Location of Antarctic ice cores. EDC=EPICA Dome C; EDML=EPICA Dronning Maud Land.

what the EPICA group called a "Stable carbon-cycle climate relationship during the late Pleistocene" (Siegenthaler et al. 2005). The Antarctic ice-core record of methane (Loulergue et al. 2008) and CO_2 (Lüthi et al. 2008) now extends back to 800 ka, covering eight orbital cycles.

In addition to temperature and greenhouse gas concentrations, atmospheric circulation and Southern Ocean sea ice using a variety of glaciochemical proxies of ocean and terrestrial aerosols have been reconstructed for the last 740 ka from the EDC core. Figure 5.9 compares aerosol proxies (sources): sodium (marine sea salt, high ssNa=cold climate, maximum sea ice), iron (terrestrial), calcium (mostly terrestrial), and sulfate (marine biogenic emissions) ion fluxes to δD and the marine $\delta^{18}O$ records (Wolff et al. 2006). Atmospheric iron and calcium in dust is controlled by many factors—in the case of the location of the EDC core in Antarctica, by Patagonian atmospheric circulation, temperature, and glacial history. The EDC record indicated a strong coupling between Southern Ocean sea ice recorded in sea-salt proxies and southern South American climate. Importantly, this led to the conclusion that there is no clear change in

climate-sea-ice feedback mechanisms that might explain the MBE shift at 430 ka.

Kawamura et al. (2007) studied a 360-ka record from the Dome Fuji ice core using a new chronology based on the ratio of oxygen-to-nitrogen molecules in trapped air, which serves as a proxy for local insolation independent of marine isotope curves. By comparison with the Vostok and marine isotope records, they addressed northern- and southern-hemisphere climate shifts during Terminations I through IV. Southern-hemisphere temperature and CO_2 concentrations clearly lagged behind northern-hemisphere temperature by several millennia during each of the last four terminations. The Dome Fuji record provided evidence that northern-hemisphere summer insolation, climate, and ice-sheet volume led southern-hemisphere climate during major climatic transitions by several millennia, at least for the last few cycles.

Tropical Sea-Surface Temperatures

Paleoceanographic tropical SST records of Pliocene-Pleistocene orbital cycles are available from Mg/Ca ratios in

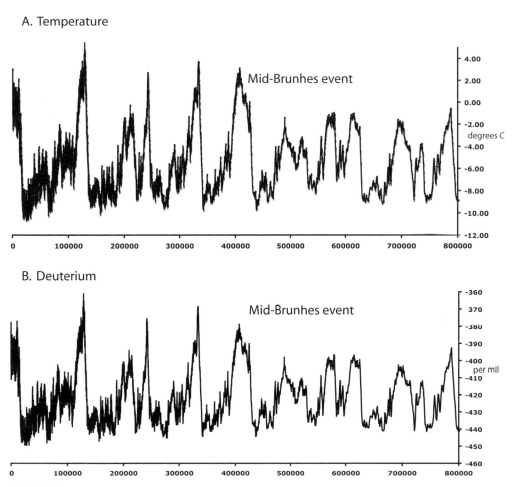

A. Temperature

Mid-Brunhes event

B. Deuterium

Mid-Brunhes event

FIGURE 5.7 EPICA Dome C (A) estimated temperature anomaly (compared to the last 1000 years) and (B) deuterium record, from Jouzel et al. (2007). Note the increase in amplitude of 100-ka cycles during the Mid-Brunhes Event.

Ice-core atmospheric CO₂

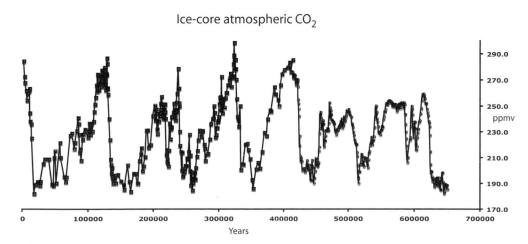

FIGURE 5.8 Vostok (open squares = 415 ka–present) and EPICA Dome C (open circles = 413–650 ka, Bern Switzerland values) atmospheric CO₂ record, from Petit et al. (1999) and Siegenthaler et al. (2005).

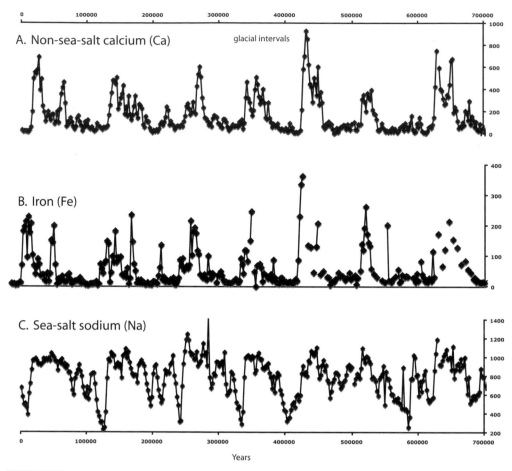

FIGURE 5.9 EPICA Dome C glaciochemical trends during the last seven glacial-interglacial cycles, from Wolff et al. (2006): (A) Non-sea-salt calcium (Ca), a terrestrial source indicator; (B) iron (Fe), mainly dust from continental terrestrial sources; (C) sea-salt sodium (Na), mainly a marine source chemical indicator. Note the high levels of both non-sea-salt Ca and Fe during glacial intervals. Wolff et al. (2006) also measured methanesulphonate and sulphate ions as indicators of biogenic emissions of dimethylsulphide. No major changes were seen during the Mid-Brunhes Event that might signify changes in sea ice or other feedbacks.

planktic foraminifers (Chaisson and Ravelo 2000; Medina-Elizalde and Lea 2005; Rickaby and Halloran 2005; Wara et al. 2005) and alkenone paleothermometry (Liu and Herbert 2004; Lawrence et al. 2006). Figure 5.10 shows eastern equatorial Pacific (EEP) SST since 1.8 Ma. In contrast to the dominant precession influence at low latitudes during the last 100 ka, during the early Pleistocene from 1.2–1.8 Ma EEP SST and productivity are dominated by 41-ka cycles, which varied, antiphased, with local annual insolation. These results suggest that high-latitude processes forced tropical SST during this interval prior to the MPT, after the tropics were more influenced more by precession.

Terrestrial Records

Two especially well studied terrestrial records of Asian climate variability come from the Chinese loess and paleosol

sequences and Lake Baikal, Russia sediment cores. Loess is windblown silt that accumulates in relatively thick deposits during drier and windier glacial periods, alternating with soils formed during wetter interglacials. Extensive Chinese loess-soil sequences cover 440,000 km², reach 200 m in thickness, span the last 2.5 Ma, and provide terrestrial evidence for orbital cycles (Heller and Liu 1982; Kukla 1987; Kukla and An 1989; Rutter and Ding 1993). Figure 5.11 shows loess magnetic susceptibility and grain size, both excellent proxies for climate and monsoonal activity for the Brunhes Chron (Liu and Ding 1998). The oscillations in loess and soils show a similar pattern to other proxy records for this period, notably much greater amplitude during the last five 100-ka cycles. Orbital and suborbital monsoon variability has also been closely linked to global ice volume.

The Lake Baikal record of biogenic silica also provides an orbital-climate-monsoon link based on paleoclimate recon-

Eastern equatorial Pacific sea-surface temperature

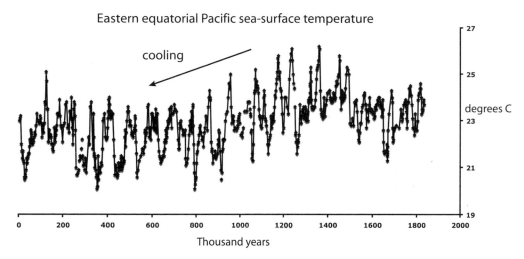

FIGURE 5.10 Eastern equatorial Pacific (EEP) sea-surface temperature (SST) based on alkenone paleothermometry for the last 1.8 Ma, from Liu and Herbert (2004). The record shows long-term cooling in EEP and 41 ka obliquity cycles from 1.8 to 1.2 Ma, a minimal precession signal. Early Pleistocene SST varied in an opposite direction to EEP annual insolation and was more influenced by high-latitude obliquity forcing.

struction and analysis of modern lake processes and their relationship to seasonal biogenic activity by diatoms (Colman et al. 1995; Williams et al. 1997; Prokopenko et al. 2001, 2006). Figure 5.12 shows the Baikal record reflecting the typical 100-ka orbital cycles during the Brunhes Chron, with additional features, such as three precession peaks in each of the last three interglacials (MIS 5, 7, and 9) and an extremely strong MIS 11 interglacial.

Another excellent record of orbital-scale variability comes from combined pollen-and-spore and oxygen isotopic analyses of the planktic foraminifer *N. pachyderma* from deep-sea sediment cores off southern Greenland (mainly ODP Site 646). De Vernal and Hillaire-Marcel (2008) showed cyclic variation for the past 1.0 Ma in terrestrial vegetation and Greenland ice volume and, although each interglacial period had a distinct pollen signature, interglacials during MIS 5e,

Chinese loess mass accumulation rate

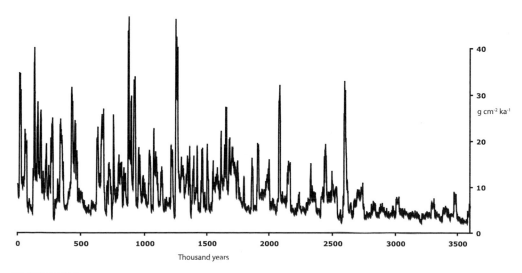

FIGURE 5.11 Chinese loess plateau mass accumulation rate from bulk density measurements for the past 3.6 Ma, from Sun and Z. An (2005). The record comes from two loess sections, Zhaojiachuan and Lingtal, and a red clay sequence. The inception of intervals of loess deposition are near 2.6 Ma at the onset of high-amplitude glacial intervals, changes in monsoons, and increased aridity related to high-latitude ice-sheet growth.

Lake Baikal biogenic silica

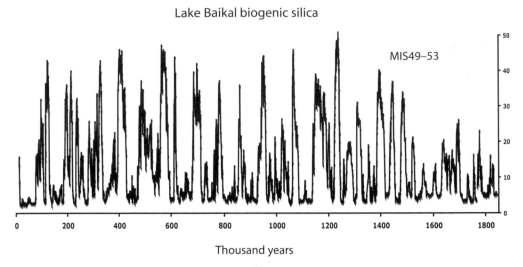

MIS49–53

Thousand years

FIGURE 5.12 Lake Baikal, Siberia biogenic silica (BiSil) record for the last 1.8 Ma showing precession frequency oscillations, from Prokopenko et al. (2001, 2006). The age model was astronomically derived by tuning BiSil to the September perihelion on the basis of seasonal ecological processes known from the Holocene section of the Baikal record. A large increase in the amplitude of cycles occurs near 1.5 Ma during Marine Isotope Stages (MISs) 49–53.

11, and 13 were particularly strong (high pollen flux) and reduced ice volume (*N. pachyerma* isotopes). A notable conclusion was that MIS 11 ice volume was less than that during MIS 5e, consistent with an extended, warm interglacial (see below).

Uncertainties, Mysteries, and Paradoxes in Orbital Theory

These and other paleoclimate records covering the last 1–5 Ma suggest an obvious, albeit qualitative and variable, relationship to the earth's orbital geometry. In this sense, paleoclimate records support at least some tenets of the general theory of orbital forcing of climate. On the other hand, longer time series and new atmospheric and tropical proxy records have raised perplexing issues regarding the changing patterns of the past few million years. This section briefly introduces some still unresolved and closely related mysteries and paradoxes in orbital theory, and is followed by a section on orbital forcing hypotheses.

Pliocene Transition at 2.8 Ma

The late Pliocene transition about 2.8 Ma is evident in the benthic oxygen isotope record already shown here in the shift into the dominant obliquity world of 41-ka cycles. This late Pliocene transition appears to represent a change in the style of orbital climate dynamics, perhaps initiated by tectonically forced boundary conditions in the ocean-atmosphere system, pushing the climate system across a threshold from sustained El Niño–like to La Niña–like conditions.

Ravelo et al. (2004) integrated high- and low-latitude paleorecords covering the last 4 Ma and related them to changes in climate sensitivity. They propose that the 2.8 Ma shift was not accompanied by a change in climate sensitivity and that no single threshold event was responsible for it. Rather, a series of climatic transitions occurred, including a gradual transition period from ~2.0–1.5 Ma characterized by equatorial cooling, the development of Walker Cell circulation in the Pacific, and greater sensitivity of high-latitude regions to orbital forcing.

What is important for orbital theory is the evidence that the 2.8-Ma transition was not accompanied by as large a shift in tropical ocean conditions. On the other hand, the East Asian monsoon system clearly experienced a large shift about 2.5 Ma marked by the change from "red clay" deposits to the loess-soil pattern in Chinese stratigraphic sequences and changes in the Asian monsoon. The stepwise shift to larger northern-hemisphere glacial periods centered on 2.8 Ma is also linked to Indonesian and Central American gateway closing (see Chapter 4 in this volume).

The 100-ka Enigma

The dominant 100-ka cycles of the late Pleistocene pose a challenge to orbital theory because the total change in insolation in an eccentricity cycle are small compared to the scale of climate change (Imbrie et al. 1993). One hypothesis is that, although eccentricity by itself cannot cause large ice ages, it can play a role by modulating the influence of precession or obliquity so that when the earth is in a circular orbit (low eccentricity), precession or obliquity influence is enhanced.

Eccentricity modulation has been associated with the asymmetric sawtooth pattern in marine isotope records, in particular the abrupt nature of terminations of the past 400 ka. At times when eccentricity and precession insolation are both weak, abrupt terminations that characterize the 100-ka cycles occur.

In fact, the term *100-ka cycle* is somewhat misleading because late Pleistocene interglacial peaks are actually spaced anywhere from 82–123 ka apart. Maslin and Ridgewell (2005), for example, discovered artifacts in spectral analyses and call the 100-ka cycle the "eccentricity myth." Various explanations have been offered to account for the irregular spacing of 100-ka cycles, their large amplitude, and their sawtooth pattern. Opinions fall into two broad categories. Some paleoclimatologists argue that precession forcing influences climate during periods of low precession. For example, Raymo (1997) and Ridgwell et al. (1999) emphasized that late Pleistocene glacial terminations occur every fourth or fifth precession cycle, which would explain the observed spacing of interglacials. Ruddiman (2003, 2006b) also emphasized the importance of precession forcing. Others focusing on ice-core records suggest that obliquity drives the 100-ka cycles (Masson-Delmotte et al. 2006). Huybers and Wunsch (2005) preferred an obliquity link suggesting that two or three obliquity cycles occurred before a termination spaced every ~82–123 ka spacing. Climate modeling has played a large role in efforts to simulate the 100-ka cycles (Imbrie et al. 1993; Tziperman and Gildor 2003; Clement et al. 1996; Muller and MacDonald 1997a; Saltzman 2002); however, there is no consensus on any single explanation (see the subsections that follow).

Mid-Pleistocene Transition

The MPT refers to the shift from the 41-ka obliquity world of the late Pliocene and early Pleistocene epochs to the 100-ka world of the last 700,000 years. The MPT, also called the "mid-Pleistocene revolution" (EPICA Community Members 2004) and "late Pleistocene transition" (Paillard 2001), signifies the final stage in the 50-million-year transition from global warmth in the Eocene to the coldest glacial periods of the late Cenozoic era. Two comprehensive review papers define the MPT as a broad climate transition from 1.25 Ma to 700–500 ka encompassing MISs 36–13 (Head and Gibbard 2005; Clark et al. 2006). Schmieder et al. (2000) characterized the MPT as a "discrete state" of ocean circulation and climate between 920 and 640 ka, based on magnetic susceptibility, carbonate concentration, and diatom assemblages from the South Atlantic following a transitional period from dominant 41-ka cycles between 1.0 Ma and 920 ka.

The most striking manifestation of the MPT occurs in the deep-sea foraminiferal oxygen isotope LR04 stack (Figure 5.5) (Lisiecki and Raymo 2005). Low-amplitude 41 ka

cycles dominate between 2.8 and 1.2 Ma (MIS 104-37), followed by a period of irregular, moderate-amplitude glacial-interglacial cycles between 1.2 Ma and 700 ka (MIS 36 17), and a final period of 100-ka cycles (MIS 7-1). The first-order benthic isotope signal represents changes in global ice volume and, secondarily, BWT. The MPT also shows up in deep-sea carbon isotopic records. Raymo et al. (1997) proposed that a large carbon isotopic excursion at the MPT signified a massive transfer of carbon from terrestrial to marine reservoirs, ultimately leading to greater air-to-sea drawdown of CO_2 and colder glacial periods after ~700 ka. Many other records show a major change centered on 700–900 ka (Clark et al. 2006).

The MPT embodies a number of related mysteries surrounding orbital theory. In addition to the 100-ka problem, why are 41-ka cycles, so regular and dominant between 2.8 and 0.7 Ma, not predicted in Milankovitch's original hypothesis? What feedbacks—such as carbon cycling and greenhouse gases, sea-ice and ocean circulation, and albedo changes from ice sheet—contribute to the timing and amplitude of these cycles, and what changed about 700 ka?

Research on the MPT and its causes has exploded in the past few years. Clark and Pollard (1998) and Clark et al. (2006) proposed the regolith-ice-dynamics hypothesis whereby prior to the MPT, northern-hemisphere ice sheets were nearly as geographically extensive as those that developed after the transition, but ice sheets during the last four or five glacial periods were much thicker, accounting for greater ice volume and lower sea level. This idea calls for the exposure of the crystalline Precambrian shield, the core of the North American continent, by repeated episodes of glacial erosion, creating a "high-friction substrate" that led to an altered response to orbital forcing. Under this scenario, greater silicate rock weathering and erosion would lead to lower atmospheric CO_2 concentrations and positive feedback toward greater cooling. As ice sheets thickened and sea level fell to extreme lows, sediments on the continental shelf and upper continental slope would be exposed, providing a source of organic carbon and nutrients.

Various hypotheses involve Southern Ocean sea-ice cover, stratification and nutrient cycling, greenhouse gas feedbacks, tropical ocean and atmosphere dynamics, or a combination of mechanisms to amplify the original insolation signal. Tziperman and Gildor (2003) proposed a sea-ice switch mechanism to explain the 100-ka cycles. This involves progressive Pleistocene cooling of the deep ocean, a critical threshold for ice-sheet ablation, and sea-ice feedbacks influencing air-to-sea heat (insulation) and carbon exchange (Gildor and Tziperman 2000). The 41-ka cycles are, in their view, "purely forced" by Milankovitch forcing, and the 100-ka cycles required the deep-sea cooling and development of extensive sea-ice. The sea-ice/carbon exchange mechanism also explains the symmetry of the 41-ka and the asymmetry

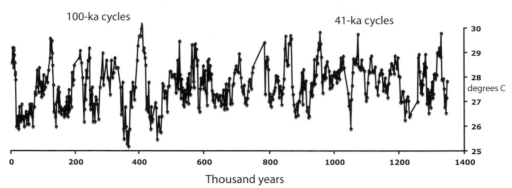

Eastern equatorial Pacific sea-surface temperature

FIGURE 5.13 Eastern equatorial Pacific (EEP) sea-surface temperature (SST) during the Mid-Pleistocene Transition (MPT), based on planktonic foraminiferal Mg/Ca ratios from Ocean Drilling Program Hole 806B, from Medina-Elizalde and Lea (2005). The EEP SST shift from 41-ka to 100-ka cycles near 950 ka preceded the shift toward large continental ice volume, suggesting a role for atmospheric CO_2 in the MPT.

of the 100-ka cycles that are characteristic of isotopic and other proxies.

Clement et al. (1996, 2004) showed that climate models were able to simulate the 100-ka cycles successfully and suggested that a tropical thermostat mechanism operating in the Pacific Ocean caused threshold-like behavior of global climate to insolation forcing. McClymont and Rosell-Melé (2005) also suggested that the tropics play an important role in the MPT, arguing that the timing coincides with a transition to strong Walker cell circulation in the tropical Pacific. Medina-Elizalde et al. (2005) also proposed that the tropical SSTs seen in their EEP record (Figure 5.13) are coupled to southern high-latitude climate, where orbital insolation forcing is amplified by outgassing of CO_2, possibly influenced by Southern Ocean sea-ice cover (see Schefuss et al. 2004).

The Mid-Brunhes Event and the MIS Stage 11 Paradox

The MBE is a period of carbonate dissolution in the oceans, centered on MIS 11, that coincides with a shift from low- to high-amplitude 100-ka climatic cycles in the middle of the Brunhes magnetic polarity epoch (Jansen et al. 1986; Berger and Jansen 1994). The MBE interval has also been called the Mid-Brunhes "dissolution interval" (Barker et al. 2006) and a CO_2 paradox (Raynaud et al. 2005) because changes in the global carbon cycle are evident from the extreme dissolution of deep-sea sediments when atmospheric CO_2 concentrations were similar to those of the Holocene and other interglacials. This apparent anomaly suggests that the widely accepted link between atmospheric CO_2 and ocean carbonate chemistry did not follow typical orbital patterns during MIS 11. The "400-ka" problem (Imbrie and Imbrie 1980), in which

this low-amplitude eccentricity frequency was absent in paleorecords (at least until recently; see Chapter 4 in this volume), may also be related to the MBE transition. Over long timescales, the 400-ka problem has been explained by frequency modulation of the 100-ka insolation signal (Rial 1999, 2004; Elkibbi and Rial 2001).

The MBE and MIS 11 are unique for several reasons. First, the largest glacial-interglacial transition in terms of temperature and sea-level amplitude occurred during Termination V during the MIS 12–11 transition at 430 ka. Second, MIS 11 is also considered by many to be the longest interglacial period of the last 750 ka and has similar, though not identical, insolation parameters to the current Holocene interglacial, making it a possible analog. Third, the MBE represents the final stage in the evolution of orbital forcing, the culmination of the MPT transition, and long-term evolution of the icehouse climate.

Age and Duration of the MBE A large literature is available on paleoclimate records of MIS 11 and the MBE (for details, see Droxler and Farrell 2000; Droxler et al. 2003; and Raynaud et al. 2007). The EPICA Dome C ice-core record epitomizes the glacial-interglacial amplitude shift after the MBE transition (figures 5.7 and 5.8) (EPICA Community Members 2004). The last five interglacial intervals (MIS 1, 5, 7, 9, 11) had more positive deuterium values (i.e., higher temperatures) and the last five glacial periods (MIS 2–4, 6, 8, 10, 12) had higher dust mass than those of the prior three cycles. A similar first-order contrast before and after the MBE is seen in EPICA CO_2 and methane concentrations (Siegenthaler et al. 2005). Another feature of MIS 11 noted by Imbrie et al. (1993) is that the amplitude, shape, or both of some glacial-interglacial cycles, as measured by various proxy records, are not scaled to the pattern of insolation changes calculated from celestial mechanics.

MIS 11 was an extremely long (20–30 ka) interglacial dated at 394–422 ka in Antarctic ice cores (EPICA Community Members 2004; Raynaud et al. 2007) and about 388–410 ka in deep-sea records (Kandiano and Bauch 2007; Hodell et al. 2000). De Abreu et al. (2005) estimated that peak interglacial conditions off the Iberian continental margin in the northeast Atlantic during MIS 11 lasted 18 ka. Ruddiman (2006b, 2007), however, maintains that the MIS 11 interglacial was far shorter, only ~6,000–10,000 years, fitting the SPECMAP timescale.

Sea Level During the MBE

The sea-level ice-volume change during Termination V at the MIS 12 to MIS 11 transition was the largest in at least the last million years and probably longer. Estimates from records of glacio-eustatic forced hypersalinity in the Red Sea (Rohling et al. 1998) and emerged shorelines (Hearty et al. 1999) show that this SLR was 25% larger than during other Terminations (160 m versus 125 m). Emerged MIS 11 shorelines and sediment deposits are known from many relatively stable coastal regions, such as the eastern United States (Cronin et al. 1981), the Bahamas and Bermuda (Kindler and Hearty 2000), and Indonesia (Howard 1997), often at elevations 13 to as high as 21 m (Olson and Hearty 2009) above present sea level, much higher than the 4–6-m elevation of MIS 5 shorelines. High MIS 11 sea level may have been the result of the collapse of the West Antarctic Ice Sheet (see Hearty et al. 1999; Scherer et al. 1998).

SST During MIS 11

The spatial and temporal pattern of SST variability during MIS 11 is also complex. In some regions of the North (McManus et al. 2003; de Abreu et al. 2005) and South (Hodell et al. 2000) Atlantic, SSTs were relatively warm but variable within the interglacial. In the Labrador and Nordic Seas, SSTs were cool (Bauch et al. 2000). In the high-latitude North Atlantic Ocean, Henning Bauch and colleagues (Bauch et al. 2000; Helmke and Bauch 2003; Kandiano and Bauch 2007) demonstrated that Termination V and MIS 11 had a complex oceanographic and climatic history distinct from Termination I and the Holocene. Differences included a longer duration for Termination V, peak interglacial warmth lasting ~10–12 ka after the onset of the interglacial in contrast to early Holocene peak warmth, and significant variability within the interglacial (see McManus et al. 2003). Bauch estimated that the MBE was preceded by a 200-ka transition between 700 and 500 ka that involved major changes in high-latitude sea-ice conditions.

Greenhouse Gas Concentrations and Carbon Cycling During the MBE

Greenhouse gas records from EPICA ice cores show that CO_2 and CH_4 concentrations during MIS 12 were 200 parts per million by volume (ppmv) and 380

parts per billion by volume (ppbv), respectively, rising to 275–280 and 680 ppbv during the MIS 11 interglacial (EPICA Community Members, 2004; Siegenthaler et al. 2005; Spahni et al. 2005). Raynaud et al. (2005) restudied the Vostok ice core and found similar CO_2 concentrations for MIS 11. Thus MIS 11 CO_2 levels remained at levels roughly equal to preindustrial Holocene values for up to 30 ka, although MIS 11 did not have an early interglacial peak in CO_2 as occurred during the following three interglacials. The EPICA Dome C record shows a strong coupling between CO_2, temperature (δD), and atmospheric circulation (dust aerosols), but no unusual carbon-cycling effects that might have led to the extensive carbonate dissolution during the MIS 12–11 interval.

In general, Antarctic ice-core records of MIS 11 are consistent with high- and low-latitude North Atlantic climate patterns (Healy and Thunell 2004; de Abreu et al. 2005). However, records from the equatorial Pacific, Indian, and North Atlantic oceans indicate that the MBE began during the glacial MIS 13, when silicate weathering from an enhanced Asian monsoon and greater tropical precipitation shifted the ocean's nutrient balance in favor of silicate-using diatoms over carbonate coccoliths. Such a silicate hypothesis has been proposed for the LGM as a way to explain greater ocean CO_2 uptake (Harrison 2000). Excessive drawdown of atmospheric CO_2 during MIS 12 would have led to an exceptionally strong glacial period and extensive carbonate dissolution. Wang and colleagues also propose that low-latitude forcing of rainfall patterns and biogeochemical cycling caused the MBE.

Holocene-MIS 11 Compared

The MBE-MIS 11 problem also has relevance for debates about rising greenhouse gas concentrations and future climatic trends. The low eccentricity during MIS 11, similar to that during the Holocene, should in theory lead to a dampening of precession forcing. This similarity, and its warm climate and high sea level, has made MIS 11 a prime candidate as a possible analog for future climate and a keystone for understanding orbital forcing, internal climatic feedbacks, and global carbon cycling.

Figure 5.14 presents deuterium profiles for the Termination V and MIS 11 transition, Termination I and MIS 1 and northern and southern high-latitude insolation (Masson Delmotte et al. 2006). Because the precise age of the ice-core MIS record is not known, two options are available to compare the phasing of insolation and paleorecords. In Figure 5.14A, obliquity is synchronized and MIS 11 and MIS 1 begin simultaneously. In Figure 5.14B, precession is synchronized and obliquity minimized. EPICA project members (2004) and Masson-Delmotte et al. (2006) favor the obliquity-driven model in Figure 5.14A.

Ruddiman (2003, 2006b) carried out similar comparisons between MIS 11 and MIS 1, favored the latter precession-dominated scenario, and concluded that the two interglacials

FIGURE 5.14 Comparison of deuterium curves for Marine Isotope Stages (MIS) 2/1 and 12/11, from EPICA Dome C ice-core and mean annual orbital insolation curves at 75°S (mainly obliquity) and mid-June insolation at 65°N latitude, from Masson-Delmotte et al. (2006), for 30,000 years into the past and future. (A) Synchronized deglaciation in phase with obliquity as proposed by EPICA Project Members (2004). (B) Diagram shows precession-dominated phasing in which obliquity and deglaciations are out of phase with obliquity (Berger and Loutre 2003; Ruddiman 2003). Courtesy of V. Masson-Delmotte and *Climates of the Past*.

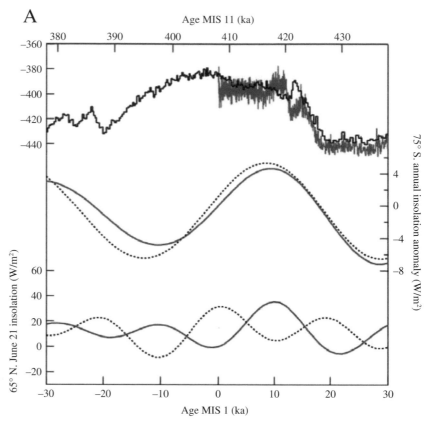

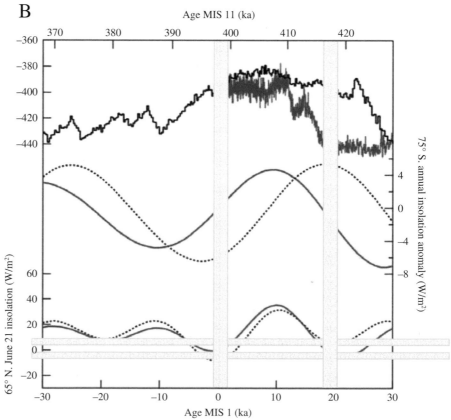

would have been much more similar had early-mid Holocene land-use activity not disrupted the natural response of the climate cycle (see Loutre and Berger 2000; Berger and Loutre 2003). Ruddiman's Holocene anthropogenic forcing theory leans heavily on orbital forcing over the last 500 ka. It holds that ice-core records indicate Holocene CH_4 peaked near 10 ka, in agreement with monsoon and boreal wetland fluxes, and fell during the early Holocene, only to rise again beginning 5000 years ago from 580 to 720 ppbv. CO_2 concentrations also steadily rose since the early Holocene, from ~260 to 280–285 ppmv prior to 20th-century industrialization. Although methane and CO_2 clearly have different sources, sinks, and feedbacks, Ruddiman argues that, had humans not influenced greenhouse through land clearance beginning in the early Holocene, atmospheric CO_2 and CH_4 concentrations would have steadily declined as they did during MISs 9, 7, and 5. In theory, CO_2 would have reached a threshold of about 240–250 ppmv, when it might have triggered glaciation (Berger and Loutre 2003; Crucifix et al. 2005, 2006). Broecker and Stocker (2006) discounted Ruddiman's hypothesis by explaining Holocene CO_2 patterns in terms of ocean carbonate chemistry.

Ocean Carbonate Chemistry A second MBE issue is carbonate dissolution and ocean acidification, a growing concern today because during the past century anthropogenic CO_2 emissions and uptake of anthropogenic carbon by the oceans have caused a decrease in the ocean's pH (see Chapter 12 in this volume). Deep-sea carbonate dissolution during MIS 11 has been attributed to ocean-wide changes in carbonate chemistry and lowered mean saturation state during the mid-Brunhes. Possible causes include SLR and a shift in carbonate production (Berger 1982), decreased influx of calcium from continents (Bassinot et al. 1994), proliferation of the coccolith *Gephyrocapsa* from evolutionary or ecological (greater productivity) factors (Barker et al. 2006), and increased low-latitude carbonate production by reefs relative to open ocean planktonic carbonate secreting organisms (Droxler et al. 1997). The MIS 11 dissolution problem is not adequately explained.

Temperature and CO_2 Phasing During Deglaciation

One important question in orbital theory is what proportion of the large amplitude of glacial-cycle temperature change is due to direct insolation forcing and how much to CO_2 amplification. Detailed analysis of the temperature-CO_2 phase relationships from Antarctic ice cores show that CO_2 concentrations lag regional and global temperature during all glacial Terminations studied in detail (e.g., Genthon et al. 1987; Barnola et al. 1987, 1991; Fischer et al. 1999; Monnin et al. 2001; Caillon et al. 2003; see also the review in Raynaud et al. 2000). Figure 5.15 shows temperature-CO_2 curves for Terminations I and II from Taylor, Vostok, and Byrd ice cores (Fischer et al. 1999). These authors estimate that CO_2 lags by 400–1000 years during the course of a glacial termination. Uncertainty stems mainly from the air-ice age difference because of different close-off times, which are a function of the snow accumulation rate. During Termination III, Fischer et al. (1999) estimated that CO_2 lagged temperature by 600 ± 200 years, an estimate refined to 800 ± 200 using argon isotopes from air as a temperature proxy of trapped air (Caillon et al. 2003). The Dome Fuji record supports the CO_2 lag seen in Vostok, EPICA, and Byrd cores and suggests that the trigger for the glacial-interglacial cycles is northern-hemisphere insolation with amplification by CO_2 (Kawamura 2003, 2007).

What Caused Low Glacial CO_2 Concentrations?

The 80-ppmv rise in atmospheric CO_2 concentrations during late Pleistocene terminations occurred over about 10,000 years; anthropogenic CO_2 emissions represent a 100-ppmv rise in CO_2 in about a century. A large proportion of anthropogenic carbon—118 out of a total of 244 Petagrams (Pg) C since the year 1800—has been taken up by the oceans (Siegenthaler and Sarmiento 1993; Sabine et al. 2004; Mikaloff Fletcher et al. 2006), leading to acidification, expansion of the $CaCO_3$ undersaturated zone, carbonate dissolution, and biological impacts on carbonate secreting organisms, among other changes to the oceans' carbonate chemistry (Feely et al. 2004; Kleypas and Langdon 2006). The fate and impacts of future CO_2 emissions is still poorly constrained and demands a better understanding of the oceans' role in CO_2 cycling during glacial cycles. In fact, understanding the mechanisms underlying the astounding regularity of the 100-ka glacial-interglacial CO_2 cycles has been called the holy grail of paleoclimatology (Sigman and Boyle 2000).

Carbon is exchanged among several reservoirs—the lithosphere (carbonate sediments, including coral reefs, deep-sea pelagic oozes, and other rocks in the earth's crust), atmosphere, riverine dissolved carbonate, the terrestrial biosphere, dissolved carbon in oceans, and the ocean biosphere—each in theory a potential sink and source during glacial-interglacial transitions. There is some consensus that the oceans were the main reservoir that stored CO_2 during the Pleistocene glacials and from which CO_2 was released to the atmosphere during terminations.

The ocean-CO_2 sequestration hypothesis goes as follows (Sigman and Boyle 2000). Excluding the lithosphere as a glacial carbon sink because of the long timescales involved with lithospheric processes, the terrestrial biosphere is also an unlikely sink for carbon because continental ice sheets covered large areas of the northern hemisphere, creating completely different biomes from those of interglacials. The glacial terrestrial biosphere was consequently a source, not a

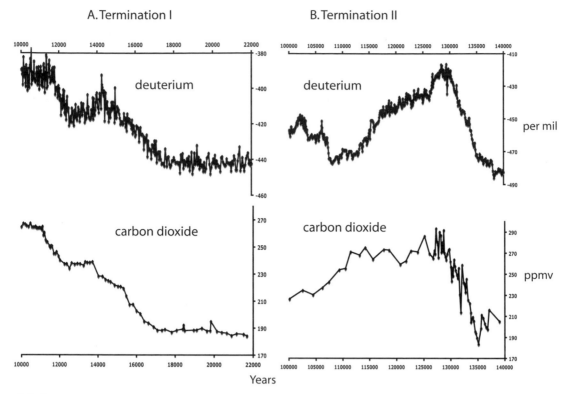

A. Termination I

B. Termination II

deuterium

deuterium

per mil

carbon dioxide

carbon dioxide

ppmv

Years

FIGURE 5.15 Comparison of deuterium isotope temperature proxy (top curve) and atmospheric CO_2 (bottom curves) for (A) Termination I from the EPICA Dome C (EDC) Ice core, from Monnin et al. (2001) and Jouzel et al. (2007), and (B) Termination II from the Vostok ice core from Petit et al. (1999) and Fischer et al. (1999). The ice-air age difference is several millennia, depending on the interval in the cores (i.e., the EDC at the Last Glacial Maximum is about 5500 years). Age models are subject to revision and precise timing of temperature, and CO_2 change is subject to uncertainty, but evidence from these cores and records from the Taylor and Byrd ice cores (Fischer et al. 1999; Indermühle et al. 1999, 2000) and Dome Fuji (Kawamura et al. 2003) show that the deglacial rise in CO_2 concentrations lags warming by about 400–800 years during Terminations I and II.

sink, of additional atmospheric carbon. Estimates from biome reconstruction are 700–1350 Pg C (Crowley 1995), those from depleted $^{13}C/^{12}C$ ratios in glacial foraminiferal $CaCO_3$ shells yielding a somewhat lower range, from 300–700 Pg C. All other things being equal, a decreased glacial-age terrestrial biosphere might add about 45 ppmv worth of CO_2 to the ocean-atmosphere system. Carbon dissolved in the ocean would be buffered by carbonate dissolution, which would partially offset this atmospheric CO_2 rise by an estimated 15 ppmv. Still, a net increase in glacial-age atmospheric CO_2 would occur if other reservoirs were not involved.

This leaves the oceans, which exchange CO_2 with the atmosphere through three primary processes known informally as the carbonate, solubility, and biological pumps. The carbonate pump refers to the precipitation and dissolution of biogenic carbonate by marine organisms, expressed as

$$CaCO_3 \text{ (calcium carbonate)} + CO_2 + H_2O \rightarrow Ca^{2+} + 2HCO_3^-$$

Berger (1982) proposed the coral-reef hypothesis, which holds that during deglacial periods when sea level rises, con-

tinental shelves are flooded and shallow-water biogenic carbonate forms largely through coral-reef skeletal growth, causing degassing of CO_2 from ocean to atmosphere. This idea is called the basin-to-shelf transfer of carbonate because carbonate deposition is shifted from the deep sea to shallow shelf regions. Deep-sea glacial-age carbonate sediment accumulation rates are generally thought to be too low to account for a CO_2 drawdown as large as the ice cores indicate (Archer and Maier-Reimer 1994).

Ocean-atmosphere exchange of carbon is governed by solubility (the partial pressure of CO_2 gas in the atmosphere and dissolved CO_2 in the ocean). CO_2 is dissolved into seawater if pCO_2 is lower than that in the atmosphere. The preindustrial ocean atmosphere is considered to have been in a steady state in terms of air-sea CO_2 exchange, with dissolution in some regions offsetting CO_2 release in others (Feely et al. 2004; Sabine et al. 2004). The solubility of CO_2 and its relationship to dissolved inorganic carbon species is expressed as

$$CO_2 \text{ (aq)} + H_2O \rightarrow H_2CO_3 \text{ (carbonic acid)}$$
$$\rightarrow HCO_3^- \text{ (bicarbonate)} + H^+ \rightarrow CO_3^{2-} \text{ (carbonate)} + 2H^+$$

In idealized circumstances (a homogeneous ocean), CO_2 is more soluble in colder water, so the mean ocean solubility of CO_2 would have been higher during glacials than during interglacials. Although more precise and geographically distributed estimates for glacial-age SST and BWT are needed, Sigman and Boyle (2000) estimated that high- and low-latitude cooling of 2.5°C and 5°C would lower atmospheric CO_2 by ~30ppmv. However, the ocean temperature-CO_2 solubility relationship is temporally dynamic over orbital and millennial timescales because of feedbacks between radiative CO_2 forcing and ocean chemistry and regional ocean temperature variability (high- versus low-latitude, surface versus deep temperature) (Archer et al. 2000). Martin et al. (2005) refers to an "enhanced solubility relationship" during deglacial periods, perhaps due to the effects of ocean alkalinity, differential cooling in high southern-latitude oceans, sea-ice feedbacks, and nutrient-related chemical and biological changes.

Another factor in the ocean's ability to store carbon is its salinity. Continental ice sheets stored large quantities of water on land, leading to elevated mean ocean salinity about 3% higher than interglacial values. Higher salinity would work in the opposite direction of decreased temperatures and might have raised atmospheric CO_2 by 6.5ppmv. Thus, the net terrestrial biosphere (+15ppmv)-temperature (−30ppmv)-salinity (+6.5ppmv) effects would cause only an approximate −8.5ppmv glacial decrease in CO_2.

The biological pump is expressed as

$$CH_2O + O_2 \rightarrow CO_2 + H_2O$$

This equation represents the production of organic matter (CH_2O) through photosynthesis carried out in the oceans with primary nutrients phosphate (PO_4^{3-}), nitrate (NO^{3-}), iron (Fe), and silicate (H_4SiO_4). Soon after the discovery of reduced glacial atmospheric CO_2 concentrations in ice cores, many hypotheses were offered to explain the ice-core data. Some involved the ocean and perhaps the biological pump as a mechanism to draw down carbon from the atmosphere and sequester it in the oceans (Broecker and Peng 1982; Knox and McElroy 1984; Sarmiento and Toggweiler 1984; Siegenthaler and Sarmiento 1993; see also Sundquist and Broecker 1985). There are now several generations of ocean, climate, and carbon-cycle models (Sarmiento et al. 1998, 2004), improved observations regarding anthropogenic CO_2 (Caldeira et al. 1999), and paleoceanographic reconstructions (Farrell et al. 1995; Francois et al. 1997; Frank et al. 2000; Moore et al. 2000; Robinson et al. 2004). Many lines of evidence point to the high-latitude Southern Ocean as the location of enhanced glacial CO_2 drawdown.

Several mechanisms invoke enhanced surface ocean productivity through greater nutrient availability or more efficient nutrient utilization (Sigman and Haug 2003). One factor that might enhance productivity would be greater airborne nutrients, in particular iron (Fe), which might "fertilize" now relatively unproductive regions of the world's oceans (Martin 1990). John Martin's iron fertilization hypothesis holds that certain regions of the world's oceans, called high-nutrient-low-chlorophyll (HNLC) because of anomalously low chlorophyll concentrations despite high concentrations of nitrogen and phosphorous, would be more productive if additional iron were available. Iron can be a limiting nutrient, and during glacial periods, which were drier and dustier with large land areas exposed to wind, aeolian sediment flux to the oceans could theoretically boost productivity. Modern HNLC regions under this hypothesis may have been more productive during glacials, and several ocean enrichment experiments have been carried out to artificially fertilize HNCL regions and simulate the response of the biological pump (Buesseler et al. 2004; Coale et al. 2004; Buesseler and Lampitt 2008). Iron enrichment does induce increased particulate organic matter, but the artificial fluxes were generally found to be relatively small compared to natural blooms and not capable of drawing down significant amounts of anthropogenic CO_2 from the atmosphere. Experiments were nonetheless limited in temporal and spatial scope and do not answer the broader question about glacial-age iron fertilization, and Pollard et al. (2009) recently found evidence for natural iron fertilization and enhanced carbon export to deep water in the Southern Ocean, supporting Martin's hypothesis.

Greater utilization of nutrients other than iron, in particular NO^{3-}, might also lead to carbon sequestration in the oceans. Broecker and Henderson (1998) proposed that iron flux from aerosol dust might increase the rate of nitrogen fixation by increasing the available stock of NO^{3-}. Nitrogen isotope ratios ($^{15}N/^{14}N$) in sediments are a proxy for nutrient utilization patterns during the last glacial period because phytoplankton take up relatively more ^{15}N during photosynthesis. In one study of circum-Antarctic sedimentary $^{15}N/^{14}N$ ratios using a new method to measure nitrogen isotopes, Robinson et al. (2004) found no large difference between glacial and modern nitrate utilization in the Atlantic sector and only 20% higher glacial values in the Indian Ocean sector.

Figure 5.16 shows one scenario for LGM circulation, carbon cycling, and nutrient utilization, from Sigman and Boyle (2000), that accounts for reduced glacial atmospheric CO_2 concentrations as compared to the modern interglacial. It features LGM deepening of the oxygen minimum zone, enhanced circulation in warmer shallower water masses, and more sluggish deep circulation, greater nutrient (NO_3^-) utilization in the Subantarctic Ocean, and enhanced atmospheric dust near the Polar Front. In essence, the LGM experienced greater organic carbon export in the Subantarctic, reduced nutrient export to low latitudes via Antarctic Intermediate and Subantarctic Mode Water, and less CO_2 flux from the Southern Ocean to the atmosphere.

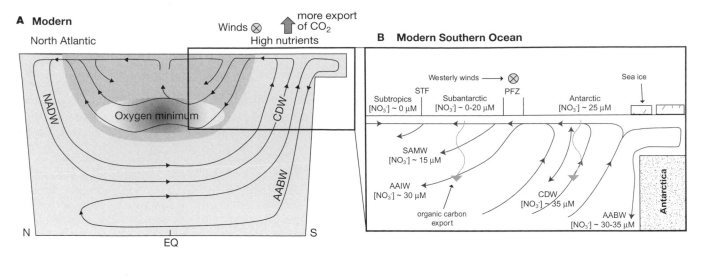

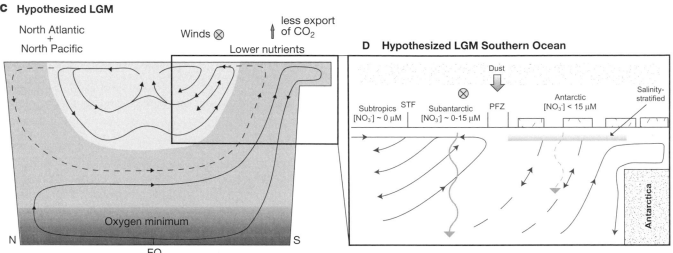

FIGURE 5.16 Comparison between modern and last glacial maximum (LGM) ocean circulation and hypothesized changes in atmosphere-ocean carbon exchange, modified from Sigman and Boyle (2000): (A) Modern Ocean; (B) enlargement of Southern Ocean; (C) LGM Ocean; (D) enlargement of Southern Ocean LGM. The Southern Ocean nutrient utilization hypothesis holds that oceans around Antarctica were more productive because of more abundant or better-utilized nutrients, or because of greater ocean stratification in the Subantarctic region, than today, and thus they sequestered more carbon, explaining lower CO_2 concentrations observed in ice cores. The hypothesis also involves changes in wind and dust deposition in the Polar Front Zone (PFZ). AABW=Antarctice Bottom Water; AAIW=Antarctic Intermediate Water; CDW=Circumpolar Deep Water; EQ=equator; NADW=North Atlantic Deep Water; SAMW=Subantarctic Mode Water; STF=Subtropical Front. Courtesy of D. Sigman.

This model is consistent with some oceanographic observations and model simulations. For example, Marinov et al. (2006) call the partitioning of biological production in the earth's Southern Ocean in two near-concentric oceanographic belts around Antarctica the "biogeochemical divide." Using ocean biogeochemical modeling simulations and chemical tracers, they found that carbon export to the deep sea, and thus ocean-atmosphere CO_2 exchange, is controlled in regions of deep-water formation near the Antarctic continent, whereas global export production is controlled by intermediate and deep-water formation in Subantarctic

Mode Water, a belt of water 40–55° south latitude, that carries nutrients to large parts of the ocean's thermocline, fueling about 75% of biological production in lower latitudes (Sarmiento et al. 2004). In other words, deep and shallow ocean circulation, as well as nutrient availability, plays an important role in carbon uptake, and low-latitude ocean regions may contribute to CO_2 uptake.

The Southern Ocean–nutrient utilization model of glacial CO_2 drawdown is one hypothesis, with still unresolved factors. Another explanation involves greater stratification in the upper layers of the Southern Ocean leading to greater sequestration of CO_2 during glacial periods (Toggweiler 1999; Sigman et al. 2003, 2004). It has also been proposed that circum-Antarctic sea ice may have been a factor in reduced ventilation in the Southern Ocean (Stephens and Keeling 2000; Keeling and Stephens 2001; Hillebrand and Cortese 2006), and some support for such a scenario exists in mid-depth paleoceanographic evidence in the eastern Pacific Ocean off Baja California (Marchitto et al. 2007).

In another important study, Anderson et al. (2009) provided strong evidence from biogenic opal in sediments that the release of CO_2 from the oceans following the LGM involved wind-driven upwelling around Antarctica. Their analyses of cores from various sectors of the Southern Ocean linked ventilation of deep water forming around Antarctica with the regulation of atmospheric CO_2 concentrations.

Another idea is that photosynthetic carbonate-secreting coccoliths compete for nutrient resources with siliceous diatoms and, under different oceanographic conditions, the concentrations of H_4SiO_4. A recent volume was devoted to the role diatoms play in ocean nutrient biogeochemical cycling (Kemp and Dugdale 2006). Diatoms produce opal, and a number of new diatom-based proxies have been applied to opal-bearing sediments (De La Rocha 2006). Silica-limiting oceanographic conditions in the subsurface equatorial Pacific highlights the connectivity between biological, nutrient, and ocean-circulation processes in controlling air-to-sea carbon flux (Sarmiento et al. 2004). Another factor is that the ultimate fate of organic matter in the oceans and its export to the deep ocean is influenced by a variety of other complex, partially known, ecological and physical processes, including top-down "grazing" mainly by copepods, and flocculation of organic material in the water column during descent. Other actively researched topics pertaining to glacial carbon export and storage in the oceans include the impact of large-scale ocean-circulation changes during glacial periods (Lynch-Stieglitz et al. 2007 and the impact of vertical diffusive mixing in lower-latitude surface-deep-ocean water mass (Archer et al. 2000).

Before proceeding, we note an important alternative to the ocean-storage hypothesis of glacial CO_2 called the permafrost hypothesis. It holds that the continents may have stored significant amounts of glacial-age carbon in frozen loess deposits (called yedoma in Siberia), other permafrost, and peatlands (Sher et al. 2005). Based on modern carbon content measurements from thick permafrost deposits, Zimov et al. (2006) estimate these three related sinks might have held 500, 400, and 50–70 Pg C, respectively. They suggest that a continental source of old carbon depleted in radiocarbon (^{14}C) might explain some of the decreases in atmospheric radiocarbon during deglaciation (see Chapter 7 in this volume). Large reservoirs of modern permafrost carbon pose a serious poten-tial positive feedback as high-latitude northern-hemisphere regions warm (see Chapter 12 in this volume).

Orbital Hypotheses and Mechanisms

Paradoxes, enigmas, and mysteries notwithstanding, many hypotheses exist to explain orbital-scale climate cycles (Table 5.1). We outline here some of the more prominent ones, placing them into six quite mutable and overlapping categories for convenience. At one extreme, some theories that hold that apparent orbital patterns in the geological record actually reflect stochastic, unforced processes and ice ages can come and go, at least in computer model simulations, without any insolation forcing. At the other extreme, many hypotheses call for orbital insolation as a trigger, with feedback mechanisms and climate thresholds—some previously mentioned here—acting as amplifiers that transmit the signal throughout the climate system. Between these extremes, one might argue that glacial cycles occur without orbital insolation forcing but that the orbital cycles control the phasing, frequency, or both of the cycles, that is, they influence when terminations and glacial inceptions will occur.

Clearly competing hypotheses are not mutually exclusive of one another, and no single mechanism can account for climate variability over the last 5 Ma. Hypotheses differ more in emphasis on particular feedbacks and thresholds in the climate system, or focus on a particular time interval. Our emphasis here is on conceptual hypotheses of reconstructed variability, some of which have been formally analyzed in climate model studies. Comprehensive discussions of the mathematics, spectral properties, and physics of climatic components and feedback mechanisms of orbital models can be found in Hinnov (2000), Saltzman (2002), and Elkibbi and Rial (2001).

Free Oscillations, Stochastic Noise

Proponents of hypotheses lumped into this category suggest that cyclic paleoclimate and associated orbital patterns do not necessarily reflect deterministic forcing from insolation, although insolation might sometimes play a role. Although in the minority, these opinions deserve consideration. One class of models holds that the earth's climate exhibits a free oscillation of the atmosphere-ocean-cryosphere and lithosphere (Le Treut and Ghil 1983; Saltzman and Sutera 1987), and attempt to simulate the interaction between orbitally forced and "free" oscillations over a certain period. The idea of free oscillations is illustrated by the potential effect of large temperate ice sheets on the climate system. Some models indicate that when ice thickness exceeds a critical value,

TABLE 5.1 Orbital Theory and Mechanisms

Mechanism Name	Reference	Aspect of Orbital	Evidence*	Processes	Other Reference
Ice-volume rate change	Roe 2006	100-ka	Oxygen isotopes, insolation	Support for Milankovitch model, secondary CO_2 role	
Carbon burial	Zeng 2003, 2007	100-ka	Modeling	Carbon-climate feedbacks	
Antarctic sea ice	Raymo et al. 2006	41-ka cycles and MPT	Oxygen isotopes	MPT shift to Antarctic marine-ice margins, inphase N/S hemisphere, stronger 23-ka cycle	
Climate continuum	Huybers and Curry 2006	Various	Modeling	Annual to orbital frequencies	
Integrated summer insolation	Huybers 2006	41-ka cycles	Climatology, energy balance	Intensity and duration of insolation controls 41-ka cycles	
Fast-ice feedback	Ruddiman 2006a, b	100- and 41-ka	Paleoclimate	Ice albedo and CO_2 feedbacks on precession and obliquity	
Late spring albedo flip	Hansen et al. 2007	Sawtooth pattern	Energy balance, paleoclimatology	Water vapor, aerosols, clouds, sea ice, snow feedbacks	
CO_2 variability	Peacock et al. 2006	100-ka CO_2 variability	Multiple	Falling sea level increases nutrients and alkalinity, ocean circulation during deglaciation	
CO_2 amplification	Medina-Elizalde and Lea 2005	MPT	Paleoceanographic	After MPT shift, greater GHG forcing	
Obliquity	Huybers and Wunsch 2005	100-ka cycles	Modeling	Termination occurs every 2d or 3d obliquity	
Tectonic/gateway threshold	Mudelsee and Raymo 2005	4–2.4 Ma transition			
Walker zonal cell intensifications	McClymont and Rosell-Melé 2005	MPT	Paleoceanographic	1.1–0.9 Ma Walker circulation intensifies	
Tropical SST, CO_2 amplification	Lea 2004	100-ka	Paleoceanographic		
Obliquity forces tropical SST	Liu and Herbert 2004	MPT	Paleoceanographic (SST, productivity)	Pre-MPT tropical Pacific shows 41-ka cycles	Liu et al. 2005
Tropical forcing during MPT	Schefuss et al. 2004	MPT	Paleoceanographic (SST)	Trpical-S, hemisphere coupling, sea-ice important	
Precessional forcing of tropics	Clement et al. 2004				
Stochastic, internal forcing	Wunsch 2003b, 2004	100-ka	Modeling, statistical	Small contribution from insolation	
Internal, free oscillation	Saltzman 2002	several	Modeling	Review of orbital models, carbon cycle, ice-sheet feedbacks	Saltzman and Maasch 1991
Global carbon-cycle amplification	Shackleton 2000		Oxygen isotopes, ice core	Ice-sheet dynamics not sufficient	

(continued)

TABLE 5.1 (*continued*)

Mechanism Name	Reference	Aspect of Orbital	Evidence*	Processes	Other Reference
Tropical role, semiprecession cycles	Rutherford and D'Hondt 2000	100-kyr	Phase demodulation	10-kyr semiprecession cycles	
Frequency modulation by 413-kyr cycle	Rial 1999				
Multiple climate state	Paillard 1998, 2001	Several	Deglacial thresholds	Ocean circulation and CO_2 feedbacks	Parrenin and Paillard 2003
	Muller and MacDonald 1997a, b				Muller and MacDonald 2000
Precession forced nonlinear ice-sheet feedback	Imbrie et al. 1992, 1993	Sawtooth and 100-kyr	Multi-proxy	Ice-sheet dynamics	
Tropical thermostat	Clement et al. 1996		ENSO climate model	Sea ice, deep-sea cooling after MPT	
Marine-based ice sheets	Raymo et al. 2006	MPT	Paleoceanographic	MPT shift from terrestrial to marine ice-sheet margins	
Regolith-ice dynamics	Clark et al. 2006	MPT	Multiproxy, glaciological	Subglacial conditions influence ice dynamics	
CO_2-tropical	Medina-Elizalde and Lea 2005	MPT	Equatorial Pacific SST	41–100-kyr shift at 950 ka, SST leads ice volume by 3 kyr, WEP, EEP coherence in SST indicates CO_2 forcing	
Sea-ice switch	Tziperman and Gildor 2003	MPT	Modeling	Ablation, accumulation changes with deep-ocean cooling	Gildor and Tziperman 2000
Ice-sheet calving, bedrock	Mudelsee and Schultz 1997	MPT	Modeling	100-kyr cycle lags ice-volume increase	
Global aridity, Terrestrial C influx	Raymo et al. 1997	MPT	Foram carbon isotopes		
	Berger and Jansen 1994	MPT	Foram oxygen isotopes		
Orbital and greenhouse feedbacks	Ruddiman 2007	MIS 11	Multiple	MIS 11 closest analog to MIS 1, which has anomalous GHG trends	
Ocean carbonate chemistry	Broecker and Stocker 2006	MIS 11	Geochemical	MIS 1 CO_2 explained by natural carbonate chemistry	Barker et al. 2006
Modeling CO_2, paleoclimate	Crucifix et al. 2005	MIS 11	Climate-ice-sheet model	MIS 11 slightly different from MIS 1	Loutre and Berger 2003
Coral-reef hypothesis	Berger 1982	Termination 1, later MIS 11	First aimed at CO_2 rise during last deglaciation	Shallow carbonate production increases during transgression	Opdyke and Walker 1992

Note: See reviews: Berger and Loutre (1991); Elkibbi and Rial (2001); Saltzman (2002). EEP = Eastern Equatorial Pacific; ENSO = El Niño–Southern Oscillation; GHG = greenhouse gases; kyr = kiloyear; MIS = Marine Oxygen Isotope Stage; MPT = Mid-Pleistocene Transition; N/S = northern/southern; SST = sea-surface temperature; WEP = Western equatorial Pacific.

*Most modeling studies use a deep-sea oxygen isotope record as a proxy.

oscillations with periods ranging from 70 to 130 ka appear (Oerlemans 1982; Ghil and Childress 1987). Saltzman and Maasch (1991) and Saltzman and Verbitsky (1993) incorporated several "slow responders" in the climate system—CO_2, isostatic adjustment of ice-covered continents and ice-calving—into several versions of a dynamic model. One conclusion offered by Saltzman is that the 100-ka pattern is a free oscillation that is paced by Milankovitch forcing; but it is not a necessary condition for the existence of the 100 ka cycle. Saltzman (2002) presents a full account of the physics incorporated into his generalized theory of global climate change.

Wunsch (2003b, 2004) presents a case that the majority of variability observed in some ice-core and deep-sea paleoclimate records can be generated as a random walk exercise, with only minor contribution (~11%) from obliquity and less from precession to total variance. The orbital contribution depends on the length of the record analyzed, but in all situations it was a relatively small percentage of total variance. Wunsch makes an analogy to unforced internal modes of variability, such as the El Niño–Southern Oscillation and North Atlantic Oscillation, which operate over annual to multidecadal timescales, despite the obvious diurnal and annually forced cycles of weather (see Chapter 10 in this volume). Until the null hypothesis is proved incorrect, Wunsch considers an "unforced" hypothesis of long-term orbital-scale climate changes as an option. More fundamentally, Wunsch (2004) points out that if insolation changes do not drive large changes in climate, i.e., if they do represent unforced "noise," then the problem of which mechanisms carry the orbital insolation signal through the climate system disappears.

Orbital Inclination and Accretionary Dust

The orbital inclination–accretionary dust model was proposed to explain the 100-ka cycles of the last 700 ka in a series of publications by Muller and MacDonald (1995, 1997a, b). The essence of their argument is that, while the dominant 41-ka cycle between 3 and about 1 Ma is consistent with orbital insolation forcing, the 100-ka pattern in deep-sea isotopic records matches changes in the inclination of the earth's orbital plane every 100 ka. The hypothesis has three parts. First, the inclination of the earth's orbital plane varies cyclically every hundred thousand years, like the 100-ka eccentricity cycle. Second, inclination cycles influence the volume of interplanetary dust particles (IDP) originating in the zodiacal cloud enveloping parts of the solar system that reaches the earth. IDP flux can reach 3×10^7 kg annually and varies over 100-ka cycles by a factor of two to three (Kortenkamp and Dermott 1998). Third, helium isotopes (^{3}He) in deep-sea sediments were considered a proxy for IDP accretion rates and supported the inclination-accretion hypothesis for the

last few hundred thousand years (Farley and Patterson 1995). Muller and MacDonald's basic conclusion is that ice-volume cycles of the past 700 ka are primarily driven by inclination and perhaps through its influence on IDP flux or meteoric clouds. Their theory does not preclude obliquity and precession forcing of other climatic cycles, such as the 41-ka ice volume of the Pliocene and late Pleistocene.

The significance of the helium sedimentary record as a proxy for IDP accretion was questioned by Marcantonio et al. (1996), whose paired helium-thorium isotopic analyses showed that helium variability in sediments was related to changes in continental sources and deep-sea sediment transport processes (see Higgens et al. 2002). Spectral analyses of isotopic records by Rial (1999) also suggested that, instead of an inclination-accretion model, a kind of frequency-modulation hypothesis could account for the 100-ka cycle. He argued that the 100-ka cycle could be explained as Milankovitch-like insolation forcing in which the 100-ka frequency is modulated by the stronger 413-ka orbital cycle. The frequency-modulation model also explained the spacing of 100-ka cycles and the lack of the 413-ka cycle, at least in late Cenozoic records, although Rial (2004) later was able to "tease out" the 413-ka signal.

Thresholds and Millennial Climate Events

Paillard (1998, 2001) presented a general, modified Milankovitch theory emphasizing thresholds in the climate system to explain abrupt glacial terminations in particular and orbital cycles in general. Drawing heavily on suborbital climate variability clearly not driven by orbital insolation (see chapters 6 and 7 in this volume), a simple temperature-ice-volume relationship is less important than climate behavior in response to critical thresholds. For example, Paillard postulated that if southern-hemisphere temperature and carbon cycling in the Southern Ocean is linked, as ice-core and other records suggest, then abrupt climate terminations will occur via hysteresis as proposed for the Cenozoic Antarctic ice sheet (De-Conto and Pollard 2003a, b) and via meridional overturning circulation during glacial terminations (Rahmstorf 1995). Paillard's hypothesis is one of several holding that greenhouse gases amplify an initial insolation signal in the southern hemisphere, at least for the past few climatic cycles.

Schultz and Zeebe (2006) introduced the "insolation canon" hypothesis to explain the timing of terminations since 700 ka. The central theme involves thresholds in timing and energy flux during high-latitude insolation maxima. They analyzed midsummer insolation—June at 65°N and December at 65°S latitude—and discovered overlapping northern- and southern-hemispheric energy maxima, with a slight southern-hemisphere lead. Late Pleistocene energy maxima surpassing threshold levels of 1000 years in duration and 0.95 TJ (10^{12} joules) m^{-2} in energy flux coincided precisely with the onset of Terminations VII through I at 632, 546, 419, 345, 252, 139, and 23 ka. The precise timing was caused by the combined interplay of both obliquity and precession, and Schultz and Zeebe concluded that this is why terminations occur in multiples of precession cycles. There is no such coincidence between obliquity and precession forcing from about 2 Ma up until the MPT. Possible feedback responses to insolation triggering probably involve land and sea ice and deep-ocean-circulation responses.

Regolith-Ice-Sheet-Carbon-Cycle Feedbacks

Ice-sheet continental bedrock interactions have long been a feature of modeling efforts such as those of Saltzman to understand late Pleistocene glacial periods. The regolith hypothesis of Clark et al. (2006) calls for the combined influence of ice-sheet erosion and thickness to explain the MPT. The first tenet of the regolith model is that progressive erosion of crystalline bedrock of the North American Precambrian shield created a high-friction interface between the basal ice sheet and the bedrock, allowing thicker ice sheets to grow since about 900 ka when high-amplitude 100-ka cycles began. Thicker ice sheets respond differently to orbital forcing, in part because of their distinct dynamics in basal layers and around their margins where they are in contact with the oceans. Second, it is estimated that a faster rate of silicate weathering of crystalline basement rocks would lead to a decrease in atmospheric CO_2 concentrations, perhaps as much as 7–12 ppmv. The weathering-CO_2 drawdown-climate cooling link described in Chapter 4 explains some aspects of long-term Cenozoic cooling. Third, there is a role for ocean circulation in the regolith model in that deep-sea isotopic evidence suggests land-to-sea organic carbon transfer during the MPT (Raymo et al. 1997). Clark's model holds that the source of this carbon was the continental shelf and upper slope, which became subaerially exposed for the first time in millions of years during the unprecedented low sea levels caused by thicker ice sheets. Thus, this theory incorporates several complex feedbacks and mechanisms including continental geology, sea-level and ice volume, ice sheet ocean dynamics, and carbon cycling and greenhouse gas forcing to explain the 41–100-ka shift in orbital forcing.

Slow and Fast Feedbacks

Several theories involve feedback mechanisms that amplify an initial orbital warming through their effects on radiative budget, albedo, and other climatic processes. We have seen that there are several "slow feedback" mechanisms in the climate system—ice sheets, terrestrial vegetation, and CO_2—that depend on inertia, biological processes, and ocean storage and circulation. Several theories involve a role for CO_2 forcing of long-term late Cenozoic cooling and amplification of the 100-ka insolation cycle or both. Shackleton (2000) postulated CO_2 amplified orbital cycles based largely on the phasing and spectral characteristics of various proxies of ice volume, northern- and southern-hemisphere climate, deep-ocean circulation, and greenhouse gas concentrations. Ruddiman (2006a, b) defended the idea that the 100-ka pattern result from complex interplay of several orbital processes, contrasting the 22-ka, 41-ka, and 100-ka orbital forcing as follows. First, a 41-ka forcing of northern-hemisphere ice sheets, which lag insolation by several thousand years, is amplified by ice-sheet and CO_2 feedbacks. This 41-ka cycle was dominant before the MPT. Summer obliquity forcing may also induce rapid response in the southern hemisphere. An ~22-ka precession forcing also influences high-latitude temperature and ice-sheet response, but the main impact of precession is an "early" response methane flux from low- and midlatitude wetlands. Methane flux potentially amplifies the insolation signal.

In contrast, the 100-ka late Pleistocene pattern originates from precession, obliquity forcing with strong CO_2 and ice-sheet feedbacks, or both. An important element in Ruddiman's hypothesis is the concept of "fast" ice-sheet-greenhouse gas (CO_2, CH_4) feedbacks in the 41-ka obliquity insolation band and greenhouse gas forcing, especially methane, during 22-ka precession forcing (Ruddiman 2006a). The details surrounding ice-sheet and greenhouse gas impacts involve various still partially understood additional factors such as atmospheric dust, deep-ocean circulation and chemistry (alkalinity and carbonate chemistry), and Southern Ocean sea ice.

Zeng (2003, 2007) proposed a view of the 100-ka cycle in which carbon from terrestrial vegetation is buried under ice sheets during glaciation but is released rapidly once ice sheets reach a critical size. If carbon release is fast enough, it cannot be taken up by the oceans and remains in the atmosphere, forcing deglaciation. The interglacial state is temporary, lasting only until atmospheric carbon is eventually stored in the oceans and a glacial inception occurs. Essentially, this hypothesis involves an extremely strong carbon-cycle feedback mechanism to amplify weak insolation forcing. A self-sustained quasi-cyclic 90–100-ka pattern of ice-sheet growth and decay was simulated in Zeng's coupled carbon-cycle-climate-ice-sheet model.

Fast-feedback mechanisms include clouds, aerosols, water vapor, sea ice, and snow. Hansen et al. (2007) hypothesized an "albedo flip" sea-ice mechanism to explain climatic cycles of the last 700 ka of climate, with emphasis on rapid terminations, radiative forcing in the tropics, and sea-ice feedbacks. In essence, rapid terminations during orbital cycles represent a critical shift from snow and ice (high reflectivity) to water (low reflectivity), and Hansen et al. proposed that spring, rather than summer insolation, was the critical variable in sea-ice, ocean-atmosphere dynamics during abrupt climate transitions. What is most important is that albedo provides a fast mechanism in contrast to many slow responders such as ice sheets.

Raymo et al. (2006) theorized that the dominant 41-ka patterns in oxygen isotope records and the lack of precession between 3 and 1 Ma signified late Pliocene-early Pleistocene ice-volume changes driven by both northern- and southern-hemisphere terrestrial ice-sheet variability and sea-level fluctuations. Citing improved constraints on Pliocene sea level and Antarctic ice-sheet volume, they argued that East Antarctic Ice Sheet (EAIS) dynamics before the MPT involved mainly a terrestrial-based ice-sheet margin. They simulated the 41-ka world in a new global, ablation-driven ice-volume model forced by summer insolation at orbital timescales, with cool summers forcing northern-hemisphere ice-sheet growth and warm summers forcing Antarctic ice-sheet decay. This out-of-phase ice-sheet behavior skillfully simulated the first-order 41-ka ice-volume pattern seen in oxygen isotope record. The MPT shift represented a change in the EAIS to a more marine-based margin, the appearance of 23-ka precession forcing, albedo and CO_2 feedbacks in amplifying the insolation signal producing the higher amplitude, and spacing of the 100-ka cycles.

The importance of sea-ice extent is central to the model of Gildor and Tziperman (2000) and Tziperman and Gildor (2003) previously discussed herein. Their hypothesis holds that thermohaline circulation cannot account for rapid climate transitions and that sea-ice changes caused by any number of forcing factors (CO_2, ocean circulation, volcanic activity) involve critical thresholds that can be simulated and can account for both the 41-ka and 100-ka cycles. The ice-bedrock model proposed by Mudelsee and Schultz (1997) to explain the MPT also involves significant ice calving around ice-sheet margins.

Paillard and Parrenin (2004) offered a conceptual model featuring an "ocean-CO_2" switch mechanism to explain certain features of orbital patterns, including the abrupt terminations. This idea is based on growing evidence for dynamic climate/ice-sheet interaction in Antarctica during orbital cycles and sea ice, brine rejection, deep-water formation, and air-sea CO_2 exchange processes in the high-latitude Southern Ocean. They maintain that glacial deep-ocean water was saltier and denser and the greater stratification allowed the oceans to store larger quantities of CO_2 at depth. The critical switch from glacial to interglacial involves the Antarctic Ice Sheet, which extends onto the continental shelf as glaciation proceeds (see Chapter 7 in this volume).

At some threshold near glacial maxima, sea ice forms over deeper water off the shelf edge. As a consequence, the formation of salty deep water ceases after several millennia, the ocean becomes less stratified, and CO_2 can be released to the atmosphere, further enhancing insolation-induced northern-hemispheric warming. Interglacial to glacial transitions are explained by northern-hemisphere Milankovitch-type forcing. The Paillard/Parrenin model explains the MPT shift as related to changes in "bottom-water efficiency" due either to gradual tectonic, erosional, or other processes around Antarctica.

In sum, there are many dynamic and conceptual models that incorporate internal feedback mechanisms related to Southern Ocean sea ice, CO_2 sequestration, and ocean transport and storage into explanations of climate response to orbital insolation. It is thus interesting that, among the many monumental discoveries about climate change over the past 800 ka from the EPICA ice-core temperature (EPICA Community Members 2004), CO_2 (Siegenthaler et al. 2005), methane (Spahni et al. 2005), and glaciochemical sea ice records (Wolff et al. 2006), there is not unequivocal support for major changes in internal feedback mechanisms such as sea ice and albedo associated with the important transition during the MBE.

Tropical Climate and Orbital Cycles

A number of studies address the role of the tropics in orbital forcing. Plio-Pleistocene tropical ocean temperature records using relatively new alkenone and Mg/Ca ratios as paleothermometers show several common features. First, they firmly establish SST variability in tropical oceans at orbital timescales. Second, there is evidence for a 41-ka pattern during the early Pleistocene and little or no evidence of typical precession forcing of low-latitude temperature for the past 500 ka. Third, a climatic transition began at 1.4–1.2 Ma and culminated near ~900–950 ka in parts of the Atlantic and ~650 ka in the southeast tropical Atlantic, suggesting a clear "nonstationarity" in orbital forcing during the Pleistocene. Fourth, some authors theorize a strong tropical-and-southern-hemisphere linkage to explain paleoceanographic records.

From the perspective of tropical climate, there is some subjectivity in defining the mid-Pleistocene transition and different interpretations of mechanisms responsible for the 41-ka early Pleistocene cycles. Liu and Herbert (2004) reconstructed EEP SSTs since 1.8 Ma, showing a dominant 41-ka cycle between 1.8 and 1.2 Ma and a change to 23- and 100-ka periods around 900 ka. Early Pleistocene glacials experienced strong tradewinds and shoaling thermocline and the obliquity signal indicated a tight high- and low-latitude coupling. Liu et al. (2005) found a similar pattern with $\delta^{15}N$ isotopes from cores in the same region and suggested that changes in denitrification might influence nitrate inventory and thus control oceanic CO_2 uptake.

In the equatorial and south Atlantic, Schmieder et al. (2000) (see also Wefer et al. 1996) measured carbonate, magnetic susceptibility, and diatom proxies and discovered a 920-ka carbonate minimum. They proposed that the MPT between 920 and 540 ka was not just a transitional period but rather a distinct climatic state characterized by unique ocean circulation, carbonate dissolution, and climate. Schefuß et al. (2004) studied ODP Sites 1077 and 1092 and other cores from the Angola Basin and the southeast tropical Atlantic and found a dominant 41-ka signal before the MPT. The 41-ka phasing shows that tropical Atlantic SSTs lag the initial tilt forcing by 5.5 years, ice volume lags by about 8–9 ka, and finally NADW by 12–13 ka. This phasing generally supports Imbrie's model. They suggest that tropical southern-hemisphere coupling was stronger before the MPT. In contrast, during late Pleistocene 100-ka cycles, NADW preceded the O-18/ice-volume signal, suggesting that NADW plays a larger role in deglaciation after the MPT.

McClymont and Rosell-Melé (2005) studied alkenone records from ODP Sites 806 and 840 in the western and eastern equatorial Pacific (WEP, EEP). Progressive EEP cooling began about 1.2 Ma, culminating near 900 ka when a near-modern temperature gradient was reached. This transition signifies the onset and intensification of Walker atmospheric circulation and more La Niña-like conditions. Cooler EEP SST could be wind driven or reflect shoaling of the thermocline and colder deep-water source for upwelled water. The implication for orbital forcing is that an intensified Walker cell transports more moisture to high latitudes, fostering ice-sheet growth.

Based on the phasing of equatorial Pacific SSTs, Antarctic ice-core CO_2, and orbital insolation seen in many records (Pisias and Mix 1997; Lea et al. 2000; Visser et al. 2003). Lea (2004) concluded that CO_2 played a major amplification role in tropical SST variability during late Pleistocene 100-ka cycles. He used the paleoclimate record to better constrain the climate sensitivity values to 4.4–5.6°C. Medina-Elizalde and Lea (2005) also analyzed equatorial Pacific SST patterns for the entire Pleistocene, spanning the 41–100-ka MPT transition (Figure 5.13). These studies showed the WEP had long-term stability and no trend in SST, but there was EEP gradual cooling beginning ~1.34 Ma until ~900 ka, superimposed on the shift in orbital cycles. The equatorial Pacific SST gradient increased by ~1.3°C, almost totally from EEP cooling. They agree that thermocline shoaling explains EEP cooling, but dispute the moisture transport hypothesis of McClymont and Rosell Melé because the 400-ka-long transition period

had such a stable high-latitude hydrology and climate. The phasing during 41-ka cycles, where SST variability of 3–4°C in both WEP and EEP leads ice volume by ~3–7 ka, suggests a strong role for CO_2 forcing during early Pleistocene 41-ka cycles. The hypothesized role of CO_2 forcing prior to the MPT does not support the idea that large northern-hemispheric ice sheets were major factors in obliquity-scale SST variability and should be tested with new ice-core data older than 700 ka.

Ashkenazy and Tziperman (2006) modeled the phasing of tropical SST and ice volume and suggested a more complex situation than tropical oceans leading northern-hemisphere ice sheets. They suggest that if SST proxies represent a warm season and not annual SST, they may lead global temperature but not northern-hemisphere ice-sheet volume changes. They also point out issues regarding temporal resolution of the SST phasing, such that ice-sheet melting might begin before changes in SST and CO_2. These parameters "catch up" and surpass ice-sheet decay, creating a positive feedback and enhanced ice-sheet melting.

Clement et al. (2004) suggest that emphasis on ice volume, global temperature, and hydrology to explain orbital cycles is misplaced; instead, in the tropics precession forcing can operate independent of high-latitude climate process changes. Their model simulations of low-latitude precession forcing show that minor insolation changes can have large hydrological impacts expressed in changes of seasonal features of large-scale Walker and Hadley circulation. Glacial and precession forcing have distinct climatic responses. However, the modeling studies by Clement and colleagues raise a more fundamental point—that the search for regional leads and lags in climate response due to various feedback mechanisms may be misguided because certain regions such as the tropics may respond independently to regional insolation variability. This basically means that both local and large-scale response to insolation forcing must be taken into account to properly assess climate dynamics of orbital cycles.

Nonlinear Phase Locking

Throughout this chapter we discussed several physical mechanisms that might transmit small insolation forcing into large glacial cycles. The concept of nonlinear phase-locking, with roots in SPECMAP and modeling, might explain late Pleistocene ice-volume cycles regardless of the internal mechanism involved. Tziperman et al. (2006) explain this as follows: "Nonlinear phase locking may determine the phase of the glacial cycles even in the presence of noise in the climate system and can be effective at setting glacial termination times even when the precession and obliquity bands account only for a small portion of the total power of an ice volume record."

This idea has several key features. First, it means that even if there is noise in the climate system, Milankovitch-type forcing by precession and obliquity pace the timing of terminations. Second, it operates regardless of which internal nonlinear feedback mechanisms (sea ice, CO_2) amplify the initial signal. Third, it is distinct from the idea that the eccentricity signal is amplified by linear or nonlinear mechanisms. As is the case with many theories aimed at orbital forcing, rigorous testing requires accurate chronological control of proxy records to assess the phasing in different parts of the climate system as well as climate modeling to assess physical mechanisms. At present, accurate chronology is limited to a small number of proxy records, and spatial coverage of orbital-scale climate variability remains spotty. Consequently, more well-dated proxy records are needed to evaluate a phase-locking mechanism in segments of the climate system other than oxygen isotopic records of ice volume.

Improvements in Interpolar Phasing

In many studies of orbital forcing and climate response, the exact phasing relationship between insolation and northern- and southern-hemisphere climate change is unclear because of age uncertainty and assumptions involved with orbital tuning, with the exception for Termination I (see Chapter 7 in this volume). Kawamura et al. (2007) overcame this problem by measuring oxygen-nitrogen ratios in Dome Fuji and Vostok ice cores as an independent indicator of insolation. The O_2/N_2 ratio is a proxy for summer solstice insolation because O_2 molecules are excluded from the trapped air more than N_2 molecules during the transition from snow to ice. Thus, Kawamura's group tuned the O_2/N_2 ratios to orbital insolation in a manner similar to orbital tuning of deep-sea oxygen isotope records. They discovered that Antarctic climate lags northern-hemisphere insolation by several thousand years over the last 360 ka. Proxies for both Antarctic atmospheric temperature and CO_2 concentrations increase in response to northern summer insolation. They also found that the tuning of ice-core and deep-sea records emphasizing $\delta^{18}O$ of the atmospheric O_2 and ocean water (from $\delta^{18}O$ of foraminifera, Shackleton 2000) produced incorrect phase relationships because of complexities of oxygen isotopic fractionation in the atmosphere and ocean.

Their results provided some of the strongest support to date for Milankovitch theory, at least for the late Pleistocene glacials, and perhaps more importantly, a means for climate modelers to simulate the role of atmospheric CO_2 amplification of insolation forcing during terminations.

Perspective

Epochal discoveries offer the scientific community new proxy records to test hypotheses about orbital forcing and internal climate feedbacks for the past 5 million years. Tom Crowley made a telling comment about the breakneck pace of research on the orbital theory and greenhouse gas forcing: "The sheer number of explanations for the 100,000-year cycle and for CO_2 changes seem to have dulled the scientific community into a semipermanent state of wariness about accepting any particular explanation" (Crowley 2002). Our brief discussion in this chapter affirms this view. However, if there is any single thread connecting recent discoveries, it might be that climatic cycles occurring at orbital timescales cannot be assessed outside the context of abrupt climate transitions, multiple thresholds, and internal ocean, cryosphere, biogeochemical, and atmospheric processes.

It can be argued that the most important question surrounding orbital climate change also pertains to an elementary issue facing climate science today. Does orbital theory predict that the earth's climate would be entering a glacial state had not humans disrupted the global carbon cycle, extracting carbon from sedimentary rocks, short-circuiting its long-term burial, and destroying large parts of terrestrial biomass, injecting CO_2 and methane into the atmosphere? Several researchers estimate that human events, unprecedented in rapidity in the geological record, when viewed from the standpoint of orbital forcing, might produce an "unnatural" extended interglacial period. Archer and Ganopolski (2005) suggest that the next glaciation may not occur as scheduled via orbital processes, even at CO_2 concentrations near 560 ppmv. Their model simulations show that the release of 5000 Pg of carbon from either fossil-fuel burning or marine methane hydrates, ignoring positive and negative carbon-cycle feedbacks and under projected future orbital configurations, would disrupt orbital cycles and possibly "prevent glaciation for the next 500,000 years," a period spanning the next two eccentricity minima. With this sobering thought, let us turn to the next chapter on suborbital climate changes.

LANDMARK PAPER Orbital Variations: Pacemaker of the Ice Ages

Hays, James D., John Imbrie, and Nicholas J. Shackleton. 1976. Variations in the earth's orbit: Pacemaker of the ice ages. Science 194:1121–1132.

The orbital theory of climate change has provided a greater understanding of climate dynamics, the causes of ice ages, ocean-climate-carbon cycling, and the biological impacts of climate than any other single aspect of paleoclimatology. Change in the geographical and seasonal insolation caused by gravity's influence in the solar system led to the growth and decay of continental ice sheets, sea-level changes exceeding 125 m, polar and tropical cooling of more than 10° and 3° C, and many other environmental changes over frequencies ranging from tens to hundreds of thousands of years during much of the earth's history.

The 1976 paper on Milankovitch orbital climatic cycles by three pioneers in paleoclimatology ranks among the most significant achievements in paleoclimatology and is a landmark application of geological and paleoclimate records to test astronomical hypotheses about external forcing of global climate. Hays, Imbrie, and Shackleton used deep-sea core records of planktonic foraminiferal oxygen isotopes as a global ice-volume proxy and a means of correlation, radiolarian assemblages to estimate sea-surface temperature, and the species *Cycladophora davisiana* as a proxy for near-surface oceanography for the past 450,000 years. They found that there are three distinct spectral peaks at periods of 23, 42, and ~100 ka that correspond to changes in the earth's orbital precession, obliquity (tilt), and eccentricity, respectively. These periods were predicted by Serbian mathematician Milutun Milankovitch in the early 20th century on the basis of astronomical calculations of temporal changes in geographical and seasonal distribution of solar radiation due to cycles in the earth's orbital geometry. Milankovitch's theory held that summer insolation at high northern-hemisphere latitudes was the primary factor in ice-age cycles, an idea generally supported in the "Pacemaker" paper.

In the ensuing decades, the Hays, Imbrie, and Shackleton paper had lasting impacts. Orbital-scale paleoclimate reconstructions have been produced from most parts of the world's oceans and many continental sites, and numerous versions of orbital theory have been proposed. Most researchers agree that orbital insolation changes are a primary causal factor for climate variability during much of the Mesozoic and Cenozoic eras, although mechanisms that translate insolation into glacial-interglacial cycles and shifts in the relative strength of precession, obliquity, and eccentricity remain actively debated. Along with the global climate reconstruction of the last glacial maximum world, the Hays et al. paper represents the most influential product of the celebrated Climate: Long-Range Investigation, Mapping, and Prediction (CLIMAP) Project.

6

Glacial Millennial Climate Change

Introduction

In this chapter we describe evidence and possible causes for what are referred to as millennial climate events during the last glacial interval between about 115,000 and 22,000 years ago, after the peak interglacial warmth of Marine Isotope Stage (MIS) 5e and before the last deglaciation. This type of climate variability is, in many ways, unique in that the earth's climate state changes rapidly, within decades, during a prolonged glacial period and persists in a new, fairly stable climate state for 500 years to several millennia. Millennial climate events have global impacts but occur at suborbital timescales and cannot be explained by the orbital processes discussed in Chapter 5, although insolation may play a role. Our focus here is on intensely studied millennial-scale climate events called Dansgaard-Oeschger (DO) and Heinrich events (H-events). Later chapters describe millennial climate reversals during the last deglacial period, 22–11.5 ka, and the Holocene interglacial period, 11.5 ka to the present.

Defining DO and H-Events

DO events were named after their discoverers, Willi Dansgaard and Hans Oeschger, who first identified periodic sharp spikes in oxygen isotopes in Greenland ice cores, suggesting abrupt changes in the atmospheric temperature of up to 15°C (Dansgaard et al. 1984; Oeschger et al. 1984; Dansgaard et al. 1993). Figure 6.1 plots the $\delta^{18}O$ record from the Greenland Ice Core Project (GRIP) core against ice-core depth—rather than age—to emphasize the thickness of the Greenland Ice Sheet (GIS) and the compaction of ice at its greatest depths. These isotopic temperature events are called Greenland "interstadials," a term referring to a relatively warm period within an otherwise cold glacial interval. DO events last about 500 to 2500 years, depending on the event, before climate returns to a colder state called a "stadial."

In contrast, H-events are ice-ocean-sediment phenomena, identified by Helmut Heinrich (1988), building on work by Ruddiman (1977) as brief but dramatic periods of detrital sediment deposition by iceberg-rafting accompanied by surface ocean cooling across a large part of the northern North Atlantic Ocean from 40 to 60°N latitude (Bond et al. 1992; Andrews et al. 1998; Hemming 2004). As we see below, H-events are now known to be associated with several larger and perhaps some smaller DO events.

The literature on millennial climate change has its own jargon. It is referred to as a kind of "flip-flopping" between two distinct, relatively stable "modes" or states of climate. Global climate behaves as though a "switch" was turned on, and then, after 1000 years or so, turned off again. Many researchers agree that these patterns signify two or three distinct, self-sustaining climatic states. Each state apparently represents a climatic equilibrium condition, which persists until a threshold is reached and the climate system is quickly knocked into the opposing stable state. The term "bipolar seesaw" is commonly used to refer to the out-of-sync (antiphased or asymmetric) pattern in which warming in the high latitudes of the northern hemisphere coincides with cooling in polar regions of the southern hemisphere and vice versa (Broecker 1994). A tropical atmospheric seesaw is a concept that has also been proposed to explain asymmetric monsoon patterns during DO events (Wang et al. 2006).

Climatic changes during millennial events are, however, not globally synchronous, equal in amplitude and pattern, or precisely periodic. Over Greenland, where first discovered, most events have an exceptionally abrupt onset, beginning within decades, high-amplitude (up to ~16°C) warming, and a saw-toothed shape, but they are more gradual, muted, and sinusoidal in the Antarctica ice-core record. Tropical temperature and precipitation were strongly affected by millennial events, but regional patterns are complex and partially understood. The spatial and temporal complexity of DO and H-events presents a significant challenge to researchers seeking to identify their causes and has spawned an explosive literature replete with a new paleoclimatic vocabulary. Table 6.1 lists common terms and abbreviations for glacial-age millennial events and what are now standard designations and accepted ages for Greenland and Antarctic ice-core climatic oscillations for MIS 4 through 2. Note that the ages given in Table 6.1 are frequently being revised with better radiometric dating, annual-layer counting, and other methods.

The most important convention is the numbering of events consecutively from youngest to oldest as follows: Greenland DO events: DO-1 through DO-25; North Atlantic

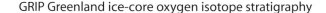

GRIP Greenland ice-core oxygen isotope stratigraphy

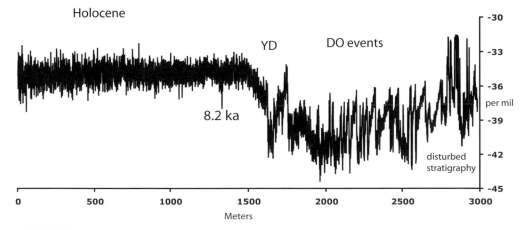

FIGURE 6.1 Greenland Ice Core Project (GRIP) oxygen ice-core record plotted against core depth (in meters) in the Greenland Ice Sheet. Major climate events (Younger Dryas [YD] and Dansgaard-Oeschger [DO], 8.2 ka) discussed in chapters 7 and 8 of this volume are shown.

TABLE 6.1 Event Stratigraphy and Terminology of Glacial Millennial Climate Events

Event	Common Useage— Abbrev.	Ice Core	Marine– N. Atlantic	Other Terms	Approx. Age (ka)*	Comment	Associated Event	Other References
Marine Oxygen Isotope Stage	MIS4				~80–59			
Marine Oxygen Isotope Stage	MIS3				~59–29			Voelker et al. 2002
Marine Oxygen Isotope Stage	MIS2				~29–10			
Dansgaard-Oeschger events		Greenland events	SST cycle	W. European Pollen		Rousseau et al. 2006		
	DO-25	GS-GIS25	NAC-NAW25	Grand Pile W event 1	117–115	Proposed standard ice-core marine–W. European terminology renumbering Greenland stadials and interstadials to match marine millennial cycles. Some papers published prior to this time might have DO event numbering offset by one event. Rousseau's scheme reconciles synchronous marine, ice-core, European millennial climate events.		Rousseau et al. 2006; Landais et al. 2006
	DO-24	GS-GIS24	NAC-NAW24	Grand pile Melisey 1	110–108			Landais et al. 2006
	DO-23	GS-GIS23	NAC-NAW23	Grand Pile M-event	105–104			Landais et al. 2006
	DO-22	GS-GIS22	NAC-NAW22	Grand Pile W event 2	97–90			Brook et al. 1996
	DO-21	GS-GIS21	NAC-NAW21	Grand Pile Melisey 2	88–85			Brook et al. 1996
	DO-20	GS-GIS20	NAC-NAW20		78–77			Flückiger et al. 2004
	DO-19	GS-GIS19	NAC-NAW19		74–73			Genty et al. 2003; Fluckiger et al. 2004
	DO-18	GS-GIS18	NAC-NAW18		69–65			Genty et al. 2003
	DO-17	GS-GIS17	NAC-NAW17		63–60		H-6	Huber et al. 2006
	DO-16	GS-GIS16	NAC-NAW16		59			Huber et al. 2006
	DO-15	GS-GIS15	NAC-NAW15		57			Huber et al. 2006
	DO-14	GS-GIS14	NAC-NAW14		56–55			Huber et al. 2006
	DO-13	GS-GIS13	NAC-NAW13		52–50			Huber et al. 2006
	DO-12	GS-GIS12	NAC-NAW12		49–47		H-5	Burns et al. 2003; Huber et al. 2006
	DO 11	GS-GIS11	NAC-NAW11		45–44			Huber et al. 2006
	DO-10	GS-GIS10	NAC-NAW10		43–42	Burns et al. 2003		Andersen et al. 2006
	DO-9	GS-GIS9	NAC-NAW9		41–40			Andersen et al. 2006

(continued)

TABLE 6.1 (*continued*)

Event	Common Useage— Abbrev.	Ice Core	Marine– N. Atlantic	Other Terms	Approx. Age (ka)*	Comment	Associated Event	Other References
	DO-8	GS-GIS8	NAC-NAW8		40–38		H-4	Andersen et al. 2006
	DO-7	GS-GIS7	NAC-NAW7		37–35			Andersen et al. 2006
	DO-6	GS-GIS6	NAC-NAW6		34.5–33.5			Andersen et al. 2006
	DO-5	GS-GIS5	NAC-NAW5		33–32			Andersen et al. 2006
	DO-4	GS-GIS4	NAC-NAW4		32–28.5		H-3	Andersen et al. 2006
	DO-3	GS-GIS3	NAC-NAW3		28–27			Andersen et al. 2006
	DO-2	GS-GIS2	NAC-NAW2		26–23	GS1, GIS1, GS2, GI2 subdivided by Björck et al. 1998	H-2	Andersen et al. 2006
	DO-1	GS-GIS1	NAC-NAW1		22–14.5		H-1 and Bølling	Andersen et al. 2006
Heinrich events	H-6–H-1 (H-0 = Younger Dryas)							Hemming 2004
	H-6				64–60 ka			Hemming 2004
	H-5.2	H-5a			56–55	Rashid et al. 2003		Hemming 2004
	H-5				49–47	See Burns et al. 2003		Hemming 2004
	H-4				40–38	See Wang et al. 2006		Hemming 2004
	H-3				32–29	See Wang et al. 2006		Hemming 2004
	H-2				26–24	See Wang et al. 2006		Hemming 2004
	H-1				17.5–16.5	See Wang et al. 2006		Hemming 2004
Antarctica millennial events	Original Antarctic cooling events							
	A-7	Byrd, Antarctica ice core			84	Antiphasing observed between DO and Antarcticic millennial events.		Blunier and Brook 2001
	A-6				73			
	A-5				68			
	A-4				59			

(continued)

TABLE 6.1 (*continued*)

Event	Common Useage—Abbrev.	Ice Core	Marine–N. Atlantic	Other Terms	Approx. Age (ka)*	Comment	Associated Event	Other References
	A-3				52			
	A-2				48–45			
	A-1				38			
Antarctic Isotopic Maxima	AIM12	EDML		A2	48	AIMs are isotopic maxima signifying warming in southern hemisphere antiphased with DO events 12–1. AIM warmings start before corresponding DO events.		EPICA Community Members 2006
	AIM11				42			EPICA Community Members 2006
	AIM10				41			EPICA Community Members 2006
	AIM9	Antarctic Isotopic Maxima			40.5			EPICA Community Members 2006
	AIM8	Antarctic Isotopic Maxima		A1	39.5			EPICA Community Members 2006
	AIM7				36			EPICA Community Members 2006
	AIM6				34			EPICA Community Members 2006
	AIM5				33			EPICA Community Members 2006
	AIM4.1				32–31			EPICA Community Members 2006
	AIM4				30–29			EPICA Community Members 2006
	AIM3				28			EPICA Community Members 2006
	AIM2				24–23			EPICA Community Members 2006
	AIM1			Precedes ACR	15			EPICA Community Members 2006

Note: Age models for ice cores, sediments, stalagtites, and other records are often revised to understand the timing of climate signal in different regions and parts of the climate system. ACR = Antarctic Cold Reversal; EDML = EPICA Dronning Maud Land Antarctic ice core site; EPICA = European Ice Core Project; GIS = Greenland interstadial; GS = Greenland stadial; NAC = North Atlantic cool; NAW = North Atlantic Warm; SST = sea-surface temperature (see Rousseau et al. 2006).

*Ages are approximate for duration or midpoint of millennial cycles, varying by author. See references for methods and age uncertainty.

H-events: H-0 through H-8 (including H-5.2); Antarctic low-frequency stadials A-1 through A-7; and high-frequency interstadials called Antarctic Isotopic Maxima (AIM1 through AIM12) Stadial-interstadial climatic variability has also been identified in glacial periods MIS 6 and 8. Although not nearly as well studied as those of the last 100 ka, they suggest that glacial-age climate is predisposed to unstable climate behavior (Pahnke et al. 2003; Martrat et al. 2004, 2007; Siddall et al. 2007).

What Makes a Millennial Climate Event?

There are four generally recognized ingredients to a millennial climate change of the past 100 ka (Broecker 2003; Alley et al. 2003). First, a trigger mechanism, a catalyst, is needed to initiate a large shift in climate from one state to another. This might be an abrupt catastrophic forcing event, such as a change in the hydrological cycle altering ocean temperature or salinity, or small changes in solar irradiance. Or it can be a nonlinear response of the climate system passing a threshold because of relatively gradual changes. Second, one or several feedback mechanisms are needed to amplify the original signal. Commonly discussed feedbacks involve atmosphere, ocean, land and sea-ice interactions. One example would be rising sea level due to melting of northern-hemisphere continental ice, which can destabilize Antarctic ice-sheet margins, leading to additional sea-level rise (SLR). Third, a mechanism is required to transmit the signal across different climatic zones and to both hemispheres. Alley et al. called this a "globilizer." In the case

of rapid climate shifts at the inception of millennial events that are synchronous in one or both hemispheres, a change in atmospheric circulation might transmit a climate signal quickly. For millennial climate changes out of phase between northern and southern hemispheres, the slower-circulating ocean is a more likely mechanism. Finally, the new climate state must be sustained for anywhere from a few centuries to one to two millennia. Broecker calls this the "flywheel" concept. The world's oceans, because of their ability to store and transport heat globally over centuries, is a primary mechanism for sustaining a climate state. In the next three sections we briefly review necessary background on the ocean's meridional overturning circulation (MOC), the concept of hysteresis, and the dating and correlation of millennial climate events.

Meridional Overturning Circulation and Hysteresis
Global Ocean Circulation

Many paleoclimatologists and climate modelers hypothesize that global ocean circulation plays an important, perhaps dominant, role in glacial-age millennial-scale climate variability. Broecker (1997) called ocean circulation the Achilles heel of the climate system because of its sensitivity to small changes in freshwater influx and made the analogy to a heat-distributing global conveyor belt. The ocean's MOC (AMOC refers to the Atlantic component—see Chapter 1) involves northward-flowing warm, salty tropical water in the North Atlantic, and the formation of deep water in the Nordic and

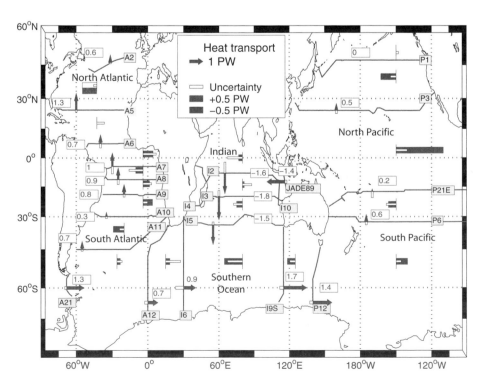

FIGURE 6.2 Map of global ocean heat transport in petawatts (PW), from the World Ocean Climate Experiment (WOCE) (Ganachaud and Wunsch 2000, 2003. Lines labeled A2, A3, etc. show hydrographic sections used in calculations. Arrows show the strength and direction of heat transport. Horizontal boxes show net heat loss (to the left of the axes) and gain (to the right of the axes) in specific regions. See the original papers for details on the calculations. Courtesy of A. Ganachaud and C. Wunsch.

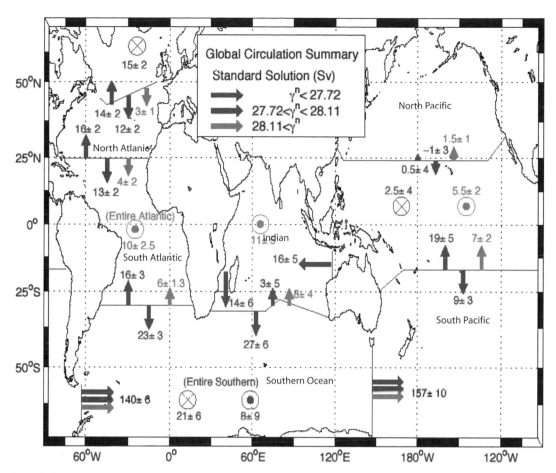

FIGURE 6.3 Map of global ocean mass transport in Sverdrups (Sv), from the World Ocean Climate Experiment (Ganachaud and Wunsch 2000, 2003). Calculations are zonally integrated mass transport values from density data from hydrographic sections. Courtesy of A. Ganachaud and C. Wunsch.

Labrador Seas, which spills over the Denmark Strait and Iceland-Faroes Rise into the deep North Atlantic and mixes with Labrador Sea Water to form North Atlantic Deep Water (NADW). Antarctic Bottom Water (AABW) formed around the Antarctic continent is colder and denser than NADW and fills the deepest abyssal plains of the Atlantic below NADW.

Ocean circulation is complex over many spatial and temporal scales and can be viewed from different perspectives such as the flux of mass, heat, salt, and various chemical properties such as carbon, oxygen, nutrients, and other chemical species (Talley 2002; Schmittner et al. 2007b). Figures 6.2 and 6.3 illustrated this point for heat and mass based on results from the World Ocean Circulation Experiment (Ganachaud and Wunsch 2000, 2003). Figure 6.2 shows that the net export of heat from low to high latitudes across the North Atlantic is 1.3 Petawatts (PW) at transoceanic hydrographic section A5. In Figure 6.3, a volume of 15 Sverdrups (Sv) of NADW is produced in high-latitude Nordic Seas and spills over the Denmark Strait and Iceland–Faroes Ridge into the North Atlantic basin. AMOC variability in heat transport is related to millennial climate changes described below be-

cause the Laurentide, Fennoscandian, and several smaller ice sheets surrounding the North Atlantic–Arctic region during the last glacial period provide a source of freshwater to the North Atlantic and thus potential disruptions to surface salinity, density, and circulation.

Hysteresis and Millennial Climate

Millennial-scale changes in global ocean circulation can be illustrated using the concept of hysteresis. Hysteresis refers to Rooth's idea (1982), refined by climate modelers (Rahmstorf 1995), that AMOC operates in two distinct, end-member equilibrium modes, one with strong deep-water formation and MOC and the other with weak or shut-off MOC. The on-off switch between the two modes is controlled to a large degree by the hydrological budget, in particular, freshwater inflow into the North Atlantic Ocean in regions sensitive to deep-water formation. MOC is strongly influenced by positive and negative feedbacks. On the ocean basin scale, one mechanism to explain large-scale circulation and abrupt transitions involves the positive feedback in which greater thermohaline circulation (THC) transports more salt from

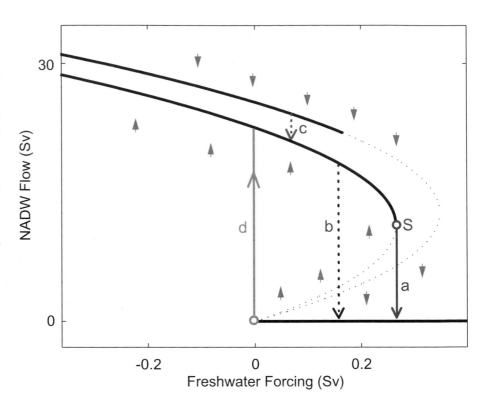

low to high latitudes, further enhancing THC. This is called Stommel's salt-advection model (Stommel 1961) and is supported by climate modeling studies (Manabe and Stouffer 1995, 1999; Rahmstorf 1995).

Figure 6.4 illustrates a hysteresis loop with NADW strength on the y-axis and freshwater forcing on the x-axis, both measured in Sverdrups. Equilibrium condition 1, a modern condition with NADW = ~30 Sv, is represented by the upper line, and condition 2 by the lower line. As freshwater is added to the ocean system, shown as increasing Sv along the x-axis, equilibrium condition 1 remains fairly stable until a threshold point is reached when NADW suddenly falls. This bifurcation point marks a transition to equilibrium condition 2, when NADW strength is reduced to zero. The theoretical extreme in condition 2 would be a complete shutdown of MOC, the opposite of the modern situation of vigorous MOC and NADW formation. The width of the hysteresis loop, that is, where the bifurcation point lies as measured in Sv of freshwater forcing, is of paramount importance because it reflects how much freshwater influx is needed to shut down, or reduce, NADW and switch climate into a different state. The loop is completed when, after an unknown time, the "off" state for MOC is switched back to "on," possibly the result of the cessation of freshwater forcing. A comparison of simulations from 11 climate modeling groups confirms the hysteresis-like behavior of MOC to a range of freshwater forcing from 0.1–0.5 Sv (Rahmstorf et al. 2005).

Hysteresis explains millennial-scale climate variability because, as we see in the next section, ice- and marine-core paleoclimate records provide strong evidence for contrasting modes of climate equilibrium states that are switched on and off again very likely by freshwater forcing. Hysteresis is also directly relevant to future climate change because climate model projections indicate that greenhouse gas forcing will enhance the global hydrological cycle, leading to greater precipitation, freshwater runoff, and ice melting in high latitudes of the North Atlantic. This freshwater forcing can tip the balance of the MOC; some researchers believe it already has. Consequently, understanding past climate changes and the sensitivity of MOC to freshwater forcing is extremely important for efforts to detect human-induced climate changes above the background of natural variability (Latif et al. 2006, 2007). The issue of whether AMOC is today undergoing large-scale human-induced changes is discussed in Chapter 12 in this volume.

Chronology for Millennial-Scale Climate

Accurate absolute dating and interregional correlation are critical elements in the study of abrupt climate transitions and global patterns of millennial climate oscillations. Chronology for paleoclimatology of the last 100 ka relies on a combination of uranium-series dating of corals and speleothems, annual layers in marine and lacustrine sediments, and ice cores, as well as radiocarbon dates calibrated to a calendar-year timescale derived from tree rings, annual sediment layers, and uranium-series radiometric dating. Each method has its strengths and uncertainty, and we briefly discuss each.

Uranium-Thorium

Uranium-thorium (U/Th) in cave speleothems and corals is the primary method for dating material older than the limit of radiocarbon dating. As we saw in Chapter 5, U/Th dating has played a central role in the analysis of orbital-scale climate variability. It is based on the decay of ^{234}U with a half-life of 245,000 years into a daughter product, ^{230}Th, which is itself is unstable with a 75,000-year half-life. The U/Th method measures the degree to which equilibrium between the two unstable isotopes has been restored, and is useful back to about 500 ka.

U/Th dating is used extensively in paleoclimatology of the glacial interval between 115 and 22 ka in two ways. First, it can be used as a direct dating method of cave speleothems and emerged coral reefs for millennial-scale climate changes beyond the range of radiocarbon dating. For example, DO events between 83 and 32 ka were dated in a speleothem from Villars Cave, France (Genty et al. 2003). Second, the U/Th method is used to calibrate the radiocarbon curve.

Layer Counting

For more than four decades, extensive efforts have been made to obtain accurate chronology for ice-core records. Chronology is critical to allow intercore comparisons of climate records and correlation to terrestrial and marine climate records (Hammer 1989; Oeschger and Langway 1989). Today, annual-layer counting remains a primary means of dating Greenland ice, using visual stratigraphy, electrical conductivity, stable isotopic, and glaciochemical measurements to identify seasonal cycles and year-to-year variability. The thickness of annual layers varies from about 2–5 cm over the last 15,000 years, covering the transition into the Bølling interstadial through the Holocene epoch, and these layers are amenable to analysis over subannual timescales.

Layer counting extends back to ~41 ka in Greenland (Svensson et al. 2006), but older ice is dated using less direct methods such as volcanic marker layers, radiogenic isotopes, and modeling of ice flow. Uncertainty stems mostly from missing precipitation, redeposition, and ice melting, depending on the location of an ice core and the time interval under investigation. Recently, Rasmussen et al. (2006) presented a new "Greenland Ice Core Chronology 2005" aimed at placing the North GRIP (NGRIP), GRIP, and Dye-3 ice-core records on common footing for the study of deglacial and Holocene climate. They estimated an error up to 2% for NGRIP and GRIP ice cores for the mid to late Holocene and 3% for the deglacial-early Holocene (14.8–7.9 ka). Additional background on Greenland ice-core chronology can be found in Alley et al. (1997), Clausen et al. (1997), Meese et al. (1997), and North Greenland Ice Core Project Members (2004).

The temporal resolution of ice-core records fluctuates greatly because of variable precipitation and snow accumulation rates across the Antarctic continent and under different climatic conditions. Accumulation rate influences the age difference between the ice itself and the air bubbles trapped in the ice. This ice-air age difference stems from the enclosure time during the transition from snow to porous firn to ice when there is exchange through molecular diffusion between the atmosphere and incipient ice bubbles (Schwander and Stauffer 1984; Joos and Spahni 2008). This process also leads to an age distribution of the gas trapped in a sample of air, resulting in a smoothing out of short-term changes in atmospheric conditions. The thickness of the porous layer can be 100 m or more. In the case of the Byrd station and the European Project for Ice Coring in Antarctic (EPICA) ice cores discussed in the section that follows, the temporal resolution is about ~100 years back to about 60 ka, based on ice-flow modeling and correlation with NGRIP cores using methane concentrations in trapped air (EPICA Community Members 2006). Ice-core chronologies are frequently revised, and it is important to consider age uncertainty in the assessment of any millennial-scale record.

Some restricted marine basins also accumulate sediment rapidly, in some cases in annual layers formed from the absence of dissolved oxygen in bottom waters, which prevents benthic organisms from inhabiting and mixing bottom sediments. Two excellent paleoclimate records of the past 100 ka come from the Cariaco Basin off Venezuela and the Santa Barbara Basin (SBB) off California. The Cariaco Basin record has temporal resolution of decades for some intervals, allowing correlation of tropical atmospheric patterns (Hughen et al. 1996, 1998; Peterson et al. 2000), sea-surface temperature (SST) (Lea et al. 2003), and radiocarbon activity, a proxy for ocean circulation (Hughen et al. 2000) to high-latitude records. The SBB off California, although not amenable to annual-layer counting, has a thick, proxy-rich sediment record of the last 60 ka (Kennett et al. 2000a) and ~1-ka long laminations coeval with Greenland interstadial warmings. The SBB record led Hendy et al. (2002) to hypothesize synchronous climate changes between the eastern Pacific and Greenland for the past 30 ka. Annually resolved sediment records are available from several other anoxic basins, including several in the Gulf of Califronia (e.g., see Barron and Bukry 2007).

Cosmogenic Radionuclide Isotopes

Cosmogenic Isotopes Heliophysics is the study of the relationship between solar activity and the earth, and one subfield involves the study of radiogenic nuclides produced in the earth's atmosphere from solar activity and galactic cosmic rays. Table 6.2 lists the characteristics of the three main nuclides used in paleoclimatology—beryllium-10 (^{10}Be), chlorine-36 (^{36}Cl), and carbon-14 (^{14}C, or radiocarbon). Atmospheric radionuclides are formed from atmospheric nitrogen, oxygen, and argon by cosmic-ray bombardment mainly by protons, and provide a means of dating and correlating millennial-scale climate records, exposure surfaces of landforms, and a long-term record of solar activity thought by many to influence the earth's climate. The production rate of these isotopes varies as a function of two primary processes—changes in the intensity of the solar magnetic field and the strength of the earth's geomagnetic dipole. The lower the solar activity, the greater the production of cosmogenic isotopes. There is no independent method by which to estimate changes in solar activity, which is known with some confidence only back to the 16th century from observations of sunspot cycles and, more recently, instruments and satellites (Lean et al. 2005; Foukal et al. 2006; Woods and Lean 2007; see also Chapter 8 in this volume). Cosmogenic isotopes therefore actually provide a means of estimating solar activity back through time if geomagnetic intensity can be determined and processes influencing the deposition and cycling of isotopes can be constrained (Muscheler and Beer 2006).

Changes in the earth's magnetic intensity shield the atmosphere from cosmic particles and modulate nuclide production. This effect is strongest at the equator and absent at the poles. The weaker the geomagnetic field, the greater the cosmogenic isotope production. Paleomagnetic intensity can in fact be reconstructed from archaeomagnetic (ceramic remains) and sediment paleointensity measurements. Standardized geomagnetic intensity curves have allowed researchers to estimate changes in the production of cosmogenic isotopes due to solar activity (Laj et al. 2000, 2002; Wagner et al. 2000). We will see below and in later chapters a growing range of applications for cosmogenic isotopes in paleoclimatology that include evaluation of solar forcing of climate, calibration of the radiocarbon dates, and correlation of polar ice-core climate records. It is worth emphasizing that changes in solar irradiance and the earth's geomagnetic field, and the production, transport, deposition, and postdepositional history of cosmogenic isotopes, involve many complex and actively researched processes, and the reader is referred to a number of excellent papers for additional discussion (Beer et al. 2002; Muscheler et al. 2005a; Bard and Frank 2006; Snowball and Muscheler 2007).

As discussed in Chapter 2, radiocarbon dating can be used for both chronology and as a proxy for changes in the global carbon cycle and ocean circulation. Figure 6.5 shows a radiocarbon-calendar-year calibration curve based on Cariaco Basin sediment radiocarbon activity (Δ^{14}C) (Hughen et al. 2004a, 2006). Large changes in Δ^{14}C activity, especially before ~10 ka, in part reflect change in the rate of carbon uptake by the ocean because of millennial-scale climate changes. This curve depicts the offset of radiocarbon and calendar-year ages previously outlined here, as well as radiocarbon plateaus, or intervals when the curve is relatively flat.

TABLE 6.2 Cosmogenic Nuclides

Nuclide	Abbreviation	Target Atom	Global Production Rate*	Atmospheric Sink	Deposition	Cycling Variability	Postdepositional Processes[†]
Carbon-14	^{14}C	Nitrogen, oxygen	2.02	CO_2 aerosols	Carbon cycle	Complex atmosphere, ocean, biosphere	Various, carbon cycling
Beryllium-10	^{10}Be	Nitrogen, oxygen	0.018		Precipitation	Atmospheric transport	Migration in ice
Chlorine-36	^{36}Cl	Argon	0.0019	HCl gas	Precipitation	Atmospheric transport	HCl migration and release from ice

Note: See Masarik and Beer (1999) and Beer et al. (2002).

*Flux (F) measured in atoms per square centimeter per second (cm^{-2} sec^{-1}). F = pca where p = density (grams per cm^3), c = concentration (atoms per gram), and a = accumulation rate (cm per year); a (accumulation rate) refers to tree-ring, sediment, or ice archive.

[†]Can alter original concentration.

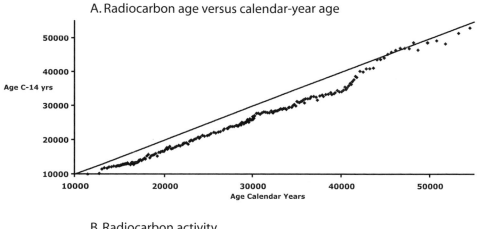

A. Radiocarbon age versus calendar-year age

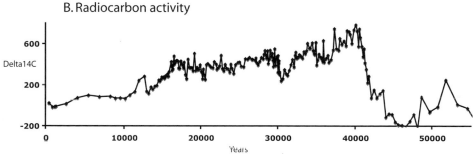

B. Radiocarbon activity

FIGURE 6.5 (A) Radiocarbon age of Cariaco Basin sediments Ocean Drilling Program (ODP) Site 1002 compared to a calendar-year timescale from Greenland Ice Sheet Project 2 (GISP2) ice core, from Hughen et al. (2004a, b). Radiocarbon is corrected for a 410-year global marine correction. (B) Cariaco Basin radiocarbon activity (Δ^{14}C) on GISP2 timescale, showing major changes during the last glacial period related to changes in the strength of ocean circulation and, to a lesser degree, radiocarbon production. Hughen et al. (2006) provide additional comparison of the Cariaco radiocarbon record on timescales derived from Hulu cave speleothem and Barbados coral-reef U/Th dating.

Climatic changes during plateaus, given constant or small changes in radiocarbon production, provide evidence for reduced uptake of newly produced radiocarbon by the ocean and a slowdown in MOC (discussed in the sections that follow). However, material from radiocarbon plateau intervals is difficult to date precisely because of the large error bar on calibrated dates.

Dansgaard-Oeschger Events

Ice-Core Records—Background, Terminology, and Correlation

Background The GIS covers a 1.7×10^6-km^2 area, reaches more than 3000 m in thickness at its summit, and holds 2.9×10^6 km^3 of ice—enough, if melted, to raise global sea level by 7.3 m. The deepest parts of the GIS are about 125,000 years old, but ice near its base that is older than ~115 ka has stratigraphic uncertainty due to ice flow. Pioneering studies

of GIS cores by Dansgaard et al. (1969, 1982, 1993), Oeschger et al. (1985), and Dansgaard and Oeschger (1989) revealed numerous interstadial events characterized by rapid swings in oxygen isotope ratios of oxygen trapped in ice between about 115 and 20 ka. Figure 6.6 shows the location of Greenland ice cores containing records of millennial climate variability.

The distinct pattern of rapid warming at the inception of an interstadial period and subsequent, more gradual return to cooler stadial conditions is now universally called a Dansgaard-Oeschger event, or a Dansgaard-Oeschger cycle. During the period 115–20 ka, high-latitude solar insolation due to orbital processes was characterized by low to moderate variability; however during DO events, the atmosphere over Greenland experienced abrupt temperature increases of 8–16°C within decades. This warming represents an increase above background glacial atmospheric temperatures over Greenland nearly equivalent to that seen over a full glacial-interglacial cycle. Early studies suggested that DO

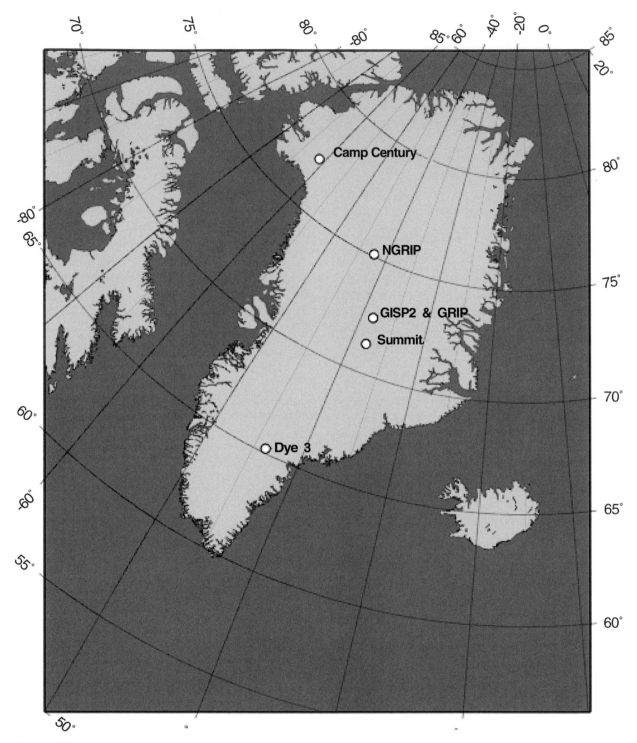

FIGURE 6.6 Location of Greenland ice cores. Greenland Ice Sheet Project 2 (GISP2) and the Greenland Ice Core Project (GRIP) are about 20 km apart. NGRIP = North Greenland Ice Core Project.

events occurred every 2–3 ka and had an asymmetric shape to the temperature cycle, with a steplike, sometimes steep warming, a plateau, and an abrupt or gradual decline (decades to centuries).

Ice-Core Programs A series of major research programs have been carried out over the past few decades to under-

stand DO events and other climate signatures preserved in Greenland ice. These include European programs (GRIP, http://www.nerc-bas.ac.uk/public/icd/grip/griplist.html, and NGRIP, http://www.gfy.ku.dk/~www-glac/ngrip/index _eng.htm) and American programs (Greenland Ice Sheet Project [GISP, GISP2], http://www.gisp2.sr.unh.edu/). A comparison between GRIP and GISP2 records (Grootes et al.

1993) showed a consistent pattern for 24 DO events, now also recognized with some differences in the amplitude of the signal at the NGRIP site shown in Figure 6.6 (North Greenland Ice Core Project Members 2004).

Comparable studies in Antarctica include EPICA introduced in Chapter 5. In addition to the orbital-scale CO_2 and climatic cycles from EPICA Dome Concordia (EDC), EPICA recovered a millennial-scale record from Dronning Maud Land (EDML) core located in the Atlantic sector of Antarctica (EPICA Community Members 2006; Fischer et al. 2007).

Terminology Rapid advances from ice-core programs and related projects in deep-sea and terrestrial paleoclimatology have resulted in major improvements in understanding millennial climate change but also in a large, potentially confusing millennial-scale terminology. Rousseau et al. (2006) proposed a standardized climatostratigraphic terminology applicable to DO events seen in Greenland ice cores, North Atlantic deep-sea cores, and European terrestrial paleoclimate records covering the period 120–10 ka. This scheme, shown in Table 6.1, is justified both on the basis of international rules of stratigraphic nomenclature and evidence for synchroneity of millennial events in the North Atlantic-western European region (McManus et al. 1994; Van Kreveld et al. 2000). Figure 6.7 shows the NGRIP ice-core record with DO events plotted against time.

The naming and numbering of millennial DO events is as follows: Greenland stadials and interstadials are called GS and GIS, respectively, and, from 120 until 60 ka, correlative North Atlantic warm and cool intervals are designated NAW and NAC. DO events are numbered consecutively from the youngest GS0 (=the Younger Dryas) to GS/GIS 25 at about 115–117 ka. H-events associated with some of the largest DO oscillations are also taken into account. To take one example, DO-12 refers to the complete DO cycle from 47–50 ka, GS 12 is associated with H-event 5 dated at ~50 ka, and GIS 12 refers to the abrupt warming after H-5 (Burns et al. 2003).

Two classes of millennial climatic oscillations are recognized in Antarctica. The first group, called A-1 though A-7, consists of large millennial-scale warming events between 38 and 84 ka, reconstructed from the Byrd Station ice core in West Antarctica. A second chronology for Antarctic millennial events was introduced by EPICA project members (2006) and covers more abrupt glacial-age events between 14 and 48 ka. It is based on oxygen isotopic and methane records from the EPICA EDML and EDC cores and previously published records from Byrd, GISP2, GRIP, and NGRIP. The EPICA group introduced the terms Antarctic Isotopic Maximum (AM1–AM12) for these warming events.

Interhemispheric Correlation of Polar Ice Cores

Greenland and Antarctic ice-core records have been correlated with one another using layer counting and a method called methane (CH_4) synchronization. Figure 6.8 shows covariance between Greenland and Antarctic methane records. Unlike global orbital-scale variations in atmospheric CO_2, which reflect mainly carbon storage and release from the earth's oceans, atmospheric CH_4 concentrations largely reflect changes in flux from tropical and high-latitude wetland sources. Methane produced by anaerobic bacteria in wetlands is a function of hydrological and temperature changes. Additional minor natural sources of CH_4 include ancient fossil (coal, lignite, natural gas) and abiotic (volcanic) sources, termites, animals, marine gas hydrates (clathrates), and permafrost. The primary sink for CH_4 is oxidation by the hydroxyl radical (OH) in the troposphere, ultimately producing formaldehyde.

CH_4 in ice cores reflects tropical and high-latitude wetland activity over orbital (Chappellaz et al. 1990) and millennial (Chappellaz et al. 1993; Blunier et al. 1995; Brook

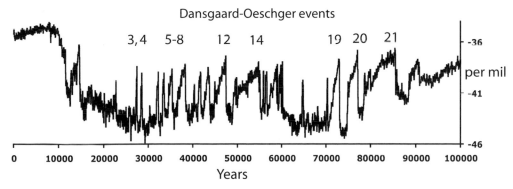

FIGURE 6.7 North Greenland Ice Core Project (NGRIP) oxygen isotope record showing Dansgaard-Oeschger (DO) oscillations, from North Greenland Ice Core Project Members (2004). Major DO events are labeled. See Table 6.1 for the ages of all DO events.

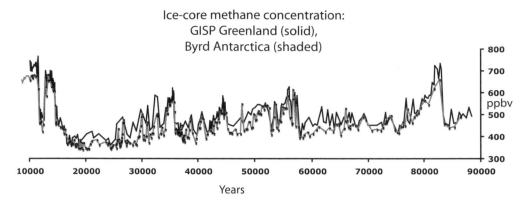

Ice-core methane concentration:
GISP Greenland (solid),
Byrd Antarctica (shaded)

FIGURE 6.8 Interhemispheric methane variability during glacial-age Dansgaard-Oeschger events. The thin solid line is Greenland Ice Sheet Project (GISP2) methane; the shaded line with circles is methane at Byrd, Antarctica, both tuned to the Greenland Ice Core Project ice-core timescale from Blunier and Brook (2001).

et al. 1996) timescales. Other things being equal, the warmer and wetter the regional climate, the greater the production of CH_4 by the earth's wetlands. Methane's short life span in the atmosphere (~10 years) and its wetland source mean that changes in atmospheric CH_4 occur quickly in response to precipitation and temperature and faster than changes in atmospheric CO_2. Consequently, methane synchronization in polar ice cores has provided a means to correlate global climate variability during DO events, measured with a host of ice-core geochemical proxies. Methane measurements are supplemented with beryllium isotopic data for correlation, including the well-known 40.4-ka beryllium spike during the Laschamps geomagnetic event, DO-9 and DO-10, and the Mono Lake event at 32 ka during DO-6 and DO-7 (Beer et al. 2002; Raisbeck et al. 2007).

Blunier et al. (1998), Brook et al. (1999), and Blunier and Brook (2001) showed that methane concentrations from GRIP, GISP2, and Byrd in West Antarctica varied during DO cycles from as little as 50 to as much as 300 parts per billion by volume (ppbv); most DO events showed changes on the order of 100–150 ppbv. They showed convincingly that large Antarctic warming events recorded in $\delta^{18}O$ measurements preceded GIS counterparts by 1500–3000 years.

Short-term Antarctic (Byrd, EPICA EDC) and Greenland records have also been correlated using timescales derived from methane synchronization. This research resulted in the first definitive correlation between short-term millennial events in northern and southern hemispheres, shown in Figure 6.9. As was the case for larger warming events A-1 through A-7, southern-hemisphere atmospheric warming between AIM1 and AIM12 also preceded their counterpart Greenland interstadials.

Greenland and Antarctic ice-core isotope records now stand on secure chronological footing and provide indisput-

able evidence that, to a first approximation, warming events in the southern hemisphere coincide with cooler events in the north and vice versa. We note, however, that air-ice age differences vary among ice cores and at different times during the last 60 ka. Improvements are continually being made to ice-core age models and methane synchronization correlations (Blunier et al. 2007). The following sections summarize current knowledge about the scale of Greenland temperature change, the duration and shape of DO events, their postulated 1470-year periodicity, and global manifestation in paleoclimate records outside Greenland and Antarctica.

Atmospheric Temperature Change During DO Events

DO events were first identified by their oxygen isotopic composition of ice ($\delta^{18}O_{ice}$) as an indictor of atmospheric temperature change. Other things being equal, the $\delta^{18}O_{ice}$ falls about 1‰ for every 1.5°C of temperature. Oxygen isotopic records of Greenland ice cores for glacial-age millennial events of the last 100 ka presented in Figure 6.10 show that DO atmospheric temperature increases over Greenland range from 8–16°C, depending on the particular event.

Whereas the temperature of precipitation near an ice-core site often dominates the $\delta^{18}O_{ice}$ signal, the temperature-$\delta^{18}O_{ice}$ relationship is complex and influenced by several factors. During the snow-firn-ice transition, thermal diffusion and gravitational processes can alter the original isotopic composition of oxygen in the ice. Moreover, the source area and seasonality of precipitation, which is subject to change over the past 100 ka, is also important (Steig et al. 1994; Cuffey et al. 1995; Johnsen et al. 1995, 2001). Thus, proxies of temperature in addition to $\delta^{18}O_{ice}$ have been developed. One useful indicator is the nitrogen isotopic ratio ($^{15}N/^{14}N$,

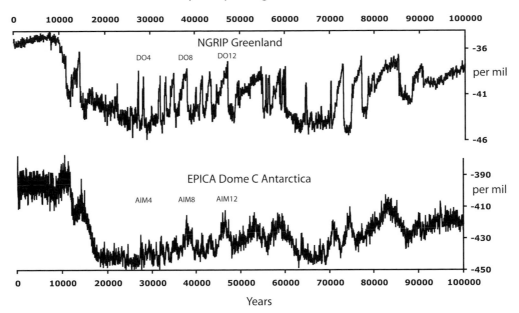

Interhemisphere phasing of millennial events

FIGURE 6.9 North Greenland Ice Core Project (NGRIP) oxygen and EPICA Dome C ice-core deuterium records showing antiphasing between northern- and southern-hemispheric millennial climate events, from North Greenland Ice Core Project Members (2004) and EPICA Community Members (2004, 2006). Several larger Dansgaard-Oeschger (DO) events and Antarctic Isotopic Minima (AIM) are labeled. DO events in Greenland have an asymmetric shape, larger amplitude, and later phasing than more dampened and symmetric AIMs.

$\delta^{15}N_{air}$) from air trapped in the ice. $\delta^{15}N_{air}$ measurements have led to refinement of estimated temperature changes during abrupt climate events including DO events (Severinghaus et al. 1998; Leuenberger et al. 1999; Severinghaus and Brook 1999; Huber et al. 2006). The $\delta^{15}N$ of has been used to calibrate the $\delta^{18}O_{ice}$ from the same samples, and the resulting estimated temperature changes during various DO events range from 8 to 15°C. Temperature change during two of the largest DO events, DO-19 ~ 70 ka, and DO-12 ~ 47–50 ka, were estimated to be 16°C and 12.5°C, respectively (see also Schwander et al. 1997; Jouzel 1999; Lang et al. 1999). In general, millennial temperature changes in Antarctica during stadial-interstadial cycles of the last glacial was of lower amplitude in relation to the temperature scale over Greenland (Watanabe et al. 2003; EPICA Community Members 2004, 2006).

Factoring in changes in atmospheric circulation and moisture source temperature during periods of rapid DO climate changes is also important because the $\delta^{18}O_{ice}$ reflects, in part, the temperature difference between the ocean surface, where water vapor formed, and the site where precipitation falls; ice cores provide additional information on source-region temperatures. Comparison of the $\delta^{18}O_{ice}$ and deuterium isotopes (δD) from the same ice-core samples provides an index called deuterium excess, which serves as a

measure of evaporation temperature of the moisture that eventually falls on ice. Deuterium excess is thus closely tied to SST and other factors such as humidity in the source area for evaporation (Stenni et al. 2001; Stenni et al. 2003). The deuterium excess method has been used extensively in Antarctic and Greenland ice-core records to analyze climate variability over a variety of timescales during the Holocene (Masson-Delmotte et al. 2004), deglacial (Morgan et al. 2002), and orbital-scale climate variability (Masson-Delmotte et al. 2005).

Jouzel et al. (2007) (see also Masson-Delmotte et al. 2005 and Landais et al. 2004a, b, 2006) compared the 3000-m long GRIP, NGRIP, and Dye ice-core records of deuterium excess to evaluate large-scale atmospheric changes in the North Atlantic moisture source region during DO and deglacial events. These studies suggest that changes in temperature over Greenland and source temperatures may be antiphased with each other, suggesting that complex hydrological changes at seasonal timescales occur during short-term glacial and deglacial events that influence the Greenland isotopic record. Stenni et al. (2003) also examined DO-event deuterium excess records from the EPICA EDC core and found contrasting temperature histories for the core site and the moisture source area. For example, temperature at the EPICA site decreased gradually up to the Last Glacial Maximum

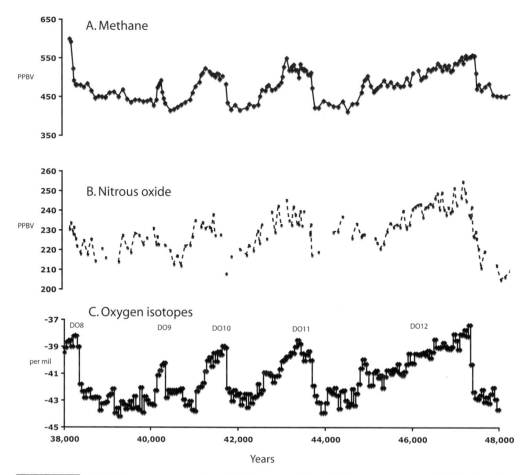

FIGURE 6.10 (A) Methane concentrations, (B) nitrous oxide, and (C) oxygen isotope ($\delta^{18}O$) values for North Greenland Ice Core Project (NGRIP) Greenland ice core for Dansgaard-Oeschger (DO) events 8–12, from Flückiger et al. (2004). The plot shows the different duration, amplitude, and shape for various DO events. The NGRIP oxygen isotope data is from S. Johnsen (North Greenland Ice Core Project Members 2004). Nitrous oxides are subject to artifacts (see text). The timescale is preliminary but does not affect the trends in atmospheric gases.

(LGM) after the warm interstadial event A1 ~38 ka, but source-area temperature increased from 41–28 ka. These studies, as well as those of the LGM (Taylor et al. 2004), exemplify the complexity of atmospheric variability throughout Greenland and Antarctica and the surrounding oceans during glacial-age millennial events.

Duration and Shape of the Atmospheric DO Cycle

DO events were first identified through visual inspection of Greenland isotopic records and it became evident that individual events varied greatly in duration from about 500 to more than 2500 years (Johnsen et al. 1992; Dansgaard et al. 1993). Many DO events also typically have a saw-toothed pattern of temperature change characterized by an initial sharp increase over only a few decades or less, followed by a gradual

decline to about half the initial value and finally a sharp drop. This three-part shape is seen in some isotopic records from ice cores having higher accumulation rates and temporal resolution such as larger DO events 8, 12, 19, and 21.

The exact shape of atmospheric change during each DO event is complex, however, and varies depending on the proxy measured (Sowers et al. 2003). Figure 6.10 from Flückiger et al. (2004) illustrates the variability in the shape of the oxygen isotope, methane, and nitrous oxide curves in several DO events, based on measurements from NGRIP and GRIP cores. The patterns show that atmospheric nitrous oxide, N_2O, rises slightly earlier than CH_4 and $\delta^{18}O_{ice}$ for longer DO events, the amplitude of the CH_4 rise generally matches summer insolation curve, and the amplitude of the N_2O signal seems to be related to the duration of the DO event. The gas age/ice age difference precludes assessment of the precise timing of $\delta^{18}O_{ice}$ in relation to greenhouse gases; however, the primary signals from CH_4 and N_2O almost certainly signify changes in their respective sources in wetlands, terrestrial soil, and the ocean.

In Antarctic ice cores, several proxies demonstrate that the shape of glacial-age millennial events is more sinusoidal in contrast to the spikiness of Greenland records (EPICA Community Members 2004, 2006; Jouzel et al. 2007). The methane-synchronized records also allowed the important conclusion that the amplitude of the Antarctic temperature rise was positively correlated with the duration of Greenland interstadials. Thus, in addition to the interhemispheric synchronization of cool-warm cycles, the relative strength of the Antarctic cycle compared to that in Greenland supports the hypothesis for an oceanic teleconnection between the North Atlantic and Antarctic via the bipolar seesaw mechanism.

Age and Periodicity of DO Variability

As mentioned above, the phasing of Greenland temperature oscillations with their Antarctic counterparts is known from the synchronization of methane records. An accurate absolute chronology is also essential to evaluate the timing of DO and Antarctic millennial events in relation to non-ice-core records. Annual ice-layer counting of Greenland ice is useful for chronology back to about 40 ka, depending on the site, but many DO events are beyond the capability of annual-layer counting and radiocarbon dating, so recent efforts to date DO events have focused on uranium-series dating of speleothem paleoclimate records. The more notable examples include Villars Cave in southwestern France for the period 83–32 ka (DO-6–20) (Genty et al. 2003) and Moomi Cave (M1–2), Socotra Island, Indian Ocean (DO-9–13) (Burns et al. 2003). The isotope record from the Villars Cave stalagmite (Vil9) had hiatuses in calcite growth called D1–D7. The three largest hiatuses were dated at 78.5–75.5 ka (D2), 67.4–61.2 ka (D3), and 55.7–51.8 ka (D4), and event D3, called the Villars Cold Phase, corresponds with H-event 6. The Moomi Cave record also revealed interstadial warming during DO-12 over a period of only 25 years. These and other records used to date DO events are summarized in Table 6.1.

The spacing of DO events and the possibility that they are cyclic in nature is the subject of a large literature because periodic climate oscillations might imply some kind of extrinsic forcing. Some early studies did indeed suggest periodic climate behavior during glacial-age millennial climate events. Mayewski et al. (1997) performed spectral analyses on a high-resolution record of GISP2 glaciogenic time series, a robust atmospheric signal of the last 110 ka. In addition to lower-frequency orbital-scale spectral peaks, they identified statistically significant 6100-year and 1450-year cycles. The longer cycle was attributed to internal ice-sheet dynamics, a

phase lag between ice sheet and orbitally driven insolation, or a combination of the two.

Whether or not DO events are periodic or irregular depends on a wide range of factors related to practical laboratory procedures and statistical methods. How closely spaced isotopic samples are in an analysis of ice or marine-sediment core will determine the temporal resolution of the record. Which ice-core timescale is used can influence the statistical results. How one actually defines an individual event is itself a subjective exercise. The criteria of a 2‰ peak-to-trough shift in Greenland ice oxygen isotopic composition over a 200-year-long interval is one definition (Alley et al. 2001, 2002). The time interval can also influence results. The outcome might be affected by whether one includes the entire interval from 115 to 10 ka, which include events after northern-hemisphere orbital insolation began to increase following the LGM, or only the interval from 42 to 10 ka, when annual-layer-counting chronology is secure. Numerous statistical methods are also available to evaluate DO patterns (Wunsch 2000, 2003a; Alley et al. 2001; Ditlevsen et al. 2007). The periodicity of DO events is discussed below in the section on causes of glacial climate variability.

Global Manifestation of DO Events

DO events have hemispheric and global climate signatures recognized in a variety of paleoclimate records. Emblematic of the widespread interest in DO events, a 2002 workshop collated published records of millennial climate records covering MIS 3 between 59 and 29 ka and DO events 17–4. The work group identified a total of 183 sites, 82 of which had a temporal resolution of 200 years or less (Voelker et al. 2002). Many more studies have since been published, and we summarize some examples of DO-scale variability in atmospheric composition, temperature, and ocean temperature and circulation.

Ice-Core Atmospheric Changes Ice cores also contain evidence for the global nature of DO events in their record of greenhouse gases CO_2, N_2O, and CH_4 trapped in the ice. We remind readers that the temporal resolution of ice-core greenhouse gas records of centennial and millennial climate events is limited by the rate of snow accumulation, the age difference between the ice (isotope temperature proxies) and the trapped air (gas proxies), and molecular diffusion in air bubbles during close-off from the atmosphere. The ice-air age difference influences the age model for the core, and diffusion the age distribution of gas within the trapped air. Slower accumulation regions can have an age uncertainty of several thousand years. Because accumulation rates vary during glacial and interglacial periods, the age uncertainty also changes through time for any particular ice-core site. These preservation issues pertain most to abrupt climatic transitions such as the inception and termination of DO events. With this caveat in mind, we see that atmospheric carbon dioxide (CO_2) varied between about 190 and 220 parts per million by volume (ppmv) during DO events between 30 and 65 ka (Neftel et al. 1988; Stauffer et al. 1998; Petit et al. 1999; Indermüle et al. 2000; Siegenthaler et al. 2005; Kawamura et al. 2007). Ahn and Brook (2007) determined that between 65 and 30 ka, atmospheric CO_2 concentrations rose just before major H-events, as well as several thousand years prior to major associated DO interstadial warming during DO events 8, 12, 14, and 17. Within the limits of the correlation, rising CO_2 concentrations at these times appeared to be coeval with large Antarctic warming events.

Ahn and Brook (2008) also examined the CO_2-climate connection between 90 and 20 ka using the high accumulation rate in the Byrd, Antarctica ice core and several Greenland ice-core records synchronized by methane. They found a positive correlation between CO_2 and temperature in Antarctic ice cores during DO-Heinrich variability. In particular, they noted that CO_2 concentrations rose rapidly coincident with Antarctic warming and synchronous cool stadials in Greenland. CO_2 declined 1–2 ka after cooling began in Antarctica. Relating the global CO_2 flux to paleoceanographic records, they found that CO_2 concentrations rose about the same time that NADW shoaled in the North Atlantic and near-surface stratification decreased in the Southern Ocean. For the most part, these patterns characterized most of the largest DO events in Greenland (DO-8, 12, 14, 17, 20, 21) and generally support a hypothesis that CO_2 is released from the ocean to the atmosphere, according to one model, when stratification decreases in the Southern Ocean (Schmittner et al. 2007a; see also Marchitto et al. 2007 and Robinson et al. 2007). The Ahn/Brook study reveals a number of other interesting points about the CO_2-climate link, interhemispheric asynchronous changes during millennial events, and differences between various DO events and those early and late in the glacial period from 90 to 20 ka.

Measurement of N_2O in trapped air also shows strong variability. For example, there was a 65-ppbv rise, from 205 to 270 ppbv, for DO-19 and DO-20 and slightly smaller amplitude changes for DO-8–12 (Flückiger et al. 1999, 2004). For comparison, N_2O concentrations rose during the transition from the LGM to the Holocene interglacial (Termination 1) from ~200 to 275 ppbv. N_2O oscillations signify changes in the flux of this greenhouse gas from soils and, to a lesser extent, the oceans (Sowers et al. 2003).

As already noted, methane gas measured in Greenland and Antarctic ice cores reflects the global dimension of DO events (EPICA Community Members 2004; Spahni et al. 2005). In one recent study, Brook et al. (2005) reconstructed

interstadial-stadial variability in methane and isotopic records of trapped oxygen (O_2) gas between 9 and 57 ka from the West Antarctic Siple Dome ice core. Although some differences exist in Antarctic interstadial cycles, they confirmed that Antarctic warming precedes abrupt warming in Greenland, as seen in other ice-core records, especially for the four largest DO events. Antarctic warming events A1 and A2 precede Greenland warming during DO-8 and DO-12 by 1.2 and 1.4 ka, respectively.

The NGRIP ice core provides an even more detailed look at atmospheric dynamics and greenhouse gas fluxes during a DO event. By calibrating NGRIP temperature variations using $\delta^{15}N_{air}$ and methane concentrations, Huber et al. (2006) showed that methane concentrations lag temperature changes over Greenland by between 25 and 75 years during DO events 9–17 (Flückiger et al. 2004). These results provide additional evidence that DO events represent a nearly synchronous (multidecadal) response by the world's wetlands to temperature changes.

Terrestrial Records Evidence for DO-type events is now known from terrestrial records. For example, Owens Lake in the Great Basin of California is sensitive to hydrological variations from precipitation and springtime runoff. Using total inorganic and organic carbon, oxygen isotopes, and magnetic susceptibility as proxies, Benson et al. (1996) discovered 19 distinct glacial advances in the Sierra Nevada between 52.5 and 23.5 ka. Most advances coincided with decreased discharge to Owens Lake, whereas intervening glacial recessions represent periods where the lake level receded below the core depth. The timing of Owens Lake desiccation cycles was every ~1500 years, close to that of DO events in Greenland ice cores.

High-resolution isotopic records from cave speleothems have added significantly to the body of evidence that DO events are global events with strong teleconnections in tropical regions. Figure 6.11 shows the pattern of DO variability in decadally resolved isotopic records from Hulu Cave, China (Wang et al. 2001) and Botuverá Cave, Brazil (Wang et al. 2006). These records reflect dynamic changes in the East Asian Monsoon (EAM) and South American Monsoon (SAM) systems, respectively. Other noteworthy studies of tropical atmospheric variability during the last glacial period include Dongge Cave, China (EAM) Yuan et al. 2004), Xiaobailong Cave, southwest China (Cai et al. 2006), and Moomi Cave, Socotra Island (Indian Ocean Monsoon) (Burns et al. 2003), Villars Cave, France (North Atlantic atmosphere) (Genty et al. 2003), Soreq and Peqiin Caves, Israel (Bar-Matthews et al. 1999, 2003), and Austria (Spötl and Mangini 2002).

The Hulu and Botuverá records suggest that the EAM and SAM systems were antiphased with one another during

Hulu Cave, China stalagmite oxygen isotope curve

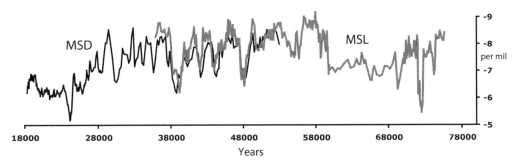

FIGURE 6.11 Oxygen isotopic record from Hulu cave stalagmites called MSD (dark line) and MSL (light line) showing Dansgaard-Oeschger-type variability mainly in monsoon-driven summer precipitation, from Wang et al. (2001).

many glacial-age millennial events. Wang et al. (2006) interpret these patterns as evidence for migration in the Intertropical Convergence Zone (ITCZ) producing an asymmetric response in Hadley circulation in the northern and southern hemispheres. They outline an idealized sequence of events during a DO cycle based on the correlation of paleorecords and climate modeling studies (Clement et al. 2004; Chiang and Blitz 2005) as follows. At times of weak MOC during stadial periods, there is a southward shift in the ITCZ, which in turn causes a warm temperature anomaly in the South Atlantic, a stronger SAM, but a weakened EAM. In effect, northern-hemisphere tropics are drier and southern-hemisphere tropics wetter during cool stadials. The opposite occurs during northern-hemisphere interstadial parts of DO events. This type of variability is called a low-latitude atmospheric seesaw, implying significant global hydrological changes originating in the tropics during DO events.

Marine Records Paleoceanographic records from the subpolar North Atlantic Ocean reveal rapid changes in SST and deposition of ice-rafted debris (IRD) attributed to DO variability (Bond and Lotti 1995; Bond et al. 1999). Rasmussen et al. (1996, 1997, 2003) identified 15 DO cycles during the past 58 ka in a marine-sediment core from the Faroe-Shetland channel. They showed a three-part structure to each DO cycle: (1) a period dominated by warm interstadial planktonic foraminiferal assemblages and inferred strong meridional overturning and NADW formation; (2) a transition into cooler SST, reduced NADW, and increased sea-ice cover, reflected in benthic foraminiferal assemblages; and finally (3) a period of peak abundance in the cold-loving foraminifera *N. pachyderma* (sinistral form) during the peak cooling of the stadial period. Magnetic susceptibility from the sediment record allowed a detailed correlation of these marine manifestations of DO cycles to the Greenland

ice-core records. Chapman and Maslin (1999) also reconstructed strong SST and sea-surface salinity (SSS) gradients between 40 and 50°N in the North Atlantic between 40 and 20 ka, which they attributed to the influence of low-latitude orbital precession on surface ocean circulation. Many other sediment-core records yield Heinrich and DO scale SST variability in the subpolar North Atlantic Ocean, revealing complex spatial and temporal patterns, including significant differences among various events (Labeyrie et al. 1999; Sarnthein et al. 2000; van Kreveld et al. 2000).

DO variability is also recognized in the low-latitude oceans. Sachs and Lehman (1999) used alkenone paleothermometry to measure a mean 3.1°C SST (range 1.7–5.3°C) change in the subtropical Atlantic Sargasso Sea during 12 DO events. Some DO events experienced SST cooling nearly equivalent to that seen during a full glacial-interglacial cycle in this region. SST cooling during two large events, DO-8 and DO-12, occurred in less than 250 years. Sachs and Lehman also reconstructed significant spatial variability from analyses of sites in the Blake and Bahama outer ridges. Other regions of the equatorial North Atlantic also exhibit SST cooling of 2–3°C during DO events (Curry and Oppo 1997; Sarnthein et al. 2001). In sediments deposited in the SBB off California, Jim Kennett and colleagues studied records of DO events in Ocean Drilling Program (ODP) Site 893. They identified 19 of 20 DO events since 75 ka (Behl and Kennett 1996) that perhaps signified changes in bottom-water oxygenation (Hendy and Kennett 2000). Additional multi-proxy analyses of ODP Site 1017 in the SBB indicate complex atmospheric and oceanographic changes related to wind, nutrient cycling, and productivity during DO events of the last 60 ka (Hendy et al. 2004). Ortiz et al. (2004) reevaluated the eastern Pacific record of DO events using cores from Magdalena Margin off Mexico. Rather than the proposed ventilation events proposed for the SBB, they instead concluded that DO events represented changes in surface ocean productivity such that during Greenland stadials, the eastern Pacific was characterized by a deep nutricline and low productivity.

The Cariaco Basin record off Venezuela provides an exceptional view of tropical ocean-atmosphere interaction during DO events. Peterson et al. (2000) used sediment color and mineral assemblages to identify all 21 DO events between 90 and 10 ka and to characterize tropical rainfall variability. Greenland interstadials were closely tied to high levels of tropical rainfall and stadials to drier conditions. Peterson et al. proposed that large tropical feedbacks involving changes in Atlantic-to-Pacific moisture export and SSS were important during DO events.

Kiefer et al. (2001, 2002) discovered an apparent correlation between Pacific Ocean warming of 3°C or more and stadial events in Greenland. Their climate modeling results suggest that such warm Pacific Ocean temperature events would lead to dramatic increases in snow accumulation over Greenland. In the Southern Ocean, SST and ice-rafting variability is generally synchronous in timing and amplitude with the Antarctic atmospheric temperature changes during the last glacial period (Kanfoush et al. 2000; Sachs et al. 2001;).

Deep-sea temperature and circulation also changed during some DO events. Using ostracode Mg/Ca paleothermometry, Cronin et al. (2000) and Dwyer et al. (2002) documented bottom-water temperature changes of 1–3°C between 30 and 90 ka at 3400 m of water depth in the North Atlantic Ocean. Martin et al. (2002) used foraminiferal Mg/Ca paleothermometry to estimate bottom-water temperature excursions of >0.5°C in the eastern tropical Pacific Ocean during MIS 3. Geochemical evidence for reduced NADW production during glacial-age climate oscillations comes from a number of sites in the North Atlantic Ocean (Keigwin and Jones 1994; Oppo and Lehman 1995). Elliott et al. (2002) analyzed $\delta^{13}C$ isotopic measurements on the benthic foraminifera *C. wuellerstorfi* for the interval 10–60 ka and found that NADW formation weakened during H-events associated with large DO events but not during smaller DO events, at least at water depths shallower than 2000 m.

These are just a few of the oceanic records of suborbital variability during MIS 4–3; additional discussion is found in Bard et al. (2000a), Sarnthein et al. (2001), Voelker et al. (2002), Labeyrie et al. (2003), and Lynch-Stieglitz (2004). We now turn our attention from atmospheric DO events to the ice-ocean phenomena known as H-events before considering the causes of these closely linked processes.

Heinrich Events

H-events are massive iceberg discharges from North American and European ice sheets during the last and probably previous Pleistocene glacial periods. Heinrich layers are 10–15-cm-thick layers of sediment deposited across a wide belt of ocean floor in the North Atlantic Ocean during H-events. These events signify the vulnerability of glacial-age ice sheets to abrupt large-scale destabilization events, including ice-sheet surges and rapid melting. Modern ice sheets also show signs of dynamic behavior, so it is especially important to fully understand glacial-age ice-sheet history in light of modern glaciological processes and to establish the relationship between Heinrich and DO events.

Heinrich layers were identified by Ruddiman (1977), who mapped their distribution across the North Atlantic. The distribution was called "cemented marls" by Heinrich (1988) in a study of Dreizack seamount sediment records, and was named as H-events by Bond and colleagues, who described them in depth in a series of papers that elevated their status as an important class of millennial climate change (Bond et al. 1992, 1993; Broecker et al. 1992; Broecker 1994; Bond and Lotti 1995). Like DO events, H-events have been intensely studied. In a comprehensive review, Hemming (2004) listed 44 different sediment cores from the northern North Atlantic region in which H-events have been studied, a testament to the attention they have since generated. Table 6.3 is a summary of basic information about H-events.

Early Studies of Heinrich Events

The canonical H-event involves the periodic collapse of the Laurentide and European Ice Sheet margins in the region of Hudson Bay and Hudson Strait and northwestern Europe (including Iceland and the British Isles) and Greenland, causing the discharge of many large icebergs into the North Atlantic Ocean. The icebergs eventually melt and deposit layers of IRD in deep-sea sediments. Heinrich layers of the last glacial period mainly consist of detrital sediment layers and are associated with planktic foraminiferal faunas that record SST and salinity changes. Figure 6.12 shows the typical oceanic IRD-foraminiferal patterns during H-events from Bond's studies.

IRD has long been recognized as evidence for iceberg activity in the subpolar North Atlantic. Ruddiman (1977) pioneered the study of IRD in North Atlantic sediment by examining the >62-µm-size fraction of sediment in 32 subpolar North Atlantic cores to establish patterns of IRD deposition during the last interglacial-deglacial cycle. IRD patterns of deposition during interglacials were distinct from those during cooler stadial climate regimes. For example, a shift in the axis of IRD deposition to an east-west zonal orientation characteristic of what are now known as Heinrich layers occurred about 75 ka. Ruddiman estimated that the total volume of sand blanketing the central and peripheral subpolar North Atlantic during the last interglacial was about 60×10^{11} grams; in contrast, the amount increased to about 220×10^{11} grams during the glacial interval. At one

TABLE 6.3 Heinrich Events

Statistics for H-events		Reference*	Comment
Duration	250–500 years	Hemming 2004; Andrews 2007	Varies by event
Freshwater volume	10^5 to 10^7 km^3	Clarke et al. 2003	Depends on discharge rate, surface layer thickness
IRD volume	100–350 km^3	Alley and MacAyeal 1994; Hemming 2004	
Sediment layer thickness	10–15 cm	Hemming 2004	
Meltwater discharge	Varies: $0.1–7 \times 10^6$ km^3	Roche et al. 2004 and refs	Varies by event, H-4 = ~2 m sea-level equivalent
Atmospheric GHG	CO_2 rises before large H-events	Ahn and Brook 2007	
Sea-level—ice sheet surge	3–11 m total ice volume change	Dowdeswell et al. 1995; Hemming 2004	15 m from Antarctica: Rohling et al. 2004
Sea level—lake outburst	0.2 m jökulhlaups	Alley et al. 2005	
IRD provenance proxies			
Weight % IRD		Ruddiman 1977; Andrews 1998	Mapped as IRD concentration mg/cm^2/kyr
Grain size abundance		Andrews 1998	No. of lithic grains/g in >63–>180 micron fraction
Mineralogy		Bond et al, 1992	Volcanic glass, detrital carbonate, hematite-stained quartz, feldspar
Rare earth element isotopes		Grousset et al. 1993	Strontium/neodymium indicates age of source material
Lead isotopes		Hemming et al. 1998	
Potassium-argon dating	$^{40}Ar/^{39}Ar$ commonly used	Hemming et al. 1998, 2000	On bulk sediment, hornblende, or feldspar grains
Magnetic susceptibility		Grousset et al. 1993; Robinson et al. 1995	Mainly to identify H-layers
Mineralogy		Andrews 2007	X-ray diffraction
Organic compounds	Lipids, others	Rosell-Melé et al. 1997; Huon et al. 2002	
Regional sources	Study region		
Laurentide IS	Hudson Strait, Labrador Sea, NA margins	Farmer et al. 2003	Hudson Strait source for IRD Belt H-1, H-2, H-4, H-5
Northeast Atlantic	Dreizack Seamounts	Heinrich 1988	IRD layers called "cemented marls"
Icelandic and E. Greenland IS	Denmark Strait	Andrews 2007	Iceland IS influence after 25 ka
British IS	Off Ireland	Peck et al. 2007	H-4 and H-5: no European precursor, BIS advance ~26.5 ka
Laurentide IS	Labrador Sea	Andrews and Tedesco 1992; Stoner et al. 1996	
Greenland, Iceland Laurentide IS	Irminger Sea, Northern N. Atlantic	Elliott et al. 1998; Van Kreveld et al. 2000	European IS precursor IRD events, Laurentide only large events
Mainly Laurentide IS	Midlatitude N. Atlantic	Bond et al. 1992; Grousset et al. 1993	
Arctic Ocean	Fram Strait	Darby et al. 2002	H-1–H-4 Arctic source precursors

Note: BIS = British Ice Sheet; GHG = greenhouse gases; IRD = ice-rafted debris; IS = Ice Sheet; kyr = kiloyear; NA = North Atlantic.
*See reference for additional papers.

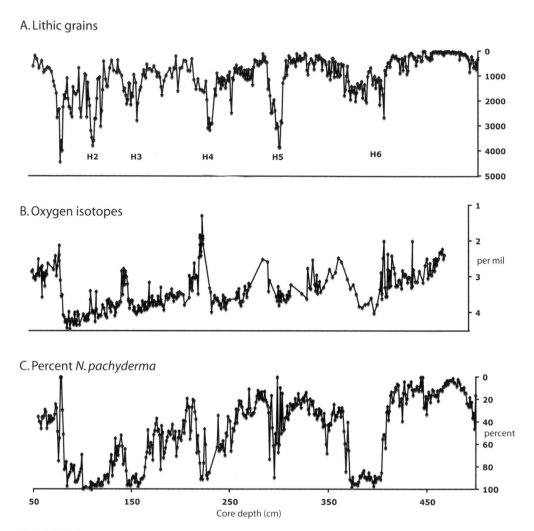

FIGURE 6.12 Heinrich events in Deep Sea Drilling Project (DSDP) Site 609, from Bond et al. (1992, 2001): (A) the number of lithic grains; (B) oxygen isotopes on *N. pachyderma* (left-coiling); (C) the percent of *N. pachyderma* (left-coiling). Heinrich events are labeled per Hemming (2004); also see Bond et al. (2001).

deposition site located under the path of icebergs exiting the Greenland sea near the Denmark Strait, IRD levels rose from 200 mg cm^{-2} during the interglacial to 800 mg cm^{-2} during MIS 4, and to >1200 mg cm^{-2} during the glacial maximum. Parallel changes occurred in the narrow regions between the Labrador Sea and the Northern Iberian Peninsula. Maximum glacial IRD deposition occurred in the subpolar North Atlantic cyclonic gyre near 50°N latitude, where southeastward-flowing Laurentide and Fennoscandian icebergs meet warm North Atlantic Drift water.

It is also significant that glacial geological research in North America and Europe accompanied the study of IRD layers in deep-sea cores. John Andrews (1998) reviewed the history of research on northern-hemisphere ice-sheet margins, especially in the Hudson Bay/Strait source area, for Heinrich icebergs during the last glacial interval. He shows

evidence that dynamic behavior such as ice-sheet surges was firmly rooted in North American glacial geology before the recognition of IRD layers in deep-sea cores. In addition, it has long been recognized from field studies that glacial ice-margin destabilization is asynchronous along various margins of northern-hemisphere ice sheets (Dowdeswell et al. 1999). This fact has large implications for interpreting patterns and evaluating causes of H-events.

Number and Age of Heinrich Events in the North Atlantic

Bond and colleagues mainly used detrital lithic grains that constitute IRD layers and planktic foraminiferal assemblages to establish the temporal and geographical extent of Heinrich layers in the subpolar North Atlantic. They named six Heinrich layers (H-6–H-1) spaced approximately 5–10 ka apart between 70 and 14 ka. These layers are dated by volcanic ashes, oxygen isotope stratigraphy, and ^{14}C dating. McManus et al. (1994) later identified two older H-events called H-7 and H-8, and more recent studies identify an event between H-6 and H-5 called H-5.2 (Rashid et al. 2003). H-event 1 actually follows the LGM, and some refer to the Younger Dryas as H-event H-O. These two deglacial events are discussed in depth in the Chapter 7 in this volume.

Chronology for H-events has been greatly improved in recent years. Younger Heinrich layers can be dated directly by radiocarbon dating of foraminifera in or near the detrital IRD layer. For example, Chapman and Maslin (1999) obtained ^{14}C dates directly from Heinrich layers at 40 and 50°N latitude in the North Atlantic. The approximate ages were 47.2 ka (H-5), 34.8 ka (H-4), 28.4 ka (H-3), 22 ka (H-2), and 15 ka (H-1). Direct dating by radiocarbon has several limitations. The first is the decreasing precision of radiocarbon dates as one reaches the limit of the method. Another is the significant age uncertainty introduced by complex changes in glacial-age carbon cycling, especially in high latitudes of the North Atlantic where sea ice can affect carbon uptake (Voelker et al. 1998). Nonetheless, some deep-sea researchers assume the standard ~400 marine reservoir correction in dating H-events in lieu of better data on local reservoir effects.

Indirect dating of high-resolution climate records outside the immediate North Atlantic region is also used. The ages for older events in Table 6.1 come largely from U/Th ages of speleothem isotopic records correlated to deep-sea and ice-core paleoclimate records. These ages are subject to revision as additional paleoclimate records are obtained, but generally show that the duration of H-events varies from centuries to over 2000 years.

Characteristic Features of Heinrich Layers

Figure 6.13 is a map showing the location of Heinrich layers in the North Atlantic Ocean and plots of magnetic susceptibility and IRD during H-events 1–5, from Hemming (2004). Sometimes this is called the IRD belt, or the "Ruddiman belt" after Bill Ruddiman, who originally mapped Heinrich layers. Deep-sea regions containing ice-rafted sediment are called far-field sites to distinguish them from regions more proximal to the ice sheets themselves. The essence of deep-sea H-events lies in the composition, distribution, and age of layers of lithic grains interbedded with more typical deep-sea sedimentation of calcareous microfossils. Figures 6.14 and 6.15 from Eynaud et al. (2007) show photographs of sediments deposited off the northwest European margin during H-event 1 in MIS 2 and MIS 6, as well as a schematic

reconstruction of how IRD is deposited. These photos show that IRD layers are associated with planktic foraminifera used to reconstruct changes in SST and SSS before, during, and after each IRD event.

Heinrich layers also contain a suite of petrologic and geochemical tracers of iceberg sources. For example, detrital carbonate from the Laurentide Ice Sheet (LIS), hematite-stained grains from the European (Greenland) Ice Sheet, and volcanic glass from the Icelandic Ice Sheet are among the most common first-order lithological indicators. Radiometric dating of individual Precambrian hornblende grains by ^{40}Ar/^{39}Ar (Hemming and Hajdas 2003), strontium and neodymium isotopic tracers (Grousset et al. 1998), and X-ray diffraction mineralogy (Andrews 2007) are also applied to provenance analysis.

Numerous studies of IRD layers by Heinrich (1988), Bond et al. (1992, 1993), Andrews and Tedesco (1992), Andrews et al. (1993), Alley and MacAyeal (1994), Dowdeswell et al. (1995), Grousset et al. (1993), Hemming et al. (1998), Hemming (2004), Peck et al. (2007), and Andrews (2007) provide important clues about H-events (Table 6.3). In the main IRD belt, most layers defined as H-events are characterized by a high percentage of detrital carbonate, high magnetic susceptibility, Precambrian potassium-argon radiometric dates, and relatively sparse planktonic foraminifera. The fine structure for H-1, H-2, H-4, and H-5 shows a sharp basal transition; these H-events are generally considered as coming from the Hudson Strait area of LIS discharge. H-4 is particularly well studied and shows consistent abundance of Precambrian detrital material across its deposition belt. H-3 and H-6 have a lower concentration and a different source of lithic grains and are distinct from Hudson Strait Heinrichs.

The volumes of IRD constituting Heinrich lithic layers are estimated to range from 100 to 350 km^3. Each event deposited approximately 5×10^{15} kg of sediment, blanketing an area of 5×10^{12} m^2. Detrital layers reach a maximum thickness of 1 m near the glacial outflow and a minimum thickness of a few centimeters in more distal regions. The average layer is about 10 cm thick. The deposition of the lithic layer itself took from ~250 to 2300 years, but on average about 500 years (Labeyrie et al. 1999; Hemming 2004).

Some H-events seem to occur as a multistep process. The first step consists of a fall in SST accompanied by a decrease in the net flux of planktic foraminiferal shells to the ocean bottom. The planktic foraminifer *Neogloboquadrina pachyderma*, a cold-water Arctic form, reached near-total dominance (80% to >90% of foraminiferal assemblages) at DSDP site 609 during H-events. Second, there is a brief, massive discharge of icebergs into the North Atlantic Ocean. The debris-laden iceberg path is traceable 3000 km in a west-to-east series of cores as a swatch of coarse-grained sediment covering the sea floor from Canada to near Portugal (Figure 6.14). The

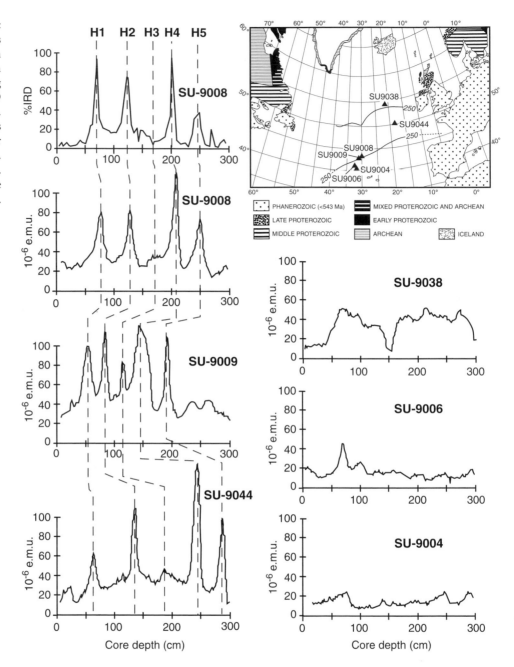

FIGURE 6.13 Spikes in magnetic susceptibility and ice-rafted debris (IRD) lithic grains for Heinrich events 1–5 in cores from the North Atlantic Ocean, and map showing general vicinity of Ruddiman IRD belt, from Hemming (2004). Patterns for continental regions show provenance regions for lithic-grain fingerprinting. E.m.u.=electromagnetic units. Courtesy of S. Hemming and the American Geophysical Union.

impact of Heinrich icebergs is felt outside the immediate belt of IRD, as shown by Bard et al. (2000a), who documented a decrease in both salinity (1–2 psu) and temperature (3–6°C) off the Iberian Peninsula.

If the source of H-event icebergs can be determined, then it might be possible to establish the mechanism causing sudden glacial discharges. Synchronous release of icebergs from all ice-sheet margins surrounding the North Atlantic implies a common forcing, perhaps related to solar radiation, rising or falling sea level, or feedbacks related to orbitally driven insolation, whereas asynchroneity implies regional forcing, internal ice-sheet dynamics, and possibly

one or several feedback mechanisms. Early geochemical analyses of IRD grains to establish provenance gave hints that not all Heinrich layers are lithologically identical and thus might have different source regions (e.g., see Grousset et al. 1993). For example, event H-3 was clearly distinct from events H-I, H-2, H-4, and H-5 (Bond et al. 1992; Gwiazda et al. 1996a, b). Event H-3 began gradually, in contrast to the abrupt beginning of other events. H-3 also had a lower detrital content compared with that of other Heinrich layers and was characterized by a depletion in foraminiferal abundance, reflecting a low-productivity surface ocean. Strontium and neodymium analyses of the silicate fraction

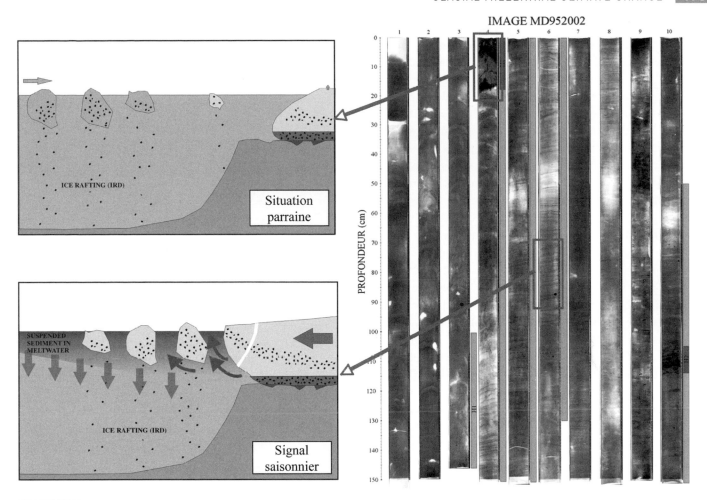

FIGURE 6.14 Photograph of deep-sea core off northwest Europe showing sediment layers and models of ice-rafting deposition (IRD) during Heinrich events. Courtesy of F. Eynaud and S. Zaragosa.

Core MD012461   Laminae inside Heinrich event layer

Porcupine Bight, North Atlantic

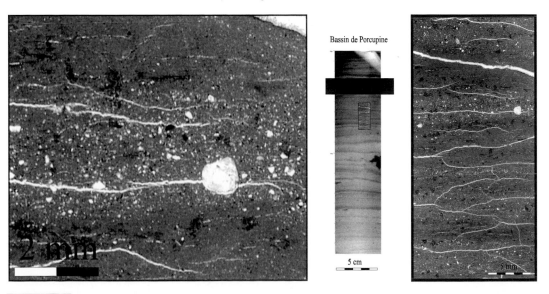

FIGURE 6.15 Photos of the Heinrich layer from Porcupine Bight, northeast Atlantic Ocean. White circles are the shells of foraminifera. Courtesy of F. Eynaud and S. Zaragosa.

show that H-3 had a different source of origin than other layers. Lead isotopes from individual feldspar grains indicated a Greenland-Scandinavian source region for the lithic grains, as opposed to a Canadian Shield source for other Heinrich layers coming from the Hudson Strait area. Most H-events also had radiometric ages on the fine fraction carbonates based on K/Ar dating that are much older than the age of material in background sediment. But in H-3, the K/Ar ages of Heinrich layer and background sediment are about the same.

Bond and Lotti (1995) fingerprinted IRD sediments by analyzing 15 different types of lithic grains from cores of North Atlantic sediments off western Ireland. They identified three main source areas: (1) Hudson Strait (Andrews and Tedesco 1992; Andrews et al. 1993; Dowdeswell et al. 1995); (2) the Gulf of St. Lawrence, draining portions of the Laurentide Ice margin; and (3) the Icelandic Sea, a source of ice from eastern Greenland, Svalbard, and the Arctic Ocean via the Denmark Strait. Hematite-coated grains from the Gulf of St. Lawrence discharge and basaltic glass traced to the volcanic regions of Iceland were particularly good provenance tracers. IRD fingerprinting suggested that discharges from Hudson Strait might have slightly lagged iceberg discharges originating from the other sources, raising the idea of "precursor" events (see "Precursor Events" below).

Several other proxies have since been used to identify the source of IRD material. These include lead isotopic composition of ice-rafted feldspar grains and organic carbon, among other methods (Rosell-Melé et al. 1997; Hemming et al. 1998, 2000; McManus et al. 1998; Huon et al. 2002). Most methods indicate a Hudson Strait source for at least some layers of H-1, H-2, H-4, and H-5. H-3 and H-6 seem to have originated from more complex sources. Grousset et al. (2000) fingerprinted H-2 using strontium and neodymium and concluded that the initial iceberg calving was from European Ice Sheet sources and was synchronous with changes in atmospheric dust recorded in Greenland ice cores. The associated H-2 LIS surge and deposition of the remaining part of the Heinrich layer did not occur until 1500 years after the initial European ice discharge. At least for this event, the idea of precursor discharges from European ice sheet margins was supported (Scourse et al. 2000).

Glacio-marine processes proximal to the ice-sheet margins control the discharge and surging of ice and thus influence the lithological composition and geographical pattern of far-field IRD deposition and, ultimately, the search for causes of H-events (Dowdeswell et al. 1995, 1999; Andrews 1998; Hillaire-Marcel and Bilodeau 2000). One important distinction is that between meltwater-sorted debris, which is common in the Labrador Sea slope region, and poorly sorted iceberg-rafted sediments common in the Nordic Seas. By using these and other sediment properties in regions close to the source of icebergs (Zaragosi et al. 2001; Peck et al. 2006; Eynaud et al. 2007), it is possible to identify the source region as well as the mechanism of IRD transport.

Oceanic Response During Heinrich Events

Changes in Atlantic surface temperature during H-events have been reconstructed using a variety of SST proxies. Bond et al. (1992, 1993) clearly showed surface ocean cooling for much of the North Atlantic using *N. pachyderma* abundances, and Sachs and Lehman (1999) and Rühlemann et al. (1999) estimated that the tropical Atlantic experienced 3–5°C decreases in SST for some events. Cortijo et al. (1997) found a 1–2°C SST decrease and a 1.5–3 psu surface salinity decrease between 40°N and 50°N, but little salinity change north of 50°N, during H-4 in the North Atlantic Ocean using foraminiferal isotopic records. Zahn et al. (1997) found complex mid-depth THC changes based on oxygen isotopes and IRD off the Portuguese margin. Bard et al. (2000a) used alkenone paleothermometry SST estimates and found SST decreases of ~4–5°C for H-2 and H-3 and up to 6°C for H-1. The amplitude of reconstructed SST changes during H-events from these and other studies suggests that glacial-age oceanic response to iceberg discharges includes significant oceanic cooling of several degrees and decreased salinity in places (Voelker et al. 2002). Rashid and Boyle (2007) also found foraminiferal oxygen isotopic evidence for increased vertical mixing of the mixed ocean layer and the thermocline in the North Atlantic Ocean, perhaps due to greater storminess, during H-0, H-1, and H-4.

Relationship Between Heinrich and Dansgaard-Oeschger Events

Large H-events occur only during the largest DO events, and the reason for this pattern, and more generally the connection between atmospheric DO variability and ice-sheet-dominated Heinrich dynamics, is still only partially understood. Initial evidence that Heinrich layers were deposited about every 7–12 ka in a quasi-cyclic pattern led to the binge-purge hypothesis (MacAyeal 1993a, b; Alley and MacAyeal 1994). This theory holds that as large continental ice sheets grew and thickened (the binge stage), they eventually reached a critical threshold when a surge discharged enough ice to thin ice sheets by as much as 1200–1460 m (the purge stage). Such an explanation is somewhat independent of any potential relationship to rapid atmospheric changes during DO events.

Bond and Lotti (1995) first showed, however, that calving events produced IRD events, albeit smaller than the largest Heinrich layers, every 2000–3000 years in North Atlantic cores, suggesting a closer link between DO warming in Greenland ice cores and H-events. A large and rapid warming followed the IRD event as subpolar oceanic temperatures reached nearly interglacial warmth. The saw-tooth SST patterns seen in this series of brief 2–3-ka-long interstadial periods, each one slightly cooler than the previous one, were bundled together over a 10–15-ka period and these were called Bond cycles by Lehman (1993). In a comprehensive review of North Atlantic H-events, Bond et al. (1999) identified 53 hematite-stained layers with an approximately 1–2-ka pacing, the largest events corresponding to the standard H-1–H-6 events. Comparing these records to ice cores, they concluded that IRD layers coincided with Greenland stadials and were caused by a common climatic forcing, which was amplified in the DO temperature oscillation. They acknowledged that the large H-events, particularly H-1, H-2, H-4, and H-5, would require massive surge-like behavior of the LIS margin in the vicinity of Hudson Strait, as proposed in the binge-purge hypothesis.

The timing of DO and H-events is now recognized from many paleoclimate records. For example, comparison of Indian Ocean Moomi and Chinese Hulu speleothem records to Greenland ice-core records shows that H-5 represents a cooling phase that precedes warming during DO-12 and that the total sequence takes only about 1000 years (from 50.4–49.4 ka) (Burns et al. 2003). Huber et al. (2006) measured nitrogen isotopes from ice cores and linked H-events to Greenland stadial intervals as follows: H-6 to DO-18/17, H-5.2 to DO-15/14, H-5 to DO-13/12, and H-4 to DO-9/8 (Genty et al. 2003). The relationship between large DO and H-events can be seen in the methane synchronized greenhouse gas record illustrated in Figure 6.16 from Ahn and Brook (2007). It shows CO_2 concentrations increase during large DO events 8, 12, 14, and 17 prior to the corresponding H-events H-4, H-5, H-5.2, and H-6. Oxygen isotopic values peak, signifying maximum warmth during Greenland interstadials, during or just after the H-event and about the time CO_2 concentrations begin to decline. Ahn and Brook's analysis also shows that CO_2 concentrations generally parallel the atmospheric temperature over Antarctica. The current understanding of H-DO pairs given in Table 6.1 is subject to refinement but for the most part is independently verified for some large events (see Ahn and Brook 2008).

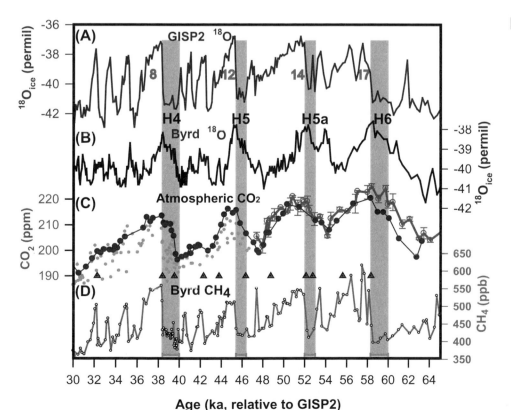

FIGURE 6.16 Comparison of northern-hemisphere temperature (Greenland Ice Sheet Project 2 [GISP2] oxygen isotopes) and southern-hemisphere temperature (Byrd oxygen isotopes) with atmospheric CO_2 concentrations from the Byrd and Taylor Antarctic ice cores during Dansgaard-Oeschger and Heinrich events, from Ahn and Brook (2007) and references cited in their paper. Interhemispheric synchronization is done using methane concentrations. CH_4 = methane; ppb = parts per billion; ppm = parts per million; permil = parts per thousand. Courtesy of J. Ahn, E. Brook, and the American Geophysical Union.

Causes of Glacial-Age Millennial Climate Events

Rapid climatic events during the last glacial period involve all parts of the climate system—ice, oceans, atmosphere, and biogeochemical cycles—in high and low latitudes, in the atmosphere and oceans, and in both the northern and southern hemispheres. The global climate system acts as a highly coupled system, with all parts responding to and, in some cases, playing a role in glacial-age climate reversals through complex, partially understood feedbacks. Climatic transitions are abrupt, switching between stadial and interstadial modes. Moreover, there is evidence that millennial-scale climate variability is a persistent characteristic of the earth's glacial-age climate system prior to the last glacial interval (McManus et al. 1999; Martrat et al. 2004; Siddall et al. 2007). Thus, it is important to consider the hypotheses generated over the past few years to explain paleoclimate records of glacial-age climate behavior.

Soon after they were first recognized as an important mode of global climate variability, a number of causes were proposed to explain millennial climate variability. These include changes in low-latitude hydrological cycles perhaps related to insolation (Chappellaz et al. 1993), internal dynamics of the LIS where basal lubrication causes periodic surges about every 7 ka (MacAyeal 1993a, b); ice-ocean-atmosphere interactions in North Atlantic Ocean circulation (Bond et al. 1993; Bond and Lotti 1995) possibly related to solar activity (Bond et al. 1997); continental ice-sheet influence of atmospheric circulation (Clark and Bartlein 1995); tropical atmospheric changes in trade winds, linked to THC (Hughen et al. 1996); atmospheric changes in water vapor (Lowell et al. 1995); and perhaps other greenhouse gases responding to orbital precession (Curry and Oppo 1997).

One overriding theme in efforts to attribute causality to millennial events is the notion of synchroneity. If millennial climate oscillations are triggered by a single external cause, such as small changes in solar irradiance, or a threshold level of greenhouse gas concentrations, their manifestation at least in the atmosphere might be synchronous although spatially complex and varying in amplitude (i.e., polar regions showing greater warming or cooling). Conversely, the presence of asynchronous, antiphased climate patterns in the two polar regions implies a strong role for MOC, at least in transferring the initial signal in one hemisphere to the other, as already discussed herein. Another possibility discussed in several papers is that DO and H-events are quasi-stochastic processes related either to the atmospheric and oceanic response to internal dynamics of large ice sheets or to the dynamics of the atmosphere-ocean system itself. Here we review processes and mechanisms, none mutually exclusive of the others, that have been proposed to explain characteristics of DO and H-events.

Bipolar Seesaw

Asynchronous High-Latitude Climate Variability

The bipolar seesaw is one widely accepted hypothesis to explain the asynchronous pattern of northern- and southern-hemispheric climate oscillations between about 100 and 20 ka. Climate events originating in the North Atlantic region would in theory alter meridional oceanic circulation, with global climatic repercussions. A related concept is the "salt oscillator" hypothesis (Broecker et al. 1988a, b, 1989, 1990b) whereby NADW formation and global MOC is vulnerable to the influence of glacial meltwater from continental ice sheets. Freshwater influx and resulting decrease in surface salinity made North Atlantic surface water less dense and slowed down NADW formation in the Nordic Seas. Without deep-water formation pulling warm water from the North Atlantic Drift-Gulf Stream system northward from low latitudes, northern Europe would be deprived of the oceanic current.

In addition to the convincing evidence for asynchronous interhemispheric atmospheric change during DO events from ice-core records as already described herein, Raisbeck et al. (2007) confirmed that DO stadial events coincided with Antarctic interstadial events, based on synchronized beryllium records from the Antarctic EPICA and Greenland ice cores. The exceptional record of DO-10, which had the least chronological uncertainty from ice-air age differences, provides particularly compelling evidence for interhemisphere asymmetry of atmospheric millennial variability.

There is now considerable evidence for interhemispheric asynchroneity during millennial climate variability in the last glacial period. H-events are accompanied by a slowdown in NADW formation and incursion of southern-source deep water into North Atlantic basins, although patterns of deep water and ice-rafting vary among different H-events (Vidal et al. 1997, 1998; Keigwin and Boyle 1999; Labeyrie et al. 1999). Vidal et al. (1999) reconstructed the timing of radiocarbon-dated H-events in the north, equatorial, and south Atlantic Ocean using oxygen isotope records and climate modeling back to 60 ka. Their evidence clearly showed antiphasing in which northern-hemisphere isotopic records of warming lag those in the south hemisphere by about 1500 years for H-5 and H-4 events. Their studies supported a slowdown of NADW formation and a freshwater forcing mechanism for H-events.

Southern-hemisphere oceanic records provide support for the bipolar seesaw concept and possible freshwater forcing of MOC. Kanfoush et al. (2000) provided one of the best records of Southern Ocean millennial variability between 20 and 74 ka derived from IRD and foraminiferal $\delta^{13}C$ measurements. They found that strong increases in NADW formation and synchronous deposition of southern-hemisphere IRD in the southeast Atlantic coincided with warming events over Greenland. Antarctic ice-sheet destabilization was inferred as the cause of IRD deposition, and two possible mechanisms to explain the observed patterns would be either northern-hemisphere SLR due to northern-hemisphere ice-sheet melting, strong export of warmer NADW water to the southern hemisphere where it eventually might influence ocean temperature around Antarctica, or both. Jennerjahn et al. (2004) identified major marine and terrestrial climate changes during H-events off the Brazilian margin between 0 and 85 ka, demonstrating a several-thousand-year lag between the ocean and atmosphere. Pahnke and Zahn (2005) showed antiphasing between periods of strong NADW (Greenland interstadials) and Antarctic Intermediate Water in a study of Southern Ocean paleoceanographic records. The asynchronous pattern was particularly clear for H-4 and H-5, and in the Southern Ocean for H 1, H-3, H-5a, and H-6. In general, they provided robust support for the bipolar seesaw hypothesis from the southern-hemisphere ocean.

Brown et al. (2007) studied a detailed millennial climate record from Lake Malawi for the last 55 ka. Evidence suggests that rapid climatic shifts coincide with DO warming events in Greenland and, although the Malawi sediment age model event does not permit detailed comparison, Brown et al. suggest that the tropical climate shifts involve southward shifts in the position of the ITCZ and that these preceded the DO warming seen in ice cores. The lake record thus also suggests an asymmetric pattern between high and low latitudes during glacial-age millennial climate events.

The issue of whether the bipolar seesaw mechanism, including changes in NADW and AABW strength, operates during more frequent DO events is not as clear. Keigwin and Boyle (1999) found evidence for SST and NADW variability on the Bermuda Rise in the Sargasso Sea region at a water depth of >4500 m between 32 and 58 ka, including clear evidence from carbon isotopes and Cd/Ca ratios for deep-water variability during DO-8. The location of the Bermuda Rise cores is extremely sensitive to changes in the relative strength of NADW and AABW. In the North Atlantic at shallower water depths (<2000 m), however, there is no clear deep-water mass variability associated with DO events, leading Elliott et al. (2002) to call DO deep-water variability more "subtle" than that during H-events. Additional high-resolution ocean records are needed to resolve details of deep circulation during DO events.

Freshwater Forcing
For the salt-switch oscillator to work, there must be a source of freshwater to lower surface ocean salinity, resulting in a slowdown of NADW forma-

tion. Clark et al. (2001) reviewed evidence for glacial lake discharge in the North Atlantic region and concluded that freshwater forcing played a major role in H-event 1 about 18 ka, as well as in other deglacial millennial events. Hall et al. (2006) postulated that a relatively small, targeted freshwater forcing from ice somewhere in the northern Europe-British Isles region was in evidence from $^{231}Pa/^{230}Th$ ratios from high-sedimentation-rate regions on the Rockall Plateau. The marine sediment record in this region is a particularly good recorder of large, rapid deglaciation events along the northwest European margin. They discovered that a large slowdown in NADW pre-dates the H-1 event by 1200 years. A series of freshwater discharge events, some lasting only 90–150 years and probably originating in localized glacial-age ice, were identified. The implication is that relatively minor freshwater discharge events during earlier H-events and DO oscillations might produce a slowdown of NADW formation and the characteristic DO temperature patterns.

Another locus of freshwater forcing is the Gulf of Mexico, where LIS meltwater drained via the Mississippi River system during the last glacial period. Hill et al. (2006) documented five periods of freshwater discharge into the Gulf of Mexico flood between 28 and 45 ka that were coeval with warming events in Antarctica but did not coincide with Greenland interstadials. They suggested that these discharge events might correspond with global ice-volume changes.

Modeling Studies
A variety of climate models are able to simulate threshold-like climate behavior, hysteresis, and abrupt changes in THC, some forced with relatively small amounts of freshwater influx to sensitive parts of the world's oceans, similar to that interpreted from some paleoclimate patterns (Tziperman 1997; Marotzke 2000; Ganopolski and Rahmstorf 2002). Some simulations show that changes in MOC might result in changes in polar climate that are approximately asymmetric about the equator—the subpolar North Atlantic Ocean and adjacent continents warm as strong MOC carries heat to high latitudes and releases it to the atmosphere; southern oceans will cool correspondingly (Manabe and Stouffer 1988; Rind and Chandler 1991; Rahmstorf 1994, 2003). Stocker and Johnsen (2003), for example, successfully modeled the bipolar seesaw behavior evident in Greenland and Antarctic ice cores during MIS 3. Intrahemispheric synchroneity is also reproduced in some models in which both polar and equatorial regions in one hemisphere would cool and warm simultaneously because of atmospheric circulation changes and teleconnections, as is seen in the correlation between the Greenland ice core and northern-hemisphere low-latitude speleothem records. Other simulations, however, suggest that changes in THC will not necessarily result in a global climate change (see Stocker 2000). We

now turn to alternative mechanisms that trigger or amplify THC to explain DO and H-events, and in Chapter 9 on geological evidence for abrupt climate change we return to the data-modeling comparisons specifically aimed at simulating the impact of freshwater discharge through different route, various volumes, and durations.

Solar Variability

Solar variability is a potential external trigger that might initiate or modulate millennial atmospheric variability. Bond et al. (1997, 2001) viewed solar forcing as one factor responsible for patterns of millennial variability during both the last glacial period and the Holocene interglacial. The solar theory hinges in part on the apparent 1470-year periodicity of millennial events, evidence from cosmogenic nuclides for past changes in solar activity, and model simulations showing small solar changes can be amplified through stratospheric feedbacks to produce large climate oscillations (Shindell et al. 1999, 2001).

A wide range of opinions exists on the significance and even the reality of the ~1470-year quasi-cycle. Some researchers argue that, if the 1470-year quasi-cyclic variability is real, confirmed by improved age models and global in extent based on multiple proxies, then it suggests some kind of external forcing of climate. If there is no 1470-year cycle, it is more likely that DO events are caused by internal processes and represent "noise" in the climate system (Ditlevsen et al. 2005).

We previously mentioned that, although 1470 years is commonly cited as the periodicity for DO events, the timing between DO events actually varies greatly. Wunsch (2000) analyzed DO variability and argued that the apparent ~1500-ka period might result from aliasing of the seasonal cycle, resulting in a kind of artifact produced by applying a 365-day year to a 365.25-day solar year. Because of insufficient sampling resolution, aliasing as a potential bias in the analysis of cycles can lead to identification of nonexistent cyclic climate behavior. Wunsch proposes that millennial DO and Heinrich variability are aperiodic and part of a continuum of stochastic variability across millennial timescales. Alley et al. (2001) evaluated whether DO events might signify stochastic resonance, a term that refers to possible maximizing of the signal-to-noise ratio in nonlinear, bistable systems such as it appears characterizes the glacial-age climate. Stochastic resonance involves both stochastic-system behavior and possible deterministic forcing that might cause the system to jump from a stadial to an interstadial state. Alley suggests that stochastic resonance might explain why, at times, DO events are "skipped"—that is, why more than 1500 years separate consecutive events.

Schulz (2002) performed statistical analyses of DO variability between 13 and 46 ka and argued that the 1470-year cyclicity applies to only three of the more than 20 DO cycles (DO-5, 6, and 7), that DO spacing varies by up to 20% around the mean 1470-year period (or multiples of it), and that changes in NADW strength do not account for the patterns. Schulz prefers to refer to the 1470-year pattern as "pacing" the DO-style of variability. Braun et al. (2005) showed through climate model simulations that the 1470-year periodicity could be produced without direct solar forcing as a result of well-known 210- and 87-year solar cycles, amplified by small freshwater forcing of MOC. Thus, the millennial events may reflect the combined influence of nonlinear dynamic response to high-frequency solar oscillations and a manifestation of long-term ocean circulation.

In contrast, Ditlevsen et al. (2005) provided a statistical argument that the apparent 1470-year cycle could simply be explained in terms of noise in the system. Using new time-scales for NGRIP and GISP2 cores, they later (Ditlevsen et al. 2007) concluded that the recurrence time for DO events could not be distinguished from stochastic behavior, except in one case in which the DO-9 event is omitted from the GISP2 core. This conclusion was generally consistent with Muscheler and Beer (2006), whose analyses of [10]Be and $\delta^{18}O$ isotopes for DO cycles 2–15 from the GRIP ice core did not find low [10]Be values before DO interstadial events, which would be expected if solar-induced cosmogenic isotopic forced the interstadial warming. Their results thus seem to discredit a proposed solar-DO linkage, although as previously herein, the reconstruction and interpretation of cosmogenic isotopes in the pre-Holocene ice is complicated by processes of [10]Be formation, transport, and deposition. Thus, Muscheler and Beer emphasize that their results do not mean that solar variability could not have influenced climate through feedbacks and amplifications by THC.

For now, there is no consensus about the hypothesis that DO-Heinrich variability is periodic and externally forced by solar variability, but further testing with better records of solar activity and climate sensitivity to small solar changes are needed.

Orbitally Driven Changes in Insolation

Although our focus here is on millennial events, glacial-age climate variability in atmospheric methane, pushed by precession-driven changes in insolation and possible monsoon and wetland activity, have long been known from ice-core and other records (Chappellaz et al. 1990). Orbitally driven changes in insolation may be influential in presetting conditions for the development of more abrupt DO and H-events. Brook et al. (1996) suggested that orbital changes are not directly responsible for observed methane changes but might modulate the millennial-scale amplitude of methane cycles. Chapman and Maslin (1999) also offered evidence for the progressive influence of precession-driven insolation in tropical latitudes from 40 ka until the LGM. The effect of insolation was to greatly increase the latitudinal SST gradient up to 6–8°C and the salinity gradient as much as 2‰ due to advection in the surface North Atlantic.

Atmospheric Greenhouse Gases

Greenhouse gas concentrations clearly fluctuate during DO and H-events. Most researchers favor wetland sources for CH_4 variability between 50 and 200 ppbv and believe that methane concentrations rise as a consequence to, not as a primary forcing agent of, abrupt DO temperature fluctuations. An alternative source of methane is deep-sea gas hydrates. Kennett et al. (2000b) proposed that rather than terrestrial wetland, methane gas hydrates into a solid form of water and methane stored in ocean sediments, particularly along continental margins, under the right pressure and temperature conditions. Changes in bottom-water temperature, sea level, submarine landslides, and other causes have been proposed to explain methane hydrate disassociation.

The methane hydrate instability hypothesis has been discounted for certain glacial intervals and the last deglacial and Holocene epoch by analysis of Greenland and Antarctic ice cores and circumarctic peatland records. Brook et al. (2000), for example, showed that the rise in methane concentrations during DO interstadial events 8 and 12 about 32 and 45.5 ka occurred over several centuries, much slower than the accompanying rise in atmospheric temperature and snow accumulation in Greenland. Similar timing was observed for the Bølling-Allerød and Younger Dryas events of the last deglaciation. MacDonald et al. (2006) demonstrated that high-latitude wetland growth contributed significantly to the postglacial and early Holocene rise in atmospheric CH_4 concentrations and suggest that wetland activity might also explain glacial-age methane variability. Schaefer et al. (2006) also argued against an oceanic methane source on the basis of analyses of $\delta^{13}C$ of carbon trapped in CH_4 in the GISP2 ice core. They concluded that changes in tropical wetlands or methane production by aerobic production could account for measured swings in methane.

As previously noted herein, CO_2 concentrations began to rise several thousand years before peak DO temperatures and the deposition of Heinrich layers. This pattern raises the possibility that oceanic uptake and release of CO_2 might play a role in millennial events. Given that some regions of the deep sea cooled ~1–3°C during millennial events of the last 60 ka (Cronin et al. 2000; Martin et al. 2002), Martin et al. (2005) used Mg/Ca paleothermometry to reconstruct deep-water temperature and evaluate oceanic controls on pCO_2 solubility during DO and H events. Although low sedimentation rates precluded precise ice-core-bottom-temperature correlations, their results indicate that deep-ocean glacial-age temperature change exerted a primary control over oceanic uptake of CO_2 from the atmosphere. In contrast to carbon-ocean cycling during the last deglacial interval, when changes in biological productivity and ocean circulation account for the postglacial rise in atmospheric CO_2 and temperature, Martin's study suggested that the pCO_2/temperature relationship during the last glacial period was different. Relatively large millennial-scale CO_2 changes seem to occur despite proportionately smaller bottom-water temperature changes. This hypothesis holds that the solubility of atmospheric CO_2 in the world's oceans during DO-Heinrich oscillations was mainly temperature-dependent.

Figure 6.16 from Ahn and Brook (2007) shows that atmospheric CO_2 began to rise several thousand years before DO events 8, 12, 14, and 17 in Greenland and that it ceased its rise at the onset of the interstadial marked by the isotopic temperature signal (see also Stauffer et al. 1998). CO_2 concentrations rose with Antarctic warming, though there was a high degree of variability among millennial events. In general, these results support a northern-hemisphere forcing of Southern Ocean leading to changes in Southern Ocean productivity and CO_2 drawdown. This interpretation is supported by climate modeling of rapid freshwater forcing in the North Atlantic (Schmittner et al. 2007b).

Finally, water vapor might influence global climate and produce a synchronous pattern of warming and cooling. Glacial records from New Zealand and South America, for example, suggest that southern-hemisphere glacial advances may have been synchronous with the northern-hemisphere ice rafting and isotopic climate events (Lowell et al. 1995; Lowell 2000; Moreno et al. 2001). If this were the case, atmospheric forcing from changes in water vapor would be one possible explanation.

Ice-Sheet Dynamics

Early Theories The LIS was the largest of the last glacial period and may have been sufficient to trigger H-events and perhaps a global climate response. MacAyeal's binge-purge hypothesis explaining internal dynamics of the LIS could account for H-events in that, when an ice sheet is thin, the temperature at its base is < 0°C (a cold-based ice sheet). As the ice sheet thickens, the basal temperature of the ice sheet increases until it reaches 0°C and basal melting occurs. Basal melting can lubricate the bed, causing ice to surge, eventually thinning and refreezing. Ice-sheet surging can lead to the discharge of icebergs into the North Atlantic surface layer, lowering the salinity and reducing deep-water formation from the conveyor belt. As THC diminishes, less heat is drawn to high northern latitudes, and this acts as a negative feedback leading to increased ice growth in the Laurentide area. A self-sustaining cycle is created.

Clark and Bartlein (1995) favored a role for the LIS, but suggested its primary impact was on atmospheric circulation. They discussed four possible mechanisms to explain synchronous millennial climate variability in the North Atlantic region and western North America. They dismiss the atmosphere and the ocean as the primary drivers, in part because general-circulation models do not predict that western North America should cool to the degree indicated by the glacial stratigraphy. Clark and Bartlein favor the hypothesis that the LIS itself exerted an influence on temperature and atmospheric circulation. Drawing on the analogous situation of the LGM, when the ice sheet depressed western North American temperatures 4–l0°C and shifted the jet stream southward (Imbrie et al. 1992), they propose that a similar situation may have occurred during millennial-scale climate change, when a second-order climatic variability was superimposed on predominant long-term orbitally driven cycles. In essence, this theory holds that short-term glacial expansion before an H-event would increase the glaciation potential in the west, and decreased ice-sheet size would cause the reverse. These ice-sheet effects acted in concert with changing levels of insolation during the past glacial period about 50–20 ka.

Precursor Events The binge-purge model might operate through basal melting and lubrication of parts of the LIS (Marshall and Clarke 1997), but it does not explain several other aspects of H-events. For example, if destabilization and iceberg calving occurred along one region of an ice-sheet margin, the resulting SLR might destabilize its other margins or other ice sheets. This notion of precursor events is widespread in the paleoclimate literature on deep-sea IRD, isotope records, and glaciological processes. The most widely mentioned example is asynchroneity between SLR associated with initial massive iceberg discharge from northern Europe's ice sheet and the subsequent drawdown of Laurentide ice feeding Hudson Strait.

The role of precursor events prior to many of the largest DO-H-events was recognized by Bond and Lotti (1995) in the dominance of *N. pachyderma* (sinistral) and hematite and glass IRD in the early phase of some H-events. Important geochemical evidence suggests a European ice-sheet source for IRD ~1.5 ka before Laurentide ice IRD for most H-events (Grousset et al. 2000). Lekens et al. (2005) also found extensive evidence for European Ice Sheet in the Norwegian Sea and Svalbard sectors ~18.6 ka, pre-dating Greenland and LIS discharges by 600–1000 years for H-1. They proposed that warm ocean water flowing into the Norwegian Sea was the trigger for the initial collapse of parts of the Fennoscandian ice-sheet margin. Complex ice-stream-induced meltwater

events from Fennoscandian ice also occurred in the Norwegian Sea during H-2–H-4, not associated with IRD events (Lekens et al. 2006).

McCabe et al. (1998) analyzed the paleoceanographic and glacial history of the margins of the ice sheet in the Irish Sea Basin for H-1 and recognized a sequence of broad deglaciation, a readvance during H-1, and finally ice-sheet collapse. Their conclusion regarding the relative timing of the Irish Sea margin and that of the LIS was in direct contrast to that of Lekens for the Norwegian Sea region. McCabe instead postulated that the cooling in the North Atlantic region caused by LIS ice discharge was responsible for Irish Sea ice readvance during H-1.

Peck et al. (2006) performed multi-proxy analyses of northeast Atlantic H-events between 26.5 and 10 ka using lithology, strontium and neodymium isotopes, and $^{40}Ar/^{39}Ar$ ages of hornblende grains to fingerprint IRD layers and their provenance. They concluded that British Ice Sheet variability occurred at a 2-ka periodicity and that its behavior was independent of LIS surges. Because the northwestern European Ice Sheet margin (Icelandic and British ice sheets combined) was mostly absent during H-3 (31.5 ka), H-4, and H-5, this ice sheet could not have served as a precursor ice-sheet source.

The behavior of ice-sheet margins as responders to and forcing agents of H-events is discussed in many papers (Labeyrie et al. 1999; Elliott et al. 2002; Hemming 2004). To bring some clarity to the issue, Dowdeswell et al. (1999) reviewed potential types of glaciological behavior during the glacial period. They point out that an ice-sheet response to a climatic event or a rise in sea level is not necessarily uniform along all ice-sheet margins. Surges, or some internal dynamics of the glacial ice itself, can extend the margin of an ice sheet away from its center, possibly making it vulnerable to ocean temperatures or other factors. A second factor relates to the fact that ice sheets equilibrate in response to long-term climate changes, such as orbital cycles, and thus IRD layers may simply reflect a delayed response to much longer-term forcing. Moreover, an ice sheet can exhibit a direct response of parts of its margin to a local or regional change in factors such as atmospheric, or more likely, oceanic temperatures. This last category seems to be important during rapid transitions at the onset of DO events. Dowdeswell et al. thus propose that random processes can lead to ice-sheet surges along Europe during the last glacial and so, to a certain degree, there is a random component to H-events, even if some external forcing agent is involved.

Heinrich Events, Ice-Shelf Collapse, and Jökulhlaups
Besides the binge-purge hypothesis, other hypotheses specifically addressing H-event ice dynamics have been proposed, and we mention two of them here. The first, by Hulbe (1997) and Hulbe et al. (2004), holds that ice shelves that

formed during stadials along eastern Canada were vulnerable to collapse if slight warming during the summer season occurred, causing surface meltwater to seep into crevasses. H-events under this scenario occur during robust ice-sheet growth when the ice shelf, which lies seaward of the ice-sheet grounding line, is the major glacial feature to collapse. The ice-shelf-collapse hypothesis draws heavily on the abrupt nature of H-events and the recent observations in modern Antarctica in which destabilization of large ice shelves in 1995 and 2002 has been attributed to a similar melt-freshwater mechanism (Vaughan and Doake 1996; Scambos et al. 2000).

In contrast, Alley et al. (2006) suggest that, as temperatures cool and ice-sheet margins advance onto continental margins, at times they form subglacial lakes in silled basins behind their grounding lines. Freshwater is trapped in lakes because the glacier's base is frozen to a sill, but some water remains liquid. Warming and destabilization of the glacial margin can cause catastrophic outbursts of overpressurized freshwater called jökulhlaups. The outburst is followed by surges in the margins of the ice sheet. Alley and colleagues suggest that most features expected from jökulhlaups are observed in geomorphological records and simulated by glaciological models and thus provide a reasonably good mechanism for freshwater forcing. Modern subglacial lakes are well known, the most famous being Lake Vostok under parts of the Antarctic Ice Sheet, but others exist elsewhere, and late Pleistocene jökulhlaups have been invoked to explain geomorphological and sediment features in formerly glaciated terrain.

Both ice shelves and jökulhlaups involve ice-sheet behavior governed by complex glacial and geomorphological controls and climate-ice dynamic linkages beyond our discussion here. Nonetheless, they are important for the consideration of H-events and, as we see in later chapters, the future of modern ice sheets.

Sea-Ice Switch

One final mechanism involving ice that is proposed to explain large DO temperature warming and Heinrich-type ice-sheet behavior during millennial climate events is the sea-ice switch mechanism of Gildor and Tziperman (2000, 2001) and Kaspi et al. (2004). According to some paleoceanographic reconstructions, large regions of the subpolar North Atlantic were periodically covered with sea ice during the last glacial period, and rapid sea ice grows seasonally in the modern Southern Ocean around Antarctica. The proposed sea-ice switch mechanism, which is successfully simulated in simple box and coupled atmosphere-ocean models, is capable of amplifying a weak response of glacial THC to relatively small freshwater discharges suggested by some paleo-

oceanographic records (Elliott et al. 2002) and accounts for both 5–10-ka Heinrich and 1–2-ka DO cycles. Sea ice acts to amplify and dampen glacial climate oscillations through its effect on albedo and as an insulating mechanism preventing air-sea heat exchange. The hypothesis holds that sea ice grows rapidly during the winter season but is also sensitive to abrupt collapse from ocean warming. Three climate states are envisioned during an idealized Heinrich cycle: (1) an ice-sheet growth phase with growing sea ice, (2) an ice-sheet collapse phase when meltwater forces weakened THC, resulting in atmospheric and SST cooling, and (3) a sea-ice melting phase caused by heat stored in the deep ocean brought to the surface by convective mixing. Rapid sea-ice melting releases heat to the atmosphere, causing the abrupt atmospheric warming seen in DO interstadials following H-events, Moreover, Elliott et al. argue that the sea-ice switch mechanism is robust whether or not ice-sheet destabilization is synchronous among various ice-sheet margins.

Sea-Level Change During DO-H-Events

Sea-level change during DO-H-events is closely linked to ice-sheet dynamics. Moreover, the mean glacial sea-level position creates a boundary condition unlike those during interglacial times such that large continental ice volume seems to produce a condition conducive to destabilization, rapid freshwater influx, and periodic jumps in sea level. However, the precise scale of global sea-level oscillations during the last glacial is subject to some uncertainty. On the basis mainly of tropical coral-reef terraces and deep-sea oxygen isotopic records, Chappell (2002) estimated sea-level changes of ~15 m during strong DO-H-events. Siddall et al. (2003) used isotopic records from the Red Sea as a proxy for salinity controlled by sea-level oscillations and the sill depth at the entrance to the Red Sea. The discovered sea level oscillated as much as 45 m during the last glacial period. Siddall et al. (2007) concluded that the range of ~30 m in sea-level variability during MIS 3 was due to about equal contributions from northern-hemisphere and Antarctic ice sources.

A number of researchers have addressed the issue of DO variability in relation to low mean glacial sea level. Schulz et al. (1999), for example, concluded that the ~1470-year pattern of stadial-interstadial variability occurs only at times when continental ice volume exceeds that required to lower global sea level by about 45 m below its current level. These conditions existed for most of the period from 60–20 ka. Antarctic ice-sheet destabilization is also implicit in studies of the temperature-sea-level phasing (Shackleton 2000), the timing of Southern Ocean IRD events (Kanfoush et al. 2000; Pahnke et al. 2003), and the ages of sea-level oscillations obtained from coral terraces (Thompson and Goldstein 2005, and Red Sea isotopic records (Siddall et al. 2003).

Tropical Climate During DO Events

Low-latitude temperature and moisture balance (Grimm et al. 1993), wind strength and oceanic biological productivity (Hughen et al. 1996; Curry and Oppo 1997; Peterson et al. 2000), ITCZ position, and monsoon strength (Y. Wang et al. 2001, 2005; Fleitmann et al. 2007) all experienced major changes during millennial-scale events. Curry and Oppo (1997) suggest that meridional THC can account neither for the degree of tropical cooling and NADW reduction nor for the synchroneity of tropical and high-latitude climate change during DO climate events. Instead, they propose that water-vapor changes originating in the tropical oceans played a role. More recent evidence from the highest-quality paleoclimate records confirms intrahemispheric synchroneity between the northern-hemisphere tropics and high latitudes during DO-H-events within the limits of dating methods (usually decades to a century). The Cariaco Basin record, when compared to records from Greenland ice cores and the low-latitude Hulu cave (Wang et al. 2001), shows a strong inphase relationship between low-latitude rainfall, Greenland temperatures, and other atmospheric components (Hughen et al. 2006). Wang et al. (2006) synthesized the relationship between Greenland and tropical-atmosphere speleothem records and found that weak MOC leads to a southward shift in the ITCZ, which in turn leads to warmer southern-hemisphere SSTs and a stronger SAM. Conversely, the East Asian and Indian monsoons weakened during stadial events. The converse occurs during interstadials. As seen in Chapter 8, similar patterns of high-latitude and tropical atmospheric variability occur during centennial and millennial climate changes during the Holocene interglacial period. In both situations, there seems to be an asymmetry to Hadley cell circulation, reflecting changes in the ITCZ (Wang et al. 2006; Fleitmann et al. 2007).

Although tropical atmospheric and oceanic changes clearly accompany glacial-age climate variability, there is still no consensus as to whether they represent a direct response to changes in North Atlantic atmospheric circulation, on the impact of THC variability on tropical surface ocean and atmosphere, or on whether the tropics actually trigger all or some millennial excursions (Vidal and Arz 2004).

Perspective

It is not yet possible to ascribe a single cause or mechanism to all DO-H-events because not all events have the same pattern and because high-resolution paleoclimate records are lacking from certain ocean and terrestrial regions. One is nonetheless drawn to several conclusions. First, millennial climate variability occurs in the northern and southern hemispheres. These patterns are a fundamental feature of the earth's climate system during the last glacial period. There are sufficient strategically located records from low latitudes and high southern-hemisphere latitudes to say that DO-H-events are global in scale, albeit of varying amplitude. The trigger for millennial reversals might be altered MOC caused by a freshwater influx from glacial lakes and ice sheets, although changes in solar irradiance, greenhouse gas concentrations, ice-sheet thickness, and sea-ice cover likely play roles as well.

Second, whatever the initial trigger that sets off the observed series of events, the initial forcing seems to be small compared to the scale of the climate response. This implies that the glacial climate state has an inherent sensitivity to relatively small and abrupt disruptions, but quantifying climatic thresholds remains a difficult task.

Third, multiple feedbacks amplify the climate response to a small initial forcing. The most likely mechanisms include rapid changes in subpolar sea-ice extent, internal ice-sheet dynamics, sea-level change and destabilization of ice-sheet margins, and tropical-to-polar atmospheric moisture and heat transport.

Fourth, the first-order interhemispheric timing of polar millennial variability indicates a large role for the world's ocean in storing and transferring heat and maintaining a sustained quasi-equilibrium state for centuries to about a few millennia before climate returns to its former state.

LANDMARK PAPERS Glacial-Age Millennial Climate Instability

Dansgaard, W., S. J. Johnsen, H. B. Clausen, D. Dahl-Jensen, N. Gundestrup, and C. U. Hammer. 1984. North Atlantic climatic oscillations revealed by deep Greenland ice cores. In J. E. Hansen and Taro Takahashi, eds., *Climate Processes and Climate Sensitivity*, pp. 288–298. Washington, DC: American Geophysical Union.

Dansgaard, W., S. J. Johnsen, H. B. Clausen, D. Dahl-Jensen, N. S. Gundestrup, C. U. Hammer, C. S. Hvidberg, J. P. Steffensen, A. E. Sveinbjörnsdottir, J. Jouzel, and G. Bond. 1993. Evidence

for general instability of past climate from a 250-kyr ice-core record. Nature 364:218–220.

Johnsen, S. J., H. B. Clausen, W. Dansgaard, K. Fuhrer, N. Gundestrup, C. U. Hammer, P. Iversen, J. Jouzel, B. Stauffer, and J. P. Steffensen. 1992. Irregular glacial interstadials recorded in a new Greenland ice core. Nature 359:311–313.

Oeschger, H., J. Beer, U. Siegenthaler, B. Stauffer, W. Dansgaard, and C. Langway. 1984. Late glacial climate history from ice cores. In J. E. Hansen and Taro Takahashi, eds., *Climate*
(continued)

Processes and Climate Sensitivity, pp. 299–306. Washington, DC: American Geophysical Union.

Millennial-scale climate variations occur at suborbital timescales and often involve abrupt onset and terminations. Dansgaard-Oeschger (DO) and Heinrich events are types of millennial climate variability discovered independently in Greenland ice and North Atlantic sediment cores.

One of the most perplexing types of millennial climate variability involves the way atmospheric temperature in high latitudes of the northern hemisphere abruptly rises or falls within decades with no apparent forcing from insolation or any external trigger. This flip-flopping between two apparently stable climate states has occurred repeatedly, but at irregular intervals and lasting from a few centuries to one or two millennia. These Dansgaard-Oeschger events dominate the climate of the last glacial period between about 100 and 20 ka and attest to extreme climate instability and the ability of climate to abruptly switch between two apparently stable states.

We list four critical papers on glacial-age climate that appeared in the 1980s and early 1990s based on analyses of Greenland ice cores by groups working mainly in Copenhagen, Denmark and Bern, Switzerland. Willi Dansgaard, Hans Oeschger, and colleagues published two in a 1984 American Geophysical Union volume; later, two seminal papers from Sigfus Johnsen (1992) and Dansgaard et al. (1993) presented high-resolution oxygen isotope records from nearly 3000 m of cored ice in the Greenland Ice Core Project at the Summit Greenland site, showing at least 21 stadial-interstadial events, each warm stadial lasting about 500–2000 years. Dansgaard and Oeschger recognized the extreme sensitivity of climate to small perturbations, the role of the oceans in millennial variability, and the value of ice-core paleoclimatology.

Bond, G., H. Heinrich, W. Broecker, L. Labeyrie, J. McManus, J. Andrews, S. Huon, R. Jantschik, S. Clasen, C. Simet, K. Tedesco, M. Klas, G. Bonani, and S. Ivy. 1992. Evidence for massive discharges of icebergs into the North Atlantic ocean during the last glacial period. Nature 360:245–249.

Ocean sediments also contain a history of abrupt glacial-age climate variability. Gerard Bond and colleagues at the Lamont Doherty Earth Observatory, building on the analysis of ice-rafted sediments in the North Atlantic Ocean William Ruddiman, recognized periodic layers of ice-rafted debris (IRD) during the last glacial caused by armadas of icebergs discharged into the Atlantic from the Laurentide and Eurasian Ice Sheets between 50 and 17 ka. They named these Heinrich events (H-events), in honor of Helmut Heinrich's 1988 paper of specific IRD sediment layers from the Dreizack Seamount. H-events are the quintessential example of the critical role ice-sheet dynamics play in abrupt climate variability and are directly relevant to efforts to understand the sensitivity of today's Greenland and Antarctic Ice Sheet margins to rapid disintegration.

Millennial Climate Events During Deglaciation

Introduction

The transition from the Last Glacial Maximum (LGM) to the Holocene interglacial is arguably the geological interval most studied by paleoclimatologists. The last deglaciation took place from ~22 ka until 11.5 ka, when the earth's mean annual atmospheric temperature rose about 5°C and there was regional high-latitude warming exceeding 10°C and smaller but significant tropical and deep-sea warming of 2–4°C and 1–2°C, respectively. Deglaciation also brought major changes in the global hydrological budget and regional precipitation patterns. Most of the world's large northern-hemisphere ice sheets in Eurasia and North America, the Patagonian Ice Sheet, and parts of the Antarctic Ice Sheet (AIS) melted, causing sea level to rise about 120 m. Ocean circulation and chemistry changed dramatically and atmospheric greenhouse gas concentrations rose from low glacial levels.

Deglaciation was neither gradual nor synchronous globally. It occurred in steplike fashion, interrupted by several climate reversals that were characterized by lower regional temperatures, reduced rates of ocean-to-atmosphere carbon dioxide (CO_2) release, slower ice-sheet retreat, altered meridional overturning ocean circulation, and other manifestations. Climate reversals began abruptly in the northern hemisphere and lasted several centuries to one or two millennia. The fall in temperature during deglacial reversals was synchronous in high and low latitudes of the northern hemisphere, but was out of phase between the polar regions. Such a pattern resembles abrupt millennial events during the glacial period described in Chapter 6, except that during deglaciation climate reversals occurred when high-latitude northern-hemisphere summer insolation was steadily rising. This chapter describes patterns of deglacial climate change and their causes.

Deglacial Terminology and the Last Glacial Maximum

We begin with a discussion of terminology that has evolved over more than a century of investigation (Table 7.1). The terms *last deglaciation*, *Termination I*, and *the marine oxygen isotope stage (MIS) 2/1* transition all refer to the period between the LGM ~22-ka and the beginning of the Holocene at 11.5 ka. Older Pleistocene deglacial periods are called Terminations II, III, and MIS 6/5, 8/7, etc. The most famous deglacial event is the abrupt change from the Bølling-Allerød (B/A) warm period to the YD, now dated in the northern hemisphere at 12.9±0.1 ka. The reader should be familiar with deglacial terms, their ages, and their abbreviations.

Last Glacial Maximum

The term *Last Glacial Maximum* is used in many different contexts and requires clarification if we are to discuss its termination and the onset of deglaciation. Defining the LGM is not a trivial exercise because it represents a critical time slice of the earth's climate characterized by global cooling. The LGM is a chronozone encompassing the last time ice sheets advanced at or near their maximum extent and maximum thickness and global atmospheric temperatures were at or near their minima. It is the culmination of the nearly 100-ka-long glacial period that followed MIS 5e. From the perspective of orbital climate oscillations of the last 700 ka, a glacial "maximum" climate state is a valuable concept serving as a direct contrast to an interglacial preindustrial Holocene climate. The view of the LGM as a cold

end-member state is extremely useful for evaluating the earth's climate sensitivity to insolation changes and for testing the skill of climate models to simulate glacial conditions (e.g., Otto-Bliesner et al. 2006). Defining the LGM as a distinct climate state has provided insight into fundamental properties of global ocean circulation under a climate state different from that of the Holocene (Duplessy et al. 1988; Curry and Oppo 2005; Lynch-Stieglitz et al. 2007).

Age of the Last Glacial Maximum

The age of the LGM depends on how one defines it—as the period with the coldest mean atmospheric temperatures, the lowest greenhouse gas concentrations, the maximum geographical coverage of continental ice, the thickest ice sheet, or the maximum ice volume and lowest global sea level. The historic "Climate: Long-Range Investigation, Mapping, and Prediction" (CLIMAP) Project began in 1971 and in many ways revolutionized the study of past global climate changes (CLIMAP Project Members 1976, 1981). CLIMAP mapped the surface of the earth during the LGM in a global "time-slice" climate reconstruction. CLIMAP used a variety of proxy methods to reconstruct sea-surface temperatures (SSTs), land vegetation, and the thickness and extent of ice sheets. CLIMAP preceded the advent of accelerator mass spectrometry (AMS) radiocarbon dating and calibrated radiocarbon timescales and the CLIMAP LGM timeslice was dated at 18 ka. In addition to a much cooler ocean and extensive sea ice in the high-latitude North Atlantic, CLIMAP found that the tropical glacial oceans were about the same temperature or slightly warmer during the LGM than they are today. Building on CLIMAP, Figure 7.1 shows a more recent reconstruction of SST in the Atlantic Ocean based largely on planktonic foraminifera and other proxies (Pflaumann et al. 2003).

A more recent international project called "Environmental Processes of the Last Ice Age, Land, Oceans, and Glaciers" (EPILOG) began in 1999 with the objective of improving our understanding of climate during the LGM (Mix et al. 2001). EPILOG was built on the foundations of CLIMAP and chose the interval 23–19 ka for the LGM because it was centered on the 21-ka age of minimum summer insolation at 65°N latitude and the approximate time that ice sheets reached their maximum volume as defined by the lowest global sea level. Figure 7.2 shows the relationship between the EPILOG glacial maximum and post-LGM millennial events as represented by Greenland ice-core stable isotope records.

Another large program called "Glacial Atlantic Ocean Mapping" (GLAMAP) reconstructed the glacial Atlantic Ocean from 275 deep-sea sediment cores (Sarnthein 1995;

TABLE 7.1 Event Stratigraphy and Terminology of the LGM and Last Deglaciation

Category	Event	Common Useage—Abbrev.	Other Terms/ Comment	Approx. Age (ka)*	Reference
General	Marine Oxygen Isotope Stage	MIS2		~29–10	Shackleton and Opdyke 1973
	Termination 1	T-1	LGM to Holocene transition	~22–10 (various)	Broecker and Van Donk 1970
	Last Glacial Maximum	LGM	Between H-2 and H-1	23–19	Mix et al. 2001, EPILOG
	Last Isotope Maximum	LIM	Marine oxygen isotopes	21.5–18	Sarnthein et al. 2003, GLAMAP
Antarctic stadials	Antarctic stadial	AIM2		24–23	EPICA Community Members 2006
	Antarctic stadial	AIM1	Precedes ACR	15	EPICA Community Members 2006
	Antarctic Cold Reversal	ACR	Antiphase with YD	14–13	Jouzel et al. 1995
N. Atlantic	Heinrich-2	H-2		26–24	Wang et al. 2006; Hemming 2004
	Heinrich-1	H-1		17.5–16.5	Wang et al. 2006; Hemming 2004
	Oldest Dryas	OD-1	OD-1, B/A, OD, YD were originally biostratigraphic units in Europe and the Alps = to pollen zones I, II, III	15.7–14.7	
	Bølling/Allerød	B/A		14.65–12.9	Stanford et al. 2006
	Older Dryas	OD-2		14.1–13.9	
	Intra-Allerød Cold Period	IACP	Brief cooling within the Bølling/Allerød	13.4–13.3	Hughen et al. 2000; Rayburn et al. 2005
	Killarney-Gerzensee	KG	Brief cooling at Allerød-YD boundary	~13	
	Younger Dryas	YD	Some call it Heinrich-0	12.9–11.5	
	Preboreal Oscillation	PBO		11.5–11.3	
	Boreal	None	Onset of Holocene	11.3–~8.5	
	Greenland stadial (GS) and interstadial (GIS) (see Chapter 5)	GIS-2		22–21	Björck et al. 1998
		GS-2	GS-2a = H-1	21–14.7	Björck et al. 1998
		GIS-1	Bølling/Allerød	14.7–12.9	Björck et al. 1998
		GS-1	Younger Dryas	12.9–11.5	Björck et al. 1998
Global sea level	Meltwater Pulse 1ao	MWP1Ao		~19	Clark et al. 2004; Gornitz 2008: 19 ka SL rise
	Meltwater Pulse 1a	MWP1A		14.1(max)–13.8	Fairbanks 1989; Stanford et al. 2006
	Meltwater Pulse 1b	MWP1B		11.5–11	Fairbanks 1989; Stanford et al. 2006
	Meltwater Pulse 1c	MWP1C		8.2–7.6	Blanchon and Shaw 1995; Cronin et al. 2007

*Ages are approximate for duration or midpoint, vary by author, and are subject to revision.

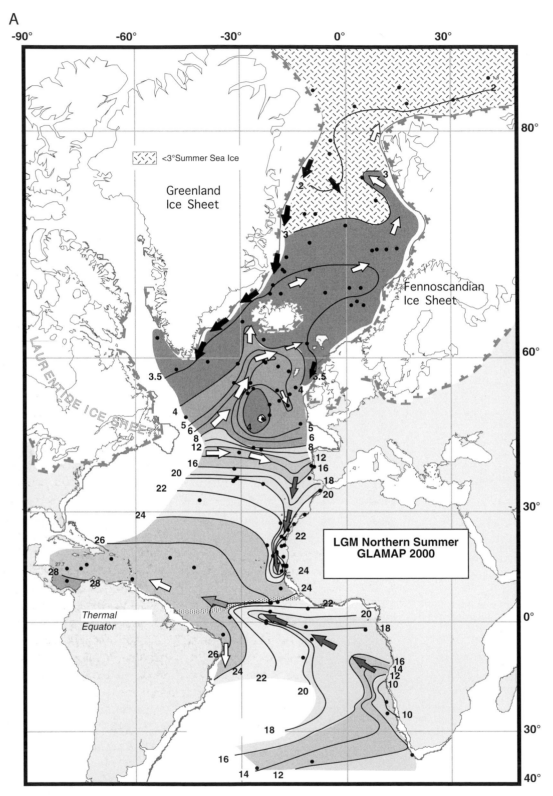

FIGURE 7.1 (A) Atlantic Ocean Last Glacial Maximum (LGM) sea-surface temperatures (SST) during northern-hemisphere summer from Glacial Atlantic Ocean Mapping (GLAMAP) Project. (B) LGM SST during southern-hemisphere summer GLAMAP reconstructions (Pflaumann et al. 2003). Dots show location of cores. Courtesy of U. Pflaumann and the GLAMAP Project.

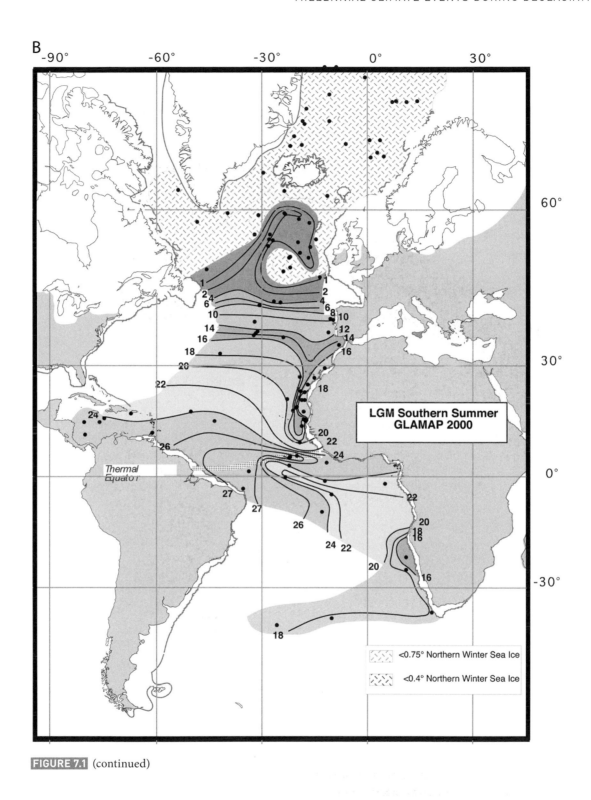

FIGURE 7.1 (continued)

Sarnthein et al. 2003a). GLAMAP's LGM chronozone was defined as 21.5–18 ka, slightly different from EPILOG's, mainly because it coincided with the nearly ocean-wide planktonic foraminiferal oxygen isotope peak called the "last isotope maximum." EPILOG and GLAMAP both used major improvements in chronology that were due mainly to AMS dating and improved SST proxies. Among the most important difference was the newly recognized cooling in tropical oceans during the LGM. Both projects and most others, therefore, place the age of the LGM between Heinrich events (H-events) 2 and 1 about 24–18 ka (Denton et al. 2005).

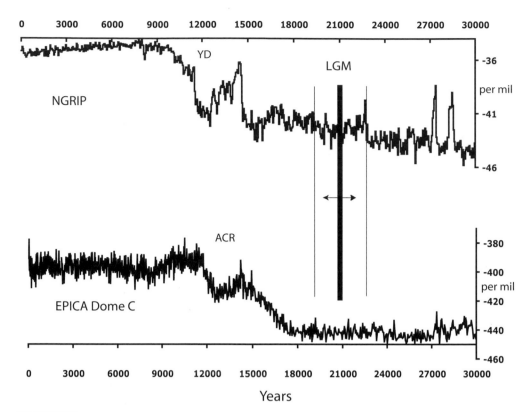

FIGURE 7.2 Last glacial maximum (LGM) timeslice at 21 ka±2 ka (arrow), defined by the Environmental Processes of the Last Ice Age, Land, Oceans, and Glaciers (EPILOG) (Mix et al. 2001) and Multiproxy Approach for the Reconstruction of the Glacial Ocean Surface (MARGO) (Kucera et al. 2005) projects, shown here against North Greenland Ice Core Project (NGRIP) ice-core oxygen isotope data (North Greenland Ice Core Project Members 2004) and EPICA Dome C Antarctic ice-core deuterium (Jouzel et al. 2007). ACR = Antarctic Cold Reversal; YD = Younger Dryas.

The Mediterranean is one of the most intensely studied regions for LGM paleoclimatology. Kuhlemann et al. (2008) synthesized a large amount of marine (SST) and terrestrial (floral, alpine-glacier equilibrium line) proxy data to decipher LGM atmospheric circulation. They discovered strong polar atmospheric incursions into the northwestern Mediterranean and steep temperature gradients in the central Mediterranean during the LGM. Importantly, they deduced that these patterns were qualitatively similar, albeit of larger amplitude, to those reconstructed for the Little Ice Age (e.g., Luterbacher et al. 2004).

Finally, the Multiproxy Approach for the Reconstruction of the Glacial Ocean surface project (MARGO) is an attempt to use multi-proxy records to reconstruct the glacial ocean (Rosell-Melé et al. 2004; Kucera et al. 2005). MARGO is the most ambitious reconstruction, covering all the world's oceans and including the following proxy methods: diatoms, dinoflagellates, planktonic foraminifera (faunal assemblage and δ18O), radiolaria, Mg/Ca ratios, and alkenone. MARGO used the same LGM timeslice as EPILOG but used one of three levels of age uncertainty for each site, depending on chronology. Several papers in *Quaternary Science Reviews* 2005, volume 24, describe the results.

Standard North Atlantic Climatostratigraphy

Björck et al. (1998) recognized the need for standardization of chronostratigraphic and climatostratigraphic terminology for the interval from 22 to 11.5 ka. They proposed a formal scheme focusing on the North Atlantic region, incorporating Greenland ice core, North Atlantic Ocean marine, and European terrestrial records (Figure 7.3). The premise for such an "event stratigraphy" was that Greenland ice cores provided a stratotype for the deglacial, an approach adopted for glacial-age Dansgaard-Oeschger (DO) events (Rousseau et al. 2006). Paleoclimatic changes across the entire North Atlantic region were synchronous within limits of dating methods. Björck followed the convention applied to DO events by numbering Greenland ice-core stadials

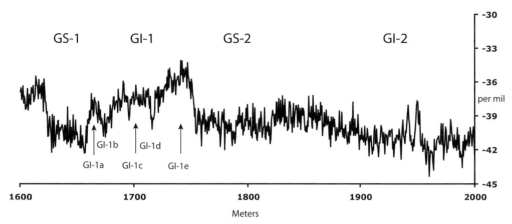

FIGURE 7.3 Oxygen isotope stratigraphy from Greenland Ice Core Project (GRIP) ice core (Johnsen et al. 1992) and deglacial terminology from Björck et al. (1998). GS=Greenland stadial; GI=Greenland interstadial; GS-1 is equivalent to the Younger Dryas; GI-1 is the Bølling-Allerød. period; GS-2 is equivalent to Dansgaard-Oeschger Event 2, which is also divided into GS2a, 2b, 2c.

(GSs) and interstadials (GISs) from top to bottom, as climatostratigraphic units. In this scheme, the onset of North Atlantic regional deglacial warming, called GI-2, begins about 21.8 ka and is followed by stadial and interstadial intervals called GS-2 (21.2–14.6 ka), GI-1 (14.7 ka), the B/A warm period, and GS-1 (12.9 ka), which is the Younger Dryas (YD) cool period. Several substages are recognized within stadial GS-2 and interstadial GI-1. These include GS-2a ~18–16 ka (H-1) and several brief (centuries) reversals called substages GI-1e through GI-1a within the B/A. Some of Björck's events correspond to events first discovered in continental Europe and designated as pollen zones, and used since in paleoclimate studies around the world. These zones include the Older Dryas, the Oldest Dryas, the Intra-Allerød Cold period, and the Preboreal Oscillation (Table 7.1). Rasmussen et al. (2006) updated the absolute ages of Björck's units using annual-layer counting in several Greenland ice cores.

Defining the end of the deglacial period and the beginning of the Holocene interglacial (the Pleistocene-Holocene boundary) is also subject to differing opinions. Its age is widely cited as 10 ka. However, we follow Björck et al. (1998), who adopt the convention that the Holocene began with warming that ended the YD stadial ~11.5 ka. Antarctic ice-core studies also refer to the Pleistocene-Holocene boundary as 11.5 ka (Masson-Delmotte et al. 2000).

Geology is inherently regional and the literature on the last glacial and deglacial intervals is replete with local lithostratigraphic, chronostratigraphic, and climatostratigraphic terms, many predating the discipline of paleoclimatology itself. Regional terms will remain in use for stratigraphic correlation, geological mapping, engineering, and other applications. There is nonetheless a need for consistency in paleoclimatic terminology, given improvements in radiocarbon dating, annually resolved records, and use of calendar-year timescales. The standard European deglacial terms and southern-hemisphere counterparts given in Table 7.1 have gained the widest use and form the primary framework for deglacial climate studies.

The Onset of Deglaciation

Background

There are many views of when the LGM ended and deglaciation began. Take the example of the major transition at the inception of the B/A about 14.6 ka. This abrupt atmospheric warming of up to 12°C over Greenland is also seen in marine and continental records throughout the North Atlantic region, and dominates regional deglacial warming. In a study linking B/A warming to the ice-sheet melting and rapid sea-level-rise event called Meltwater Pulse 1a, Weaver et al. (2003) suggested the B/A onset "marked the termination of the last glacial period." The widely reproduced Greenland ice-core oxygen isotopic records shown in Figures 7.2 and 7.3 give the impression that this was the most significant period of warming between 22 and 11.5 ka.

Most agree, however, that more gradual warming prior to the B/A event marks the end of the glacial. Table 7.2 lists papers describing this LGM-deglacial transition. The ages given represent the inflexion point in various proxy records. Choosing an inflexion point is a critical decision for interpretation of the phasing of climatic events, but it is a subjective exercise

TABLE 7.2 Onset of Post-LGM Warming and Deglaciation

Climate System	Climate Indicator	Region/Site	Deglacial Onset (ka)*	Comment	Source
Atmosphere	Atmospheric CO_2	EPICA Dome C	17	Also Law, Byrd, Vostok core	Monnin et al. 2001
	Antarctic temperature	EPICA Dome C	17.8	Varies at other Antarctic sites	Monnin et al. 2001
	Antarctic temperature	Taylor Dome	17	3000 yrs after Byrd	Steig et al. 2000; Grootes et al. 2001
	Antarctic temperature	Vostok	17		Petit et al. 1999
	Antarctic temperature	Dome Fuji	19–18		Kawamura et al. 2007
	African tropical temperature	W. Africa	17.5	Land-sea seasonal contrasts	Weijers et al. 2007
	Greenland temperature	Multiple records	22	Orbital forcing 65°N	Alley et al. 2002
	Greenland temperature	GISP2	20	Gradual, accelerates ~14.6 ka	Meese et al. 1997
	Greenland temperature	Greenland ice cores	21	Multiple sites	Johnsen et al. 1992, 2001
	Atmospheric methane	GISP2, Byrd	17.5		Blunier and Brook 2001
	Atmospheric methane	Antarctica, Greenland	~17	Compares Vostok, Byrd, GRIP CH_4	Blunier et al. 1998
	Northern-hemisphere methane	Circumarctic peatlands	16, 14.5	Accelerates at 11.5 ka	MacDonald et al. 2006
	Atmospheric methane	North GRIP		Emphasis on pre-20 ka	Flückiger et al. 2004
	SE Asian temperature, monsoon	Dongge Cave, China	16		Yuan et al. 2004
	SE Asian temperature, monsoon	Hulu Cave, China	16		Wang et al. 2001
	Greenland N_2O	Greenland	16 or older	N_2O rises with temperature	Flückiger et al. 1999
Surface ocean	SST	SE Pacific	17.5		Lamy et al. 2004
	SST	W. Equatorial Pacific	17 and 21		Visser et al. 2003
	SST	Cariaco Basin, Caribbean	16 and 21		Lea et al. 2003
	SST	W. Equatorial Atlantic	16–19		Rühlemann et al. 1999
	SST	N. Atlantic	~16	Multi-proxy comparison	Elderfield and Ganssen 2000
	SST	Santa Barbara Basin	19–21 and 16.5–17		Hendy et al. 2002
	SST	Sunda Shelf	19	Also sea-level rise	Hanebuth et al. 2000; Steinke et al. 2003

(continued)

TABLE 7.2 (*continued*)

Climate System	Climate Indicator	Region/Site	Deglacial Onset (ka)*	Comment	Source
	SST	S. China Sea	19	Synchronous w/Greenland	Kienast et al. 2001
	SST	S. China Sea	18 and 14.5	North and south different amplitude	Pelejero et al. 1999
	SST	E. Equatorial Pacific	18 (17–20)	N/S gradient	Koutavas et al. 2002, 2004
	SST	E. Equatorial Pacific	16		Lea et al. 2000
	SST, SSS	Off Indonesia	18	Temp rise at 18 ka H-1	Levi et al. 2007
	SSS	Columbia Basin, Caribbean	16 and 21	Salinity and temperature	Schmidt et al. 2004
	SSS	SE Pacific	18–15.5	From 35–31 ppt	Lamy et al. 2004
Deep, intermediate ocean	Atmospheric ^{14}C	Lake Suigetsu, Japan	~18–20	Younger, Older Dryas plateaus	Kitagawa and van der Plicht 1998
	Atmospheric ^{14}C	Cariaco Basin, other	See curve	Beck et al. 2001	Hughen et al. 2000, 2006
	NADW	S. Atlantic	18		Piotrowski et al. 2005
	Deep-sea temperature	E. Equatorial Pacific	18 (22?)		Martin et al. 2005
	Deep-sea temperature	W. Tropical Pacific	19	Precedes SST warming	Stott et al. 2007
	Deep-sea circulation	Bermuda Rise			Boyle and Keigwin 1987
	Deep-sea circulation	Bermuda Rise			Keigwin et al. 1991
	Deep-sea circulation	Bermuda Rise	19.5	No data pre-20 ka	McManus et al. 2004
	Deep and intermediate N. Atlantic	Little Bahama Banks	18–19 and 14–15	Deep, intermediate water contrast	Marchitto et al. 1998
	Intermediate Atlantic	SW Atlantic, off Brazil	18–19 and 14–15	Complex N. Atlantic water geometry	Came et al. 2003
	Intermediate Atlantic	SE Atlantic, off Angola	20.5 and 18.5	ODP 1078C, other sites	Rühlemann et al. 2004
	S. Atlantic	General	Deglaciation		Wefer et al. 2004
Sea-level rise	Global	Irish Sea Basin	19	SL rose 10–15 m	Clark et al. 2004
	Global	Bonaparte Gulf, Australia	19		Yokoyama et al. 2000
	Global	Sunda Shelf	19		Hanebuth et al. 2000
	Global	Barbados	19–20	Accelerates 15–16 ka	Fairbanks 1989; Peltier and Fairbanks 2006
	Global	ICE-5G Model	~20		Peltier 2004
	Global		~19–20		Lambeck and Chappell 2001
	Global	Various	Various	Review of sea level	Pirazzoli 1996

(*continued*)

TABLE 7.2 (*continued*)

Climate System	Climate Indicator	Region/Site	Deglacial Onset (ka)*	Comment	Source
Ice retreat[†]	Laurentide		19–20	LGM = 21.5, Laurentide, Cordilleran Innuitian not in phase	Dyke et al. 2002
	SE Laurentide		23–24		Lowell et al. 1999; Balco et al. 2002
	British-Fenoscandian	Bay of Biscay marine	21.5 and 17–19	Reactivation of European rivers	Menot et al. 2006
	Scandinavian Ice Sheet	Sweden and Baltic region	19 and 14.6	Southern margin of SIS	Rinterknect et al. 2006
	Barents–Kara Sea Ice Sheet	LGM minor compared to early and mid-Weichselian			Svendsen et al. 2004
	British and Irish Sea Ice Sheet		21.4 and 17.4		Bowen et al. 2002
	Arctic Ocean	Eurasian Basin	16–18	Spatially and temporally variable	Nørgaard-Peterson et al. 2003
	Alpine glaciers	Tropical Andes	21 and 15	Mountain and ice sheet asynchroneity	Smith et al. 2005
	N. American alpine glaciers		21–22.5		Clark 2002
	Tropical Andes glaciers		<21		Seltzer et al. 2002
	Midlatitude alpine glaciers	S. and N. America, New Zealand	Mean = 17.2	[10]Be dating moraines ~16–18-ka range	Schaefer et al. 2006
	Patagonian Ice Sheet	S. America	17.5		Denton et al. 1999; Lamy 2004; Kaiser et al. 2005
	W. Antarctic Ice Sheet	Not Including Antarctic Peninsula	15–12		Anderson et al. 2002
	E. Antarctic Ice Sheet	Deglaciation varies by sector	various	W. and E. Antarctic Peninsula asynchronous	Anderson et al. 2002
	Antarctic Ice Sheet	Antarctic continental shelves	14–8	See Denton and Hughes 2002 for dimensions	Ingólfsson 2004
	Global Ice Sheets				Ehlers and Gibbard 2003
	Norwegian sector	Norwegian Sea	17.5–18	Retreat from shelf	Mangerud 2004

Note: EPICA = European Ice Core Project; GISP = Greenland Ice Sheet Project; GRIP = Greenland Ice Core Project; LGM = Last Glacial Maximum; NADW = North Atlantic Deep Water; N/S = North/South; ODP = Ocean Drilling Program; SIS = Scandinavian Ice Sheet; SL = sea level; SSS = sea-surface salinity; SST = sea-surface temperature.

*Age for warming/deglaciation onset is from figures or text in papers—see the original papers. Age model revisions are common, resolution and chronology vary.

[†]See Ingolfsson (2004) and Domack et al. (2005) for Antarctica.

for several reasons. Studies vary greatly in sampling design, temporal resolution of climate proxies, error bars on the measured parameter (i.e., SST and sea-surface salinity [SSS]), and the stratigraphic position of chronological control tie points (i.e., ages from radiocarbon, cosmogenic, uranium-thorium, varves or other dating methods). Moreover, many studies were designed to investigate specific deglacial millennial climate reversals, such as the B/A–YD transition, and not to identify the onset of warming.

With these factors in mind, the summary in Table 7.2 leads to the conclusion that most parts of the climate system begin to change between 22 and 18 ka. Two opposing views about the phasing of deglaciation have emerged. The first holds that the southern hemisphere leads the northern hemisphere with the following sequence of events:

1. South Pole temperature rises, 19 ka (Dome Fuji) (Kawamura et al. 2007).

2. Pacific deep-sea temperature rises, 19–18.5 ka (Stott et al. 2007).

3. South Pacific SST rises, 19–18 ka (Kaiser et al. 2005).

4. Global sea level rises, 19 ka (Hanebuth et al. 2000; Clark et al. 2004; Peltier and Fairbanks 2006).

5. Northern-hemisphere ice sheet initial decay occurs, 19 ka.

6. Atmospheric CO_2 concentrations rise, ~18–17 ka (Monnin et al. 2001; Kawamura et al. 2007).

7. Midlatitude alpine glaciers retreat, 17.2 ka (Schaefer et al. 2006).

8. Methane concentrations rise, ~17 ka.

This sequence suggests that deglaciation began in the southern hemisphere with surface and deep-water warming, followed by tropical SSTs and with CO_2 concentrations rising 1000–2000 years later (e.g., Stott et al. 2007; Wolff et al. 2009).

A second view is that northern-hemisphere deglaciation leads the southern hemisphere:

1. European Ice Sheet margins retreat, 21–18 ka (Bowen et al. 2002; Ménot et al. 2006; Rinterknect et al. 2006).

2. North American alpine glaciers retreat, ~21 ka (Clark et al. 2002).

3. Low-latitude glaciers retreat, ~21 ka (Clark et al. 2002; Seltzer et al. 2002; Smith et al. 2005).

4. Global sea level rises, 20–19 ka (Hanebuth et al. 2000; Clark et al. 2004; Peltier and Fairbanks 2006).

5. Low-latitude SST rises and subsequent CO_2 rises, 19–17 ka (depends on the study and interpretation of SST proxy).

6. Methane concentrations rise, ~17 ka.

In this scenario, high-latitude northern-hemisphere summer insolation is the trigger for ice-sheet decay and sea-level rise (SLR) (Alley et al. 2002; Clark et al. 2004). In both cases, CO_2 concentrations rise after the initial phase of warming, ice-sheet melting, or both, but not in the major phase during the B/A in the north. A third view, held in the minority, is that deglaciation was synchronous in both hemispheres. The main point is that an idealized atmosphere-ocean-ice-sheet sequence for the inception of deglaciation is not available and is subject to diverging interpretations. A closer look at several key proxy records shows why this is the case.

Atmospheric Records

Middle and Low Latitudes Alpine glaciers and snow lines are sensitive monitors of atmospheric temperatures. In a comprehensive study of midlatitude, high-elevation moraines from North and South America and New Zealand dated by cosmogenic isotopic dating, Schaefer et al. (2006) obtained an age of 16–18 ka and a mean age of 17.2 ka for moraines. Ice recession from moraines thus suggests a later glacial recession compared to higher-latitude regions. Porter (2001a) synthesized records of low-latitude (23.5°N–23.5°S) alpine glaciers and snow lines during and since the LGM, showing that the maximum alpine glacier and snow line extent was highly diachronous, ranging from 31.3 to 19.0 ka. Under certain assumptions about the temperature-glacial-mass-balance relationship, he showed that following the LGM, alpine snow lines began a 1000-m rise in elevation, signifying a rise in temperature of 5.4°C (~4.7°C if one accounts for a 120-m sea-level drop). Denton et al. (1999) also studied the complex chronology of South American Llanquihue moraines of Chile, with the youngest moraines dated at roughly 17.5–17.0 ka. Seltzer et al. (2002) argued for early deglaciation ~21 ka for the tropical South American Andes, on the basis of paleolimnological records from Lake Titicaca, but Clark (2002) disagreed, suggesting a revised Titicaca-Greenland chronology and interpretation of the Andean lacustrine records that did not require early alpine glacial retreat.

High-Latitude Records Many polar records contain evidence for atmospheric change between 22 and 19 ka such as warming in northern (Alley et al. 2002; Clark 2002) and southern high latitudes (Monnin et al. 2001; Morgan et al. 2002). Taylor et al. (2004) produced a deuterium-based temperature record from Siple Dome, Antarctica that illustrates the difficulty pinpointing a single age for the inception of deglaciation. Siple Dome is situated near the Ross Sea sector of Antarctica and provides a good indication of temperatures in the high-latitude Pacific region. They documented a sharp 6°C temperature rise within only a few decades at ~22 ka, one of the earliest signs of postglacial warming in any record. As seen in Chapter 6, millennial temperature variability in Antarctica is typically subdued compared to the large spikes in Greenland, but the Siple event resembled abrupt Greenland-type interstadials. In contrast, Taylor et al. identified a more gradual warming beginning about 20–21 ka at the Byrd, Antarctica site, located only 500 km away from Siple.

Alley et al. (2002) identified initial deglaciation from the Greenland ice core by filtering millennial-scale climate "noise" from possible orbitally induced warming during the last deglacial. Their conclusion that deglaciation in Greenland began about 21–22 ka, just after summer insolation reached a minimum at 65°N, is consistent with orbital theory as discussed in Chapter 5 and indicates early temperature rise, albeit low in amplitude, in the northern hemisphere. These studies show that there is spatial variability in the scale and timing of initial warming in atmospheric records for the interval 22–17 ka.

Marine Records

The onset of deglaciation in the world's ocean must be viewed against a backdrop of zonal and meridional surface and vertical deep-water variability in the LGM ocean, the low resolution in most marine sediment records, chronological uncertainty due to radiocarbon reservoir effects, and the relatively small scale of the initial signal.

Atlantic Ocean The surface LGM Atlantic Ocean, by far the most intensely studied ocean, was more spatially and seasonally complex in terms of temperature anomalies than today's (Pflaumann et al. 2003). For example, glacial summer SSTs were as much as 12°C cooler in eastern North Atlantic at 45°N latitude, and 6°C in parts of the Nordic Seas. LGM SST cooling in parts of the subtropical gyre was only ~0.5–1.5°C. This spatial variability in and of itself makes identification of the first warming subject to errors associated with the proxy method and temporal resolution of the sediment record. Nonetheless, post-LGM North Atlantic Ocean warming occurs during the H-1 to B/A transition and at the end of the YD. This pattern is generally consistent with Greenland ice-core records despite radiocarbon reservoir effects (Waelbroeck et al. 2001). SST variability between 18 and 22 ka before H-1 is relatively minor compared to warming during these large events.

Pacific Ocean In the Pacific Ocean, Stott et al. (2007) found that deep-sea warming of 2°C between 19 and 17 ka preceded equatorial warming by ~1000–2000 years. SSTs warmed at the same time as Antarctic surface temperatures (~17.8 ka), and just before the initial rise in global atmospheric CO_2 concentrations (~17–17.3 ka). This phasing led to the conclusion that southern-hemisphere insolation, probably amplified by sea-ice feedbacks around Antarctica, caused the early onset of warming. Lamy et al. (2004) also identified an early 3°C warming from 19.2–17.4 ka off Chile in the southeastern Pacific.

Indian Ocean Rashid et al. (2007) used Mg/Ca ratios and oxygen isotopes from planktic foraminifers to examine the LGM-deglacial interval in the Andaman Sea in the Indian Ocean. They recognized a 3°C cooling during the LGM and high $\delta^{18}O_{seawater}$ values from foraminiferal values during the early deglacial (19–15 ka) and the YD, indicative of higher ocean salinity. In contrast, $\delta^{18}O_{seawater}$ values and lower salinity characterized the B/A and early Holocene (10.8–5.5 ka). They interpreted these patterns as evidence for decreased evaporation-precipitation, increased riverine influx due to a stronger Indian Ocean and East Asian Monsoon (EAM) during the B/A and early Holocene, or both.

Deep-Ocean Circulation Significant changes in deep- (Marchitto et al. 1998; McManus et al. 2004) and intermediate-ocean circulation occurred during deglaciation (Came et al. 2003). High-resolution records of the LGM-deglacial transition are sparse, and those that are available show a small initial deglacial signal compared to that during H-1 and the YD. For example, by using protactinium/thorium ratios ($^{231}Pa/^{230}Th$), McManus et al. (2004) showed a minor strengthening of North Atlantic Deep Water (NADW) from ~19.5 to 19 ka followed by a slowdown in NADW during the H-1 event. The 19.5–19-ka NADW change was accompanied by minimal SST warming in the subpolar North Atlantic (Bard et al. 2000a) and over Greenland. A complex surface, intermediate-, and deep-water Atlantic deglacial pattern is also reflected in the rapid and intense intermediate-depth warming that accompanied the slowdown of thermohaline circulation (THC) during the H-1 event between 19 and 16 ka in the western North Atlantic (Rühlemann et al. 2004).

Sea Level, Ice-Sheet Extent, and Volume

Minimum LGM sea level ranged from 115 to 135 m below its present position (Pirazzoli 1996; Lambeck and Chappell 2001; Peltier 2004). Uncertainty comes from a variety of factors—ice-sheet thickness, vertical movement of submerged LGM shorelines, depth ranges for and transport of shoreline indicators (coral reefs, marsh, and mangrove peat), mantle viscosity, and other geophysical characteristics used in modeling glacio-isostatic adjustment, as well as complications imposed by diachronous ice-sheet margin behavior already discussed herein. On the basis of new uranium-thorium dating of Barbados corals, Peltier and Fairbanks (2006) estimated a minimum sea level of −120 m lasting several thousand years between 26 and 20 ka, followed by a slow rise beginning about 20 ka. An initial post-LGM SLR of ~10–15 m dated at ~19 ka by Clark et al. (2004) from sediment cores off the British Isles is consistent with submerged coral and coastal sediment ages from several regions. Still, the exact age and magnitude of the first few meters of post-LGM SLR is extremely difficult to determine on the basis of paleo-shorelines (Hanebuth et al. 2000; Yokoyama et al. 2000). Because sea level is closely linked to ice-sheet decay, we review the status of the advance and initial retreat of LGM ice sheets.

North American Ice Sheets   Dyke et al. (2002) reviewed the history of the Innuitian, Cordilleran, and Laurentide sectors of the North American Ice Sheet (NAIS) complex on the basis of advances in AMS dating and new geological mapping (Figure 7.4). The NAIS was the largest, ~74 m of sea-level equivalent (Peltier 2004). Dyke concluded that the Laurentide Ice Sheet (LIS) advanced ~27 ka towards its maximum position by ~22–19 ka and then began to retreat. LIS advance and retreat were influenced by hemispheric climate changes. Conversely, the Cordilleran Ice Sheet west of the continental divide advanced and retreated several thousand years after the LIS because of regional atmospheric circulation and precipitation. The smaller Innuitian Ice Sheet, where relict ice still exists on Ellsemere, Axel Heiberg, and other Canadian Islands, was governed by isostatic rebound,

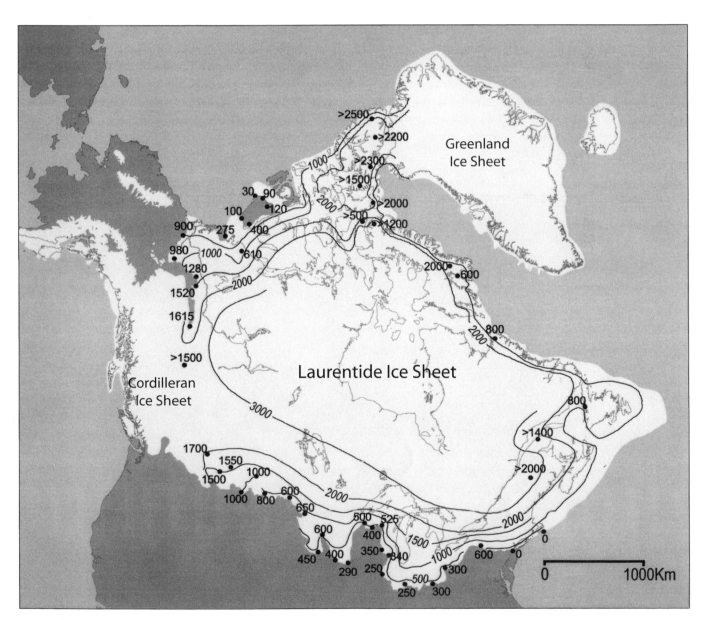

FIGURE 7.4 North American Ice Sheet during Last Glacial Maximum showing contour lines of ice-sheet thickness; the numbers refer to high elevation points near the ice-sheet margin used to constrain its thickness. The figure is modified from Dyke et al. (2002). Courtesy of A. Dyke and the Canadian Geological Survey.

topography, and complex drainage patterns through channels. NAIS dynamics are thus regionally diachronous and controlled by climatic and nonclimatic factors, and maximum ice extent spans several millennia. The NAIS probably contributed to early SLR between 19 and 16 ka during H-1 through large-scale thinning rather than marginal recession. It should be kept in mind that the thickness of the NAIS, particularly in eastern Canada, is still widely debated (e.g., Miller et al. 2001).

Eurasian Ice Sheet The second largest ice sheet—the Eurasian Ice Sheet (EIS)—is composed of the Fennoscandian, Barents-Kara Sea, and British Isles Ice Sheets and was the subject of an international project called the Quaternary Environments of the Eurasian North (QUEEN) (Svendsen et al. 2004). Figure 7.5 shows QUEEN EIS reconstructions for the LGM Weichselian stage and the penultimate glacial called the Saalian equivalent to Marine Isotope Stage (MIS) 6. It is clear that the LGM ice sheets were less extensive than those of MIS 6. The QUEEN Project also discovered that the Barents-Kara Sea Ice Sheet was more extensive during the early and middle Weichselian (100–50 ka) than during the late Weich-

selian (25–15 ka). As with the NAIS, there was no simple synchronous advance-and-retreat pattern among sectors of the EIS. Along northern Eurasia, for example, its history is partly governed by onshore topography and bathymetry in the Barents-Kara Sea, regional precipitation, and temperature. These factors combine to produce a regionally complex and varied pattern of thickening and thinning during the LGM and the deglacial interval. The QUEEN Project also carried out numerical glaciological model studies of the EIS forced by global sea-level oscillations and orbitally controlled solar insolation. The model simulations provide reconstructions of the thickness and extent of EIS in accord with empirical field evidence from glacial geological mapping.

Several syntheses are available for sectors of the EIS (see Ehlers and Gibbard 2003). For example, Jan Mangerud (2004) emphasized the diachronous nature of ice-sheet advance and retreat for that portion of the Fennoscandian Ice Sheet located along the coast of Norway. Evans et al. (2005) concluded that the British Isles Ice Sheet reached maximum extent asynchronously. Sejrup et al. (1994) discovered that the Fennoscandian and British Ice Sheets were connected in the North Sea region, at least during some periods of the last glacial.

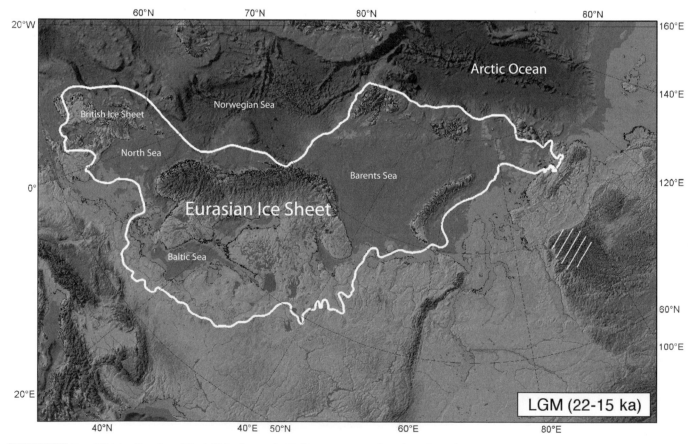

FIGURE 7.5 Late Weichselian Last Glacial Maximum (LGM) Eurasian Ice Sheet based on the reconstruction of the Quaternary Environments of the Eurasian North (QUEEN) Project (Svendsen et al. 2004). Ice-sheet distribution comes also from Landvik et al. (1998) and Ehlers and Gibbard (2004). Courtesy of J. Svendsen, M. Jakobsson, the QUEEN Project, and Elsevier.

In addition to diachronous ice-sheet behavior is the link between the EIS and large-scale climate and sea-level events. Using beryllium and chlorine cosmogenic isotopes to date moraines from the southern EIS in Scandinavia, Rinterknecht et al. (2006) found that early retreat at 19 ka was coincident with SLR (Clark et al. 2004). Ménot et al. (2006) compared the continental record of the western EIS with a sediment record of river discharge into the northern Bay of Biscay southwest of the English Channel. During the LGM, the Channel River system drained eastward from the southern margins of the British Ice Sheet, integrating the modern Rhine, Thames, Seine, and Maas catchment areas, a measure of organic material transported by rivers and several marine proxies. Ménot discovered a rise in branched and isoprenoid TetraEther index (BIT) values began to increase at 21.5 ka, accelerating rapidly at ~18.5 ka, reflecting a large fluvial discharge from Europe and the British Isles during early deglaciation. Reactivation of northern Europe's river systems was coeval with warming over Greenland and parts of Europe. Fluvial discharge ceased abruptly at ~17 ka as sea level rose and the Channel River mouth shifted location. The Channel River record exemplifies the steplike nature of deglaciation near LGM ice-sheet margins (Bowen et al. 2002).

Arctic Ocean

Paleoceanographic history of the Eurasian Basin of the Arctic Ocean is related to the behavior of the northern margin of the EIS (Knies et al. 2001; Nørgaard-Pedersen et al. 2003; Spielhagen et al. 2004, 2005; Wollenburg et al. 2004; Darby et al. 2006). There was extensive LGM sea ice in parts of the Eurasian Basin but minimal deglacial meltwater and ice-rafted debris (IRD) deposition except near the Siberian continental margin. In general, the Arctic Ocean record supports the 100-ka-long EIS history synthesized in the QUEEN project, where there was ice-sheet expansion in the early and mid-Weichselian and less ice during the LGM.

Ice gouging and other seabed erosional features up to 1000-m in water depth on the Lomonosov Ridge indicate thick ice in the central Arctic Ocean at times during the last glacial period. Erosion may have been caused by a seaward extension of the EIS into the central Arctic in the form of ice shelves (Polyak et al. 2001; Jakobsson et al. 2005), or by deep-draft icebergs draining from the Barents-Kara Sea region through the St. Anna Trough (Vogt et al. 1994; Kristofferson et al. 2004). Polyak et al. (2007) argued that ice-shelf grounding occurred during the LGM and an earlier episode occurred on the Chukchi margin off North America in the western Arctic Basin. However, most seabed erosional features in the Arctic Ocean are not well dated.

Antarctic Ice Sheet

AIS history is equally complex in terms of the timing of maximum advance and deglacial retreat. George Denton and Terry Hughes, editors of the CLIMAP volume *The Last Great Ice Sheets* (Denton and Hughes 1981), updated their views on the extent of LGM Antarctic ice, concluding that it accounts for about 14 m of sea level, equivalent to the total ~120 m glacial low sea level. Huybrechts (2002) estimated Antarctic and Greenland LGM ice volume based on glaciological modeling, indicating that these ice sheets were larger than those of today by 13–21 m and 2–3 m sea-level equivalent, respectively. Huybrechts's "best guess" of 17.5 m for the AIS at the LGM is about 29% larger than the modern AIS. According to his glaciological model, this maximum was not reached until 10 ka, in contrast to the ice volume maximum on Greenland at 16.5 ka. Most additional LGM ice was located in the West Antarctic Ice Sheet (WAIS) on the Antarctic Peninsula and in the Ross and Weddell Sea regions, where it may have been as much as 2000 m thicker. The East Antarctic Ice Sheet (EAIS) may have been only 50–100 m thicker, on average.

J. Anderson et al. (2002) synthesized the AIS on the continental shelf along various sectors of the Antarctic continental margin on the basis of geophysical and sediment core studies. They concluded that the WAIS extended to the outer continental shelf during the LGM and has been retreating since, continuing even into the late Holocene epoch for some regions. Ages for maximum ice extent range from 13–18 ka in the Ross Sea sector to 12–14 ka on the Antarctic Peninsula and in the Weddell Sea region. Initial retreat from the WAIS began ~12–15 ka but later in some regions. The larger but less studied EAIS extended to midshelf position during the LGM, and in some places it appears it was located near its present location. The more limited dating of maximum EAIS ice and the onset of retreat ranges from 9–25 ka but may have been as old as 30 ka in Lutzow-Holm Bay. Ingólfsson (2004) also showed that AIS deglaciation was spatially variable, spanned the period 14–8 ka, and in some regions continued into the mid- to late Holocene epoch. Because of poor chronology on the continental shelf, there is no comprehensive correlation between AIS history and atmospheric records from ice cores.

Sea-ice history in the Southern Ocean around Antarctica is also important for understanding glacial-interglacial climate change and carbon cycling (Sarnthein et al. 2003a; Gersonde et al. 2005). Enhanced Southern Ocean productivity during the LGM may have contributed to the drawdown of atmospheric CO_2 (François et al. 1997; Elderfield and Rickaby 2000; Sigman and Boyle 2000; Stephens and Keeling 2000). Shemesh et al. (2002) reconstructed deglacial sea-ice history in the Atlantic sector of the Southern Ocean and its relationship to atmospheric pCO_2 increase using diatom isotopes, IRD, and other proxies. Each proxy showed a slightly different timing and rate of deglacial change. Sea ice and ocean-nutrient cycling began to change at 19 ka, about 2000 years before the postglacial rise in atmospheric pCO$_2$. Combined with other Southern Ocean records, they concluded that this delay contradicts the hypothesis that the Southern Ocean played a dominant role in glacial-interglacial ocean-atmosphere exchange. Their study highlights the need for precise chronology of sea-ice retreat, surface ocean productivity, CO$_2$ degassing from the ocean, and deglacial warming for the period from 22 to 16 ka.

Summary of Last Glacial Maximum–Deglacial Transition   This summary was meant to underscore the highly diachronous, spatially variable, relatively gradual, and low-amplitude nature of climate change between 22 and 18 ka. Many factors contribute to uncertainty in the relative timing of the onset of deglaciation: ice-sheet dynamics, the effects of SLR on ice-sheet margins, the small amplitude of temperature change, and the likelihood that feedbacks from sea ice, SLR, and greenhouse gas concentrations were just beginning to amplify insolation forcing. Improved proxies of sea ice, more mapping and dating of submerged glacial-age shorelines, better calendar-year chronology, quantification of regional reservoir effects, and high-resolution marine records from poorly known parts of the ocean are needed.

Millennial Climate Reversals During Deglaciation

Understanding abrupt millennial climate reversals poses one of the great challenges in paleoclimatology. This section describes background and then presents detailed proxy records from critical parts of the climate system.

Early Studies of Deglacial Millennial Events

The YD is the most celebrated deglacial climate reversal. Its name comes from fossil pollen from *Dryas octopetala*, an Arctic flower that lives today in cold regions in the northern hemisphere. Spikes in *Dryas* pollen in deglacial sediments from northern Europe and the Alps were recognized decades ago as part of a standard European pollen zonation that includes three *Dryas* zones (Oldest, Older, and Younger), alternating with the Bølling, Allerød, and Preboreal periods (Table 7.1) (Jensen 1935; Mangerud et al. 1974).

The YD was more than a regional event. The Camp Century Greenland ice-core isotopic evidence showed that atmospheric temperature rose 5–10°C in only a century (Dansgaard et al. 1989). Quaternary fossil insects indicate rapid climate change on land (Coope 1977, 1994; Elias 1994). Rapid changes punctuated North America and Europe during the Wisconsinan and Weichselian glacial stages. For example, glacial stratigraphy in the eastern Great Lakes and St. Law-

rence Valley of the United States and Canada reflects a dynamic, oscillating LIS margin, with alternating glacial and nonglacial deposits like loess, lake sediments, and peat (Dreimanis and Karrow 1972). The Port Talbot, Plum Point, Erie, Mackinaw, and North Bay Interstadials reflect warmer periods within the glacial period. Complex LIS, glacial lake and regional environmental changes occurred during the YD in the Great Lakes and the St. Lawrence River (Dyke and Prest 1987; LaSalle and Elson 1975; Cronin 1977; Hillaire-Marcel and Occhietti 1980). In the southern hemisphere, early studies laid the groundwork for a global search for the YD age and other millennial-scale events (Suggate 1990).

John Mercer (1969) was among the first to directly address the question of whether the Allerød-YD transition was local or indicative of global climatic cooling. Knowing that the Allerød ended abruptly in Europe, Mercer estimated that the YD lasted about 650 years in Europe and proposed that it was related to sea-ice export from the Arctic Ocean and Norwegian Sea into the North Atlantic Ocean. He argued that 6000 km^3/yr of ice exported from the Arctic would cool Europe, weaken the North Atlantic Drift, and cover much of the North Atlantic within a 30-cm-thick freshwater layer annually. Mercer recognized that these events might have worldwide repercussions, but concluded that they diminished as one moves farther from the North Atlantic region.

Jim Kennett and Nick Shackleton investigated deglacial SST changes in the Gulf of Mexico (GOM). The GOM was important because if deglacial LIS meltwater draining southward through the Mississippi River system were rerouted through another outlet, it might be evident in GOM temperature and salinity records. Kennett and Shackleton (1975) found isotopic and foraminiferal faunal evidence in the Orca Basin for meltwater discharge during early deglaciation and reduced SST nearly equal to those during the LGM. In a series of more highly resolved records from the Orca Basin in the GOM, more definitive evidence for altered freshwater discharge patterns were discovered (Leventer et al. 1982; Kennett et al. 1985; Flower and Kennett 1990).

Despite these early studies, before the 1980s millennial-scale climate variability had been overshadowed by research on orbital-scale cycles. In their volume *Abrupt Climatic Change*, Berger and Labeyrie (1987) noted how little was known about abrupt climate change and the causes of the YD event. This changed when Broecker et al. (1988a, 1989) proposed that the retreat of the LIS margin diverted glacial meltwater from the Mississippi River–Gulf of Mexico to a St. Lawrence Estuary–North Atlantic drainage routing. Rerouting caused catastrophic glacial lake drainage that lowered North Atlantic salinity enough to influence density-driven deep-water formation, slow down meridional ocean circulation, and cause YD cooling. Broecker et al. (1990b) suggested this phenomenon operated like a salt oscillator that might deprive Europe's atmosphere of its heat source. The trigger might be the abrupt influx of glacial lake meltwater from the LIS. Broecker (1998) proposed the bipolar seesaw theory of millennial-scale climatic events, which held that a full or partial shutdown of North Atlantic THC during YD-like events could explain the asymmetric temperature patterns in two hemispheres.

These discoveries raised several questions addressed in the rest of this chapter: Did THC shut off during the YD? Was the YD a global event, and if so, how strong was the signal outside the North Atlantic? Was YD cooling synchronous in the northern and southern hemispheres? In the tropics of both hemispheres? Were there other deglacial climate reversals and were they synchronous? What was the role of ice sheets, meltwater, and icebergs in thermohaline variability in millennial climate change?

Atmospheric Change During Deglaciation

Atmospheric Composition   The Vostok and European Project for Ice Coring in Antarctic (EPICA) Maud Dronning Land (MDL) ice cores show that the concentrations of atmospheric CO_2 rose about 80–100 ppmv during each of the last four terminations and about 50–60 ppmv for the preceding three terminations. Methane and nitrous oxide concentrations also rose. Table 7.3 summarizes the variability in greenhouse gases based on ice-core records.

Atmospheric CO_2 during the last deglaciation was first studied in ice cores from Byrd, Antarctica by Neftel et al. (1982, 1988) and later from Taylor Dome (Fischer et al. 1999; Indermühle et al. 1999, 2000) and EPICA Dome Concordia (EDC) (Monnin et al. 2001). Figure 7.6 shows a period of relatively invariant CO_2 concentrations with a mean value of 189 ppmv during the last glacial interval 18.1–17 ka, followed by a gradual rise to 220 ppmv between 17 and 15.4 ka, and a distinct slowdown in the rate of increase until about 14 ka.

An abrupt rise of >10 ppmv occurs about 13.9–14 ka, the onset of the B/A, followed by a plateau at 240 ppmv until 12.2 ka, and then a rapid rise to 265 ppmv by ~11 ka.

The relationship between CO_2 concentrations and temperature is important for interpreting mechanisms of climate change. The EDC record of Monnin et al. (2001) confirmed that CO_2 lagged temperature, as discussed in Chapter 5. Temperature, as measured by deuterium (δD), began to rise 17,800±300 years, and CO_2 concentrations rose at 17,000 ± 200 years. When age uncertainty from the ice-air close-off time during snow-firn-ice transformation is considered, the most conservative estimate for the lag time is 800 ± 600 years, but it is probably slightly less. Fischer et al. (1999) compared Vostok, Byrd, and Taylor Dome CO_2 concentrations, noting some age discrepancy (~1000 years) in identifying the onset of the postglacial CO_2 rise. This uncertainty was caused by different ice-accumulation rates and age models for the cores and by minor uncertainty in the Antarctic-record-to-Greenland-ice-core age scale. There is also uncertainty from the age distribution of CO_2 within individual air-bubble samples, where the enclosure process during the firn-ice transition produces a smoothing of the actual variability. This uncertainty is about 200 and 20 years for high- and low-resolution Antarctic ice cores, respectively, and is usually greater for glacial-age samples because of slower rates of snow accumulation.

Despite these uncertainties, there is no major discrepancy between ice-core records for the temporal relationship between CO_2 and temperature for the interval between 10 and 15 ka. CO_2 concentrations definitely lagged temperature by about 600–1000 years. There was a 300–500-year lag when CO_2 fell ~10 ppmv after the Antarctic Cold Reversal (ACR) ~13–14 ka and up to 1000 years before the YD began. Fischer also showed that CO_2 concentrations lagged temperature by 600–1000 years during the two prior deglaciations (Terminations II and III), which is confirmed by studies of Dome Fuji ice cores (Kawamura et al. 2003, 2007).

To place these changes in the modern context, the rates that atmospheric deglacial CO_2 additions vary are from as

TABLE 7.3 Glacial and Holocene Greenhouse Gas Concentrations

Gas	Glacial-Interglacial Range	Holocene Range	Holocene Range as % of Glacial-Interglacial Range	Radiative Forcing (W/m²)*
CO_2	76 (189–265) ppmv	20 (260–280) ppmv	26	1.66 +/ −0.16
CH_4	306 (368–674) ppbv	112 (562–674) ppbv	37	0.48 +/ −0.05
N_2O	70 (200–270) ppbv	10 (258–268) ppbv	14	.16 +/ −0.02

Sources: References: Blunier and Brook (2001); Monnin et al. (2001), Flückiger et al. (2002), Sowers et al. (2003); Siegenthaler et al. (2005); Spahni et al. (2005); Forster, Ramaswamy et al. (2007b).

Note: ppmv = parts per million by volume; ppbv = parts per billion by volume.

*Since pre industrial times.

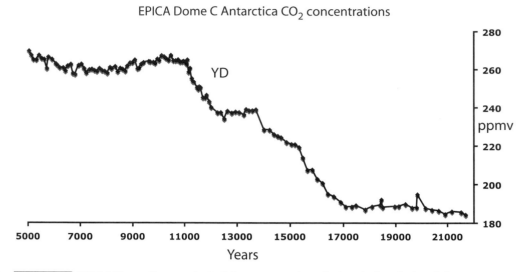

FIGURE 7.6 EPICA Dome C atmospheric CO$_2$ concentrations during the last deglacial, from Monnin et al. (2001). Note that the initial deglacial rise begins at 17 ka and a plateau occurs at 12.5–13.5 ka. YD=Younger Dryas.

low as 1 ppmv to as high as 20 ppmv per 1000 years. In stark contrast, anthropogenic activity has caused CO$_2$ concentrations to rise at rates exceeding 1 ppmv per year, and this rate is expected to increase in the next few decades (Hansen and Sato 2004). Human-induced rates, therefore, are many times greater than those during even the most rapid deglacial warming. Based on analyses of all available ice-core data, Joos and Spahni (2008) concluded that the 20th-century CO$_2$ increase and associated radiative forcing was at least an order of magnitude faster than "any sustained change during the past 22,000 years." They argue that it is highly unlikely that a peak in CO$_2$ comparable to that of the 20th century occurred during the last 50,000 years.

Methane concentrations from ice cores are used to reconstruct changes in global methane budget, to correlate Greenland and Antarctica records, and to identify methane source regions (Blunier and Brook 2001; EPICA Community Members 2006; Blunier et al. 2007). Unlike CO$_2$, methane concentrations reflect the impact of climate on high-latitude and tropical wetland sinks and sources. Methane has a shorter residence time in the atmosphere (9–10 years), and any changes in methane production or sequestration are rapidly mixed in the global atmosphere.

Blunier et al. (1998) showed that methane concentrations increased in an irregular fashion between 18 and 10 ka in Greenland Ice Core Project (GRIP), Byrd, and Vostok ice cores (see also Monnin et al. 2001; Brook et al. 1996, 1999). Figure 7.7 shows a typical pattern from the Greenland Ice Sheet Project 2 (GISP2) and Taylor cores. Mean concentrations rose from near 368 ppbv during the LGM to 674 ppbv at the end of the deglaciation (Table 7.3). Like CO$_2$, methane rises slowly at the beginning of deglaciation ~17 ka, but the patterns for the

two gases are different after that. At the onset of the B/A interstadial (~14.6 ka), methane concentrations increase rapidly by almost 200 ppbv, before falling almost as much during the YD. Severinghaus and Brook (1999) showed that methane rose about 140 ppbv in only 50 years at the Bølling transition and lagged temperature by only 20–30 years, which was probably the time it took for terrestrial ecosystems to respond to temperature and hydrological changes.

The dramatic B/A warming and coeval events in Antarctica have received intensive focus. Based on methane synchronization, it was synchronous within the North Atlantic region and nearly correlative with the cool ACR. Morgan et al. (2002) analyzed Law Dome, East Antarctic ice-core temperature history using methane concentrations for correlation with EPICA Dome C and Vostok records. They confirmed that the ACR was a 1000–1500-year-long pause in deglaciation but that its onset may have been as old as 15 ka in the southern hemisphere. No matter what age model they used, the ACR began prior to the onset of the Bølling in the northern hemisphere. They also showed that at the end of the ACR, southern-hemisphere warming began 1000 years before the onset of the YD. The end of the YD ~11.5 ka in the northern hemisphere did not coincide exactly with counterpart Antarctic cooling, although there is spatial variability throughout Antarctica for this interval. These results suggest that although first-order millennial climate reversals are antiphased in northern and southern high latitudes, there is not a simple multicentury delayed response in the south and the bipolar seesaw mechanism is too simplistic to explain centennial-scale patterns.

The Interpolar Methane Gradient is influenced by changes in latitudinal sinks and sources of methane and distin-

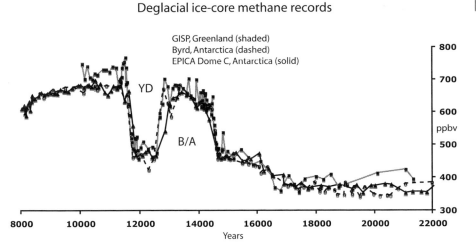

Deglacial ice-core methane records

GISP, Greenland (shaded)
Byrd, Antarctica (dashed)
EPICA Dome C, Antarctica (solid)

FIGURE 7.7 Deglacial methane records from Greenland Ice Sheet Project (GISP) (grey line, squares) and Byrd (dashed line, open circles) (Antarctica) ice cores on Greenland Ice Core Project age scale, from Blunier and Brook (2001) and EPICA Dome C ice core (solid line, triangles), from Monnin et al. (2001) and Spahni et al. (2005). B/A = Bølling-Allerød (BA); YD = Younger Dryas.

guishes between tropical and boreal sources of methane during deglaciation (Chappellaz et al. 1997). Brook et al. (2000) combined empirical ice-core measurements and box modeling of methane cycling to show that both low- and high-latitude sources contributed to the rapid rise in methane concentrations at the Bølling outset and at the end of the YD, as well as to a decrease in high northern-latitude sources during YD cooling. Schaefer et al. (2006) analyzed carbon isotopes in methane ($\delta^{13}C$, a proxy for the source area) in Greenland ice cores during the late deglacial period. Their $\delta^{13}CH_4$ trends showed a rapid rise in tropical methane release about 11.5 ka. There is also evidence for explosive growth of northern-hemispheric peatlands from 12–8 ka as ice sheets rapidly retreated and exposed large land regions where peat-forming habitats could grow (MacDonald et al. 2006).

Atmospheric Temperature An extremely large literature exists on atmospheric temperature during the YD. Rind et al. (1986) reviewed 80 European and 78 North American palynological and lake-level studies and found definitive evidence for cooling throughout Europe and less conclusive but now confirmed evidence for cooling from eastern parts of North America. YD cooling ranged from −2 to −10°C in various regions. Terrestrial paleoclimate records of the YD are also well known from maritime Canada (New Brunswick and Nova Scotia) (see Stea and Mott 1998). Pollen and chironomid midge larva indicate YD regional cooling of up to 12°C when a forest-tundra ecosystem returned to eastern Canada. Levesque et al. (1993) also discovered palynological evidence for a pre-YD cooling of about 8°C, which they named the Killarney oscillation. Peteet (1993, 1995) showed that continental records from Europe, Asia, North, Central, and parts of South America, Africa, and New Zealand all show evidence for a YD climate reversal, but chronology in many regions needed improvement (Lowell and Kelly 2008).

We now know that global and regional temperature oscillations during the YD and other deglacial events were often abrupt. Severinghaus et al. (1998) and Severinghaus and Brook (1999) used nitrogen and argon isotopes and methane concentrations from trapped air in Summit, Greenland ice cores, with oxygen isotopes from ice, to show that temperature rose ~9°C in just a few decades at the beginning of the Bølling. They also showed that warming after the YD occurred within a few decades. The timing of temperature and methane changes indicated that northern high-latitude warming preceded tropical methane release by only 20–80 years.

As discussed in Chapter 6 for glacial-age events, Antarctic and Greenland ice-core records show two primary interhemispheric differences between northern- and southern-hemisphere temperature patterns during these rapid events. First, atmospheric temperature oscillations were out of phase such that at least parts of the warm B/A were coeval with the ACR and parts of Antarctica were warm during the cold YD. Second, temperature oscillations were more muted in Antarctica. For example, the atmosphere over parts of Antarctica warmed by only 1°C during the YD, in contrast to much greater simultaneous cooling over Greenland. To many researchers, these patterns support the antiphased bipolar seesaw mechanism of interhemispheric oceanic transmission of temperature changes.

Atmospheric Circulation Close inspection of ice-core records reveals complex deglacial changes in atmospheric circulation. Mayewski et al. (1996) compared isotopic (mainly temperature) and glaciochemical (calcium, chloride) records of atmospheric circulation from GISP 2 and several Antarctic ice cores. The GISP 2 curve showed steep increases in calcium concentration during all major millennial stadials—the Oldest, Older, Younger Dryas, and the Intra-Allerød cold period just prior to the YD. The YD event was by far the largest. From the onset of the Bølling to the end of the YD, Taylor Dome Antarctic calcium concentrations were relatively invariant; chloride concentrations rose during the YD. Both ice-core records show high-frequency changes in atmospheric circulation at decadal to centennial timescales.

Stenni et al. (2001) examined the relationship between temperature over the EDC site in Antarctica, at the moisture source region in the Indian Ocean, and atmospheric circulation patterns using deuterium excess and glaciochemical measurements (sodium). The total glacial-early Holocene temperature rise at the EPICA site was 8–9°C but only ~3°C in the Indian Ocean. Moreover, they discovered an abrupt Oceanic Cold Reversal about 800 years after the ACR. By calculating the difference between temperature at the core site and at the moisture source area, Stenni et al. showed a strong latitudinal temperature gradient between 14 and 12.5 ka during parts of the Allerød and YD. Both the isotopic and sodium records from EPICA Dome C core indicated that enhanced atmospheric circulation occurred during this period.

Steffensen et al. (2008) sampled the North GRIP ice core at an unprecedented subannual resolution for the B/A, YD, and Preboreal to examine atmospheric changes during abrupt deglacial transitions. They found generally similar paleo-atmospheric patterns during the two major warming intervals at the beginning of the B/A and Preboreal at 12.9 and 11.7 ka, respectively. Both involved warming of about 10°K, although B/A warming occurred over only one to three years in contrast to the 60-year-long warming at the YD-Preboreal transition. The onset of the YD experienced an extremely rapid (one to three years long) shift in deuterium excess, suggesting hydrological or temperature changes or both at the source region or a shift in the location of the moisture source itself. If—as the authors suspect on the basis of correlation to their dust, glaciogenic, and accumulation-rate proxies—this shift signified a change in the moisture source region for Greenland, this means the YD onset involved a large change in polar atmospheric circulation patterns in only one to three years. This shift was followed by a more gradual 50-year decline in temperature at the Greenland core site.

In addition to ice cores, lakes, marine, and speleothem records show dynamic low-latitude atmospheric changes between 22 and 10 ka. Hughen et al. (1996, 2000, 2004a), Peterson et al. (2000), and Haug et al. (2001b) studied the Cariaco Basin off Venezuela, near 10°N latitude, a region sensitive to shifts in the Intertropical Convergence Zone (ITCZ). Figure 7.8 shows the Cariaco Basin radiocarbon and grayscale sediment color records during the end of H-1, the transition to the B/A, and the YD. The grayscale record shows major reversals in surface upwelling and wind intensity over only a few decades. Radiocarbon activity, related to atmospheric ^{14}C production, shows near-synchronous changes in deep-ocean ventilation. Other Cariaco proxies show (1) drier conditions during the YD due to a southward shift in the mean ITCZ position, (2) a 50-year lag of tropical vegetation behind climate, and (3) synchroneity between tropical deglacial changes and those in high latitudes of the northern hemisphere. Significant atmospheric changes are also evident in Cariaco proxies for the Preboreal Oscillation ~11.3–11.4 ka. Nakagawa et al. (2005) also showed a vegetation-climate lag of 250–400 years during the YD using pollen and proxies from Lake Suigetsu, Japan. However, they suggested that the Suigetsu record is out of phase with the Greenland record by several centuries during the deglacial interval.

Asian speleothem oxygen isotope values reflect insolation forcing of the Asian monsoon (Fleitmann et al. 2007) and abrupt oscillations coincident with climate reversals of the last deglaciation in southeast Asia and in the Indian Ocean region. Two spectacular new records from Hulu and Dongge Caves in southern China shed light on the deglacial atmosphere (Wang et al. 2001; Yuan et al. 2004; Dykoski et al. 2005). Figure 7.9 shows abrupt, large-amplitude oxygen isotope oscillations in Hulu and Dongge speleothems at the onset of the B/A and YD events. These are in phase with each other and with high-latitude northern-hemisphere temperature. Although multiple factors influence the $\delta^{18}O$ of cave calcite, the similarity between the two sites, which are 1200 km apart from each other, and the large amplitude of the oscillations suggest changes in isotopic composition of precipitation in source regions in the Indo-Pacific and over southeast China. Wang et al. (2001) argued that the Asian monsoon system was in phase with the North Atlantic region and that stadials such as the YD were characterized by more zonal atmospheric flow. Strong zonal flow during the YD was also in evidence from marine sediment core records off the Amazon (Maslin and Burns 2000). Importantly, the steep decline in the monsoon at the B/A onset occurred over less than 200 years. The Hulu and Dongge cave records provided strong evidence for tropical atmospheric instability, abrupt climate transitions, and first-order synchroneity with the high-latitude North Atlantic during the deglacial.

Nakagawa et al. (2003) found evidence for cooler temperatures and higher winter precipitation during the YD in Lake Suigetsu cores. They speculated that this was due to changes in Siberian air masses reflecting an enhanced winter season Asian monsoon. Yancheva et al. (2007) reconstructed monsoon records from Lake Huguang Maar in southeastern China and also inferred deglacial changes in both the position of the ITCZ and the Asian monsoon. As was the case with the Suigetsu record, they found evidence for a strong YD Asian winter monsoon but not so for the summer monsoon that is dominant today.

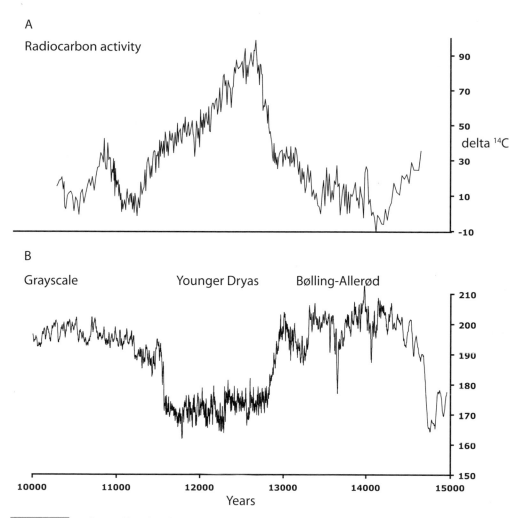

A

Radiocarbon activity

delta ^{14}C

B

Grayscale   Younger Dryas   Bølling-Allerød

Years

FIGURE 7.8 Bølling-Allerød and Younger Dryas (YD) climate record from Cariaco Basin: (A) Radiocarbon activity (Δ^{14}C); (B) grayscale, from Peterson et al. (2000) and Hughen et al. (1996, 2000). Note the abrupt onset and termination in grayscale and the abrupt rise and more gradual decline in radiocarbon activity during the YD event.

Deglacial atmospheric variability (lake levels, temperature) over tropical Africa is another well-studied topic (Gasse 2000). For example, Powers et al. (2005) used the TetraEther index (TEX$_{86}$) paleothermometer method to reconstruct temperature from Lake Malawi (9–14°S latitude). The TEX$_{86}$ index of organic compounds with tetraethers that have 86 carbon atoms is used in paleolimnological and marine records (Schouten et al. 2002). Powers et al. showed a total glacial-Holocene warming of 3.5°C beginning ~21–20 ka. They also discovered a 2°C cooling coincident with the ACR near the end of the Allerød and early part of the YD and a 2°C cooling during the 8.2-ka event. Aridity and shifts in wind patterns and the position of the ITCZ have also been thoroughly documented from records in tropical Africa (Barker et al. 2004; Castaneda et al. 2007).

Weijers et al. (2007) identified a coupling between the tropical African atmosphere and the SSTs off West Africa in a steplike pattern during deglaciation and during an inphase relationship with the rest of the North Atlantic region. Using organic geochemical methods called the Methylation Index of Branched Tetraethers and the Cyclization Ratio of Branched Tetraethers on a marine core from the region off the Congo River Basin, they estimated a total 4°C (21–25°C) warming since the LGM, beginning about 17 ka. The deglacial onset occurs at the same time, as seen in other highly resolved records such as the EPICA cores and the ^{10}Be-dated moraines of mid-latitude alpine glaciers. The Congo Basin record also showed much larger glacial-age cooling on continents than in the tropical ocean, a pattern predicted by LGM climate models. This land-sea thermal gradient rises rapidly after about 15 ka and again in two steps from 12 ka until 9.5 ka. The Congo record is a good example of the tropical hydrological response to climate through atmospheric feedbacks distinct from those in midlatitude and high-latitude processes.

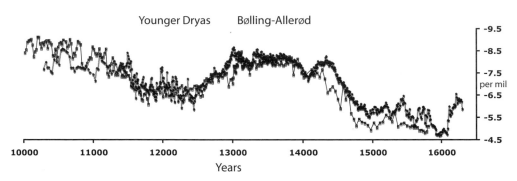

Deglacial speleothem record, China

Hulu Cave (shaded)
Dongge Cave (solid)

FIGURE 7.9 Deglacial oxygen isotopic record from Hulu (shaded line diamonds) and Dongge (thin line, squares) cave speleothems, China, from Wang et al. (2001), Yuan et al. (2004), and Dykoski et al. (2005). The Asian monsoon weakened during the Younger Dryas and strengthened during the Bølling-Allerød.

Deglacial Surface Ocean Temperature

Since CLIMAP's ice-age reconstruction, improved proxy methods have led to complete revision of tropical and subtropical SST variability (Hastings et al. 1998; Roselle-Melé et al. 2004). We now know that the tropical oceans cooled from 2–4°C during the LGM. LGM-modern SST contrasts are 1.5–3°C for the eastern tropical Pacific (Lea et al. 2000; Koutavas et al. 2002; Koutavas and Lynch-Stieglitz 2003), 3.3°C for the western pacific off Borneo (Visser et all. 2003), 2.3°C in the Sulu Sea (Rosenthal et al. 2003), 3–4°C in the South China Sea (Pelejero et al. 1999; Kienast et al. 2001), 2.5°C in the Colombian Basin, Caribbean (Schmidt et al. 2004), 2°C on the Sunda shelf off Indonesia (Steinke et al. 2003), 2°C in the Indian Ocean off Madagascar (Roselle-Melé et al. 2004), and 3.5–4°C off Granada in the Caribbean (Rühlemann et al. 1999), >5°C in the Gulf of Mexico (Flower et al. 2004), and ~3°C off west Africa (Weijers et al. 2007).

SST also changed significantly during deglacial climate reversals. Ruddiman and McIntyre (1973) and Ruddiman et al. (1977) documented winter and summer YD cooling from 8.2°C to 1.8°C and from 14.3°C to 7.4°C, respectively, in the North Atlantic (40°N and 65°N). At the end of the YD, winter and summer SSTs rose from 1.8 to 10.1°C and from 7.4 to 15.2°C, respectively. Duplessy et al. (1981) also found oxygen isotopic evidence for YD cooling and ice-volume changes that interrupted the deglaciation. In the Norwegian Sea, YD cooling was identified in a number of studies (Jansen and Bjorklund 1985; Jansen 1987; Lehman and Keigwin 1992; Koç and Jansen 1994). Jansen (1987) used three proxies—foraminifera, radiolarian, and stable isotopes—to estimate a major cooling in the southeastern Norwegian Sea from 7°C above modern during the B/A to near-glacial temperatures during the YD. Lehman and Keigwin (1992) and Koç-Karpuz and Jansen (1992) were among the first to document rapid SST changes up to 9°C at the YD-Preboreal transition in <50 years. Using long-chain alkenones of *Emiliana huxleyi* and planktonic foraminifera, Sikes and Keigwin (1994) estimated a 3–6°C temperature drop and a meltwater-induced decrease in surface salinities during the YD. Koç-Karpuz and Jansen (1992) and Koç and Jansen (1994) used a diatom transfer function to show that the YD cooling in the Norwegian Sea was synchronous with European continental and Greenland cooling but complex in terms of Nordic Sea ocean-temperature gradients (Koç et al. 1996).

Tropical SST also varied during stadial events. The alkenone-based SST reconstruction for the tropical west Atlantic off Grenada of Rühlemann et al. (1999) (see also Flower et al. 2004) shows a 1.5°C rise in temperature at 19–16 ka, a plateau at 15–13 ka, a 1.2°C warming about 12 ka, and progressive rise after 10 ka. Warming occurred in this part of the western tropical Atlantic during H-1 and the YD and thus was out of phase with the subpolar North Atlantic and the eastern tropical Atlantic, but in phase with Antarctic temperature. Their explanation is that NADW slowdown during H-1 and the YD reduces the export of heat to high latitudes but leads to a concomitant increase in SST in the source region near their core site.

Figure 7.10 shows significant cooling of 3–4°C in the Caribbean during the YD coincident with a southward shift in ITCZ position and a drop in Greenland temperatures and atmospheric methane concentration (Lea et al. 2003).

Cariaco Basin deglacial sea-surface temperature

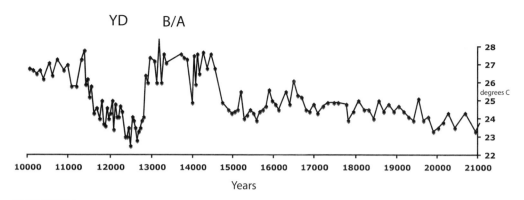

FIGURE 7.10 Sea-surface temperature (SST) from Cariaco Basin showing steep warming at the Bølling-Allerød (B/A) and 5°C cooling at the Younger Dryas (YD), from Lea et al. (2003). Cariaco SST shows near-synchroneity with high-latitude northern-hemisphere climate records.

Smaller temperature oscillations (~1°C) are seen in the tropical and South Atlantic (Arz et al. 1999, Rühlemann et al. 1999). Kienast et al. (2001) measured a sharp 1°C warming at the Bølling and a 0.4–0.8°C cooling during the YD in the South China Sea. L. Wang et al. (1999) discovered ten century-scale events between about 20 and 10 ka in high-resolution marine cores from the South China Sea. They interpreted these as abrupt shifts in the EAM interrupting its glacial-to-interglacial transition. During some events, the summer monsoon shift seemed to precede climatic and sea-level changes in the Atlantic region, and they postulated that the monsoons have contributed to the deglaciation.

Parts of the subpolar North Atlantic cooled during the glacial by up to 12°C (Pflaumann et al. 2003), and Waelbroeck et al. (2001) identified a significant two-step 7–12°C SST increase from the LGM to the Holocene epoch at the H-1-Bølling transition and the end of the YD. The southeastern Pacific Ocean off Chile at Ocean Drilling Program (ODP) Site 1233 near 41°N latitude cooled from 9 to 15°C, with shorter-term temperature oscillations synchronous with those seen in Antarctic ice cores (Lamy et al. 2004).

SSS changes have also been measured for the deglacial period. Duplessy et al. (1992) noted that salinity fell during H-1 and the YD, and Lamy et al. (2004) found salinity varied between 31 and 35 psu off Chile. Schmidt et al. (2004) estimated that salinity in the Caribbean during the LGM was 2.3–2.7 psu higher than that today. The largest salinity event in the Caribbean was a sharp decrease at the beginning of the Bølling, which coincided with rapid northward ITCZ migration, greater rainfall in the Caribbean, and enhanced advection of warm, salty Caribbean water to high latitudes at the inception of stronger B/A THC. Rashid et al. (2007)

showed links between the Indian Ocean hydrology and salinity during the YD and the strength of the EAM.

Deglacial Deep-Sea Bottom Temperature

Oxygen isotopic (Labeyrie et al. 1987) and Mg/Ca paleothermometry (Dwyer et al. 1995, 2000) suggest that deep-sea bottom-water temperature (BWT) cooled in the North Atlantic during the glacial periods as much as 2–4°C, depending on water depth. Martin et al. (2002) showed that BWT in the low-latitude Atlantic and Pacific fell by 2–4°C and 2–3°C, respectively, during glacial periods of the last 300 ka. Stott et al. (2007) compared tropical Pacific SST (Figure 7.11), deep Pacific BWT, and Antarctic records of deglacial climate. The results show low-amplitude cooling in both SST and BWT near the onset of the Bølling at 14.5–14 ka, confirming a connection between south polar and tropical Pacific climate phasing during the last deglacial.

Deglacial Intermediate and Deep-Ocean Circulation

Large-scale changes in deep and intermediate-water masses during deglacial millennial climatic events are reconstructed using six major proxy methods. Because of their importance, it is useful to discuss these methods briefly.

Ocean Circulation Proxy Methods The first two methods are $\delta^{13}C$ (Oppo and Lehman 1993, 1995) and cadmium/calcium (Cd/Ca) ratios of benthic foraminifers (Boyle and Keigwin 1982; Lehman and Keigwin 1992). These "nutrient" tracers were among the first lines of evidence

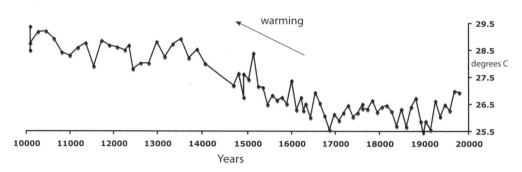

FIGURE 7.11 Sea-surface temperature from western equatorial Pacific Ocean showing ~3°C warming beginning about 17 ka, from Stott et al. (2007).

that deep-ocean-circulation changes accompanied millennial events such as the YD in the eastern (Sarnthein et al. 1994) and western Atlantic (Curry and Oppo 2005). Carbon isotopes are also used to contrast large-scale differences in the earth's oceans during the LGM and the Holocene interglacial (Curry and Oppo 2005; Lynch-Stieglitz et al. 2007). The ratio of coprecipitated cadmium to the primary calcium ion of the shell in the benthic foraminifera *Cibicidoides wuellerstorfi* correlates positively in most of the earth's oceans with the nutrient content of seawater, notably phosphorous, and serves as a paleonutrient tracer (Boyle 1988, 1992; Duplessy et al. 1988). Both carbon isotopic and Cd/Ca approaches rely on biogeochemical properties of modern deep-water masses such as the contrast between nutrient-rich, isotopically depleted Antarctic Bottom Water (AABW) and nutrient-poor, isotopically enriched NADW (Boyle 1990).

Carbon isotopes are influenced by biological, ecological, and geochemical processes in seawater or porewater (McCorckle et al. 1990, 2008; Mackensen et al. 1993; Fontanier et al. 2006) and atmospheric temperature and air-sea carbon exchange, and conflicting interpretations arose between Cd/Ca and $\delta^{13}C$ and other methods for the strength of deep-sea circulation during the LGM and the Holocene. Additional proxy methods were developed as more-direct "dynamic" tracers of ocean circulation. One method uses the ratios of protactinium and thorium isotopes ($^{231}Pa/^{230}Th$) in ocean sediments (Yu et al. 1996; Marchal et al. 2000). Both radionuclides are produced naturally in seawater as byproducts of radioactive decay of ^{235}U and ^{234}U, respectfully, at a ratio of 0.093. They are then adsorbed onto particles and deposited in sediments. Because of the different reactivity behavior of the two radionuclides, the residence time of ^{231}Pa in the North Atlantic is 100–200 years whereas that of ^{230}Th is 20–40 years. Consequently, changes in the $^{231}Pa/^{230}Th$ ratios can be used to estimate changes in ocean circulation. Other factors being equal, slower MOC results in less ^{231}Pa

export and a higher $^{231}Pa/^{230}Th$ ratio in sediments. Complete MOC shutdown would in theory result in a ratio of 0.093 (Henderson and Anderson 2003).

A fourth approach utilizes the fact that ^{14}C decays after uptake into seawater and the calcareous shells of foraminifers and deep-sea corals and thus the ^{14}C to ^{12}C ratio of shells (radiocarbon activity, $\Delta^{14}C$) is a tracer for the age of ancient water masses. Radiocarbon activity is used in several ways to reconstruct circulation changes. The radiocarbon age difference between paired planktonic and benthic foraminiferal specimens from the same deep-sea core sample, for example, can be used to calculate the age difference between surface and bottom waters and to estimate the ventilation rate of deep water during important climate periods such as the LGM and YD (Broecker et al. 1990a, b; Keigwin 2004). Another approach is to measure radiocarbon activity in annually deposited sediments such as the Cariaco Basin. Cariaco $\Delta^{14}C$ records indicate deep-ocean ventilation changes during deglacial stadial events and are also used to construct the radiocarbon age calibration curve (Hughen et al. 2004a, b). Radiocarbon activity in deep-water corals is another approach. Unlike hermatypic corals that dominate warm shallow waters, some coral genera live in water up to 2500 m deep and can live more than 100 years. These corals can be dated by uranium/thorium decay and their skeletal chemistry provides annually resolved records of deep circulation.

A fifth method called sortable silt (SS) is based on measurements of the fine fraction of deep-sea sediment from 10–63 microns. SS is derived mainly from terrigenous sources, and the weight and mean size of the silt fraction of deep-sea sediment are a function of the strength of deep-sea currents. SS thus provide a direct means to estimate the strength of circulation (McCave et al.1995; McCave and Hall 2006).

The sixth method is neodymium isotopes ($^{143}Nd/^{144}Nd$), which vary in the modern ocean and in ferromanganese oxide precipitates found in sediments (Goldstein and Hemming 2003). Ferromanganese oxide Nd isotopes have been applied to higher-frequency paleocirculation changes during the late Quaternary (Rutberg et al. 2000) because they are not strongly influenced by vital effects, low-temperature geochemical processes, or changes in carbon cycling that pose complications for nutrient proxies.

Deglacial Ocean Circulation Application of these proxies to deep-sea cores has led to major discoveries regarding intermediate and deep-sea circulation during the last deglacial. Changes in $\Delta^{14}C$ in Cariaco Basin sediments are evidence for reduction in deep-water formation and weaker THC in the North Atlantic during H-1 and the YD (Hughen et al. 1998). These MOC changes are synchronous with atmospheric temperature changes between 15 and 10 ka (Hughen et al. 2000). The onset and termination of the YD

are characterized by abrupt (< 100-year) transitions that cannot be accounted for by solar activity estimated from ^{10}Be records. Centennial-scale changes in $\Delta^{14}C$ before and after the YD event include a possible Older Dryas event at 14.1 ka and another during the Intra-Allerød Cold Period at 13.3 ka. The unprecedented high resolution and sensitivity of the Cariaco proxy record demonstrated convincingly that low- and high-latitude northern-hemisphere atmospheric changes were nearly synchronous (within decades) and coincided with large changes in deep ventilation rates. In addition, $\Delta^{14}C$ measurements from corals from 1700 m water depth dated at ~15.3 ka also show that deep-water ventilation can change extremely rapidly, over decadal timescales (Adkins et al. 1998).

Evidence for changes in deep-ocean ventilation comes from $^{231}Pa/^{230}Th$ ratios from a core at 4500 m water depth on the Bermuda Rise (Figure 7.12) (McManus et al. 2004). There

Deglacial radiochemical and isotope record of deep-ocean circulation

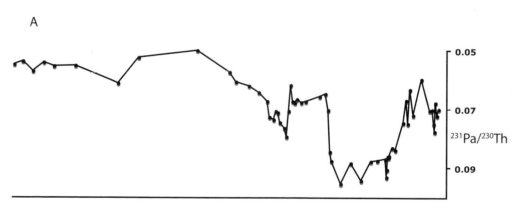

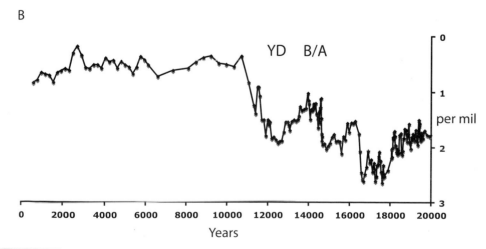

FIGURE 7.12 Radiochemical and isotopic record of deglacial (A) $^{231}Pa/^{230}Th$ deep-circulation tracer (based on the uranium 232 correction method) and (B) oxygen isotopes on planktonic foraminifer *Globigerina inflata* from OCE326 GCG5 Bermuda Rise core, from McManus et al. (2004). Note the strong decrease in circulation strength at this site during Heinrich event 1 and the smaller decrease during the Younger Dryas (YD). B/A = Bølling-Allerød.

appears to be a nearly complete shutdown of MOC during H-1 within centuries of large iceberg discharges and freshening of subpolar North Atlantic surface layers. The resumption of strong MOC at the end of H-1 near 15 ka was extremely rapid, on the order of several decades. Keigwin and Boyle (2008) note, however, that ocean-productivity changes and enhanced diatom flux to the Bermuda Rise might have influenced the flux of protactinium, leading to an overestimation of NADW slowdown from $^{231}Pa/^{230}Th$ ratios during H-1. Their evidence from radiocarbon activity in bivalve shells and other proxies suggest that NADW slowdown during H-1 was not as great as that during the LGM. McManus et al. (2004) also document an MOC slowdown from $^{231}Pa/^{230}Th$ ratios during the YD that was smaller than that during H-1.

Piotrowski et al. (2004, 2005) used Nd isotopes from the southeastern Atlantic to show that NADW first strengthened after the LGM at 17–18 ka and then experienced a series of oscillations during the Bølling, Allerød, and YD. NADW strength was forced by northern-hemisphere insolation and modulated by high-latitude sea ice, which influenced the location of deep-water formation. Piotrowski et al. outlined a first-order sequence of deglacial events that included initial ice-sheet retreat, changes in the global carbon budget, and finally changes in THC.

Intermediate-ocean circulation also changed during the deglacial (Marchitto et al. 1998; McCave et al. 2002). During the LGM, North Atlantic Intermediate Water (NAIW) penetrated far into the south Atlantic along the Brazilian Margin (Oppo and Horowitz 2000). Came et al. (2003) used benthic foraminiferal Cd/Ca from an intermediate-depth site in the western South Atlantic to show that the NAIW contribution to the South Atlantic began to decrease at ~14.5 ka, at the inception of the B/A. This was the most notable shift in ocean circulation during the entire deglacial. High cadmium values in the intermediate South and deep North Atlantic during the YD suggest reduced export of deep and intermediate water at this time. Rühlemann et al. (2004) also showed that THC slowdowns during H-1 and the YD were accompanied by intermediate-depth warming in the tropical Atlantic Ocean due to reduced ventilation by colder water and downward mixing of warmer water from the thermocline.

Robinson et al. (2005) measured $\Delta^{14}C$ activity in deepwater corals (500–2500-depth habitat), paired planktonic-benthic foraminifers, and constructed a vertical transect of oceanic circulation from 28–10 ka. During H-1, radiocarbon-rich water overlaid radiocarbon-poor water, a pattern interpreted as signifying more AABW southern-source deep water than for NADW. At 16.3–16.2 ka an abrupt change in $\Delta^{14}C$ activity occurred coincident with other climate and oceanographic indicators at the end of H-1. They also identified a large change in $\Delta^{14}C$ near 15.4 ka representing an increase in southern-source water coeval with a plateau in

the atmospheric temperature. At the end of the 15.4-ka event, ocean radiocarbon $\Delta^{14}C$ activity is inverted, that is, radiocarbon-poor water overlies radiocarbon-rich water. This is consistent with Southern Ocean records for strong intermediate-water formation during deglaciation (Pahnke and Zahn 2005). Robinson found elevated $\Delta^{14}C$ activity and strong Holocene-like MOC during the B/A interstadial period but an inverted $\Delta^{14}C$ profile during the YD. They concluded that deep-water changes during H-1 and the YD coincided with reduced NADW strength, but that large changes in intermediate-depth $\Delta^{14}C$ activity supported synchronous interhemispheric variability.

Labeyrie et al. (2005) found rapid, almost simultaneous intermediate-water changes during the deglacial in isotopic data from the Indian, Pacific, and Atlantic Oceans. There were notable changes at H-1 and the YD. In contrast to patterns at intermediate depths, deep-water isotopic shifts were more progressive, lagging surface and intermediate-depth changes by up to ~1.5 ka. Labeyrie proposed that the deglacial deep-intermediate-water decoupling is related to strong brine formation around Antarctica, causing vigorous Antarctic Intermediate Water variability.

Other biogeochemical and biological changes occurred in the deep and intermediate-depth ocean during deglaciation. These include intensification in nitrification in the oxygen minimum zone along continental margins whereby steplike changes in nitrification and dissolved oxygen levels occurred during H-1, at the onset of the B/A, and during the post-YD interval (Kienast et al. 2001; Altabet et al. 2002; Thunell and Kepple 2004). Yasuhara et al. (2008) discovered catastrophic changes to deep-sea benthic communities and species diversity during all major deglacial climate reversals in the North Atlantic Ocean.

The Arctic Ocean is a source of deep and surface water and sea-ice export to the Greenland Norwegian Seas and probably plays a role in deglacial climate. Duplessy et al. (1996), for example, proposed that sea-ice exported from the Arctic via the east Greenland Current might be responsible for the YD, and Tarasov and Peltier's (2005) model simulations indicated the importance of Arctic freshwater export during the YD. High-latitude paleorecords reveal enormous spatial and temporal complexity between 20 and 10 ka that was related to sea ice, surface ocean productivity, ice rafting, and deep-water formation (Nørgaard-Pedersen et al. 2003; Birgel and Haas 2004; Spielhagen et al. 2004; Wollenburg et al. 2004).

Ice Volume and Sea-Level Change During Deglaciation

Deglacial Sea-Level Curves Global SLR was about 120 m from its LGM lowstand ~21–26 ka to its present level by about 5–6 ka. The global sea-level curve in Figure 7.13 is

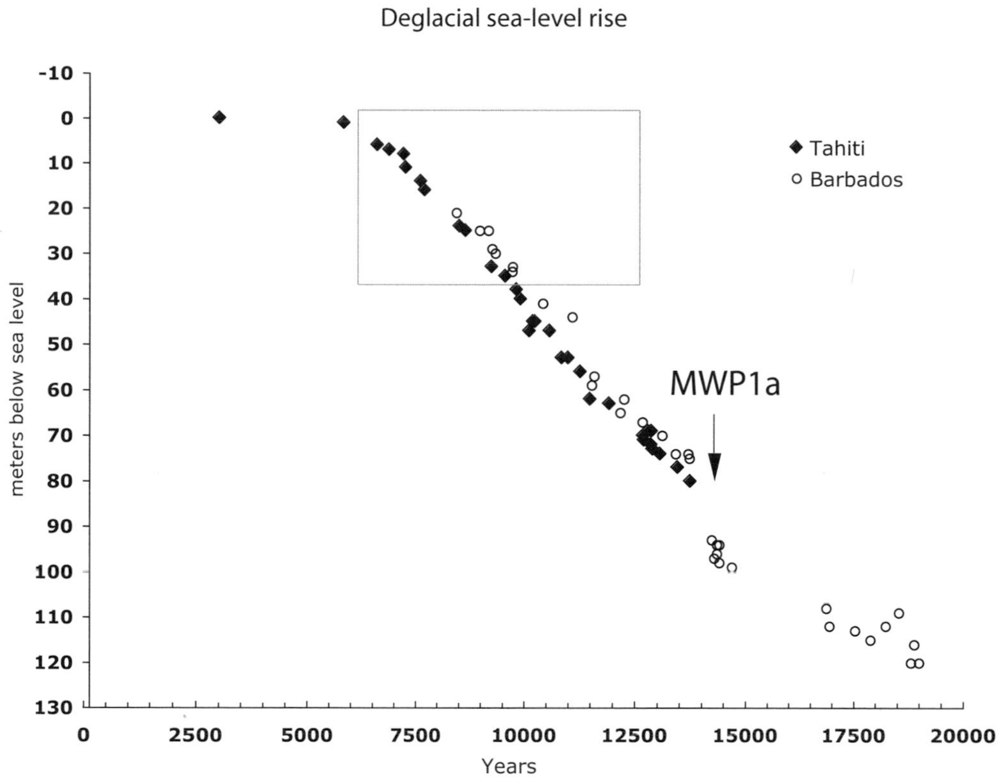

Deglacial sea-level rise

FIGURE 7.13 Deglacial sea-level curves from submerged coral reefs in Barbados (Fairbanks 1989) and Tahiti (Bard et al.1996). The depths are corrected for local uplift and subsidence. Caribbean coral *Acropora palmata* depth habitats is ~7 m, and slightly greater for Tahiti corals. Additional coral ages for the glacial period older than 19 ka are given in Peltier and Fairbanks (2005). MWP1a=inferred Meltwater Pulse 1a. Box shows interval of final sea-level rise.

based on Tahiti and Barbados coral reefs (Fairbanks 1989; Bard et al. 1996). These data are integrated with much other sea-level data into the glacial isostatic adjustment ICE-5G model of Richard Peltier (2004). The most commonly dated proxies of deglacial shorelines are the coral *Acropora palmata,* which is considered a reliable indicator of water depths less than 5–7 m based on its modern depth habitats in the Caribbean, and basal peats formed in tidal marshes.

The ICE-5G model is the culmination of decades of research on the isostatic response of the earth's upper mantle and surface to changing distribution of mass during ice-sheet decay. Its design is such that it gives sea-level positions in discrete 1000-year step increments, so that unlike many high-resolution paleorecords discussed herein, it does not model submillennial sea-level variability during abrupt climate transitions. This is in part due to the response rate of geophysical processes in the earth's viscoelastic upper mantle to changes in mass and limitations of coastal sea-level records. During periods of modest SLR, from 1 to 10 mm yr^{-1}, many coral reefs and tidal marshes "keep up" with rising sea level, growing at nearly the same rate, other factors being equal. During rapid SLR, reefs and marshes cannot sustain their growth. These intervals, called reef- or marsh-drowning events, are usually recog-

nized by a gap in U/Th- or radiocarbon-dated corals or basal peat, and signify a rate of SLR too rapid for corals or marshes to accrete.

Figure 7.13 shows the post-LGM steady SLR, which includes four periods of rapid rise at 19 ka, ~14.5 ka, 11.5 ka, and 8 ka. The most widely recognized is Meltwater Pulse 1a (MWP1a), named by Fairbanks (1989) for the hypothesized influx of deglacial meltwater from ice sheets. Rates of SLR during MWP1a exceeded 40 mm yr^{-1}, more than 10 times the modern rate of SLR. Three other possible rapid SLR events are at 19 ka (Clark et al. 2001), at MWP1b (Fairbanks 1989), and at 8.2–7.6 ka (Blanchon and Shaw 1995). Blanchon and Shaw called these Catastrophic Reef Events (CREs). CRE1 and 2 were linked with MWP1a and 1b; CRE3 was identified as an event about 7.6–8 ka. Bard et al. (1996) did not detect CRE3 in the Tahiti sea-level record, and Blanchon and Shaw's hypothesized third event received some criticism (Clark et al. 1995). However, recent studies suggest an early Holocene period of rapid SLR with rates of at least 10–12 mm yr^{-1}. Yu et al. (2007) identified an event culminating at 7.6 ka in the Baltic Sea region, Cronin et al. (2007) discovered a two-step period of marsh drowning in the Chesapeake Bay during abrupt cooling from 8.5–8.2 ka, and Carlson et al. (2007b) identified an LIS event at about 7.6–7.2 ka.

Sea-level reconstructions based on coral records rely on a firm understanding of the biology of various taxonomic groups as well as the stratigraphy and sedimentology of reef sequences. Reef growth is governed by tectonics, nutrients, light, turbidity, and the rate of SLR, to mention a few factors. In a review of Indo-Pacific deglacial reefs, Montaggioni (2005) counted 684 cores that contribute to the rich tropical-reef history of SLR. Forty-four reefs recovered in 140 cores have been studied in Australia's Great Barrier Reef alone (Hopley et al. 2007). Montaggioni was able to identify four periods of reef growth, called Reef Generation (RG) periods. During RG0 from 23–19 ka, there was a period of slow (1 mm yr^{-1}) SLR, consistent with many other paleoclimate records, for a gradual initial period of deglaciation. RGI, RGII, and RGIII were dated at 17.5–14.7, 13.8–11.5, and 10 ka, respectively. In Montaggioni's chronology, RGII followed MWP1a and coincided with surface ocean warming that lasted until the onset of the YD.

Steinke et al. (2003) and Hanebuth et al. (2000) studied SST and sea level on the Sunda continental shelf and slope off Indonesia. Their multi-proxy records showed that the shelf was inundated in steplike fashion beginning about 18–19 ka following the LGM. After a modest initial SLR during the interval 19–16.5 ka, sea level rose from 16.5–14.5 ka, until a rapid 16-m SLR occurred at 14.6–14.3 ka, which they correlate with MWP1a and the B/A warming. The final stages of the postglacial record showed decelerated rates of SLR between 14.3 and 12.5 ka and into the Holocene.

In sum, there is little question that the rate of SLR during MWPs and CREs exceeded 10 mm yr^{-1} and probably was as fast as 50 mm yr^{-1} at times when global sea level jumped 10 to 20 m in 500 years or less.

Uncertainty in Sea-Level-Climate Correlations Three main problems arise when we try to correlate the postglacial SLR with climate reversals already discussed herein. First, there has long been debate about the ice-sheet source for each meltwater-sea-level event (Clark et al. 1996; Peltier 2002). This issue is exemplified in the identification of the source of meltwater during MWP1a, dated by various authors at 14.0–14.5 ka. Reversing a long-standing opinion that southern LIS meltwater caused MWP1a, Clark et al. (2002) suggested instead that a large part of this event may have originated in melting Antarctic ice (see Weaver et al. 2003). This opinion generated significant discussion (e.g., Peltier 2005) because, as discussed above, there is substantial evidence for northern-hemisphere deglaciation at this time and later deglaciation of at least parts of Antarctica.

In one study linking SLR to climate, Stanford et al. (2006) showed that MWP1a did not coincide with the onset of B/A warming but instead correlates with cooling during the Older Dryas about 14 ka. They also found that during peri-

ods of NADW slowdown during H-1 and the YD, there was not a significant influx of northern-hemisphere ice-sheet glacial meltwater. Their argument for decoupling between the rate and magnitude of meltwater influx and THC suggests that either the location where small meltwater pulses reach the surface ocean is more critical than their rate or magnitude, or that other factors, such as sea-ice extent, are important in explaining large-scale climate and ocean-circulation changes.

A second issue revolves around the contribution to SLR directly from melting ice versus that from catastrophic discharge of large volumes of glacial lake water. Törnqvist et al. (2004), for example, compiled estimates of SLR due to drainage of the largest glacial lake, Agassiz-Ojibway, near 8.2 ka. This large discharge would generate only a 1-m SLR, yet it clearly had a major impact on global climate, possibly catalyzing catastrophic large-scale ice-sheet or ice-shelf destabilization (Kanfoush et al. 2000; Cronin et al. 2007). Chapter 9 addresses questions of the rate, magnitude, location, and timing of freshwater lake discharges in more depth.

The third complication arises from the complexity of ice-sheet and ice-shelf decay in terms of internal ice dynamics. Asynchronous melting of various ice-sheet margins, in some cases unforced by climatic factors, means that submillennial-scale sea-level variability must ultimately be obtained and linked to regionally specific glaciological records of deglaciation. Such temporal and spatial resolution is available only for a few regions at present, but it will be required to integrate sea level into annual and decadal climatic records.

Mechanisms to Explain Deglacial Climate Changes

At the beginning of this chapter, we showed that there are two basic schools of thought about the timing of deglacial climate changes. The first holds that northern and southern-hemisphere changes are asynchronous and that either northern-hemisphere climate leads changes in the southern hemisphere, or vice versa. An alternate view is that northern- and southern-hemispheric deglacial events are synchronous or at least do not show the proper lead-lag relationships consistent with northern-hemispheric forcing (Steig et al. 1998, but see Piotrowski et al. 2005). This dichotomy is of course an oversimplification of abrupt climate reversals during deglacial periods. Nonetheless, we briefly summarize these viewpoints.

Within-Hemispheric Synchroneity

The synchroneity between high- and low-latitude northern-hemisphere climate change during the YD now seems to be thoroughly documented. In the polar and subpolar regions of the North Atlantic, climatic events are clearly synchronous within decades (Björck et al. 1998; Alley et al. 2002). Low-latitude tropical SST (Lea et al. 2003), tropical Atlantic trade winds (Hughen et al. 1996), eastern Pacific Santa Barbara Basin paleoceanography (Hendy et al. 2002), and Hulu and Dongge cave speleothem records of EAM variability (Y. Wang et al. 2001; Yuan et al. 2004) all indicate synchronous changes within the decadal limits of various age models.

Efforts to establish whether low and high latitudes in the southern hemisphere are synchronous with each other for certain deglacial millennial events show that warming in Antarctic ice cores generally began simultaneously ~17.5–17 ka (EPICA Community Members 2004). The ACR varies slightly in its duration and amplitude across Antarctica (Morgan et al. 2002). Excellent records come from South African speleothems (Holmgren et al. 2003), Patagonian ice-sheet melting (Hulton et al. 2002), eastern tropical Atlantic SSTs (Schefuß et al. 2005), and tropical African atmospheric temperatures (Weijers et al. 2007). Southeastern Pacific Ocean SSTs also indicate synchroneity between Antarctic ice core and other southern-hemisphere climate records (Lamy et al. 2004).

Interpolar Asymmetry

The recognition of interpolar asynchronous millennial climate changes goes back to efforts to identify a global YD signal and establish its causes (Sowers and Bender 1995). For example, Markgraf (1991, 1993) argued that midlatitude pollen records from Chilean lakes and peat from Tierra del Fuego and southern Patagonia reveal complex vegetation shifts that did not support the idea of a simple climatic reversal during the YD. In higher southern latitudes, there is marked short-term variability in vegetation, but the timing of changes in specific plant taxa did not match each other. This asynchroneity led Markgraf to suggest that the changes signified vegetative response to local events such as fire disturbance.

The YD in New Zealand has also received considerable attention. New Zealand's Waiho Loop, a large important moraine of Franz Josef Glacier, has been at the center of the debate. Radiocarbon dating indicated a YD-age glacial advance, and regional cooling in New Zealand suggested synchronous climate changes in both hemispheres (Denton and Hendy 1994, reviewed in Lowell et al. 1995). Singer et al. (1998), however, concluded on the basis of pollen evidence that there was no YD cooling in New Zealand. Barrows et al. (2007) recently re-dated the Waiho Loop moraine using [10]Be and [36]Cl cosmogenic isotopes and determined that it was actually several thousand years older than previously believed and not synchronous with YD cooling. Their marine SST record from the Tasman Sea showed a steady oceanic warming beginning early in the deglaciation of 19 ka right through the YD into the Holocene epoch. A similar controversy exists in the dating and interpretation of the Southern Patagonian ice field, recently re-dated by Ackert et al. (2008), who concluded that these glaciers advanced after the YD cooling episode in the northern hemisphere.

Definitive evidence for first-order asynchroneity between high northern and southern latitudes during millennial climate reversals comes from ice cores. The bulk of evidence comes from temperature records synchronized by methane from Greenland and Antarctica (Blunier et al. 1998,; Blunier and Brook 2001; Morgan et al. 2002). These show that the ACR coincided with the B/A warming and that Antarctica experienced low-amplitude warming during the YD.

The paleoceanographic record sheds light on the asymmetry of climate changes. Lea et al. (2003) found that Cariaco Basin SST changed in phase with Greenland temperature, but several others have found the opposite pattern for tropical Atlantic sites in the nearby Tobago Basin (e.g., Rühlemann et al. 1999; Hüls and Zahn 2000), the Gulf of Mexico (Flower et al. 2004), and the western tropical Atlantic in the North Brazilian Current (Weldeab et al. 2006).

Interhemispheric Asymmetry— Low Latitudes

There is growing evidence, mainly from U/Th-dated cave records, for an asymmetric pattern of deglacial climate changes between the tropical regions in the northern and southern hemispheres (Y. Wang et al. 2001, 2005; Yuan et al. 2004; Dykoski et al. 2005). These patterns are consistent with high-latitude evidence for an asymmetric climate pattern at millennial timescales between the two hemispheres.

Firestone et al. (2007) presented an extensive array of proxy records from archaeological sites and the Carolina Bays in the eastern United States, suggesting that an extraterrestrial impact about 12.9 ka caused the extinction of some large North American megafauna, the end of the Clovis point culture, and the onset of YD cooling. They cited the occurrence of magnetic grains and microspherule, high concentrations of the elements iridium and nickel (possibly extraterrestrial in origin), charcoal, carbon spherules (biomass burning, wildfires, climate changes), and other geochemical and sedimentological features at the 12.9 stratigraphic horizon at multiple archaeological sites as evidence for this event. There is no identifiable impact crater for the proposed YD event, so they hypothesis it may have involved prior fragmentation and airbursts in the atmosphere, perhaps over the retreating LIS, leaving negligible evidence. The YD-extraterrestrial impact theory for the cause of the YD event requires additional testing through field stratigraphic studies and modeling.

Older Terminations

YD-like events occur in older glacial terminations, although few studies have explicitly searched for them. For example, Sarnthein and Tiedemann (1990) discovered that, during several of the past six glacial terminations, apparent reversals toward glacial-like climate characterized the deep-sea isotopic record of foraminifera from Deep Sea Drilling Project (DSDP) Site 658 off West Africa. Haake and Pflaumann (1989) and Sarnthein and Altenbach (1995) discovered a reversal in the $\delta^{18}O$ of *N. pachyderma* from the Nordic Seas during Termination II that looked similar to a YD pattern. Seidenkrantz et al. (1996) named a YD-type event during Termination II the "Zeifen-Kattegat oscillation." This rapid event appeared in stable isotopes in speleothems, biostratigraphy, and pedostratigraphy in soil sequences, as well as in ice-core climate records from 24 lacustrine and marine sites. In the North Atlantic, Oppo et al. (1997, 2001) studied Termination II using faunal and isotopic records from several sites. They found evidence for surface ocean SST and ice-rafting oscillations, including minor cooling in the middle to late stages of deglaciation. SST remained cool for most of Termination II, rising 8–10°C relatively late in deglaciation. Oppo points out a link between surface ocean conditions and deep-water formation in the Nordic Seas during warm phases and in the northern North Atlantic during deglacial stadial events, similar to the LGM situation (see Cortijo et al. 1999; Rasmussen et al. 1999, 2003). Pahnke and Zahn (2005) found carbon and oxygen isotopic evidence for climatic reversals during Terminations I, II, III, and IV in the southwest Pacific Ocean at intermediate water depth. Siddall et al. (2006) discovered a 30-m sea-level reversal during Termination II in Red Sea isotopic records, strongly suggesting millennial-scale variability such as the YD.

Perspective

The most startling feature of deglacial millennial climate events is how quickly they begin and end. Like DO events, large regional temperature, atmospheric, and oceanic changes occur over only a few decades, reversing the general deglacial trend. Terrestrial ecosystems also respond rapidly, within decades, dispelling a long-held view that they lag climate change by centuries to millennia. Even deep-ocean-circulation changes abruptly affecting the composition and diversity of deep-sea ecosystems change within centuries during deglacial abrupt events. As such, deglacial climate variability is a further testament to the extreme sensitivity of the climate system to relatively small forcing, changes in boundary conditions, or both. The direct causes of abrupt millennial events are still under intense investigation; the most commonly cited trigger—the catastrophic discharge of large proglacial lakes—is discussed further in Chapter 9.

LANDMARK PAPER Deglacial Sea-Level Rise: The Barbados Record

Fairbanks, R. G. 1989. A 17,000-year glacio-eustatic sea level record: Influence of glacial melting rates on the Younger Dryas event and deep-ocean circulation. Nature 342:637–642.

The possibility of rapid sea-level rise in the future is one of the most serious potential environmental changes confronting society. Just how fast can sea level rise during periods of climatic warming?

Sea-level curves reconstructed from fossil coral reefs and coastal marsh peat provide many answers. The Barbados Sea Level curve is one of the most recognizable paleoclimate reconstructions for its significance as a historical record of global ice volume during the last glacial maximum (LGM) and the deglacial interval. Richard Fairbanks of Columbia University's Lamont Doherty Earth Observatory recovered cores from off Barbados and dated fossil corals of *Acropora palmata* from them with radiocarbon and later uranium-thorium methods. Fairbanks's Barbados sea-level paper provided a means to quantify the volume of deglacial meltwater discharge to the world's oceans, calibrate the marine foraminiferal oxygen isotope curve for its ice-volume component, and examine meltwater's impact on ocean salinity and deep-ocean circulation. It showed that sea level was, at a minimum, about 121 m below modern sea level during the LGM. Even 20 years later, this estimate is within meters of those of more recent studies.

Moreover, Fairbanks discovered an intermittent pattern of sea-level rise, suggesting that there were two periods of massive ice-sheet decay and rapid rates of global sea-level rise exceeding 40 mm per year, more than 10 times the current rate. These periods, called Meltwater Pulses 1a and 1b, are benchmark discoveries in paleoclimatology. Generally supported in later studies of Tahiti and the Sunda Shelf by Bard et al. (1996) and Hanebuth et al. (2000), the Barbados sea-level curve is a standard against which deglacial paleoclimate records are compared and glacio-isostatic geophysical models are tested (Peltier and Fairbanks 2006).

Thanks to the Barbados, Tahiti, and other sea-level records, we now know that the rates of sea-level rise during rapid climatic warming reach anywhere from 3 to 10 times the current rate.

8

Holocene Climate Variability

Introduction

This chapter deals with the challenging subject of climate change during the Holocene interglacial period covering the past 11.5 ka. The Holocene epoch poses unique problems for paleoclimatology. One unique aspect of the Holocene is that the past few thousand years constitute the benchmark against which climatologists differentiate 20th-century human-induced climate change from natural variability. Compared to global climate changes over orbital and millennial timescales, the signal-to-noise ratio for Holocene climate variability is relatively small. For example, global mean temperature has oscillated by only 0.5–1°C during the past few millennia, whereas there has been a 5°C or more mean global cooling during glacial intervals. Some plausible forcing mechanisms to generate global temperature variability <1°C within an interglacial period are volcanism, solar irradiance, atmospheric-oceanic interactions, atmosphere-vegetation feedbacks, and atmospheric trace gases.

Consequently, a related problem is that Holocene climate change on hemispheric and global scales has been difficult both to detect with proxy methods and to correlate from region to region. Even today, one sees the Holocene called a relatively stable climate state when it is compared to the unstable climate of the last glacial and deglacial periods. In this same vein, Holocene climate change was a fledgling field just a few years ago. Pioneering studies of terrestrial (Bradley and Jones 1992; Diaz and Markgraf 1992; Hughes and Diaz 1994a) and ice-core records (e.g., Meese et al. 1994; O'Brien et al. 1995) yielded no consensus about interhemispheric climate patterns or their causes.

Viewing Holocene climate as stable or invariant would be totally incompatible with available evidence. Detailed paleoreconstructions from sensitive parts of the climate system—glaciers and ice caps, the tropical oceans, regional atmospheric systems such as monsoons, the Intertropical Convergence Zone (ITCZ) and El Niño–Southern Oscillation (ENSO)—reveal dynamic climate variability during the Holocene (Fischer et al. 2004). Rapid advances in proxy methods; the recovery of high-resolution tree-ring, ice-core, speleothem, and ocean and lake sediment records; and climate modeling have allowed great strides in Holocene paleoclimatology. Coastal records show that since about 6000 years ago, sea level varied within a small envelope of one to two meters, although the influence of local land movements complicates sea-level reconstructions (Figure 8.1).

Another unique aspect of Holocene climate is that it influenced human cultural evolution. The literature is full of studies linking climate change to the decline of or stress on civilizations (deMenocal 2001). Examples include the collapse of the classical Central American Maya (Hodell et al. 1995; Haug et al. 2003), the North American Anasazi and other native North American cultures during Medieval Warm Period (MWP) mega-droughts (Benson et al. 2007), indigenous cultures in the American southwest since 4 ka (Polyak and Asmerom 2001), Little Ice Age (LIA) 15th-century Norse colonies in western Greenland (Dansgaard et al. 1975), ancient Egyptian (Krom et al. 2002; Nicoll 2004) and other Asian and African old-world civilizations (Drysdale et al. 2006), and even early English and Spanish colonists in North America (Stahle et al. 1998a). Not all civilizations are linked to large climate change (Enzel et al. 1999), but it is still fair to say that modern society, as it confronts future climate change, can benefit from an understanding of these historical events.

A final issue introduced in Chapter 5 is whether preindustrial human activity actually changed the course of Holocene interglacial climate through impacts on carbon cycling long *before* industrialization. This view, espoused by Bill Ruddiman of the University of Virginia, holds that agriculture affected mid-Holocene methane flux, preventing an interglacial climate from falling into the depths of the next glacial age (Ruddiman 2007). Ruddiman's early Anthropocene hypothesis is widely debated and reflects the importance of Holocene climatic patterns described in this chapter.

Several large international projects have been designed to understand Holocene climate variability. The first effort was called the Cooperative Holocene Mapping Project

FIGURE 8.1 Emergent middle Holocene *Porites* coral reef about 1.5 m above modern sea level from the Sunda Strait. See also figure 2-21 in the *World Atlas of Holocene Sea-Level Changes* (Pirazzoli 1991). The reef is emergent because of glacio-isostatic adjustment and possibly also slightly lower-than-modern continental ice volume. Courtesy of P. Pirazzoli.

(COHMAP) (COHMAP Members 1988). COHMAP incorporated proxy-based reconstructions of global climate at 12, 9, 6, and 3 ka, as well as climate modeling, to understand causal mechanisms (Kutzbach and Guetter 1986; Wright et al. 1993). Since 1991, an international project called PAGES (Past Global Changes), carried out under the auspices of the International Geosphere-Biosphere Programme (IGBP), has fostered Holocene climate research. Under the PAGES umbrella project, the Pole-Equator-Pole (PEP I, II, and III) projects are north-to-south paleoclimate reconstructions along transects spanning the Americas, East Asia-Australasia, and Europe and Africa, respectively. These efforts have provided an enormous source of new paleoclimate time series and time slices. The International Marine Global Change (IMAGES) and Ocean Drilling Programs (ODP) are complimentary efforts to understand global climate changes in the world's oceans. The Paleoclimate Modeling Intercomparison Project (PMIP) is a major program that combines paleoreconstructions and climate modeling to understand global climate at 6 ka and during the Last Glacial Maximum (LGM) (Braconnot et al. 2007). Recently, PMIP data-model simulation studies have demonstrated the importance of southern-hemisphere sea-ice feedbacks and Atlantic circulation changes during the LGM (Otto-Bliesner et al. 2007). Several major Greenland and Antarctic ice-coring programs introduced earlier provide high-resolution Holocene atmospheric records. In this chapter we draw on these and other efforts to outline relatively cohesive, geographically robust patterns of Holocene climate change and to discuss the large challenges that lie ahead in attributing cause.

Holocene Paleoclimatology: Terms

Terms and concepts introduced during the 19th and early 20th centuries form the foundation of Holocene paleoclimatology, and many are still widely used today (Table 8.1). Some early concepts of Holocene climate are historically and regionally important, but there is often confusion about their exact meaning or climatic significance. For example, entire volumes have been devoted to questions such as "Was there a Medieval Warm Period?" This is not a trivial question—the existence of preindustrial warm periods fuels debates about the degree to which humans have caused 20th-century warming. So it is worth understanding the origins and meanings of these terms.

The Holocene Epoch

The Holocene epoch, called the Flandrian interglacial in Europe, was first named in 1885 at the International Geological Congress to designate the "wholly recent" geological period (Bowen 1978). The Holocene is traditionally distinguished from the Pleistocene glacial period as a time of relatively warm interglacial climate, and until recent decades had been viewed as a time of stable climate relative to that of

TABLE 8.1 Holocene Paleoclimate Terminology*

Name	Abbreviation	Approximate Age	Reference	Other Names
Holocene Stage	N/A	11.5 ka to present		Flandrian
Neoglacial	N/A	About 5 ka–A.D. 900	Porter and Denton 1967	
Medieval Warm Period	MWP	A.D. 800–1500	Lamb 1965	Medieval Climate Anomaly
Little Ice Age	LIA	A.D. 1500–1900	Matthes 1939; Grove 1988	
Maunder Minimum	N/A	A.D. 1645–1710	Eddy 1976	
Hypsithermal		Early to mid-Holocene	Deevey and Flint 1957; Davis 1984	
Altithermal		mid-Holocene	Meltzer 1999 and Antevs publications therein	
Early Holocene Thermal Maximum			Various	
Dalton Sunspot Minimum		A.D. 1800–1860	Hoyt and Schatten 1997	
Wolf Sunspot Minimum		A.D. 1290–1390	Wolf 1868	
Spörer Sunspot Minimum		A.D. 1450–1540	Spörer 1889	

Source: http://www.geo.arizona.edu/palynology/geos462/holobib.html.
*Many other regional terms exist, especially in pollen and peat stratigraphy literature. See Owen Davis's compilation (1984).

the last glacial. The definition of the base of the Holocene has a long history (Bowen 1978; Davis 1984; Roberts 1998), but today paleoclimatologists place it either at 10 ka or at 11.5 ka, coincident with the end of the Younger Dryas (YD) and the beginning of the Preboreal period in Europe (Björck et al. 1998). Based on global climate patterns described in the sections that follow and in Chapter 7, we adopt an age of 11.5 ka.

There is no universally accepted time stratigraphic subdivision of the Holocene into formal substages. Most other authors use the terms early, middle, or late Holocene informally with reference to regional paleoclimatic patterns. Roberts (1998) refers to the early Holocene as the period between 11.5 and 5 ka, the middle and late Holocene from 5 ka up to about 500 years ago, and the last 500 years as modern times. Major climate changes during the middle Holocene ~5–6 ka might make this period a convenient boundary. However, the nature of the mid-Holocene transition—either gradual or abrupt—is still debated, and evidence exists for other climatic transitions such as one near 3.5 ka. Consequently, a formal climatostratigraphic Holocene nomenclature, such as that applied to glacial and deglacial intervals, is not recommended at this time and the terms early, mid-, and late Holocene should be used informally.

An absolute timescale is preferred for correlating and interpreting Holocene climatic patterns because an increasing number of paleorecords achieve annual or decadal resolution and reveal high-frequency climate variability. However, there is no universal convention on a common "zero" point from which we count back (Rose 2007; Wolff 2007). Some authors use the terms B.C. and A.D., and these can be useful, particularly in discussing historical climate changes. The year A.D. 1600 is easier to understand than 408 yr BP, 0.408 ka. The most obvious drawbacks are the practical inconvenience of having to subtract the current year from the absolute age and the difficulty merging a B.C./A.D. convention with ka used for older records.

The older literature is full of uncalibrated radiocarbon dates, usually referred to in "yr BP," or years before present, with "present" meaning the year 1950. But confusion arises because tree-ring, ice-core, coral, and speleothem researchers using methods other than radiocarbon dating for age control count back from the year the archive was sampled or some other reference time. One option suggested by Wolff (2007) is to simplify the issue and count back from the year 2000 using the abbreviation "b2k" for before 2000. The term b2k is already in use in some recently published papers (e.g., Rasmussen et al. 2006), and the international community should decide in the coming years if this option is acceptable. Here we use ka for the Holocene timescale and calendar years for historical periods younger than 2000 years (see chapters 11 and 12 in this volume).

Early Holocene: The Altithermal, Hypsithermal, and Little Climatic Optimum

The period between about 11.5 and 5 ka has traditionally been characterized as a time of climate warmth called the altithermal (Antevs 1928, 1955), the Atlantic warm period, the Little Climatic Optimum in Europe, the postglacial hypsithermal (Deevey and Flint 1957), the early Holocene thermal optimum, and even a cultural "golden age." Early studies of palynological, marine, tree-line, and sea-level records mainly from Europe and the Americas indicated that temperatures and sea-level positions exceeded those of the late Holocene (Fairbridge 1961). Many low-latitude alpine glaciers were at their minimum extent before an abrupt advance about 5.2 ka (Thompson et al. 2006a). The hypsithermal was also characterized by enhanced precipitation in some areas.

Our current understanding of Holocene climate renders terms such as *altithermal*, *hypsithermal*, and *Little Climatic Optimum* of limited use and a potential source of confusion. For example, there is significant diachroneity in the timing of Holocene climatic temperature changes even at the regional scale. In a study of 11 Antarctic ice cores with multidecadal resolution, Masson-Delmotte et al. (2000) showed that temperature maxima occur between ~11.5–9 ka and 8–6 ka in the Ross Sea region of West Antarctica, 6 and 3 ka in central Antarctica, and as late as 3 ka in other regions of Antarctica. The Arctic Ocean marine record is not as well dated as those from the Antarctic continent, but it has equally complex spatial variability in early Holocene ocean-surface conditions (Cronin et al. 1995; Nørgaard-Pedersen et al. 2003; Spielhagen et al. 2004; Wollenburg et al. 2004; Polyak et al. 2007). Spatial and temporal complexity also characterizes tropics and midlatitude atmospheric and ocean temperatures and precipitation (see the subsections that follow). Consequently, viewing the early Holocene as a period of global warmth is far too simplistic.

Neoglaciation and the Little Ice Age

Neoglaciation refers to a period of advancing alpine glaciers during the Holocene first identified by Matthes (1939) in studies of glacier activity and moraines in the Sierra Nevada of the western United States. Matthes coined the term *Little Ice Age* to refer to "an epoch of renewed but moderate glaciation which followed the warmest period of the Holocene." Porter and Denton (1967) renamed the Holocene interval of North American glacial readvances the "Neoglaciation." Neoglacial alpine glacial history recognized in pioneering studies by Denton and Karlén (1973) are now recognized as major features of Holocene climatology (Mayewski et al. 2004). The term *Little Ice Age* is now most often used to refer to the period from about the 15th century through the end of the 19th century, when atmospheric temperatures fell and alpine glaciers advanced in many regions (Grove 1988).

In addition to northern-hemispheric Neoglacial cooling, a late Holocene cooling trend is also evident in east Antarctic ice-core temperatures (Mosley-Thompson 1996), the Agassiz ice cap, Ellesmere Island (Fisher et al. 1995), the Penny ice cap, Baffin Island (Fisher et al. 1998), summer melt layers in the Greenland Ice Sheet Project 2 (GISP2) ice core (Alley and Anandakrishnan 1995), southern-hemisphere glacial history (Clapperton 1993; Porter 2000a), and planktic foraminiferal isotope (Keigwin 1996) and bottom-water temperature (Dwyer et al. 2000) records. Late Holocene cooling is not, however, a simple unidirectional trend but instead is punctuated by complex temperature and precipitation oscillations discussed in the subsections that follow. Thus, the term *Neoglacial* should be used cautiously to characterize the first-order Holocene trend toward glacial advances up until the 20th century.

Medieval Warm Period

The 9th through the 14th centuries are often referred to as the Medieval Warm Period (Lamb 1965). A more narrowly defined period, A.D. 1100 to 1250, is sometimes called the Medieval Climate Anomaly, or Medieval Warm Epoch, which coincides with the Medieval Solar Maximum reflecting relatively high solar activity (Jirikowic and Damon 1994). Medieval warmth has traditionally been contrasted with the cooler conditions of the subsequent LIA from the 14th to the end of the 19th century.

During the past 20 years the MWP has emerged from relative obscurity to prominence as the lightning rod for controversy surrounding 20th century and future climate changes. Hughes and Diaz (1994b) asked, "Was there a Medieval Warm Period, and if so, where and when?" finding no sustained globally coherent MWP-LIA climate sequence, but some evidence for warmth during the parts of the 12th through 14th centuries. Crowley and Lowery (2000) asked, "How warm was the Medieval Warm Period?" Broecker (2001) asked, "Was the Medieval Warm Period Global?" and concluded that it was part of natural global-scale millennial-scale Holocene climate variations related to ocean circulation and heat transport.

The question persists: How much of the observed warming over the last century is due to natural climate variability, such as that during the MWP and LIA, and how much is human-induced change? There are two very different approaches to this question. One is to view the MWP from Broecker's perspective of millennial variability that characterizes an interglacial period such as the Holocene. This topic is addressed in the current chapter. The second considers climate variability of the last millennium, as a baseline against which instrumental records of climate can be assessed. This topic is addressed in chapters 11 and 12 of this volume.

Solar Insolation and Tropical Atmospheric Processes

Holocene orbitally driven insolation and variability in solar irradiance are two important forcing mechanisms of Holocene climate. The ITCZ and global monsoon system—two closely linked tropical climate features sensitive to relatively small forcing—have a strong influence on global climate. This section briefly describes these aspects of Holocene paleoclimatology. High-latitude processes such as sea ice, deep-ocean circulation, and freshwater forcing also play a role in Holocene climate; relevant background was given in earlier chapters.

Orbital Insolation

Orbitally induced changes in seasonal and geographical solar insolation provide the broad context for Holocene climate evolution. Figure 8.2 shows orbitally driven insolation changes in high and low latitudes of the northern and southern hemispheres, from the calculations of Berger and Loutre (1991). We see that boreal summer insolation peaked near 11 ka at 60°N latitude and the equator and has declined since. Boreal winter insolation was lowest between 12 and 10 ka. At the equator and at 60°S latitude, insolation increased during the Holocene. These small but critical changes in insolation operating through partially understood feedback mechanisms are believed to induce significant changes in global climate, as described in the subsections that follow.

Solar Irradiance

Solar forcing of Holocene climate is an actively researched and controversial topic. Solar irradiance—the amount of solar energy reaching the top of the earth's atmosphere—varies over different timescales because of complex, still partially understood solar processes (Foukal et al. 2006). On the basis of observations of sunspots, other solar activity, and recent satellite monitoring, historical changes in irradiance have been identified at several distinct frequencies (Table 8.2) (Hoyt and Schatten 1997; Pap et al. 2004; Benestad 2005). Frequencies most commonly encountered in Holocene paleoclimatology are the 11-year Schwabe sunspot, 22-year Hale double sunspot, 88-year Gleissberg, 200-year de Vries (also called Suess), and the so-called 1500-year Bond cycles. The Holocene Bond cycle should not be confused with that

Holocene orbital insolation

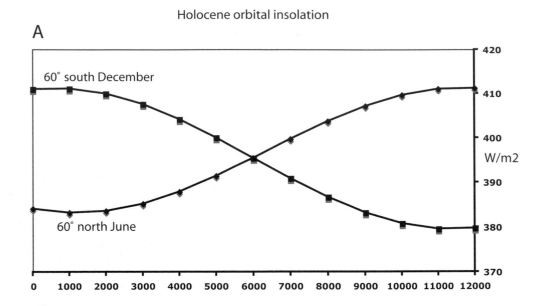

A

60° south December

60° north June

420
410
400   W/m2
390
380
370

0   1000   2000   3000   4000   5000   6000   7000   8000   9000   10000   11000   12000

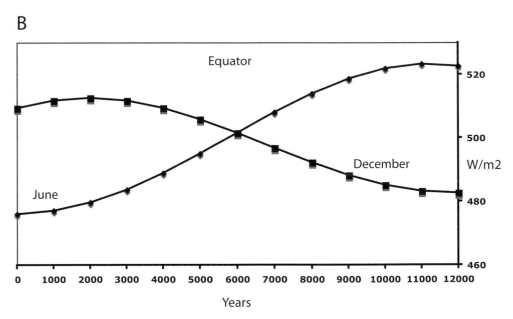

B

Equator

June

December

520
500
480   W/m2
460

0   1000   2000   3000   4000   5000   6000   7000   8000   9000   10000   11000   12000

Years

FIGURE 8.2 Holocene orbitally driven insolation: (A) 60° S December and 60° N June; (B) insolation at equator in December and June. From Berger and Loutre (1991).

given the same name for bundles of Dansgaard-Oeschger (DO) events during Marine Isotope Stage (MIS) 3 (see Chapter 6 in this volume). The Holocene "1500-year" pattern is at best quasi-periodic.

Solar variability during the Holocene has also been inferred from a rapidly growing literature on cosmogenic isotope (radionuclide) and geomagnetic paleointensity reconstructions (Table 8.2). For example, tree-ring, sedimentary, and ice-core evidence for the de Vries cycle can be found in Sonett et al. (1992), Stuiver and Braziunas (1993), Cook et al. (1996), Ram and Stoltz (1999), Yu and Ito (1999), Mc-

Cracken et al. (2001), and Peristykh and Damon (2003), to mention a few.

Radionuclide concentrations measured in tree rings, ice cores, and sediments represent the combined effects of heliophysical and geomagnetic processes that govern their production rate in the atmosphere and their transport and deposition after formation. In a series of classic studies, Stuiver and Braziunas (1993; Stuiver and Quay 1980; Stuiver and Braziunas 1989; Stuiver 1993) quantitatively analyzed a 10-ka ^{14}C record from five fir trees from the Pacific coast of the United States. The apparent periodicity in ^{14}C suggested several

TABLE 8.2 Solar Cycles

Cycle Name	Frequency*	Comment	Reference
Schwabe	10.4-yr, 11-yr		Schwabe 1844
Hale	22.3-yr	Double sunspot cycle	
Yoshimura	55-yr		
Gleissberg	80–90-yr, usually 88-yr		Gleissberg 1966
de Vries	200–210-yr	Also called Suess	de Vries 1958
Bond cycles—Holocene[†]	1470-yr, 1500-yr	Holocene	Bond and Lotti 1995; Bond et al. 1997, 2001
Bond cycles—glacial	10–15 ka packages of DO events		Bond et al. 1992, 1993; Lehman 1993

*Slightly higher and lower periodicities can be found in the literature, depending on the proxy record under study.

[†]Some consider Holocene 1500-year Bond cycles to be of solar origin. Glacial Bond cycles are distinct from those of the Holocene and are associated with Dansgaard-Oeschger and Heinrich events, and are unlikely of solar origin.

solar-ocean-atmosphere linkages. For example, the 11-year cycle appeared to be associated with external forcing by solar modulation of cosmic-ray flux. In contrast, multidecadal variability seemed to be tied to changes in thermohaline circulation and the rate of carbon storage in the oceanic carbon reservoir, as discussed earlier for glacial and deglacial periods. A ~200-year solar cycle was also evident.

Solanki et al. (2004) reproduced sunspot variability from an 11,400-year-old tree-ring ^{14}C record to determine if 20th-century sunspot activity was unusual (Figure 8.3). Numerous sunspot minima are evident, including those during the LIA, between 5.5 and 8 ka, and several times during the early Holocene between 9 and 11.5 ka. Solanki concluded that 20th-century solar activity is unusual in the sense that similar short-term excursions occurred only 10% of the time over the past 11,400 years. They also concluded that solar forcing of late 20th-century climate played a secondary role to greenhouse gas forcing.

Solanki's sunspot curve is a useful model but it is not the final word on Holocene solar variability. For example, Muscheler et al. (2005b) commented that radiocarbon records indicate several episodes of strong solar magnetic wind during the last few centuries such as those observed today. In addition to radiocarbon, beryllium, and chlorine isotopes used in the Holocene, solar-climate studies improve controls on past solar variability (Muscheler et al. 2005a). By combining beryllium and radiocarbon records, it is theoretically possible to separate the effects of carbon cycling and ocean circulation from the cosmogenic production rate and solar activity. For example, Hughen et al. (2000) compared the Cariaco Basin ^{14}C record to ice core ^{10}Be for the deglacial period, showing especially large changes in ^{14}C activity during the YD. These patterns clearly reflected changes in deep-ocean circulation rather than solar activity. Hughen et al. also found evidence for the 206-year cycle observed by Stuiver and Braziunas and a 2–6-year cycle related to ENSO ocean-atmosphere dynamics.

Holocene reconstructed sunspot activity

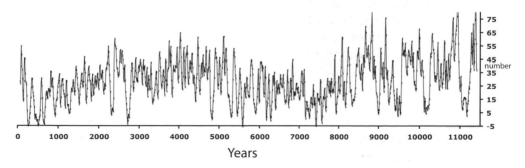

FIGURE 8.3 Reconstructed sunspot activity based on dendrochronologically dated radiocarbon concentrations, from Solanki et al. (2004). The curve shows that recent sunspot activity is relatively high compared to that over the past 8000 years; however, reconstructing past solar activity remains a challenge (see Muscheler et al. 2004, 2007).

Snowball and Muscheler (2007) integrated Holocene records of sunspots, cosmogenic production of beryllium and chlorine isotopes, and magnetic data from archeological and sediment records (see, among others, Yang et al. 2000; St. Onge et al. 2003; Solanki et al. 2004; Muscheler et al. 2005a). Their major conclusions are that geomagnetic reconstructions vary greatly depending on the region and source of measurements, that equally complex obstacles surround reconstruction of long-term modulation of solar-energy flux by solar wind and other processes, and that these factors preclude a simple quantification of millennial and short-term solar irradiance changes from radionuclide records.

Consequently, although many researchers hypothesize solar irradiance as a primary cause of Holocene climate variability (see the subsection "Solar Irradiance" in this chapter), a great deal of caution is needed when interpreting paleoclimate records before A.D. 1600 because reconstructed radionuclide records are not direct measures of solar activity.

The Intertropical Convergence Zone

The ITCZ is the global belt of atmospheric moisture convection manifested in extensive cloud systems that form near the rising limb of Hadley cell circulation. Figure 8.4

FIGURE 8.4 A Geostationary Operational Environmental Satellite (GOES). Satellite photograph of Intertropical Convergence Zone (ITCZ) taken August 27, 1999 (moon in lower left corner). The ITCZ is a ring of cloud systems in the Pacific Ocean north of the equator. Courtesy of D. Chester and the National Aeronautics and Space Administration (NASA).

shows an image of the modern ITCZ, with its northern-hemisphere bias located near the southern boundary of the northeast trade-wind belts. The ITCZ shows up as a ring of convective clouds over the tropical Pacific Ocean. During the boreal summer, the ITCZ reaches its northernmost position, shifting southward to near the equator during the winter. Seasonal migrations of the ITCZ are tightly coupled with changes in equatorial sea-surface temperatures (SSTs), which reach 5°C annually in the eastern Pacific Ocean cold tongue. This coupling is related to stronger atmospheric winds when the ITCZ is in a northerly position, causing more intense ocean upwelling off northwestern South America. Conversely, when the ITCZ is in a southerly position, winds are weaker and upwelling diminishes. In addition to seasonal changes, the ITCZ varies over short-term (40–60-day, Madden-Julian Oscillation), interannual (ENSO—see Chapter 10 in this volume) and, as seen below, centennial and millennial timescales during the Holocene.

Monsoon Systems

The other major climate pattern in the tropics is the development of seasonal monsoons. The term *monsoon* connotes a giant sea breeze, but it actually involves the entire tropical atmosphere and complex atmospheric-ocean links. Trenberth et al. (2000) characterize the modern monsoon as "a global-scale persistent overturning of the atmosphere throughout the Tropics and subtropics that varies with the time of year." Monsoon-driven precipitation occurs over extensive low- to midlatitude land areas in association with changes in SSTs and upwelling in the adjacent ocean.

The conventional view is that modern monsoons develop in response to strong seasonal contrasts in ocean (warm) and continental (cool) temperatures. Seasonal sensible heating of the Asian landmass combine with latent heat flux from the tropical Pacific and Indian Oceans. Some modeling studies suggest that monsoons can also develop with high SSTs and in the absence of a continent (Chao and Chen 2001). Papers on monsoons and their relationship to climate can be found in Tyson et al. (2002).

Two important aspects of monsoon variability are the intensity of wet season rainfall and the timing of seasonal climate extremes. Figure 8.5 depicts seasonal wind and precipitation patterns for the Asian Monsoon system, which is subdivided into the Indian or South Asian Monsoon west of 105°E longitude and the East Asian Monsoon east of 105°E (P. Wang et al. 2005). The figure shows the seasonal shift in wind direction from southwestern wind flow in winter to an onshore northeastern direction in summer. These shifts are accompanied by summer precipitation on the Indian subcontinent and in southeastern Asia. Monsoon systems also influence regions of west and east Africa, southwestern North America, and Australia.

The study of paleomonsoons is a relatively new sub-discipline of paleoclimatology. Some of the first evidence for large-scale, long-term monsoon variability came from classic reconstructions of Arabian Sea temperature and upwelling (Prell and Kutzbach 1987), Chinese loess and paleosol stratigraphy and sedimentology (Kukla and An 1989; Ding et al. 1994), and African paleolimnology (Street-Perrott and Perrott 1990, 1993). Rapid progress has since been made deciphering the behavior of monsoon systems over million-year (Prell and Kutzbach 1992; Sun et al. 2006), orbital (Clemens et al. 1991), and millennial timescales (Porter 2000b; An and Porter 1997; Leuschner and Sirocko 2000). We see in the next section that the Holocene monsoon has also been especially unstable over decadal to millennial timescales.

Holocene Records of Atmospheric Composition and Circulation

This section focuses on some of the best proxy records of Holocene atmospheric variability reconstructed from ice cores, lakes, peats, and other archives. Many atmospheric patterns are tightly coupled with oceanic and glaciological processes described in the following subsections.

Greenhouse Gases

Figures 8.6 and 8.7 show records of carbon dioxide (CO_2) and methane (CH_4) from several Antarctic ice cores (Etheridge et al. 1996, 1998; Indermühle et al. 1999; Monnin et al. 2001; Flückiger et al. 2002). Following an early Holocene peak of 265–270 parts per million by volume (ppmv) that occurred ~10–11 ka, CO_2 concentrations fell to about 260 ppmv by about 8 ka before rising gradually to 280 ppmv during the preindustrial period. Minor differences in the CO_2 records from different ice cores can be attributed to age differences between cores. Partly on the basis of measurements they made on isotopes of the carbon in the CO_2 gas, Indermühle et al. (1999) proposed that the progressive increase in Holocene CO_2 concentrations represents a decrease in terrestrial biomass during the Holocene epoch. The interpretation of Broecker et al. (1999a, b, 2001) is that this atmospheric CO_2 rise represented compensation by the ocean's carbonate chemistry.

Wagner et al. (1999, 2002) also reconstructed Holocene atmospheric CO_2 from a stomatal index from fossil leaves from Sweden. Following an early Holocene minimum similar to that in ice cores, they observed a rise between 5000 and 2800 years ago from 290–320 ppmv, values slightly higher than those from Antarctic ice cores. They also noted a brief decrease in CO_2 concentrations near 3.7 ka, probably related to changes in terrestrial biomass.

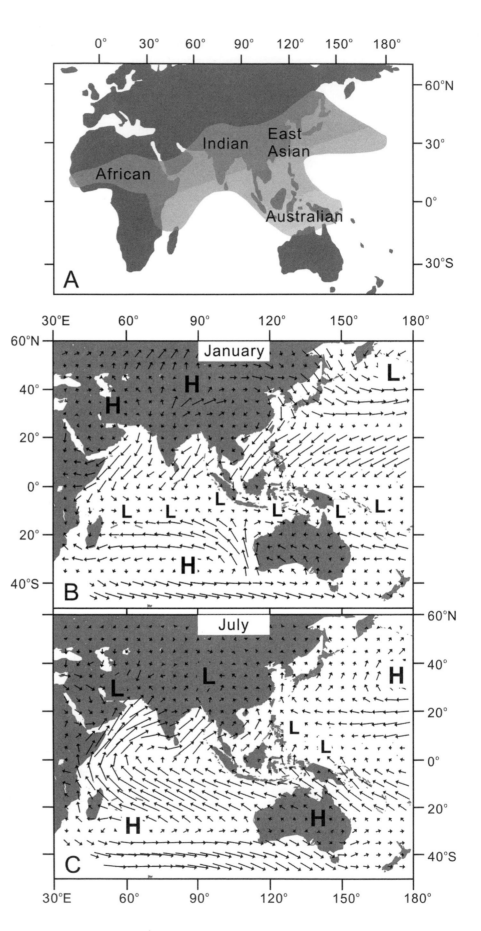

FIGURE 8.5 General location of (A) Asian monsoon, (B) atmospheric pressure, and (C) wind direction and strength during January and July, from Pinxian Wang et al. (2005). Courtesy of P. Wang and Elsevier.

Holocene ice-core atmospheric
CO_2 concentrations

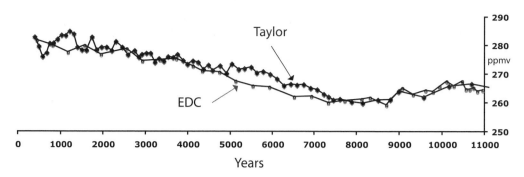

FIGURE 8.6 Holocene atmospheric CO_2 concentrations from Taylor (Indermühle et al. 1999) and EPICA Dome C (EDC) (Monnin et al. 2001). Antarctic ice cores are plotted against original timescales that were slightly revised by Monnin et al. (2004). Note the progressive rise from a minimum near 260 ppmv to preindustrial levels near 280 ppmv.

The methane curve is quite different from that of CO_2. It shows a steep decline from maximum concentrations of ~690–700 parts per billion by volume (ppbv) at 9.5–10 ka to mid-Holocene minimum values below 575 ppbv and then an increase to about 675 ppbv at preindustrial levels. As described in Chapter 7, atmospheric methane has several sources such that the interhemispheric gradient in methane concentrations changes, especially over periods of rapid climate change. MacDonald et al. (2006) showed that the early-to mid-Holocene methane pattern parallels the massive development of northern-hemisphere boreal peatlands from 12–8 ka following final retreat of northern-hemisphere ice sheets. They also argued that the late Holocene increase in methane since about 5–6 ka cannot be explained by methane flux from northern-hemisphere peatlands (Blunier et al. 1995) because late Holocene peat formed in ombrotrophic bogs that produce relatively small amounts of methane.

Methane buildup also coincides with changes in terrestrial vegetation in Africa (Johnson et al. 1996; deMenocal et al. 2000). The late Holocene methane rise is probably due either to anthropogenic activity (Chappellaz et al. 1990; Etheridge et al. 1998; Ruddiman 2003) or terrestrial sources other than boreal peatlands.

Nitrous oxide (N_2O) also varies during the Holocene. There is a maximum concentration during the early Holocene and ~2.5–2.0 ka, with a broad minimum centered on 8 ka and preindustrial values near 270 ppbv. Nitrous oxide measurements have relatively large analytical uncertainty considering the low amplitude during the Holocene that was equivalent to only 14% of that during a glacial-interglacial cycle (Flückiger et al. 2002). The sources of atmospheric nitrous oxide are soils and oceans, and its atmospheric lifetime is also influenced by ultraviolet radiation and atmospheric circulation. In spite of these complications, Flückiger et al. (2002)

Holocene methane concentrations,
EPICA Dome C, Antarctica

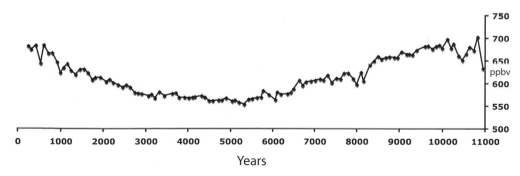

FIGURE 8.7 Holocene methane concentrations from the EPICA Dome C, Antarctica ice core, from Flückiger et al. (2002). Note that the pattern is different from the CO_2 curve in Figure 8.6 because of terrestrial sources and sinks in methane.

concluded Holocene variability in N_2O reflects changes in soil and possibly ocean sources.

Tropical Atmospheric Circulation

There is a substantial body of evidence for tropical atmospheric circulation changes during the Holocene (Koutavas and Lynch-Stieglitz 2004; Cane et al. 2006; Koutavas et al. 2006; Rashid et al. 2007). Table 8.3 lists several exceptional paleoclimate records from low latitudes. The most signifi-

cant first-order Holocene trends involve a progressive southward shift in the mean position of the ITCZ; decreased strength in the Asian, African, and North American monsoons; and greater El Niño activity.

ITCZ Variability   Figure 8.8 exemplifies the southward migration of the ITCZ in the Indian Ocean, south Asia, region based on the Qunf speleothem record from Oman (Fleitmann et al. 2003) and the Cariaco Basin titanium record of Haug et al. (2001b). Progressively more positive oxygen

TABLE 8.3 Holocene Atmospheric and Oceanic Records

Region	Proxy	Conclusion	Reference
Taiwan—Retreat Lake	Total organic carbon	Monsoon peak 8.6–7.7 ka, the insolation-induced decrease	Selvaraj et al. 2007
African lakes	Various	ITCZ shift and ENSO-like variability	Russell and Johnson 2003a, b, 2005a, b
British Columbian lakes	Diatoms, sediments	1220-yr cycles, forcing unclear	Cumming et al. 2002
Tibetan Plateau, China	Hongyuan peat isotopes	Monsoon-N. Atlantic IRD coupling	Hong et al. 2003
Dongge Cave, China	Speleothem isotopes	3.55-ka abrupt monsoon weakening	Dykoski et al. 2005
Cariaco Basin	Iron, titanium, river influx	Wetter early Holocene shift to drier climate and more southerly ITCZ	Haug et al. 2001b
Dongge Cave, China	Speleothem oxygen isotopes	Progressive monsoon weakening and centennial events	Y. Wang et al. 2005
N. Africa	Dust at ODP Site 658C	Southward expansion of Africa desert	deMenocal et al. 2000
Oman	Stalagtite 18O	Southward shift in ITCZ	Fleitmann et al. 2003, 2007
Lake Titicaca, Bolivia	Lacustrine geochemistry	Mid-Holocene dry to wet transition, southward shift in ITCZ, antiphased with northern tropics	Seltzer et al. 1998; Baker et al. 2001a, b, 2005
Lake Malawi, Africa	Biogenic silica, dust	Mid-Holocene southward ITCZ shift	Johnson et al. 2002
Tropical Pacific	Ecuador lake sediments	Weak early- to mid-Holocene ENSO	Moy et al. 2002
Tropical Pacific	Foraminiferal shell chemistry	Mid-Holocene northward ITCZ, weakened El Niño	Koutavas et al. 2006
W. Tropical Pacific	Foraminiferal shell chemistry	Progressive decrease in SST and SSS in Holocene	Stott et al. 2004
Ecuador	Lacustrine sediments	Modern ENSO established 5 ka	Rodbell et al. 1999
Galapagos	Foraminiferal shell chemistry	Minor SST change in Holocene	Lea et al. 2006
Great Barrier Reef	Coral geochemistry	Mid-Holocene SST and hydrological changes	McGregor and Gagan 2004; Gagan et al. 1998, 2004

Note: ENSO = El Niño–Southern Oscillation; IRD = ice-rafted debris; ITCZ = Intertropical Convergence Zone; ODP = Ocean Drilling Program; SSS = sea-surface salinity; SST = sea-surface temperature.

Holocene ITCZ and monsoon variability

A. Cariaco Basin titanium

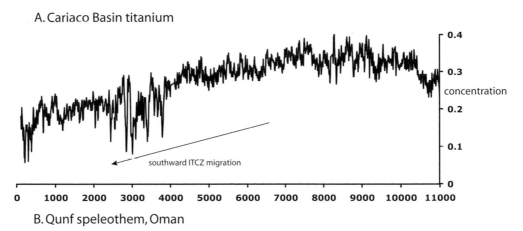

B. Qunf speleothem, Oman

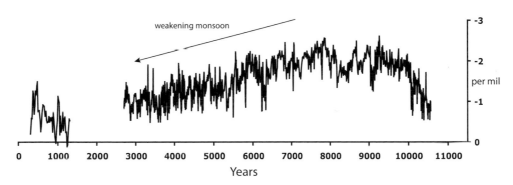

Years

FIGURE 8.8 Holocene Intertropical Convergence Zone (ITCZ) and monsoon migration from (A) titanium concentrations from the Cariaco Basin, Caribbean, from Haug et al. (2001b) (3-point smoothing) and (B) the oxygen isotope record from Qunf Speleothem, Oman from Fleitmann et al. (2004, 2007). Note the gradual southward migration of the ITCZ and the weakening of the Asian monsoon during the Holocene, with many short-term excursions.

isotope values signify a southerly retreat of major monsoon rainfall since about 7.5–8.0 ka. Numerous other records of Asian rainfall reflect this general trend (Kudrass et al. 2001; Fleitmann et al. 2004, 2007; P. Wang et al. 2005). In addition to the Cariaco record, comparable paleoprecipitation records are available from the northern tropics, including the Caribbean (Haug et al. 2001b; Tedesco and Thunell 2003), the Gulf of Mexico and southwestern United States (Poore et al. 2003, 2005), ocean regions off West Africa (SST and terrigenous influx) (deMenocal et al. 2000), and African lake records (Gasse 2000; Russell et al. 2003a, b). The southward migration of the ITCZ led to drier mean late Holocene conditions in the northern tropics and expanded deserts in low northern latitudes of Africa. The opposite occurred in southern-hemisphere low latitudes, where late Holocene climate became progressively wetter (Finney et al. 1996; Seltzer et al. 1998, 2000; Maslin and Burns 2000; P. Baker et al. 2001a, b, 2005; Castanada et al. 2007).

There remains an issue of whether the northern-hemisphere mid-Holocene (~5 ka) shift in summer ITCZ position and Indian, Asian, and North American monsoon precipitation intensity was an abrupt event (Morrill et al. 2003) or a gradual response to decreasing low-latitude precession (Fleitmann et al. 2003, 2007). Most evidence favors a more gradual transition, but there is not a sufficient density of records to preclude abrupt regional climate shifts as well (see Kröpelin et al. 2008).

Monsoon Variability The Holocene monsoon also exhibits long-term Holocene changes evident in lake, peat, speleothem, and ocean sediment records. Figure 8.9 shows a representative record of the Holocene Asian monsoon from the Dongge Cave, China speleothem (Y. Wang et al. 2005). It shows gradually more positive oxygen isotopic values increasing from an early Holocene minimum near 6.8–7.0 ka to an isotopic maximum a few centuries ago. Pollen, lake level, and loess records show that these monsoonal shifts were time-transgressive, with the zone of maximum precipitation migrating from northwest to southeast across southeast China between 10 and 3 ka (An et al. 2000). This pattern

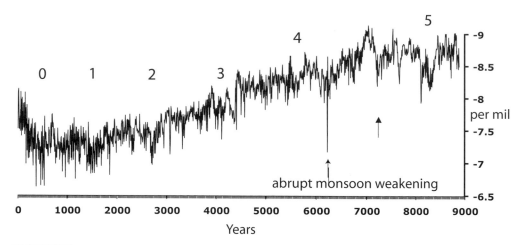

Holocene Asian monsoon variability,
Dongge Cave, China speleothem

FIGURE 8.9 Holocene oxygen isotope record from Dongge Cave, China DA speleothem, from Y. Wang et al. (2005), showing evolution of Asian monsoon over millennial and shorter timescales. The numerals 0 through 5 refer to one- to five-century-scale periods of weakened monsoon corresponding to the North Atlantic Ocean ice-rafting Bond events of Bond et al. (2001). Arrows indicate two other abrupt weak monsoon events.

was interpreted as evidence for insolation-forced time-transgressive weakening of monsoon systems.

Fleitmann et al. (2007) reviewed 22 paleorecords of the Holocene Asian and Indian monsoons and the ITCZ from China, Asia, and the Indian Ocean that were based on a variety of proxy records (planktonic foraminifera, terrigenous sediment, speleothem oxygen isotopes, and marine nitrogen isotopes). The evidence shows strong early Holocene Indian Ocean summer monsoons, progressively weakening during the Holocene in association with reduced oceanic upwelling and increased SST. Enzel et al. (1999) showed that following a mid-Holocene (~5 ka) wet period, lake levels controlled by the Indian monsoon fell, indicating drought conditions at about the same time that dune stabilization occurred. Carbon isotopic data from tree cellulose and specific plant types from the Hongyuan Peat Bog on the Tibetan Plateau also show a stronger Indian Ocean monsoon during the early Holocene, weakening gradually since about 5.5 ka (Hong et al. 2003).

The Holocene monsoon has also been reconstructed from a growing number of paleorecords from Africa (Gasse 2000; Barker et al. 2001; Weldeab et al. 2005), the Red Sea and eastern Mediterranean Sea (Arz et al. 2003), and the North Atlantic Ocean off West Africa (deMenocal et al. 2000). The primary trend is toward monsoon weakening and the end of what is called the early Holocene humid period in Africa. Weijers et al. (2007) produced an exceptional multi-proxy geochemical record from marine sediments off tropical west Africa, a region influenced by the Congo River drainage

system. They showed enhanced early Holocene precipitation, progressive large-scale drying, and a complex ocean-atmosphere linkage during the Holocene. Variability in the southwest North American monsoon is also evident in terrestrial records from the southwestern United States and in marine records from the Gulf of Mexico (Poore et al. 2003).

Submillennial Variability Significant submillennial variability superimposed on the long-term Holocene trend is also evident in high-resolution records shown in Figures 8.8 and 8.9, among others (Halfman et al. 1994; Haug et al. 2001b; Hong et al. 2003; Dykoski et al. 2005; Y. Wang et al. 2005; Russell and Johnson 2005a; Fleitmann et al. 2007). Centennial-scale variability reflects short-term changes in monsoon strength that in some cases can be correlated with episodes of ice rafting and ocean cooling in the subpolar North Atlantic Ocean and with terrestrial atmospheric records from other continents. The Dongge Cave record, in particular, provided compelling evidence that abrupt changes in the East Asian monsoon correspond with large-scale changes in global climate (Figure 8.9). Six 10–500-year-long isotopic excursions correspond precisely to six North Atlantic ice-rafted debris (IRD) events (see Yu et al. 2006). Haug's Cariaco Basin sediment record also reflects submillennial tropical climate variability (Figure 8.8). High rainfall characterized the Caribbean during the "thermal maximum" at 6–10 ka, much drier conditions during the late Holocene, and centennial-scale changes are evident during periods such as the LIA.

ENSO Variability   One of the most striking aspects of Holocene climate variability is the progressive development of ENSO over the past few thousand years. Figure 8.10 is a sediment record from Laguna Pallcacocha, Ecuador interpreted as a record of tropical storms caused by El Niño variability (Rodbell et al. 1999; Moy et al. 2002). The record shows subdued or even absent El Niño activity during the early Holocene and increasing ENSO activity during the late Holocene, peaking 1200 years ago. A similar first-order Holocene pattern has been inferred from geoarchaeological data from northwestern South America (Sandweiss et al. 1996) and terrestrial and marine records from New Zealand (Gomez et al. 2004). In the western Pacific region, Gagan et al. (1998, 2004) also found that ENSO variability was absent in the Great Barrier Reef, Australia around 5.4 ka. Tudhope et al. (2001) analyzed oxygen isotopes from New Guinea fossil corals and found that Holocene ENSO variance was lowest at 6.5 ka, whereas it was 2–3 ka for both the late Holocene and the modern age

Clement et al. (1999, 2000) simulated ENSO variability using the Zebiak-Cane model originally designed to analyze modern ENSO activity (Zebiak and Cane 1987; see also Chapter 10 in this volume). Their conclusion that Holocene changes in solar insolation influenced the intensification of ENSO variability during the last 10 ka was consistent with a growing body of evidence from paleoclimatology and climate modeling for stronger ENSO activity after 5 ka, coincident with the southward ITCZ shift, and peak activity about 3–1 ka (Cane 2005; Cane et al. 2006).

Despite the general patterns in Holocene ENSO activity, there is, nonetheless, a great need for better geographical coverage of annually resolved Holocene tropical atmospheric and SST records.

Midlatitude Regions

Midlatitude regions also experienced large changes in Holocene precipitation and temperature. Forman et al. (2001) documented extensive dune activity between 7 and 5 ka and extensive loess deposition that occurred intermittently between 10 and 4 ka, reflecting windier, more arid early- to mid-Holocene climate in the central parts of North America. Off the western United States, marine records indicate that the California Current was characterized by a longer and less vigorous upwelling season, with decreased seasonal contrast, prior to the mid-Holocene (Diffenbaugh et al. 2003; see also Barron et al. 2003, 2004).

Millennial variability is also evident in many Holocene terrestrial records. Diatom assemblages from Big Lake in British Columbia, Canada indicate a 1200-year quasi-cyclic behavior of moisture variability that is positively correlated with glacier advance-and-retreat patterns (Cumming et al. 2002). Viau et al. (2002, 2006) conducted a comprehensive analysis of North American temperature and also found strong evidence for millennial variability during the Holocene. Along the eastern United States, Holocene variability in rainfall over millennial timescales is antiphased with that observed in the northern tropics (Willard et al. 2005; Cronin et al. 2005). Midlatitude regions during the early Holocene experienced much drier conditions than the progressively wetter conditions of the late Holocene.

Holocene climate in Europe is also regionally and temporally complex. Some notable first-order European Holocene trends come from Ammersee ostracode shell chemistry, which showed a close link to Greenland ice-core atmospheric temperature over various timescales (von Grafenstein et al. 1998, 1999). The oxygen isotope record from a speleothem

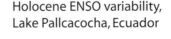

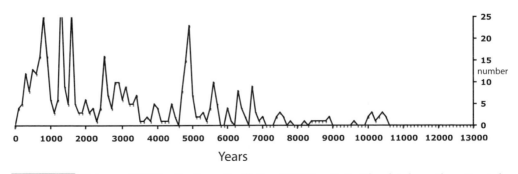

FIGURE 8.10   Holocene El Niño–Southern Oscillation (ENSO) activity. The plot shows the estimated number of El Niño events during the last 13,000 years, based on sediment analysis of Lake Pallcacocha, Ecuador, from Rodbell et al. (1999) and Moy et al. (2002). The early Holocene experienced little or no ENSO-like activity.

from Spannagel Cave, Austria also showed major regional cooling following an early Holocene northern-hemisphere climatic optimum ~9–6 ka (Vollweiler et al. 2006). This trend was consistent with many other lake and alpine records. Magny et al. (2003) and Magny (2004) inferred extremely unstable European Holocene precipitation patterns based on 26 lake-level reconstructions from radiocarbon, tree-ring, and archaeological records that came from regions in the Alps, the Swiss Plateau, and adjacent areas. They postulated solar forcing as a causal factor in European rainfall. Davis et al. (2003) integrated 500 pollen records into a detailed spatial network of Holocene European temperature patterns. They identified a summer thermal maximum that occurred in Northern Europe coincident with relatively cool temperatures in Southern Europe and intermediate temperatures in Central Europe. Mean annual temperatures in Europe increased gradually until about 7.8 ka, after which there remained fairly stable conditions.

European pollen records have also been integrated with climate modeling to understand patterns and causes of Holocene climate in the PMIP 1 project (Harrison et al. 1998; Prentice et al. 1998; Masson et al. 1999; Bonfils et al. 2004). The latest generation of data-model studies, called PMIP 2, integrated 400 pollen records into a data-modeling study of moisture availability, temperature during the coldest month, and growing season at 6 ka (Brewer et al. 2007). Twenty-five general-circulation models were generally successful in simulating the direction, and in some regions the magnitude, of Holocene climate. The PEP III project revealed an enormous amount of regional paleoclimate variability in Europe and adjacent regions (Battarbee et al. 2004; Verschuren et al. 2004).

High-Latitude, High-Elevation Atmospheric Variability

Greenland ice cores provided convincing evidence for Holocene atmospheric variability in the high-latitude North Atlantic region. A series of cooling events is evident in ice-accumulation rates (Meese et al. 1994) and glaciochemical records of dust and sea salt (O'Brien et al. 1995) dated at >11, 8.8–7.8, 6.1–5.0, 3.1–2.4, and 0.6–0 ka. Paleorecords other than ice cores provide an increasingly complex picture of Holocene climate in the high-latitude North Atlantic region. These include centennial-scale temperature oscillations over west Greenland (Willemse and Törnqvist 1999) and multiple cold, windy episodes in Iceland (Jackson et al. 2005), among others. Masson-Delmotte et al. (2000) demonstrated an Antarctic-wide early Holocene temperature maximum but regional variability in mid-Holocene warming events across the continent. Aperiodic millennial temperature oscillations were also evident.

High-elevation, low-latitude ice cores from the Quelccaya ice cap, Peru (Thompson et al. 1979, 1985, 1986), the Huascarán ice cap of Peru (Thompson et al. 1995b), and the Dunde (Thompson et al. 1989) and Guliya ice caps on the Qinghai-Tibetan Plateau in China (Thompson et al. 1995a, 1997) yield invaluable information on low-latitude climate change during the Holocene. Thompson et al. (2002) studied cores from the ice cap on Mount Kilimanjaro near the equator in east central Africa. The Kilimanjaro oxygen isotopic and glaciochemical records show abrupt climate shifts at 8.3, 5.2, and 4 ka (Figure 8.11). These patterns were consistent with trends deduced from African lake level and pollen records (Street-Perrott and Perrott 1990; Gasse and Van Campo 1994; Peyron et al. 2000),

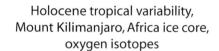

Holocene tropical variability,
Mount Kilimanjaro, Africa ice core,
oxygen isotopes

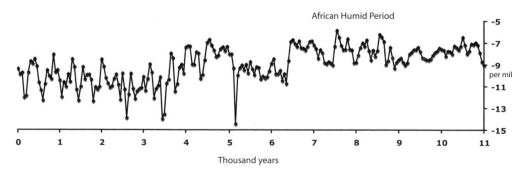

FIGURE 8.11 Oxygen isotope record from ice cores from Mount Kilimanjaro, Africa showing abrupt climate changes at 5.2 and 4–3.5 ka, marking the end of the early Holocene African Humid Period. The record is a composite from NIF2 and NIF3 ice cores, from Thompson et al. (2002).

South American tropical glaciers (Thompson et al. 1995b), Greenland ice cores, and a variety of other paleoclimatic and historical records. Barker et al. (2001) also proposed large Holocene precipitation changes from sediment records from Lake Kenya. Periods of strong rainfall at 11.1–8.6, 6.7–5.6 ka, 2.9–1.9 ka, and <1.3 ka correlated with North Atlantic SSTs.

Alpine glaciers are among the most sensitive indicators of atmospheric temperature and precipitation changes. Mayewski et al. (2004) compared alpine glacial records from North America, Asia, and Europe to 50 Holocene paleoclimate records from around the world to evaluate the timing of glacial advances in relationship to possible causal factors such as atmospheric CO_2, cosmogenic records of solar activity, and freshwater glacial meltwater. They found that nearly synchronized global glacial advances were centered on 8.5, 5.5–6, ~4, 2–3, 2, and 1.0 ka and during the LIA beginning ~600 years ago. With the exception of the LIA, their timing coincides with polar cooling, tropical aridity, and large-scale reorganization of global atmospheric circulation.

Summary of the Holocene Global Atmosphere

The main features of Holocene atmospheric climate variability are as follows:

1. The ITCZ migrated southward toward its present mean position slightly north of the equator.

2. Early Holocene monsoon activity was more intense and progressively weakened after the early Holocene thermal maximum and the end of the African humid period.

3. There was a mid-Holocene shift from minimal ENSO activity to more frequent or intense El Niño events (or both), peaking about 3–1 ka.

4. Centennial-scale variability is common at various frequencies in many records; some suggest synchronous low- and high-latitude centennial climate changes.

5. The onset and termination of some centennial events are abrupt.

Ocean Variability and Climate

Surface Ocean Temperatures and Ice Rafting

Midlatitude to High-Latitude North Atlantic Region

Millennial-scale variability evident in multi-proxy records from the subpolar North Atlantic Ocean reflects quasi-periodic behavior in the Holocene ocean-ice-atmosphere sys-

tem (Bond et al. 1997, 1999, 2001; Sarnthein et al. 2003b). Bond et al. (2001) analyses of petrologic tracers, planktic foraminifera, and stable isotopes as proxies for ice-rafting and SST resulted in what is now widely referred to as the 1500-year Holocene climate pattern. Ice-rafted tracers include percentages of Icelandic glass, detrital carbonate, and hematite-stained grains representing their source regions in the Iceland, Norwegian-Greenland, and Labrador Seas. Eight maxima in the three petrologic tracers mark major millennial-year ice-rafting events. Bond linked Holocene millennial events to continental glacial readvances (Denton and Karlén 1973) and to reconstructed atmospheric radionuclide production rates, a proxy for solar activity. Interestingly, Bond and colleagues were somewhat cautious in their interpretation of Holocene climate, calling these apparent cycles "enigmatic" and "at best quasi-periodic."

The hypothesized 1500-year cycles have generated enormous attention during the past few years. Additional evidence for significant variability in IRD and SST has been discovered in the Nordic Seas but not necessarily at a 1500-year periodicity (Jennings et al. 2002; Schulz and Paul 2002; Risebrobakken et al. 2003). Because of their importance as a hypothesized solar-global climate link, we outline several sources of uncertainties in Holocene marine records.

First, some SST proxies are biased toward summer temperatures, and a more complete picture emerges when marine records are integrated with those on continents and from ice cores and when changes in winter and mean annual temperature are also accounted for. Moros et al. (2004) demonstrated the importance of seasonality in an integrated study of Holocene steplike shifts in SST and ice rafting in the Norwegian Sea and Reykjanes Ridge areas. They identified four distinct phases of Holocene climate, including a Holocene summer thermal maximum lasting until 6.7 ka and an extremely unstable climatic period at 3.7 ka. They proposed that decreased seasonality might explain the coincident with the 3.7-ka ocean event, possible regional winter season warming, and coeval Neoglaciation in northern Europe.

A second issue is that marine records typically do not have the annual or decadal resolution of ice cores, varves, and tree rings. Marine sediments provide excellent records of millennial and sometimes centennial-scale variability, but it is difficult to compare them to more detailed terrestrial climate records.

Third, ice rafting from the Greenland Ice Sheet and other sources should not necessarily be synchronous, even when there is a common forcing present, because of internal ice-dynamic response to atmospheric temperature, sea-level changes, and local geology and oceanography. This complication was emphasized in earlier chapters for glacial periods when large ice sheets surrounded much of the high-latitude North Atlantic region, but it applies also to Holocene glacial dynamics of alpine glaciers and the Greenland Ice Sheet (see Bischof 2000).

Another complication is that the Nordic Seas sit between the subpolar North Atlantic and the Arctic Ocean. This region—characterized by strong, spatial ocean gradients in SST, sea ice, and salinity—experiences complex changes in currents, sea-ice cover, productivity, deep-water convection, and surface- and deep-water exchange with adjacent regions (Nees et al. 1997). The primary first-order Holocene paleoceanographic pattern is progressive Holocene surface-water cooling in the high-latitude North Atlantic linked to mid-latitude SST cooling of 2°C off west Africa near 31°N latitude (Kim et al. 2007). But significant spatial and temporal variability exists as well (Koç and Jansen 1994; Eiríksson et al. 2000; Marchal et al. 2002; Moros et al. 2004; Hald et al. 2004; Giraudeau et al. 2004). Andrews et al. (2003), for example, noted Holocene quasi-cyclic variability at several solar frequencies (88, 125, and 200 years) in proxy records from a core off northern Iceland.

Transects of marine cores show that oceanographic changes in the Nordic Seas were time-transgressive during rapid climate transitions. Hald et al. (2004, 2007) studied the YD–early Holocene (Preboreal) transition in the Norwegian Sea. SST increased at a rate of 0.1°C per year in the south, but more slowly in the north. Several Holocene cool events centered on 9–8, 7.5, 6.5, 5.5–3.0, and 1 ka were also documented in the northern region, probably because of the influence of cold water originating in the Arctic Ocean. Interestingly, Hald attributed temperature variability in the northern Norwegian Sea to orbital insolation changes but in the south to other factors such as deglacial meltwater influx, changes in the strength of meridional overturning circulation, atmospheric changes in the westerlies, and polar amplification due to snow from sea-ice feedbacks.

In addition to the complexity of the surface ocean, the three-dimensional structure of the high-latitude northern-hemisphere ocean circulation is poorly constrained for the Holocene. Only recently have mid-depth paleoceanographic records with good chronology become available, and these indicate complex changes in ocean temperature structure in the Iceland shelf region during the mid to late Holocene (Kristjánsdóttir et al. 2007).

Northern Pacific Holocene paleoceanography is less studied than the Atlantic version, but recently Kim et al. (2004) used alkenone paleothermometry to compare Pacific and Atlantic records. They identified an interocean antiphasing pattern in which Holocene warming in the North Pacific since 7 ka coincided with cooling in the North Atlantic. This dipole seesaw behavior was attributed to interoceanic teleconnections perhaps similar to climate behavior seen during opposite phases of the modern Pacific North Atlantic and North Atlantic Oscillation modes of internal variability.

Low-Latitude Surface Oceans   A growing number of sediment-core records using alkenone and Mg/Ca paleothemometry are available for the tropical oceans. These include records from the eastern equatorial Atlantic (deMenocal et al. 2000), the eastern (Lea et al. 2000, 2006; Koutavas et al. 2002, 2006) and western (Stott et al. 2002, 2004; Rosenthal et al. 2003; Visser et al. 2003) equatorial Pacific, and the Indian Ocean (Jung et al. 2004). Typically, small (<1°C) fluctuations occur over millennial and shorter timescales but they have important implications for Holocene climate dynamics. The SST minimum between about 5 and 8 ka in what is today the modern eastern equatorial Pacific cold tongue reflects suppressed trade winds, diminished upwelling, warmer surface waters, and reduced west-to-east temperature gradient consistent with a more La Niña–like mean condition before 5 ka. Significant surface salinity variability accompanies Holocene millennial SST changes in the tropical Pacific, Atlantic, and Indian Oceans (Stott et al. 2004; Oppo et al. 2007). In general, equatorial SST records are in line with the atmospheric evidence for southward shifts in the ITCZ during the Holocene.

The relationship between low- and high-latitude ocean and atmospheric patterns is also a widely discussed topic. DeMenocal et al. (2000) reconstructed Holocene tropical eastern North Atlantic equatorial SSTs and African rainfall from ODP Site 658C (Figure 8.12). They proposed a strong coupling between temperature over Greenland, low-latitude Atlantic SSTs, and basin-wide climatic shifts near the 1500-year frequency described by Bond. Notably, they also suggested that the transitions between cool and warm conditions occurred over only decades or centuries and signified large-scale ocean-atmosphere reorganizations. Subsequent studies have confirmed the first-order timing between low- and high-latitude Holocene variability in both marine and atmospheric records (Weldeab et al. 2005).

There is still much uncertainty regarding the relationship between Holocene oceanic and atmospheric variability because few ocean sediment records have decadal to centennial resolution, regional ocean variability is especially large in critical regions, and uncertainty surrounds some SST proxies (Mix 2006). Greater spatial and temporal coverage is needed to evaluate the phasing between atmospheric and ocean patterns during key Holocene climatic transitions.

Deep-Ocean Circulation

Research on Holocene deep-ocean circulation variability is still in its infancy compared to studies of deglacial ocean circulation (Marchitto et al. 1998; McManus et al. 2004). Keigwin and Boyle (2000) reviewed the sparse evidence then available for altered Holocene deep-ocean circulation from

Holocene SSTs off West Africa ODP 658C

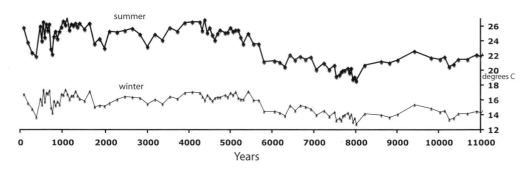

FIGURE 8.12 Summer and winter sea-surface temperatures (SSTs) off West Africa from Ocean Drilling Program (ODP) Site 658C, from deMenocal et al. (2000). SST increase in the mid-Holocene about 5.5. ka was coincident with a large change in seasonality, greater terrigenous windblown influx, African aridity, and the end of the African Humid Period.

Bermuda Rise isotopes (Keigwin and Jones 1989), sedimentary evidence from sortable silt records (Bianchi and McCave 1999), and geochemical evidence for deep-water circulation changes during the LIA (Keigwin 1996). More recent evidence has come from deep-water temperatures (Dwyer et al. 2000; Marchitto and deMenocal 2003), isotopic (Oppo et al. 2003b; Keigwin 2004), and faunal proxies (Yasuhara et al. 2008). The bottom-water temperature decrease during the Holocene was large, ~4°C in the mid-depth Atlantic and about 1°C near 4500 m water depth (Dwyer et al. 2000). Submillennial-scale records of deep-sea circulation and temperature are scarce.

Ocean Chemistry and Carbon Cycling

The ocean's carbonate chemistry is closely linked to increasing Holocene CO_2 concentrations already described herein. Carbonate ion (CO_3^{2-}) concentration is inversely proportional to atmospheric CO_2 concentration, such that a rise in CO_2 concentration of 20 ppmv (20 μmol/kg) would be equivalent to a decrease in carbonate ion concentration from 6–8.3 μmol/kg (Berelson et al. 1997; Broecker et al. 1999a). Various methods have been used to reconstruct Holocene carbonate ion history (Broecker et al. 2001). Broecker and Clark (2003) used foraminiferal shell weight, a shell fragmentation index, and $CaCO_3$ concentrations from tropical Pacific Ontong-Java cores as measures of Holocene ocean carbonate chemistry. They estimated that the decline of 10–15 μmol/kg in the ocean's Holocene carbonate ion concentration is larger than the expected 6 μmol/km decrease due to a 20 ppmv rise in atmospheric CO_2.

How is this discrepancy explained? Broecker holds that the larger-than-expected Holocene carbonate ion decrease

may be related to stronger thermohaline circulation or enhanced ocean biological pump at about 8 ka, and that either or both processes subsequently weakened over the remainder of the Holocene. Uncertainty in reconciling the atmospheric and ocean records stems mainly from disagreement in the carbonate ion proxy estimates, the influence of pore-water carbonate chemical processes, and poor constraints on the strength of Southern Ocean Holocene deep-water formation and biological uptake.

Holocene Sea Level and Ice Volume

Background

The lack of a global sea-level-ice-volume curve for the Holocene ranks among the most serious deficiencies in paleoclimatology's quest to establish a firm climate-ice-volume link for the interglacial climate state. Its significance stems from the fact that, although Holocene sea-level oscillations were small in amplitude compared to those during glacial-interglacial cycles, even a small sea-level change of a few meters during an interglacial represents a major environmental concern from the standpoint of future coastal habitats. Sea-level records of prior interglacial periods also present challenges to paleoclimatology. It is important that Rohling et al. (2008) showed that rates of sea-level rise during the last interglacial period (MIS 5) ranged from about 6–25 mm yr[-1], about two to eight times higher than rates during the last century.

Uncertainty about Holocene sea level and ice-volume history does not reflect a lack of effort. Radiocarbon- and U/Th-dated Holocene peat, coral, and other shoreline proxies from around the world number in the thousands (Chappell 1983; Pirazzoli 1996). Regional sea-level records form the foundation for the ICE-5G sea-level/glacio-isostatic adjustment (GIA) model of Peltier (2002, 2004) discussed in Chapter 7. The ICE-5G model is designed to study sea level, ice volume, and glacio-isostatic adjustment at discrete 1000-year-long increments and jumps in sea level exceeding ~5–7 m (the depth range of U-series-dated corals). Thus, it is not suitable for analysis of high-frequency, low-amplitude Holocene sea-level variability. In this section, we discuss the major trends, disputes, and sources of uncertainty in regional and global sea-level reconstruction for the last 8000 years.

All local or regional reconstructions are called relative sea-level (RSL) curves because they represent the combined effects of glacio-eustatic (ice-volume) changes and vertical changes in the land. Vertical movement along coasts is caused by several processes. The global redistribution of load from melting ice sheets following the LGM—glacio-isostatic and hydro-isostatic adjustment—leads to a regionally variable response of the earth's lithosphere. Regions are grouped into near-field, intermediate-field, and far-field categories on the basis of their proximity to the northern-hemisphere ice sheets. Near-field sites under or near the margin of LGM ice sheets experienced glacio-isostatic uplift following ice retreat. As a consequence, near-field coastal regions underwent Holocene emergence. Intermediate-field sites located in unglaciated regions adjacent to LGM ice sheets bulged up during maximum ice advance and thickness but subsided following ice retreat. Far-field sites across much of the Pacific Ocean and parts of the southern hemisphere in theory experienced mid-Holocene coastal emergence, the amplitude of which was a function of the size of an island, its proximity to a continent, and local lithospheric properties.

In addition to glacio-isostatic adjustment, co-seismic displacement, sedimentation (progradation), sediment compaction, thermosteric ocean expansion, and regional upwelling can also influence an RSL curve. Uncertainty in the depth range for various shoreline proxy indicators introduces another source of error into regional sea-level reconstruction. It is also noteworthy that the deep-sea oxygen isotope curve is not useful in reconstructing Holocene sea level because of the small amplitude of isotopic variability due to ice-volume. An isotopic shift of 0.1‰, for example, equals on average about a 10-m change in ice volume, much larger than the known Holocene sea-level amplitude. The net result is that it is difficult to integrate RSL curves from various areas into an integrated glacio-eustatic Holocene sea-level curve. With these factors in mind, we review Holocene sea-level and ice-volume records from coastal regions, Antarctica, and the northern hemisphere.

Coastal Records

RSL curves constructed from coastal records produce conflicting views of Holocene ice-volume history. On the one hand, some records show progressive submergence of coastlines during the last 6–7 ka, suggesting small but significant ice-sheet melting long after the main phase of deglaciation. Such a typical "smooth" RSL curve was developed in the Gulf of Mexico (Shepard 1963, 1964) and south Florida (Scholl and Stuiver 1967; Scholl et al. 1969).

Other records suggest that sea level reached or even exceeded modern levels during the middle Holocene, after which it fell, either progressively in a smooth, regular fashion or in an irregular, oscillating fashion (Figure 8.1). Short-term Holocene sea-level oscillations of only 1–2 m are a signature feature of the curve constructed by Fairbridge (1961). Lambeck (1993) postulated that emergent mid-Holocene shorelines on many islands in the Pacific Ocean represented a hydro-isostatic response to the 125-m sea-level equivalent of water added to the world's ocean basins from ice-sheet melting. Another possibility is that there was 1–3 m less global ice volume during the mid-Holocene, and thus the absolute elevation of global sea level at this time remains a major point of contention (Peltier 2002).

Disputes about the shape of Holocene sea-level curves even permeate interpretations of regional stratigraphic and geomorphological records. In the well-studied Gulf of Mexico area, two sea-level models exist: (1) progressive regional submergence, seen in records from Florida, the Bahamas, and parts of the Caribbean (Scholl et al. 1969; Toscano and McIntyre 2003; Gonzalez and Törnqvist 2006; Cronin et al. 2007a), and (2) high early- to mid-Holocene sea level and subsequent emergence evident along the Texas coasts (Morton et al. 2000; Blum et al. 2001; see also Stapor and Stone 2005). Blum et al. (2001) suggested that sea level rose from −9 m near 7.8 ka to +2 m by 6.8 ka, a remarkable amplitude, but one that is inconsistent with mangrove and coral Holocene records from around the Caribbean and southern Gulf of Mexico (Toscano and McIntyre 2003).

Integrated coastal geomorphological sea-level reconstructions combined with regional glacio-isostatic modeling have improved our understanding of Holocene sea level for some regions. In the British Isles and surrounding areas, Ian Shennan and colleagues (Shennan and Hamilton 2006; Shennan et al. 2002, 2005; Shennan and Horton 2002) decoupled the RSL curve from global ice-volume changes due to Antarctic and European ice-sheet melting since 16 ka. They assembled more than 1200 radiocarbon dates from the Holocene alone. Their hypotheses about the source and rate of ice melting, as well as postglacial uplift rates, can be tested in other coastal areas having different isostatic responses. However, few regions have received such intensive scrutiny.

Baker and Haworth (2000a, b; R. Baker et al. 2005) considered smooth versus oscillating hypotheses for Holocene sea level, focusing on southern-hemisphere Holocene records from nonglaciated, tectonically passive coasts. They concluded that an oscillating model was a possibility but that additional records from "fixed" shoreline proxies from nonglaciated regions were necessary to fit a quantitative curve to the sea-level data.

Dickinson (2001) conducted a comprehensive review of Pacific Island records, concluding that "the dominant pattern of relative sea-level change, where not overprinted by local tectonism or lithospheric flexure, was a uniform early Holocene rise in eustatic sea level followed by a regionally variable late Holocene hydro-isostatic drawdown in sea level." Montaggiani's (2005) review of Pacific coral-reef records also shows regionally variable Holocene RSL curves with emergent reefs of different ages throughout the region (see Camoin et al. 1997, 2007). The basic pattern shows that the mid-Holocene high stand differs in age and elevation throughout the Pacific region. The temporal and spatial variability is attributed to a hydro-isostatic response of the Pacific region to glacial melting and isostatic response (Mitrovica and Peltier 1991; Lambeck 1993; Peltier 2002).

In sum, several factors complicate efforts to reconstruct a single glacio-eustatic sea-level curve from paleoshorelines, precluding a comparison between Holocene ice-volume and other paleoclimate records at submillennial timescales.

Antarctic Ice Volume

Glaciological studies around Antarctica can provide direct evidence for Holocene ice-volume changes. The ICE-5G GIA model holds that most glacial melting and sea-level rise during the early Holocene ~9–7 ka came from an Antarctic ice source (Peltier 2004). As for the LGM and the onset of deglacial ice-sheet retreat, however, the Holocene history of the Antarctic Ice Sheet is subject to great uncertainty. Ingólfsson and Hjordt (1999) and Ingólfsson (2004) reviewed the terrestrial record of Antarctica's glacial and climatic history, showing a regionally complex pattern of glacial advance and retreat in both East and West Antarctica. Debate about the timing of glacial activity even surrounds regional records. Taking the example of the Vestfold Hills region of East Antarctica, Ingólfsson points out that one view holds slow progressive deglaciation after 10–8 ka, while another holds that deglaciation was already advanced by 8.6 ka. There is also evidence that Holocene climate trends around Antarctica lagged those in the northern hemisphere, such as the relatively late mid-Holocene warming at 4.5–2.5 ka in several parts of Antarctica.

J. Anderson et al. (2002) found similar regional complexity in continental shelf regions of Antarctica, acknowledging

data gaps in many regions. Regional atmospheric temperature variability around Antarctica is also evident from ice-core proxies (Masson-Delmotte et al. 2000); however, ice-core proxies record atmospheric trends at high elevations on the interior of the ice sheet, and these are not necessarily similar to those at its margins near sea level or over the adjacent ocean.

Some evidence nonetheless indicates that Antarctic ice contributed to postglacial sea-level rise between 14 ka and 6 ka (Ingólfsson 2004). There is also substantial evidence for dynamic changes in Holocene ice-margin positions that were very likely sources of submillennial glacio-eustatic sea-level variability on the order of a few meters. Nonetheless, better Holocene Antarctic glaciological records are needed to quantify the contribution of the Antarctic ice sheet to Holocene sea-level variability.

Northern-Hemisphere Ice Volume

Most glacial-age ice in North America and European ice sheets had melted by about 8.5–8.0 ka, when glacial Lake Agassiz-Ojibway drained into Hudson Strait (see Chapter 9 in this volume). Licciardi et al. (1999) and Dyke et al. (2002) showed that the Laurentide Ice Sheet had only a few meters of ice volume equivalent remaining by about 8 ka. Carlson et al. (2007b, 2008) postulated that sea level abruptly rose several meters at about 9 and 7.6 ka because of the final stages of Laurentide Ice Sheet melting in the Labrador and Hudson Bay regions. After this event, the Holocene ice-volume record of the Greenland Ice Sheet and remaining ice caps in the Canadian Arctic is poorly constrained and not yet integrated into a Holocene sea-level curve.

Causes of Holocene Variability

We divide hypotheses about the causes of Holocene climate variability into three broad categories of external forcing mechanisms—orbital insolation, solar irradiance variability, and volcanic activity—and several feedback mechanisms related to terrestrial vegetation, atmospheric greenhouse gases, and high-latitude ocean-ice processes.

Orbital Insolation

ITCZ Migration The southward migration of the ITCZ during the Holocene has been linked to low-latitude precession insolation in several studies (Haug et al. 2001b; Fleitmann et al. 2003; Tedesco and Thunell 2003; Koutavas and Lynch-Stieglitz 2004; Y. Wang et al. 2005). Stott et al. (2004) interpreted higher early Holocene SST and salinity in the western equatorial Pacific as consistent with external forcing by insolation of the basin-scale hydrological cycle. Kouta-vas and Lynch-Stieglitz (2004) suggest that using August-September rather than July insolation may account for the spatial and temporal variability in tropical Holocene rainfall records of the ITCZ migration. September insolation peaked between 9 and 6 ka, and this pattern matches some records showing maximum rainfall at this time (Baker et al. 2001b; Haug et al. 2001b). Partin et al. (2007) reconstructed tropical Pacific Holocene climate from speleothems from Borneo that supported the hypothesis of low-latitude summer and fall precession leading to the southward migration of the ITCZ. However, they point out that the precise response mechanism in the tropical Pacific region remains unclear and that zonal changes better explain an anomalous precipitation maximum at 5 ka in Borneo.

Monsoon Activity   Changes in orbital insolation are also considered responsible for the long-term evolution of Holocene monsoon activity, which itself is tightly coupled with the ITCZ migration. Overpeck et al. (1996) postulated that the peak in monsoon intensity near 9–7 ka seems to lag maximum northern-hemisphere solar insolation at 11 ka, possibly because of the lingering effects of the remaining northern-hemisphere ice sheets. They noted that monsoon intensity and insolation are more closely linked since about 6 ka. Some high-resolution records also show a 1–2-ka lag of hydrological changes after the early Holocene insolation peak (Fleitmann et al. 2003, 2007; Haug et al. 2001;)

COHMAP (COHMAP Members 1988) incorporated proxy-based reconstructions of global climate at 12, 9, 6, and 3 ka and modeling studies to understand causal mechanisms (Kutzbach and Guetter 1986). Among its many contributions, COHMAP discovered that Asian and African monsoon systems respond to precession forcing. The greater amplitude of the northern-hemisphere seasonal cycle during the early to mid-Holocene (~5–9 ka) is linked to enhanced continental warming, stronger Asian monsoon intensity, and higher rainfall than during the late Holocene (Kutzbach and Street-Perrott 1985).

Additional support for the insolation-monsoon link comes from more recent studies. An et al. (2000) combined paleomonsoon reconstructions based on pollen, lake levels, and soils with climate modeling and showed that the Holocene migration in peak monsoonal strength across Asia, with maximum intensity at 10–8 ka in the northeast and 3 ka in the south, was likely a response to seasonal changes in monsoon strength, caused by progressive weakening of the summer insolation anomaly. Aerts et al. (2006) analyzed global patterns of Holocene river discharge and found that steadily decreasing discharge from large rivers located in monsoon regions was driven by northern-hemisphere summer insolation. In contrast, southern-hemisphere dis-

charge, typified by the Amazon River, increased during the Holocene because of rainy winter-season insolation forcing.

The precise mechanisms translating insolation changes into global climate anomalies are still unclear. PMIP examined mid-Holocene climate at 6 ka and during the LGM, resulting in a number of model simulations (Hall and Valdes 1997; Hewitt and Mitchell 1998; Braconnot et al. 2007), paleo-data compilations (Harrison et al. 1996; Cheddadi et al. 1997), and integrated data-modeling studies (Harrison et al. 1998; Guiot et al. 1999; Joussaume et al. 1999; Masson et al. 1999; Kohfeld and Harrison 2000). One successful application of PMIP was the use of model-simulated output in conjunction with the BIOME vegetation-growth-physiology model (Prentice and Webb 1998; Prentice et al. (2000). This synergy led to the conclusion that climate model simulations produce smaller-amplitude signals than those suggested by paleoclimate records, at least for some regions (see Harrison et al. 1998 and Forman et al. 2001).

Shin et al. (2006) simulated mid-Holocene climate and found a better match than PMIP for explaining the paleoclimate evidence for dry central North American climate and more humid African tropics at 6 ka. They suggested that, in addition to orbital insolation and terrestrial vegetation feedbacks, La Niña–like conditions in the Pacific Ocean, especially cool eastern equatorial SSTs, might explain the amplitude of the continental paleoclimate patterns. Oppo et al. (2007) also analyzed ocean-atmosphere dynamics during the Holocene through reconstructed low-latitude Pacific Ocean salinity using isotopic and Mg/Ca data and climate modeling. She found that mid-Holocene tropical Pacific Ocean salinity was higher because of water vapor transport changes (less water vapor transported from the Atlantic to the Pacific and more from the Pacific to the Indian Ocean), changes in surface ocean advection, and a stronger monsoon, leading to enhanced precipitation over the Asian continent and less over oceanic regions.

ENSO Variability   In addition to its influence on the ITCZ and monsoons, orbitally driven insolation may be responsible for changes in the strength of ENSO during the Holocene. Clement et al. (2000) applied an anomalous heat flux due to precession-driven in-seasonal insolation to the Zebiak-Cane ENSO model. Their model produced patterns similar for the paleo-ENSO patterns already described herein for suppressed mid-Holocene ENSO activity (fewer and less intense El Niño events). They argue that this is the result of the earth passing its perihelion (closest annual point to the sun) in the summer during the mid-Holocene, in contrast with the earth passing the modern perihelion near the beginning of the calendar year. In effect, there is diminished forcing by northern-hemisphere winter insolation during the mid-Holocene, affecting ENSO ocean-atmosphere dynamics.

Renssen et al. (2005a) also suggested that orbital forcing drives Holocene climate in the southern hemisphere. Their model simulations showed a summer thermal optimum in the mid-Holocene between 6 and 3 ka when atmospheric temperatures exceeded mean preindustrial values by as much as 3°C. Autumn temperatures increased gradually during the early Holocene. In contrast, winter ocean and spring continental temperatures peaked 3–3.5°C above preindustrial temperatures at 9 ka, after which progressive cooling occurred. Southern-hemisphere temperature trends were interpreted as showing a slightly lagged (months) response to orbital forcing and oceanic processes.

High Latitude Response   Insolation may have played a direct role in North Atlantic surface ocean cooling and deep-water convection during the last 11,000 years. Renssen et al. (2005b), for example, found that insolation influenced surface and deep-ocean processes in the Nordic and Labrador Seas. Sachs (2007) documented large SST decreases of 4–10°C in the western North Atlantic off eastern North America, noting that these coincided with a decline in summer insolation between 20 and 65°N latitude that was equivalent to 36 and 48 W m^{-2}. Mean annual insolation in these latitudes fell by 7–9%. A progressively stronger SST gradient could have led to the intensified westerly wind, cooler slope waters, and stronger Labrador Sea deep-water formation in evidence from paleoceanographic records (de Vernal and Hillaire-Marcel 2001 2006; Hillaire-Marcel and Bilodeau 2000).

Western-Hemisphere Continental Climate   There is a large literature on early- to mid-Holocene climate of the Americas, including several generations of model simulations aimed at understanding mechanisms (Bartlein et al. 1998; Webb and Kutzbach 1998). Harrison et al.'s (2003) comprehensive modeling study of the mid-Holocene at 6 ka in North America incorporated extensive moisture and precipitation records (lake levels, dunes, pollen, etc.) from 237 total sites—mainly from North, Central, and South America—into the BIOME vegetation model (Prentice and Webb 1998; Prentice et al. 2000). Two coupled ocean-atmosphere model simulations produced a consistent picture of atmospheric and terrestrial vegetation dynamic response to orbital forcing characterized by greater summer monsoonal precipitation over the American southwest, Central America, and northern parts of South America and an arid midcontinental North America. Mid-Holocene land-surface warming dominates the seasonal signal, although oceanic feedbacks may also have been involved.

Diffenbaugh et al. (2006) ran global and nested model simulations to evaluate North American mid-Holocene precipitation patterns. The simulated climate generally matched North American paleoprecipitation records from western lakes and pack-rat middens (Bartlein et al. 1998), Holocene sand dune reconstructions (Forman et al. 2001), mid-Atlantic rainfall reconstructions from pollen (Willard et al. 2005) and estuarine salinity (Cronin et al. 2005; Saenger et al. 2006), and pollen proxies from the northeastern United States (Shuman et al. 2002). Diffenbaugh concluded that summer insolation forcing led to drier-than-modern Mid-Holocene summer aridity in central North America and to climate wetter than today along the Atlantic coast. The mid-continent climate reflected upper-level atmospheric moisture transport in enhanced mid-Holocene anticyclonic and cyclonic circulation, respectively, and reduced low-level moisture over the Gulf of Mexico and southcentral United States. Despite these advances, feedbacks between forest, land surface, and climate are complex and can involve both amplification and dampening of the initial climate forcing (Bonan 2008).

High- and Low-Latitude Coupling   A number of studies show tightly coupled low- and high-latitude atmospheric climate variability during the Holocene. For example, Hong et al. (2003) reconstructed Indian Monsoon variability from the high-elevation (3466 m) Hongyuan peat carbon isotopic record from the Tibetan Plateau and linked it to North Atlantic IRD layers for the last 6 ka. The Dongge Cave, China speleothem record showed a coeval shift in the Asian Monsoon and the high-latitude North Atlantic at centennial and millennial timescales (Dykoski et al. 2005). There are also similarities in centennial-scale Holocene climate records at Dongge Cave and in South America. Porter and Weijian (2006) found a correspondence between the eolian-paleosol records of China and the North Atlantic IRD record, as did Russell and Johnson (2005b) for Holocene sediments from Lake Edward, Africa. Selvaraj et al. (2007) correlated periods of weakened East Asian monsoon activity to reduced North Atlantic Deep Water (NADW) and low atmospheric methane at 8.2, 5.4, and 4.5–2.1 ka. They attributed these patterns to both orbitally driven insolation and solar irradiance and thus propose one of the more all-encompassing hypotheses linking the sun-ocean-monsoon-NADW global system.

Solar Irradiance

Hypothesized solar-induced cyclic climatic variability during the Holocene constitutes an exhaustive literature. Table 8.4 lists a few representative studies attributing proxy records to solar irradiance over a range of timescales. Most

TABLE 8.4 Selected Holocene Records Possibly Related to Solar Cycles*

Record	Proxy	Parameter	Quasi-cycles (years)[†]	Reference
Irish bogs	Tree ring-14C	Precipitation	800	Turney et al. 2005
North Atlantic IRD-SST	Multiple	Ice-rafting, SST	~1470	Bond et al. 2001
Chesapeake Bay pollen	Palynology	Precipitation and temperature	~1500	Willard et al. 2005
Midwestern N. American lake	Ostracode shell chemistry	Lake chemistry, salinity	100–400	Yu and Eto 1999
Scottish bog	Peat	Precipitation	1100	Langdon et al. 2003
African lake	Multiple		725	Russell et al. 2003a, b
Gulf of California	Diatoms, other	Productivity, wind, temperature	MWP through present	Barron et al. 2003; Barron and Bukry 2007
Dongge Cave, China	Oxygen isotopes	Asian monsoon strength	512, 206, 148	Y. Wang et al. 2005
Dongge Cave, China	Oxygen isotopes	Asian monsoon strength	208, 86, 11.6	Dykoski et al. 2005
Hoti, Qunf Oman Caves	Oxygen isotopes	Indian Ocean Monsoon	~207, 88, 11.6	Neff et al. 2001; Fleitmann et al. 2003
Tibet Plateau Qinghai Lake	Sediment color	Asian, Indian Monsoon	~200, others	Ji et al. 2005
Lake Titicaca	Carbon isotopes	S. American rainfall	100–600	Baker et al. 2005
Lake Chichancanab, Yucatan Peninsula	Sediment geochemistry	Central American rainfall	206	Hodell et al. 2001
Gulf of Mexico	Planktonic foraminifers	ITCZ, surface oceanography	180, 512, other centennial	Poore et al. 2003, 2005
Arolik Lake, Alaska	Biogenical silica	Limnology	6 peaks from 135–950-yr	Hu et al. 2003

Note: IRD = ice-rafted debris; ITCZ = Intertropical Convergence Zone; MWP = Medieval Warm Period; SST = sea-surface temperature.

*This list is a small sample of paleoreconstructions attributed to solar forcing.

[†]See the original papers for the authors' opinions and statistical analyses.

propose a solar-climate link on the basis of a match in spectral peaks between proxy time series and solar activity such as Bond, de Vries, Gleissberg, and Schwabe cycles with periodicities of ~1500, 200, 88, and 11 years, respectively.

Two influential studies linking climate and solar activity were those of Stuiver et al. (1995, 1997), who linked Greenland ice-core climate records to solar activity, and Bond et al. (2001), who integrated cosmogenic isotope, North Atlantic SST and IRD, and Greenland ice-core records. Stuiver et al. stated (p. 259): "The timing, estimated order of temperature change, and phase lag of several maxima in ^{14}C and minima in δ^{18}O are suggestive of a solar component to the forcing of Greenland climate over the past millennium. The fractional climate response of the cold interval associated with the Maunder sunspot minimum (and ^{14}C maximum), as well as the Medieval Warm Period and Little Ice Age temperature trend of the past millennium, are compatible with

solar climate forcing, within an order of magnitude of solar constant change of −0.3%" (1997, p. 259).

Bond hypothesized that Holocene 1500-year climate cycles evident in ocean, terrestrial, and solar proxy records were similar to high-amplitude glacial age DO events, albeit at dampened amplitudes, and might signify a solar-climate link at millennial timescales during the interglacial.

In addition to the influence of these studies, the growing popularity of the solar-climate hypothesis might be due to several other factors. First, evidence for changes in solar variability from cosmogenic variability in radiocarbon production in the atmosphere is now fairly convincing (Stuiver and Braziunas 1993; Muscheler et al. 2000—see the subsection "Solar Irradiance" above) despite the many uncertainties (Snowball and Muscheler 2007). This point is especially relevant because many consider, or at least imply, that Holocene changes in carbon cycling and ocean circulation are less important (although not insignificant) than solar activity, at least compared to full glacial cycles.

Second, climate modeling suggests that even small solar irradiance changes can at least indirectly induce multidecadal to millennial-scale climate variability such as that associated with the North Atlantic Oscillation and centennial changes during the LIA through internal feedbacks. A change in mean irradiance as small as 0.2–0.35%, like that during the LIA-Maunder Minimum (Maunder 1922; Lean 2000), or 0.1% during a modern 11-year solar cycle was enough to produce surface temperature change through effects on stratospheric ozone and other atmospheric feedbacks (Shindell et al. 1999, 2001; see also Chapter 11 in this volume). Consequently, the magnitude of solar irradiance changes inferred from Holocene beryllium and radiocarbon proxies are considered by some sufficiently large to have potentially had a large influence on climate.

Third, many paleoclimate records yield spectral peaks in proxy data at frequencies that match observational solar cycles. A qualitative assessment of the literature suggests that the ~200-year de Vries cycle and the 1500-year cycle are two of the most commonly encountered patterns. However, one also finds what seems to be an extreme number of other putative solar-related cycles based solely on single-site paleorecords. The study of biogenic silica, organic carbon, and pollen records from a sediment core from Arolik Lake in Alaska by Hu et al. (2003) is a case in point. They identified spectral peaks at 135, 170, 195, 435, 590, and 950 years that were significant above the 90% confidence interval and a 1500-year peak significant at the 80% level. Taking into account age model uncertainty, they noted that most of these correspond to peaks in solar activity inferred from cosmosgenic records, and the authors supported Bond's hypothesis of the solar forcing of Holocene climate. In an even more extreme case of what Hoyt and Schatten (1997) called "Cyclomania," Hong et al. (2001) found no fewer than 19 purported decadal to centennial periodicities in Chinese carbon isotopic records from peat, some linked to solar activity.

As striking as the number of spectral peaks might be, the variety of proxy methods and archives that have yielded quasi-cyclic decadal to millennial-scale patterns of variability is equally remarkable. These run the gamut from annually resolved Irish bog tree rings showing an 800-year periodicity (Turney et al. 2005), North Atlantic planktonic foraminifera (Chapman and Shackleton 2000), lake levels in the European Alps (Magny 2004), bog and peat deposits in the British Isles (Langdon et al. 2003; Charman et al. 2006), and various records from arid regions of China (Feng et al. 2006).

Speleothem records provide intriguing lines of evidence for solar cycles because of their well-constrained proxies (usually oxygen isotopes) and high temporal resolution that allow correlation to high-latitude annually resolved tree-ring and ice-core records. Speleothem records from Oman (Neff et al. 2001; Fleitmann et al. 2003), and China (Dykoski et al. 2005; Y. Wang et al. 2005) show a number of spectral peaks at solar periodicities.

The 1500-year Bond cycles have received the most attention, with various authors supporting or refuting Bond's solar hypothesis. Clemens (2005) presented a modified view of millennial-scale cycles by comparing the Hulu Cave, Greenland Ice Core (GRIP), and GISP2 $\delta^{18}O$ ice-core records. Instead of a dominant 1500-year periodicity, he identified two strong peaks centered on 1667 and 1190 years, as well as a weaker ~1500-year cycle like that proposed by Bond. These three millennial-scale periodicities corresponded to sums or tones of centennial solar cycles of atmospheric ^{14}C activity and a solar-climate link was supported.

Renssen et al. (2006) modeled solar irradiance from cosmogenic isotopic proxy records and climate variability and compared simulation output to paleoclimate evidence. They supported the link between solar irradiance such that negative irradiance anomalies coincided with deep-water formation in the Nordic Seas, where sea-ice feedbacks may have amplified the signal. They found sustained periods of reduced irradiance at 5.2, 2.7, and 0.5 ka, each corresponding with large IRD and cooling events.

In contrast, Turney et al. (2005) evaluated Bond's North Atlantic climate-solar link using Irish tree-ring records of atmospheric moisture from bogs. Because both moisture and radiocarbon activity are reconstructed from the same tree-ring records, the Irish bogs preserve the proxy of both solar activity and atmospheric conditions on a common absolute timescale for the past 9000 years. Holocene changes in moisture variability were found to be coincident with ocean SST and IRD, but they did not coincide with solar cycles. Viau et al. (2006) also found that millennial variability in North American temperature occurred with a 900–1100-year cycle and was consistent with terrestrial, glacial, and marine records. They hypothesize that this coherence might signify a common forcing during the Holocene. Russell and Johnson (2005b) also discovered a 725-year periodicity in the records of Lake Edwards, Africa, which they could not attribute to any known solar periodicity but suggested instead a possible link to changes in ENSO dynamics.

Centennial- and multidecadal scale variability during the Holocene is particularly problematic in terms of solar forcing because of still unknown patterns of internal climate variability (see Chapter 10 in this volume). Renssen et al. (2007) addressed this issue using a coupled climate model forced by solar irradiance, internal unforced processes, and freshwater forcing of ocean circulation to study abrupt cooling events. They found that short-term Holocene cooling reflected a mode of internal variability characterized by southward location of deep-water convective cells and weakened meridional overturning circulation. Potential solar forcing from abrupt solar flux events and freshwater forcing, especially for the 8.2-ka event, influenced the behavior of this internal mode of variability. Although preliminary, these results signify a new horizon in climate research in that they reflect an attempt to evaluate unforced internal variability intermediate in frequency between glacial-age DO events and annual to multidecadal modes well known from instrumental records.

The solar-climate link has attracted researchers not only from paleoclimatology but also solar physics, meteorology, archaeology, and climate modeling, to mention a few disparate fields. We cannot review all purported and disputed solar-climate cycles found throughout the literature, but it is fair to say this topic will attract continued attention during the next few years. Rind (2002) captured the flavor surrounding Holocene solar-climate debate when he asked, "Is the sun the controller of climate changes, only the instigator of changes that are mostly forced by system feedbacks, or simply a convenient scapegoat for climate variations lacking any other obvious cause?"

There is as yet no satisfactory answer to this question as it pertains to millennial and submillennial Holocene climate variability. Tinsley (2003) likened the status of the solar-climate hypothesis to "that of 'continental drift' half a century ago; evidence from sea floor spreading could not be ignored, and the field was data-rich and theory-poor." Foukal et al. (2006) and Snowball and Muscheler (2007) caution that uncertainty due to long-term total solar irradiance and geomagnetic intensity lead to large uncertainty in simply computing Holocene solar intensity.

Some prerequisites for improving theory and testing Holocene solar-climate hypotheses include more annually resolved paleorecords, spatially robust paleomagnetic intensity curves, long-term proxy records of solar activity, improved understanding of Holocene deep-sea circulation and carbon cycling, the role of stratospheric processes, low-latitude ocean-atmosphere interaction, and high-latitude sea-ice feedbacks in amplifying small irradiance changes. In addition, multiple causal mechanisms—including reconstructed (i.e., not simulated) unforced decadal to centennial internal variability and volcanic activity—must be integrated with solar proxy records into model simulations. Until these topics are more fully investigated, it is premature to attribute causality to paleoclimate spectral peaks, particularly those limited to a single lake or other archive, that appear to be similar to solar cycles themselves deduced from still poorly understood long-term processes.

Volcanic Activity

Volcanic eruptions have large impacts on climate, but usually individual stratospheric eruptions lower the mean hemi-spheric or global temperature only 0.2–0.3°C and last only one to three years, depending on several factors (Rampino et al. 1985, 1988; Robock 2000; Zielinski 2000). The volcanic-climate link has been most intensely studied for the last 2000 years because of the critical need to distinguish volcanic from the solar and greenhouse gas forcing of recent climate changes (see Chapter 11 in this volume). Excluding what are called "mega-eruptions" (Rampino and Self 1992), extended periods of more frequent stratospheric eruptions might in theory alter climate over longer timescales. Holocene records of volcanic activity from sulfate aerosols formed from H_2SO_4 are used as a proxy for volcanic activity in Greenland (Clausen and Hammer 1988; Zielinski et al. 1994, 1996, 1997; Clausen et al. 1997) and Antarctic (Cole-Dai et al. 1997) ice cores. However, no definitive link between the observed climate variability discussed in this chapter and volcanic activity is yet available. Complicating the situation is the hypothesis that rapid climatic changes actually force increased volcanic activity (reviewed in Zielinski 2000). For example, increased activity during deglaciation between 13 and 7 ka corresponds to global-scale shifts in loading as northern-hemispheric ice sheets melted, with possible effects on volcanic activity in active tectonic regions. The lack of a detailed and quantitative volcanic index for the entire Holocene hampers the analysis of volcanic forcing and its relationship to solar and insolation processes and reflects a still unknown possible mechanism to explain at least some events of the last 11.5 ka.

Feedback Mechanisms

Vegetation and Land Cover   One coherent theme in Holocene paleoclimatology is that orbitally driven changes in solar insolation are modulated by vegetation (land cover) and albedo feedbacks. Climate-vegetation feedbacks can operate on fine or continental spatial scales largely through changes in convective activity and soil moisture. Continental vegetation feedbacks triggered by orbital insolation, often modulated by tropical SSTs or high-latitude atmospheric or other processes, are considered possible amplification mechanisms for Holocene climate changes in North America (Bartlein et al. 1998; Kohfeld and Harrison 2000; Harrison et al. 2003; Diffenbaugh et al. 2006), Africa (Kutzbach et al. 1996; Irizarry-Ortiz et al. 2003; Weldeab et al. 2005; Russell et al. 2003a; Renssen et al. 2006; Notaro et al. 2007), the Mediterranean (Arz et al. 2003), and globally (Prentice et al. 2000 and the PMIP references therein; Liu et al. 2006).

The vegetation-climate feedback link is complicated, however, as more paleorecords and more sophisticated climate model simulations have become available. For example, Liu et al. (2007) used transient simulations from the atmosphere-ocean-vegetation "Fast Ocean Atmosphere Model" (FOAM) climate model to show a mismatch between a rapid vegetation shift in Africa at 5 ka and more gradual changes in regional precipitation. The implication of this study is that the collapse of African vegetation was not a strong, direct positive feedback on insolation-driven rainfall but instead a nonlinear response to some sort of a precipitation threshold. SST feedbacks from the nearby ocean have also been shown to enhance mid-Holocene monsoon variability during transient climate changes (Kutzbach and Liu 1997).

In addition to climate-driven changes in land surface and vegetation, even anthropogenic land clearing activities have been shown to have impacts on regional climate. Two notable examples are land-use changes during the MWP in Europe (Goosse et al. 2006) and in the 20th century in peninsular Florida (Marshall et al. 2004).

Sea Ice   In prior chapters, sea-ice feedbacks were considered important amplifying mechanisms during abrupt climate transitions over orbital and millennial timescales (Tuenter et al. 2005). Sea ice also played a possible role during the Holocene interglacial. Renssen et al. (2005b) used a coupled atmosphere-sea-ice-ocean-vegetation model forced by orbital insolation and atmospheric CO_2 and CH_4 concentrations to investigate early Holocene climate. Simulated mean annual, continental, and oceanic temperatures during the early Holocene matched those reconstructed from proxy records, but positive feedbacks related to albedo and insolation from sea ice amplified high-latitude northern-hemisphere temperature changes. Deep-water convection in the Nordic and Labrador Seas was also affected by insolation-greenhouse-gas-forced changes.

Despite Renssen's encouraging results, when compared to the literature on continental land-surface-vegetation feedback, paleoceanographic reconstruction of Holocene sea-ice trends is still inadequate for both northern and southern hemispheres to fully evaluate its response to and amplification of small solar changes from orbital-scale and solar irradiance changes.

Greenhouse Gases

There is reasonable agreement that the Holocene CO_2 and CH_4 patterns already described herein signify the response of the ocean, biosphere, and carbon cycle to climate and that these gases were not the major forcing factor in Holocene climate variability. The "regeneration" hypothesis of Broecker et al. (1999a) holds that the rise in Holocene CO_2 concentrations from an early Holocene low level was a response of the earth's oceans in compensating for approximately 500 Gigatons (Gt) C sequestered in terrestrial forest regrowth and soil formation following the final retreat of the midlatitude

ice sheets (Broecker and Stocker 2006). Mechanisms involving the oceans as a source for increasing Holocene atmospheric CO_2 include outgassing due to changes in SST, with a resulting impact on dissolution of CO_2 uptake, and changes in calcite compensation (Broecker et al. 2001).

MacDonald et al. (2006) suggested that the explosive growth in northern-hemisphere peatland between 12 and 8 ka would have sequestered between 29 and 58 Gt C if circumarctic peat at 8 ka was on average 0.5–1 m thick and covered only one-quarter the area that modern peatlands cover. Such terrestrial storage might explain up to 7 ppmv of the early Holocene CO_2 minimum. Y. Wang et al. (2005) ran modeling experiments to explain the observed Holocene pattern of CO_2. They showed that a decrease in Saharan (70 Gt C) and an increase in Southern Ocean storage (40 Gt C) occurred during the Holocene. Factors such as the retreating Laurentide Ice Sheet, which trapped carbon, and vegetation-albedo feedbacks caused the increase in carbon stored in terrestrial environments between 8 and 6 ka. They surmised that the initial 10-ppmv increase in atmospheric CO_2 came from the earth's oceans. Another 5–7 ppmv (~68–95 Gt C) may have been released from carbon storage on land since 6 ka. They estimated that this would raise atmospheric CO_2 by 5–7 ppmv (Joos et al. 2004).

As discussed in Chapter 5, Ruddiman (2003, 2007) contends that the mid- to late Holocene methane trend reflects, in part, agricultural practices of early civilizations and that CO_2 concentrations would have fallen during the mid- to late Holocene as they did during other interglacials, had not humans altered the global carbon cycle. Part of Ruddiman's theory revolves around the fact that orbital theory of insolation-driven glacial interglacial cycles, manifested in ice-core CO_2 records, would predict a progressive Holocene decline in atmospheric CO_2 concentrations instead of the observed rise. The Ruddiman hypothesis merits serious consideration regarding the very essence of how one views anthropogenic modification of global climate (see chapters 10 and 11 in this volume).

Perspective

Holocene paleoclimatology is in a state of flux. Traditionally studied in high-latitude, high-elevation terrestrial systems, high quality Holocene paleoreconstructions from oceans and low-latitude continental regions have become available only during the past few years. These provide first-order trends in atmospheric and oceanic temperature and in tropical atmosphere circulation and precipitation, but there is still limited spatial coverage and little is known about ocean circulation and sea-ice trends during the Holocene. Reconstructions of solar irradiance using cosmogenic isotopes and geomagnetic proxies have improved substantially, but hypotheses about solar forcing of Holocene climate require further testing. Moreover, Holocene climate model simulations have emphasized time slices, such as those in PMIP projects. Transient model simulations with climate forced by multiple agents, when compared to spatially robust networks of paleoreconstructions produced from multiple proxy methods, will provide a clearer picture of patterns and causes of interglacial climate variability.

LANDMARK PAPER Unstable Holocene Climate

Denton, George H. and Wibjörn Karlén. 1973. Holocene climatic variations—Their pattern and possible cause. Quaternary Research 3:155–205.

The notion that the Holocene interglacial period, when the earliest human societies developed, was characterized by a stable climate persisted for many years. In fact, an extremely unstable Holocene climate was evident from early studies of alpine glacial history by George Denton from the University of Maine and Wibjörn Karlén of Stockholm University. Their 1973 paper included an in-depth analysis of glacial history in the St. Elias Mountains of the Yukon Territory, Canada and Alaska, where Denton and Karlén discovered glacial expansion about 3300–2400 years ago and during the Little Ice Age and five periods of Holocene glacier recession, including the post-1920s. They also reviewed evidence from 40 glaciers in Swedish Lapland and other regions, conclud-

ing that there were multiple periods of glacial expansion lasting about 900 years and contraction up to 1750 years. Glaciers are sensitive monitors of temperature, and Denton and Karlén hypothesized a possible link between solar activity deduced from radiocarbon records and glacial activity. They wrote two other papers, in 1975 and 1977, on Swedish and Alaskan-Yukon Territory glaciers.

The impact of their studies of glaciers around the world has led to a much fuller grasp of interglacial climate instability. Their evidence for millennial-scale climate variations are today among the most informative for understanding recent glacial retreat and global warming. By using newer high-resolution ocean, speleothem, tree-ring, and ice-core records, we now know that climate exhibits an extraordinary level of variability at centennial and millennial timescales. It is against this variability that we must compare current trends in climate.

9

Abrupt Climate Events

Introduction

We devote a full chapter to the important topic of abrupt climate change because the scientific community, as well as society in general, must—through choice or necessity—confront the issue of whether today's climate is experiencing an abrupt climatic transition due to human activities. At its simplest, the question is this: Are rapidly rising greenhouse gas concentrations and large-scale land use disturbing the global hydrological cycle, pushing global climate into a new state, or will such a change happen soon? Stocker and Marchal (2000) phrased the question another way: Where is the current climate state situated on the hysteresis loop in which small freshwater forcing leads to large-scale oceanic and climate changes? Is modern climate transitioning between bistable climate states, and what will the new climate look like?

Earlier chapters described widespread evidence that instabilities punctuate the earth's climate during glacial, deglacial, and interglacial periods. The Younger Dryas (YD) episode, Dansgaard-Oeschger (DO) events, and some Holocene climate transitions attract attention from scientists, the media, policy makers, and the public at large, because they begin and end abruptly, within a human lifetime or less. This chapter probes more deeply into their ultimate causes. How strong is evidence from geology, geomorphology, glaciology, and paleoceanography to support the hypothesis that freshwater from ice sheets, glacial lakes, rivers, or precipitation, when transferred to the oceans, caused global climate shifts in the past? Are there alternative explanations for climate transitions other than freshwater forcing? Can we place quantitative constraints on the "what, when, where, and how" about triggers of abrupt events?

To answer these questions, the focus will be recent efforts to quantify the volume, timing, and duration of freshwater forcing events built on concrete field evidence for late Quaternary ice-sheet retreat, glacial lake drainage history, and modern digital elevation analyses. We also address evidence for ocean hydrological and temperature response along those continental margins located near former ice sheets and freshwater outlets. In addition to well-publicized ice-core and deep-sea records and model-predicted climate scenarios, the Quaternary geology of continents remains a pillar underlying the hypothesis that abrupt climate change is triggered by hydrological perturbations to ocean circulation and climate, not just from glacial lakes but from any origin. It is hoped this chapter will inform the value and complexity of Quaternary geology, perhaps underappreciated in modeling efforts to quantify, express in terms of parameters, simulate, and predict climate variability.

This chapter traces the topic of abrupt climate change from its theoretical roots and simulated impacts to hard physical evidence for catastrophic triggers and to downstream impacts on ocean salinity and temperature. It begins with background sections on the definition of abrupt climate change and climate model simulations of freshwater forcing of climate. We then describe geological and geomorphological records of glacial lake history for North America and Eurasia in the context of the abrupt deglacial climate reversals described in Chapter 7. Next, evidence is presented for dilution of ocean salinity and sea-surface cooling in marginal seas and the open ocean due to freshwater discharge from continents. The final section reviews paleoclimate evidence that some abrupt climate transitions are linked to tropical atmospheric processes rather than to high-latitude freshwater forcing.

Defining Abrupt Climate Change

Many descriptive terms characterize abrupt climate transitions: collapses, jumps, shocks, flips, triggers, thresholds, catastrophes, flickers, switches, and surprises. Can abrupt climate change be formally defined? The onset of the Paleocene-Eocene Thermal Maximum in a few thousand years certainly qualifies as abrupt in the context of the Cenozoic greenhouse-to-icehouse climatic transition. The meteor impact at 65 Ma at the Cretaceous-Tertiary boundary and the massive eruption of the Toba, Sumatra volcano at 74 ka were, geologically, nearly instantaneous events, each having global climatic repercussions. Nonetheless, we concentrate our discussion on abrupt climate events during the past 100,000 years because they are most pertinent to understanding future climate under the modern ocean-basin-continent configuration and sustained high greenhouse gas concentrations.

Catastrophic Lake Drainage Events

In response to indisputable paleoclimatic evidence for the abrupt climate reversals described in chapters 6 and 7, the U.S. National Academy of Sciences National Research Council published a report called "Abrupt Climate Change: Inevitable Surprises," addressing the topic of abrupt climate change during the last 100,000 years (National Research Council 2002). The report provided two useful definitions. The first refers to a climate change in which the scale of the response exceeds that of the original forcing. This definition refers to a situation in which a small perturbation triggers a much larger change in climate because a threshold is crossed, climate feedbacks act as amplifiers, and the earth's climate enters a new fairly stable equilibrium state (Rahmstorf 1996; Manabe and Stouffer 2000). Lying at the heart of this definition is the concept of hysteresis. Stadials and interstadials of the last glacial period are prime examples of where one climate state persists in a self-sustaining mode until an abrupt forcing is applied and a threshold is crossed, knocking it into another state.

One commonly cited cause for this type of climate behavior is small changes to the hydrological balance in high latitudes of the northern hemisphere. As described earlier, changes in freshwater influx into the North Atlantic and Arctic Oceans can in theory lead to changes in meridional overturning circulation (MOC), ocean-heat transport, and climate changes. The argument goes that the volume of freshwater from glacial lakes and ice sheets is relatively small compared to that of the ocean, but its impact can be huge if freshwater enters into ocean regions quickly, at the right time and place.

Freshwater sources might be river discharge or precipitation, but the most common example from the geological record involves the catastrophic release of freshwater stored in large proglacial lakes and ice sheets. Glacial lake outbursts, also called floods, superfloods, or mega-floods (Saint-Laurent 2004), can involve a nearly geologically instantaneous event occurring over only a month to a year (Teller et al. 2002). A closely related phenomenon called a rerouting (or routing) event involves a more sustained flow of freshwater from either glacial meltwater or a change in the outlet route for lake drainage (Clark et al. 1999, 2001; Meissner and Clark 2006). Climate modeling simulations show that the subtle difference between an outburst and a routing event may be of utmost importance in terms of its impact on ocean circulation.

Potential outlet routes for North American and European glacial lake drainage include the Mackenzie River to the Arctic; the Mississippi River into the Gulf of Mexico (GOM); the Kategat into the North, Norwegian, and Greenland Seas; the English Channel; and the St. Lawrence and Hudson Rivers and the Hudson Strait into the North Atlantic Ocean. Freshwater from the margins of the Antarctic Ice Sheet can also influence ocean circulation and climate. The direct impact of freshwater influx into sites of deep-water formation, such as the Greenland-Norwegian or Labrador Seas, would be to lower surface salinity and ocean density and reduce the strength of deep-water formation. In the extreme case, deep-water formation and thermohaline circulation (THC) are completely shut down.

Adaptation to Climate Change

The second definition given by the National Research Council describes abrupt change as a shift in climate so large that natural ecosystems and society itself might not adapt fast enough. This definition focuses on the impacts of abrupt change rather than on its causes and dynamics. *Adaptation* is a term traditionally found in evolutionary theory to refer to the way species, through their phenotype (morphology, physiology, behavior, and geochemistry), evolutionarily "adjust" to their environment. Today, a large community comprising ecologists, social scientists, and others dealing with the potential impacts of climate change uses the term *adaptation* to refer to the various ways ecosystems and human societies can respond—through policy, management, behavior, and technology—to future climate change. Adaptation thus pertains to ecosystems, communities of interacting organisms, and on a larger scale, entire biomes and their sensitivity to climate. Ecosystem adaptation to climate change is beyond the scope of this book, but the Fourth Assessment by the Intergovernmental Panel on Climate Change (IPCC) Working Group (WG) II is a useful introduction to issues of impacts, mitigation, and adaptation to climate change (IPCC WG II 2007).

Regime Shifts

Another way of looking at abrupt climate change is as a large-scale reorganization in regional or global climate—often described as a "regime shift"—rather than as a single catastrophic event. A growing number of researchers believe that the tropics play a pivotal role in global climate in general and in abrupt climate-regime shifts in particular. A change in tropical climate can alter global hydrological balances and in theory shift the earth's climate into a different equilibrium state. Tropical atmospheric forcing of climate—called the "sleeping dragon" by Pierrehumbert (2000)—is like freshwater forcing of high-latitude oceans in that, like the hypersensitive THC, the tropical atmosphere is extremely "tippy" to minor external forcing. If a threshold is reached, relatively small changes in tropical processes might tip the climate system out of balance, and such a transition may actually be occurring in El Niño–Southern Oscillation (ENSO) patterns documented over the past few decades. Clement et al. (1996) argued for a tropical "thermostat" mechanism like that described by Ramananthan and Collins (1991) and others, whereby tropical Pacific Ocean dynamics, particularly small changes in sea-surface temperatures (SSTs), regulate tropical climate.

Major Holocene changes in monsoon patterns and intertropical convergence zone (ITCZ) migrations seen in Chapter 8 are global climate-regime shifts driven by relatively small changes in total solar irradiance output, orbital insolation, land-atmosphere feedbacks, or a combination of factors. Although not universally accepted, many researchers view low-latitude paleoclimate reconstructions as evidence that the mid-Holocene tropical climate change occurred abruptly.

In addition to mid-Holocene tropical climate changes, several historical climate transitions can be considered regime shifts caused by relatively minor external forcing. Changes in solar irradiance during the Maunder Minimum in the late 17th century caused relatively small changes in energy absorbed at the top of the earth's atmosphere—only 2–3 Watts (W) m^{-2}, or 0.2% of total 1365 W m^{-2} (Lean 2000)—yet these changes are suspected as a cause of the Little Ice Age. The Atlantic-Arctic ocean-climate system underwent what are called regime shifts during the 20th century. One large-scale ocean-atmospheric event—the "Great Salinity Anomaly" in the high-latitude North Atlantic Ocean from 1968 to 1982 (Dickson et al. 1988)—has generated a great deal of attention because it signifies a complex link between freshwater influx, ocean salinity anomalies around the subpolar surface gyre, and deep-ocean circulation. Belkin (2004) reviews evidence for two additional salinity anomalies, and it is now believed that large ocean salinity anomalies are closely linked with the multidecadal internal climate variability known as the North Atlantic Oscillation (Hurrell et al. 2001; Hurrell and Dickson 2004). The dynamics of North Atlantic salinity anomalies are complex (Wadley and Bigg 2006), but they nonetheless represent a change in hydrological budget and basin-wide climate. Moreover, the anomalous behavior of the North Atlantic Oscillation since about 1960 remains difficult to explain by natural climate variability (Hurrell et al. 2006), and modeling and observation of Atlantic salinity and MOC forms part of efforts to detect whether human beings have actually initiated an abrupt climate-regime shift. The climate-regime shift in 1976 in the North Pacific Ocean region is another large-scale transition of unknown origin, at least in the context of long-term variability (see Chapter 10 in this volume).

These examples illustrate the variety of perspectives found in paleoclimatology, oceanography, climatology, and climate modeling on what constitutes abrupt climate changes.

Background Climate State

Much of our focus will be on evidence for proglacial lake drainage between the Last Glacial Maximum and the early Holocene. Abrupt climate events occurred at times when ice sheets and glacial lakes provided an obvious source of freshwater, and thus offer an exceptional opportunity to understand abrupt-climate dynamics for several reasons. First, abrupt climate changes do in fact punctuate relatively warm interglacial climate states like the Holocene and the last interglacial (McManus et al. 1994, 1999; Adkins et al. 1997; Overpeck and Webb 2000; Sprovieri et al. 2006). Prior chapters described ample evidence that large glacial lakes are not a prerequisite for abrupt climate transitions.

Second, large glacial lakes are not the only potential source of freshwater to the world's oceans. There is growing evidence that changes in key parts of the global hydrological cycle represent a response to greenhouse gas forcing. Trends include freshwater discharge from the world's two remaining large ice sheets, Greenland (Hanna et al. 2005, 2006) and Antarctica (DeAngelis and Skvarca 2003; Howat et al. 2007), retreat of alpine glaciers (Oerlemanns 2005; Meier et al. 2007), changes in ocean salinity (Curry and Mauritzen 2005), and altered global precipitation patterns (Treydte et al. 2006). Some of these changes can be simulated by coupled atmosphere-ocean general-circulation models designed to study the climatic response to future greenhouse gas concentrations (Cubasch et al. 2001). However, many uncertainties remain. For example, it is not clear whether additional freshwater forcing from melting Greenland ice would significantly affect MOC in the next century (Jungclaus et al. 2006). These topics are discussed in chapters 11 and 12 in this volume, but for our purposes here, they highlight the importance of understanding freshwater forcing whatever the source.

Third, climate models play a central role in the assessment of future climate change, and some models are evaluated in terms of their ability to successfully simulate past climate conditions or hydrological events reconstructed by paleoclimatologists. More accurate quantitative estimates on the volume, age, duration, location, and impacts of past lake and ice-sheet freshwater fluxes are required to better constrain the baseline conditions for model simulations of past and future climate changes.

Finally, during the past century, atmospheric greenhouse gas concentrations have risen, and continue to rise, at rates that are unprecedented when compared to those derived from direct measurements of greenhouse gases in ice cores for the past 700,000 years and from other atmospheric CO_2 proxies for the past 600 million years. There is almost universal agreement that increasing greenhouse gas concentrations will lead to an enhanced global hydrological cycle, including greater freshwater flux to high-latitude regions where deep-water masses form. Potential anthropogenic disruption to the earth's hydrological balance and ocean circulation magnifies the importance of understanding the freshwater-abrupt climate link.

Models of Freshwater Forcing of Climate

In this section, we address four critical factors regarding freshwater forcing of ocean circulation and climate: the timing, volume, location, and duration of freshwater discharge events.

Timing of Freshwater Discharge

Certain climate states, like the last glacial period when large ice sheets surrounded the North Atlantic region, might be more conducive to freshwater forcing than others. Ganopolski and Rahmstorf (2002) simulated Dansgaard-Oeschger (DO) and Heinrich events using the CLIMBER-2 model designed to simulate long-term glacial-age climate changes. They first applied relatively small amounts of freshwater (0.015–0.045 Sverdrups, or Sv) to the Nordic Seas under glacial boundary conditions and then removed the freshwater source. Their simulations with freshwater injections equivalent to 0.03–0.045 Sv resulted in a characteristic DO-type climate signal with abrupt THC strengthening, increased sea-surface salinity, and a rapid rise in atmospheric temperature over Greenland. In contrast, their simulation with smaller forcing did not result in a DO event. Their Heinrich event simulation injected a much larger freshwater forcing into the Nordic Sea region, 0.15 Sv, an amount considered representative of a large surge in the Laurentide Ice Sheet (LIS). This forcing produced a shutdown in THC, which did not return to more vigorous convection until after the freshwater forcing was turned off.

Rahmstorf et al. (2005) conducted an intermodel comparison study under modern climate and ocean conditions in which 11 separate climate models of intermediate complexity were forced with freshwater influx into the North Atlantic at a rate of 0.05 Sv. They found a wide range of MOC and climate sensitivity but qualitative agreement among the models supporting the idea of hysteresis behavior and the salt-advection model of Stommel (1961). The bifurcation points on the hysteresis loop—the point at which modern North Atlantic Deep Water (NADW) formation cannot be sustained—fell at different degrees of cumulative freshwater influx, ranging from <0.1 to >0.5 Sv. They postulate that more complex general-circulation models used in future climate assessments would also exhibit a similar style of MOC sensitivity. Another aspect of the timing of freshwater discharges is the preexisting climate state shown by LeGrande and Schmidt (2008) to be important for the impact of early (8.2 ka) Holocene-like discharge.

Volume of Freshwater Forcing

Climate modeling by Rahmstorf (1995), Ganopolski and Rahmstorf (2002), and Manabe and Stouffer (1995, 1997) raised the disconcerting possibility that a relatively small volume of freshwater can have significant impacts on the strength of MOC. The experiments of Rahmstorf (1995, 1996, 2002) included a variety of forcing flux values and locations for freshwater injections into the oceans. One widely cited result was that a flux as small as 0.06 Sv into high latitudes could shut down NADW and push the climate into a new equilibrium state. Weaver (1995) called this "the point of no return" for MOC.

The impact of different volumes of freshwater flux is also evaluated in what are called freshwater "hosing" experiments. Stouffer et al. (2006) compared the response of THC and global climate using 14 different coupled climate models subjected to continual 0.1 Sv and 1.0 Sv forcing for 100 years spread evenly over the North Atlantic Ocean between 50 and 80°N latitude. The 0.1-Sv volume was chosen because this is the amount of increased freshwater flux predicted by model simulations of a climate state with atmospheric carbon dioxide concentrations four times those of preindustrial times. The 1.0-Sv value, although considered unlikely due to human-induced climate shifts, is comparable to freshwater forcing events from glacial lake drainage already discussed in this chapter. For comparison, the 0.1 Sv equals about the modern Amazon River flow, and the 1.0 Sv equals the total global river input, the equivalent of about a 9-m rise in global sea level. The 0.1-Sv experiments yielded a mean 30% shutdown in THC and an intermodel range of 9–62%. A 30% reduction equals a drop in meridional overturning of 5.6 Sv. The 1.0-Sv experiments all led to complete NADW shutdown, significant cooling in the North Atlantic, and a southward shift in the position of the Atlantic's ITCZ. Simulated

recovery of NADW after the shutoff of freshwater hosing was quite variable among different models.

Location of Freshwater Influx

The importance of the location of freshwater influx has been investigated in a number of independent studies using different climate models. Using a coupled atmosphere-ocean model forced with 0.1 Sv over 500 years, Manabe and Stouffer (1997) contrasted the effects of freshwater influx into the northern and subtropical North Atlantic to simulate a YD-like climate change. The results showed that THC weakened in response to freshwater forcing and the response was four to five times as large in the northern experiment.

Tarasov and Peltier (2005) investigated the drainage of freshwater through the Mackenzie River system into the western basin of the Arctic Ocean. They argued that glacial Lake Agassiz drained into the Arctic through the Mackenzie River at the onset of the YD and Meltwater Pulse 1a (MWP1a) (see Fisher et al. 2006). This hypothesis contrasts with those calling for a St. Lawrence River routing, as originally proposed by Broecker et al. (1989). Tarasov and Peltier's simulations suggests that export of freshwater from the Arctic, eventually through pack ice export through the Fram Straight, is a more likely forcing of YD cooling.

LeGrande et al. (2006) modeled the ocean-atmosphere response to glacial lake drainage into the North Atlantic during the 8.2-ka event using 2.5 and 5 Sv of freshwater forcing for 0.5- and one-year simulations. They generally found consistent patterns in model simulations and proxy data that were based largely on tracing oxygen isotopic composition through the entire hydrological cycle (glacial lake, seawater, precipitation, ice sheet) and evidence from other proxies. In particular, they found evidence for a brief but distinct (about 50%) slowdown in MOC. LeGrande and Schmidt (2008) found reduced Atlantic Ocean heat transport, greater atmospheric heat transport, and greater low cloud and sea-ice cover during the 8.2-ka event and, at least locally, cooling was as much as 3°C.

Duration of Freshwater Events

The duration of a freshwater event—the period over which freshwater flux is sustained—is one of the most challenging topics for the geological community because it requires annually or decadally resolved geological records. This topic has been addressed in modeling experiments designed to simulate the impact of early Holocene Lake Agassiz-Ojibway outburst on THC and climate during the 8.2-ka event, a subject we return to in the section "Abrupt Change in the 8.2-ka Event" later in this chapter. Renssen et al. (2001) and Wiersma and Renssen (2006) used 4.67×10^{14} m^3 as the volume of freshwater discharged during this event, a value that is near the higher end of estimates (von Grafenstein et al. 1998; Barber et al. 1999; Törnqvist et al. 2004). They varied the discharge rate from 0.03 to 1.5 Sv over 10, 20, 50, and 500-year intervals. One simulation showed that a freshwater flux of 0.75 Sv into the Labrador Sea sustained over 20 years caused a weakening of THC for 320 years and general surface cooling around the North Atlantic. This scenario was considered to be the most realistic given the geological data on the 8.2-ka event. The impact of a 10-year pulse at 1.5 Sv was even more dramatic, shutting down THC completely for 1000 years. Conversely, injecting the freshwater over a 500-year interval at a rate of 0.03 Sv produced a very small response.

Summary

This brief discussion does not do justice to advances in model simulations of abrupt climate change and freshwater forcing. Many fine points about model parameterizations, intermodel discrepancies, and simulation-instrumental data comparisons can be found in the original publications. Moreover, our understanding of large-scale features of the ocean's THC and its potential response to perturbation remains rudimentary in many respects (Marotzke 2000; Wunsch and Ferrari 2004). Nonetheless, for our present discussion, we note simply that many modeling studies reveal an Atlantic MOC and THC system that is highly sensitive to small freshwater forcing, highlighting the importance of the paleoclimate evidence for the location, volume, and duration of past freshwater discharges under known climate boundary conditions. Our objective now is to describe direct field evidence for deglacial and Holocene freshwater forcing, starting with geological evidence for glacial lakes in North America and Eurasia, moving "downstream" to nearshore ocean salinity variability as well surface and deep-ocean circulation changes.

Continental Records of Glacial Lake Drainage

This section summarizes the spatial extent, configuration and volume, and age and drainage routes of major proglacial lakes formed during the last deglaciation between approximately 22,000 and 7000 years ago. The last deglacial period was characterized by extensive paleo-lake systems. The Lake Bonneville and Lake Lahontan Systems, which covered large parts of what is today Utah and Nevada some distance from the LIS, are among the most famous. Bonneville was an impressive lake—its shorelines lay several hundred meters above the relict modern Great Salt lake, it covered almost 52,000 km², and it drained more than 4000 km³ of water catastrophically about 15 ka through the Snake River.

Regions around the Aral and Caspian Seas were also covered by immense late Quaternary lakes, the one in the Caspian basin reaching 424×10^3 km² (Mangerud et al. 2001). Our main emphasis, however, is on lakes dammed by ice-sheet margins that are believed to have had large climatic impacts.

Terminology and History

The term *proglacial lake* (also called glacial lakes) refers to lakes that have part of their shoreline formed by the retreating ice-sheet margin. Table 9.1 summarizes information about major glacial lakes with drainage episodes linked to specific abrupt climate events recognized in paleoclimate records from ice cores, oceanic, speleothem, and other archives. The table gives the estimated age, volume, flux measured in Sverdrups, and associated abrupt climate event. Both the hydrological budget of proglacial lakes and their drainage history are strongly influenced by the behavior of the ice-sheet margin. Proglacial lakes varied greatly in area and volume. The volume of the largest, glacial Lake Agassiz, reached 163,000 km³ during its final Kinojévis phase, when it merged with glacial Lake Ojibway. Various phases of Lake Agassiz resulted in drainage outbursts, described in the subsection "Glacial Lake Nomenclature," ranging from 2500–163,000 km³ during the final Agassiz-Ojibway event (Teller and Leverington 2004).

Both North America and Europe have a long tradition of investigations of glacial lakes extending back to the 19th and early 20th centuries. Some classical studies include those of glacial Lake Hitchcock, Agassiz (Upham, 1895; Leverett, 1932), Vermont-Albany (Woodworth 1905; Chapman 1937), the Great Lakes region (Leverett and Taylor 1915), and the Baltic Ice Lake (see Björck 1995). Citations to other early studies can be found in Teller et al. (2005), Karrow and Occhietti (1989), Lewis and Anderson (1989), Franzi et al. (2007), and Jakobsson et al. (2007b). The serious student of glacial lake history and abrupt climate change would be remiss not to read the insightful and ingenious comments of these pioneers.

Glacial Lake Nomenclature

Figures 9.1 and 9.2 show the location of major North American and European proglacial lakes formed mainly from melting water from the North American Laurentide and Fennoscandian Ice Sheets. These maps, while based on extensive field observations, are synoptic representations not meant to capture the entire history of a particular lake or the complex hydrological interconnections among lakes. Their purpose is to illustrate the source and possible outlet routes of freshwater discussed later in the chapter.

TABLE 9.1 Ice-Dammed Lakes and Discharge Events Linked to Abrupt Climate Changes

Lake Name— North American	Region	Stage/ Phase*	Potential Flood Vol. (km³)	Freshwater Forcing (Sverdrup)	Age (ka)	Route†	Climate Event	Comments
Agassiz	W. North America	Herman	9500	0.3	12.9	S/E		1-yr long event; Herman Phase ~100 yr
Agassiz	W. North America	Norcross	9300	0.29	11.7	S/E; N/W		Conflicting interpretations of routing
Agassiz	W. North America	Tintah	5900	0.19	11.2	S/E		
Agassiz—2 Campbell stages	W. North America	Campbell	7000/3700	.22, .12	10.4–10.6	East (N/W)		Several stages
Agassiz—4 stages	W. North America	McCauleyville, Hillsboro, Burnside, The Pas	7600 total	0.17 total	9.2–10.3	East		
Agassiz-Ojibway	W. North America	Kinojévis	163,000	5.2	8.4	North-Hudson Strait	8.2 ka event	Over 6 months to 1 year
Agassiz	W. North America		21,000	0.5	11.3	Mackenzie-Arctic	Preboreal Oscillation	1.5–3 yr, through Clearwater-Athabasca spillway and Lake McConnell
Agassiz—18 lake outlet events	W. North America	Herman, Tintah, Kinojévis	Various	Various	Various	Various	YD, PBO, 8.2	Three of 18 total Agassiz events linked to Greenland Ice Cores
Algonquin	Midwestern North America	Kirkfield			~13.5	South		Low Stanley Phases between Algonquin and Mattawa, and after Mattawa
Algonquin	Midwestern North America	Main Algonquin		Up to 0.12‡	11.1	East	Meltwater pulse 1B	
Algonquin	Midwestern North America	Mattawa		Up to 0.9‡	8.5–8.9	East	Not known	
Algonquin	Midwestern North America							
Algonquin	Midwestern North America							
Vermont—Iroquois	Ontario–St. Lawrence–Champlain		700		13.3	Hudson Valley		Two-stage event

(continued)

TABLE 9.1 (*continued*)

Lake Name— North American	Region	Stage/ Phase*	Potential Flood Vol. (km³)	Freshwater Forcing (Sverdrup)	Age (ka)	Route†	Climate Event	Comments
Vermont— Iroquois	Ontario–St. Lawrence– Champlain		2500	0.14–0.18	13.1	Hudson Valley		Sv based on 6-month flood event
Vermont— Candona-St. Lawrence	Ontario–St. Lawrence– Champlain		1500	0.08–0.11	12.9–13.0	St. Lawrence Estuary	YD	
Iroquios	Ontario		N/A	N/A				
Labrador— Ungava lakes	Labrador	30 lake events	Total all events 6000	Variable, some > 0.015	7–8.4 ka	Labrador Sea, Ungava Bay, Hudson Strait		30 different meltwater pulses

Lake Name– Europe								
Baltic Ice Lake	Baltic Sea	final phase	7800	0.12–0.25	11.5–11.6	North Sea/ North Atlantic		1–2-yr flood duration
Lake Komi	Pechora River region, Russia		2400		80–90 ka	Probably west between White and Baltic Seas		1400-km length, 100 m ASL
White Sea Basin	White Sea		15,000		80–90 ka	Probably west through Baltic		
West Siberian Plain	Siberia		15,000		80–90, 50–60 ka	Probably east		
Lake Taimyr	Taimyr Peninsula region				80, 60 ka			

References: Lewis and Anderson (1989); Pair and Rodrigues (1993); Rea et al. (1994); Barber et al. (1999); Leverington et al. (2000); Moore et al. (2000); Fisher et al. (2002); Teller et al. (2002); Clarke et al. (2003, 2004); Fisher (2003); Jansson and Kleman (2004); Teller and Leverington (2004); Rayburn et al. (2005, 2007). ASL = above sea level; PBO = Preboreal Oscillation; YD = Younger Dryas.

*Terms are stages or phases of glacial lakes. See the references for the names of strandlines, beaches, outlets, and other lakes.

†Includes Algonquin, Iroquois, and S/E margin of Laurentide Ice Sheet.

‡East = St. Lawrence-Champlain-Hudson system; South = Mississippi system; N = Arctic via Mackenzie River system.

Lake Agassiz is the most thoroughly studied proglacial lake. For about 5000 years, it inundated the plains in the Canadian provinces of Manitoba, Ontario, and parts of Saskatchewan, as well as parts of Minnesota and North Dakota in the United States. Agassiz's complex history was recounted in papers by James Teller, Harvey Thorleifson, and colleagues (Teller and Thorleifson 1983; Thorleifson 1996; Fisher et al. 2002; Teller and Leverington 2004; Teller et al. 2005). Thorleifson's (1996) comprehensive summary of the history of the study of Lake Agassiz is a valuable source of information on its geology and geomorphology.

Lake Agassiz formed from glacial meltwater and runoff filling a northward-tilted low-elevation region in the central part of North America. At the time of its formation, the Hudson-Atlantic drainage divide separating the northern part of the continent's drainage had been abandoned by the retreating ice sheet but remained isostatically depressed from the weight of the ice sheet. The northern margin of Lake Agassiz consisted of the southwestern lobe of the Hudson Bay dome of the LIS margin as it receded toward western Hudson Bay. The southeast margin abutted against the continental divide northwest of modern Lake Superior. The

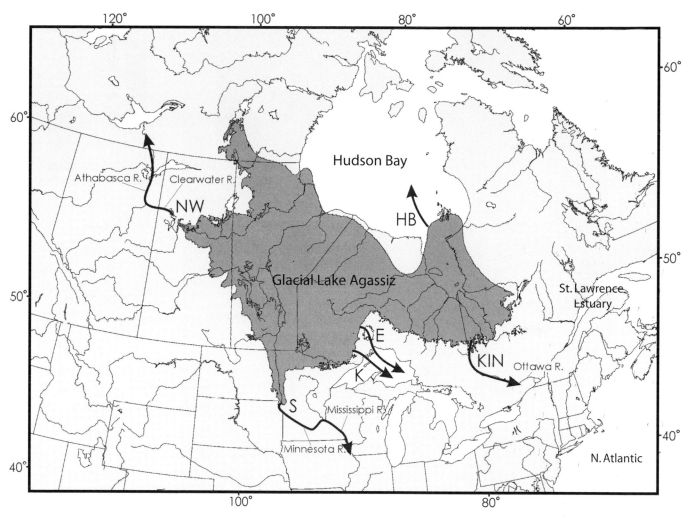

FIGURE 9.1 Map of proglacial Lake Agassiz in North America showing major outlets for lake drainage, from Teller et al. (2002) and Teller and Leverington (2004). E=Eastern (Nipigon Basin); HB=Hudson Bay; K=Eastern (Thunder Bay); KIN=Kinojévis; NW=Northwest (Clearwater-Athabasca-Mackenzie River); S=South (Mississippi). Other lakes occupied the modern Great Lakes, St. Lawrence–Champlain Valleys (see text). Courtesy of J. Teller and D. Leverington and the Canadian National Research Council.

southwestern shoreline was located in the central high plains regions.

There are at least 24 preserved Agassiz paleoshorelines found at successively lower elevations, which represent high lake levels, most of which were followed by a large fall in lake-level during drainage (Figure 9.3). Drainage events occur when outlet routes open because of changes in the position of the retreating LIS margin—i.e., retreat (or advance) of the ice-sheet margin opens (or closes) outlet routes—or because of isostatic uplift of recently deglaciated areas. The five phases, their most likely outlet routes, and the latest age estimates are (1) Lockhart Phase–GOM (13–13.6 ka), (2) Moorhead Phase–North Atlantic Ocean (11.6–13 ka), (3) Emerson–Arctic Ocean (10.7–11.6 ka), (4) Nipigon–North Atlantic (8.6–10.7 ka), and (5) Ojibway–Hudson Bay (8.4–8.6 ka). In the case of the Nipigon Phase, multiple eastward drainage events occurred in the modern Lake Superior Ba-

sin, flowing through proglacial Lake Kelvin (modern Lake Nipigon). Leverington and Teller (2003) propose that 10 recorded lake-level drops ranging from 9 to 58 m in elevation change had equivalent freshwater volumes between 1900 and 8100 km^3. These estimates might be halved or doubled depending on the actual position of the LIS margin (Leverington et al. 2002a).

Smaller glacial lakes occupied the modern Great Lakes region east of Lake Agassiz. From the standpoint of the potential impact on climate, the most important was Lake Algonquin (the modern Lake Huron basin) because all eastward drainage of Agassiz water had to flow through this basin before it exited the continent through either the St. Lawrence or the Hudson River systems. The six major phases of Lake Algonquin history are called the Kirkfield, Main Algonquin, early Stanley, Middle Stanley, Main Mattawa, and late Stanley. These phases encompass periods characterized

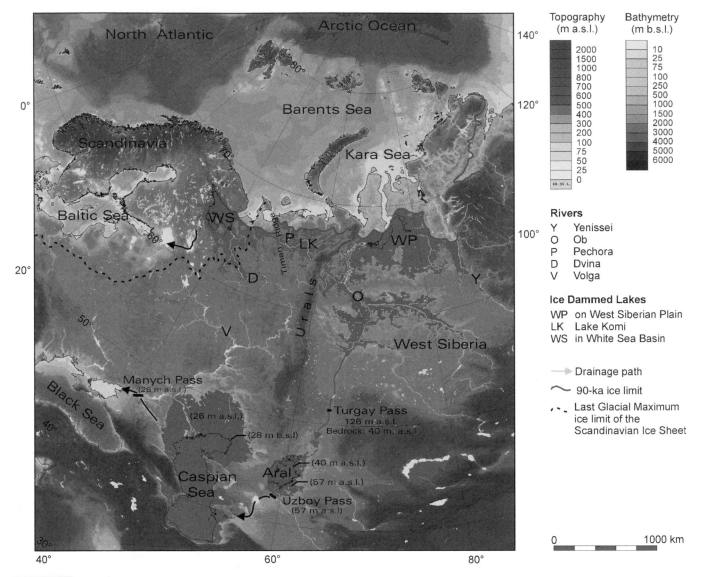

FIGURE 9.2 Location of late Pleistocene (Weichselian) lakes in Eurasia, modified from Mangerud et al. (2001a). Note that the northern Eurasian lakes, in contrast to the Caspian and Aral Sea basins, were ice dammed. Courtesy of J. Mangerud and M. Jakobsson, the Quaternary Research Association, and John Wiley & Sons.

by both high (main Algonquin) and low (early and middle Stanley) lake levels. Many studies in the Lake Superior Basin, immediately adjacent to Lake Agassiz at times during deglaciation, apply stratigraphic or lake terminology (or both) from either Lake Agassiz or Lake Algonquin. Breckenridge et al. (2006) gives a useful review of glacial Lake Superior and its relationship to LIS and Lake Agassiz drainage events between 10.7 and 8.9 ka, based on its varve record.

East of Lake Algonquin, Lake Iroquois occupied the modern Lake Ontario Basin, which at times was connected to lakes in the modern St. Lawrence estuary and the Champlain Valley of New York and Vermont. In these regions, proglacial lakes go by various names: Lake Candona and Lake St. Lawrence in Canada (Pair and Rodrigues 1993; Parent and Occhietti 1999), and Lake Vermont in New York and

Vermont (Chapman 1937; Rayburn et al. 2005, 2007; Franzi et al. 2007).

In Europe, the Baltic Ice Lake that formed in the modern Baltic Sea Basin ~12.5–1.5 ka was about 50–60 m above present sea level and is one of the most important proglacial lakes in terms of its drainage history (Figure 9.2). This lake was located between the retreating Fennoscandian Ice Sheet (part of the Eurasian Ice Sheet, or EIS) in Sweden and Finland on the northwest and deglaciated coastal regions of Germany, Estonia, Latvia, and Lithuania, and of Poland on the southeast. As we see in the subsection "Ice-Sheet Buttresses," the drainage of the Baltic Ice Lake is linked to major deglacial ocean circulation and climate changes.

Late Quaternary ice-dammed lakes in northern Europe and Siberia have been inferred from ice-sheet configura-

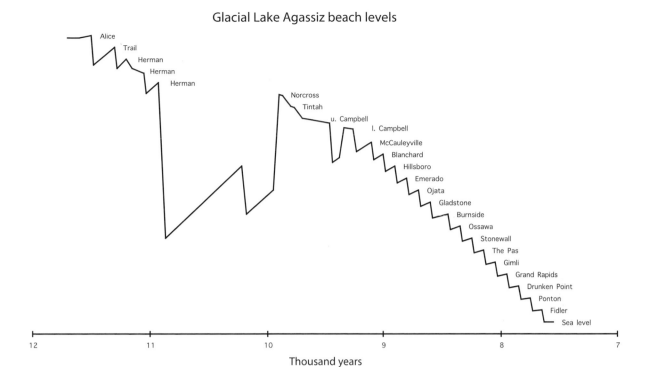

FIGURE 9.3 Beach levels of glacial Lake Agassiz, formed between 12 and 7.5 ka, from Thorleifson (1996) and Teller and Thorleifson (1983). See the tilted beaches in Figure 9.7. Ages are in uncorrected radiocarbon years. Courtesy of H. L. Thorleifson.

tions, lacustrine sediments, and paleoshorelines. Much of this work was done by Jan Mangerud and colleagues (Mangerud et al. 2001a, b, 2004) and others involved with the Quaternary Environments of the Eurasian North (QUEEN) Project (Svendsen et al. 2004). Like North American lakes, proglacial lakes in Europe and Siberia were constrained by the margins of various segments of European and Siberian parts of the EIS. However, various segments of the EIS did not reach maximum extent simultaneously and the glacial lake history in northern Europe is quite distinct from that in North America. For example, the large proglacial Lake Taimyr formed in the early Weichselian ~70–80 ka when the Barents–Kara Sea Ice Sheet was large enough to block its drainage through rivers flowing north, east, and west. In contrast, during the late Weichselian, the Barents–Kara Sea Ice Sheet was not as large and did not extend far enough southward to block river drainage into the Arctic Ocean.

Mangerud et al. (2001a, 2004) synthesized the literature on eastern European and western Siberian Quaternary lakes that formed at roughly 90–50 ka. These lakes had a combined area of 907×10^3 km^2 and freshwater volume equaling 32.4×10^3 km^3. Their significance lies not only in the large quantities of water they held, but in the fact that they drained to the south and west because the EIS lay to the north, blocking drainage into the Arctic Ocean—the opposite of large modern river drainage patterns in northern Eurasia.

Many other smaller lakes formed during the last deglacial interval have been named in North America and Europe. Stea and Mott (1998) list as many as eight glacial lakes from Nova Scotia alone. The paleolimnological and varve records of glacial Lake Hitchcock in the Connecticut River Valley have been extensively studied (Ridge et al. 1999; Rittenouer et al. 2000). Other lake sediments deposited before the last deglacial interval (early- to mid-Wisconsin) are known, such as the St. Pierre Beds of Quebec, the Massawippi Formation deposited between glacial till layers, and several others in the Great Lakes region (Karrow et al. 2000). The freshwater volumes and drainage history for older North American lakes are poorly constrained because sediment and geomorphological records were destroyed by late Wisconsinan ice-sheet advance and retreat. Thus, it remains unclear if they were ice dammed or a source of catastrophic outbursts.

Glacial Geology and Geomorphology Applied to Abrupt Climate

Quaternary geologists have a large array of geological and geomorphological features from which to reconstruct glacial lake history. These features provide critical clues surrounding the timing and volume of freshwater discharge to the world's oceans. We summarize those that pertain to glacial lake outbursts and routing as a background for discussion on specific glacial lakes.

Gorges and Outlet Channels

Geomorphological evidence for catastrophic lake drainage events comes from several sources. The Gulf, a 50-m-deep gorge shown in Figure 9.4, exemplifies one of the most convincing lines of evidence for abrupt freshwater discharge. It was formed when the LIS margin moved north of Covey Hill, the northernmost topographical high of the Adirondack Mountains, allowing Lake Iroquois to flow southeastward into the Champlain Valley and eventually down the Hudson River (Rayburn et al. 2005; Franzi et al. 2007). The Gulf's steep sides, 1.5-km length, 45-m depth, and plunge pool

formed by a former waterfall all attest to a rapid erosional event. Mapping by several generations of geologists (see Denny 1974 and Franzi et al. 2007) show that it was formed by the drainage of the Frontenac Phase of Lake Iroquois southward into the Champlain Valley, lowering the Frontenac lake level 76 m and discharging 2500 km³ of freshwater.

The Gulf is one of many erosional features related to lake drainage, including spillways and meltwater channels cut into bedrock and exposed bedrock surfaces located in formerly glaciated terrain scoured of overlying glacial action and carved by lake drainage events. Other outlet channels and related erosional features in North America are associated with lake outbursts: two at the northern end of modern Lake Superior; the Ouimet and Rabbitt Canyons, which drained two Lake Agassiz levels into the Lake Nipigon; Devil's Crater, where the Nipigon phase of Agassiz drained; Big

FIGURE 9.4 Photo looking west of "The Gulf," a gorge on the New York State-Quebec, Canada border near Covey Hill formed from catastrophic Lake Frontenac discharge from the west (left in photo) in the modern Lake Ontario Basin about 13 ka. The Gulf is ~1.5-km long and has a maximum depth of ~45 m cut into Cambrian Potsdam sandstone bedrock, one of many geomorphological features documenting catastrophic lake drainage (Franzi et al 2007). The drainage event lowered lake levels in the St. Lawrence Lowlands by ~76 m and released ~2500 km³ through the Hudson River system (Rayburn et al. 2005). Courtesy of D. Franzi.

Stone Moraine in Minnesota, draining into the Mississippi River; and the Athabasca-Clearwater spillway in the northwestern region of Lake Agassiz, which connected the lake to the Arctic via the Mackenzie River system. Breaching of the Athabasca-Clearwater spillway has been linked to the short-lived climate reversal known as the Preboreal Oscillation (PBO) (Fisher et al. 2002). Lowell et al. (2005) summarized the geological and geomorphological characteristics of the spillway as follows: "The geomorphic evidence is stunning: a wide, linear channel with numerous feeder channels at its eastern end, and a large delta at its downstream end. Catastrophic flood deposits contain wood giving a maximum age of 9860 ± 230 ^{14}C years B. P. [Fisher et al., 2002]. Strandlines (water-plane indicators such as beaches, spits, or escarpments) near the head of this system are discontinuous and are covered by boreal forest, with the only known Agassiz strandline projecting to the base of the spillway. Evidence that Lake Agassiz existed at the head of the spillway is based on the distribution of scattered high-elevation strandlines, lacustrine sediment, and radiocarbon age-dated flood gravels."

Gorges, spillways, channels and related features provide direct evidence for massive erosion that took place during lake drainage. One important aspect of outlet channels is that their geometry can be used to estimate freshwater discharge rate with a high degree of accuracy using the following approach taken from basic hydrology. Discharge (Q) is a function of the cross-sectional area of the channel (A) and the velocity of the flow (V), which itself is estimated from several methods.

$$Q = A \times V$$

Fisher et al. (2002) calculated that the most likely velocity V for Lake Agassiz drainage through the Athabasca-Clearwater Spillway was 12 m sec^{-1} and the spillway area A was 180,000 m^2 (channel width = 1800 m × depth = 100 m). These values result in a peak discharge of 2.16 Sv. Rayburn et al. (2005, 2007) used similar methods to calculate discharge volume for two Lake Vermont freshwater discharge events through the Fort Ann, New York outlet and one through the Gulf of St. Lawrence near Quebec City. They showed that the background discharge volume of 56,000 m^3 sec^{-1} for the region was overwhelmed by the three catastrophic lake drainage events, with volumes totaling 700 km^3, 2500 km^3, and 1500 km^3.

Ice-Sheet Buttresses

During deglaciation, lakes form when large volumes of glacial meltwater accumulate in low-lying regions that were isostatically depressed from the weight of the ice sheet because drainage outlets through paleoriver systems cannot accommodate the large volume of water. The water is essentially dammed and, consequently, lakes fill up to limits controlled by topography and the ice-sheet margin. In some cases, there is a topographical point against which a part of

the ice-sheet margin is lodged. Retreat from these buttress points causes abrupt release of freshwater, and geologists have identified several such critical points.

One example comes from the Baltic Ice Lake (Figure 9.5). Mount Billigen in south-central Sweden was a key topographical feature in the history of the Baltic Ice Lake. Rising 305 m above sea level, it served as a pinning point for the ice sheet and a constraint on the geographical extent of the Baltic Ice Lake. More than a century of geological studies of the Baltic Ice Lake has led to an in-depth understanding of its drainage history. Chronology for the Baltic Ice Lake comes from radiocarbon dating (Björck 1995), varve chronology (Andrén et al. 2002), and optically stimulated luminescence (Kortekaas et al. 2006). The many features documenting its history and catastrophic drainage include glacial lake varve stratigraphy, scoured bedrock surfaces, large (8 m^3) transported boulders gouged from Mt. Billingen bedrock, and coarse sediment carried downstream by floodwater and deposited on scoured bedrock surfaces. In a recent synthesis, Jakobsson et al. (2007b) used a Digital Terrain Model that incorporates ice-sheet positions, lake paleoshorelines, and isostatic uplift rates to quantify Baltic Ice Lake bathymetry, volume, and hypsometry. These studies showed that, as the Fennoscandian Ice Sheet retreated north of Mount Billigen at the end of the YD ~11.7 ka, lake level fell 25 m and led to 8000 km^3 of freshwater influx into the North Sea (Björck and Digerveldt 1989). This discharge was one of the largest deglacial lake drainage events recorded.

Mount Billigen is one of several major buttressing points that temporarily prevented glacial lake drainage into the oceans. Others include Covey Hill at the Canadian-U.S. (New York) border (Clark and Karrow 1984; Franzi et al. 2007), the highlands surrounding Hudson Strait (Andrews et al. 1995; Barber et al. 1999; Dyke et al. 2002), and various pinning points near the continental divide separating glacial Lake Agassiz to the west and the Great Lakes basins to the east (Thorleifson 1996). They are obviously a key link in the chain connecting freshwater lake drainage to ocean circulation and climate.

Paleoshorelines and Lake Volumes

Channel dimensions are the main source of information for calculating lake discharge rate, but beaches and other shoreline features provide the most tangible field evidence for glacial lake geometry and a means to estimate lake volume prior to drainage. Proglacial lake shorelines can be identified and mapped in the field and from aerial surveys on the basis of constructional deposits, such as emergent beaches and spits, and erosional features, such as scarps. The locations of glacial Lake Agassiz phases as defined by paleoshorelines are shown in the map view in Figure 9.6 and as a cross-section in Figure 9.7. The approximately south-to-north differential

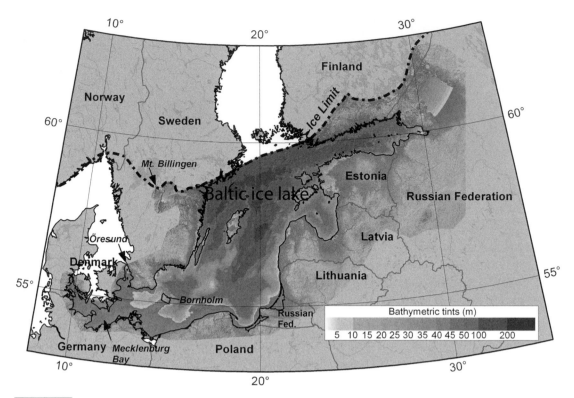

FIGURE 9.5 Map of Baltic Ice Lake showing location of buttress point at Mount Billigen, from Jakobsson et al. (2007b). Retreat of the ice-sheet margin north of this point released freshwater to the North Sea, producing oxygen isotopic excursions, shown in Figure 9.11. Courtesy of M. Jakobsson.

isostatic uplift of progressively younger (lower) shorelines causes the shape of each Agassiz beach profile.

Lake volumes are calculated from shoreline elevation data and geographical extent in the following way (Leverington et al. 2000, 2002a, b; Leverington and Teller 2003). First, building on the shoreline mapping of Teller and Thorleifson (1983, reviewed in Thorleifson 1996), the elevations of mapped shorelines are corrected for regional isostatic uplift from isobases, which are contour lines of equal isostatic rebound constructed from the elevations and ages of emerged shorelines in a region. Paleotopography of the lake region can be constructed by subtracting the modern shoreline elevation from the amount of isostatic uplift if the ages of shorelines are known. The resulting topographical map would include the paleobathymetry of the ancient lake phase. Modern digital elevation modeling is then applied to the paleotopographical data at spatial scales of a few tens to hundreds of meters. Such an approach has been applied to European lakes as well (Jakobsson et al. 2007b).

Shorelines are found in most places a glacial lake existed for an extended period, controlled by a balance between rates of inflowing ice-sheet meltwater and isostatic uplift. The literature on glacial lake shorelines is too large to review here but some of the notable shoreline compilations are those for Lakes Agassiz (Teller 1990; Thorleifson 1996; Leverington et al. 2000),

Algonquin (Lewis et al. 1994), Iroquois (Pair and Rodrigues 1993), Vermont-Candona (Chapman 1937; Parent and Occhietti 1999; Occhietti and Richard 2003; Rayburn et al. 2005, 2007), and the Baltic Ice Lake (Björck and Digerfeldt 1989; Stromberg 1992). Well-dated shorelines are not only important for reconstructing lake drainage but are also an essential ingredient in the development of global paleotopographical models of postglacial isostatic adjustment, such as the ICE-5G model of lithsperic response to ice loading and unloading (Peltier 2004) that is used extensively in sea-level studies.

In addition to shorelines formed during the last deglaciation, paleoshorelines are known from the early to mid-Weichselian in Europe and Siberia. One example is a large lake 80 m above sea level ice-dammed against the Barents-Kara Sea ice sheet, a portion of the northern margin of the EIS at ~60 ka. This was one of several lakes in the region formed during the last 90 ka that was identified from geological mapping and stratigraphic analysis of glacial, marine, and lacustrine deposits (Mangerud et al. 2004). These older lakes, although not treated in depth here, are receiving greater attention because their drainage appears to be recorded in paleoceanographic records. Spielhagen et al. (2004), for example, identified nine distinct meltwater events from marine isotopic records during the past 180 ka in the Arctic Ocean linked to glacial history.

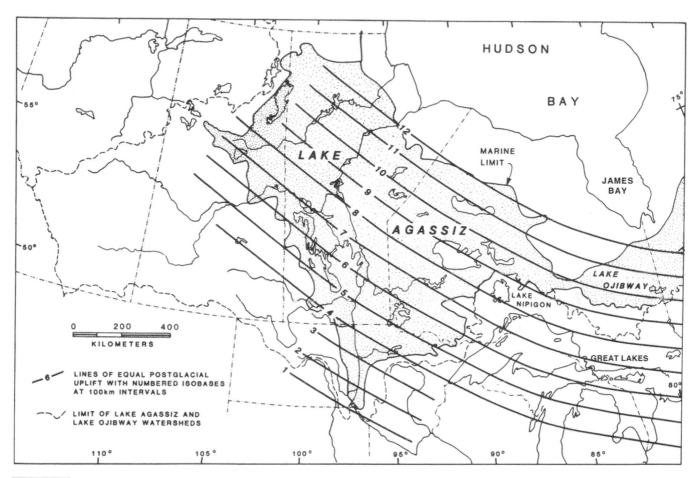

FIGURE 9.6 Isobases across midcontinent North America where glacial Lakes Agassiz and Ojibway inundated isostatically depressed regions during the deglacial interval after the Laurentide Ice Sheet retreated. Twelve curved lines indicate areas of equal isostatic uplift that influenced the timing, direction, and rate of glacial lake outflow, from Teller and Thorleifson (1983), Thorleifson (1996), and early studies. Courtesy of H. L. Thorleifson and J. Teller.

Quantitative estimates of glacial lake volume based on mapped shorelines provide a critical link between the field geologist and climate modelers attempting to simulate freshwater impacts on ocean circulation.

Lake Sediment Records

Geomorphological features like beaches provide precise elevations for past lake levels and volumes but they are "snapshots" of a particular time and do not provide information on low stands in lake level nor a continuous record of regional lake discharge and ice-sheet meltwater history. For more complete proglacial lake records of rapid drawdown and rerouting events, including their drainage, we must turn to sediment deposition within glacial lake basins and in surrounding areas where outflow deposits sediment during lake drainage.

There are many types of sedimentological evidence for glacial lake drainage events. The most dramatic are fluvial boulder beaches and sands deposited relatively close to the main outlet channel, signifying high-energy fluvial systems. Fluvial boulder beds are known from eastern drainage outlets for Lake Agassiz (Teller and Thorleifson, 1983, as reviewed in Leverington and Teller 2003), the Mackenzie River channels in Athabasca-Clearwater (Fisher et al. 2006), Lake Iroquois-Vermont (Franzi et al. 2007), and the Baltic Ice Lake.

Proglacial lakes often deposit thick, fine-grained sediment sequences that are useful in reconstructing the paleolimnological and drainage history of the lake. Many modern lakes are relics of the last glacial cycle, and their basins contain deglacial and Holocene sediments studied in several coring projects. These include the modern Great Lakes Basin (Colman et al. 1994; Rea et al. 1994; Moore et al. 2000), Lake Manitoba in the western North American prairies (Last et al. 1994), and Lake Champlain in New York and Vermont (Hunt and Rathburn 1988; Cronin et al. 2008).

Lithological proxies used in glacial lake sediments include changes in varve thickness, sand layers, rip-up clasts

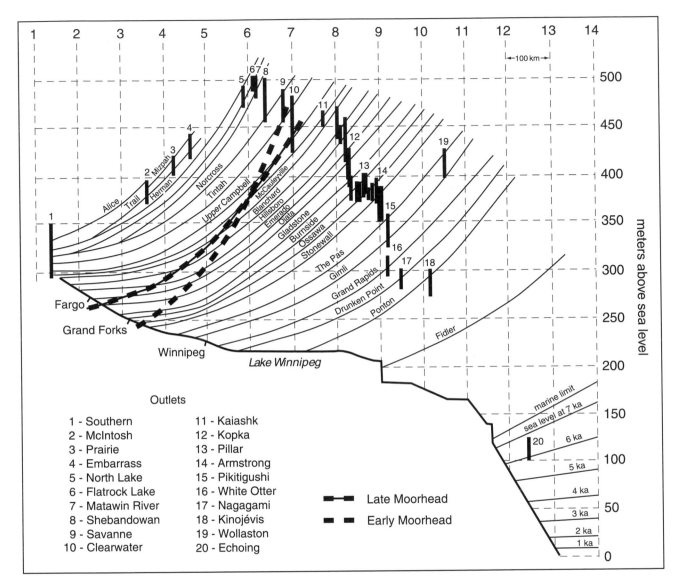

FIGURE 9.7 Lake Agassiz beaches formed during the last deglacial, showing uplift and tilting from differential isostatic adjustment during and after Laurentide Ice Sheet retreat, from Teller and Thorleifson (1983) and Thorleifson (1996). The numbered vertical bars are isobases, shown in Figure 9.6. The black vertical bars are outlet channels. Courtesy of H. L. Thorleifson and J. Teller.

from scouring and erosion, and ice-rafted debris (IRD). A significant decrease in varve thickness, for example, between early and late phases of glacial Lake Vermont-Candona coincides with major changes in lake dimensions due to the lake-level fall from the Coveville to Fort Ann Phase (Rayburn et al. 2005). Changes in the position of the ice margin and other factors can also lead to changes in varve thickness. In Sweden, changes in the color, lithology, and diatom content of varves may also be a reflection of Baltic Ice Lake drainage. Andrén et al. (2002) showed that a tripling in varve thickness began within decades after the lake drained near the onset of the PBO.

Thick sand beds, interlayered with varves, some containing rip-up clasts of material transported from known prov-

enance, can also indicate a major change in sedimentation due to a lake discharge event. Rayburn et al. (2007) found sand layers in Lake Vermont sediments, signifying the downstream influence of the Lake Frontenac outburst that cut the Gorge and deposited boulder beaches closer to the outlet channel.

Paleohydrological reconstructions of glacial lake sediments also provide clues on lake drainage history using a suite of geochemical methods including oxygen and carbon isotopic analyses of ostracodes (Lewis et al. 1994; Remenda et al. 1994) and sedimentary cellulose from phytoplankton (Last et al. 1994; Buhay and Betcher 1998; Birks et al. 2007). Isotopic records provide a means of reconstructing changes in the source of lake water from ice-sheet meltwater, precipitation, river inflow, or lake water draining from an adjacent basin (Moore et al. 2000).

Marine Incursions

Along low-lying continental margins, catastrophic lake drainage occurrence is usually followed by rapid marine inundation of glacio-isostatically depressed areas. Semienclosed glacio-marine bodies include the Tyrell Sea around Hudson Bay, the Champlain Sea in the upper St. Lawrence-Champlain Valleys, the Goldthwait Sea in the outer St. Lawrence Estuary, and the Yoldia Sea in the Baltic (Björck et al. 2002), to mention some important ones. Many coastal areas in formerly glaciated regions were also inundated for at least a few centuries to millennia. Some have been given formal geological names, such as the Presumpscot Formation in coastal Maine and the Everson interstade in Puget Sound (Amundsen et al. 1994; Dethier et al. 1995). Weddle and Retelle (2001) provided a comprehensive volume on postglacial marine deposits and sea-level change along the coasts of the northeastern United States and eastern Canada. In Spitzbergen and Franz Josef Land, Forman et al. (2004) documented a number of emergent marine shorelines. In some regions, marine deposits are referred to simply as glacio-marine sediments assigned to the Wisconsinan or Weichselian.

Glacio-marine deposits are important for several reasons. First, if they cover a region formerly occupied by a glacial lake, they provide indisputable evidence that the lake drained at the onset of marine deposition. Second, some marine incursions left well-preserved shoreline, which is useful for calculating isostatic uplift rates. Third, as we see in the subsection "St. Lawrence River–Gulf of St. Lawrence," their sediments provide a means of reconstructing paleosalinity and hydrology that might reflect freshwater outflow from later glacial lake drainage events. Fourth, they provide paleoclimate proxies (i.e., marine foraminifera) and chronology (material for radiocarbon) that is unavailable in lake sediments.

Chronology for Glacial Lake History

The primary methods used to date deglacial history are radiocarbon dating, varve counting, cosmogenic isotopic dating of exposed bedrock or glacial erratic surfaces, and paleomagnetic records. Early field studies pre-dated the development of radiocarbon dating in the 1950s, and many used varve counting to estimate the age of deglaciation. Glacial lake varves are coarse- to fine-grained sediment couplets deposited during summer melt and winter freeze seasonal cycles. Before radiocarbon dating, varve sequences were used as a means of dating the retreat of ice sheets following the Last Glacial Maximum and for regional correlation using varve thickness. The foundation of varve chronology stems from the work of Swedish geologist Gerard De Geer, who used the Swedish term *varv* to refer to sediment couplets, culminating in classic papers (De Geer 1912; see Wohlfarth 1996 for review) that form the foundation of modern varve chronologies from Sweden, Scotland, Iceland, the Americas, and other regions. De Geer's student Ernst Antevs continued varve research in North America, especially glacial Lake Hitchcock in the Connecticut River Valley (Antevs 1922, 1928; see Ridge et al. 1999 for review).

In the 1960s and 1970s, glacial geologists applied radiocarbon dating of plant, wood, bone, bulk carbon, mollusk, and other material for chronology, filling the literature with thousands of radiocarbon ages on glacial lakes and related glacial features. There are several limitations when using radiocarbon dates from the older literature. First, these dates pre-dated the development of calibration programs used to convert radiocarbon dates to calendar-year timescales, and many do not have the $\delta^{13}C$ isotopic measurement on dated material used in calibration. Second, the geographical or stratigraphic provenance (or both) of the material is sometimes unknown or ambiguous, making it difficult to establish the precise relationship between the radiocarbon ages and glacial lake events. Third, early radiocarbon ages were determined using a beta-counting method, which requires a large amount of material and has larger errors associated with counting carbon-14 atoms. Today samples as small as a few hundred micrograms are dated using accelerator mass spectrometry. Finally, global and local carbon reservoir effects and other sources of old carbon (groundwater, sediment pore-water, bedrock) can alter the real age of a sample. For these and other reasons, some researchers even today choose to cite new dates in radiocarbon years rather than calendar years for comparison with older dates. This can make it difficult to compare paleo-lake history to the calendar-year age models almost universally employed for ice-core, speleothem, and many deep-sea paleoclimate records.

Because of the need for calendar-year chronology of deglacial history and abrupt climate events, there is a renewed interest in varve chronology as a means of achieving annually resolved continental paleoclimate records and calibrating the terrestrial radiocarbon timescale to calendar years. In Europe a large effort has been made to synchronize radiocarbon and varve chronologies (Björck and Möller 1987; Wohlfarth et al. 1995). Wohlfarth and Possnert (2000) reviewed late glacial and Holocene radiocarbon and varve chronology back to almost 11 ka; they were based on glacio-lacustrine varves from more than 1000 varve diagrams. Andrén et al. (1999, 2002) used Swedish varves to correlate continental deglacial history to paleoclimate records from Greenland ice cores for the YD-Preboreal interval. Ridge and colleagues have developed an integrated 6000-year-long New England varve and paleomagntic intensity chronology for New England in the northeastern United States (Ridge et al. 1999). These studies have resulted in a varve-magnetic calibration of the regional radiocarbon age chronology, permitting a reevaluation of the retreat of the LIS, the age of postglacial marine inundation, and the dating of abrupt glacial Lake Vermont drainage (Rayburn et al. 2007).

In sum, improvements in radiocarbon age calibration and the application of varve chronologies have helped pinpoint the age of several glacial lake drainage events discussed in the remainder of the chapter.

Ice-Sheet Meltwater

The Laurentide and European Ice Sheets are the ultimate source of most glacial lake water. In addition to digital mapping of lake volumes, quantitative reconstruction and modeling of the volume of the LIS and its runoff history were developed by Licchiardi et al. (1999) on the basis of revised ice-sheet-retreat history (Dyke et al. 2002). Dyke's ice-sheet margin reconstructions are given for 500-year increments for the main phase of deglaciation. Incorporating precipitation and evaporation into compilations, Licchiardi concluded that LIS cumulative baseline runoff maintained a fairly constant rate of about 0.3 Sv from about 21 ka until about 9.5 ka. Modeling studies by Marshall and Clarke (1999) generally supported this estimate. In Eurasia, the QUEEN and projects focused on the British Isles, Scandinavia, Iceland, Greenland, and other regions (see Chapter 7 in this volume) produced similarly detailed ice-sheet reconstructions of ice-sheet retreat. These provide information on the background rates of meltwater discharge exceeded to varying amounts during lake drainage events.

Glacial Lakes and Abrupt Climate Events

This section summarizes efforts to correlate glacial lake drainage events to global climate reversals recorded in Greenland ice-core oxygen isotopes, deep-sea carbon isotopes, and Cariaco Basin grayscale records. Clark et al. (2001) presented an all-encompassing model relating deglacial freshwater discharge from LIS melting to paleoclimate records from Greenland Ice Sheet Project 2 (GISP2) oxygen isotopes, radiocarbon activity, marine carbon isotopes, IRD in the North Atlantic, and the percentage of carbonate. They argued that periods of freshwater flow into the North Atlantic coincided with millennial climate reversals between 17 and 8 ka, including the YD cooling. They called these rerouting events and numbered them R1 through R8 (Figure 9.8). Event R1 coincided with the 8.2-ka event (Barber et al. 1999), R2 was a brief event near 10.2 ka (perhaps the PBO), R3 equaled the YD, R4–R6 were three brief events during the Bølling-Allerød (B/A) (possibly the Intra-Allerød Cold Period, the Killarney-Gerzensee cooling, and the Older Dryas), and R7 and R8 occurred just after and before Heinrich event 1 (H-1). In their reconstruction, most meltwater discharged through the Mississippi River system between 22 and 17 ka, after which the association between rerouting events and climate oscillations becomes apparent. Clark notes that a midsize ice sheet, such as the Laurentide during deglacial retreat, might have a larger impact on ocean circulation and

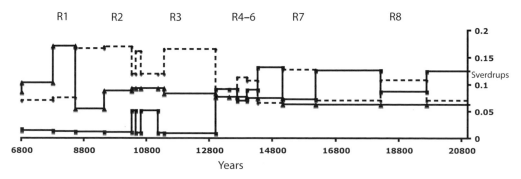

North American glacial lake drainage routing events

FIGURE 9.8 Freshwater flux to ocean from glacial lake routing events during the last deglaciation, measured in Sverdrups, from Clark et al. (2001). Routing events R1–R8 are labeled. Three major drainage routes are shown: (1) The dashed line with circles combines the Hudson and St. Lawrence Rivers; (2) the gray line with squares indicates the Mississippi River; and (3) the solid line with triangles indicates Hudson Strait. Mississippi River flow dominates the early deglacial, whereas Hudson Strait dominates the 8.2-ka event. The precise timing and route of drainage events is still actively researched; other routes include the Mackenzie River during the Younger Dryas, 13–11.5 ka. Data courtesy of P. Clark.

climate because freshwater exits the continent via eastern routes as opposed to a southern route through the GOM.

Teller and Leverington (2004) proposed a correlation between Agassiz discharge and the Greenland ice-core oxygen isotope record (Figure 9.9). This compilation differs from that of Clark in that it shows drainage volume for outburst events, assuming a one-year duration for each flood, although they acknowledge that the precise duration of each

event may have lasted anywhere from a month to several decades. It also shows baseline flow from Lake Agassiz, total LIS background flow, and the drainage routes for each Agassiz phase (Ojibway, Nipigon, Emerson, Moorhead, and Lockhart). The 19 lake outbursts increased total freshwater volume significantly above baseline Agassiz flow, from less than 2,500 to 10,000 km³, except for the final event near 8.5 ka, which reached 163,000 km³. Teller's major proposition—that

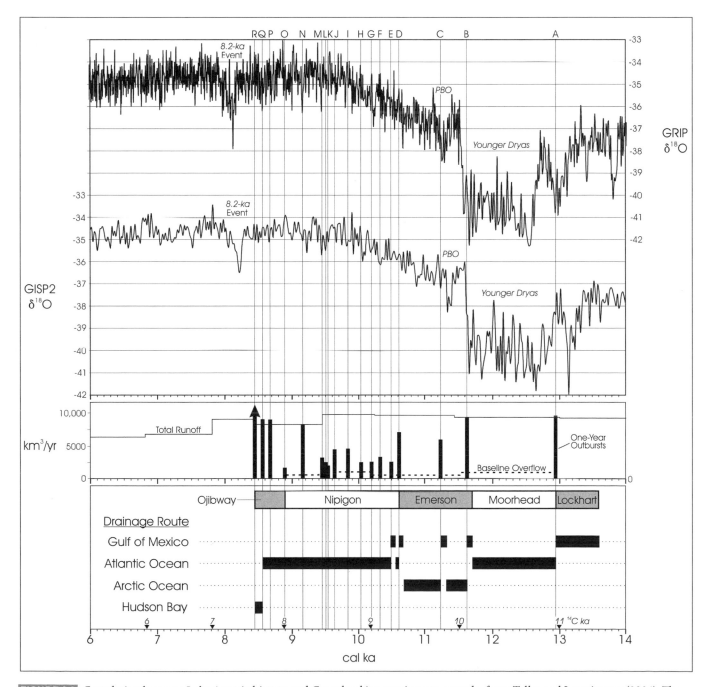

FIGURE 9.9 Correlation between Lake Agassiz history and Greenland ice core isotope records, from Teller and Leverington (2004). The top two panels are Greenland Ice Core Project (GRIP) and Greenland Ice Sheet Project 2 (GISP2) oxygen isotope records; the lower two panels are glacial lake discharge estimates and drainage routes for glacial Lake Agassiz. Courtesy of J. Teller and D. Leverington and the Geological Society of America.

major outburst events can be correlated with ice-core isotopic excursions during the YD, PBO, and the 8.2-ka event, with varying degrees of certainty—is supported by a substantial amount of field evidence. However, recent field evidence suggests alternative routing for YD-age glacial lake discharge out through the Mackenzie River (Lowell et al. 2005; Broecker 2006) and other studies suggest that an extraterrestrial impact coincided with the onset of the YD (Firestone et al. 2007). Future work is needed to definitively link lake outbursts with Greenland temperature changes and explore alternative theories.

There are still unresolved issues regarding the routing of each event shown in Figures 9.8 and 9.9. For example, the large drainage at the end of the Lockhart Phase is shown as draining through the GOM, but it may have had an eastward (Broecker et al. 1989) or northwestward route (Tarasov and Peltier 2005; Teller et al. 2005). Agassiz's Moorehead Phase ~11.6 and 11.2 ka, which followed a quiescent period of baseline flow during the low lake level, may have discharged via the Mackenzie River into the Arctic Ocean in one or two events near the PBO (Fisher et al. 2002). These may have occurred in only one-and-a-half to three years and equaled 0.5 Sv. Other Preboreal-age lake drainage events from the Great Lakes via the St. Lawrence (Cronin et al. 2008) and the Baltic (Björck et al. 1996) have also been postulated, raising the possibility of simultaneous discharges into three separate high-latitude ocean regions.

Lewis et al. (1994) and Moore et al. (2000) synthesized Lake Algonquin (modern Lake Huron basin) discharge history between 8 and 14 ka based on geomorphological evidence, stable isotopic analyses of ostracode shells from sediment cores, and modeling freshwater outflow. The Algonquin record suggests that maximum flow from the lake Huron Basin preceded the onset and succeeded the end of the YD cooling episode and coincided with the large MWP1b inferred from the Barbados sea-level record. However, considerable uncertainty exists from the radiocarbon chronology, the lack of suitable ostracode material during certain lake phases, and poorly constrained estimates for the volume of inflowing freshwater from Lake Agassiz and other lakes. Nonetheless, these studies demonstrate a still untapped potential for paleohydrological analysis of the interplay between interconnected proglacial lakes necessary to refine their discharge histories.

In Europe, there is strong evidence that the Baltic Ice Lake coincided with the inception of the PBO (Björck et al. 1996, 1997; Andrén et al. 2002). Among the strongest arguments for a lake-drainage-climate linkage is the exceptionally robust documentation of the lake's extent, its relation to the buttress point at Mount Billigan, and its drainage history. The precise chronology provided by varves is especially important because of the plateau in the radiocarbon curve at the time

the Baltic Ice Lake existed. In addition, detailed mapping of ice-sheet positions and lake shorelines, lithological and geochemical changes in varve sediments, and precise dating of the post-Baltic Lake marine incursion called the Yoldia Sea (Andrén et al. 2002) make this drainage event exceptionally well documented. As we see in the next section, support for a catastrophic freshwater event from the Baltic Ice Lake comes from paleoceanographic reconstructions in the North Sea.

Paleoceanographic Changes in Marginal Seas

Estuaries, bays, and ocean regions adjacent to continental margins, especially near outlet routes for glacial lake drainage, provide a critical link between the abrupt lake discharge observed on continents and the broader paleoclimate record from ice cores, speleothems, and deep-sea cores. This is because a massive flow of freshwater should in theory reduce salinity in estuaries, although the hydrodynamic response must have been extremely complex, with no historical analog except perhaps extreme hurricane or storm events. In this section we review evidence for meltwater events in sediment cores from regions around North America and Europe to better constrain the location, timing, and extent of catastrophic discharge events.

A number of sediment proxies are used to infer lake discharges. These include distinctive "red" sediment layers (Kerwin 1996); dinoflagellate (de Vernal et al. 1996), benthic foraminiferal (Corliss et al. 1982; Rodrigues and Vilks 1994), and ostracode (Cronin 1989; Hunt and Rathburn 1988) assemblages; mineralogy and composition of IRD (Andrews et al. (1999); barium/calcium ratios (Hall and Chan 2004b); and magnetic susceptibility. By far the most widely used and diagnostic proxy method in ocean sediments, however, is the oxygen isotope chemistry of planktic foraminiferal shells. Living in the upper surface layers of the ocean, planktic foraminferal $\delta^{18}O$ values should record periods of changing hydrology from freshwater discharges. This approach is based on the principle of mixing fresh and marine water with very different isotopic composition. The isotopically depleted freshwater end-member $\delta^{18}O$ can range from −35 to −16‰ (standard mean ocean water), depending on the isotopic composition of source water from melting ice, precipitation, lakes and rivers, evaporation, and other processes. In contrast to heavily depleted freshwater, ocean water $\delta^{18}O$ is enriched, usually ranging from ~−1 to +2‰. Assuming that other factors, such as temperature, global ice volume, and biological processes during shell secretion remain constant, the closer a marginal marine site is to the freshwater source, and the larger the freshwater discharge event, the greater the isotopic depletion. Thus, the oxygen isotope proxy is a measure of changing hydrology from source water and mixing, not a direct measure of ocean margin salinity. We review here some evidence for freshwater discharge events.

St. Lawrence River–Gulf of St. Lawrence

The St. Lawrence Estuary is a critical outlet for freshwater, originally thought to have caused the YD when Lake Agassiz drainage shifted from its early deglacial southward-flowing route to the GOM to an eastern routing (Broecker et al. 1989). Eastward drainage from North American lakes would have first influenced the salinity of the Champlain Sea (13–9 ka), an inland extension of the Atlantic that inundated the isostatically depressed St. Lawrence–Champlain Valleys. More distally, the modern Gulf of St. Lawrence Estuary should have also been affected.

A number of attempts have been made to identify a salinity signal in the St. Lawrence system, and evidence about the timing and impact of freshwater is still unclear. Decreased salinity changes during the middle of the Champlain Sea episode led Rodrigues and Vilks (1994) to conclude that meltwater from the continent did not flow into the Atlantic at the onset of the YD, as originally proposed by Broecker. More recently, improved radiocarbon dating of wood indicates that the onset of the Champlain Sea marine episode actually coincided with the beginning of the YD (Occhietti and Richard 2003). This means that glacial Lake Vermont in the Champlain Valley and Lake Candona in the St. Lawrence lowlands drained at this time. There is also evidence for abrupt freshening in the Champlain Sea only decades after its formation. This means there might have been a double pulse of freshwater—first lake Vermont-Candona, and then water originating from another western lake source, perhaps an early phase of Lake Agassiz. Another freshening event in the Champlain Sea occurred ~11.4 ka, also hypothesized to come from the Great Lakes region, probably glacial Lake Algonquin (Cronin et al. 2008).

On the basis of paleoceanographic analyses to the east of the Champlain Sea in the St. Lawrence Estuary region, De Vernal et al. (1996) challenged the idea that meltwater flow out of the St. Lawrence caused the YD reversal. They interpreted dinoflagellate and isotopic data from a transect of cores from the Cabot Strait, Laurentian Channel, and Northwest Atlantic Ocean as showing an initial lowering of salinity in the Cabot Strait region coincident with the YD inception, but concluded that the salinity record indicated only a modest meltwater event that apparently influenced only the upper surface layers and not the more oceanic sites in the Laurentian Channel. In essence, the impact of any meltwater signal seemed to dissipate near the shelf break and was too small to cause the YD under current ideas of North Atlantic sensitivity to freshwater forcing.

More recently, Carlson et al. (2007a) restudied the outer St. Lawrence Estuary region using new proxies $^{87}Sr/^{86}Sr$, U/Ca, and Mg/Ca as tracers of source water from the interior of North America, as well as stable isotopes to reconstruct paleosalinity (Figure 9.10). In this way, they tried to "fingerprint" the source of freshwater exiting the St. Lawrence, as well as to reconstruct its impact on salinity. They concluded that discharge from source water from the western Canadian Plains occurred at the start of the YD. Their salinity calculations suggest that the initial pulse ~12.9 ka was equivalent to a flux of 0.06 ± 0.02 Sv, and that this was followed by an additional meltwater event equivalent to an additional 0.06 Sv ending ~12.3 ka.

Hudson River System

Thieler et al. (2007) presented new geophysical analyses and insights into the processes that formed the present Hudson Shelf Valley. They identified geomorphological evidence interpreted as signifying a major flood caused by the failure of the terminal moraine, which was damming water to the north, at the Narrows between Long Island and Staten Island in the New York City area. Such a failure would allow glacial lakes Iroquois, Vermont, and Albany to drain rapidly southward, eventually reaching the Hudson continental shelf. Their estimate for the age of this event was 13.35 ka, coincident with the Intra-Allerød Cold period. At present, no direct evidence for paleosalinity changes is available from regions off the Hudson Estuary, and these will be required to confirm these inferences.

Gulf of Mexico

The GOM holds a unique place for LIS meltwater events because evidence from the 400-km^2 anoxic Orca Basin on the GOM continental slope led Broecker et al. (1989) to propose that diversion of southward-to-eastward drainage of meltwater triggered the YD. Emiliani et al. (1975), Kennett and Shackleton (1975), and Leventer et al. (1982) established that a link exists between GOM salinity—reconstructed from several proxies, notably isotopic records of planktonic foraminifera—and midcontinent deglaciation; this was later confirmed by Flower and Kennett (1990). Maximum flow occurred during the B/A, indicated by more negative (depleted) $\delta^{18}O$ values for *Globigerinoides ruber* (Broecker et al. 1989; Flower et al. 1990). High percentages of Cretaceous reworked calcareous nannofossils also indicate large river inflows during the B/A and even suggest a brief period of decreased flow within the B/A during the Older Dryas event ~14 ka (Marchitto and Wei 1995). Following the B/A, reduced flow is inferred from a sharp decrease in $\delta^{18}O_{foram}$ and reworked nannofossils during the YD. Marchitto and Wei also

FIGURE 9.10 Multiproxy record of Younger Dryas event from core HU90031-044, St. Lawrence Estuary: (A) Oxygen isotope seawater; (B) Mg/Ca (western Canada runoff); (C) U/Ca ratio; (D) strontium isotope ratio (ice sheet/bedrock, sea-surface temperature [SST]); (E) Sea-surface temperature (SST) fall during Younger Dryas. Magnesium, uranium, and strontium data are used as tracers of provenance, from Carlson et al. (2007a). The SSTs are derived from dinoflagellates and oxygen isotopes on *N. pachyderma*, from de Vernal et al. (1996). Note that the eastward drainage of Lake Agassiz water near 13 ka lowered salinity and temperature, increased transported Mg/Ca, and slightly later caused uranium and strontium changes. Data courtesy of A. Carlson.

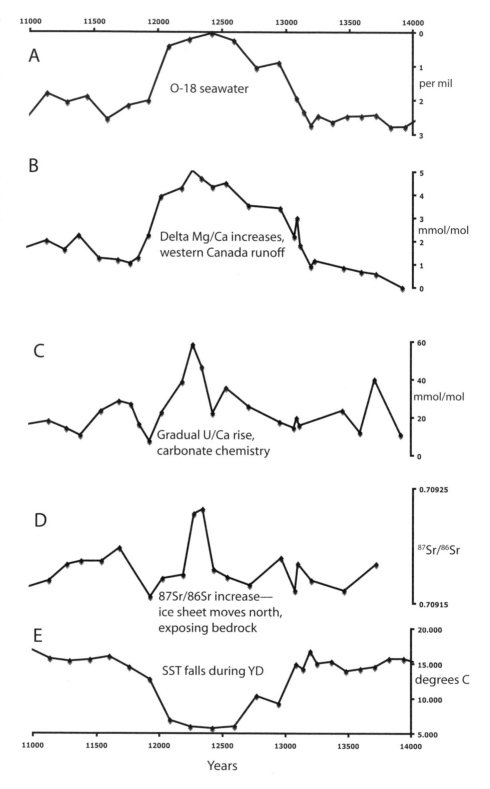

argued that following the YD freshwater discharge, meltwater did not return to the GOM.

Brown and Kennett (1998), Brown et al. (1999), Poore et al. (2003), and Aharon (2003) supported the GOM midcontinent link for the YD and other events during the deglacial interval. For example, Poore et al. (2003) concluded that a post-YD positive isotopic excursion at 9.0 and 7.8 ka radiocarbon years correlates with rerouting events R2 and R1 of Clark et al. (2001). Aharon's (2003) study is particularly interesting because it attempted to quantify freshwater flux to the world's oceans via the GOM for most of the deglacial period between 16 and 8.5 radiocarbon years. Aharon interpreted isotopic excursions in $\delta^{18}O$ of *G. ruber* in terms of meltwater floods that were separated by "pauses." Aharon developed an $\delta^{18}O$-based mixing model of fresh and marine end-member water to compute freshwater volume during the floods. He calculated that the floods were as brief as <100 years and as long as 1200 years. Three of the isotopic excursions signified discharges eight times the volume of the largest historic Mississippi River floods. "Superfloods" characterized the period before the YD and smaller events occurred after ~11.5 ka. Aharon recognized the PBO, YD, Older Dryas, Oldest Dryas, and H-1 stadials in his GOM isotope curves. Importantly, he also concluded there was an antiphasing between the GOM isotope depletions (i.e., Pause P-4 coincided with the YD), consistent with the idea that flow was rerouted through a more northerly route. Floods MWP-4 and 5 bracketed the YD indicating a return to GOM after the YD.

The GOM clearly contains some of the strongest evidence for large meltwater floods and periods of reduced flow as glacial lake and ice-sheet meltwater flowed through outlets other than the Mississippi. What remains an issue is the chronology and duration of floods. It is difficult to correlate the radiocarbon-based GOM sediment record with highly resolved paleoclimate records from Greenland ice cores and Cariaco Basin sediments. Sediments deposited at 20 cm per 1000 years, even when sampled at 1-cm intervals (Poore et al. 2003), yield at best a 50-year resolution without taking into account bioturbation. Moreover, the glacial geology of the midcontinent is complex, and radiocarbon dating there is subject to relatively large errors. Thus, it remains uncertain (1) whether meltwater events are coeval with ice margin advances (midcontinent stadials) or retreats (interstadial) and (2) whether each event was multidecadal or centennial in duration, as suggested by Aharon, or one-month to one-year events, as postulated by Teller.

Hudson Strait–Labrador Sea

Hudson Strait and Ungava Bay lie between Baffin Island on the north and Labrador on the south and occupy a key location for iceberg release during Heinrich events. This region is also critical as a potential route for the largest and final catastrophic deglacial meltwater event—the drainage of the merged final phase of glacial Lake Agassiz (Kinojévis lake phase of Teller et al. 2002) and Lake Ojibway about 8.5–8.2 ka. Barber et al. (1999) compiled several lines of evidence on the final phase of deglaciation from earlier studies (Andrews et al. 1995), and proposed a link between the drainage event and the 8.2-ka cooling known from Greenland ice cores (Alley et al. 1997), European lakes (von Grafenstein et al. 1998), and other paleorecords (Rohling and Pälike 2005). They estimated that the 2×10^{14} m^3 of glacial lake water, if released over 1, 10, or 100 years, would increase freshwater flux out through Hudson Strait and into the Labrador Sea by 6, 0.6, and 0.06 Sv, respectively.

The discovery of the 8.2-ka event in ice-core records (Alley et al. 1997) and glacial lake (Barber et al. 1999) records led to a number of investigations in search of its salinity signal across a large part of the eastern North American continental margin from the Hudson Strait all the way to the Hatteras slope of the mid-Atlantic United States. On the Cartwright Saddle on the Labrador Shelf, Andrews et al. (1999) discovered a large 1‰ depletion in planktonic foraminiferal $\delta^{18}O$ at ~8.8 ka. In contrast, they noted that isotope records from the shelf off Nova Scotia and in the Labrador Sea did not record this event. De Vernal and Hillaire-Marcel (2006, and their references) reviewed extensive literature from the northeast North Atlantic and found a possible reduced salinity excursion near 8.2 ka on the Orphan Knoll in the Labrador Sea.

In regions southward along the coast of eastern Canada, de Vernal et al. (1996) and Keigwin et al. (2005) found evidence for substantial surface ocean cooling in both the Laurentian Fan, off the mouth of the St. Lawrence, and the distal part of Cabot Strait. Sarnthein et al. (2003b) and Knudsen et al. (2004, 2008) found surface ocean cooling in the Barents and Iceland Seas, respectively, approximately coincident with the 8-ka event. Keigwin's study included multiple proxy methods with high temporal resolution. On the Laurentian Fan, they identified a distinct two-step cooling and a large drop in salinity beginning at 8.5 ka and lasting about 700 years. They also identified a sharp 1‰ isotopic depletion in $\delta^{18}O$ of *Neogloboquadrina pachyderma* (dextral) farther to the south off the Hatteras Slope. In reviewing the paleoceanographic evidence for the entire eastern North American region, they concluded that the Laurentian Fan and the Hatteras Slope showed a clear 8.2-ka signal but that the signal seemed to go unnoticed in the Labrador Sea.

The evidence for freshwater influx from off northeastern North America is thus somewhat mixed regarding its influence on salinity between about 8.5–8.2 ka. In addition to chronological uncertainty from carbon reservoir effects, there is also the possibility that the discharge event was so abrupt that it cannot be distinguished in ocean sediments using isotopes or dinoflagellate assemblages as proxies, or that the sampling interval or sedimentation rate was insufficient to capture it.

North Sea Region

Uncertainty from low sedimentation rates, sampling intervals, and salinity proxies are not a problem in the North Sea region. Bodén et al. (1997) provide some of the best oxygen isotopic evidence for rapid discharge of the Baltic Ice Lake as the EIS margin retreated northward from Mount Billigen (Figure 9.11). Their very high sedimentation-rate marine records reveal two events called BILL-1 predating the YD and BILL-2 less than 10 years later, superimposed on a general trend of decreasing background meltwater influx. This marine isotopic signature of a lake drainage event is unique in its temporal resolution and provides strong evidence for the abrupt nature of Baltic Ice Lake discharge.

Norwegian-Greenland-Barents Seas

Jansen and Veum (1990), Veum et al. (1992), Sarnthein et al. (1995), and Bauch and Weinelt (1997) all identified deglacial meltwater spikes from marine isotopic records that were distinct from those reflecting changes in SST and sea-ice cover and, linked with varying degrees of confidence, to Fennoscandian ice-sheet retreat (see Sarnthein et al. 2003a). Bauch and Weinelt (1997), for example, used foraminiferal oxygen isotopes to show that complex surface oceanographic changes occurred in the Norwegian Sea between 9.0 and 7.5 ka. They related isotopic excursions to surface-water stratification in response to the final stage of deglaciation of the Fennoscandian ice sheet, but a precise temporal correlation to glacial events is not yet established.

Arctic Ocean

The Arctic Ocean is also an important potential source of freshwater for several reasons. First, as discussed above, revision of Lake Agassiz history suggests that some drainage

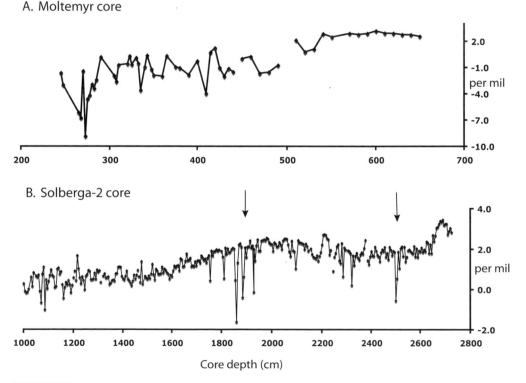

Baltic Ice Lake drainage events

FIGURE 9.11 Oxygen isotopic record from benthic foraminifera *Elphidium excavatum*, from (A) Moltemyr and (B) Solberga-2 cores from the North Sea, showing the hydrographic impact of Baltic Ice Lake Drainage events, from Boden et al. (1997). Arrows show two lake outbursts called BILL-1 and BILL-2. Data courtesy of J. Wright.

events may have been routed through the Mackenzie River system. Second, improved paleoceanographic records from the western Arctic Ocean suggest some deglacial meltwater influence on Arctic Ocean salinity. Third, freshwater export from the Arctic Ocean is important for deep-water formation in the Nordic Seas and, more broadly, for global MOC. Finally, modeling studies of North American isostatic history support the notion of a Mackenzie River–Arctic drainage route for the YD event (Tarasov and Peltier 2005).

In contrast to marginal marine regions adjacent to the continents, most paleoceanographic studies of the Arctic have focused on deglacial records from cores taken from submarine ridges and plateaus, where sedimentation rates average about 1 cm ka^{-1} (Backman et al. 2004). Nonetheless, Stein et al. (1994), Nørgaard-Petersen et al. (1998), and Poore et al. (1999) all show evidence for foraminiferal δ^{18}O depletions roughly coinciding with the onset of the YD. Along the Barents Sea margin, Wollenburg et al. (2004) also reconstructed surface ocean productivity changes during the YD from benthic foraminiferal records.

In one study of the Beaufort Sea margin, Andrews and Dunhill (2004) analyzed high-resolution cores from 405 m water depth along the North American margin of the Arctic Ocean. The core locations—west of the Mackenzie River delta in a water depth near the top of an inflowing, warm Atlantic water layer—makes these records potentially sensitive to both inflowing warm water and to freshwater discharged from North America. Stable isotopic values (δ^{18}O and δ^{13}C) indicated a distinct low-δ^{18}O event, captured in both the benthic and planktonic isotope records, possibly signifying the glacial Lake Agassiz outburst flood associated with the PBO cold event (Fisher et al. 2002). Benthic foraminiferal changes also suggest changes in mid-depth oceanography coincident with the 8.2 cold event.

Polyak et al. (2007) presented new IRD, faunal, and isotopic data from cores from the Chukchi margin, as well as a synthesis of evidence from other regions of the Arctic Ocean, including the Mendeleyev Ridge (Poore et al. 1993) and the Beaufort Sea (Andrews and Dunhill 2004). They argued there were at least two major meltwater events, one ~13 ka coincident with the YD onset and another ~11 ka coincident with the PBO-discharge event of Fisher et al. (2002). The multi-proxy evidence from several sites is fairly convincing that there was in fact significant influx of fresh meltwater from North America. However, it is still not possible to estimate the duration of these discharge events or to establish precise correlation to extra-Arctic paleoclimate records because of relatively low sedimentation rates and reservoir effects on radiocarbon dates.

In sum, there have been impressive gains in reconstructing the paleoceanographic response to freshwater discharges into marginal oceanic regions. It is clear that there is extraordinary potential for dating the timing and scale of freshwater discharges in high-sediment-rate regions near the freshwater sources.

Abrupt Change During the 8.2-ka Event

The response of surface and deep-ocean temperature, salinity, and circulation during abrupt glacial and deglacial climate transitions such as the YD was discussed in chapters 6 and 7 in this volume. We focus our discussion here on the quintessential abrupt climate event forced by glacial lake discharge, the 8.2-ka event, for several reasons. It occurred during the Holocene interglacial several thousand years after the northern-hemisphere insolation maximum. The volume of water discharged during the release of glacial Lake Ojibway-Agassiz is well constrained and much larger than that of any other North American deglacial lake event. Moreover, the age of this event is well constrained from both onshore and offshore geological studies in the region of Hudson Bay and Hudson Strait.

The atmospheric patterns surrounding the 8.2-ka event were established by Alley et al. (1997) from the GISP2 ice core, focusing on five proxies: δ^{18}O$_{ice}$ (air temperature), chloride (Cl$^-$, sea salt), calcium (Ca^{2+}, continental dust), methane (CH$_4$, moisture availability), and ammonium (wildfire-produced). All proxies indicated a dry, dusty, century-long cooling about half the magnitude of the YD event. Moreover, they identified the 8.2 event in the Dye, Greenland Ice Core Project (GRIP), and Camp Century ice cores (see Alley and Agustsdottir 2005)

In the years since Alley's paper, the literature on the 8.2 event has exploded. Intense scrutiny shows that the 8.2 event was actually a more complex series of events than simply a brief cooling. Rohling and Pälike (2005) statistically analyzed 17 high-resolution paleoclimate records spanning the early Holocene. They discovered that the 8.2-ka event was actually a sharp spike embedded within a geographically widespread cooling period that started at 8.6 ka and lasted between 400 and 600 years and that this event was manifested in low-latitude hydrological changes (see LoDico et al. 2006). Thus, the 8.2-ka spike is superimposed within a longer-term cooling interval. Another implication is that caution is necessary when interpreting paleoclimate reconstructions far from the North Atlantic region and immediate impact of the freshwater discharge. Abrupt events forced by lake discharges must be viewed in the context of Holocene climate variability such as documented by Bond et al. (1997, 2001) from North Atlantic SST and ice-rafting cycles, attributed to solar variability. Furthermore, Leverington et al. (2000) (see also Teller et al. 2002 and Teller and Leverington 2004) suggest the drainage of glacial lakes may have actually occurred in two pulses separated by only a century or two.

In a large data-modeling study, Wiersma and Renssen (2006) and Renssen et al. (2001) compiled 121 sites from the paleoclimate literature on the 8.2-ka event. By comparing the reconstructed atmospheric and oceanic conditions to model simulations forced by the drainage of glacial Lake Agassiz-Ojibway, they found that the reconstructed pattern fit that expected from the model. Thomas et al. (2007) re-studied the Greenland ice-core record of the 8.2-ka event in exceptional detail in four Greenland ice cores. They supported the 8150-ky [before 1950 to match radiocarbon chronologies] age estimate of Muscheler et al. (2004), concluding that the age of the oxygen isotopic minimum was just younger than 8200 years. The entire 160-year event encompasses about 15 m of Greenland ice in the composite depth scale, from 1325–1340 m below the surface. The duration and nature of the event based on the age model (Rasmussen et al. 2006) and the oxygen isotopic signature included a 160.5-year-long isotopic decrease within which there was a 69-year-long interval in which the $\delta^{18}O$ values were, minimum, more than one standard deviation from the mean. This study included many other proxy measurements for the 8.2-ka event, including glaciochemical records suggesting (although atmospheric temperature fell significantly over Greenland for ~160 years) that there was a relatively minor change in atmospheric circulation. For this reason, they concluded that the 8.2-ka event was not in these respects similar to changes in the atmosphere during the YD cooling, also attributed to large freshwater influx.

Whether MOC was affected by 8.2-ka lake drainage remains an unresolved issue. Bianchi and McCave (1999) noted increased sortable silt reflecting Iceland-Scotland Ridge overflow and stronger NADW, but during the entire Holocene, not just at 8.2 ka. In terms of the scale of the deep-ocean response, Keigwin and Boyle (2000) noted that deep-water circulation was reduced more during the Little Ice Age than it was by the 8.2-ka event. Oppo et al. (2003b) analyzed $\delta^{13}C$ values from Ocean Drilling Program (ODP) Site 980 and found reduced NADW formation at 9.3 ka and 8.0 ka. Hall et al. (2004) analyzed both $\delta^{13}C$ and sortable silt and identified only a small decrease in NADW during the 8.2 event, concluding that it was a fairly indistinct event. Muscheler et al. (2004) used ice core $\Delta^{14}C$ activity to infer that there was no major change in NADW during the 8.2-ka event.

By contrast, Ellison et al. (2006) provided evidence for a large drop in SST and IRD nearly simultaneous with the 8.2-ka event in a high-resolution core from the Gardar Drift. Their record included a 400-year-long slowdown of MOC marked by a decrease in sortable silt (Figure 9.12). This pattern indicated a slightly lagged and sustained MOC response following the freshwater forcing event. Kleiven et al. (2007) measured eight different proxies of surface and bottom circulation in a core from off the southern tip of Greenland, and confirmed a reduced density and perhaps flux of meridional overturning at the 8.2-ka event, which they estimated lasted about only about 100 years, with an abrupt beginning and a more gradual end. Their results, like those from other paleoclimate records, support model simulations (Clarke et al. 2004; Wiersma and Renssen 2006; LeGrande et al. 2006). In addition to decreased MOC, Cronin et al. (2007) identified a rapid two-step sea-level rise coincident within centuries of the changes in sortable silt and other deep-circulation proxies. This coincidence implied a delayed response of ice-sheet melting, possibly in Antarctica, to the initial freshwater forcing.

In sum, the largest Holocene climate anomaly involved significant cooling over the North Atlantic region, diminished deep-water ventilation of the North Atlantic by NADW for a century or so, and rapid sea-level rise within centuries. The event is almost certainly forced by an abrupt drainage of glacial Lake Ojibway-Agassiz in one or perhaps two pulses, but this lake outburst happened to occur during a multicentury-long cooling period similar to others during the Holocene interglacial. The global climatic response to the 8.2-ka event was more subdued than that in the North Atlantic region.

Tropical Forcing of Abrupt Events

We conclude this chapter with a brief discussion of the issue of regime shift in climate during the middle Holocene. Evidence outlined in Chapter 8 in this volume leaves little doubt that a major shift in the tropical climate system occurred about 5 ka, manifested by changes in ENSO variability, ITCZ positions, monsoon patterns, and extratropical climate. The main issue regarding the mid-Holocene shift is whether it represents an abrupt, semipermanent change in climate state, a threshold-like shift in the tropical atmosphere-ocean system, or just a step in a more gradual, secular progression of century-scale climate excursions related to either orbital forcing or solar variability. Unlike the situation of high-latitude freshwater forcing of MOC, the tropics do not have such an obvious trigger as massive proglacial lakes to induce such an abrupt change.

Nonetheless, several climate model simulations suggest that such an abrupt change can occur in the tropics. For example, one argument for a tropical mechanism causing abrupt climate reversals has two parts (Cane and Clement 1999; Clement et al. 2001). The first is that high-latitude freshwater forcing is not sufficient to explain the mismatch between

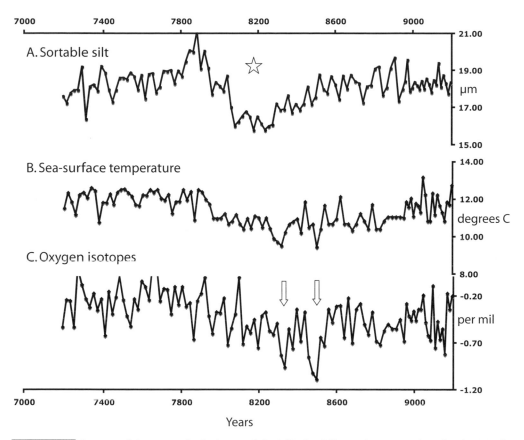

8.2-ka event in North Atlantic Ocean

A. Sortable silt

B. Sea-surface temperature

C. Oxygen isotopes

Years

FIGURE 9.12 Impact of abrupt 8.2-ka drainage of glacial Lake Ojibway-Agassiz in the subpolar North Atlantic, from Ellison et al. (2006): (A) Mean size of sortable silt, a proxy for deep-circulation strength; (B) reconstructed summer sea-surface temperature (SST); (C) $\delta^{18}O$ of seawater, a function of sea-surface salinity (SSS). Note the double spike in decreased SST and SSS (arrows), followed within one to three centuries by a drop in sortable silt (star).

some paleorecords and model simulations. Specifically, they cite inconsistencies between the timing of glacial lake discharge and the YD climate reversal. In particular the oceanic and climate response to YD-type freshwater forcing shows a primary impact in the North Atlantic region (e.g., Manabe and Stouffer 2000), not in the southern hemisphere, where some evidence exists for YD cooling. Certainly not all researchers would agree that such a discrepancy exists, given the antiphasing during deglacial millennial events, including the Antarctic Cold Reversal, B/A, and YD events.

The second part of the argument is that short-term climate variability in the tropics—ENSO as the prime example—has rapid global teleconnections that might occur over longer timescales. Building on the Zebiak and Cane (1987) ENSO model, Clement and colleagues simulated changes in ENSO during orbitally driven changes in tropical seasonal insolation for the last 500 ka. They found that ENSO

variability "flickers" on and off for century-long periods during regime shifts forced by the strength of the seasonal insolation signal. The tropical climate system essentially locks into a seasonal cycle, producing a climate state with no clear ENSO oscillation. This phenomenon of ENSO shutdown occurs when the earth's perihelion takes place during either the winter or summer season, which conveys an 11-ka periodicity to the abrupt changes, and eccentricity is weak. The duration might last from 30 to 309 years. By changing what is called the model's drag coefficient, a means of calculating the wind stress that affects the strength of atmosphere-ocean coupling, the model produced shutdowns as long as 1000 years. Thus, the simulated pattern is not unlike some reconstructed abrupt climate events during the deglacial and Holocene and demonstrates that the tropical Pacific can generate an abrupt response to gradual forcing by orbital insolation changes.

Pierrehumbert (2000) points out that the modern tropical air mass, consisting of the interconnected Walker and Hadley cell circulation systems, is buoyantly neutral so that small changes in moisture content or temperature in key regions might affect atmospheric convection. Relatively small changes in convection can in theory be amplified by water vapor feedback mechanisms like those during ENSO events. The equatorial Pacific thermocline, which is strongly affected in ENSO dynamics, is coupled to the world's ocean circulation over long timescales and might contribute to abrupt transitions emanating in the tropical upwelling. In this sense, concern about whether anthropogenic activity has upset ENSO patterns has a deeper significance because the major Pacific region climate shift around 1976 could be related to changes in the relative strength of Antarctic Intermediate Water.

There are two opposing views on the abrupt or gradual nature of the mid-Holocene change. Morrill et al. (2003) compiled 36 Holocene paleoreconstructions of the Asian monsoon, using stringent criteria imposed on the chronology and methodology of each study. Seventeen records had a sampling resolution from 100 to 200 years, fourteen from 10 to 80 years, and five were annual to decadal. They evaluated the widely recognized mid-Holocene monsoon shift statistically to judge whether it was abrupt or gradual. Their conclusion was that this shift constituted an abrupt climate change. The causes are still unclear and may reside in the high-latitude North Atlantic region ENSO-forced monsoon shift, perhaps from orbital insolation changes on seasonal ENSO dynamics (Clement et al. 2000), or in terrestrial vegetation-atmospheric-ocean interactions. Regardless, what distinguishes this view of abrupt change is that, unlike the situation for the 8.2-ka event, where many records indicate annual resolution, *abrupt* here refers to a change occurring over about a century or more.

Fleitmann et al. (2007) addressed the same topic of mid-Holocene changes in the Asian monsoon, reviewing 22 continental reconstructions, many of them highly resolved speleothem isotopic records. They came to the opposite conclusion, arguing that the mid-Holocene climate experienced a continuous southward migration in the ITCZ and decreasing monsoon precipitation due to decreasing solar insolation. They also found century-scale abrupt shifts in monsoon precipitation superimposed on this gradual shift. Kuper and Kröpelin (2006) also found a gradual mid-Holocene transition in Africa. Such a gradual climatic transition punctuated by frequent abrupt centennial-scale reversals is observed in the Indian monsoon (Hong et al. 2003), the Caribbean ITCZ (Haug et al. 2001b), and many other segments of the climate system (see Chapter 8 in this volume). The study of tropical climate variability is a relatively new field, and additional records should shed light on the patterns and causes of climate changes there.

Perspective

Concrete evidence for "abrupt" geological, glaciological, and hydrological events is pervasive in the geological record, no matter how one chooses to define the term *abrupt*. Large freshwater discharge events with well-constrained volumes entered the North Atlantic Ocean from proglacial lakes multiple times during the last deglacial and early Holocene. Some affected regional ocean salinity and probably surface ocean temperature, deep circulation, and ultimately high-latitude atmospheric temperatures, with varying impacts in other regions.

Abrupt climate change from tropical processes is a more complex matter. The issue is not whether extensive atmospheric and oceanic tropical changes occurred during abrupt climate transitions, as they certainly are now well documented. The questions are what critical thresholds were crossed, pushing climate into a new state, and what if any were triggering mechanisms. Uncertainty surrounding these issues may simply reflect the greater maturity of research on high-latitude glacial records, going back more than a century, and extensive climate modeling of freshwater forcing, but it does not diminish the potential for large-scale changes to emanate in the tropics.

LANDMARK PAPER Abrupt Climate Change from Catastrophic Glacial Lake Drainage

Broecker, Wallace S., James P. Kennett, Benjamin P. Flower, James T. Teller, Sue Trumbore, Georges Bonani, and Willy Wolfli. 1989. Routing of meltwater from the Laurentide Ice Sheet during the Younger Dryas cold episode. Nature 341:318–321.

How would society respond if global precipitation patterns changed abruptly, in only a few decades, affecting the world's agriculture and food supply? Or if the rate of sea-level rise increased tenfold over the current rate of 3–4 mm per year?

Definitions vary, but abrupt climate change refers to a situation where a threshold is crossed in a sensitive part of the climate system, such as the ocean's meridional overturning circulation, that is amplified rapidly through feedback mechanisms, with resulting global effects.

Many papers by Wallace Broecker of Columbia University's Lamont Doherty Earth Observatory on ocean chemistry, isotope dating, and geochronology could be considered landmark contributions to paleoclimatology. Among his most influential

(continued)

papers was the 1989 paper on catastrophic rerouting of glacial Lake Agassiz water from a drainage route along the Mississippi–Gulf of Mexico to one along the St. Lawrence Estuary–North Atlantic about 13 ka, causing the abrupt (within decades) inception of the Younger Dryas cooling episode. This paper was followed by a series of others in which Broecker and colleagues hypothesized that millennial-scale climate reversals were triggered by freshwater forcing of thermohaline circulation, the slowing down of what is called the global "conveyor belt" of ocean circulation, and the maintenance of a quasi-stable climate state for about 1000–2000 years through a mechanism called the bipolar seesaw.

The impact of the 1989 paper was large and immediate. It generated a worldwide search for evidence for the Younger Dryas and other abrupt climate reversals. More generally, it shifted emphasis from research on orbital-scale climate dynamics to suborbital timescales, especially the abrupt onset and termination of millennial events, which remain directly relevant to today as a reminder about the vulnerability of the climate system to abrupt changes.

10

Internal Modes of Climate Variability

Introduction

Quantifying the human fingerprint on climate and predicting future climate changes are two of the greatest environmental challenges facing society. There is an urgency to distinguish climate change caused by human activity, such as fossil-fuel burning and land-use change, from that caused by natural processes. In fact, there is an entire branch of climate science devoted to detection and attribution (D&A) of recent climate change (International Ad Hoc Detection and Attribution Group 2005). Detection deals with the identification of a trend and attribution addresses its causes.

As seen in Chapter 1 of this volume, climatologists divide causes of climate change into two categories: (1) external mechanisms, including greenhouse gases, aerosols, volcanic activity, and solar irradiance, which all affect the earth's radiative budget, and (2) internal processes that do not involve changes to net radiative balance. Internal climate variability refers to what is frequently called "unforced" climate changes, or "modes" of variability. These modes represent climate anomalies that originate in a particular region, such as the tropical Pacific Ocean, but often have significant hemispheric or global impacts.

Internal unforced variability raises serious problems for the D&A of climate change because, unlike changes to the earth's radiative balance, internal processes involve complex ocean-atmosphere dynamics interacting over various spatial and temporal scales that are only partially understood and, according to many experts, stochastic and unpredictable. In this sense, climate variability interferes with D&A efforts, making forecasting climate change due to external forcing even more difficult. This chapter describes instrumental records and paleoreconstructions of major modes of internal climate variability. The topics covered here serve as background for the next two chapters on climate changes over the last 2000 years, including those caused by human activity.

Approaches to the Study of Internal Climate Variability

Let us first look at internal climate variability from two perspectives. Climatologists who study the modern system refer to climate variability as fluctuations generated by dynamic interactions between the ocean-atmosphere system (see the discussion of the World Climate Research Programme [WCRP] in Busalacchi and Palmer 2006). The tropical Pacific's El Niño–Southern Oscillation (ENSO) is the prime example of an internal mode of variability originating in one region and having widespread global impacts or "teleconnections" (Bjerknes 1969). This type of interannual-to-decadal climate variability lies at the intersection of weather and more profound transient changes in mean climate state, described in earlier chapters.

Hunt (2006) called natural internal climate variability "that which occurs solely due to internal interactions within the climate system, without any external forcing whatsoever." Some researchers view at least certain modes of internal variability seen in instrumental records as red noise and unpredictable, what has been called the "climate noise paradigm" (Hurrell et al. 2003b). Others believe it is critical to attempt to model and predict even part of the variance associated with internal climate processes.

By contrast, from a paleoclimatological point of view, climate variability represents the high-frequency end of a continuum caused by any number of factors, such as tectonic, geological, or hydrological events (glacial lake drainage) (Dima and Lohmann 2004). The contrast in perspectives between modern climatologists and paleoclimatologists is important for several reasons. For example, as we see in the sections that follow, some internal modes involve oce-

anic thermohaline circulation (THC) changes as well as atmosphere and surface ocean interactions (Dickson et al. 2003; Hurrell et al. 2006). THC by its nature involves large-scale changes to freshwater budgets and occurs over decadal, centennial, and longer timescales. As a consequence, high-frequency variability blends interannual-to-decadal variability with lower-frequency modes of climate variability described in prior chapters.

The recognition of the ocean's role in internal climate variability is evident from observational programs, such as the World Ocean Circulation Experiment (WOCE) (Ganachaud and Wunsch 2000) and the Estimating Ocean Circulation and Climate Program (ECCO) (Wunsch and Heimbach 2006, 2007). ECCO addresses climate and sea-level variability over decadal timescales (Wunsch et al. 2007). Furthermore, paleoreconstructions show that modern internal modes do not persist over long periods, changing in amplitude and frequency—and even whether or not they exist—over the last millennium and longer. This point was illustrated in Chapter 8 for the case of Holocene ENSO, but applies also to other modes discussed later in this chapter. It is thus worthwhile to further examine some challenges surrounding internal climate variability that most pertain to paleoclimatology before discussing each particular mode.

Climate Variability Issues

A common set of complicating issues surrounds the study of various modes of internal climate variability. The first stems from the need to establish whether a certain climatic pattern is stochastic or whether there is an identifiable regularity, or periodicity, to atmospheric or oceanic variability. Wunsch (1999) defines stationarity as "having statistics that do not change in time," and the concept of stochastic behavior has formed the foundation for much of statistical hydrology. Ironically, it has recently been recommended that the concept of stationarity be abandoned in hydrology because of the overprint of human-induced climate change (Milly et al. 2008).

Stationarity is important in part because a central theme of climate research is to predict future climate behavior. Whether or not a mode of climate variability is stochastic is often approached through analyses of observational records and climate model simulations, though often with ambiguous results due to incomplete understanding of mechanisms, relatively short instrumental records, and stochastic climate behavior. Periodicity is most relevant to lower-frequency modes such as the 60–90-year pattern in the Atlantic Multidecadal Oscillation (AMO) (Schlesinger and Ramankutty 1994; Enfield et al. 2001) because these timescales intersect those of human-induced greenhouse gas forcing. As noted by Delworth and Mann (2000), from a statistical standpoint, observed ocean-atmosphere variability over the last few centuries includes some spectral peaks that sometimes exceed those expected from stochastic red noise.

Paleoclimatology of internal variability involves a search for what we refer to informally as persistence—that is, whether a particular mode of variability has been typical of Holocene interglacial climate both prior to and since instrumental records have been kept. Taking the example of ENSO, it represents an interannual process with quasi-cyclic El Niño events occurring every two to eight years, but decadal and multidecadal ENSO variations also occur and are also important for future climate patterns. Persistence of ENSO and other modes is important because the instrumental record is relatively short and paleorecords cover a period before the climate system was contaminated by human-induced greenhouse gas emissions and land-use changes.

Climate modelers have developed approaches to detect human influence on climate using patterns of internal variability generated artificially by computer simulations in conjunction with external forcing of climate. Simulated time series of unforced variability, called pseudo-proxies (Osborn and Briffa 2004), are used in lieu of empirical proxy data to analyze climate change, for several reasons, including the perceived inadequate spatial coverage of paleorecords, errors associated with proxies, and the mismatch between some proxy-based temperature reconstructions and expected patterns produced by some climate model simulations. In one study of surface air temperature, pseudo-proxy temperature time series generated by model simulations that were forced by external agents were used to evaluate empirical, but "noisy," tree-ring and other proxy-based temperature data (Mann and Rutherford 2002; von Storch et al. 2004). In the von Storch study, the discrepancy was large—tree-ring-based temperature reconstructions appeared to underestimate expected centennial-scale variations.

Another example involves the concern that human-induced warming may have raised sea-surface temperatures (SSTs) in tropical regions of cyclogenesis, leading to anomalous hurricane activity during the past few years. In an attempt to evaluate this problem, Santer et al. (2006) stated that the lack of a control experiment without human-induced greenhouse gas forcing means that experiments to quantify climatic noise "can be performed only with numerical models of the climate system." They generated unforced SST patterns in Atlantic and Pacific regions of cyclogenesis using 22 climate models and concluded that observed SST variations were larger than those produced by model-simulated unforced SST. They therefore rejected the hypothesis that "natural" internal variability "as simulated in current climate models" caused higher tropical SSTs. Using a complex D&A approach, Barnett et al. (2008) similarly concluded that human activity was responsible for up to 60% of the hydrologi-

cal changes in the western United States during the last few decades.

A related complication is the more general question of how external forcing influences internal variability and vice versa (Hunt 2006; Hunt and Elliott 2006). The importance of this issue should already be apparent from paleoclimate-modeling studies cited in prior chapters where attempts are made to simulate past climate conditions under different boundary conditions of geography, orbital insolation, solar irradiance, sea-ice area, atmospheric carbon dioxide, and other factors. Separating internal and external agents of climate change is also relevant to efforts to detect human influence on regional climate. In a study of the impact of the North Atlantic Oscillation (NAO) on European climate under elevated greenhouse gas conditions, Scaife et al. (2008) concluded that "great care is needed to assess changes due to modes of climate variability [like the NAO] when interpreting extreme events on regional and seasonal scales."

Another issue in the study of internal climate variability is that interactions between different modes of variability must be sorted out to achieve a better understanding of dynamic mechanisms and predictive capability. For example, the Pacific North American (PNA) mode is a decadal atmospheric oscillation emanating in the North Pacific Ocean and having large impacts on North American climate and weather. The PNA is influenced by processes originating in the tropical Pacific, so that in some cases, the positive phase of the PNA coincides with El Niño events and the negative phase with La Niñas (McCabe and Dettinger 1999). The PNA has been called the extratropical arm of ENSO (Hurrell et al. 2003b).

The interactions between the troposphere, where most internal variability modes are measured as indices of atmospheric pressure variations, and the overlying stratosphere and the surface ocean also pose complications. The stratosphere, for example, plays a role in short-term variability related to internal modes such as the Arctic Oscillation (AO) (Shindell et al. 1999) and is influenced by explosive volcanic eruptions through effects on radiative properties. Similarly, the ocean has a larger heat capacity and slower physics than the atmosphere, and is thus especially relevant to THC and NAO variability (Dickson et al. 2003). Both troposphere-stratosphere and troposphere-surface ocean interactions add complexity to multidecadal climate variability that cannot be addressed solely from short-term instrumental records.

In sum, D&A of recent climate change has, with a few exceptions, emphasized using climate model simulations, external radiative forcing time series, observational records, and computer-generated patterns of internal variability. It represents one approach to understanding climate change. It is worth noting the intense focus on external forcing rather than on empirical records of unforced climate variability.

Hunt (2006) states: "This reluctance amongst observationalists to acknowledge a contribution from internal climatic variability has resulted in an undue emphasis on the role of external forcing." In at least some segments of the community there is a sense that enhanced spatial and temporal coverage of proxy records of preindustrial internal variability, coupled with climate modeling, is essential to assess internal climate variability, in part because we cannot assume that the brief period covered by instrumental records is sufficient. With these points in mind, we now provide a brief survey of modern modes of climate variability before examining paleoreconstructions.

Indices and Terminology

Climatologists and oceanographers have a large lexicon of terms, abbreviations, and indices—calculated from atmospheric and oceanic measurements—to characterize internal modes of climate variability originating in various regions. Table 10.1 lists the most common indices, climatic patterns associated with each, and seminal papers on each internal mode. At the outset, it should be stressed that there is no "correct" way to define any particular mode of variability and that several options are available for each. Most indices are designed to capture regional climatic oscillations from measurements of surface atmospheric pressure (SAP), often expressed as the pressure difference, in millibars (hPa), between two regions that have different mean climatology. These regions are often called the "centers of action" for a particular mode, and long-term station records are used to compute the index. Sometimes midtroposphere pressure readings are used, but these do not extend as far back as SAP does.

In addition to station records, multivariate statistical analyses (usually principal components analysis) of atmospheric data generate what are called Empirical Orthogonal Functions (EOFs) (Hannachi et al. 2007). The leading EOF explains the most variance and is often used to study a particular mode. EOFs have the advantage over station records in their better spatial coverage of complex climate processes, but have the disadvantage of shorter periods of record, geographical complexity, and statistical issues regarding the orthogonality of principal components. However defined, climate anomalies originating in one region often influence weather and climate in other areas, including changes in atmospheric heat and moisture transport, wind strength, and the frequency of storms, among other things.

Climate anomalies due to internal variability often have their strongest manifestation and impacts during a particular season. For example, the NAO measures variability in the relative strength of atmospheric high- and low-pressure systems over the Azores and Iceland, respectively (Hurrell 1995). The Azores, located in the subtropical North Atlantic, typically experience high pressure and dry climate, the opposite of Iceland, but the strength of these patterns is reversed during periods when the NAO index is negative. The NAO index, expressed in mean sea-level pressure (SLP), is primarily a northern-hemisphere winter-season phenomenon, so winter atmospheric conditions are most commonly used to calculate the NAO index. NAO variability has a strong influence on European climate particularly during winter and spring seasons. Most indices plotted in figures here represent the particular season in which the mode of variability is most dominant.

As already mentioned in this chapter, most station records of SAP and SST go back less than 100 years; satellite records extend back only decades. SAP and SST records also diminish in spatial coverage before 1950, and even records covering the last few decades can have large spatial gaps. In some cases, SAP measurements from regions other than those originally designated can be used to extend instrumental records. Jones et al. (1999), for example, extended the NAO time series back to 1821 using observations from Gibraltar and southwest Iceland, and Luterbacher et al. (1999) used several sources of data back to 1675.

The limited temporal coverage of instrumental records means that they are by themselves sufficient neither to characterize decadal and multidecadal patterns nor to distinguish natural variability from human-induced changes. This situation has led to the need for developing paleo-indices of internal modes of climate variability. Table 10.2 summarizes several paleo-indices selected from a growing literature. Most studies develop a proxy index tied to modern modes of variability, either directly as paleo-indices of NAO, ENSO, and so forth are calibrated to instrumental records, or indirectly by virtue of their location in or near centers of action where a proxy might be sensitive to climate variability. The following sections outline each major mode of internal climate variability and long-term patterns in evidence from some of the more illustrative paleorecords.

El Niño–Southern Oscillation

The ENSO mode of variability originates in the tropical Pacific Ocean through dynamic interaction between the ocean and atmosphere causing an El Niño "warm" event about every two to eight years, with global climate repercussions. During El Niño events, anomalously warm SSTs occupy the central and eastern equatorial Pacific (EEP) and the thermocline in the EEP deepens, decreasing the normal east-to-west tilt (Figure 10.1). La Niña "cool" events are the opposite of El Niño events, occurring in years when winter-season SST warming is unusually weak, relatively cool ocean temperatures from upwelled water characterize the EEP, and the thermocline tilt is steeper than normal. La

TABLE 10.1 Internal Modes of Climate Variability*

Name—Index	Abbreviation	Location	Calculation	Frequency, Pattern	Instrumental Record*	Comment	References
Southern Oscillation Index	SOI	Tropical Pacific	MSLP, Tahiti, or Easter Island minus Darwin, Australia	2–8-yr	mid-1800s	Atmospheric ENSO index	See text, many
North Atlantic Oscillation	NAO	N. Atlantic Ocean	Monthly or seasonal SLP, Iceland-Azores	Decadal, multidecadal	Early-mid 1800s	Mainly winter influence	Wallace and Gutzler 1981; Hurrell 1995
Pacific Decadal Oscillation	PDO	N. Pacific Ocean	NPI = MSLP in Gulf of Alaska, an index of N. Pacific SST N. 20°N	multidecadal	Since 1900	Also NPI	Trenberth and Hurrell 1994; Mantua et al. 1997
Southern Annular Mode	SAM	High southern-hemisphere latitudes	Difference in MSLP in southern hemisphere middle (~45°S) and high (~65°S) latitudes or amplitude of leading EOF S of 20°S			Also Antarctic Oscillation and Annular Mode	Thompson and Wallace 2001; Marshall 2003
Arctic Oscillation/ N. Annular	AO	Arctic-high northern-hemisphere latitudes	Leading EOF of winter monthly northern hemisphere MSLP north of 20°N	Decadal	Since 1899	Also Northern Annular mode	Thompson and Wallace 1998; Quadrelli and Wallace 2004
Quasi-Biennial Oscillation	QBO	Pacific	Stratospheric zonal winds				Holton and Lindzen 1972; Rasmussen et al. 1990
Pacific N. American	PNA	N. Pacific	Mean normalized 500hPa height, at 20°N, 160°W and 55° N, 115°W–45°N, 165° W and 30°N, 85°W				Wallace and Gutzler 1981; Halpert and Smith 1994
Atlantic Quasi-Decadal mode	QD	Atlantic					Deser and Blackmon 1993; Dima et al. 2001
Atlantic Multidec-adal Oscillation	AMO	Atlantic Ocean	SST	Multidecadal	Since 1856		Kaplan et al. 1998; Enfield et al. 2001
NINO-1-4	NINO-1-4	Tropical Pacific	SST	3–7-yr	NINO-3/4 since 1854	Ocean ENSO index	Smith and Reynolds 2004
Interdecadal Pacific Oscillation	IPO	Midlatitude Pacific, S. Pacific Conver-gence Zone	SST			Linked to ENSO	Power et al. 1999; Folland et al. 1999, 2002
El Niño-Southern Oscillation	ENSO	See SOI and NINO-1-4					

Source: Bridgman and Oliver 2006 for review. See http://iri.columbia.edu/climate/ENSO/background/monitoring.html and http://www.cdc.noaa.gov/Pressure/Timeseries/.

Note: EOF Empirical Orthogonal Function; MSLP = mean sea-level atmospheric pressure; NPI = North Pacific Index; SST = Sea-surface temperature. NINO1+2 (0–10°S, 80–90°W): SST warming early in an El Niño event. NINO3 (5S–5°N; 150W–90°W): largest variability in SST associated with ENSO variability. NINO3.4 (5°S–5°N; 170°W–120°W): large SST variability during ENSO events is often used as the ocean index of ENSO. NINO4 (5°S–5°N: 160°E–150°W): SST reaches 27.5°C.

*Records extend up to 1990s or 2000s depending on study; older intrumental records are available from some regions but do not reveal spatial patterns.

TABLE 10.2 Paleo-Indices of Internal Modes of Variability*

Mode	Time Interval (A.D.)[†]	Proxy	Pattern, Notes	Reference
AO	1650	Tree rings, temperature, and SLP		D'Arrigo et al. 2003
NAO	1648	Greenland ice accumulation rate		Appenzeller et al. 1998
NAO	140	Tree rings		Cook et al. 2002
NAO	1000	Greenland ice core glaciochemical	10.4, 62-yr, weak in LIA	Fischer and Mieding 2005
NAO	1200	Cariaco planktonic foraminifera	60–70-yr	Black et al. 1999
NAO	1500	Multiple	Seasonal and annual	Luterbacher et al. 1999, 2002
NAO	1860	Greenland ice accumulation		Mosley-Thompson et al. 2005
NAO	1890	Coral geochemistry		Rosenheim et al. 2005
AMO	600	Tree rings	40–120-yr, 60–70-yr	Gray et al. 2004
PDO	1768	Tree rings and corals	Corals	Evans et al. 2001a, b; Linsley et al. 2000b
PDO	1100/1358–2001	Tree rings		D'Arrigo et al. 2001, 2005a; D'Arrigo and Wilson 2006
PDO	1661	Tree rings	ENSO decadal variability	Biondi et al. 2001
PDO	1840	Tree rings and corals	20 and 50–70-yr	Gedalof and Mantua 2002
PDO	1840	Coral oxygen isotopes		Urban et al. 2000
PDO	1565	Tree rings		D'Arrigo and Wilson 2006
PDO/ENSO	1893/1726	Corals		Linsley et al. 2000a, b
ENSO	1408–1978	Tree rings for NINO-3 SSTs	2–8-yr, low amplitude during LIA	D'Arrigo et al. 2005
ENSO	1000	Corals		Cobb et al. 2003
ENSO	1880–1995	Mg/Ca ratios	5–6-yr, decadal	Cronin et al. 2003
ENSO	1607–1990	Coral oxygen isotopes	Changing ENSO frequency	Evans et al. 2002
ENSO	1600	Coral oxygen isotopes		Dunbar et al. 1994
ENSO	1708–1884	Coral oxygen isotopes		Linsley et al. 1994
ENSO	1657–1992	Coral oxygen isotopes		Quinn et al. 1998

(continued)

TABLE 10.2 (*continued*)

Mode	Time Interval (A.D.)[†]	Proxy	Pattern, Notes	Reference
ENSO	1635–1957	Coral oxygen isotopes and radiocarbon		Druffel and Griffin 1993
ENSO	>100 years	Coral geochemistry		Gagan et al. 2000
ENSO	1893–1989	Coral oxygen isotopes		Cole et al. 1993
ENSO	~1400	Tree rings-NINO-3/4 SST	Multiple interannual peaks	Cook unpublished; D'Arrigo et al. 2005b
SOI	1706–1977	Tree rings	Winter season	Stahle et al. 1998b
ENSO	1400/1650–1980	SST-NINO-3		Mann et al. 2000, 2005
IPO	1726, 1780	Coral geochemistry	Major reorganization ~1880	Linsley et al. 2004

Note: AMO = Atlantic Multidecadal Oscillation; AO = Arctic Oscillation; ENSO = El Niño–Southern Oscillation; IPO = Interdecadal Pacific Oscillation; LIA = Little Ice Age; NAO = North Atlantic Oscillation; PDO = Pacific Decadal Oscillation; SOI = Southern Oscillation Index; SLP = sea-level pressure.

*See each paper for a detailed discussion of patterns of variability. Some studies attempt to separate external forcing from internal variability—see Chapter 12.

[†]The record extends from the date indicated until about 1980 to 2005, depending on the study.

Niña events have also been called El Viejo. Neutral years are those in which neither strong El Niño nor La Niña conditions develop.

History of ENSO

South American fishermen named the annual arrival of the warming of ocean currents off northwestern South America during Christmas as El Niño, or La Coriente del Niño. This phenomenon reflected suppressed upwelling of cold, nutrient-rich subsurface waters due to weaker northeastern trade winds. Every few years, ocean warming was more intense, earlier than usual, or both; upwelling was suppressed; and warm, nutrient-depleted surface ocean waters reached coastal regions, disturbing pelagic marine ecosystems and affecting fish harvests. This oceanic event came to be called El Niño.

The Southern Oscillation is an atmospheric phenomenon discovered through observations of weather anomalies that also affected society. At the turn of the 19th century into the 20th, Great Britain's Gilbert Walker investigated the periodic catastrophic breakdown of the Asian summer monsoon and drought conditions that affected the Indian

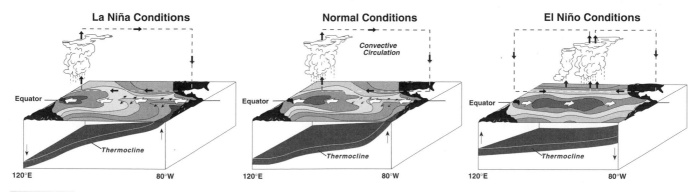

FIGURE 10.1 Ocean and atmosphere in the tropical Pacific during La Niña, Neutral, and El Niño conditions, from the Pacific Marine Environmental Laboratory, National Oceanographic Atmospheric Administration. Courtesy of M. McPhaden.

subcontinent's agricultural production (Rasmussen and Carpenter 1982; Philander 1990). In typical years, the summer monsoon brings extensive rainfall to the Indian subcontinent; conversely, high surface pressure typically dominates the summer atmosphere in the EEP and off the Pacific coast of South America.

Walker discovered a periodic flip-flopping of sea-level pressure in the tropics when typical weather patterns like the Indian monsoon reversed themselves (Walker 1924; Walker and Bliss 1932). High pressure typical of the eastern Pacific moved westward, and low-pressure systems over Africa and Australia and including the Indian subcontinent shifted eastward (Figure 10.1). During the 1899–1900 El Niño, the Indonesian-Asian monsoon low-pressure system was suppressed, leading to drought, while eastward migration of low pressure brought heavier-than-usual rains to coastal South America. Walker's discovery is known as the Southern Oscillation.

Bjerknes (1966, 1969) established the link between the ocean's El Niño and the atmosphere's Southern Oscillation. He recognized the dynamic connection between ocean and atmosphere that was the result of changes in wind stress, which influenced equatorial upwelling and thermal ocean structure. Bjerknes named the zonal tropical atmospheric pattern Walker Circulation and concluded that the normal pattern of Walker cell circulation weakened or collapsed during El Niño events, and precipitation diminished in the western Pacific and increased in the eastern Pacific. Decreased oceanic upwelling from weaker trade winds led to high SSTs in the EEP as surface ocean waters that had been pushed into the western equatorial Pacific flowed eastward (Wyrtki 1975). Walker and Bjerknes also both recognized that interannual climatic anomalies influenced extratropical climate through teleconnections. Additional historical aspects of the discovery of ENSO and its dynamics can be found in Philander (1990).

Measurements of ENSO

There are many ways to measure ENSO climate variability. The two most common are the Southern Oscillation Index (SOI) and the NINO ocean SST index (sometimes called Niño, with lowercase letters). The SOI is the difference between atmospheric pressure around Darwin, Australia and either Tahiti or Easter Island in the central Pacific Ocean. Negative SOI values designate a tendency toward El Niño events, positive values La Niña, and values near zero are neutral conditions. The SOI successfully captures large-scale seasonal and interannual changes in the strength of zonal Walker cell and associated precipitation anomalies. SST measurements from regions in the tropical Pacific called NINO-1, 2, 3, and 4 (from east to west) are available from an integrated network of buoys and satellite monitoring systems. NINO-3.4 refers to the eastern and western halves of NINO-4 and NINO-3 regions in the central tropical Pacific from 5°S to 5°N latitude, where SST variability is evident.

ENSO Patterns and Processes

ENSO Events The generic El Niño event begins in the summer or early fall and develops fully during the boreal winter season and usually subsides the following spring and summer (Philander 1983). This is why an El Niño or La Niña event is usually designated by two years, such as the 1982–1983 event. Typical events last a total of roughly 18 months. During the onset phase, the Intertropical Convergence Zone (ITCZ) migrates, bringing heavy rains to regions that are typically arid regions, and SSTs in the eastern Pacific increase during the late summer or fall season. At its peak, precipitation is heavy in the central equatorial Pacific during November through January, northeast trade winds weaken, warm western Pacific surface water migrates eastward, upwelling weakens, and the thermocline loses its steep west-east gradient as the eastern Pacific thermocline deepens. Sea-level height varies significantly across the Pacific during ENSO events, reflecting low-frequency, low-amplitude eastward-propagating Kelvin and westward-propagating Rossby equatorial waves (Wyrtki 1973, 1975). The demise of an El Niño event occurs as SSTs fall and strong trade winds resume in the eastern Pacific and neutral conditions return to the entire Pacific Ocean.

Each El Niño event is also distinct in detail from others. This may reflect the fact that ENSO climate anomalies occur within the context of a continuum of weather-climate variability, from high-frequency seasonal and Madden-Julian Oscillations to low-frequency centennial-scale changes. Another factor is that defining an El Niño and La Niña event is a subjective exercise resting on whether one uses the atmospheric SOI or the oceanic NINO-1-4 index and what particular season one analyzes (Rasmussen et al. 1990; Halpert and Ropelewski 1992).

ENSO Teleconnections ENSO originates as a tropical ocean-atmosphere disturbance that leads to global weather anomalies (Ropelewski and Halpert 1986, 1987), schematically diagrammed in Figure 10.2. In addition to the failure of the Asian monsoon and the production of a warmer, wetter central and eastern Pacific, ENSO-driven anomalies include summer drought in parts of Australia, warm winters in parts of North America, wetter winters in the southeastern United States, and summer warmth in parts of South America, to mention a few.

ENSO Temporal Patterns The temporal patterns and frequency of El Niño events are the subject of an extensive

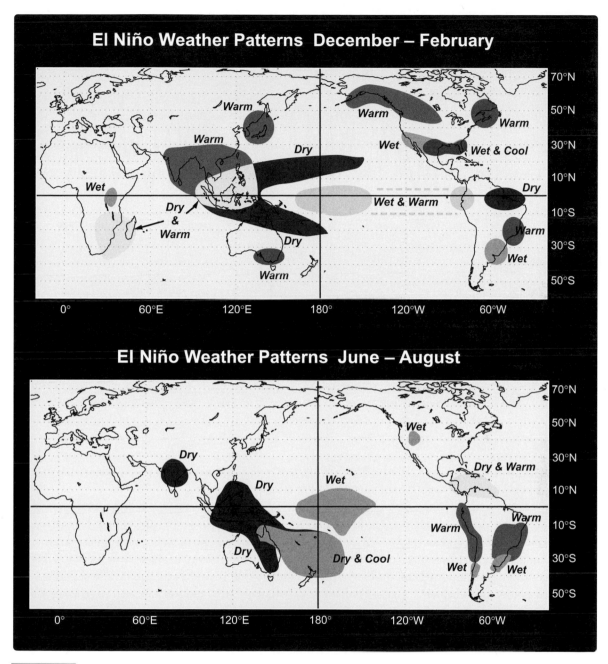

FIGURE 10.2 Global impacts or teleconnections of El Niño events during northern-hemisphere winter and summer season. Courtesy of the Pacific Marine Environmental Laboratory, National Oceanographic Atmospheric Administration, and M. McPhaden.

literature. Figure 10.3A shows the NINO-3.4 index back to the late 19th century. Simple inspection of this curve reveals a quasi-periodic pattern, with El Niño events occurring every few years. Figure 10.4A shows variability in the SOI index back to the 1860s. Both records show the classic two- to eight-year quasi-periodicity of ENSO extremes in both atmospheric pressure and ocean temperature. A trend toward more intense ENSO variability since the 1960s is also evident in both the NINO3.4 and SOI indices, as are the strong individual 1982–1983 and 1997–1998 events, which caused

global temperature and precipitation anomalies, flooding, drought, anomalous fires, disease, billions of dollars in property damage and agricultural losses, and the loss of many lives (McPhaden 2004).

ENSO Causes The causes of modern ENSO climate variability stem from internally generated dynamics between the ocean and the atmosphere (Cane 1986; Trenberth and Shea 1987). Although some studies have suggested a possible relationship between ENSO and external forcing mechanisms,

ENSO sea-surface
temperature variability

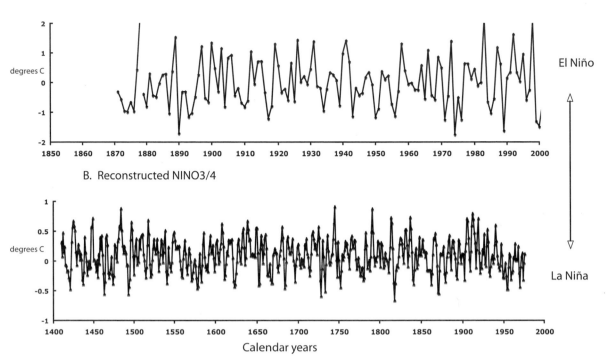

FIGURE 10.3 (A) NINO-3⁄4 sea-surface temperature (SST) index for the central tropical Pacific, from Rayner et al. 2003, Hadley Met Office. (B) Reconstructed NINO-3⁄4 index from tree rings, from Cook (2000, data available in NOAA World Datacenter for Paleoclimatology) and D'Arrigo et al. (2005a). Warm (cool) SSTs occur during El Niño (La Niña) events. Arrows point to strong 1982–1983 and 1997–1998 El Niño events.

as currently understood, ENSO dynamics are not directly triggered by external forcing mechanisms. Volcanic, solar, and greenhouse gas forcing might nonetheless influence patterns of ENSO variability (see chapters 11 and 12 in this volume).

The variability seen in SOI, NINO-3.4 indices, and documentary records raises the question of whether there is a random element to ENSO variability, due either to chaos associated with the deterministic aspects of ENSO or to noise in the climate system. This topic has been widely discussed for years. Diaz and Kiladis (1992), for example, stressed that the occurrence of ENSO events can be quite irregular, such as the absence of El Niño during much of the 1940s. From 1990 through 1994, El Niño conditions persisted for several consecutive years. It continues to be studied today in the context of anthropogenic forcing of ENSO (Trenberth et al. 2002; Mann et al. 2005a). We see below that ENSO varies over decadal and centennial timescales in addition to its irregular interannual recurrence interval.

ENSO Prediction   Modeling and prediction of ENSO climate variability pioneered by Cane (1986) and Zebiak

and Cane (1987) has since grown into a large, active field of applied climatology (Trenberth et al. 2002). Barnston et al. (1999) and Landsea and Knaff (2000) discuss the performance of ENSO models in predicting the large 1997–1998 El Niño, and Chagnon (1999) provides an in-depth look at its impacts.

ENSO climate variability thus encompasses more than the simple interaction of the tropical atmosphere-ocean system. McPhaden et al. (2006) commented that "ENSO is unique among climate phenomena in its strength, predictability, and global influence, projecting beyond the tropical Pacific," emphasizing ENSO impacts on biological and societal systems, as well as its physics.

Proxies of ENSO

Coral skeletal geochemistry and tree rings are the primary methods used to reconstruct paleo-ENSO. Three aspects of coral biology make them useful to study ENSO: (1) their colonial, sedentary life history allows them to record decades to centuries of ocean conditions; (2) coral skeletons have annual growth bands, sometimes with seasonal ocean signals;

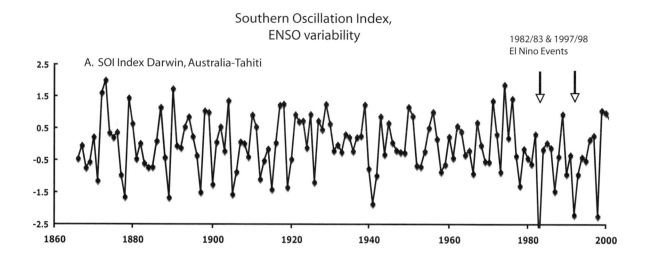

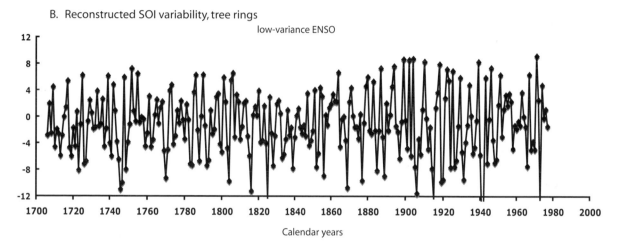

FIGURE 10.4 (A) Southern Oscillation Index (SOI) from National Oceanographic Atmospheric Administration as calculated by Ropelewski and Halpert (1987). (B) Tree-ring reconstruction of SOI-related El Niño–Southern Oscillation (ENSO) activity, from Stahle et al. (1998b). Note the decadal and lower-frequency patterns.

and (3) coral skeletal chemistry is affected by environmental factors influenced by ENSO activity (Quinn and Sampson 2002; Corrège 2006).

Oxygen isotopes ($\delta^{18}O$) of corals are valuable for recording SST and sea-surface salinity (SSS) (Fairbanks and Dodge 1979; Druffel 1985; McConnaughey 1989; Carriquiry et al. 1994; Dunbar et al. 1994; Linsley et al. 1994; Leder et al. 1996; Wellington et al. 1996; Cole et al. 2000). Coral skeletal $\delta^{18}O$ is controlled by temperature and isotopic composition of the seawater as well as biological factors. If interannual SST variability is strong and seawater $\delta^{18}O$ variability minimal, then $\delta^{18}O$ as a paleothermometer is most valuable. Gagan et al. (1994) reported an isotope-temperature relationship of 0.18‰ per 1°C (see also Gagan et al. 2000). In studies on the genus *Porites* from Caño Island, Costa Rica, Carriquiry et al. (1994) found that a strong isotopic depletion characterizes skeletal isotope values from El Niño events,

including the 1982–1983 El Niño. In contrast, interannual SST variability in the Gulf of Chiriqui off Panama is only 1–2°C but $\delta^{18}O$ variability reflects seasonal ITCZ and rainfall variability (Dunbar and Wellington 1981; Linsley et al. 1994). In addition to SST and SSS, biological factors are important in coral colonial growth (McConnaughey 1989; Lough and Barnes 1990, 1992; Wellington et al. 1996).

Carbon isotopes in coral skeletons ($\delta^{13}C$) yield a more ambiguous signal for ENSO variability (Swart et al. 1996b). Coral $\delta^{13}C$ is related to endogenous (coral ecology, physiology) and exogenous (e.g., light intensity; see Carriquiry et al. 1994) factors. Photosynthetic activity of the zooxanthellate algae that live symbiotically in the corals is one important factor (Fairbanks and Dodge 1979; Swart 1983; De Villiers et al. 1995; Wellington and Dunbar 1995). Because solar insolation is partially a function of sunlight and clouds, it can affect carbon isotopic values through its impact on photosynthesis and can thus affect coral tissue and skeleton $\delta^{13}C$. Shen et al. (1992) showed an inverse correlation between the $\delta^{18}O$ and $\delta^{13}C$ in corals from Punta Pitt in the eastern Galapagos Islands, and the carbon isotopes were considered related to cloud cover during rainy and dry seasons. Turbidity due to runoff, resuspension, upwelling, and plankton blooms may also affect coral carbon isotopes variability (Wellington and Dunbar 1995; Carriquiry et al. 1994).

Elemental concentrations of strontium (Beck et al. 1992; Cohen et al. 2002; Linsley et al. 2004), cadmium (Shen and Boyle 1988; Reuer et al. 2003), manganese (Shen et al. 1991; Delaney et al. 1993), barium (Lea et al. 1989; McCullough et al. 2003), uranium (Shen and Dunbar 1995; Hendy et al. 2002), and magnesium (Mitsuguchi et al. 1996) in coral skeletons have also been used as proxies in ENSO studies (see Sinclair 2005). The literature is too extensive to review here and progress is rapid (Lea et al. 2003), but suffice it to say that coral skeletons are most useful for ENSO reconstruction, using combined Sr/Ca ratios $\delta^{18}O$ records sometimes augmented by radiocarbon activity ($\Delta^{14}C$) (Gagan et al. 2000). Multiple factors, including biological processes, present challenges to the development of quantitative ENSO indices (Sinclair et al. 2006; Grottoli and Eakin 2007). In addition to coral and tree-ring proxies (see the subsection that follows), laminated sediments such as those found in ocean-margin basins with laminated sediments can also be useful in paleo-ENSO studies (Bull et al. 2000).

Paleoreconstructions of ENSO

Coral records of ENSO variability over the last century are found in a number of classic studies by Shen et al. (1991), Cole et al. (1993), Linsley et al. (1994, 2000a), Dunbar et al. (1994), and Quinn et al. (1998), among others. Several review papers on coral records of ENSO have been written over the last 15 years. As of 1993, Dunbar and Cole (1993) counted about 92 different coral records of varying length. A few years later, Gagan et al. (2000) listed 15–20 coral records that extended back to the mid-1800s. Grottoli and Eakin (2007) counted 27 coral $\delta^{18}O$ and 26 coral $\Delta^{14}C$ records extending back 30 and 20 years, respectively, with a common overlap period of 1962 to 1979. They concluded that the influence of ENSO activity on SST, SSS, or both is reflected in coral $\delta^{18}O$ records and that skeletal $\Delta^{14}C$ activity has an unrealized potential to help researchers understand oceanic carbon uptake during ENSO events.

The study by Dunbar et al. (1994) of the Urvina Bay, Galapagos *Pavona* coral oxygen isotope record was an influential reconstruction because the Galapagos Islands are located in the center of action of ENSO activity, the $\delta^{18}O$ recorded mainly SST variability, and the coral colonies extended back to the year 1587. Dunbar's group discovered

that coral $\delta^{18}O$ patterns generally match the historical reconstruction of Quinn et al. (1987), exhibit interannual SST variability of 1–2.5°C, and document cool intervals during the Little Ice Age in the 1600s and 1800s and a primary 4.6-year periodicity with other frequencies from 3.3 to 34 years. Many studies (e.g., Urban et al. 2000) since have shown that decadal ENSO variability is typical in corals, reflecting its nonstationarity.

Tree-ring records, sometimes combined with coral reconstructions, have also been useful in ENSO reconstruction. Figures 10.3B and 10.4B show ENSO variability from tree-ring proxy records of SST and precipitation from Ed Cook (unpublished 2000; see also D'Arrigo et al. 2005a and Stahle et al.1998b). The NINO-3 reconstruction in Figure 10.3B is based on 175 tree-ring records (mainly of tree-ring width) from the southwestern United States and Mexico. In addition to the typical two- to eight-year periodicity, the tree-ring record shows low-amplitude ENSO activity during much of the Little Ice Age. With the exception of the mid-19th century (Mann et al. 2005a), the general coincidence between the tree-ring and other ENSO records from around the world suggests a first-order persistence of ENSO teleconnections over the past few centuries, although the frequency and intensity of ENSO events varies. Cook, D'Arrigo, and colleagues also supported the hypothesis that external radiative forcing can influence ENSO activity, as was suggested by paleoclimate (Cobb et al. 2003) and modeling studies (Mann et al. 2005a). Stahle's reconstruction in Figure 10.4B reflects the large influence of ENSO on the western United States, clearly showing low variance during the mid-19th century, increased ENSO activity during the 20th century, and interannual variability like that in the instrumental record shown in Figure 10.4A.

Long-Term ENSO Records

We saw in Chapter 8 that evidence from lakes and marine-sediment cores suggest that major changes in ENSO occurred during the last 11,000 years of Holocene climate evolution. Many other studies infer "ENSO-like" behavior over longer timescales (Sandweiss et al. 1996; Rodbell et al. 1999). Some examples include deglacial-Holocene loess-soil carbonate carbon isotopes (Wang et al. 2000), corals from the last interglacial (Hughen et al. 1999; Tudhope et al. 2001), cold glacial-age stadial periods (Stott et al. 2002), last glacial maximum SST (Koutavas et al. 2002), Eocene climate modeling (Huber and Caballero 2003), deglacial precipitation in New England (Rittenour et al. 2000), and mid-Pliocene global warmth (Molnar and Cane 2002; Federov et al. 2006), to mention a few. Although past patterns sometimes resemble those of modern ENSOs, uncertainty about the dynamics of long-term variability and its causes means that they do not necessarily reflect a deterministic internal mode of variability and that much additional research is needed on ENSO prior to the last few centuries (Trenberth and Otto-Bliesner 2003; Cane 2005).

ENSO from Documentary Records

Documentary evidence for ENSO events suggests nine strong and 14 moderate El Niño events over the past century and 82 events between 1607 and 1953 (Quinn et al. 1978, 1987; Quinn 1992). Whetton and Rutherford (1994) and Whetton et al. (1996) analyzed eastern-hemisphere teleconnections from Javan tree rings, Nile River flooding, droughts and famine in India, and rainfall in northern China. Although the confidence in the record for specific years declines rapidly before the late 1700s, Quinn's records remain the standard against which paleoclimate records can be compared (Whetton et al. 1996). In general, their results supported the SOI and NINO-3.4 instrumental records of variability in ENSO the past few centuries.

Pacific Decadal and Interdecadal Pacific Oscillations

Pacific Decadal Oscillation Instrumental Record

Another mode of climate variability is the Pacific Decadal Oscillation (PDO). The PDO was first identified, through the relationship between decadal SSTs and oceanic ecosystem variability (Alaskan salmon production), by Mantua et al. (1997). Mantua notes that graduate student Steven Hare originally coined the term *PDO* in 1996 (Hare et al. 1999). The PDO is defined in the North Pacific Ocean by the leading EOF of SST north of 20°N latitude, and PDO-related SST variability occurs in concert with atmospheric pressure changes in the North Pacific. Trenberth and Hurrell (1994) named this decadal-to-interdecadal atmospheric mode the North Pacific Index (NPI) and defined it as mean sea-level pressure in the Gulf of Alaska region where the Aleutian low-pressure system is centered.

Figure 10.5 shows the Mantua-Hare PDO index back to 1900. The most characteristic features are the 20–30-year variability and the notable change from negative to positive values in the PDO around 1976–1977. This change is considered by many to signify a major regime shift in regional and perhaps hemispheric or global climate, perhaps related to a change in ENSO variability or other processes (Trenberth and Stepaniak 2001; Mestas-Nuñez and Miller 2006). In addition to the 1976–1977 transition, Hare and Mantua (2000) also identified a regime shift in Pacific ecosystems in 1989 (see Grebmeier et al. 2006; Litzow 2006).

The 20th-century PDO mode has some similarities to ENSO in the sense that they both have an impact on North American climate. They differ in that (1) the instrumental record of the PDO's predominant timescale is 20–30 years, whereas it is ~18 months for ENSO; and (2) the primary PDO impacts occur in the midlatitude to high-latitude Pacific region as opposed to ENSO's main impact on the tropical Pacific. The PDO is also linked to Pacific-wide interhemispheric Interdecadal Pacific Oscillation (IPO) variability (see the subsection "PDO and IPO").

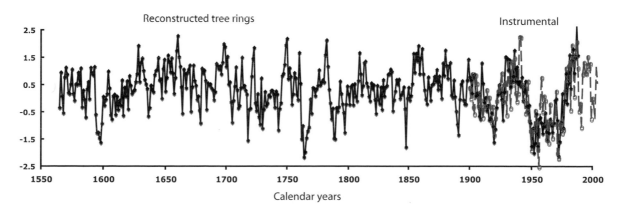

Spring Pacific decadal oscillation

FIGURE 10.5 Plot of Pacific Decadal Oscillation (PDO). The solid line is reconstructed from tree rings, the dashed line from instrumental records (D'Arrigo and Wilson 2006). The dashed line is the measured PDO, from Mantua et al. (1997). The shift at 1976–1977 is the best-documented climate transition in the Pacific region for the 20th century (Mantua et al. 1997).

Paleo-PDO Reconstructions

Figure 10.5 shows a long-term spring PDO reconstruction from Asia back to 1565 based on 17 tree-rings records from D'Arrigo and Wilson (2006). There is generally good agreement between the paleo-index and the instrumental record, notably low PDO values during the 1920s, 1950 to the 1960s, and the 1976–1977 transition. There is also a good correlation to the independent 400-year reconstruction of the NPI based on high-latitude tree rings surrounding the North Pacific region and a possible connection between the PDO and the Asian monsoon (D'Arrigo et al. 2005b).

A number of other excellent PDO reconstructions are available from the Pacific region extending back to A.D. 1840 to A.D. 1000 (Linsley et al. 2000a, b, 2004; Urban et al. 2000; Biondi et al. 2001; Evans et al. 2001a; Gedalof and Smith 2001; Gedalof and Mantua 2002). MacDonald and Case (2005) documented significant centennial-scale variability in the PDO going back more than 1000 years. In addition, Shen et al. (2006) reconstructed PDO-related variability from a drought-flood index from eastern China back to A.D. 1470, and argue that the PDO patterns suggest solar forcing on multidecadal timescales. Verdon and Franks (2006) proposed a relationship between the PDA and ENSO for the last 400 years in which the positive (negative) phase of the PDO is linked with more (less) El Niño events.

PDO and IPO

One key to understanding internal climate variability in the Pacific region was establishing the relationship between dynamic processes in the northern and southern hemispheres and establishing a connection between the PDO in the North Pacific and similar modes in the South Pacific. The South Pacific Convergence Zone (SPCZ) is a region of convective clouds and rainfall running from near the equator off northern Australia and New Guinea, southeast to 30°S latitude near French Polynesia in the eastern South Pacific. Power et al. (1999) and Folland et al. (1999, 2002) showed that SPCZ variability is related to both interannual ENSO and the lower-frequency variability now called the IPO. Their analyses demonstrated "the equivalence of the IPO and PDO in describing Pacific-wide variations in ocean climate." Nonetheless, the link between ENSO and extratropical Pacific variability since 1900 and the exact mechanisms controlling SST and atmospheric conditions associated with IPO, PDO, and ENSO modes are still incompletely understood (Deser et al. 2004; Schneider and Cornuelle 2005).

Several paleoreconstructions have supported the hypothesis that internal variability in the North Pacific Ocean is linked to processes in the low-latitude South Pacific. Linsley et al. (2000b, 2004) and Evans et al. (2001b) concluded

that the PDO reflects basin-wide processes embodied in IPO, which Evans et al. called Pacific Decadal Variability. These studies suggest a common SST pattern across the Pacific at least back to the late 18th century and perhaps longer. Another important observation was that there was reduced PDO activity during the mid-19th century. Cobb et al. (2001) also observed 12–13-year decadal variability in tropical Pacific coral SST records from Palmyra Island, which they linked to similar patterns in the Atlantic and Indian Oceans.

Decadal and multidecadal variability in the Pacific should in theory impact North American climate. One study that examined this connection was by Tian et al. (2006), who reconstructed 3100 years of drought from Steel Lake, Minnesota sediments. They identified 15–25-year and 50–70-year spectral peaks in precipitation, but it is not clear if these represent PDO-type modes of variability or, as speculated by Tian, solar forcing.

Atlantic Multidecadal Oscillation

Understanding multidecadal patterns of climate variability is particularly important for the detection of anthropogenic climate change. SST in the Atlantic Ocean varies over 65- to 90-year timescales; this pattern has come to be known as the AMO. The climate dynamics of the AMO, an example of extratropical atmosphere-ocean interaction, have been the subject of intensive research over the last decade (see Kushnir et al. 2002). The discovery of the AMO is somewhat confusing. Richard Kerr (2000) informally introduced the term *AMO* in a summary article in *Science*, referring to a number of ice-core observations and model simulations analyzing recent behavior in the climate system. Kerr's discussions with a number of specialists in multidecadal climate patterns, including Thomas Delworth, Michael Mann, Thomas Knutson, Yochanan Kushnir, John Marshall, Axel Timmermann, Mojib Latif, and Christopher Folland, led to near-universal agreement of the importance of multidecadal patterns in the search to detect human influence on climate. Some commentators, such as climate modeler Andrew Weaver, suggested that THC played a role in AMO oscillations, and paleoclimatologist Wallace Broecker noted that, despite the existence of AMO-like behavior, it did not mean that anthropogenic influence on climate was not overriding the type of natural variability during the last few decades.

Schlesinger et al. (2000) wrote a letter to *Science* pointing out that their group had already recognized a global mean temperature pattern of about a 65–70 year frequency that resulted from a 50–88-year oscillation in the North Atlantic Ocean and adjacent regions (Schlesinger and Ramankutty 1994; Andronova and Schlesinger 2000). Today, there are many more papers on multidecadal variability that show frequencies as short as 30 years to as long as 90 years. Frequencies in the range of 60–90 years seem to be the most common patterns (Delworth and Mann 2000; Cronin et al. 2003), but there is no consensus on which frequencies dominate, and the lower frequencies are interestingly close to the 88-year Gleissberg solar cycle discussed elsewhere in this book.

With this background, Figure 10.6A shows the instrumental SST record from Kaplan et al. (1998) and Enfield et al. (2001), who argued that the AMO was a 65–80-year quasi-cyclic variability of SST with a 0.4°C temperature range. Warm intervals characterize the intervals 1860 to 1880, 1940 to 1960, and since the 1990s; cool SST prevailed between 1902 and 1925 and 1960 until 1990. There was a strong statistical correlation between AMO-related SST variability and rainfall in the southeastern United States and Mississippi River flow. Warm SSTs correspond with relatively low rainfall over much of the United States and mid-

western droughts, except for parts of south Florida. McCabe and Palecki's (2006) global analysis of the Palmer Drought Severity Index revealed a connection between continental rainfall and the AMO.

Some evidence suggests that the AMO signifies a global mode of climate variability. Knight et al. (2005) analyzed global surface temperature back to the 19th century and simulated 1400-year temperature records. They concluded that "the AMO is a genuine quasi-periodic cycle of internal climate variability persisting for several centuries, and is related to variability in the oceanic thermohaline circulation." Knight et al. also note that several non-SST climate measurements—such as rainfall in northeastern Brazil and in the African Sahel, Atlantic hurricane patterns, and summer climatology in North America and Europe (Sutton and Hodson 2005)—are manifestations of the AMO.

Understanding the mechanisms driving the AMO is critical for future prediction under rising greenhouse gas concentrations. Some researchers indicate a link between Atlantic

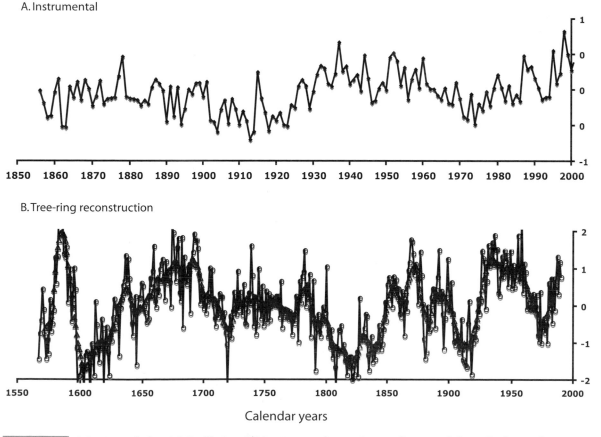

FIGURE 10.6 Atlantic Multidecadal Oscillation: (A) instrumental annual anomalies record, from Kaplan et al. 1998 and Enfield et al. 2001; (B) tree-ring reconstruction, from Gray et al. 2004. Reconstruction for sea-surface temperature anomalies is for 0–70°N latitude.

decadal variability and changes in THC (Delworth and Mann 2000), a topic discussed in the subsection "North Atlantic Oscillation" in reference to the NAO. McCabe et al. (2004) concluded that the combined influence of the AMO and PDO accounted for 52% of the low-frequency drought variability in the conterminous United States. Sutton and Hodson (2005) proposed that the AMO variability is driven by the Atlantic's THC and had a strong influence on European summer climatology. R. Zhang et al. (2007) modeled the AMO and concluded that SST variability is sufficient to influence northern-hemispheric temperature patterns. More generally, the AMO is now routinely evaluated in the context of large-scale climate phenomena that include hurricane patterns, droughts, and floods, among others, in addition to external forcing.

Paleo-AMO Records

A number of paleoclimate records from the North Atlantic region document multidecadal variability in midlatitude temperature (Cronin et al. 2003) and regional precipitation

(Cronin et al. 2005) but do not provide a direct AMO index. Gray et al. (2004) used 12 tree-ring records from eastern North America, western Europe, Scandinavia, and the Middle East to reconstruct an index of AMO activity back to the mid-16th century (Figure 10.6B). In addition to capturing the instrumental trends in the 19th and 20th centuries, they identified a clear low-frequency 60–100-year pattern in SST across a wide region (equator to 70°N) that is evident at least for the last 500 years. The existence of such a pattern reinforces the need to more fully understand AMO-type variability over longer timescales, its relationship to higher-frequency modes, and how anthropogenic activity will affect it.

North Atlantic Oscillation and Pacific North American Mode

Several modes of variability dominate the high-latitude northern hemisphere, including parts of North America, Europe, and the North Atlantic and Arctic Oceans.

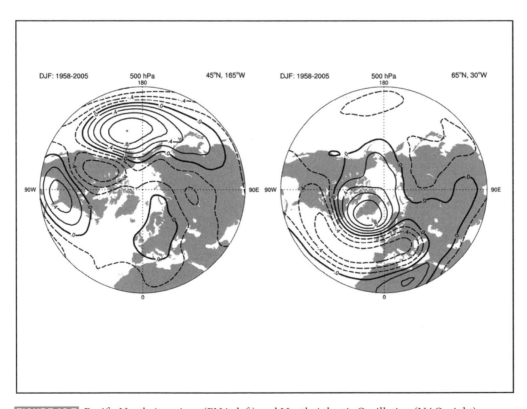

FIGURE 10.7 Pacific North American (PNA, left) and North Atlantic Oscillation (NAO, right) teleconnection patterns, shown as one-point correlation maps of 500 hPa geopotential heights for boreal winter (DJF) over 1958–2005. In the left panel, the reference point is 45°N, 165°W, corresponding to the primary center of action of the PNA pattern, given by the + sign. In the right panel, the NAO pattern is illustrated on the basis of a reference point of 65°N, 30°W. Negative correlation coefficients are shown as dashed lines, and the contour increment is 0.2. Both the figure and caption are from figure 3.26, IPCC chapter 3, "Observations: Surface and Atmospheric Climate Change" (Trenberth et al. 2007), Cambridge University Press (original figure in Hurrell et al. 2003b). Courtesy of the Intergovernmental Panel on Climate Change.

North Atlantic Oscillation

The NAO is the dominant decadal mode of climate variability in the North Atlantic region, identified by oscillations in atmospheric pressure between the Icelandic subpolar low-pressure system and the subtropical high pressure centered over the Azores (Hurrell 1995; Hurrell and Van Loon 1997). During positive NAO, the Icelandic and Azores pressure is relatively lower and higher than mean conditions, respectively (Figure 10.7). NAO is primarily a winter-season index having a strong impact on European rainfall and temperature whereby a high NAO index is associated with warmer and wetter winters and a low NAO index the opposite. The eastern United States is also affected by these oscillations.

The volume edited by Hurrell et al. (2003a) contains the most comprehensive treatment of the NAO mode of variability, with chapters on the history of NAO research (Stephenson et al. 2003), ocean-atmosphere coupling (Czaja et al. 2003), ocean response to NAO (Visbeck et al. 2003), the NAO-AO relationship (Thompson et al. 2003), possible anomalous 20th-century NAO behavior (Gillett et al. 2003), and other topics. Additional papers on the NAO and North Atlantic salinity and deep-ocean circulation by Dickson et al. (1996, 2003), 20th-century NAO behavior (Delworth and Dixon 2000), NAO prediction by Latif et al. (2004), and long-term NAO variability by Luterbacher et al. (2002) provide details.

Pacific North American Mode

The PNA mode of decadal variability emanates in the North Pacific region, having impacts on North American climate (Wallace and Gutzler 1981; Kushnir and Wallace 1989). Figure 10.7 shows the atmospheric pattern for the PNA. Like the NAO, the PNA is mainly a northern-hemisphere winter-season index of atmospheric pressure. Although the PNA is a distinct mode of variability, its behavior is linked to tropical ENSO variability (Trenberth et al. 1998; Hurrell et al. 2003b).

NAO Instrumental Records

The NAO can be defined in several ways, such as instrumental records of the monthly or seasonal SAP difference between Iceland and the Azores or EOFs from a larger spatial network of atmospheric conditions (20–70°N latitude, 90°W–40°E longitude) (Hurrell et al. 2003b). Figure 10.8 shows Hurrell's winter NAO index back to the 1860s. The plot shows clear interannual and decadal variations to the NAO index. Although there appears to be some variance at the 8–10-year frequency, it is not statistically significant and Hurrell et al. (2003b) concluded that there was no preferred timescale of NAO variability, at least based on instrumental records, which now extend back to 1823 with measurements from Gibralter and southwest Iceland (Jones et al. 1997). In addition to its atmospheric expression, observational evidence suggests the NAO is linked to surface, intermediate, and deep-water circulation in the North Atlantic Ocean, the Labrador and Norwegian-Greenland Seas, and the Arctic Ocean (Hurrell and Dickson 2004; Dickson et al. 2002, 2003).

Paleo-NAO Indices

A growing number of paleo-NAO indices have been developed from tree rings (Cook et al. 1998; Glueck and Stockton 2001), speleothems (Proctor et al. 2000, 2002), ice cores (Appenzeller et al. 1998), sclerosponges (Rosenheim et al. 2005), and multiple proxies (Cullen and deMenocal 2000; Luterbacher et al. 2002, 2004; Pauling et al. 2003, 2005). Figure 10.8 shows one of these from Appenzeller et al. (1998) showing a

North Atlantic Oscillation

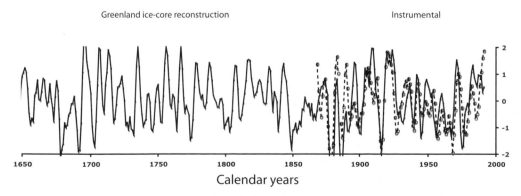

FIGURE 10.8 North Atlantic Oscillation (NAO). The solid line depicts the reconstructed annual NAO from the west Greenland ice core, from Appenzeller et al. (1998); the dashed line depicts the instrumental NAO index of atmospheric pressure in Iceland and Azores, from Hurrell (1995).

match to Hurrell's index and NAO variability back to the year 1650. Figure 10.9 shows four other NAO reconstructions from Moroccan and Finland tree rings and the Greenland ice core called ARKOA (Afraskou, Ich Ramouz, Karhunpe, GISP2 [Greenland Ice Sheet Project 2] δ^{18}O, GISP2 accumulation rate) (Glueck and Stockton 2001); a tree-ring reconstruction from the North Atlantic region (Cook et al. 2002); and combined instrumental and proxy data on monthly and seasonal western Eurasian time series (Luterbacher et al. 1999, 2002). The dense network of European records shows that NAO variability has opposing influences on southern (Mediterranean) and northern European regions. These records show positive NAO values during the early and late 1900s, as seen in the instrumental records and significant multidecadal variability before that. Luterbacher's monthly NAO index shows a subdued level of variability before the mid-18th century.

Arctic Oscillation

The AO is the leading interannual-to-decadal mode of variability defined for the region from 20°N to the North Pole (Thompson and Wallace 1998, 2001; Thompson et al. 2003). It is also called the Northern Annular Mode because of the distinctive zonal concentric pattern to atmospheric pressure gradients when viewed from a polar projection. AO variability is strongly linked with troposphere-stratosphere coupling (Baldwin and Dunkerton 1999; Rind et al. 2005). The region defined for the AO overlaps with that for the NAO, and the AO and NAO are also considered to represent the same mode of climate variability, although there are differences in the timing and spatial structure of the two.

The nature of AO variability and impacts is especially important today because polar regions are thought to amplify greenhouse gas forcing (the polar amplification hypothesis) and, given that the AO strongly influences Arctic Ocean sea ice (Rigor et al. 2002; Rothrock and Zhang 2005), it raises the issue of to what extent recent drastic decreases in Arctic sea-ice area and thickness are due to human influence, AO variability, or both. The Goddard Institute of Space Studies (GISS) climate model incorporates stratospheric and tropospheric processes into model simulations, and GISS simulations suggest that recent trends in the AO and in Arctic climatology can be confidently attributed to greenhouse gas forcing (Shindell et al. 1999). This topic is discussed again in Chapter 12 of this volume.

Instrumental AO Records

The AO is computed from monthly mean anomalies in atmospheric pressure (usually 500 or 1000 hPa) using EOFs or SAP north of 20°N. Some instrumental time series are relatively short for the Arctic Ocean region, and the AO is extended back to the year 1950 by some authors (Higgins et al.

2002; Zhou et al. 2001). However, David Thompson, who proposed the AO index with John Wallace, extended the index back to 1899 (Figure 10.10A). The winter AO index shows positive values from 1899 through 1939 and 1989 through 2002, negative values for 1940 through 1988, and positive values for 1989 to the present. Since 2002 the AO has been carefully monitored in efforts to determine its role in disappearing Arctic Ocean sea ice.

Paleo-AO Reconstructions

Figure 10.10B and C shows the reconstructed AO evident in tree-ring records as proxies for summer sea-level pressure air-temperature anomalies (D'Arrigo et al. 2003). Important highlights are the decadal and multidecadal variability in both pressure and temperature, with periods of low temperature and pressure in the 19th century followed by a steep rise in both parameters in the early 20th century.

Southern Hemisphere Annular Mode

One pattern of variability especially important for Antarctic climatology is the Southern Hemisphere Annular Mode (SAM) (Thompson and Wallace 2000; Hall and Visbeck 2002). Like the AO in the northern hemisphere, SAM variability is relevant for decoupling natural from anthropogenic variability in snowfall, temperature, atmospheric pressure, sea ice, and other parameters. Some evidence suggests that SAM may be strengthening because of anthropogenic forcing from ozone and greenhouse gases (e.g., Marshall et al. 2004; Shindell and Schmidt 2004). At present there are no specific paleoclimate reconstructions of SAM variability.

Perspective

Research on modes of internal climate variability reconstructed from proxy methods tells us the following:

- Some internal modes identified in 20th-century instrumental records have basin-scale (hemispheric) and global expressions—e.g., the AMO and IPO.
- Most modes experienced times of low-amplitude patterns during the past few centuries, in contrast with what was seen in instrumental records; the issue of persistence remains a fundamental issue to be addressed.
- There is interannual, decadal, centennial, and millennial-scale variability in ENSO.
- Multidecadal patterns suggest that ocean-circulation changes, as well as those in surface air pressure and temperature, characterize some internal modes of variability.

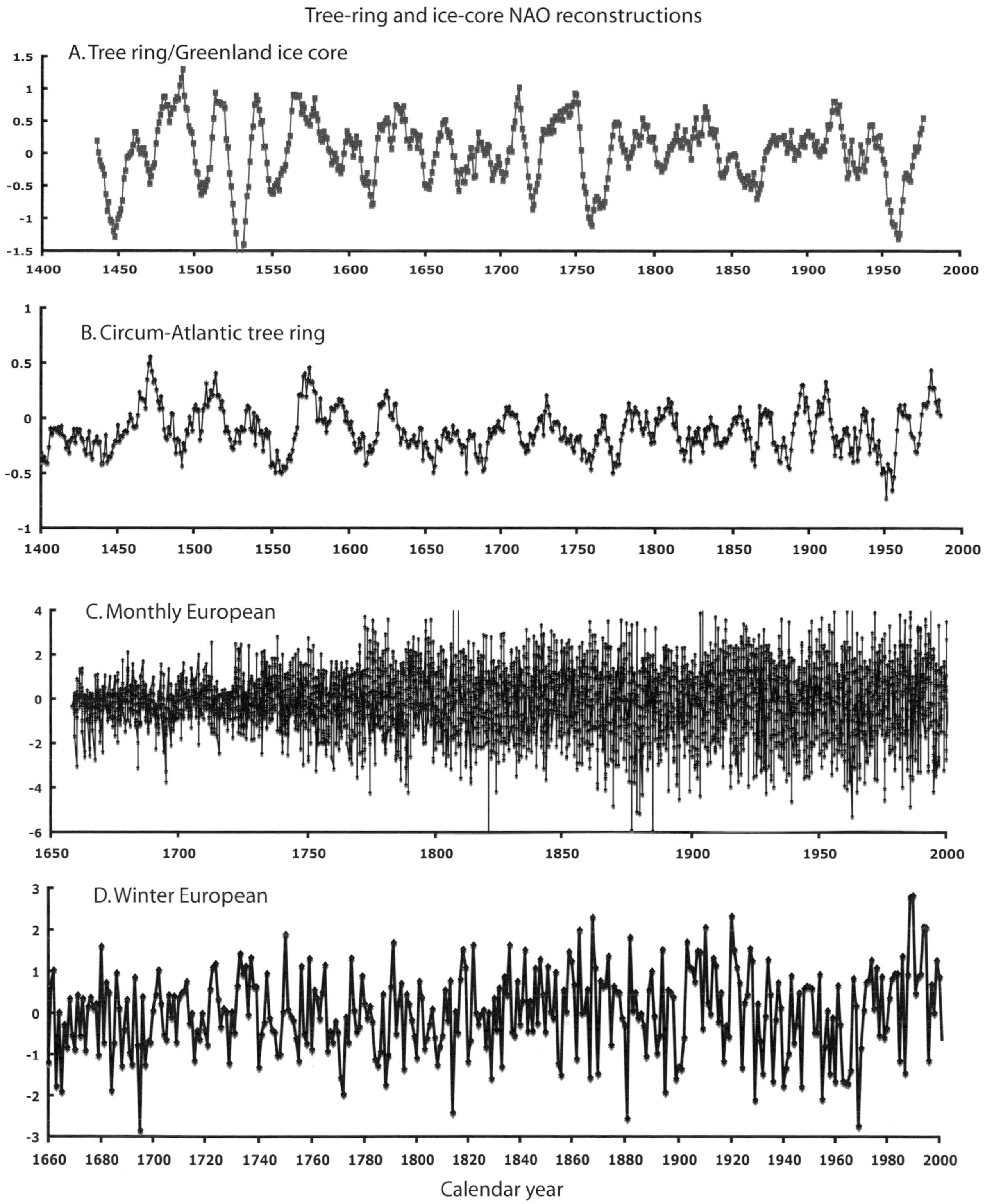

FIGURE 10.9 Four North Atlantic Oscillation (NAO) reconstructions: (A) Winter, based on Moroccan and Finland tree rings and Greenland ice core, called ARKOA (Afraskou, Ich Ramouz, Karhunpe, Greenland Ice Sheet Project 2 [GISP2] d180, GISP2 accumulation), from Glueck and Stockton (2001); (B) panel from Cook et al (2002); (C) monthly western European NAO time series, based on instruments and proxy data from Luterbacher et al. (1999) (the seasonal NAO goes back to A.D. 1500, as found in Luterbacher et al. 2002); (D) winter NAO, from Luterbacher. Most records show positive NAO during the early and late 1900s and significant multidecadal variability before that time.

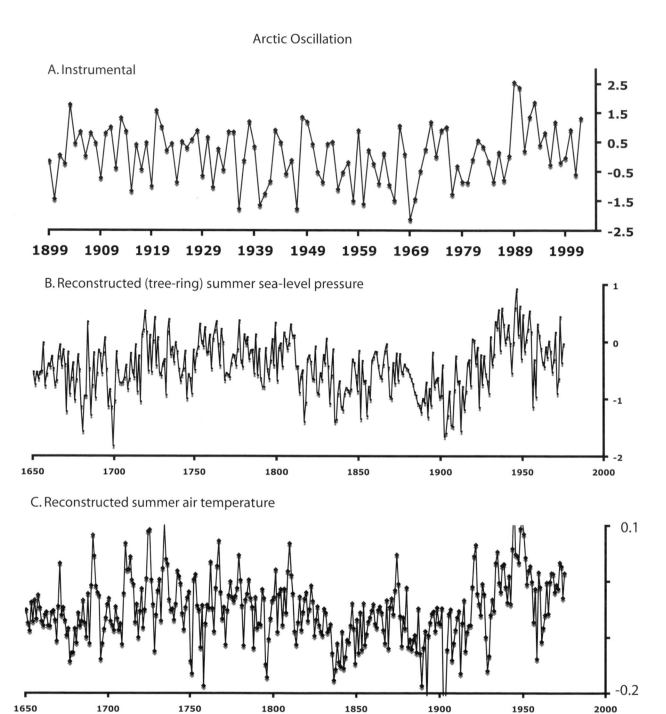

FIGURE 10.10 (A) Instrumental record of Arctic Oscillation. Reconstructed Arctic Oscillation based on tree-ring records from D'Arrigo et al. (2003). (B) Summer sea-level pressure. (C) Summer air-temperature anomalies. Complied by and courtesy of D. W. J. Thompson and Colorado State University.

A final comment about proxy reconstructions is particularly germane to this chapter. Trenberth and Otto-Bliesner (2003) commented on paleoreconstructions from observationalist and modeling viewpoints. They called for improved calibration of proxies, better spatial coverage, and integration of proxy records with climate modeling to gain a better

understanding of the physical mechanisms driving climate variability. Wunsch and Heimback (2008) also raised concern about the ability of proxy methods to skillfully reproduce past ocean-circulation changes. These comments reflect the apparent mismatch between paleorecords and more spatially robust, but shorter, instrumental records and model simulations. They serve as a challenge to develop more widespread paleoreconstructions using improved proxy calibrations of temperature and precipitation (e.g., the TetraEther index, or TEX_{86}, and deuterium), better characterization of seasonal variability, and integration with monitoring data and climate modeling. Currently available paleoclimate records cannot capture the long-term, spatial aspects of climate variability, especially over oceanic regions and polar latitudes. On the other hand, paleorecords expose our rudimentary understanding of the long-term persistence of internal modes of variability and raise the question of whether modern AMO, NAO, and other patterns are fundamental, long-term features of pre-anthropogenic Holocene climate.

LANDMARK PAPER Paleo–El Niño

Dunbar, R. B., G. M. Wellington, M. W. Colgan, and P. W. Glynn. 1994. Eastern Pacific sea surface temperature since 1600 A.D: The $\delta^{18}O$ record of climate variability in Galapagos corals. Paleoceanography 9:291–315.

Internal modes of climate variability pose especially difficult challenges for climate scientists attempting to distinguish human influence on climate from natural variability. The El Niño–Southern Oscillation (ENSO), for example, represents one of the most powerful climate phenomena recorded by historical observations and instrumental methods. Originating in the tropical Pacific Ocean, El Niño events affect global weather patterns, and the largest events, such as those in 1982–1983 and 1997–1998, cause billions of dollars in damage, loss of life, and ecosystem disturbance. But scientists still do not know ENSO's long-term variability or whether more intense and frequent ENSO activity over the past few decades is natural or human-induced.

At a time when coral records were rapidly gaining acceptance in tropical paleoceanography, the paper by Rob Dunbar and colleagues represented a breakthrough because of the length of the record (back to A.D. 1586), the sensitivity of the proxy (oxygen isotopes calibrated to sea-surface temperatures, or SSTs), and the location (Isabela Island in the Galapagos Islands) in the Eastern Equatorial Pacific (EEP). This region is a "center of action" for modern ENSO activity. Dunbar's paper showed that, when compared to historical ENSO records, corals provided a more complete record of ENSO activity, especially activity before the 19th century. Their most important conclusions were that EEP SST variability was greater during the 1600s and early 1800s but lower in amplitude during the 1700s, long-term changes occur in decadal and interannual (three to eight years) SST, and relatively minor SST warming occurred during the early 20th century.

The Dunbar paper and the 1997–1998 El Niño event kindled new research projects in coral paleoclimatology. Today, there are dozens of coral records that, together with tree-ring and ice-core records, provide a much better picture of modes of tropical and extratropical climate variability.

CHAPTER 11

The Anthropocene I: Global and Hemispheric Temperature

Introduction

In the year 2000, atmospheric chemist Paul Crutzen of the Max Planck Institute for Chemistry and biologist Eugene Stoermer of the University of Michigan assigned the term *Anthropocene* to designate the period of time since the late 18th century when human beings began to have an impact on the earth's atmospheric concentrations of carbon dioxide (CO_2) and methane (CH_4) (Crutzen and Stoermer 2000). Although Crutzen and Stoermer recognized that the date they chose for its onset was somewhat arbitrary, the Anthropocene nonetheless roughly coincided with the initial rise in atmospheric greenhouse gas concentrations recorded in ice cores, the invention of the steam engine in 1784, widespread ecosystem degradation, and a host of other environmental changes accompanying human population growth, industrialization, and urbanization. Ruddiman proposed that the Anthropocene actually began much earlier, in fact as early as 7000 years ago, when early human civilizations began large-scale agricultural practices affecting the global carbon cycle (Ruddiman 2003, 2007). The idea that early human agriculture had influenced global methane cycling is not new; it stemmed from discoveries pertaining to ice-core records of methane (Chappellaz et al. 1990; Etheridge et al. 1996, 1998). Recognizing the human impact on the earth's environments, Gradstein et al. (2004) inserted the term *Anthropocene* into the most recent timescale proposed by the International Commission on Stratigraphy.

The Anthropocene is a lightning rod for controversy about the impacts of human greenhouse gas emissions and land-use changes on global climate during the last century. For several years, an intense debate has taken place in the scientific literature and often in the media and public domain, revolving around what caused global surface air temperatures (SATs) to rise ~0.6°C during the past 150 years: natural external forcing from solar or volcanic processes, fossil-fuel burning, and cement production? Internal climate variability? Is recent warming a natural transition from the cool Little Ice Age (LIA) between A.D. 1400 and 1900? Or if all these factors played a role, how much did each contribute?

This chapter addresses the small but important segment of paleoclimatology dealing with SAT over the past 1000 years, a topic that fits under the rubric of "global warming." The 1000-year SAT curve reconstructed from tree rings, ice cores, and lake and marine sediments is the most direct application of paleoclimatology to human-induced climate change. The following topics are covered: development of SAT proxy records, the LIA, external forcing agents of late Holocene temperature, the construction of and limitations to proxy-based hemispheric temperature reconstructions, and selected climate model simulations designed to detect the causes of reconstructed temperature patterns. Chapter 12 is an assessment of parts of the climate system other than temperature and of whether paleoreconstructions provide evidence for human perturbation of climate.

The main theme in these two chapters is the analyses of SAT and the relative contributions from greenhouse gas as well as volcanic and solar forcing in order to evaluate recent climate trends. Small mean SAT changes of tenths of a degree Celsius are important from the standpoint of radiative forcing and global energy budgets at the planetary scale. The Anthropocene, however, experienced complex variability in regional temperature patterns, natural internal decadal to centennial precipitation, ice dynamics and glaciers, ocean circulation and chemistry, thresholds and nonlinear feedback systems, and other processes. Growing evidence from paleoclimatology, observations, and model simulations suggests that there is "anomalous behavior" in some segments

of the climate system, but great uncertainty surrounding others. Climate response over the past millennium is regionally variable in scale and sign (warming in one region, cooling in another) and often asynchronous because of feedbacks (Goosse et al. 2004; Mann 2007). We begin with a brief look at the temperature record.

Surface Air Temperature

The search for human influence on climate has emphasized trends in global SAT. This is obviously due to the expected impact on temperature from rising greenhouse gas concentrations. It is also, however, due to practical issues of data availability. Instrumental records of temperature extend back 150 years, longer than records of ocean temperature (40–60 years) (Levitus et al. 2000), atmospheric pressure and precipitation (30–100 years) (Trenberth et al. 2007), sea ice (30–40 years) (Serreze et al. 2007), tide gauges (50–60 years, except for a few gauges) (Woodworth and Player 2003; Bindoff et al. 2007; Jevrejeva et al. 2008), and a few decades for satellite records covering most of the ocean-atmosphere-ice system.

SAT has been carefully reconstructed from proxy and instrumental records. Figure 11.1 is a 1000-year record of northern-hemisphere SAT anomalies (departure from mean) produced from studies of late Holocene temperature from Mann et al. (1999) as compiled by Crowley and Lowery (2000). The proxy record is spliced with instrumental records for the last 150 years that is shown in Figure 11.2, the global near-surface instrumental record from England's Met Office Hadley Centre for Climate Change, updated from Brohan et al. (2006). The instrumental record has relatively good spatial and temporal coverage back to about 1850 and shows a characteristic stepwise 20th-century warming (Jones et al. 2001; Jones and Moberg 2003; Brohan et al. 2006).

The striking pattern seen in the northern-hemisphere SAT curve in Figure 11.1 became known as the "hockey stick" curve because of its distinctive shape. The large warming in the last century exceeds any prior warming events during the last 1000 years. A number of proxy temperature curves have been published in the last few years, showing differences in the amplitude of centennial-scale temperature fluctuations before the 20th century but still the same general pattern (Crowley 2000; Esper et al. 2002a; Mann and Jones 2003; Moberg et al. 2005; Osborn and Briffa 2006; Hegerl et al. 2007; reviewed in Jones and Mann 2004). Differences stem from the choice of proxy records, calibration, scaling and statistical methodology, and the objectives of each study. All curves have a greater uncertainty in temperature estimates before the instrumental period and again before about A.D. 1600, when fewer tree-ring records are available. All records, however, show cool temperatures during several periods of the LIA between 1400 and 1900, periods of

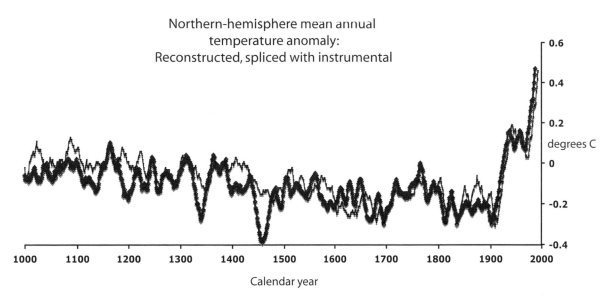

Northern-hemisphere mean annual
temperature anomaly:
Reconstructed, spliced with instrumental

Calendar year

FIGURE 11.1 Northern-hemisphere mean and annual temperature records for the last millennium (A.D. 1000–1993) based on paleo-reconstructions from Mann et al. (1999) and Crowley (2000, modified from Crowley and Lowery 2000). The Crowley record is spliced into the 11-point smoothed record of Jones et al. 1999. Two sigma standard deviations (not shown) plus/minus are about 0.3°C (A.D. 1000–1600) and about 0.25°C (A.D. 1600–1850), i.e., the beginning of the instrumental record.

Global Average Near-Surface Temperatures, 1850–2007

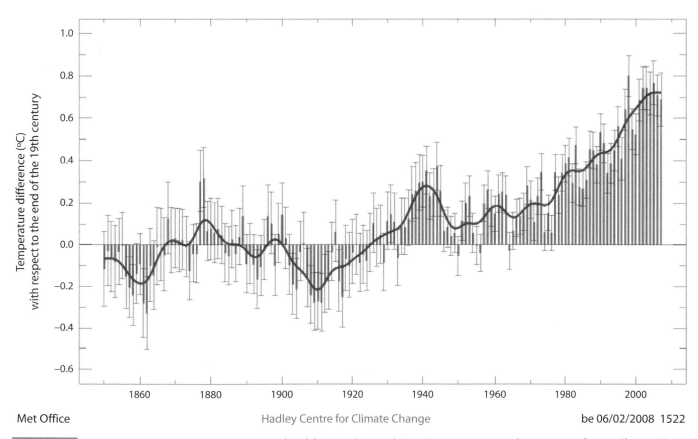

Met Office   Hadley Centre for Climate Change   be 06/02/2008 1522

FIGURE 11.2 Atmospheric temperature since 1850, updated from Brohan et al. (2006). Crown Copyright, courtesy of Met Office Hadley Centre for Climate Change, P. Brohan and P. Jones.

warmth within the Medieval Warm Period (MWP, or Medieval Climate Anomaly) (~A.D. 800–1400), and recent warming. Some reconstructions cover the last 2000 years, but spatial coverage decreases greatly before A.D. 1000 (Mann 2007; Mann et al. 2008).

Since first published by Mann et al. (1999), the statistical methods, reliability of temperature proxies, spatial averaging techniques, role of non-greenhouse-gas forcing agents, and other aspects of the hockey stick curve have been scrutinized by the scientific community. A special committee of the U.S. National Academy of Sciences (NAS 2006) evaluated the Mann et al. temperature curve and the Intergovernmental Panel on Climate Change (IPCC) reviewed a number of curves (Jansen et al. 2007). Before proceeding, proper assessment of recent temperature history requires some background on LIA climate.

Before the Anthropocene: The Little Ice Age

The LIA is the multicentury period (A.D. 1400–1900) of relatively cool climate and alpine glacial advances recognized in many regions of the world. The end of the LIA coincides with the industrial revolution and the onset of the stepwise rise in global temperatures. Thus, a problem immediately arises in distinguishing human-induced warming from a "natural" end to the LIA. Many believe that understanding LIA climate and its causes represents a critical element for understanding current and predicting future climate trends.

At the heart of the LIA climate dilemma is the most basic question: Did it actually exist? That is, is there a distinct multicentury climatic state from A.D. 1400 to 1900 recognizable globally that can be designated as the LIA? Jones and Mann (2004), echoing earlier studies, propose that the bulk of the evidence suggests the LIA concept is outdated and that there was not a climate anomaly that can be identified globally. On the other side of the argument, many studies, introduced in Chapter 8 and discussed in this chapter, suggest that millennial- and submillennial-scale variability during and within the LIA was part of typical Holocene climate patterns—in essence, the culmination of the mid- to late Neoglacial interval.

Historical Development

A few decades ago, most research on late Holocene climate involved historical sources of climatic data including diaries, flood records, official documents, farm records, ships' logs, and, since about the 16th century in Europe and later in other areas, instrumental data (Ladurie 1971; Eddy 1976; Lamb 1977, 1995; Grove 1988; Bradley and Jones 1993; Hoyt and Schatten 1997). Climate has long been linked to a host of economic factors—the price of a bushel of wheat and

sunspot numbers by Sir William Herschel in 1801, cool temperatures and the demise of English vineyards, smaller wine harvests in France, and the advance of glaciers in the Alps. Many early workers recognized a period of cooling in Europe and elsewhere before the 20th century, and this was called the LIA by English climatologist Hubert Lamb (1977).

Early field evidence for an LIA came from North America's Sierra Nevada Mountains where Matthes (1939) documented several periods of Holocene glacial advances. Glacial advances can be caused by internal ice mechanics and not necessarily climatic factors, but evidence from both observations of glaciers and field studies of moraines suggested that climate might be involved. The landmark study of Greenland ice-core oxygen isotopes by Dansgaard et al. (1969) provided some of the first direct paleoclimate evidence for a cool LIA climate. Historical and paleoclimate records of preindustrial climate have continued unabated, and excellent reviews are given in Jones and Mann (2004), Bradley et al. (2003), and Mann (2007).

Perspectives on the Little Ice Age

Recent studies of LIA climate have taken two equally important approaches leading to differing opinions on its climatic significance. These can be called, for want of better terms, the geological and geographical approaches. The geological approach is "forward-looking," starting from the perspective of long-term Holocene climate variability. It is the realm of paleoclimatologists who specialize in glacial geology, paleoceanography, and paleolimnology. The geographical approach "looks back" at the LIA from the standpoint of the 20th and 21st centuries and is adopted mainly by specialists in geography and climatology, including tree-ring experts.

Geological Approach to LIA Climate The geological approach views the LIA as one of several global millennial-scale Holocene climate anomalies described in Chapter 8. In a global compilation of Holocene glacier advance-and-retreat patterns, Mayewski et al. (2004) showed that the LIA was the last of several glacial cycles that to a first approximation coincided with climatic fluctuations seen in ocean, ice-core, and continental paleorecords. Mayewski's compilation is impressive in that it includes results gathered over more than a century of glaciological studies, many with greatly improved chronology and refined proxy methods over those available just a few years ago. By itself, this overview presents compelling evidence for a global LIA climate anomaly.

Regional studies also provide context for the LIA. Alaska is home to some of the world's most thoroughly studied glacier complexes. Wiles et al. (2008) synthesized the history of southern Alaskan glaciers whose terminus lies on land.

Land-based glaciers are sensitive monitors of atmospheric temperature and, unlike tidewater glaciers, are less likely to experience major surges due to internal dynamics. Their 2000-year reconstruction dated with tree rings, lichenometry, and radiocarbon ages showed that major cool periods occurred from A.D. 0 to 800 (especially A.D. 650) and during the LIA. Periods of glacial advance roughly corresponded to ocean cooling events discussed in the paragraphs that follow. There was also evidence for medieval warmth from A.D. 900 to 1200, when soil formation and tree growth resumed in the absence of glacial activity. In addition to millennial-scale variability, Wiles also correlated glacier advances with Wolf, Spörer, Maunder (the largest glacial advance), and Dalton sunspot minima. Many additional high-latitude glacial and lake records from Alaska (Hu et al. 2001; Daigle and Kaufman 2008), Canada (Thomas and Briner 2009), Greenland (Geirsdóttir et al. 2000), and areas across the North Atlantic indicate a complex pattern of LIA climate variability (Andersson et al. 2003; Andresen et al. 2005; Eiríksson et al. 2006; Hebbeln et al. 2006; de Vernal and Hillaire-Marcel 2006).

Greenland ice and marine-sediment cores also suggest the LIA is the most recent in a series of quasi-cyclic cool climatic periods. Changes in Greenland ice accumulation rates (Meese et al. 1994) and glaciochemical proxies of dust and sea salt (O'Brien et al. 1995) are dated at > 11,000, 8800–7800, 6100–5000, 3100–2400, and 600–0 years ago. In the North Atlantic Ocean, Holocene sea-surface temperature (SST) and ice rafting fluctuate on millennial timescales, including the LIA (Bond et al. 1999, 2001). Keigwin (1996) recognized LIA cooling in the Sargasso Sea in the western subtropical Atlantic Ocean from deep-sea cores on the Bermuda Rise and linked it to a North Atlantic Oscillation–like behavior. Keigwin and Pickart (1999) found evidence for 2°C *warming* in the subpolar North Atlantic off Newfoundland, reflecting regional variability in LIA SSTs. DeMenocal et al. (2000) provided convincing evidence that LIA SSTs off west Africa were 3–4°C lower than today and that tropical Atlantic cooling was synchronous with that in high latitudes. Cronin et al. (2003) documented multiple LIA temperature minima in Chesapeake Bay in the eastern United States.

SST in the tropical ocean also fell several degrees during the LIA. Winter et al. (2000) estimated a 2–3°C cooling in the Caribbean from coral records; Watanabe et al. (2001) analyzed oxygen isotopes and Mg/Ca ratios in corals from off southwest Puerto Rico showing a 2°C cooling and significant short-term variability within the LIA interval. Regional decreases in tropical SST are also known for at least some intervals during the LIA in the Pacific and the Indo-Pacific (Newton et al. 2006). Richey et al. (2007) estimated a 2–2.5°C cooling during the span of the LIA, on the basis of magnesium-calcium paleothermometry from cores from the Pigmy Basin in the Gulf of Mexico.

Tropical atmospheric circulation changes substantially during the LIA. Haug et al. (2001b) discovered southward migration of the ITCZ during the LIA from titanium and iron records obtained from Cariaco Basin cores in the Caribbean. They also identified three precipitation minima during 1600 to 1840 coincident with the Greenland ice-core cooling. LIA climate variability over tropical Africa was regionally complex but consistent with the Cariaco record. For example, east Africa experienced a relatively humid LIA, according to Verschuren et al. (2000), but Russell et al. (2007) also found a strong west-east regional precipitation gradient in east Africa. Brown and Johnson (2005) and Russell and Johnson (2005a, b) hypothesized broad synchroneity between tropical African and high-latitude LIA climate patterns related to a southward shift in the ITCZ during periods of high-latitude cooling.

Significant climate change also occurred in southeast Asia and the Indo-Pacific region. The East Asian monsoon weakened during the LIA, according to speleothem (Y. Wang et al. 2005) and peat (Hong et al. 2003) paleoclimate records. In the Great Barrier Reef, coral oxygen isotope records suggest that western tropical Pacific sea-surface salinity was higher during the LIA (Hendy et al. 2002). Cobb et al. (2003) found that LIA SSTs were actually warmer in some tropical regions perhaps related to increased El Niño activity during parts of the LIA.

Changes in meridional overturning circulation (MOC) very likely influenced LIA climate. Broecker et al. (1999a) and Broecker (2000) proposed that post-LIA 20th-century slowdown of Southern Ocean deep water combined with glacial advances and warming in the northern hemisphere might indicate north-south antiphasing due to the bipolar seesaw discussed in Chapter 6 in this volume. Keigwin and Boyle's (2000) benthic foraminiferal carbon isotopic and sediment grain size records from the North Atlantic suggest a slowdown in North Atlantic Deep Water (NADW) during the LIA, perhaps due to diminished overflow from the Iceland-Scotland Ridge. They noted that evidence for NADW slowdown from Cd/Ca ratios from foraminifers was still unclear. Marchitto and deMenocal (2003) found a coincident pattern of increased IRD and decreased deep-bottom water temperature during the LIA in deep-sea cores from 1800 m water depth on the Laurentian Slope off Newfoundland.

The Gulf Stream feeds warm water to the high-latitude North Atlantic, influencing MOC. Lund et al. (2006) reconstructed LIA Gulf Stream temperature, salinity, and volumetric flow using cores from the Florida Strait region. They show that the density gradient and vertical current shear in the Florida Straits were lower during the LIA and that the Gulf Stream transported 10% less volume than it does today. This evidence supports the hypothesis that, whatever triggered cool periods during the Holocene as with the LIA, there was in all likelihood a diminished flow of warm salty water to high latitudes and probably a decrease in the strength of MOC.

If deep-ocean circulation played a role in LIA climate, one would expect an asynchronous response in the high latitudes of northern and southern hemispheres. Proxy records from the southern hemisphere provide some support for interhemispheric asymmetry in LIA climate, but the picture is somewhat ambiguous. For example, southern-hemisphere warming was seen in ice-core records from the Antarctic Peninsula at times when northern-hemisphere climate cooled (Mosley-Thompson et al. 1996). However, other regions of Antarctica may have been relatively cool during parts of the LIA, and austral summer and winter temperature trends differ from one another over the past 700 years (Morgan and van Ommen 1997). Glacial advance on the Shetland Islands seems to be in phase with the classical northern-hemisphere LIA (Hall 2007), but new core results from the European Project for Ice Coring in Antarctic (EPICA) suggest a complex regional LIA climate signal (Masson-Delmotte et al. 2004) and marine records from the Antarctic Peninsula indicate reduced productivity and cooling during the LIA (Domack and Mayewski 1999) that was fairly robust regionally (Domack et al. 2003). But there is also evidence for multidecadal-to-centennial-scale climate variability in the Antarctic Peninsula (Leventer et al. 1996; Domack et al. 2001) and still ambiguous IRD patterns during the Neoglacial and LIA (Heroy et al. 2008).

Ice-core records also indicate changes in LIA atmospheric CO_2 concentrations of about 7 parts per million by volume (ppmv) (Etheridge et al. 1996). These variations may reflect significant carbon cycle feedbacks in response to LIA cooling. According to Scheffer et al. (2006), the climate-CO_2 relationships suggest that greenhouse gas-induced future temperature rise will influence carbon cycling and might lead to enhanced warming of 15–78% over a century. Regardless of the ultimate causes, LIA CO_2 concentrations were lower than those in the immediately succeeding preindustrial period, and this represents a real greenhouse gas anomaly.

In sum, when viewed from the geological standpoint of Holocene climate variability, the LIA clearly stands out as one of several widespread Holocene cooling events, with concomitant changes in many parts of the climate system reflecting deep-ocean circulation, atmospheric, carbon cycling, surface ocean, and glaciological variability (Wanner et al. 2008).

Geographical Approach The geographical approach to LIA climate stems from efforts to understand whether its end was all or in part human-induced, and to what degree it reflects natural Holocene variability. The LIA is viewed from the perspective of historical temperature trends of the last century or so, recorded by instrumental records of varying length integrated with annual or decadal paleoclimate records covering the past 1000–2000 years.

This approach reveals a distinct feature of LIA climate—a complex of temperature and precipitation patterns at submillennial timescales. Decadal and centennial variability within the LIA led some early workers to dismiss its importance as a distinct climate state. For example, North American glaciological records led Bradley and Jones (1992) to object to the notion of the LIA as a sustained period of global cooling. Luckman (2000) suggested that the concept of the LIA as multicentury-long cooling was an oversimplification of glacier advance-and-retreat records in the Canadian Rockies. Hughes and Diaz (1994a, b) reached a similar conclusion in their review of the MWP and LIA. Alpine glacial history also substantiated the view that the LIA glacier maxima were diachronous in Scandinavia, the Alps, and other regions of Europe (Grove 1988). In a comprehensive recent review, Jones and Mann (2004) suggest that the LIA concept has limited applicability because of complex spatial and temporal SAT and precipitation variability. This is not necessarily at odds with the idea that the LIA is one of several global Holocene millennial cooling events and that external forcing agents and internal climate processes operate over short timescales with complex regional expressions. One should thus not necessarily expect LIA climate changes to be regionally synchronous or equal in magnitude (Gillespie and Molnar 1995). Many records show two or three temperature minima and complex precipitation patterns during the 17th through 19th centuries. This is one reason why the detection of anthropogenic climate impacts is so difficult. No one disagrees that the end of the LIA coincided roughly with the onset of the industrial revolution and atmospheric warming. The dilemma is what combination of factors caused the transition.

Forcing Agents of Temperature During the Late Holocene

Four primary external forcing factors influenced global and regional SAT during the late Holocene: greenhouse gas concentrations, volcanic activity, solar variability, and sulfate aerosols. These are called external forcing agents, and the strength of the first three—measured in watts per square meter ($W m^{-2}$)—over the past 1000 to 2000 years has been reconstructed, modeled, or both by Crowley (2000), Bertrand et al. (2002), Mann and Jones (2003), Jones and Mann (2004), and Mann et al. (2008). These forcing agents are superimposed on progressively decreasing low-latitude summer insolation because of orbital processes discussed in earlier chapters. Tropospheric aerosols produced by human activity are an important negative forcing during the post-1950s but are not considered in this chapter. In addition to external factors, internal modes of variability related to the ocean-atmosphere-ice system cause climate to vary over the last 2000 years. These are considered in chapters 10 and 12 of this volume.

Solar Variability

The sun is an obvious potential source of climate variability, and solar variability is critical to any assessment of climate change. In their chapter "Cyclomania," Douglas V. Hoyt and Kenneth H. Schatten (1997) listed almost 40 separate solar-induced cycles, ranging in frequency from 6.6 days to 334 years, that supposedly influence climate or climate-related processes like thunderstorm frequency and precipitation. Among the notable solar-climate cycles introduced in Chapter 8 of this volume, those most relevant to climate variability during the past 2000 years are the 11-year Schwabe sunspot cycle, the 22-year Hale double sunspot, 88-year Gleissberg sunspot envelope-modulation cycle, and the ~200-year de Vries/Suess cycle.

Historical Studies of Solar Variability   Solar observation was an important activity during the European Renaissance and during alpine glacial advances and retreats in the 14th through 19th centuries. The solar-climate hypothesis gained momentum with Eddy's (1976) proposal that the

Maunder sunspot minimum (A.D. 1645–1715) may have caused cool European winters during the LIA. Eddy's paper sparked a resurgence of solar-climate research found in many volumes (McCormac 1983; Castagnoli 1988; Pecker and Runcorn 1990; Sonett et al. 1992; Nesme-Ribes 1994) (see the review by Eddy 1983). Since Eddy's paper, several reconstructions of solar variability have been produced. Figure 11.3 shows estimated solar forcing since A.D. 1000, based on work by Lean et al. (1995), Bard et al. (2000b), and Crowley (2000) using satellite, observational (sunspot activity), and cosmogenic isotopes. In addition to the Maunder Minimum, it shows that historical periods of reduced sunspots include the Spörer (A.D. 1400–1510) and Dalton (A.D. 1800–1860) minima and the medieval solar maximum. Using the Goddard Institute for Space Studies (GISS) climate model, Rind et al. (2004) estimated that Maunder Minimum cooling could be accounted for by lower greenhouse gas levels and solar variability, which, they estimated, accounted for 40% of observed cooling.

Solar Processes   The sun experiences a number of processes that potentially influence its total irradiance output: sunspots, faculae, solar flares, and coronal mass ejections, among others. Sunspots are dark, cool regions that form on the face of the sun through magnetic processes. Faculae are bright regions surrounding sunspots that emit a greater amount of energy than the sunspot regions. Both sunspots and faculae affect the solar constant, the sun's total solar irradiance. In an excellent review of the solar-climate link,

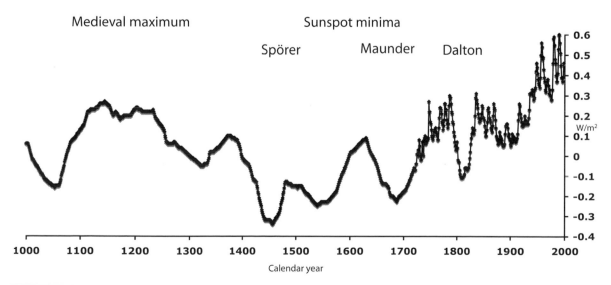

FIGURE 11.3 Estimated forcing from solar irradiance. Values since A.D. 1600 are from observed solar activity (sunspots) from Lean et al. (1995) spliced with beryllium 10 cosmogenic isotopic reconstruction from Bard et al. (2000b), by Crowley (2000). Maunder, Spörer, and Dalton sunspot minima as well as the Medieval maximum are shown.

Beer et al. (2000) point out that total solar irradiance varies because of energy transport through the sun's interior, emission through the outer layers of the sun, solar rotation, radiation from different wavelengths, and other processes. Solar magnetic fields are important modulators of outgoing solar irradiance.

The solar "constant" actually varies such that radiative forcing fluctuates from ~1–3 W m^{-2} over short timescales (11-year sunspot cycles) around a mean value of approximately 1366 W m^{-2}. Figure 11.4 shows the last two sunspot cycles based on the ACRIM (active cavity radiometer irradi-

ance monitor) project. The 1–3 W m^{-2} range is equivalent to a change of about 0.1% total solar irradiance during an 11-year cycle. Contemporary solar irradiance variability has been measured in several projects (Willson and Mordvinov 2003) and the Solar Radiation and Climate Experiment (SORCE) (Lean et al. 2005).

Cosmogenic Isotopic Records of Solar Activity

Space-based solar monitoring and historical observations do not record the full range of solar variability over long timescales, so cosmogenic radionuclides must be used as proxies of

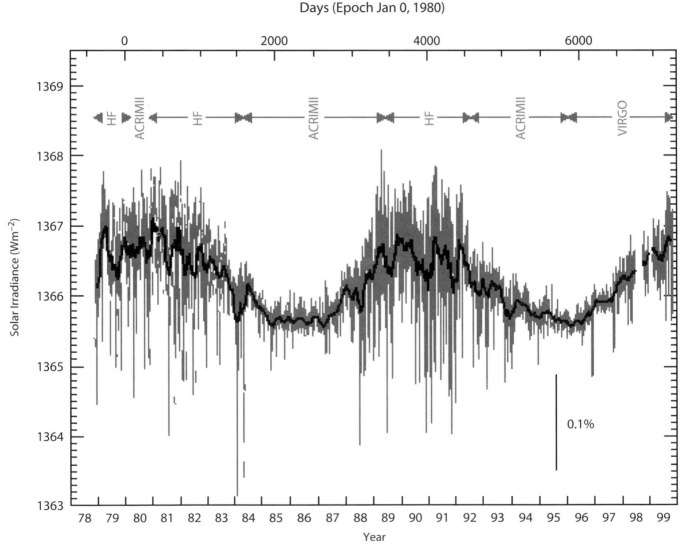

FIGURE 11.4 The composite curve is based on the active cavity radiometer irradiance monitor (ACRIM-I and II) and Virgo records. Courtesy of National Geophysical Data Center, National Oceanographic Atmospheric Administration. Before and during the gap between ACRIM-I and II, the corrected data was inserted by shifting the level to fit the corresponding ACRIM data over an overlapping period of 250 days on each side of the ACRIM sets. In early 1996 the VIRGO data took over, again shifted to agree with ACRIM-II. Finally, the composite record was adjusted via ACRIM-II to the Space Absolute Radiometer Reference to allow the comparison of different space experiments. The data from Earth Radiation Budget Experiment and ACRIM-III and empirical modeling are used for comparisons and for internal consistency checks. Compiled by C. Frohlich and J. Lean, courtesy of E. Erwin and the National Geophysical Data Center; see World Radiation Center (http://www.pmodwrc.ch/).

long-term solar trends. Cosmogenic radiogenic isotopes of carbon (^{14}C) (Stuiver 1965), beryllium (^{10}Be) (Beer et al. 1988), and chlorine (^{36}Cl) (Baumgartner et al. 1998) are produced in the atmosphere by high-energy cosmic rays modified by the strength of the earth's geomagnetic intensity field. The ^{14}C production rate is computed as

$$Q = C * A_{geo} * A_{sol}$$

where C is the production rate by unattenuated cosmic rays and A_{geo} and A_{sol} are geomagnetic and solar attenuation factors. Classic studies of ^{14}C, mainly from tree cellulose, by de Vries (1958), Stuiver (1965), and Suess (1965, 1968), and ^{10}Be measurements from ice cores by Beer et al. (1988, 1994) demonstrated the potential of cosmogenic isotopes for solar studies (see Beer et al. 1988, 1990; Stuiver and Braziunas 1989; Raisbeck et al. 1992). However, we saw in earlier chapters that there are many pitfalls using cosmogenic isotopes as direct proxies for solar activity (Muscheler et al. 2004; Snowball and Muscheler 2007). Cosmogenic isotopes trace production rate changes in the atmosphere because of irradiance changes modulated by solar magnetic processes and the earth's geomagnetic field. Long-term solar magnetic variability is poorly constrained, and geomagnetic intensity must be reconstructed independently from sediment and ice-core records and is subject to its own sources of uncertainty.

Once produced, carbon and beryllium isotopes behave differently in the atmosphere. ^{14}C is oxidized into ^{14}CO$_2$, which has a longer atmospheric lifetime (~50–200 years) than beryllium. Radioactive carbon circulates between the oceans and the biosphere, incorporated into organic material via the global carbon cycle, where processes internal to the climate system (ocean circulation, photosynthesis, nutrients) influence the ^{14}C concentrations in atmospheric, ocean, and biological systems. ^{10}Be is removed from the atmosphere within weeks of formation and generally records changes in atmospheric production. Short-term atmospheric processes, such as wind transport, can influence the deposition of ^{10}Be in polar ice and complicate the primary ^{10}Be signal resulting from solar output and geomagnetic and solar modulation. Beer et al. (1988) concluded that atmospheric mixing and transport had minor effects on ^{10}Be concentrations, that atmospheric production rate dominated the signal, and that ^{10}Be appeared to vary over 11-year sunspot cycles and possibly at 80- to 90-year periods (Gleissberg cycles).

Figure 11.5 shows two 1000-year solar modulation function curves (Φ, given in MeV), one based on a composite

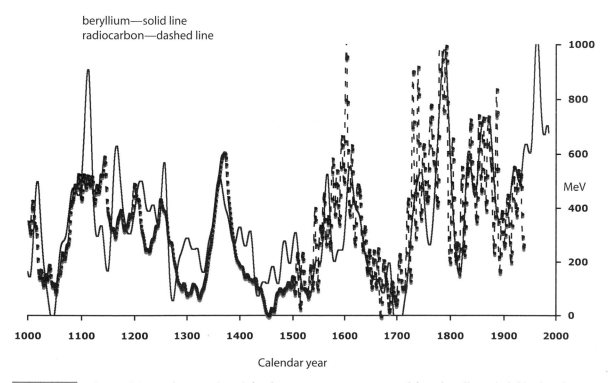

Solar modulation from cosmogenic isotopes

beryllium—solid line
radiocarbon—dashed line

FIGURE 11.5 Solar modulation function (MeV) for the past 1000 years, estimated from beryllium (solid line) and radiocarbon (dotted line with circles) cosmogenic isotopes, from Muscheler et al. (2007), normalized to the geomagnetic field intensity record of Yang et al. (2000). Data courtesy of R. Muscheler, U.S. National Center for Atmospheric Research.

beryllium-radiocarbon record from several Greenland and Antarctic ice cores normalized for geomagnetic intensity changes, and the other on radiocarbon measurements and carbon cycle modeling (Muscheler et al. 2004, 2007). Muscheler's exceptional record can be used to evaluate solar modulation of and climatic influences on cosmogenic isotope production. In general, the two cosmogenic isotopes vary in tandem for most of the past millennium, providing strong evidence for solar modulation of radionuclide production rates. Low Φ values are centered on A.D. 1300, 1450, and 1700. The authors also showed that ^{10}Be concentrations in ice are influenced by local climate variability as well as solar irradiance, and noted a mismatch around A.D. 1780 between reconstructions of solar activity from sunspots and the radionuclide record. Their study reinforces the view that despite cycling processes and geomagnetic effects, radionuclides can record changes in solar activity, but they are not precise, one-to-one proxies of solar magnetic, sunspot, or faculae activity.

Long-Term Solar Variability The likelihood that past solar variability exceeded the variability during 11-year solar cycles has implications for the interpretation of LIA and recent climate trends (Fröhlich and Lean 2004; Foukal et al. 2006) and for the assessment of human-induced and volcanic forcing using climate models (Crowley 2000; Bertrand et al. 2002; Ammann et al. 2003, 2007). Solar variability during the Maunder Minimum is linked to LIA cooling on the basis of both empirical and climate modeling research. There is a range of estimates for the solar radiative during the Maunder Minimum (Solanki and Fligge 1999; Solanki et al. 2000; Y. Wang et al. 2005; Forster et al. 2007b). Most suggest a lower radiative solar forcing at somewhere from 0.2–0.35%, or about 2–5 W m^{-2} as compared to modern sunspot minima. A best-guess Maunder Minimum estimate is ~0.24% (Lean et al. 1992; Crowley and Kim 1996; Lean 2000; Shindell et al. 2001). For comparison, Willson's (1997) ACRIM data shows an upward trend in the total solar irradiance at a rate of 0.036% per decade between the minima of sunspot cycles 21 and 22 (1980–1990s). These measurements imply that a rate of irradiance change in the opposite direction sustained over 200 years would be sufficient to account for LIA cooling of about 0.4–1.5°C.

Solar Forcing and Paleoclimate Records Reconstructed cyclic climate variability from proxies is often attributed to solar variability, and some of the best records come from highly resolved tree-ring (Stuiver and Braziunas 1987; Damon and Sonett 1991; Jirikowic and Damon 1994) and ice-core (Stuiver and Braziunas 1993; Stuiver et al. 1997; Thompson et al. 2002) records. On the basis of tree-ring and ice-core evidence for solar influence, Stuiver and Braziunas concluded

that there was a solar component to Greenland's climate variability over the past 1000 years.

The ~200-year de Vries cycle is a common frequency found in non-tree-ring paleorecords. Poore et al. (2003) recognized ~200-year cycles in Gulf of Mexico planktonic foraminiferal isotopic records of sea-surface conditions, and Lund and Curry (2004, 2006) found multidecadal and centennial cycles in Holocene isotopic records of SST changes off Florida. Lund's comparison of low- and high-latitude records from the Atlantic (Bond et al. 2001) and the Caribbean (Peterson et al. 1991) indicate that both solar variability and internal processes (NAO) may play a role in Gulf Stream strength, but a conclusive solar-oceanic SST link is not yet possible.

In North America, terrestrial records suggest a solar influence on late Holocene climate. These include the Moon Lake diatom record of drought in the Great Plains (Laird et al. 1996), the Rice Lake, North Dakota drought record (Yu and Ito 1999), fire history and prairie droughts (Brown et al. 2005), and the Steel Lake, Minnesota 3100-year oxygen isotope record of drought (Tian et al. 2006). Fluctuations in Holocene alpine glaciers in the North American cordillera have been linked to solar activity by a number of workers. Luckman and Wilson (2005) reconstructed temperatures in the Canadian Rocky Mountains using spruce tree rings, concluding that solar and perhaps volcanic activity were mainly responsible for observed patterns. This study showed that multidecadal and centennial tree-ring records appear to be coherent with other northern-hemisphere paleoclimate reconstructions. Koch and Clague (2006) reviewed glacier fluctuations of the North American cordillera and compared them to the decadal sunspot number record, showing some striking matches over the 1000 years and the entire Holocene. Periods of low sunspot numbers and low solar insolation correspond to times of glacial advances.

Monsoon rainfall in the American southwest (Asmerom et al. 2007) and East Asia (P. Wang et al. 2005) is linked to solar variability. High solar activity corresponds with decreased rainfall in southwestern North America and vice versa for the Asian monsoon. Asmerom's suggests that solar activity influences El Niño–Southern Oscillation (ENSO) and Pacific Decadal Oscillation (PDO) patterns. Hodell et al. (2001) reconstructed drought history from Lake Chichancanab sediments from the Mexican Yucatan Peninsula for the past 2600 years. A clear 208-year cycle in oxygen isotopes and gypsum deposition was evident, suspiciously close to the well-known 206-year solar cycle inferred from cosmogenic isotope records. They attributed cyclic drought and patterns of cultural development, including the ultimate demise of the Mayan cultures in Central America, to solar forcing.

Miyahara et al. (2004) analyzed tree-ring and other climate records for the past few centuries and discovered that during the Maunder Minimum, instead of a 22-year periodicity like the modern double sunspot cycle, a Maunder cycle of 28 years suggested long-term changes in irradiance patterns since the LIA.

Solar-Climate Summary Hoyt and Schatten (1997) commented that "recent studies make a good case that the sun's radiant output varies over decades and longer time scales and that these variations are playing a significant role in climate change." Qualitatively, this view is supported by cosmogenic and some paleoclimate records. Nonetheless, long-term solar variability and mechanisms translating it into climate changes are still poorly understood even for the past few centuries. Briffa (1994) could not find evidence from Scandinavian tree rings for postulated 11-year, 22-year, and 80–90-year cycles and was skeptical of the role of solar activity and climate. Friis-Christensen and Lassen (1991) argued that instead of sunspot number, the length of the sunspot cycle provides a more valid means to examine the solar-climate linkage. High solar activity is associated with short sunspot-cycle length and vice versa. Friis-Christensen and Lassen demonstrated a statistical correspondence between sunspot cycle length and northern-hemisphere surface land temperature for the past 130 years and concluded that their results supported, but did not prove, the idea that solar activity influences the earth's surface temperatures. Damon and Peristykh (1999, 2005) reanalyzed solar-cycle length for the past 300 years using the Mann et al. (1999) reconstruction as a boundary condition. They concluded that solar forcing accounted for 25% of hemispheric warming up until 1980 and

15% until 1997, whereas Beer et al. (2000) attributed an estimated 40% of hemispheric warming to solar processes since about A.D. 1860.

Rind (2002) argued that the sun's role in climate remains ambiguous because the expected magnitude of observed or reconstructed climate changes far exceeds that expected from observed or reconstructed solar activity and because the feedback mechanisms that translate the small changes in solar irradiance into large climate changes are still unclear. Lean et al. (2005) emphasized that any purported solar forcing of short-term climate requires mechanisms to amplify the irradiance signal, perhaps through its effect on stratospheric temperature, humidity, atmospheric convection, or low-latitude Hadley cell circulation. She points out that climate models are as yet unable to simulate recent solar-climate relationships with sufficient skill, let alone long-term changes over centuries or millennia. Other causes of climate variability compete with solar activity as causal factors, as amplifying mechanisms, or both for small changes in solar input. These include decadal-to-centennial changes of climate due to internal dynamics of the ocean-atmosphere system, some involving land- and sea-ice feedbacks and MOC.

Volcanic Activity

Volcanic eruptions can send large quantities of sulfate gases into the atmosphere, where they convert to particulate sulfate aerosols and influence radiative forcing (Robock 2000; Simpkin and Siebert 1994). Figure 11.6 shows estimated volcanic forcing based on Crowley's (2000) compilation. Note

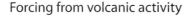

Forcing from volcanic activity

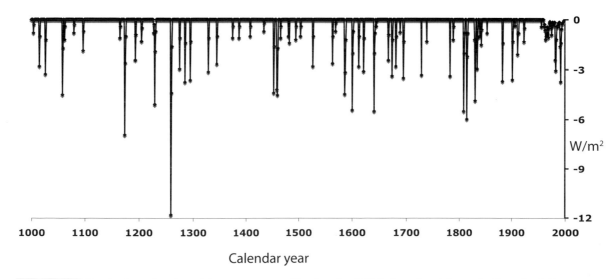

FIGURE 11.6 Forcing by volcanic activity, reconstructed by Crowley (2000), based mainly on the Greenland Crete and Greenland Ice Sheet Project 2 (GISP2) ice-core records of volcanism derived from sulphate measurements in ice (see Hammer et al. 1980; Zielinski et al. 1994, 1997).

that eruptions produce a negative forcing (cooling), plotted as negative values along the y-axis. Volcanic aerosols usually have a residence time in the atmosphere of only one to three years, depending on the size of the eruption. One primary effect is a general cooling of the surface of the earth by scattering radiation back to space. With important exceptions, eruptions from low-latitude volcanoes generally seem to have a more "global" effect, influencing climate in both northern and southern hemispheres, while higher-latitude events affect more climate in the hemisphere where they occur. In addition to directly affecting radiative balance of the earth, volcanic events have more complex effects on other parts of the climate system: stratospheric heating and ozone depletion, internal climate processes such as ENSO and the Arctic Oscillation (Self 2006; Stenchikov et al. 2002), and seasonal impacts over Europe (Luterbacher et al. 2004). Indirect effects occur over different spatial and temporal scales; some typically occur over less than a year and are difficult to identify in paleorecords.

Historical Eruptions Several famous historical events illustrate the climatic impact of volcanic eruptions (Rampino and Self 1992). One was the 1815 eruption of the Indonesian volcano Tambora, which led to the famous "year without a summer" in Europe in 1816. The Tambora eruption can be identified by a sharp spike in sulfate and dust content in ice cores from Peru, Greenland, and Antarctica. Two smaller eruptions were the 1883 eruption of Krakatau in Indonesia and the 1982 El Chichón event in Mexico. More recently, the eruption of Mount Pinatubo in the Philippines in 1991 gave atmospheric scientists an unprecedented chance to observe in detail the global effects of a giant eruption. Among its many impacts, Pinatubo reduced direct radiation at Mauna Loa, Hawaii from 520–390 W m^{-2} (Russell et al. 1996). Minnis et al. (1993) showed significant effects of Pinatubo on aerosol size, distribution, and radiative forcing. In addition to direct effects on albedo in cloud-free regions, they showed substantial variability of Pinatubo's impacts on a broad spatial scale between 40°N and 40°S latitude.

Zielinski (2000) reviewed the volcanic-climate link over geological timescales, focusing on ice-core records. He cautioned that instrumental and anecdotal records of historical eruptions are too limited to capture the full range of climatic impacts for Holocene and older volcanic events. Multiple eruptions can in fact influence lower-frequency (decadal to multidecadal) climate patterns, and some extremely large eruptions known from their widespread ash layers can have even longer climatic effects. Another limitation of ice-core records of volcanism is the difficulty in distinguishing those eruptions that reach the earth's stratosphere from those that remain in the troposphere from ice cores, on the basis of ice-core sulfate concentrations. Stratospheric eruptions occur

every few years, can have long-term radiative impacts, and might be identified using new sulfate isotopic analyses (Savarino et al. 2003).

Volcanic Indices Volcanic activity reconstructed from a combination of historical records of past eruptions yield indices that quantify the impact of volcanic activity on the atmosphere (Robock and Free 1995, 1996). These include optical dust derived from the dust veil index (DVI/Emax) (Lamb 1970, 1983), the volcanic explosivity index (VEI) (Newhall and Self 1982), and the ice-core volcanic index (IVI). Indices are typically plotted as optical depth (τ, at $\lambda = 0.55\,\mu m$) (e.g., Sato et al. 1993). The El Chichón and Pinatubo events are seen as dramatic, short-lived events. These measurements, combined with ice-core records of volcanism (Zielinski et al. 1994; Robock and Free 1996; Castellano et al. 2004), constitute volcanic forcing indices used to isolate and quantify nongreenhouse gas forcing such as those in Figure 11.6 (Robock and Free 1996; Crowley 2000).

Volcanic Impacts Volcanic eruptions have direct and indirect impacts on atmospheric radiative properties, resulting in complex spatial and temporal climatic patterns. Seasonal impacts of eruptions also vary, depending on the geographical origin of the eruptions. For example, Oman et al. (2005) showed that high-latitude volcanic eruptions produce mainly winter season cooling. Fischer et al. (2007) demonstrated that during the past 15 years European seasonal and mean annual temperature exhibited a complex response in the year following 15 known tropical volcanic eruptions. In general, these events produced colder summers and warmer winters, with mean surface temperature changes in the range of about −0.5 and +0.75°C, respectively.

Volcanic eruptions can influence internal variability, but there is still much uncertainty about the effects of other processes. For example, Robock (2000) concluded that evidence was not strong that volcanic activity influenced ENSO variability, but other studies suggest a link between volcanism and interannual ENSO pattern (Adams et al. 2003). Mann et al. (2005a) examined volcanic and solar forcing of ENSO activity using the model of Zebiak and Cane (1987) and found at least statistically a tendency for the Pacific region to experience El Niño (warm) conditions following explosive volcanic events in low-latitude volcanoes. Moreover, they were able to reproduce this response in the Zebiak-Cane model and to correlate the ENSO-volcanic-solar relationships with Pacific coral records of ENSO history. In addition to possible impacts on ENSO, analyses of 600 years of tropical volcanism uncovered multidecadal patterns in ice-core sulfate signatures, supporting the view that multiple volcanic events can have a cumulative effect on climate (Ammann and Naveau 2003).

In sum, when viewing the volcanic forcing indices used in climate models, it is wise to keep in mind the poorly known indirect impacts of eruptions, as well as the various methods used to capture volcanic activity at different timescales.

Greenhouse Gas Forcing

Greenouse gas concentrations influenced by human activity can warm the atmosphere near the earth's surface because they absorb long wavelengths of terrestrial (i.e., thermal) radiation reemitted from the earth. Sweden's Svante Arrhenius (Arrhenius 1896) implicated atmospheric CO_2 from volcanic processes as an explanation for glacial episodes, proposed that fossil-fuel burning might influence climate, and estimated that it might warm the earth by 4°C. American geologist T. C. Chamberlin (1899) also recognized atmospheric CO_2 as an important factor in climate. He proposed that higher weathering rates and reduced marine calcium carbonate production would combine to reduce atmospheric CO_2 levels. Chamberlin also believed that volcanoes played a role in elevating CO_2 concentrations in the atmosphere, reasoning that periods such as the Cretaceous had high CO_2 concentrations and warm climate. He posited that low continental elevations, high marine carbonate production, and low weathering rates of continental rocks were all factors controlling the global carbon budget. During the early part of the 20th century, efforts were made to estimate past CO_2 concentrations, but a direct means to measure ancient CO_2 concentrations was not yet available (Revelle and Suess 1957).

Charles D. Keeling published landmark papers showing a startling rise in atmospheric CO_2 concentrations from 312 to 330 ppmv at Mauna Loa, Hawaii, between 1958 and 1972 due to human activities (Keeling 1973; Keeling et al. 1976). At that time, scientists had limited knowledge about preindustrial atmospheric CO_2 content, but it was clear the 20th-century rise in CO_2 was due mainly to fossil-fuel emissions, cement production, and deforestation. Since Keeling's discoveries, research efforts accelerated to gain a better of understanding of natural and human-induced greenhouse gas forcing. In 1980, unambiguous evidence emerged from Antarctic ice cores that preindustrial, 19th-century concentrations of atmospheric CO_2 were significantly below modern levels. The ice-core record showed that CO_2 concentrations were ~280–290 ppmv during preindustrial times, about 100 ppmv below current levels. This represents an ~36% increase in CO_2 concentration over natural interglacial levels. Since 1960 the annual rate of carbon addition to the atmosphere has been about 1.4 ppmv. From 1999 to 2005 concentrations increased at an annual rate of 6.5–7.8 Gt. Hansen and Sato (2004) predicted future rates would be about 1.8–2 ppmv yr^{-1}.

Figure 11.7 shows IPCC estimates of greenhouse gas radiative forcing from CO_2, methane, nitrous oxide, and other factors. Radiative forcing is usually measured in watts per square meter (W m^{-2}). A change in radiative forcing called adjusted forcing refers to mean changes in the earth's troposphere after equilibration occurs in the stratosphere. Hansen et al. (2005, 2007) also use the term *effective forcing* (Fe), a measure of total radiative forcing that takes into account the different warming potential of the three primary greenhouse gases. Net changes in radiative forcing that correspond to warming and cooling are indicated by positive and negative values. The change in near-surface temperature, $\Delta T_{surface}$, due to a change in radiative forcing is expressed in the linear relationship (Ramaswamy et al. 2001).

$$\Delta T_{surface} = \lambda \, RF$$

As discussed in earlier chapters, CO_2, methane, and nitrous oxide concentrations reconstructed from ice-core records show complex variability governed by changes in their sources and sinks. In Chapter 8 of this volume we summarized information on atmospheric CO_2, methane, and nitrous oxide concentrations from instrumental data (Forster et al. 2007b) and ice-core records. Current concentrations significantly exceed those of the past 700,000 years and the current rate of addition of CO_2 is 100 to 1000 times faster than rates during the last deglacial interval.

Climate Sensitivity $\Delta T_{surface}$ is one general measure of the climate sensitivity of the earth that tries to capture the new equilibrium mean annual temperature after an initial radiative forcing and subsequent operation of climate processes and feedback mechanisms. One conventional way to define equilibrium climate sensitivity is the change in mean annual, global surface temperature in response to the doubling of preindustrial CO_2 levels of 280 ppmv.

As introduced in Chapter 1, refining climate sensitivity to greenhouse gas forcing has been a persistent problem in climatology. One approach is to use paleotemperature reconstructions for the last 1000 years under the assumption that forcing agents such as greenhouse gases and solar and volcanic variability dominate the temperature signal (Hegerl et al. 2006). Orbital insolation forcing influences surface albedo and can also provide information on climate sensitivity (Hansen et al. 2005, 2007). Other estimates of sensitivity come from instrumental records and climate models (e.g., Annan and Hargreaves 2006).

Climate sensitivity values usually range from 1.5–4.5°C, with "best guess" estimates of ~2 or 3°C. Climate sensitivity can also be expressed mathematically as a probability distribution with a peak near 3°C and a tail of decreasing probabilities toward larger temperature changes up to 5°C or more (Bony et al. 2006). The wide range of climate sensitivity estimates stems from uncertainty in climate processes (clouds, internal feedbacks) and stochastic "noise" in the climate

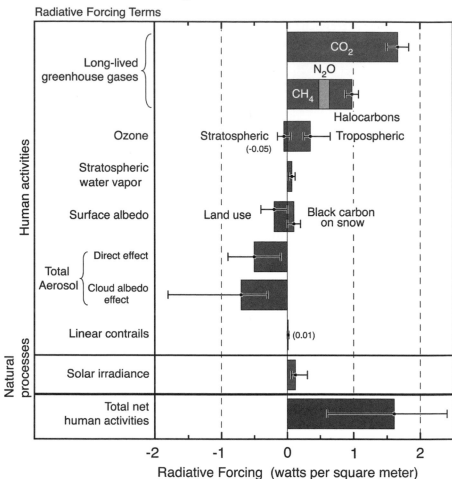

FIGURE 11.7 Summary of principal components of the radiative forcing of climate change. All these radiative forcings result from one or more factors that affect climate and are associated with human activities or natural process, as discussed in the text. The values represent the forcings in 2005 in relation to the start of the industrial era (about 1750). Human activities cause significant changes in long-lived gases, ozone, water vapor, surface albedo, aerosols, and contrails. The only increase in natural forcing of any significance between 1750 and 2005 occurred in solar irradiance. Positive forcings lead to climate warming and negative forcings lead to cooling. The thin black line attached to each colored bar represents the range of uncertainty for the respective value. The figure and caption are from FAQ 2.1, figure 2, IPCC chapter 2, "Changes in Atmospheric Constituents and in Radiative Forcing" (Forster, Ramaswamy et al. 2007), Cambridge University Press. Courtesy of the Intergovernmental Panel on Climate Change.

system. According to Roe and Baker (2007), the probability distribution is a consequence of the nature of the climate system itself and cannot be improved upon by better integration of climate processes. Complicating the climate sensitivity issue, as we saw in earlier chapters, is that the earth's climate sensitivity is itself subject to change, depending on background climate state.

The IPCC estimated that the combined increase in greenhouse gases since preindustrial times due to human activity has so far led to a globally averaged $+2.07-+2.53\,\mathrm{W\,m^{-2}}$ increase in radiative forcing (Forster et al. 2007b). Hansen et al. (2005) estimated the total human-induced radiative forcing since 1880 caused by well-mixed gasses to be $2.75\,\mathrm{W\,m^{-2}}$. Warming has been partially offset by cooling caused by a variety of aerosol particulates that contribute to a direct negative forcing between -0.9 and $-0.1\,\mathrm{W\,m^{-2}}$ and indirect cooling through changes in clouds and albedo of -1.8 to $-0.3\,\mathrm{W\,m^{-2}}$. In addition to the influence of greenhouse gases on mean temperature, they also appear to influence the

phase and amplitude of the annual cycle or surface air temperature (Stine et al. 2009).

Climate Feedbacks and Temperature Variability

In addition to external forcing and internal climate variability, feedbacks from atmospheric circulation, chemistry, water vapor, ocean circulation, and land-surface conditions modulate natural climate variability and anthropogenic greenhouse gas forcing. Natural vegetation-climate feedbacks, modification of monsoons, and high-latitude albedo changes from snow-vegetation-cover all played a role in Holocene climate. Negative feedbacks related to human-induced land-cover changes have been taking place on a large scale for several centuries. These changes can influence radiative forcing as well as global carbon cycling (Houghton 2003, 2007; Foley et al. 2005). Land-surface changes contributed ~-0.1 to $-0.2\,\mathrm{W\,m^{-2}}$ forcing during the past century (Figure 11.7);

however, Brovkin et al. (2006) estimated that this factor might be as high as 0.5 W m⁻².

Land-use changes in Europe probably influenced climate during the MWP and LIA (Goosse et al. 2006). Using the best available historical evidence for European deforestation, continental-scale proxy temperature reconstructions (Luterbacher et al. 2004), and local records, Goosse et al. (2006) applied a general-circulation model (GCM) of intermediate complexity called the ECBILT-CLIO-VECODE model to simulate the combined effects of solar, volcanic, greenhouse gas, and deforestation on surface temperatures. The distinctly cooler LIA climate, compared to relative warmth in the early part of the millennium (0.3°C winter and 0.4°C summer cooling), could be attributed to land-use changes in Europe. They also concluded that the difference between mean temperature in A.D. 1976–2000 and the MWP in A.D. 1026–1050 could not be attributed to solar or volcanic forcing. In contrast to hemispheric temperature reconstructions, climate forcing caused by regional land-use changes made it difficult to detect an anthropogenic impact on European temperature. Regional land-use change in south Florida has also been implicated as a factor in 20th-century regional drying of the Everglades wetland system (Marshall et al. 2004). Although a full discussion is beyond our scope here, the integration of land-surface processes into assessment of temperature trends is one of many issues that complicate efforts to detect and quantify a temperature impact from greenhouse gas forcing.

The Development of Surface Air Temperature Reconstructions

The reconstruction of the earth's surface temperatures for the last 1000 years has been carried out on a number of fronts, most notably in the studies of Michael Mann, Ray Bradley, Malcolm Hughes, Phil Jones, and colleagues. Splicing records mainly from tree rings, ice cores, and sediments with instrumental records, they developed the most spatially and temporally robust temperature curve existing for the evaluation of 20th-century warming. A convenient place to begin is the Bradley and Jones (1993) reconstruction of temperatures from tree rings. Building on experience with late Holocene climate (Bradley and Jones, 1992), the 1993 paper was actually a survey of LIA summer temperatures reconstructed from tree rings and ice cores. They concluded that both an absence of major volcanic activity and rising greenhouse gas concentrations were "probably related" to the observed 20th-century warming. This was one of the first, if not *the* first, comprehensive paleoreconstruction to explic-

itly attribute warming to human activities and to distinguish it from natural temperature trends. It became the foundation of more recent paleotemperature reconstructions and the source of intense scrutiny.

Building on the Bradley-Jones survey and a growing database of global tree-ring and ice-core records, Mann et al. (1995) studied 35 records covering the past 600 years, 12 records having long timescales at decadal resolution, 21 going back to A.D. 1615, and the others to A.D. 1730. Their objectives were to evaluate growing evidence for multidecadal (15–35-year) and centennial (50–150-year) variability in climate. They concluded that there was convincing evidence for persistent SAT variability at these timescales that could not be evaluated using only instrumental records. They postulated that temperature variability might be associated with ENSO, NAO, or even lower-frequency thermohaline circulation changes. Notably, they did not address the issue of 20th-century warming and did not even cover the post-1960s period, when the largest warming has occurred.

Later, Mann et al. (1998) expanded their analysis to incorporate more than 100 time series records from multiple proxy sources into a mean annual surface temperature curve spanning the last 400 years. The focus was on statistical methods used to calibrate proxy methods to instrumental records over 1902–1995 and then to examine the "skill" of the methods by verifying the temperatures with a few long-term instrumental records from 1854 to 1901. This latter procedure is also called "cross-validation." Calibration and verification of proxy data has become commonplace in studies of Holocene climate. Mann et al. found that northern-hemisphere temperature was a good "diagnostic" of global climate and, when compared to CO_2, solar, and dust veil indices of climate forcing mechanisms, they came to what would turn out to be a controversial conclusion—greenhouse gas forcing was the dominant agent of temperature change over the past century but solar forcing played a partial role in 20th-century temperature, and both solar and volcanic forcing played important roles over the past 400 years. Their results were consistent with early-model simulations suggesting that 20th-century warming was anomalous and influenced by anthropogenic greenhouse gas forcing.

Using 12 long temperature records, three of which were actually principal components produced mathematically by multivariate analyses of tree-ring records, Mann et al. (1999) extended the temperature record back to A.D. 1000. They concluded that the 20th century was "nominally the warmest of the millennium" but that during the 11th and 12th centuries, part of the MWP, temperatures were within a standard deviation of 20th-century temperature. Decadal temperature averages indicated that the decade from 1989 to 1998 was the warmest in the proxy and instrumental record and the last few years of the 1990s were the warmest of the millennium. Mann et al. also concluded what had long been suspected and was becoming evident from other studies. Regional temperature changes were often greater than those of the hemispheric mean, e.g., the MWP in Europe (Crowley and Lowery 2000; Goosse et al. 2006).

Jones et al. (2001) also analyzed 20th-century temperature and concluded that the average SAT for the period 1970 to 2000 was the warmest in the past 1000 years, about 0.2°C warmer than temperatures during the 11th and 12th centuries. The estimated rate of warming based on instrumental records back to 1850 was 0.6°C per century (0.8°C in winter, 0.4°C in summer). Warming was most pronounced from 1920 to 1945 and from 1970 to present; during the 1950s and 1960s, temperature was affected by the cooling effect of sulfate aerosols. Mann and Jones (2003) extended the SAT record back 2000 years, an important advance given that marine and coastal records show evidence for regional warmth at about A.D. 700–1000 (deMenocal et al. 2000; Cronin et al. 2003; Richey et al. 2007). Mann and Jones concluded that the 20th century experienced unprecedented warmth for the past 2000 years but conceded that southern-hemisphere data was too sparse to reach a conclusion. Osborn and Briffa (2006) also evaluated 20th-century conditions, focusing on the period between A.D. 800 and 1995, using 14 annual-to-decadal smoothed temperature records from 800 to 1995. Again, the conclusion was that the 20th century was anomalous in terms of temperature for the period studied.

Limitations to Atmospheric Temperature Reconstructions

The basic conclusion reached by most researchers analyzing patterns of mean northern-hemisphere temperature for the last 400–1000 years is that the 20th century, or parts of it, was the warmest period in the last millennium. A second conclusion, offered with varying degrees of confidence, is that 20th-century warming is anomalous in terms of natural external forcing and internal variability and is in part human-induced. The scrutiny, statistical reanalyses, and reinterpretation of the temperature curve have been substantial. Controversy revolves around relatively trivial methodological aspects; however, such attention seems to reflect the community's need to detect relatively small mean decadal temperature changes of only tenths of a degree Celsius. Our goal in this section is to summarize the response of the scientific community in three particular areas—methodological and statistical aspects, proxy limitations, and natural Holocene climate variability.

Methodological Issues

All hemispheric temperature reconstructions dominated by tree rings suffer from a large drop-off in the number of records available before about 400 years ago. This weakness led two large review panels to adopt a conservative view regarding a long-term perspective of 20th-century warming. The hypothesis that the 20th century was the warmest in the past 400 years was reaffirmed by the analysis of the temperature curve by the U.S. National Academy of Sciences (National Academy of Sciences 2006). However, the report concluded that a comparison with temperatures of the past 1000 years was premature mainly because of the relatively few annually resolved records older than 400 years and the lack of spatial resolution, especially in the southern hemisphere. In another group assessment, the IPCC considered the statistical confidence level for anomalous temperatures to be only 66–90% certainty (Jansen et al. 2007), a considerably wider range of uncertainty than the 90–95% confidence limits seen in the original papers.

Compilation and Calibration Methods Reanalysis of temperature curves has been based on methodological and statistical arguments surrounding the analyses of the paleodata. One aspect is the distinction between composite-plus-scale and climate field reconstruction (CFR) calibration proxy techniques (see Rutherford et al. 2005). The former method attempts to simplify SAT curves by first assembling a composite of proxy records by averaging a set of time series of the proxy (i.e., tree-ring width) and then scaling the composite into the target parameter, in this case northern-hemisphere temperature for an overlap period. The measured proxy is related (or so it is assumed) to a specific climate parameter such as temperature, or to a climate variable such as ENSO. In contrast to the composite-plus-scale method, the CFR produces a large-scale field from the multivariate correlation of proxy time series and regional climate variables, without assuming any direct correlation between the proxy and temperature. CFR methods sometimes focus on regional observational and proxy data sets aiming to identify the leading modes of climate variability. For example, one field approach uses reduced-space objective analysis procedures to evaluate evolving tropical climate fields over time. Evans et al. (2002) performed such analyses on coral oxygen isotopic records and was able to identify a primary ENSO mode superimposed on a more long-term trend.

Pseudo-Proxies As discussed in the last chapter, von Storch et al. (2004) suggested that empirically based paleo-proxy-time series from CFRs were unable to skillfully reproduce low-frequency climate trends because of "noise" in the proxy data. The pseudo-proxies used in ocean-atmosphere model simulations serve as a source of background climate variability. The underestimation of centennial-scale variability by paleo-proxies was to von Storch a major impediment to interpreting low-amplitude temperature of the last 1000 years. Mann et al. (2005b) responded to this issue by performing a comparison between composite-plus-scale paleotemperature CFR methods and found relatively minor differences between the two methods when applied to northern-hemispheric temperature patterns (Rutherford et al. 2005).

Sample Selection Criteria The criteria used to select proxy records are also subject to interpretation. Rather than a composite reconstruction based on numerous tree-ring records, Osborn and Briffa (2006) used 14 long-term records, not necessarily annually resolved, but each going back about 1200 years and positively correlated with regional instrumental records. They found the "most significant and longest duration feature during the last 1200 years is the geographical extent of warmth in the middle to late 20th century." Their chosen set of paleoclimate records also exhibited positive and negative anomalies during the MWP and LIA. Bürger (2007) challenged their time series selection process as introducing bias into their analysis of 1200 years of temperature. Based on statistical grounds and reanalysis of the data correcting for the proposed bias, Burger concluded that the 20th-century warming, in fact, no longer appears to be anomalous. Osborn et al. (2007) defended their selection of 14 records, noting there were very few records from which to choose that met their criteria. As already noted above, tree-ring records longer than 400 years are scarce and high-sediment-rate lacustrine and marine temperature records are widely scattered.

Calibration Period Calibration procedures are subject to differing opinions, in part because most reconstructions are based on an empirically derived proxy-temperature calibration for a selected period of overlap between the proxy and target instrumental data. In addition to practical decisions made during field collecting and laboratory analyses, subjective choices must be made regarding the target calibration data set (land, ocean, or land and ocean temperatures, station versus latitudinal mean temperatures, etc.), and the length of the calibration period (1850–1990, 1900–1980, etc.). Esper et al. (2005a) analyzed several different scenarios, varying the target instrument data and the calibration period for temperature records of the past 150 years. They found a remarkably wide range (0.51–1.31°C) of reconstructed temperature changes between the warmest and the coldest decades that depended on the target data set and the calibration period. Moberg et al. (2005) also found decadal mean

differences ranging from 0.4 to 1.0°C, depending on methodology. These levels of uncertainty may be small compared to those during glacial-interglacial cycles and Eocene climate extremes, but they are a persistent concern in the late Holocene considering the total mean annual 20th-century warming of only ~0.5–0.6°C.

Statistical Scaling The statistical method used to derive the calibration itself is also in dispute because different methods can yield radically different temperature estimates. The scaling method involves a procedure called the equalization of the mean and standard deviation values for proxy and instrumental data. In contrast, a least-squares linear-regression method takes into account the correlation coefficient. The least-squares method tends to reduce the amplitude of the temperature signal proportional to the correlation value. Esper et al. (2005a) illustrate the effect of scaling versus regression using the proxy data of Jones et al. (1998), covering the last 150 years. The difference between the warmest and coldest decades for the period of record was 0.62°C for scaling but only 0.34°C for the regression method.

Principle Components Analysis Criticism of the proxy SAT curve on statistical grounds came from McIntyre and McKitrick (2003), who argued that the Mann et al. (1998, 1999) reconstruction involved a selection bias in the multivariate-analysis procedure. In effect, by choosing the bristlecone pine—a tree species dominating the first principal component for North America for the period 1400 to 1450—in the standardization procedure, they postulated that Mann's analysis would produce a hockey-stick-shaped-curve regardless of what the real temperature history was. This idea was refuted by several authors (e.g., Jones and Mann 2004). Additional criticism from McIntyre and McKitrick (2005) suggested that Mann's tree-ring-based statistical normalization procedures introduced a bias producing a hockey stick–like shape. This possible bias, however, had fairly small impacts on the overall temperature reconstruction (von Storch and Zorita 2005), and the normalization and error estimate methods of McIntyre and McKitrick exaggerated the bias of Mann's methodology (Huybers 2005). Thus, as with any paleoclimate reconstruction, choices must be made that can introduce subjectivity, but, in this case, it seems that statistical considerations do not warrant revision of the general shape of the curve.

Methodology Summary The very brief discussion of methodological and statistical aspects of the temperature-curve debate only touches the surface of a highly complex topic. Clearly, future work is needed to reduce uncertainty in temperature time series for the past few thousand years, the inclusion of pseudo-proxies (von Storch et al. 2004) and variance adjustments to proxy records to reduce artifacts (Frank et al. 2007), the use of longer time series for calibration (Frank and Esper 2005), and the evaluation of different frequencies of climatic variability (Moberg et al. 2005). Such improvements, when combined with modeling, the use of multiple proxies, and the addition of longer non-tree-ring records (see Chapter 12 in this volume), will lead to a much more realistic picture of the complexity of late Holocene temperature patterns and processes.

Proxy Limitations—Divergence and Segment Length Curse

Divergence refers to the apparent reduced sensitivity of growth in several tree species to changing temperatures, noted mainly in high-latitude northern-hemisphere tree-ring records for the past few decades (Figure 11.8) (Briffa 1998a, b). Wilson et al. (2007) summarized the divergence problem stating that no current tree ring-based reconstruction of northern hemisphere temperatures that includes the 1990s is able to capture the range of late 20th century warming seen in instrumental records. This means that instrumental records show warming, but reconstructed temperatures from trees show cooling or no change. Divergence is

particularly important because human-induced climate change is expected to first impact high latitudes, where climate changes appear to be already occurring.

Briffa et al. (2004) examined the divergence problem and found that maximum latewood density tree growth is influenced by mean April-September temperature. Their reconstructions of six centuries of gridded summer temperatures at various spatial scales yielded "quasi-hemispheric" temperature histories reflecting volcanic eruptions and greater century-scale temperature variability than some other reconstructions. They noted that divergence in many tree-ring chronologies precluded the claim that unprecedented hemispheric warming has occurred during recent decades, at least on the basis of tree-ring records.

R. Wilson et al. (2005) considered the divergence problem in the European Alps. They make the important point that the construction of documentary and tree-ring data sets was not originally designed with the study of large regional or hemispheric summer temperature response in mind. They suggested that divergence might be circumvented by using old violins and other string instruments made from "predivergence" wood for calibration.

D'Arrigo et al. (2008) point out that there are still a number of possible causes of divergence: drought stress from high temperatures; other physiological responses to recent

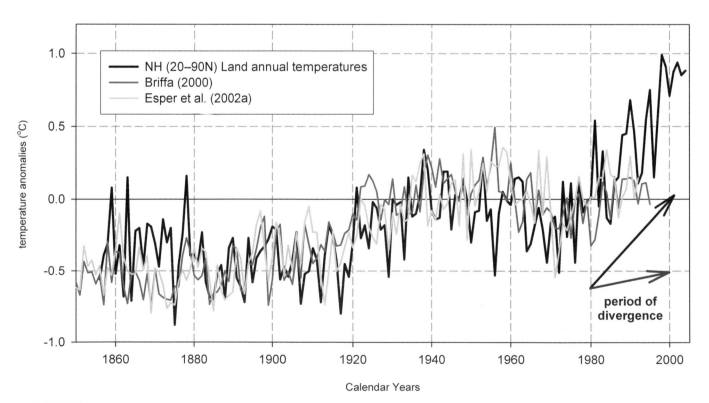

FIGURE 11.8 Divergence in tree-ring records, modified from D'Arrigo et al. (2008). NH is the northern-hemisphere surface air temperature. Other temperature reconstructions are from Briffa (2000) and Esper et al. (2002a). Divergence between tree-ring-reconstructed and instrumental temperatures begins in the 1980s. Courtesy of R. Wilson and Rosanne D'Arrigo.

warming; stratospheric ozone; snowmelt or seasonality changes; uncertainty about the influence of yearly maximum, minimum or, mean temperature on tree growth; and global dimming or decreased levels of solar radiation. It is difficult to identify causes because the environmental changes covary over the 20th century and the response of trees to some environmental parameters may be delayed, making it more complex. Another perplexing aspect is that divergence varies by tree type and some trees do not exhibit divergence. In general terms, divergence seems to be attributed to some kind of anthropogenic influence on the trees' environments, and advances in ecophysiology of various tree species is especially important (Vaganov et al. 2009).

Another problem with tree-ring records is called the "segment length curse" (Cook et al. 1995). This phenomenon refers to the fact that annual growth rings are influenced by environmental conditions, such as temperature, and also vary with the age of the tree. A tree-ring chronology for a particular region is developed from multiple trees and thus subtraction of age-related ring variability can potentially eliminate real environmental signals in the rings. Various methods have been developed to analyze the segment length curves, most commonly what is called the Regional Curve Standardization (RCS) procedure, in which a composite record is derived from multiple trees and smoothed to produce a single growth curve for the area that is not influenced by climate variables (Briffa et al. 1992, 2001; Esper et al. 2005b). The differences between RCS-derived and large-scale composite reconstructions tend to increase farther back in time.

Twentieth-Century Climate in the Context of the Holocene

Throughout this book we have seen unequivocal evidence for large-scale, often abrupt changes in climate, due to many factors. Some authors have taken such changes as evidence that recent warming is an expression of natural variability and not human-induced change. One example is the paper by Soon and Baliunas (2003). These authors examined 1000-year proxy records and concluded that "the 20th century is probably not the warmest nor a uniquely extreme climatic period of the last millennium." They suggested that methods used to reconstruct temperature from various proxies were different in nature, and therefore they could not be used to construct a global or hemispheric composite. In addition, they suggested that if there were an objectively discernible climatic anomaly during the 20th century, defining such an anomaly as a 50-year period of sustained warmth (cool), wetness (dryness), or both was an arbitrary procedure. Their literature search led them to state that there were indeed periods of warm and cool climate during the MWP and LIA, though regionally variable in their expression. The essence

of their argument is that natural climate variability can account for much of observed 20th-century warmth.

Mann et al. (2003, 2005b) addressed the methodological criticism by verifying the compositing method with the field reconstruction method. Other papers cited flaws in the Soon and Baliunas arguments (Jones and Mann 2004; Osborn and Briffa 2006). For example, Soon and Baliunas did not differentiate tree-ring paleoclimate records from those recovered from other parts of the climate system. The majority of studies cited were terrestrial records, most based on ice cores, tree-rings, or glaciological data. Many lacked the requisite chronology, proxy development, or both needed to evaluate the 20th century. Most studies cited were designed to investigate pre-20th-century intervals and not to test the hypothesis of warming in a particular region or hemispheric or global trends. In the cases of oceanic records, one would not expect sufficient temporal resolution to evaluate 20th-century trends, because of low sedimentation rates. Soon and Baliunas did not critically address the errors associated with temperature estimates from proxy methods, the age models and chronology, or regional and hemispheric variability due to internal climate processes.

Nonetheless, an important issue remains about human-induced warming: what is the nature of long-term Holocene climate variability, including the LIA and older Holocene climate cooling events, and to what degree has climate variability been disrupted during the last century? Despite many studies on LIA and MWP climate, the unevenness in the quality of the records echoes a persistent theme that additional long-term (>1000-year) temperature records are needed to fully understand the context of climate during the Anthropocene.

Regional Paleotemperature Reconstructions

Tree-ring research has produced a number of sophisticated regional paleotemperature reconstructions that address many limitations of hemispheric composite reconstructions and offer insights into anthropogenic influence on background climate variability. Esper et al. (2002a) (see also Cook et al. 2004a) synthesized northern-hemisphere extratropical land temperature from 14 tree-ring records, six extending back to the year A.D. 831. Sites were widely scattered throughout Canada, the western United States, Fennoscandinavia, central Europe, and Siberia. They used an RCS (Briffa et al. 1992), which involves complex averaging and statistical analysis effective for evaluating lower-frequency variability. Four conclusions were drawn from their new RCS: (1) the late 20th century was the warmest of the 1162-year period; (2) LIA temperature minima were centered on 1450–1460, 1600–1650, and the early 1800s; (3) an MWP maximum occurred about

1000 years ago; and (4) internal variability must be considered when interpreting any long-term tree-ring compilations. Firm conclusions about comparisons of MWP temperatures with those of the last 30 years were hampered by the scarcity of records and the divergence problem.

D'Arrigo et al. (2005a) compared 400 years of tropical North Pacific tree rings using the North Pacific Index (NPI), a measure of winter conditions of the Aleutian low-pressure system similar to the Pacific-North American Index discussed in Chapter 10. The value of this particular study is that it compares the NPI with 11 carefully selected tropical records of coral oxygen isotopes, which are mainly a proxy of tropical SSTs. Instrumental and coral records suggest that tropical decadal climate variations in the Indo-Pacific region, such as the regime shift to warmer and wetter conditions in 1976, influence higher latitudes in the North Pacific climate (Deser et al. 2004; Linsley et al. 2004). D'Arrigo identified at least 10 regime shifts similar to the 1976 event, and showed that this recent shift was not unique. They also concluded that tropical processes strongly influenced ocean-wide decadal climate variability that was possibly related to decadal or long-term ENSO variability.

D'Arrigo et al. (2005b) analyzed 14 white spruce chronologies from Seward Peninsula, Alaska dating back to the 14th century and found evidence for LIA cooling. A separate composite record of springtime tree-ring width covering the period from A.D. 978 until 2001 showed links to Bering and Chukchi Sea SST patterns and the PDO. Their RCS reconstruction of Seward temperature showed periods of warmth similar to those of the 20th century and attributed much of the Seward pattern to ocean-atmosphere links rather than to broad temperature changes seen in high-latitude northern-hemisphere continental regions (e.g., Overpeck et al. 1997).

Luterbacher et al. (2004) constructed a European network of monthly and seasonal temperatures using multiple proxies reconstructed back to A.D. 1500. In addition to unraveling complex seasonal changes during the LIA-20th-century transition, they came to the conclusion that the late 20th–early 21st century was the warmest interval for that region in 500 years. Since LIA cooling between 1500 and 1900, winter temperatures in Europe warmed by 0.5°C and mean annual temperatures by 0.25°C. Their results were in general agreement with those for the northern hemisphere (e.g., Mann and Jones 2003), but like others (Cook, D'Arrigo, Esper, Wilson) emphasized the importance of regional reconstructions for evaluating 20th-century warming.

D'Arrigo et al. (2006) reanalyzed the MWP, LIA, and 20th-century warming by quantifying the differences between composite-plus-scaling and RCS methods. They concluded that the latter method is better for observing low-frequency trends and that North American and Euro-

pean RCS series match each other well before compositing. This means there may be a common forcing for MWP and LIA climate anomalies. Moreover, they calculated that the MWP was 0.7°C cooler than the late 20th century, and that mean temperature increased a total of ~1.14°C from the coldest decade (1600–1609) to the warmest (1937–1946). Their major conclusion was that it is possible to separate the spatial heterogeneity of MWP climate variability from a more homogeneous global fingerprint for the 20th century.

A number of other studies have identified decadal and lower-frequency regional temperature variability in temperature records of the past few centuries or longer. These include low-latitude, high-elevation records for central Asia (Esper et al. 2002b, 2007), the southwestern United States (Gedalof et al. 2002), and North American Pacific region combined tree rings and coral records (Cook 2003). Trouet et al. (2009) identified a persistent positive North Atlantic Oscillation pattern in several proxy records from Europe and surrounding regions during the Medieval Climate Anomaly between about A.D. 1050 until 1300. The evidence from these and other studies indicates that, apart from hemispheric and global trends, high-amplitude regional temperature variability is commonplace at various frequencies over the past millennium, but these can be successfully detected using regionally based calibrations and proxy records.

Climate Modeling and Proxy Reconstructions

Climate model simulations can be used to test proxy-reconstructed temperature patterns whereas SAT records, when combined with time series of radiative forcing, provide insight into how well models simulate late Holocene climate.

Little Ice Age Modeling

The LIA has received considerable attention because there is great interest in how such a small preindustrial solar forcing can generate the relatively large cooling seen in many proxy records. Shindell et al. (2001) used the GISS GCM, incorporating stratosphere and ozone, to show that solar irradiance changes during the Maunder Minimum could account for much of the observed LIA cooling, although they also detected significant regional variability and a volcanic signal in temperature. Their model showed that the decadal mean temperature around A.D. 1680 was 0.34°C cooler than for A.D. 1780, a difference close to those from most proxy records. The simulations also suggested that cooling was concentrated over continents (Eurasia, North America), whereas some ocean regions actually warmed slightly. They proposed a mechanism to explain LIA cooling in which processes involving stratospheric ozone and stratosphere-troposphere interactions amplified the small irradiance changes through positive feedbacks. Shindell et al. (2003, 2004) showed that solar forcing is generally more important than volcanic forcing, which can be damped by dynamic and radiative processes, and leads to complex geographical response similar to that seen during climate variability caused by the NAO.

Yoshimori et al. (2005) also used GCMs to separate externally forced variability from internal processes during the Maunder Minimum. They achieved an excellent reproduction of short-term (decadal), large-scale volcanic forcing of mean annual temperature seen in proxy records, but a more ambiguous pattern of internal variability. Their temperature simulations were more reliable than those for precipitation, but the authors acknowledged that improvements are needed in long-term solar and volcanic forcing indices.

Modeling External Forcing and Temperature

Tom Crowley conducted a series of studies comparing energy-balance model simulations to a multi-proxy time series of the past 1000 years to detect temperature response to external forcing. Crowley (2000) used decadally smoothed mean annual surface temperature from Mann et al. (1999) and Crowley and Lowery (2000), which included ice-core, coral, and tree-ring records, suggesting that volcanic and solar forcing is responsible for between 41 and 64%, or reconstructed temperature variability, generally consistent with the conclusions of Free and Robock (1999). After removing solar and volcanic forcing, Crowley found that residual temperature patterns produced by model simulations included 20th-century and future temperature anomalies consistent with greenhouse gas forcing suggested by more complex GCMs. Crowley et al. (2003) later incorporated ocean temperatures and updated volcanic, solar, and greenhouse gas time series into energy-balance modeling and was able to successfully simulate observed oceanic warming manifested during the second half of the 20th century in heat content (Levitus et al. 2000) and sea level (Church et al. 2004). They also simulated decreased oceanic heat content during the LIA, which is consistent with paleoceanographic evidence for reduced SSTs. Crowley confirmed that late 20th-century ocean warming was forced by greenhouse gas concentrations as proposed by Levitus et al. (2001).

Bauer et al. (2003) used a model of intermediate complexity called CLIMBER-2 to simulate external forcing during the last 1000 years, but they added an additional factor—global deforestation estimated from historical cropland statistics for the period from 1700 to 1992. They found a positive correlation between simulated and proxy temperatures, confirming Crowley's results. They also discovered that widespread land-use change and relatively cool atmospheric temperature during the late 19th century confirmed the importance of deforestation on cooling. Hegerl et al. (2003) used a linear-regression method called fingerprinting in an energy-balance model to compare several proxy time series of hemispherically averaged temperature. They discovered that known forcing accounted for 49–67% of observed temperature variance between A.D. 1000 and 1960 and that volcanic forcing was consistent with expected radiative response.

Incorporating Unforced Variability

"Unforced" climate variability has been a persistent issue in attempts to explain atmospheric temperature patterns over the past 1000–2000 years. Goosse et al. (2005) found that regional temperatures are dominated by internal variability but that hemispheric and global patterns can be seen as forced by external radiative processes. Internal variability influences the amplitude of regional temperature response, and in some regions the sign. Their results demonstrate the need for improved regional reconstructions and longer (pre-A.D. 1000) proxy time series to more fully understand late Holocene temperature.

Hegerl et al. (2007) performed a detection and attribution experiment aimed at separating the influence of low-frequency, high-amplitude internal variability not easily detected in annually resolved proxy time series from externally forced changes. They applied a new "total least-squares" calibration method to a 1500-year temperature time series derived from surface temperature proxies (Crowley and Lowery 2000; Moberg et al. 2005) and 500-year-long borehole temperature records (Harris and Chapman 2001; Beltrani 2002). Their data sets included 12 decadal records to A.D. 1505, a longer 1500-year record constructed from seven records to A.D. 946, and five records to A.D. 558. Using complex detection and attribution statistics, their basic conclusion was that large hemispheric-scale temperature compilations "detect" a human fingerprint over and above the influence of volcanic forcing and natural variance in climate. By incorporating proxies, external forcing from an energy-balance model, and internal variability generated by a coupled atmosphere-ocean GCM, they concluded that decadal variability is largely externally forced (60–75% of the proxy temperature signal), a volcanic signal is clear, and greenhouse gas forcing overrides unforced variability. Human-induced greenhouse gas forcing accounted for one-third of early 20th-century warming and most of the warming in the late 20th century.

Summary of 1000-Year Temperature Patterns

Several generalizations emerge about the patterns and causes of temperature variability over the past millennium.

- Regardless of which proxies, observations, and methods are used, understanding late Holocene climate variability requires much more than simple statistical comparisons between two decades (or multidecadal comparisons), one warm, the other cold.

- Solar variability affected decadal and centennial climate before and, to a lesser extent during, the 20th century. The 17th-century Maunder Minimum has proven to be an important testing ground for solar forcing on temperature.

- Volcanic activity causes complex regional temperature anomalies, but these are generally associated with short-term cooling. Since ~1925 fewer large eruptions, increased solar irradiance, and greenhouse gas concentrations have led to warming; after ~1960 volcanic and sulfate aerosol forcing remained negative, but this was overwhelmed by greenhouse gas forcing after about 1975.

- Data-model comparisons have been able to detect anthropogenic greenhouse gas forcing during the 20th century over and above solar and volcanic forcing; the role of internal climate variability remains poorly understood.

Perspective

The debate about patterns and causes of late Holocene SAT variability has been constructive. Hands-on experience with field and laboratory methods and an appreciation for principles, practices, and complexities of stratigraphy, geology, geochemistry, tree physiology, glaciology, physical oceanography, climate dynamics, and other disciplines are required to interpret recent temperature trends. Limitations of paleotemperature reconstructions do not reside in, or at least are not dominated by, statistical methods and inferences, but rather are identified by how well researchers understand underlying processes influencing proxies, how good a chronology is, and how spatially robust a particular signal is.

Researchers actively involved with late Holocene paleotemperature reconstruction, climate dynamics, and model simulations appreciate such issues, and there is some consensus about the major features of large-scale temperature change over the last 1000 years. There is firm evidence for a warm but regionally complex MWP, multiple temperature minima during the LIA, and steplike warming during the 20th century. Reconstructed northern-hemispheric SAT curves for the past 400 to 2000 years show generally similar century-scale patterns, though the amplitude of decadal oscillations varies among reconstructions (Esper et al. 2002a, 2005b; Bradley et al. 2003; Jones and Mann 2004; Moberg et al. 2005; Juckes et al. 2007; Mann et al. 2008). It is also generally agreed that solar and volcanic forcing has been overtaken by greenhouse gas forcing the past few decades. Acute future needs include

- Proxy records back to 2000 years from a dense network of terrestrial and continental sites

- High-resolution paleoceanographic proxy records of ocean temperature and circulation

- Reduced proxy errors associated with statistics and calibration

- Long-term reconstructions of solar irradiance and volcanism

- Paleoreconstructions of low frequency and unforced variability to verify model-based pseudo-proxy time series

- Analysis of phasing of temperature response to combined internal and external forcing

- Analysis of the impacts of external forcing on natural modes of variability

---

LANDMARK PAPER The Anthropocene

Bradley, R. S. and P. D. Jones, eds. 1993. "Little Ice Age" summer temperature variations: Their nature and relevance to recent global warming trends. The Holocene 3:367–376.

Human influence on global climate is one of the most intensely debated and misunderstood issues in contemporary natural sciences, spilling over into virtually every corner of the sociopolitical landscape. What caused rising atmospheric temperatures during the past century lies at the heart of this debate.

Ray Bradley of the University of Massachusetts and Phil Jones of the University of East Anglia pioneered efforts to splice instrumental temperature records with those from proxy data to provide a temporal context for 20th-century warming trends. They reconstructed a 600-year-long temperature history for high-latitude northern-hemisphere regions using historical, ice-core, and tree-ring records, with an emphasis on the period known as the Little Ice Age (LIA), or about A.D. 1500 to 1900. In the last 15 years, Michael Mann and Malcolm Hughes, along with

(continued)

LANDMARK PAPER (*continued*)

Bradley, Jones, and colleagues, greatly expanded the density and duration of the proxy network. This work has culminated today in an unprecedented mean annual temperature reconstruction that extends back over the last 2000 years of the late Holocene. The late Holocene temperature reconstruction, known to many as the "hockey stick" curve because of the upturn in temperature above background levels during the last century, became the center of widespread controversy. Although this controversy has taken many statistical and methodological detours, at its core it revolves around distinguishing natural warming following the LIA from that caused by fossil-fuel emissions.

Overshadowed at times by public controversy, the late Holocene temperature curve catalyzed unprecedented research efforts by paleoclimatologists and climate modelers, leading major strides in understanding the causes of temperature variability over annual-to-centennial timescales. The Bradley-Jones analysis deserves credit for much of this progress and, more broadly, the successful application of paleoclimatology to modern climate issues.

12

The Anthropocene II: Climatic and Hydrological Change During the Last 2000 Years

Introduction

This chapter addresses trends in parts of the climate system other than surface air temperature for the past ~2000 years. We focus on atmospheric records of precipitation, drought, and tropical cyclone activity; ocean records of temperature, salinity, circulation, and chemistry; polar sea ice; and alpine glaciers, ice sheets, and sea-level change. Emphasis is mainly on regional climate variability, its context in the global system, and evidence for human influence on climate. We begin with a brief discussion of the scientific community's attempt to address the issue of human-induced climate change, its manifestations, and impacts through detection and attribution studies introduced in Chapter 10 of this volume.

International Ad Hoc Detection and Attribution Group Project

One international effort, called the International Ad Hoc Detection and Attribution Group (IDAG), began in 1995, aimed in part at "assessing and reducing uncertainties in the detection of climate change" over the past millennium (International Ad Hoc Detection and Attribution Group 2005). Most IDAG emphasis is on the last century, where experts draw on an extensive literature on all aspects of the climate system, including but not limited to global and regional atmospheric temperature, circulation, ocean-heat content, sea level, and tropopause height. In a 2005 review paper on its status, IDAG concluded that external forcing from volcanic, solar, and greenhouse gases is evident in parts of the climate system, that observed changes are unlikely to be caused by unforced internal variability, and that "a large fraction of the warming over the last 50 years can be attributed to greenhouse gas increases." The primary paleoclimate context against which IDAG compared instrumental records and model simulations is the hemispheric surface air temperature reconstruction described in Chapter 11 of this volume, with additional minor discussions of trends in sea level. Attribution of causality to precipitation patterns is more complex than those of temperature, and much uncertainty still exists.

Intergovernmental Panel on Climate Change

Another international effort is the Fourth Assessment Report by Working Group I (WG I) of the Intergovernmental Panel on Climate Change (IPCC). In 2007, IPCC WG I, which is focused on the physical basis of climate change, published its Fourth Assessment Report, which included multiauthored chapters on "Changes in Snow, Ice and Frozen Ground" (Lemke et al. 2007), "Oceanic Climate Change and Sea Level" (Bindoff et al. 2007), "Palaeoclimate" (Jansen et al. 2007), and "Coupling Between Changes in the Climate System and Biogeochemistry" (Denman et al. 2007), among others, including several on climate models and forecasts. These chapters are exceptionally useful syntheses of climate processes, observations, and model simulations, with emphasis on scientific advances made since the 2001 IPCC Third Assessment Report. They should be consulted by anyone involved with climate-related issues. The paleoclimate chapter is a concise, up-to-date, and valuable discussion of climate extremes pertinent to future climate.

Working Group II of the Fourth IPCC Assessment Report focused on the impacts of climate change from observational data on physical and biological systems. This multichapter volume surveyed a total of 80,000 data series from 577 published studies. Of these, 29,000 data series from 75 studies ending in the year 1990 or later and covering at least 20 years were used to assess climate trends and human impacts on climate. Most were from land areas in the northern hemisphere. Data series were categorized by region and each was judged on whether the data were "consistent with warming." One conclusion was that the majority (94% globally) of observational data series already show the impacts of human-induced climate change.

Similarly, there is growing evidence that, as more and better instrumental records become available, anthropogenic forcing of various parts of the climate system appears to be progressing at a faster rate than climate models had forecast even just a few years ago. Such a situation pertains to ocean acidification (Feely et al. 2008), Arctic Ocean sea ice (Serreze et al. 2007), and western North American hydrology (Cayan et al. 2001; Knowles et al. 2006), among others. Importantly, the interpretation of the causes of all observed trends strongly depends on still poorly known decadal and longer patterns of internal climate variability (Pacific Decadal Oscillation [PDO], Arctic Oscillation [AO], etc.).

These publications raise an important issue. With the notable exception of surface air temperature reconstructions described in Chapter 11 of this volume, the vast majority of scientific research into the detection and attribution of human-induced climate impacts has been carried out with little or no baseline variability over long timescales. Despite the Herculean effort to distinguish human-induced climate change from natural forced and unforced (internal) variability, the lack of consideration of late Holocene proxy records in these and other influential climatic compilations is symptomatic of either a major gap in paleoclimatology, a view that baseline climate data from proxy records are not important or reliable, or both.

This chapter begins to fill this gap in our understanding of preindustrial Holocene natural climate variability from proxy records. In each section, following a brief overview of recent trends from instrumental records, we review paleorecords meeting at least one of three criteria: (1) well-calibrated proxies spliced with or calibrated to instrumental measurements; (2) temporal or geographical coverage, or both, sufficient to evaluate trends for a segment of the climate system; and (3) a conclusion, albeit usually preliminary, on the detection of anomalous climate behavior during the Anthropocene. We will see that records from tree rings, tropical ice cores, corals, ocean and coastal sediments, geomorphology, speleothems, and glacial deposits, although not formally used as tools for detection and attribution, provide a long-term context for instrumental records and climate model simulations. In some instances, compelling evidence exists for anomalous 20th-century climatic patterns. In others, preindustrial decadal-to-centennial-scale variability complicates any effort to quantify human-induced impacts.

Atmospheric Records of Climate Change

We concentrate on four aspects of the atmosphere that might reflect the combined impacts of internal climate variability and greenhouse gas forcing. These are (1) global precipitation, (2) midlatitude drought, with emphasis on North America during the past 1200 years, (3) tropical atmospheric precipitation, and (4) tropical cyclonic activity.

Global and Regional Precipitation

Background Warmer air holds more water vapor, and thus average global atmospheric moisture is expected to increase as warming occurs, following the first-order relationship expressed in the Clausius-Clapeyron equation for saturation water vapor:

$$d \ln e_s / dT = L / R_v T^2$$

where e_s is saturation water vapor, R_v is the water vapor gas constant, L is the latent heat of vaporization, and T is temperature. Moisture increases at a rate of approximately 7% for each degree Kelvin and the corresponding increase in global rainfall is expected to be muted, about 1–3%. As a consequence of these relationships, climatic warming should be accompanied by an enhanced hydrological cycle and changing precipitation patterns.

Instrumental Records of Precipitation On the global scale, estimates of 20th-century land-area precipitation trends vary greatly. Trenberth et al. (2007), for example, summarized six global precipitation data sets for the period from 1951 to 2005 that showed a range of −7 to +2 mm yr^{-1} each, with large error bars. The discrepancies reflect the complexity of the precipitation response as well as different methodologies. Huntington (2006) reviewed evidence for intensification of the global hydrological cycle and concluded, despite data gaps and regional variability, that the weight of the evidence indicates an "ongoing intensification of the water cycle." Wentz et al. (2007) showed that for the past 20 years, evaporation and rainfall increased 961 mm yr^{-1} and 950 mm yr^{-1}, respectively, suggesting that the potential for increased rainfall due to rising greenhouse gas concentrations may be greater than previously thought, but that inter-

nal variability might influence this short-term record. X. Zhang et al. (2007) distinguished anthropogenic alteration of global precipitation patterns over land areas from background variability in an intermodel comparison study of 14 general-circulation models. They attributed greater midlatitude precipitation, tropical-subtropical drying in the northern hemisphere, and wetter conditions in the subtropics and tropics of the southern hemisphere to human influence and concluded that observed and modeled changes were greater than prior model simulations had suggested.

Regional precipitation should also respond to greenhouse gas forcing of large-scale patterns of evaporation and atmospheric circulation. One observed pattern is greater mid- to high-latitude rainfall since 1900, and reduced tropical rainfall, especially since 1970, reflecting a trend toward greater poleward moisture flow (Trenberth et al. 2007). Although atmospheric moisture and rainfall are expected to increase, another atmospheric response is a weakened atmospheric overturning convection, most notable in the zonal Walker Circulation (Held and Soden 2006; Vecchi et al. 2006). These and other processes produce a trend in which wet regions get wetter and dry regions drier.

Paleoprecipitation Unlike efforts to construct hemispheric and global temperature from proxy and instrumental records, there has been no comparable paleoprecipitation reconstruction. However, Treydte et al. (2006) recently reconstructed precipitation trends in the Karakorum region of central Asia using oxygen isotope records from tree-ring cellulose covering the period A.D. 826 to 1998 (Figure 12.1). The Karakorum record showed that significant increases in snowfall in Pakistan began in the late 19th century. When Treydte compared the Karakorum record to records from western North America (Cook et al. 2004b), northeast China (Sheppard et al. 2004), the southeast Asian monsoon (D. Anderson et al. 2002), and southern Germany (R. J. S. Wilson et al. 2005), they found a corresponding increase in precipitation in northeast China and parts of Europe, a stronger southwest Asian monsoon, and a drier southwestern United States beginning in the late 19th and continuing in the 20th century. Greater 20th-century precipitation in central Asia is consistent with widespread rainfall increases across high elevations in India, Tibet, and Mongolia and the reconstruction of increased 20th-century temperature from measurements of air content from Himalayan ice cores (Hou et al. 2007). Secular intensification of precipitation in central Asia and elsewhere during the last century was unprecedented for the last millennium and represents a new line of evidence for human influence on large-scale precipitation patterns generally consistent with expectations from modeling.

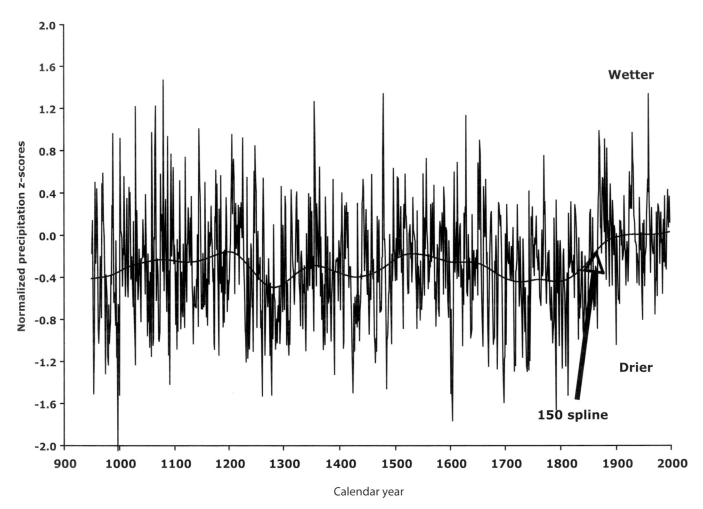

Reconstructed precipitation, Karakorum, Pakistan Central Asia

FIGURE 12.1 Winter precipitation reconstructed from oxygen isotopes in high-elevation tree rings from juniper trees in Karakorum, Pakistan, Central Asia, from Treydte et al. (2006). Normalized Z-scores (regional mean precipitation obtained by averaging standardized values) are from multiple tree-ring records; the low-frequency pattern is shown by the 150-spline curve. The precipitation increase since about A.D. 1800 is the largest in the last millennium and is also seen in paleorecords from northeast China, southwest Asian monsoons, and southern Germany, at least at low-frequency (centennial) timescales. Courtesy of K. Treydte.

Midlatitude Precipitation and Drought

In addition to greater global atmospheric moisture and an altered hydrological cycle, changes in regional rainfall regimes might result from higher greenhouse gas concentrations, but they must be distinguished from those caused by internal climate processes. Precipitation in western North America, particularly drought history, provides an important baseline for 20th-century patterns.

North American Precipitation As seen in Chapter 10 of this volume, the PDO influences hydrometeorology over western North America. Dettinger and Cayan (1995), for example, linked a post-1940 trend toward earlier spring river

runoff from the Sierra Nevada Mountains, where melting of winter snows feed river systems, to changes in PDO variability. Stewart et al. (2005) showed that the region affected by this trend is larger than first believed and that the immediate cause was higher winter and spring temperatures, rather than precipitation. They concluded that observed patterns are only partly explained by the PDO, and that a distinct spring warming trend "spans the PDO phases."

Decoupling the PDO influence from anthropogenic forcing using statistical methods showed that the PDO could not account for observed secular trends. Specifically, the PDO shift from a warm to a cooler phase in 1999 was not accompanied by a shift in North American stream flow or temperature. A similar conclusion was drawn from studies showing a decreasing ratio of snow to rain in the western United States in which PDO-induced variability was superimposed on a broad, secular warming trend since the early 20th-century (Knowles et al. 2006). Evidence from regional tree-ring records also supports the view that late 20th-century spring temperatures are relatively warmer (Luckman and Wilson 2005).

North American Paleoprecipitation   Precipitation and temperature reconstructions from western North America place recent trends in a long-term context. The existence of North American "mega-droughts" during the past few centuries, some lasting several decades, has been known from tree-ring records for several years (Stahle et al. 1998a; Woodhouse and Overpeck 1998). Cook et al. (1999, 2004b, 2007) analyzed 1200 years of temperature and drought from a network of tree-ring records in the western United States. Figure 12.2 shows Cook's compilation of drought derived from quantitative reconstruction of the Palmer Drought Severity Index. The reconstructions show a period of warmth centered on A.D. 1000 and an extended period of mega-drought during the Medieval Warm Period (MWP) between about A.D. 900 and 1300, including extremely dry episodes that were centered at A.D. 936, 1034, 1150, and 1253. This extended mega-drought ended near the inception of the Little Ice Age (LIA). Wetter conditions characterized the latter 19th and early 20th century until about 1920; the rest of the

20th century experienced drought during the 1930s (Dust Bowl), 1950s, and the unusually long (four-year) early 21st-century interval.

MacDonald and Case (2005) used tree-ring records from California and Alberta, Canada to reconstruct the last millennium of PDO-driven precipitation variability. They found 50–70-year periodicity for the last 200 years and for A.D. 1000 to 1200 and A.D. 1300 to 1500 but an absence of such frequencies between A.D. 1600 and 1800. Tian et al. (2006) also analyzed North American droughts from a 3100-year-long sediment record from Steel Lake, Minnesota. They identified PDO and low-frequency ENSO forcing of low-frequency drought variability, including large MWP dry intervals, and concluded that droughts more intense than those of the 20th century characterized 90% of the last 3100 years. Herweijer et al. (2007) confirmed the existence of mega-droughts lasting 20–40 years and showed that drought years of the past few decades were equal to prehistorical mega-droughts in severity and spatial extent, although droughts during the MWP lasted longer.

The mechanisms causing North American droughts are evident from paleomodeling studies. Herweijer et al. (2006) used paleoreconstructions in climate model simulations of three late 19th-century decadal droughts affecting much of North America, comparing them to medieval droughts. They found that 19th-century North American droughts occur during La Niña–like events in the tropical Pacific region

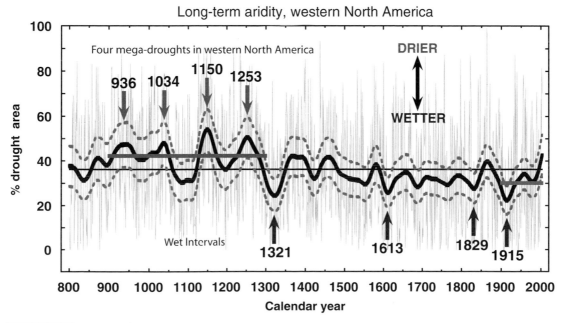

FIGURE 12.2 North American drought reconstructed from network of tree-ring records, from Cook et al. (1999, 2004b). Mega-droughts during the Medieval Warm Period exceeded the current drought in severity. See the Lamont-Doherty Earth Observatory Drought Atlas for details. Courtesy of E. Cook (http://iridl.ldeo .columbia.edu/SOURCES/.LDEO/.TRL/.NADA2004/.pdsi-atlas.html).

and proposed that tropical sea-surface temperature (SST) forcing explains precipitation patterns. Kitzberger et al. (2007) also found periods of drought during the past 550 years in an assessment of forest fire activity that has plagued the western United States since the late 1990s. The fire scar records revealed complex forcing by Pacific Ocean variability from El Niño–Southern Oscillation (ENSO), PDO, and Atlantic Multidecadal Oscillation (AMO) variability over longer timescales. The warming trend in the AMO over the past few decades corresponds to intensified recent fire activity, which may portend greater activity in the future.

Placing North American MWP droughts in a global context, Herweijer et al. (2006) evaluated paleoclimate records covering all or part of the A.D. 900–1300 interval. Most evidence indicates that the MWP was linked to persistent La Niña–like conditions in the Pacific Ocean. Cook et al. (2007) reviewed patterns and causes of late Holocene North American drought, its relationship to tropical oceanic variability (see the following sections), and how it compares to 20th-century droughts. They concluded that La Niña–like cool SST in the eastern tropical Pacific resulted in large-scale continental droughts, and that these conditions might be caused by relatively small increases in solar irradiance during the MWP and feedbacks from internal processes. Although decadal Pacific Ocean atmosphere-ocean processes, including protracted ENSO-type variability, will continue to have a large influence on North American rainfall, it remains uncertain whether the current North American drought will persist as long or be as intense as those during the MWP, or how much greenhouse gas forcing will enhance it.

Other paleoreconstructions of midlatitude regional temperature and rainfall have been compared to patterns during the 20th century. The mid-Atlantic region of eastern North America is influenced by the North Atlantic Oscillation (NAO) mode of variability at decadal-to-multidecadal timescales, such that a high (low) NAO index leads to warm (cool) and wet (dry) winter conditions. Using oxygen isotopes and Mg/Ca ratios, reconstruction of late Holocene salinity and temperature of Chesapeake Bay in the eastern United States reveals that NAO-type decadal and multidecadal variability is characteristic of these midlatitude regions (Cronin et al. 2003, 2005; Saenger et al. 2006). Results suggest that some of the largest extremes in both bay temperature and regional precipitation estimated from bay salinity occurred during the 20th century, supporting the hypothesis that NAO behavior the past few decades is extreme compared to the 150-year instrumental record (Hurrell et al. 2003a, b; Gillett et al. 2003).

In general, North American paleoclimate records reveal complex patterns that highlight the problem of confidently quantifying the degree to which late 20th-century trends in regional rainfall, temperature-driven spring snowmelt, and river runoff are unusual and evidence of global warming impacts.

Tropical Atmospheric Change

Monsoon Reconstructions   Changes in the tropical atmosphere during the last millennium have been reconstructed using tree-ring, ocean, and ice-core records of temperature and monsoon precipitation. D. Anderson et al. (2002) reconstructed Asian monsoon strength from *Globigerina bulloides*, a species that dominates planktic foraminiferal assemblages during periods of intense wind-driven upwelling in the Arabian Sea. A secular increase in wind strength is seen starting about 400 years ago, followed by a weakening in the early 1800s and another rise beginning near 1850. It is not certain if part of the post-1850 enhanced monsoon activity is related to anthropogenic forcing.

Bräuning and Mantwill (2004) analyzed a 350-year climate and monsoon record from 22 high-elevation (3700–4500 m) sites in the Tibetan Plateau using spruce and larch maximum latewood density (MLD) tree-ring measurements. MLD is a proxy for cloudiness and precipitation, although temperature can also be important. The Tibetan Plateau and surrounding regions have experienced summer and winter warming of between 0.09 and 0.32°C per decade from 1955 to 1996, but instrumental records are lacking prior to about 1950. They interpreted a decrease in MLD as indicative of an increase in the Indian Ocean monsoon beginning in 1980 that was unprecedented over the past 350 years and consistent with other high-elevation northern-hemisphere records.

Low-Latitude Ice-Core Records   The late 20th-century Indian monsoon also appears to show decoupling from background variability in records of Himalayan snow accumulation from the Dasuopu ice core (Duan et al. 2006). Between 1920 and 1995, annual snow accumulation at Dasuopu decreased by 500 mm, while at the same time northern-hemisphere temperatures were rising. This was interpreted as a decrease in monsoon-related water vapor from the Indian Ocean to the Himalayan Plateau due to decreased land-sea thermal contrast.

Ice-core research by Lonnie Thompson and Ellen Mosley-Thompson has focused on climate records from low-latitude ice caps and glaciers in the Tibetan Plateau, South America and several sites in Greenland and Antarctica (Mosley-Thompson 1996; Mosley-Thompson et al. 2001; Mosley-Thompson and Thompson 2003). The oxygen isotopic composition of ice has been a primary tool in these studies. To a first approximation, temperature influences the oxygen isotopic composition of precipitation at midlatitudes and high latitudes, but the amount of precipitation is also an influence on isotopic composition especially in lower latitudes (Rosanski et al. 1993). Consequently, both temperature and precipitation contribute to low-latitude ice-core records.

The timescale in question—daily, seasonal, interannual, or longer—is one potentially important aspect for interpretation of oxygen isotopic variability as it pertains to natural climate variability and anthropogenic influences. Hoffmann et al. (2003), for example, concluded that, at least for the 20th century, precipitation variability over the Amazon Basin regions controlled Andean ice-core $\delta^{18}O$ variability. Vuille et al. (2003, 2005) proposed that glacial isotopic records integrate several aspects of atmospheric circulation in addition to temperature. It is generally agreed that additional work on factors contributing to the isotopic signal is needed (Mosley-Thompson et al. 2006).

With this caveat in mind, a global perspective of atmospheric change can be seen in three ice-core isotopic records covering the past 1000 years, shown in Figure 12.3 from the work of Thompson and Mosley-Thompson (Thompson et al. 2006a). There is evidence for isotopic enrichment from warming and possibly precipitation changes. In other regions, coeval isotopic shifts beginning in the late 19th and 20th centuries in the South American Andes, the Tibetan Plateau, Greenland, other northern-hemisphere ice caps, and parts of Antarctica, whereas the records from Siple Station on the Antarctic Peninsula and Plateau Remote, East Antarctic show relatively flat or decreasing isotopic trends the past century (Mosley-Thompson et al. 2003). The extreme values reached during the post-19th-century interval in low and high northern-hemisphere latitudes must be considered anomalous compared to those over the last few centuries to 2000 years.

Thompson et al. (2006a) compared the 2000-year-long paleoclimate record from South American and Tibetan tropical ice cores to instrumental and reconstructed temperatures from Jones and Moberg (2003) and Jones and Mann (2004). The most obvious similarities are the more positive $\delta^{18}O$ values over the past century. Warming and glacial retreat is based not only on multiple ice-core proxy records from varying locations and elevations, but also on observations made during a return to one of the most famous sites, Quelccaya, first cored in 1983 and later in 2003. In addition to replicating the early Quelccaya record with improved methods, Thompson's group reports a trend seen in other low-latitude ice caps toward rapid retreat during the more than two decades of study, evident also in photographic and historical archives. Multiple lines of evidence suggest that a distinct and apparently abrupt climate transition is occurring. When viewing it in the context of the 10-ka-long Holocene history of high-elevation glaciers based on ice-core isotopic, glaciological, and paleobotanical evidence, Thompson notes that 20th-century glacial recession is unprecedented since Neoglacial advances began about 5.5–5.0 ka following a warmer early Holocene.

Tropical Cyclone Activity

Causes of Recent Hurricane Activity Tropical Atlantic hurricanes (tropical cyclones) have increased in frequency, intensity, and number of landfalling storms since 1970 following a relatively quiescent period during the 1950s and 60s. Hurricanes have been especially active since 1995, including the devastating hurricanes in 2004 and 2005, spurring discussion about whether these trends are part of long-term variability (Bell and Chelliah 2006) or a response to human-induced ocean warming (Emanuel 2005) (see Landsea 2005, 2007; Pielke et al. 2005; Mann and Emanuel 2006; Holland and Webster 2007; and Mann et al. 2007 for viewpoints). Two mechanisms that might enhance hurricane formation and development are thermodynamic processes (e.g., increased SSTs) (Emanuel 2005) and decreased vertical wind shear (Goldenberg et al. 2001). Both SST and wind shear are influenced directly or indirectly by climate variability such as ENSO, as well as by secular external forcing, making attribution of cause difficult. Trenberth (2005) suggests that, in theory, ocean warming should increase the intensity but not necessarily the frequency of storms.

There are several ways to measure annual hurricane activity. The total number of storms and number of category 4 and 5 storms are two simple metrics. Atlantic hurricane historical records back to the 19th century and aircraft records beginning near 1950 provide information on the number of hurricanes. The power dissipation index (PDI) is a more sophisticated measure that is highly correlated with SST. The PDI for the Atlantic Ocean correlates well with SST from the Hadley Centre for sea ice and SST (Emanuel 2005). The trend clearly shows a strong association of PDI and SST and an upsurge since the early 1990s. Nonetheless, no instrumental records extend back far enough to resolve the debate about long-term trends and human perturbation of hurricane patterns (Landsea et al. 2006).

Recent hurricane activity has prompted researchers to try to distinguish, using coupling observations and climate modeling, the impact of human-induced ocean warming on hurricanes from natural variability (Knutson et al. 2007). Trenberth and Shea (2006), for example, placed the record-setting 2005 hurricane season in the context of decadal and interannual variability due to the AMO and ENSO. During 2005, tropical North Atlantic SSTs were 0.9°C above the 1901 to 1970 mean value. Trenberth and Shea modeled AMO-driven SSTs and attributed the 0.1°C of the 0.9°C temperature anomaly to AMO variability, 0.2°C to ENSO, and almost half (0.45°C) to anthropogenic warming. Mann and Emanuel

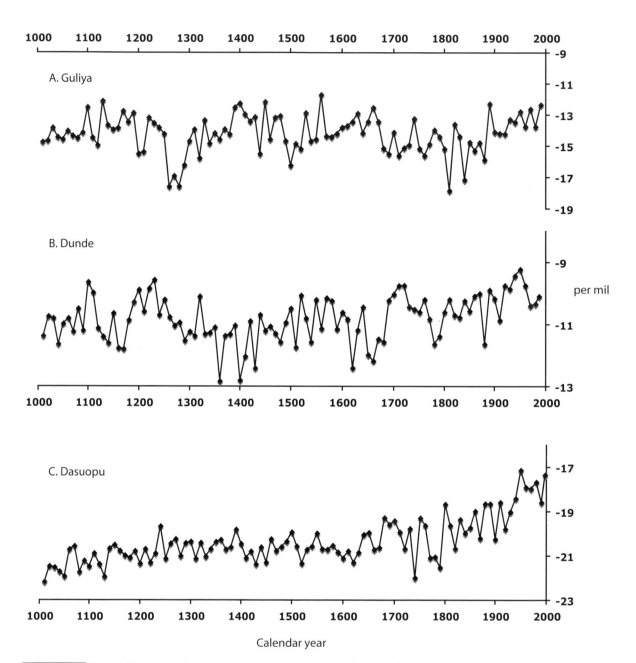

Tibetan-Himalayan ice-core oxygen isotope records

FIGURE 12.3 Decadally averaged oxygen isotope records from low-latitude, high-elevation ice cores: (A) Guliya, China; (B) Dunde, China; (C) Dasuopu, Tibet; from Thompson et al. 1997, 2003, 2006b). Note that post-19th-century isotopic enrichment is due in part to warming temperatures and at times in some records to precipitation changes. Isotopic depletion during the Chinese drought occurs from A.D. 1790 to 1796. See text.

(2006) and Mann et al. (2007) also attributed greater hurricane activity over the past few decades to rising Atlantic Ocean SSTs. Santer et al. (2006) conducted a 22-model comparison study of regions of tropical Atlantic and Pacific cyclogenesis forced by greenhouse gases, aerosols, volcanic emissions, and solar irradiance. Their main conclusions were that there was an 84% chance that external forcing, rather than internal variability, caused up to 67% of SST warming in cyclo-genesis regions, that greenhouse gas forcing was predominant in the Pacific during the 20th century, and that volcanic forcing played a role in the Atlantic region.

Paleohurricane Studies Paleotempestology, the study of past storms, hurricanes, and tsunamis using the geological record, has emerged as an important field in efforts to reconstruct past hurricane activity and place recent patterns in a long-term context. There are several proxy tools used to identify past hurricanes. Hurricanes cause large amounts of sediment flux from rivers to estuaries and deposit washover fans on barrier islands and marshes from storm surges. The impact of historical hurricanes on coastal sedimentation, in fact, has received considerable attention before the recent upsurge in activity. Hurricane Agnes, which hit Chesapeake Bay and its watershed in the eastern United States in June 1972, caused large disturbance to bay ecosystems and deposition of large quantities of sediment in the northern bay (Davis and Laird 1977). Twentieth-century hurricanes are also recognized in the sediment records from Florida Bay from radiogenic isotopic measurements (^{210}Pb, ^{137}Cs). Swart et al. (1996a) used an exceptionally stable isotopic record from a coral in Florida Bay and identified seven category 3 or greater storms between 1910 and 1948 followed by a relatively quiescent period from 1948 to 1986. Based on stratigraphic studies of overwash deposits, Donnelly et al. (2001a, b) identified two intense storms that hit New Jersey in the last 700 years and at least four storms—in 1954, 1938, 1815, and 1635 or 1638—that struck Succotash Marsh in Rhode Island.

Holocene hurricane records show that intense prehistorical hurricane activity has been common during the past 5000 years. Liu and Fearn (2000) studied lake records in northwestern Florida and discovered a "hyperactive" hurricane period between 1000 and 2000 years ago that was five times as great as activity during the last 1000 years. Nott and Hayne (2001) analyzed ridges of coral material deposited above high tide level and terraces formed by catastrophic cyclones that hit the Great Barrier Reef and Gulf of Carpentaria, Australia during the past 5000 years. They argued that superstorms hit Australia about every 200–300 years, a higher frequency than previously believed.

Donnelly and Woodruff (2007) found that hurricane activity in Laguna Playa Grande, Vieques, Puerto Rico increased around the year 1700 and continued into the 1800s (Figure 12.4). Their sedimentary evidence was supported by colonial historical records and implied that greater activity coincided with cooler SSTs during the latter part of the LIA. The 5000-year-long Vieques record also indicated that multicentury-long intervals of weak El Niño activity (sometimes La Niña–like conditions) coincided with less Caribbean hurricane activity, supporting the idea that a subdued El Niño leads to reduced wind shear and conditions conducive to hurricane development. By calibrating overwash deposits from sediment cores from the New York City area to historical 17th–19th century storms, Scileppi and Donnelly (2007) also found an increase in hurricanes after A.D. 1700 following a period of fewer storms. In addition, they discovered a period of active hurricane activity from 900–2200 years ago and concluded that these multicentury patterns were characteristic of the greater North Atlantic–Caribbean–Gulf of Mexico region.

Nyberg et al. (2007) conducted a comprehensive study in a Caribbean region of cyclogenesis for the last 270 years (Figure 12.5). They analyzed coral luminescence as a proxy for wind shear and SST from the Dominican Republic, Mona Island, and Puerto Rico plus G. bulloides from sediment cores from the Cariaco Basin. High percentages of G. bulloides are

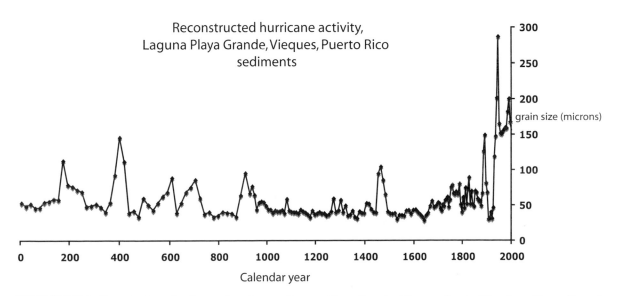

FIGURE 12.4 Bulk grain size of sediment from lagoon, Laguna Playa Grande, Vieques, Puerto Rico, from Donnelly and Woodruff (2007). Coarser sediments generally indicate greater hurricane activity. The record shows that long-term variability is influenced by the El Niño–Southern Oscillation and African monsoon atmospheric dynamics. Warm sea-surface temperatures as we see today are not necessary to cause high hurricane activity.

Reconstructed Caribbean hurricane activity

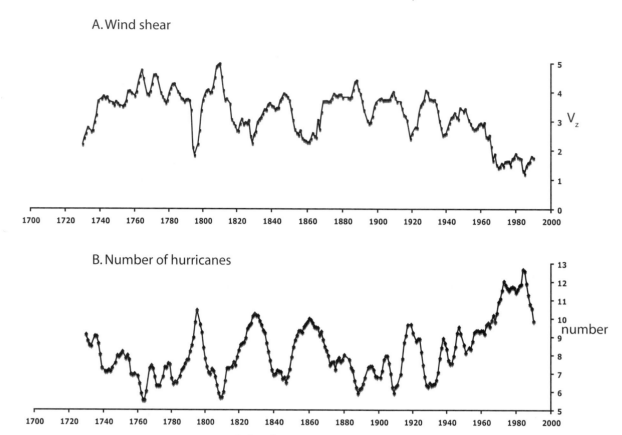

FIGURE 12.5 (A) Reconstructed vertical wind shear (Vz), and (B) reconstructed number of major hurricanes based coral and marine-core proxies from cyclogenesis regions in Caribbean-tropical Atlantic, from Nyberg et al. (2007). Records suggest that high hurricane activity in the 1990s represents a return to more typical patterns before low activity in the 1970s and 1980s.

a reliable index of more intense upwelling caused by intensified trade winds and higher wind stress. Figure 12.5 shows their two main indices for the number of hurricanes and wind shear (V_z) as compared to reconstructed solar irradiance and the AMO. At decadal timescales, reduced hurricane activity coincided with lower AMO values; peaks in solar irradiance coincided with periods of higher major hurricane activity. Nyberg et al.'s main conclusion was that post-1995 hurricane activity was not necessarily unusual in the context of the last 270 years, and they hypothesize that increasing SSTs are offset by increased wind shear, which suppresses cyclogenesis in the Caribbean region. These important reconstructions must be verified with additional paleorecords and model simulations, but they show the potential for evaluating tropical storm activity and its causes using a multi-proxy approach.

In sum, paleohurricane reconstructions, though mainly qualitative, regional, and temporally discontinuous at present, reveal that periods of hyperactivity, superstorms, mul-

ticentury intervals of greater activity (at least regionally and possibly basin-wide), and complex forcing from several mechanisms are inherent aspects to the Atlantic region and some Pacific regions. Rather than clarify the issue of whether the upsurge in hurricanes the past 40 years is anthropogenic in origin, these records raise an additional concern about extended periods of intense tropical storm activity and even larger storms than those seen recently, as well as a complex temperature-hurricane relationship at centennial timescales.

Oceanic Changes

Ocean Temperatures

Observations Global ocean-temperature data sets have been compiled by a number of groups back to the 19th century. One of the more comprehensive is England's Met Office Hadley Centre SST data set, known as HadSST

(Folland et al. 2001; Rayner et al. 2003, 2006). This compilation combined ship-based, buoy, and satellite data with multivariate analysis and smoothing methods to analyze the data. The latest update of the HadSST2 goes back to the 1850s and indicates progressive surface warming of the global ocean of 0.52±0.19°C, with slightly more (0.59±0.20°C) and less (0.46±0.29°C) warming in northern and southern oceans, respectively.

Levitus et al. (2000) also analyzed ocean temperature between the 1950s and 1990s, using the World Ocean Database. They found a warming of 0.31°C in the uppermost 300 m of the ocean, but also spatial (both depth-related and geographical) and temporal (decadal) variability in temperature across the world's oceans. These changes are comparable to an increase in ocean-heat content since the 1950s totaling 18.2×10^{22} Joules (Levitus et al. 2001, 2005). Recently, Harrison and Carson (2007) cautioned about assuming trends from relatively short-term time series, and because of methodological differences in ocean-temperature measurements different opinions also exist about the amplitude of ocean warming. Corrections to instrument biases made by Gouretski and Koltermann (2007) suggest that the total heat content changes since the 1950s estimated by Levitus and other groups may be overestimates.

Nevertheless, the consensus regarding the causes of oceanic warming from observations and climate modeling is that it is partly attributable to human causes (Levitus et al. 2005; Barnett et al. 2005). AchutaRao et al. (2006) came to a similar conclusion in an evaluation of simulations from 13 climate models with improved indexes of volcanic forcing and internal decadal variability, as did Santer et al. (2006), who found that between 0.32° and 0.67°C of increased SST cannot be attributed to internal climate variability.

SST Records   Despite paleoceanography's contributions to many aspects of the earth's climate history, most sediment records provide only a millennial-scale resolution because marine sedimentation rates in most regions are too low to resolve decadal patterns. Noteworthy studies discussed in earlier chapters show a post-LIA warming in several regions of the Atlantic Ocean (Keigwin 1996; deMenocal et al. 2000), but the exact age for the end of LIA cooling and inception of anthropogenic influence cannot be determined from deep-sea sediments. Sediments along coastal regions deposited at much higher rates are beginning to provide paleo-SST information. Fjord sediments, for example, have the potential to aid the understanding of late Holocene oceanic variability (Jensen et al. 2004; Lassen et al. 2004). Estuaries are also sediment traps containing high-resolution temperature records. Cronin et al. (2003) documented MWP, LIA and 20th-century temperature variability from Chesapeake Bay, noting warm conditions comparable to mean 20th-century temperatures between A.D. 500 and 1000. Tem-

peratures during some 20th-century warm extremes exceeded those of this "early" MWP. As high-sedimentation-rate records of ocean temperatures become more common it will possible to evaluate oceanic response to anthrpogenic greenhouse gas forcing as well as natural variability (see Andersson et al. 2003; Eiríksson et al. 2006; Sicre et al. 2008).

Two detailed ocean sediment records are those from the Cariaco Basin, Caribbean (Black et al. 2007) and the Florida Current-Gulf Stream reconstructed from sediments off the Straits of Florida, Dry Tortugas, and the Bahamas (Lund and Curry 2004, 2006; Lund et al. 2006). Black's SST record (Figure 12.6) shows temperature variability consistent with instrumental records for the past century, reflecting basin-wide trends. A divergence from the general pattern began in 1970, when the temperature rise was of larger magnitude and more rapid than any over the past 800 years. Lund showed that the strength of the Gulf Stream off Florida was as much as 10% weaker during the LIA (A.D. 1200–1850) than it is today. Gulf Stream variability was linked to meridional overturning circulation (MOC), possible Intertropical Convergence Zone (ITCZ) migrations, and solar forcing, but a detectable 20th century anthropogenic influence was not apparent.

Interannual and decadal reconstructions of tropical SSTs are more common because of the abundant coral records. Most paleostudies using tropical corals were designed to analyze changes in SST and, to a lesser extent, salinity and nutrients, associated with ENSO (Cole et al. 1993; Dunbar et al. 1994), or changes in $\Delta^{14}C$ inventory in the world's oceans as tracers of ocean circulation (Druffel 2002). In addition to ENSO, there is significant decadal periodicity in tropical climate variability, complicating efforts to distinguish human-induced from natural variability. Cobb et al. (2001) discovered that the amplitude of decadal SST variability in a 112-year-old coral from Palmyra in the central Pacific Ocean was 0.3°C and showed a 12–13-year periodicity. The Palmyra record was coherent with other coral and sediment records from the equatorial Atlantic and Indian Ocean records, and 20th-century decadal variability was evident even under increasing atmospheric carbon dioxide (CO_2) concentrations.

Gagan et al. (2000) tallied 15–20 coral records extending back to the mid-1800s and 20–30 others covering the last 50–100 years. Few coral studies explicitly consider the anomalous 20th-century behavior of tropical SSTs or internal variability, but the potential for such analysis is clear. Several examples include studies by Linsley et al. (2004, 2006) and Bagnato et al. (2005) for Rarotonga and Fiji, and Cobb et al. (2003) for Palmyra atolls. There was a warming trend in Fiji corals beginning around A.D. 1870–1880, when a major reorganization of decadal variability occurred. In addition, Goodkin et al. (2005) showed that growth rate influences the Sr/Ca ratios in coral skeletal calcite and, if not

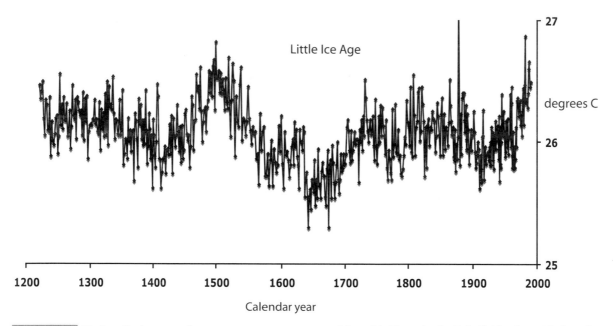

Cariaco Basin, Caribbean sea-surface temperature

FIGURE 12.6 Cariaco Basin sea-surface temperature reconstructed from Mg/Ca ratios in *G. bulloides*, from Black et al. (2007). These temperatures are generally similar to decadal and centennial variability in the Atlantic except for a divergence since 1970, when they show the largest and most rapid increase in 800 years.

corrected for, can produce an exaggerated warming after the LIA.

Wilson et al. (2006) constructed a composite coral temperature record for the tropical Indo-Pacific region back to 1750 that compared well with instrumental temperature during the period of overlap (Figure 12.7). Calvo et al. (2007) reconstructed the past few centuries of tropical surface ocean SST and salinity using Sr/Ca ratios and $\delta^{18}O$ records from corals from Flinders Reef on Australia's Great Barrier Reef in the Coral Sea (Figure 12.8). They compared this record with those from Palmyra, Abraham Reef, Fiji, New Caledonia, and Vanuatu. The combined evidence indicates progressively warmer SSTs superimposed on the well-known decadal patterns caused by internal variability. The inception of secular warming began at various times during the 19th century, depending on the region. Tropical salinity remained relatively invariant since a freshening occurred about 1800. Notably, there is a fairly coherent two-step structure in 20th-century tropical SSTs, with a rise around 1920 to 1925 and again at 1965 to 1975; the latter increase may be linked to the mid-1970s shift in ENSO and PDO variability discussed in Chapter 10 of this volume.

The prominent 20th-century warming seen at Flinders and Palmyra and in Wilson's composite record might represent an important line of evidence for anthropogenic impact on tropical SST, but we still cannot separate decadal and lower-frequency variability from secular warming, changes in precipitation and evaporation, or both. These studies suggest, nonetheless, that ENSO-type variability occurs over longer timescales and that extended coral records before the late 18th century, improved proxy methods, and additional marine-sediment records are needed to assess the last 2000 years of tropical climate.

Ocean Salinity and Circulation

Observations Sea-surface salinity is an important component of MOC and global climate over various timescales. Freshwater forcing of MOC, which might be manifested in decreased surface and near-surface salinity, is significant because past freshwater discharges into sensitive regions of the North Atlantic Ocean altered circulation and climate, and climate models predict greater freshwater influx due to greenhouse gas forcing.

Dickson et al. (1988) called the trend in decreased salinity during the 1960s in parts of the North Atlantic Ocean the "Great Salinity Anomaly." The Great Salinity Anomaly originated in the Arctic Ocean and could be traced through the Fram Strait, along the eastern coast of Greenland, through the Denmark Strait and into the Labrador Sea. Recently, oceanographic changes related to freshwater budget including other salinity anomalies (Belkin 2004) have been observed in the Arctic, Nordic Seas, North Atlantic, and Labrador Seas. Dickson et al. (2003) summarized these trends:

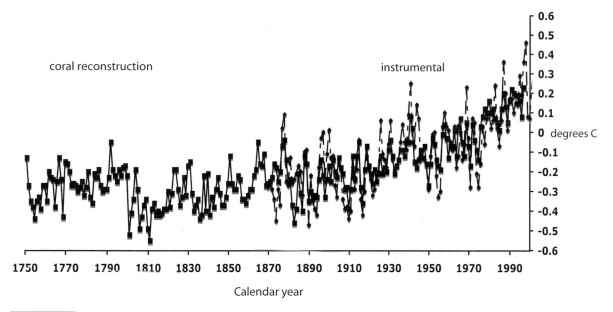

Tropical Pacific-Indian Ocean sea-surface temperature anomaly

FIGURE 12.7 Compiled winter tropical Pacific and Indian Ocean sea-surface temperature (SST) from 16 coral records (solid curve) compared with instrumental SST (dashed curve), from Wilson et al. (2006).

(1) long-term freshening of the upper 1500 m of the Norwegian Sea coincident with a shift in the NAO index toward positive, (2) increased sea-ice export from the Arctic Ocean, (3) NAO-related increased precipitation over the Nordic Seas and decreased winter sea-ice formation, and (4) increased temperature and decreased salinity of Nordic Sea deep water via the Denmark Strait between Greenland and Iceland and the Faroe-Shetland Channel between Iceland and Scotland. This latter trend is traceable across more than 2500 km from the Fram Strait to the core of North Atlantic Deep Water (NADW) flowing over the Denmark Strait and into the abyssal plain of the Labrador Sea. The salinity decrease was rapid, about 0.01 psu per decade (Dickson et al. 2002), and signified the addition of between 14,000 and 23,000 km³ of fresh water at a maximum rate of 2000 km³ per year in the 1960s (Curry and Mauritzen 2005). These salinity events might correspond to a decrease in Norwegian Sea Deep Water formation (Dickson et al. eds. 2008).

Satellite altimetry of the surface ocean from the TOPEX/Poseidon mission suggested a decline in the 1990's of subpolar circulation in parts of the North Atlantic (Häkkinen and Rhines 2004). Häkkinen and Rhines concluded that changes in buoyancy and wind stress were probably related to warming in the subpolar gyre and not to the widely varying NAO pattern. According to Curry and Mauritzen's calculations of the rate at which freshwater exported from the Arctic is retained in the Nordic Seas, it would take 100 years to signifi-

cantly influence the exchange of Nordic Sea and North Atlantic basis and thus slow down deep-water formation and 200 years to shut it down.

Salinity has also increased 0.4–0.5 psu in parts of the North Atlantic over the past 40–50 years (Curry et al. 2003), reflecting reduced freshwater influx, change in evaporation-precipitation, or change in the strength of salinity of low-latitude source waters (Rosenheim et al. 2005). Freshening in parts of the North Atlantic system may be related to freshwater export from the Arctic Ocean. Petersen et al. (2002) demonstrated increased river flow to the Arctic Ocean during the past few decades and later (Peterson et al. 2006) showed that in addition to riverine flux, sea ice and glacial melting in high northern latitudes also contributed to freshwater export and decreased high-latitude ocean salinity. The recent trends are thought to represent an amplification of the NAO-AO system.

Others have concluded that MOC circulation has decreased during the 20th century. For example, Bryden et al. (2005) analyzed records taken since 1957 from an oceanographic transect across 25°N latitude in the Atlantic Ocean. Their results generated considerable interest because they found a marked 50% strengthening of thermocline water and a 50% decrease in deeper (3000–5000 m) NADW strength, which contrasted with a generally constant surface Gulf Stream transport. Their data suggest a change in geostrophic flow and transport since 1957, and they concluded that these

Flinders Reef, Queensland, Australia tropical records

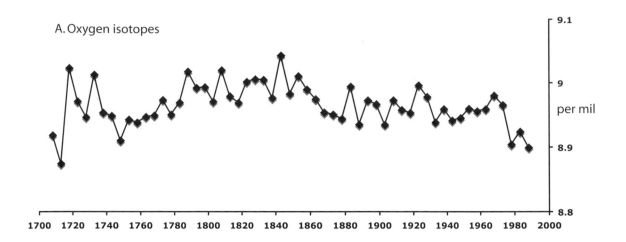

A. Oxygen isotopes

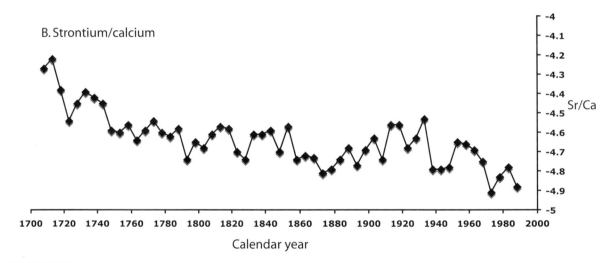

B. Strontium/calcium

Calendar year

FIGURE 12.8 (A) Oxygen isotope and (B) Sr/Ca ratios from coral from Flinders Reef, Queensland, Australia showing trends in precipitation and sea-surface temperature. Decadal trends reflect the influence of the Pacific Decadal Oscillation. When corrected for salinity variability, the Sr/Ca curve shows evidence for late 20th-century western tropical Pacific warming. From Calvo et al. (2007).

patterns provided evidence of late 20th-century MOC slow-down resulting in a net decrease in northward heat transport by the Gulf Stream system. Some evidence from the deep western boundary current in the North Atlantic do not necessarily support this view (e.g., Meinen et al. 2006).

Detecting a trend in the rate of ocean circulation is difficult because of various temporal scales of mixing of a turbulent fluid. Wunsch and Heimbach (2006) state that in turbulent fluid flow, such as MOC, variability is expected and what they call a "spectral gap" has never been observed. Consequently it is premature to conclude on the basis of long-term data from a single latitudinal transect that any long-term shift in ocean circulation has occurred. In their own analysis of MOC and heat flux at 26°N between 1993 and 2004, Wunsch and Heimbach found a slight drop in volume flux of 0.19±0.05 Sverdrups (Sv), but large spatial and temporal variability precluded any conclusion about a long-term trend and there was no statistically significant decrease in heat flux. Much larger spatial- and longer temporal-scale records are necessary to evaluate whether a climate-related trend in MOC is occurring.

Paleosalinity Records Paleosalinity records provide one long-term perspective of 20th-century changes. Rosenheim et al. (2005) analyzed salinity and temperature variability since 1890 from sclerosponge isotopic and Sr/Ca records from the Bahamas. This region is sensitive to changes in the strength of tropical water that is a source for the high-latitude surface ocean. Decadal salinity variability is also related to NAO forcing because of high wind stress during high NAO intervals. They found that a secular trend of increasing salinity changes and regional SSTs was consistent with ocean-wide warming (Levitus et al. 2005) and might reflect thermohaline circulation "adjusting to climate perturbations."

Moses et al. (2006) reconstructed 20th-century salinity in the eastern North Atlantic from coral isotopic records from the Cape Verde Islands. Salinity in this region is influenced largely by the position of the ITCZ and easterly wind-driven upwelling. They found evidence for a late 20th-century salinity increase of about 0.5 psu, similar to the increase found by instrumental records, and a link between salinity, SST, and the NAO decadal variability for some periods. The Cape Verde coral record showed evidence for a change in the NAO-salinity relationship around 1930–1931, which is not seen in the Bahaman records, suggesting the Cape Verde region may at times lie outside the influence of NAO forcing.

Ocean-atmosphere interactions can also reflect anomalous climate behavior. McGregor et al. (2007) analyzed alkenone-derived paleotemperature records from sediment cores from off northwest Africa. This region experiences strong wind-driven upwelling, such that increased wind forcing results in greater upwelling and cooler SSTs. Figure 12.9 shows that SST during the last century has declined abruptly to levels unprecedented in the past 2000 years, coincident with increasing northern-hemisphere atmospheric

temperatures and elevated greenhouse gas concentrations. The alkenone SST reconstruction of ocean and wind forcing represents one of the largest amplitude anomalies in any paleoclimate record spanning the late Holocene and the anthropogenic period. The authors interpret this trend as a response of the ocean-atmosphere system to greenhouse gas forcing, a hypothesis requiring additional paleoceanographic records.

Ocean Chemistry

Ocean carbonate chemistry involves a complex set of equilibrium reactions controlled by total dissolved inorganic carbon concentrations and total alkalinity (concentration of bases). Carbonate dissolution and secretion/precipitation is expressed as

$$CaCO_3(calcite) + CO_2 + H_2O \Leftrightarrow 2HCO_3^- \; Ca^{+2}$$

The saturation of ocean water with respect to calcite is expressed as the concentration of calcium times that of carbonate ions divided by the solubility product (K_{calc}):

$$\Omega_{calc} = [Ca^{+2}] \times [CO_3^{2-}] / Ksp_{calc}$$

Elevated atmospheric greenhouse gas concentrations significantly alter the ocean's carbonate chemistry and carbonate-secreting marine organisms, including mollusks, corals, foraminifera, and photosynthetic calcareous coccoliths. Increasing partial pressure from rising CO_2 concentrations decreases the carbonate saturation state of the world's

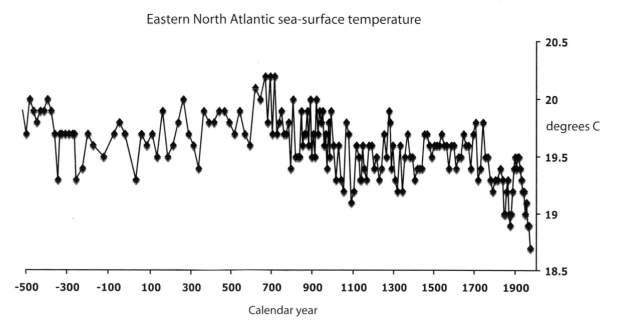

Eastern North Atlantic sea-surface temperature

FIGURE 12.9 Eastern Atlantic tropical sea-surface temperature (SST) reconstructed from alkenone records from off northwest Africa, from McGregor et al. (2007). The cooling SST trend during the 20th century signifies enhanced surface winds and oceanic upwelling unprecedented in the past 2500 years.

oceans (though not with spatial uniformity), drives the ocean toward greater acidity and carbonate dissolution, and has negative consequences for calcitic organisms and marine ecosystems.

A panel of experts on "ocean acidification" found that as atmospheric CO_2 concentrations increased from 280 to 380 parts per million by volume (ppmv), calcite saturation Ω_{calc} decreased from 5.32–4.46 ppmv (Kleypas et al. 2006; see also Feely et al. 2004). This change is comparable to that estimated for a glacial-interglacial cycle (6.63–5.52) ppmv. Likewise the oceans' mean pH is estimated to have fallen from 8.16 to 8.05 pH units, a change equal to about 70% of that during a glacial-interglacial cycle (8.32–8.16). It is estimated that by the time atmospheric CO_2 concentrations double the preindustrial values, calcite saturation will fall further to 3.52 ppmv. The problem of ocean acidification is complicated, and to a degree exacerbated, along coastal regions where additional carbonate chemistry changes over the past three centuries are caused by river influx of nutrients and organic material (Andersson et al. 2005, 2006). Nonetheless, the impact of acidification can already be identified in the eastern North Pacific Ocean off western North America (Feely et al. 2008).

Paleoreconstructions of the ocean carbonate system are extremely important for understanding climate-atmosphere-ocean biogeochemical fluxes over millennial, orbital, and longer timescales. Reconstructions of ocean acidity and carbonate saturation for the past millennium are rare. Pelejero et al. (2005) used boron isotopes ($\delta^{11}B$) to reconstruct the pH history of the Great Barrier Reef system off eastern Australia. They showed that over multidecadal (~50-year) timescales, pH varied by 0.3 pH units (Figure 12.10). This amount is three times the estimated ocean-wide change in pH that is attributable to uptake of anthropogenic CO_2. They also found that the $\delta^{11}B$ and inferred pH variability are consistent with forcing by the Interdecadal Pacific Oscillation (IPO) mode of variability. Mechanisms to explain pH variability include reef calcification processes and regional flushing time of ocean water, but regardless of the mechanism, both the strong evidence for large-scale interdecadal variability in pH and the evidence for region-specific acidification signals adds a new dimension to the already difficult task of predicting the effects of increasing greenhouse gas concentrations on ocean acidification and marine biological systems.

Patterns of Internal Climate Variability

El Niño–Southern Oscillation

Has ENSO variability exhibited anomalous behavior (i.e., more frequent or intense El Niño events, or both) during the last few decades, and if so, can it be attributed to anthropogenic influence? These questions, reflecting in part enhanced ENSO variability since the mid-1970s—including major El Niño events in 1982 to 1983 and 1997 to 1998, have generated enormous interest but no broad consensus. Federov and Philander (2000) offered the view that random climate disturbances, decadal variability, or anthropogenic warming could all be responsible for late 20th-century behavior of ENSO. More recently, the IPCC concluded that there still appears to be no consensus on whether greenhouse gas forcing has perturbed ENSO or decadal tropical climate variability (Bindoff et al. 2007; Trenberth et al. 2007). This uncertainty is partly due to the scarcity of long-term records, the influence of volcanic and solar forcing, and the recognition that there is multidecadal and centennial variability superimposed on year-to-year ENSO-related events. At a 2008 workshop, a consensus emerged that despite great progress, it was still not yet possible to construct a 500-year-long ENSO record from paleoreconstructions (Diaz and Tourre 2008).

The idea that centennial and decadal trends in ENSO occur, perhaps influenced by solar irradiance, was proposed by Anderson et al. (1992), whose historical reconstructions showed that ENSO events occurred in cycles of about 90, 50, 24, and 22 years since A.D. 622. Solar modulation of ENSO activity, according to Anderson, may have produced significant impacts on regional climate in the Pacific region. Mann et al. (2000) conducted a geographically comprehensive analysis of the past few centuries of ENSO activity using a large number of paleotemperature and other proxy reconstructions and the historical record of ENSO. They identified changing strength in ENSO variability and regional teleconnections for the preinstrumental and postinstrumental periods at various timescales. One conclusion was that 20th-century ENSO-related warming is small compared to the mean global surface air temperature rise. In addition, they noted a 20th-century increase in the incidence of large warm and cold ENSO episodes and a lower frequency of extended warm ENSO intervals, but suggest that this pattern may be deteriorating in the past few decades. If such a situation were confirmed as a response to changes in external forcing, future ENSO variability might resemble the 19th-century period of reduced ENSO activity.

Evans et al. (2002) compiled 13 coral records to evaluate how late 20th-century ENSO and other Pacific tropical variability compared to prehistorical periods. They found that periods of intense ENSO activity such as that during the past few decades also occurred in the past (i.e., 1820–1860) and seem to be correlated with relatively warm tropical SSTs. Their results mean that elevated ENSO activity of the late 20th century may reflect, in part, low-frequency tropical climate variability.

Reconstructed ocean pH proxies,
Flinders Reef, Australia

A. Boron isotopes

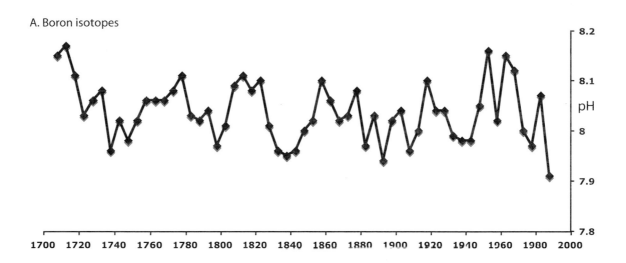

B. Carbon isotopes

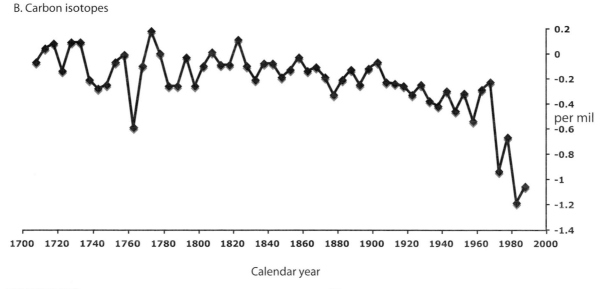

Calendar year

FIGURE 12.10 (A) Paleo-pH reconstructed from boron isotopes ($\delta^{11}B$), and (B) carbon isotopic record from from *Porites* coral from Flinders Reef, off Queensland, Australia, from Pelejero et al. (2005). Decadal trends in pH generally track multidecadal variability in the Pacific Decadal Oscillation. More negative carbon isotopic values are also evident in records from Vanuatu and Jamaica.

Mann et al. (2005a) and D'Arrigo et al. (2005a) also directly addressed ENSO behavior over the last 1000 and 600 years, respectively. Mann's study combined the Zebiak-Cane ENSO model with solar and volcanic forcing indices and concluded that ENSO dynamics are strongly influenced by radiative forcing from volcanic and solar processes (see Adams et al. 2003). D'Arrigo et al. developed a new low-latitude tree-ring ENSO index calibrated to tropical Pacific SSTs. Their record indicated low ENSO activity during the LIA, a pattern matched by some other tree-ring records. However,

their comparison to coral paleo-ENSO reconstructions and model simulations for the Pacific region showed a wide range of temporal variability in the strength of ENSO, such that a simple robust multidecadal pattern of low-frequency ENSO variability has not yet been identified. Some evidence, for example, shows that periods of decreased solar irradiance and high volcanic forcing experience increased ENSO variance. ENSO variance measured by instruments for the last few decades exceeded that seen for most intervals covered by the reconstructed ENSO records. One mechanism to account for long-term ENSO variability is the tropical "thermostat" mechanism in which ENSO intensity is influenced by radiative forcing in the tropics (Clement et al. 1996, 2000).

Fowler (2008) and Gergis and Fowler (2006) used tree-ring widths from kauri conifer trees in New Zealand to reconstruct ENSO activity from 1580 to 2002. ENSO activity exhibited low-frequency, multidecadal patterns (55–80-year periods), peaked in the early and late 20th century with an intervening midcentury muted signal, and showed high and low activity levels in the middle of the 18th century and near the beginning of the 19th century, respectively. The mid-18th- and 20th-century peaks in ENSO activity were consistent with reconstructions of Mann and D'Arrigo, although the relative strength in ENSO during the 20th century varied somewhat, perhaps due to local ENSO teleconnections. In general, however, there was an association between strong ENSO activity and periods of warming. The existence of high ENSO activity during preindustrial periods precludes a definitive link between recent trends and anthropogenic warming but raises concerns should warming continue.

Rodbell et al. (1999) and Moy et al. (2002) provide a long-term perspective for ENSO variability based on 2000-year-long lake sediment records from Ecuador (Figure 12.11). The most obvious feature in their record is the peak in ENSO activity about 1000–1200 years ago, after which there was minimum activity, at least relative to the entire late Holocene.

In sum, ENSO has exhibited large spatial and temporal variability during the last few centuries and millennia, possibly related to solar forcing (Emile-Geay et al. 2007). Understanding the spatial expression of low-frequency ENSO variability, as well as the contributions of various forcing

mechanisms, requires additional paleoreconstructions covering at least the last 1000 years and model simulations to fully understand the impact of greenhouse gas forcing.

Arctic and North Atlantic Oscillations

As discussed in Chapter 10 of this volume, the AO and NAO are modes of interannual and decadal winter season atmospheric variability responsible for large-scale climate changes over northern-hemisphere polar regions, the North Atlantic Ocean, and adjacent continents (Hurrell 1995; Thompson and Wallace 1998). Variability in the AO is highly correlated to the NAO, as well as to North Atlantic and Nordic Sea salinity and ocean circulation (Hurrell and Dickson 2004), and thus they are treated together here.

AO variability is linked to a number of important climate variables sensitive to anthropogenic forcing, including Arctic temperature, atmospheric circulation and wind field, and sea ice (Rigor et al. 2002; Rigor and Wallace 2004; Lindsay and Zhang 2005; Serreze et al. 2007). For most of the past three decades, the AO index has risen in concert with decreasing Arctic sea-ice thickness (e.g., Yu et al. 2004) and area (Comiso 2006). Since 1995, the AO index has been relatively neutral but sea-ice cover has declined (Overland and Wang 2005), leading some to postulate a decoupling of the AO-sea-ice connection (see "Arctic Sea Ice" later in this chapter) (Serreze and Francis 2006; Serreze et al. 2007). Another interpretation of this pattern is a delayed response of sea ice to NAO forcing (Rigor and Wallace 2004). Several authors have also concluded on the basis of instrumental or model simulations, or both, that NAO variability during the later part of

ENSO activity reconstruction,
Lake Pallcacocha, Ecuador

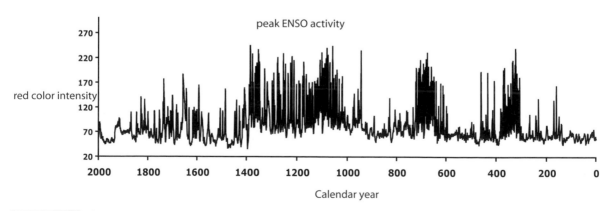

FIGURE 12.11 El Niño–Southern Oscillation (ENSO) activity the last 2000 years, from color intensity of lake sediments deposited in Lake Pallcacocha, Ecuador, from Rodbell et al. (1999) and Moy et al. (2002). Note peak El Niño activity from 600–1000 years before the present.

the 20th century is unusual in relation to that during the rest of the instrumental record (Gillett et al. 2003).

Instrumental records of the AO and NAO extend back about 50 and 150 years, respectively, and the AO-NAO mode varies over decades, so it is important to use paleoreconstructions to evaluate late 20th-century patterns. Unfortunately, there are no direct paleo-proxies calibrated to the AO index, leaving a large gap in our ability to assess the long-term context of declining Arctic sea ice. Conversely, several NAO indices extending back 300–500 years provide insight into the North Atlantic system (Appenzeller et al. 1998; Luterbacher et al. 1999, 2002; Cook 2003). Cook (2003) reviewed evidence for NAO variability in the North Atlantic region and concluded that 20th-century NAO variability was "unusual" but not unique when it was compared to reconstructed variability for the past 1000 years.

Interdecadal Pacific Oscillation

The IPO, the ocean-wide expression of the PDO of the North Pacific, exhibits quasi-cyclic variability occurring over decadal timescales, but long-term instrumental records of the IPO are limited to about the last 50 years (Power et al. 1999; Folland et al. 2002). Paleoclimate records suggest that the IPO exhibits complex behavior over the past few centuries. Linsley et al. (2004) presented Sr/Ca and $\delta^{18}O$ results from Rarotonga and Fiji (15–20°S latitude), located in the center of IPO variability. They documented a clear correspondence between the coral paleorecords from these and other tropical Pacific regions and instrumental measures of the IPO since 1880. They also found that from the late 1700s to 1880, there was interregional asynchroneity, a pattern consistent with tree-ring and oceanic paleoclimate records from the Pacific region. Apparently, interdecadal IPO variability was regionally suppressed during the mid-1800s. Perhaps more important, they postulate that there was a shift toward a modern-like IPO pattern about 1880, preceding anthropogenic atmospheric warming in the southern hemisphere by 30 years, arguing against an anthropogenic forcing of this late 19th-century regime shift.

Polar Regions and Sea Ice

Sea ice is an important climate variable for detecting anthropogenic influence on climate because of the sensitivity of polar regions to climate warming. Sea-ice cover and freshwater export from the Arctic Ocean are closely linked to the Nordic Sea and North Atlantic hydrography and circulation. Similarly, sea ice around Antarctica is dynamic over seasonal, interannual, and longer timescales. We consider north and south polar sea-ice records in the context of long-term variability.

Arctic Sea Ice

The distribution and thickness of Arctic sea ice is governed by several factors including winds, atmospheric and oceanic temperatures, and freshwater runoff, especially from large Siberian rivers. Winds tend to break up ice (causing rapid melting), push it into ridges, or cause floes to drift to regions of warmer temperatures. These processes lead to significant year-to-year and regional variability. Interannual and decadal variability in the Arctic linked to the AO (Thompson and Wallace 1998; Thompson et al. 2003) also influences seasonal wind patterns. High AO index values are associated with strong winds, which make ice more vulnerable to spring and summer melting (Rigor et al. 2002). Secular warming of Arctic surface air temperatures is also a factor in sea-ice development.

Parkinson et al. (1999) and Rothrock et al. (1999) established that the extent and thickness of summer Arctic sea-ice was decreasing since the 1970s. Recently, evidence shows a number of disturbing trends: a 20% decline in perennial sea ice (Stroeve et al. 2005, 2007), dramatic thinning of underformed ice since the 1960s (Rothrock and Zhang 2005), declining ice thickness in the central Arctic Ocean due to AO variability and thermal forcing (Yu et al. 2004), increasing surface temperatures from 1981 to 2006 for regions covered by sea ice and regions north of 70°N by 0.65 and 0.73°C per decade, respectively.

As mentioned above, while sea-ice and AO variability have been tightly linked since the mid-20th century, there was an apparent decoupling in the AO-sea-ice connection beginning around 1995 (Rothrock et al. 2003; Francis et al. 2005; Overland and Wang 2005; Serreze et al. 2007). Despite a neutral AO-NAO index since the mid-1990s, sea ice has continued to decline rapidly, leading some researchers to invoke greenhouse gas forcing to explain the trend. Peterson et al. (2002, 2006) came to a similar conclusion regarding enhanced river runoff into the Arctic Ocean the past several decades, seen in freshening of the Beaufort Sea. Comiso (2006) showed that winter sea ice also declined in 2005 and 2006, especially in the Eastern Arctic Basin.

In spite of these trends and the need to know the long-term sea-ice perspective, there are unfortunately no detailed late Holocene sea-ice reconstructions from Arctic Ocean sediment cores. This data gap reflects relatively low sedimentation rates (~1 cm yr^{-1}) (Backman et al. 2004), the complexity of surface ocean proxies (Spielhagen et al. 2004) and IRD reconstructions (Darby et al. 2002), and a research focus on orbital and millennial timescales (Darby et al. 2006). The most comprehensive multi-proxy reconstruction of the Arctic is that of Overpeck et al. (1997), who compiled tree-ring, lacustrine sediment, ice-core, and instrumental records into a single 400-year-long paleotemperature curve (Figure 12.12).

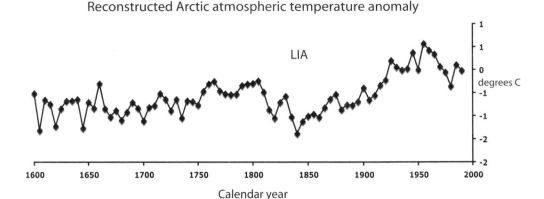

Reconstructed Arctic atmospheric temperature anomaly

FIGURE 12.12 Arctic-region temperature anomalies since 1600 based on compilation of instrumental, ice core, tree-ring, and sediment records, from Overpeck et al. (1997). LIA = Little Ice Age.

It shows evidence for 20th-century warming of more than 1°C since the mid-19th century but no direct evidence regarding sea-ice history.

Sea-ice proxies are challenging because many factors influence organisms that dwell in the ocean surface and because ice-rafted debris sources are often ambiguous. There is nonetheless a growing literature on paleo-sea-ice indicators, which hold promise for future study. In addition to provenance study from IRD lithic grains (see Chapter 6 in this volume), the most widely applied sea-ice proxies are diatoms (Gersonde et al. 2003, 2005; Leventer et al. 2002; Armand and Leventer 2003), dinoflagellates (de Vernal et al. 2001), and calcareous nannofossils (Gard 1993). De Vernal's dinoflagellate transfer function can be used to compute the number of months of sea-ice cover based on an extensive modern database. Bauch and Polyakova (2003) used diatom assemblages as a proxy for freshwater influx to reconstruct a 2800-year sea-ice cover in the Laptev Sea and identified a major sea-ice shift beginning 300 years ago.

Antarctic and Southern Ocean Sea Ice

Sea ice around Antarctica is extremely variable seasonally, covering $4 \times 10^6 \, km^2$ in summer and $20.1 \times 10^6 \, km^2$ in winter. Some regions around the Antarctic Peninsula have warmed several degrees during the past century, as opposed to a mean 0.6°C warming globally, and sea ice in the Southern Ocean has declined in the past few decades, although there is limited long-term instrumental data.

One highly publicized event was the disintegration between 1995 and 2002 of 12,500 km² of the Larsen B ice shelf off the northeastern Antarctic Peninsula, part of the West Antarctic Ice Sheet (WAIS) (Shepherd et al. 2003; Scambos et al. 2004). It may have been caused in part by regional atmospheric warming of about 2°C, observed since the 1960s,

which was responsible for surface melting, ponding of meltwater, and seepage through crevices in the ice, and in part by possible basal melting, eventually causing the breakup of the ice shelf into smaller pieces. Shepherd et al. (2004) suggested that ocean warming was an important factor in regional West Antarctic atmospheric and ice-shelf trends. Regional warming has been linked to changes in decadal variability of the Southern Annular Mode that some climate model simulations suggest is related to anthropogenic forcing over the past few decades (Marshall et al. 2004; Fogt and Bromwich 2006). Following the Larsen B ice-shelf disintegration, there has been accelerated glacial recession and a glacier surge for some glaciers that had been buttressed by the ice shelf. This has led to a drawdown of land-based glacial ice that fed the Larsen ice-shelf system and a small sea-level rise (SLR) (De Angelis and Skvarca 2003; Rignot et al. 2004).

The disintegration of the Larsen Ice Shelf highlights the persistent question surrounding the stability of the entire WAIS, which holds about 6 m of sea-level equivalent. Is the disintegration of parts of the WAIS a temporary short-term response of regional glaciers or the beginning of a long-term trend? Goosse and Renssen (2005) used model simulations of sea ice around Antarctica for the past 150–1000 years to evaluate the adequacy of historical records that preceded modern instruments. They concluded that early explorer and whaling records going back to the 19th century are too ambiguous to be used in modeling studies and assessment of 20th-century trends. Their modeling results suggested that the 20th-century sea-ice decline around Antarctica cannot be explained by volcanic and solar forcing and may reflect, at least in part, anthropogenic influence, but noted the lack of late Holocene paleorecords.

Domack et al. (2005) reconstructed Holocene paleoceanography in the Antarctic Peninsula region using multiple proxies from six sediment cores. They produced one of the most focused, long-term perspectives of a single glacial event attributed by some to anthropogenic warming. They concluded that there had been long-term Holocene thinning of the ice shelf, the recent ice-shelf disintegration was due to sustained 20th-century warming, and recent disintegration was unprecedented in the Holocene history of the region. Paleoceanographic reconstructions from sediment cores from other regions of the Antarctic margin are scarce and lack the temporal resolution to evaluate late Holocene submillennial-scale patterns.

Other ingenious methods can be applied to Antarctic sea-ice reconstructions. Curran et al. (2003) used a novel, indirect approach by measuring methanesulphonic acid (MSA) concentrations from an ice core from Law Dome, Antarctica dated back to 1841. MSA concentrations are partly a function of marine phytoplankton, specifically sea-ice algae, which produce dimethylsulphide in ocean regions of sea ice. MSA concentrations thus reflect to a first approximation sea-ice cover. Curran et al. found a 20% decline in sea-ice extent around East Antarctica since 1950, but this trend cannot be extrapolated to other margins of the Antarctic Ice Sheet (AIS).

In light of Goosse and Renssen's conclusions and the climate-sea-ice-ocean reconstructions by Domack and Curran, it is obvious that there is both a great need and untapped potential for paleoreconstructions using various proxies of sea-surface conditions around the Antarctic (Taylor and McGinn 2001; Denis et al. 2006).

Sea Level, Ice Sheets, and Glaciers

Understanding SLR over the past century is an actively researched topic because of its potential impact on coastal regions. Distinguishing a global sea-level trend from local sea-level changes along a particular coast requires comparison of observed rates of SLR from tide gauges (about a century) and satellites (the past 15 years) against the source contributions to sea level. SLR is due mainly to thermal expansion of the oceans (thermosteric effect) and melting ice from glaciers and the Greenland and Antarctic Ice Sheets, counterbalanced by impoundment of water in reservoirs. The sea-level literature is large, expanding rapidly, and potentially confusing because of the rapid pace of advances and the many processes that contribute to sea level. In this section we summarize the salient aspects of 20th-century SLR and geological records of sea-level patterns from several coastal regions.

Trends in Sea Level and Their Causes

Table 12.1 summarizes quantitative estimates of observed rates and source contributions to historical sea-level change taken from several sources. The major contributions to SLR from thermal expansion and melting ice are complicated by glacio-isostatic adjustment (GIA) (Peltier 2004). We have seen in prior chapters that GIA influences relative sea level (RSL) in that the earth's lithosphere responds to the redistribution of mass on the earth's surface that accompanies deglaciation and the melting of glacial ice sheets. GIA vertical movements vary greatly, depending on the proximity of coasts to former ice sheets, the width of the continental shelf, local sedimentation, and other factors. For example, GIA adjustments range from 0.1–0.2 mm yr^{-1} along low-latitude coasts far from Last Glacial Maximum ice sheets to 1 mm yr^{-1} near former ice-sheet margins. Estimates of SLR rates in Table 12.1 are corrected for GIA, which introduces a degree of uncertainty but does not negate the broad decadal-to-centennial changes in sea level described here.

One conclusion from Table 12.1 is that the mean rate of SLR was ~1.7–1.8 mm yr^{-1} during the last century and 3.1 mm yr^{-1} during the past two decades, when satellites provided exceptional spatial and temporal coverage of the oceans (Cazenave and Nerem 2004; Miller and Douglas 2004; Church and White 2006). Thermosteric expansion due to oceanic warming and the melting of glaciers and Greenland and Antarctic ice are the causes. A few years ago the scientific community faced a major dilemma because SLR from thermosteric effects and melting ice was smaller than the observed rise from tide gauges and satellites. Munk (2002) estimated that 12 cm of the total 21 cm of SLR over the past century could not be attributed to thermosteric effects and ice melting (Douglas and Peltier 2002; Meier and Wahr 2002). It was not clear whether this difference was due to steric or glacio-eustatic processes, changes in the earth's rotation, length of day and other geophysical processes (Munk 2002), unreliability of tide gauges (Cabanes et al. 2001), or large error bars on estimated mass changes attributable to the melting of land-based ice.

It is not surprising that little is known about the long-term trends in the world's 160,000 glaciers, because continuous monitoring data is unavailable for all but a few glaciers (Dyurgerov and Meier 2005). As a consequence, assessment of mass balance changes in glaciers and ice sheets has been focused on shorter satellite-based data. For example, Shepherd and Wingham (2007) compiled 14 satellite-based studies of ice-sheet and glacier mass balance estimates published since 1998. Mass balance estimates ranged from a large loss totaling 366 Gt (~1.0 mm yr^{-1} sea-level equivalent) to a small gain of 53 Gt (~0.15 mm yr^{-1}), equating to a net fall in sea level. Each study covered a different period, surveyed different regions of ice sheets, and used different methods, so a divergence of estimates is expected. The range of mass balance estimates also reflects the limitations of short-term records to establish clear trends.

TABLE 12.1 Observed Rates and Sources of Sea-Level Rise

Method	Rate (mm yr⁻¹)	Error	Period	Source	Comment
Observations					
Tide-gauge SLR	2.03	0.35	1904–1953	Holgate 2007	9 long-term gauge records
Tide-gauge SLR	1.45	0.34	1954–2003	Holgate 2007	
Tide-gauge SLR	5.31	N/A	1980s	Holgate 2007	Highest rate of 20th century
Tide-gauge SLR	1.74	0.16	20th century	Holgate 2007	Mean rate of 20th century
Tide-gauge SLR	1.8	N/A	~1920s–1990s	Douglas 2001	Few long-term gauges
Tide-gauge SLR	1.5–2.0	N/A	20th century	Miller and Douglas 2004	Few long-term gauges
Tide-gauge SLR	1.7	0.4	1948–2002	Holgate and Woodworth 2004	Stable coastline
Tide-gauge SLR	1.8	0.3	1950–2000	Church et al. 2004	177 gauge records
Tide-gauge SLR	1.7	0.3	20th century	Church and White 2006	
Generally accepted value for observed SLR	1.8	0.5	20th century		
Satellite altimetry—post-1993 SLR	3.1	0.7	1993–2003	Cazenave and Nerem 2004	Summary of literature
Sources of SLR					
Ocean steric SL change 0–3000 m	0.4	0.09	1955–1998	Antonov et al. 2005; Levitus et al. 2005	
Ocean steric SL change 0–700 m	0.33	0.07	1955–2003	Antonov et al. 2005; Levitus et al. 2005	
Ocean steric SL change 0–700 m	0.36	0.06	1955–2003	Ishii et al. 2006	
Ocean steric SL change 0–700 m	1.2	0.5	1993–2003*	Ishii et al. 2006	
Ocean steric SL change 0–700 m	1.2	0.5	1993–2003	Antonov et al. 2005; Levitus et al. 2005	
Oceanic steric	0.35	0.1	1950–1998	Plag 2006	Empirical model estimate
Generally accepted value	0.42	0.12	1961–2003	Bindoff et al. 2007	
Generally accepted value	1.6	0.5	1993–2003	Bindoff et al. 2007	
Melting ice—alpine glaciers, small ice caps[†]	1.1	0.24	Year 2006	Meier et al. 2007	Glacier SLR by 2100 totals 240+/−140 mm[‡]
Melting ice—alpine glaciers, small ice caps[†]	0.5	0.18	1961–2003	Lemke et al. 2007	Potential SLR 0.15–0.37 m
Melting ice—alpine glaciers, small ice caps	0.77	0.22	1993–2003	Cogley 2005; Dyurgerov and Meier 2005	Range reflects uncertainty in volume

(*continued*)

TABLE 12.1 (Continued)

Method	Rate (mm yr^{-1})	Error	Period	Source	Comment
Melting ice—Greenland Ice Sheet	0.05	0.12	1961–2003	Ohmura 2004; Lemke et al. 2007 est. higher	Inventory of 160,000 glaciers
Melting ice—Greenland Ice Sheet	0.21	0.07	1993–2003	Lemke et al. 2007	
Melting ice—Greenland Ice Sheet	0.1	0.05	1950–1998	Plag 2006	Empirical model estimates
Melting ice—Antarctic Ice Sheet	0.39	0.11	1950–1998	Plag 2006	Empirical model estimates
Melting ice—Antarctic Ice Sheet	0.14	0.41	1961–2003	Lemke, Ren et al. 2007	
Melting ice—Antarctic Ice Sheet	0.21	0.35	1993–2003	Lemke, Ren et al. 2007	
Melting ice total	0.69		1961–2003	Bindoff et al. 2007	
Melting ice total	1.19		1993–2003	Bindoff et al. 2007	
Total sources (steric + ice melt) Melting Greenland and Antarctic ice	2.79		1993–2003		
Balancing the Budget					
"Missing" SL Equivalent (observed, sources)	~0.7		1961–2003	Munk 2002; Meier and Wahr 2002; Douglas and Peltier 2002	
"Missing" SL Equivalent Mass (Observed, Sources)	~0.3		1993–2003	Bindoff et al. 2007; Rahmstorf 2007	
Annual Ice Sheet Budget (gain/loss in gigatons, Gt) since early 1992s, summarized by Shepherd and Wingham 2007 from 14 studies					
Greenland ice	Loss of 227 Gt to gain of 11 Gt/year				
West Antarctic ice	Loss of 47 to 59 Gt/year				
East Antarctic Ice	Loss of 1 Gt to gain of 67 Gt/year				

Note: SL = sea level; SLR = sea-level rise.

*Other estimates are as high as 1.6–1.8 mm yr^{-1}.

†Does not include ice caps around Greenland or West Antarctica.

‡This is Meier's estimate for A.D. 2100, assuming current acceleration. 360 Gt of ice is equivalent to 1 mm SLR per year.

The past few years have seen revision of the rates and sources of observed SLR. For example, satellite altimetry indicates a rate of 3.1 mm yr^{-1} (~2.8 mm yr^{-1} uncorrected for GIA) for combined sources for the post-1993 period (Nerem and Mitchum 2002). Jevrejeva et al. (2008) examined tide-gauge records back to A.D. 1700 and found that the rate of SLR accelerated in the late 19th century by an amount roughly 0.01 mm yr^{-1}. Quasi-periodic (60–65-year) oscillations were superimposed on the long-term preanthropogenic trend. They noted that if these trends continued, global sea level might rise 34 cm by the year 2100, at the high end of IPCC estimates (Bindoff et al. 2007; Meehl et al. 2007).

Several studies have estimated Greenland and Antarctic Ice Sheet mass balances (Hanna et al. 2005; Kaser et al. 2006). Plag (2006) developed an empirical model to separate the sources of SLR since 1950, showing that steric effects contributed ~0.35 mm yr^{-1} and Antarctica and Greenland contributed 0.39 and 0.10 mm yr^{-1}, respectively. There is now a better match between observations of sea-level change and potential sources to explain them. The IPCC compilations for 1993 to 2003 suggest that thermal expansion contributed 1.6±0.5 mm yr^{-1} and melting ice 1.2±0.7 mm yr^{-1}. Combined, these are fairly close to the observed 3.1-mm-yr^{-1} rate of SLR (Bindoff et al. 2007). These estimates exclude poorly constrained contributions from the global hydrological budget, including water extracted from porous rocks, reservoir storage, and land-use changes. Recent modeling of the ocean's mass budget suggests a rate of SLR of 3.34 mm yr^{-1}. This rate compares well with general-circulation modeling estimates constrained by TOPEX/Poseidon and Gravity Recovery and Climate Experiment (GRACE) topography data. The observed rate from 1993 to 2003 is 3.37 mm yr^{-1} and is similar to the modeled total of 3.34 mm yr^{-1}, of which 2.50 yr^{-1} is due to thermosteric and 0.74 yr^{-1} to eustatic factors (Wenzcl and Schröter 2007).

Uncertainty about the causes of SLR does not only reflect uncertainty in ice mass balance changes. Chao et al. (2008) compiled data on water impoundment since 1900, finding a total of 30 mm of sea-level equivalent has been stored in reservoirs at a mean rate of about −0.55 mm yr^{-1} for at least the past 50 years. This evidences leads to a revision in the global mean rate of sea level rise downward to about 2.5 mm yr^{-1}, implying that the observed rate of 3.4 mm yr^{-1} includes a greater contribution from human or natural factors than previously believed.

Decadal Variability

Superimposed on the general trend of rising sea level is temporal (interannual, decadal) and spatial variability in rates and patterns of SLR during the past century, complicating efforts to determine if and when global SLR rates acceler-

ated. Temporal and spatial variability in sea level is associated with internal ocean-atmosphere climate processes such as ENSO and the NAO. Wakelin et al. (2003), for example, demonstrated an NAO-sea-level relationship off the northwestern European coast since 1955. Moreover, the high rate of SLR of 3–4 mm yr^{-1} observed over the past few decades has also been observed for other decades during the last century, precluding a simple conclusion about accelerating rates (Holgate and Woodworth 2004; Church and White 2006; Holgate 2007).

Glacier and Ice-Sheet-Mass Balance

From the standpoint of eustatic sea-level changes, there is no cohesive picture about ice-volume changes over the last 2000 years, but several regional studies are informative. Haeberli and Holzhauser (2003) concluded that 20th-century glacial retreat in the European Alps exceeds those experienced during the late Holocene (see Hoelzle et al. 2007). Mosley-Thompson et al. (2006) came to a similar conclusion about low-latitude mountain glaciers. Arendt et al. (2002) studied long-term records of 67 glaciers from Alaska, representing about 20% of the state's total area covered by ice. They showed that Alaskan glaciers, which constitute a small portion of the world's total, might have contributed as much as 0.14 mm yr^{-1} to SLR since the 1950s and as much as 0.27 mm yr^{-1} since the 1990s. These numbers compare with a total of 0.5 mm yr^{-1} for post-1961 and 0.77 mm yr^{-1} since 1993 (Bindoff et al. 2007). Similar results seem to confirm rapid alpine glacial melting during the past few decades. One example was analysis of comprehensive satellite data on European alpine glaciers suggesting a much faster retreat since 1985 than had previously been thought (Paul et al. 2004; Zemp et al. 2006). Wiles et al. (2008) also studied land-terminating glaciers in southern Alaskan Wrangell and Coastal mountains and found rapid retreat during the MWP and following the LIA, continuing during the 20th century. They suggest these fluctuations were due to late Holocene millennial-scale temperature variability perhaps from solar forcing, but there was no clear anthropogenic signal.

Oerlemans (2005) compiled a 300-year-long atmospheric temperature record from 169 well-monitored glaciers. The results show a precipitous decline in the length of glaciers from Norway, Svalbard, New Zealand, Canada, and Jan Mayen Land, indicating the widespread nature of alpine ice retreat. Dyurgerov and Meyer (2005) and Kaser et al. (2006) analyzed alpine glaciers back to 1960, also documenting the recent decline in glaciers. Meier et al. (2007) provided a quantitative global-scale analysis of glacier and small ice cap trends and their contribution to sea level. They calculated that 60% of the observed nonsteric historical SLR is due to glaciers and small ice caps, that alpine glacial melting is accelerating, and that by the year 2100 glaciers and small ice caps will contribute a total of 0.1–0.25 m to SLR, a wide range due to uncertainty about actual rates of accelerated melting. Their main conclusion, nonetheless, is of significance for future SLR: alpine glacial melting will be the primary nonsteric contribution to SLR during the 21st century, after which it is expected that Greenland and Antarctic ice melting will play a greater role.

There is also progress in understanding Greenland and Antarctic Ice Sheet mass balance. Rignot and Kanagaretnam (2006) and Howat et al. (2007) used satellite measurements to document extremely rapid reversals in the rate at which ice discharged from Greenland's tidewater outlet glaciers. Rignot and Kanagaretnam's study found that the rate of thinning in some of Greenland's glaciers increased up to 100%, revising estimates of Greenland ice contribution to recent SLR by up to 0.25 mm yr^{-1}. Howat's study focused on two of Greenland's largest outlet glaciers during the period between 2000 and 2006. The observed calving and mass loss doubled in the year 2004, but this trend reversed itself in 2006. These short-term changes are superimposed on what appear to be a long-term net mass loss but, like other satellite datasets, these cover a limited period. Holland et al. (2008) recently identified warming ocean waters as a trigger for accelerated velocity and thinning of Jakobshavn Isbrae, one of Greenland's largest outlet glaciers.

In contrast to rapid wasting in most alpine and some Greenland glaciers, a positive mass balance, thus a net sea-level fall, has been documented in some studies of Antarctica. Davis et al. (2005) analyzed spatial patterns of snow accumulation on the grounded part of the AIS between 1992 and 2003, using remote sensing satellite data to calculate short-term elevation changes. This study supported the opinion that an enhanced hydrological cycle under climatic warming might lead to a net gain in AIS mass as more snow accumulates in the interior of the ice sheet, counterbalancing melting along its margins. Davis et al. found that a conservative estimate of 0.12±0.02 mm yr^{-1} of sea-level equivalent had accumulated during the 10-year study interval.

Shepherd and Wingham (2007) evaluated three different methods used to estimate recent sea-level changes from ice sources: (1) mass balance estimates from snowfall and melting processes, (2) satellite altimetry of ice elevation, and (3) gravitation methods. They concluded that the East Antarctic Ice Sheet is gaining 25 gigatons (Gt) of mass per year, the WAIS is losing twice that much, and Greenland is losing 100 Gt per year. There is a trend of decreasing elevation at four particularly well-studied Antarctic sites since 1992. Shepherd and Wingham estimate that the combined contribution from Greenland and Antarctic Ice to global sea level is equivalent to ~0.35 mm yr^{-1}. This estimate may be conservative because it might not account for rapid loss of ice mass

due to nonlinear ice-dynamic processes, which we now address.

Dynamic Changes in Ice Sheets

In addition to uncertainty surrounding the historical mass balance history of the world's ice, perhaps the greatest concern regarding future SLR pertains to recent discoveries about ice-sheet dynamics (Truffer and Fahnestock 2007). The term *Ice-sheet dynamics* refers to various subglacial processes and ocean-ice-sheet-margin interactions that, in contrast to ablation of glaciers and ice sheets in direct response to surface warming, can lead to abrupt, nonlinear behavior of ice sheets (Zwally et al. 2002; Chen et al. 2006). Discoveries in Greenland and Antarctica include, but are not limited to, glacier surging in West Antarctica (DeAngelis and Skvarca 2003), subglacial water lubrication beneath West Antarctic ice streams (Fricker et al. 2007), tidal, sea-level, and subglacial control of ice stream discharges (Bindschadler et al. 2003), and accelerated discharge from Greenland's tidewater glaciers (Rignot and Kanagaretnam 2006).

Ice dynamics are critical because climate models used to assess future SLR quantify changes in ice-sheet vertical sheer and surface ablation due to warming but do not yet have the ability to factor in abrupt, nonlinear ice-sheet response to rising regional ocean temperatures, sea-ice albedo feedbacks, basal lubrication of ice, and other processes. Shepherd and Wingham's (2007) summary reflects a widely held opinion among glaciologists that the rate of mass loss is already, or may soon be, increasing in some ice-sheet margin regions, and some of this mass loss or melting may be due to nonlinear dynamic processes. This means that the contribution of ice sheets to future SLR is poorly constrained in models but very likely to be more significant, rapid, and unpredictable than was thought only a few years ago. In addition to the scale of eustatic sea-level rise due to melting Antarctic ice, Mitrovica et al. (2009) showed that, due to gravitational effects, the "fingerprint" of West Antarctic ice sheet decay will vary greatly along coastlines of the world, in many places exceeding the mean eustatic sea-level rise. With this background, we consider records of late Holocene sea-level change.

Paleo-Sea-Level Records

In contrast to sea ice, Holocene sea-level history has been studied along many low-lying coasts, using stratigraphic records of basal peat formed in tidal marshes (Shennan and Horton 2002), and less commonly, historical and archaeological records (Lambeck et al. 2004a), often linked to historical rates from tide gauges and satellites. Figure 12.13 compares sea-level curves constructed from radiocarbon-dated basal peat for the past 1000 years (Thomas and Varekamp

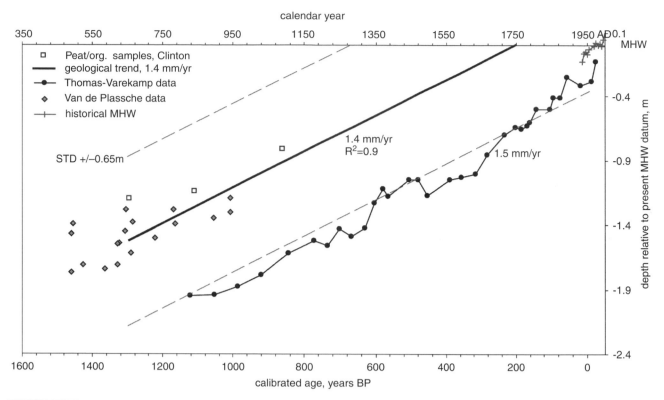

calendar year

FIGURE 12.13 Sea-level rise the last 1600 years based on basal peat (Clinton, Guilford, Connecticut, United States) and historical mean high water (MHW) from tide gauges (Rhode Island). A 1 sigma (σ) error bar is shown. The approximate intervals for the Medieval Warm Period and Little Ice Age are shown. Modified from Larsen and Clark (2006); data from Bloom and Stuiver (1963), Nydick et al. (1995), and van de Plassche et al. (1998). Courtesy of C. Larsen and I. Clark.

1991; Varekamp and Thomas 1998) and tide-gauge records along the east coast of the United States (Larsen and Clark 2006). It shows that despite claims of accelerated SLR over the past century, the 20th-century trend fits within the 2-sigma statistical error bar for the last millennium. This does not mean that the rate of SLR has not or is not accelerating. It reflects the fact that quantifying RSL history over decades to centuries is particularly difficult. When the errors associated with each method are included, it becomes difficult to identify an inflexion point that signifies a definitive acceleration in the rate of SLR.

Sea-level reconstruction is complex for several reasons. First, measuring a sea-level position within a few centimeters tests the limits of paleoshoreline biological or lithological indicators or both (Pirazzoli 1991). Tidal-marsh-dwelling benthic foraminifera are one of the most widely used groups, but even these indicators have complex ecological and preservation issues that introduce complexity into the interpretation of faunal and sea-level changes (Culver and Horton 2005; Culver et al. 2006). Uncertainty also comes from coastal uplift and subsidence because of isostatic and tectonic processes, errors associated with radiocarbon-dates (Törnqvist et al.

1998), dewatering or peat compaction or both (Kaye and Barghoorn 1964), spatial variability in marsh accretion rates (Kearney 1996), and difficulty in separating decadal sea-level variability from secular trends (Larsen and Clark 2006). It is therefore axiomatic to sea-level researchers that no single "eustatic" sea-level curve can be constructed and that all regional sea-level curves are relative curves reflecting a combination of eustatic sea-level changes, vertical land movement, sediment processes, and errors from several sources.

RSL curves from peat and other coastal sediment records, however, can show whether acceleration (or slowdown) occurred in RSL for a particular region before, during, or—when spliced with tide-gauge records—after industrialization. Several finely sampled sea-level records demonstrate this approach. Coasts along eastern North America and the Gulf of Mexico, as well as some coasts in Europe, have received most attention because of negligible tectonic effects, low tidal amplitudes, and high rates of modern SLR due in part to glacio-isostatic subsidence (Table 12.2). Radiocarbon-dated basal peats are by far the most commonly used sea-level indicators. Peat accumulates in coastal marshes as a land surface is transgressed and organic material accretes.

TABLE 12.2 Rates of Relative Sea-Level Rise During the Past Millennium

Region	Preindustrial (mm yr⁻¹)*	20th Century (mm yr⁻¹)	Current (mm yr⁻¹)	Comment	Reference
Narragansett Bay, Rhode Island	1	2.5	2.7	Late 19th-century acceleration; 20th-century rates vary decadally	Donnelly and Bertness 2001
Chesapeake Bay	0.56	3	3.0–3.9	LIA regression	Kearney 1996
Chesapeake Bay	N/A	3	>3	Varies greatly by method	Kearney 1996
Connecticut— Hammock River (Clinton, CT)	1.1	2.2	2.1–2.2	Decadal variability during LIA, MWP; most recent acceleration began ~A.D. 1700	Varekamp et al. 1992
Connecticut	1.4	Variable		Increased rate A.D. 1630	Varekamp and Thomas 1998
Connecticut— Barn Island	1	2.5–3.0	2.8	Acceleration 1893–1921	Donnelly et al. 2004
Connecticut— Hammock River	Variable	Accelerating	2.8	1400-yr record of variability, no rise in 1800s	van de Plassche et al. 1998; van de Plassche 2000
Maine	Stable 800–1300	1.5–2 (since 1800)		Unprecedented 20th-century rate	Gehrels et al. 2002
Delaware	2	>3	2.6	Rapid rise 1750–~1900	Varekamp and Thomas 1998; Larsen and Clark 2006
Delaware	2		2.6		Nikitima et al. 2000; Larsen and Clark 2006
Gulf of Mexico	0.55	NA		Sediment loading, compaction important; rates vary for last millennium	Gonzales and Törnqvist 2006; Törnqvist et al. 2008
Chezzetcook, Nova Scotia	1.7	1.6 (19th century)	3.2	Acceleration began 1900–1920	Gehrels et al. 2005

Note: Error bars for rates can be found in the original sources.

*Preindustrial usually denotes long-term mean over ~1000–1500 years and might include isostatic adjustment.

Holocene peat-derived sea-level curves have a long history going back to the earliest use of radiocarbon dating for geological records (Bloom and Stuiver 1963; Scholl and Stuiver 1967; Milliman and Emery 1968; Scholl et al. 1969).

Various authors cited in Table 12.2 came to opposing conclusions regarding 20th-century SLR. One group identifies decadal and centennial sea-level variability over the past 500–1500 years and, in some cases, finds that the onset of accelerated SLR occurred during the 18th or 19th century and pre-dated late 19th-century industrialization. Decadal variability is evident in many records, especially those along the well-studied east coast of the United States (Varekamp et al. 1992; Varekamp and Thomas 1998; van de Plassche et al. 1998; van de Plassche 2000).

The second group concludes that an acceleration of the rate of SLR occurred at the end of the 19th century and in the first few decades of the 20th century, roughly coincident with the industrial revolution. Studies by Donnelly et al. (2004) of Barn Island, Connecticut and Gehrels et al. (2005) of Chezzetcook, Nova Scotia are the most notable. Gehrels's record was especially informative because it used multi-proxy environmental records from a site in Nova Scotia situated near the Atlantic Ocean, in contrast to other studies focused on enclosed basins or estuaries. Evidence for an acceleration from ~1.6 mm yr⁻¹ to 3.2 mm yr⁻¹ between 1900 and 1920 was interpreted as evidence for human-induced warming. The similarity of the Chezzetcook record to that of Donnelly et al. (2004) for the Connecticut marsh is quite striking. An increased rate of SLR from a historical mean rate of 1.5–2.0 mm yr⁻¹ to the current ~3.1 mm yr⁻¹ is consistent with evidence from tide-gauge records (Miller and Douglas 2004; Church and White 2006).

These opposing results raise the question of whether stratigraphic marsh records, particularly those using benthic foraminiferal species as proxies of marsh habitats, provide reliable information about low-amplitude, high-frequency glacio-eustatic oscillations on the scale of 10–20 cm. Gehrels et al. (2005) emphasized the difficulty in recognizing decadal and centennial variability during the MWP and LIA, not only because of processes already mentioned here, but also because of changes in carbon cycling and radiocarbon "plateaus" in ^{14}C calibration curves. Kelley et al. (2001) and Larsen and Clark (2006) also pointed out difficulties in overcoming the small signal-to-noise ratio in short-term sea-level reconstructions from stratigraphic records.

Estimates of preindustrial background Holocene SLR rates have also been derived from combined paleo-sea-level records and GIA modeling of millennial-scale RSL. Estimates of long-term mean RSL rise for the Holocene range from <0.5 m (Lambeck and Bard 2000), 0–2 m for the last 4000 years (Jansen et al. 2007), 3–5 m (Fleming et al. 1998), 3 m since about 6 ka (Lambeck 1997), and as little as 0 m for some sites since about 7 ka (Pirazzoli 2005). Sea-level reconstructions on well-studied coasts of the British Isles (Peltier et al. 2002) and Australia (Lambeck 2002) suggest that the contribution of ice volume to global late Holocene sea-level change is small, between 0 and 0.2 mm yr^{-1}. Along these same lines, Lambeck et al. (2004a) presented archaeological evidence from the Mediterranean on Holocene sea level for the past 4000 years. They estimated that, when corrected for isostatic adjustment, sea level may have fallen 0.13 m ± 0.09 m since the Roman Period ~2000 years ago, and that the modern SLR began 100 ± 53 years ago (see Lambeck et al. 2004b). Gehrels et al. (2004, 2006) supported the absence of a detectable ice-volume contribution to SLR for the last 5000 years. These studies hypothesize a flat glacio-eustatic component of the sea-level curve for at least the last 2000 years, possibly back 4000 or 5000 years ago, and thus a negligible glacio-eustatic component to background sea-level rates.

This hypothesis that late Holocene ice volume was relatively constant—at least within the bounds of detection from combined archaeological, paleoshoreline, tide-gauge, and GIA model methods—is quite significant, although not at all confirmed. The IPCC 2007 report did not attempt to factor in rates of RSL from submillennial-scale proxy records from coastal marsh records already cited here, but instead used the Peltier and Lambeck GIA model as a background against which to view 20th-century SLR (Jansen et al. 2007). In another comprehensive analysis, no apparent acceleration of the rate of SLR was evident during the 18th to the 20th century, according to a recent volume published by the Royal Society of London (Woodworth 2006).

Holocene sea-level curves provided by GIA models are useful for millennial-scale climate and glaciological analysis

of large ice sheets, but are not designed to address decadal and centennial variability critical for interpreting 20th-century RSL. The contribution of melting ice during the past few millennia is poorly constrained at short timescales and there is still no submillennial-scale sea-level model that attempts to reproduce the ice-volume portion of sea-level changes during the last 2000 years. Consequently, despite strides made in the modeling of 20th-century rates of sea-level change, the long-standing hypothesis that the rate of 20th-century SLR is larger than the average rate of the past few millennia has remained basically unchallenged and untested for years.

In sum, sea-level proxy indicators, marsh accretion rates, glacio-isostatic modeling, and better dating methods suggest that paleo-sea-level reconstructions from stratigraphic sequences along modern coasts offer a potential source of information about decadal and centennial sea-level trends and their relationship to broader natural and anthropogenic climate changes. There remains a great need for an integrated data-model analysis of sea level for the past 2000 years, similar to that achieved by the GIA model for sea-level change since the last deglaciation and for surface air temperature discussed in Chapter 11 of this volume.

Perspective

Even this abbreviated synthesis of late Holocene paleoclimatology demonstrates a level of regional and temporal complexity not anticipated a decade ago. Several conclusions are evident. The first is that some evidence suggests 20th-century climate behavior that can be called anomalous or unusual in relation to that reconstructed from proxies over centuries or longer. Intensive efforts focusing on regional climate, such as tree-ring networks of drought and of low-latitude and polar ice-core atmospheric records, support conclusions based on instruments and modeling about at least some human influence on nontemperature parts of the climate system. Second, other records show significant decadal-to-centennial climate variability caused by internal modes of climate variability and external forcing. The amplitude of preindustrial variability can equal or exceed that seen in the instrumental record. In some records, emergence from 19th-century LIA climate began before major industrialization. Third, climate variability during the LIA and the MWP is especially large, but its causes are only partially understood. Reconstructions of precipitation, ocean temperature, sea ice, glaciers, ice sheets, and sea level, incomplete as they are, render efforts to distinguish natural and anthropogenic impacts more challenging than simply the detection and attribution of a human-induced trend in atmospheric temperature.

The future will see new and improved reconstructions of sea level from coastal regions, precipitation from lake sediments, atmospheric circulation from ice and ocean cores, tropical temperature and circulation from corals and other proxies, deep-ocean circulation from deep-sea corals and high-sediment-rate cores, and polar sea ice from marine cores. The success of coordinated programs focused on the Last Glacial Maximum (CLIMAP, EPILOG, GLAMAP), past global warmth (PETM, PRISM), and data-modeling comparisons (PMIP) reflects the value of spatially robust, integrated analyses over single-site studies in addressing the detection and attribution of climate changes.

LANDMARK PAPERS Low-Latitude Alpine Ice Cores and Climate Change

Thompson, L. G., S. Hastenrath, and B. Morales Arnao. 1979. Climatic ice core records from the tropical Quelccaya ice cap. Science 203:1240–1243.

Thompson, L. G., E. Mosley-Thompson, J. F. Bolzan, and B. R. Koci. 1985. A 1500-year record of tropical precipitation in ice cores from Quelccaya ice cap, Peru. Science 229:971–973.

Thompson, L. G., E. Mosley-Thompson, W. Dansgaard, and P. M. Grootes. 1986. The Little Ice Age as recorded in the stratigraphy of the tropical Quelccaya ice cap. Science 234:361–364.

In addition to rising atmospheric temperatures, climate change involves retreating glaciers, shifts in regional and global precipitation, an altered carbon cycle, ocean acidification, rising sea level, and many other trends. Alpine regions are among the most sensitive to climate changes, but they pose obvious limitations for paleoclimatology because of the logistical difficulty of obtaining records from lake sediments, ice caps, glacial geology, or tree rings. The 5670-m-high Quelccaya, Peru ice cap was the first low-latitude ice-core paleoclimate record obtained by Lonnie Thompson and Ellen Mosley-Thompson of Byrd Polar Research Center, Ohio State University. Quelccaya had an excellent Holocene chronology and yielded information on a variety of topics—the Little Ice Age tropical atmosphere, El Niño climate variability, and the evolution of Holocene climate.

Since these pioneering studies, Thompson, Mosley-Thompson, and colleagues have reconstructed climate records using stable isotopes and other proxies from low-latitude ice caps from the Tibetan Plateau, Mt. Kilimanjaro in Africa, and other South American Andean sites located 5300–6200 m above sea level. The interpretation of alpine glacial isotopic records is complex and revolves around whether temperature, precipitation, or moisture sources are the dominant controls. Nonetheless, ice-core records generated by Thompson and colleagues are unique archives of low-latitude climate variability and a baseline for anthropogenic climate change. In addition to the proxy records, the alarming rate at which high-elevation glaciers themselves have thinned and retreated, even in the decades since Thompson and Mosley-Thompson first cored them, has raised public awareness of accelerated climatic changes. To some researchers, alpine glaciers offer some of the strongest evidence for anomalous climate behavior over the last century. The Thompson-Mosley-Thompson studies epitomize today's much larger community of researchers trying to decouple natural from anthropogenic influence in all other parts of the climate system.

Epilogue

Natural processes in the oceans, atmosphere, ice, organisms, and lithosphere left us an archival history of the earth's climate, and, thanks to the geoscientists committed to unraveling these archives, we now know an awful lot about it. Ice sheets and glaciers catastrophically melt, sea level rises at rapid rates, temperature spikes in a lifetime or less, ocean circulation alters global heat transport, biogeochemical cycles undergo drastic extremes, persistent droughts affect human cultures, and ecosystems and biomes are often totally rearranged. This collective body of geological knowledge, hardly imagined a few decades ago, proves beyond a reasonable doubt that the earth's climate is extraordinarily unstable over all timescales, provides enormous insight into contemporary and potential future climate change, and should raise awareness, if not a clarion call, for action on several fronts.

APPENDIX

Paleoclimate Proxies

Climate Parameter*	Proxy Method	Primary Archives	Common Application	Comments	Reference
Surface air temperature	$\delta^{18}O$, deuterium (δD)	Ice core ice	Quaternary		Johnsen et al. 1995; Jouzel 1999; Jouzel et al. 2007
	Deuterium excess (d)	Ice core ice	Quaternary	$d=\delta D-8^*\delta^{18}O$, temperature, precipitation	Jouzel et al. 2007
	$\delta^{15}N_{air}$	Ice core air bubbles	Quaternary		Severinghaus and Brook 1999
	$\delta^{40}Ar_{air}$	Ice core air bubbles	Quaternary		Severinghaus and Brook 1999
	$\delta^{29}N_{air}$	Ice core air bubbles	Quaternary		Lang et al. 1999
	$\delta^{18}O$ in speleothems	Caves	Quaternary	Complex hydro-logical processes	McDermott 2004
	Dendroclimatology	Tree rings	Quaternary		Mann and Jones 2003; D'Arrigo et al. 2006
	Noble gases in fluid inclusions	Speleothems		Gases: Ne, He, Ar, Kr, Xe	Kluge et al. 2008
Precipitation	Titanium and iron	Marine sediment		Precipitation-driven river discharge	Haug et al. 2001b
	$\delta^{18}O$ in calcite	Speleothems			McDermott 2004; Mickler et al. 2006
	$\delta^{18}O$ in forams	Estuarine and marine sediment		Precipitation-driven river discharge	Cronin et al. 2005
	$\delta^{18}O$ in ostracodes	Nonmarine ostracodes	Mainly Quaternary	Paleolimnology, also temperature	von Grafenstein 2002

(continued)

Climate Parameter*	Proxy Method	Primary Archives	Common Application	Comments	Reference
	Mg/Ca, Sr/Ca in ostracodes	Nonmarine sediment	Mainly Quaternary	Paleolimnology	Ito 2002
	Dendroclimatology	Tree rings	Late Quaternary		Cook et al. 2007; Treydte et al. 2006
	Fire scars in trees	Tree rings	Holocene	Precipitation, temperature	Swetnam 1995
	δD in terrestrial plant matter	Lake sediment	Mainly Quaternary	Lake environmental change	Huang et al. 2004
Precipitation and runoff	δD, $\delta^{13}C$ in terrestrial plant matter	Nearshore marine sediment	Various	Various organic biomarkers	Pancost and Boot 2004
Snow acculation	^{36}Cl, ^{10}Be, paleomagnetic intensity	Ice cores	Greenland, Antarctica	Cosmogenic isotopes	Wagner et al. 2001
Precipitation-evaporation	$\delta^{18}O$ in forams	Mainly tropical oceans	Quaternary		Rohling and Cooke 1999; Rohling 2006
Atmospheric temperature-moisture	Organic acid fluorescence	Speleothems	Various	Complex soil-groundwater processes	McGarry and Baker 2000
Atmospheric circulation	Wind, dust	Marine sediment	Mainly Quaternary	Iron transport	Rea 1994; Murray et al. 1995
	Sulfate aerosols	Ice cores	Holocene		Duan et al. 2007
	^{10}Be, cosmogenic isotope	Ice cores	Quaternary	Atmospheric production, snow accumulation	
	Soluble glaciochemical	Ice core paleoatmosphere	Quaternary	HNO_3, HCL, H_2O_2, NH_3	Various
	Insoluble glaciochemical	Ice core sea salt, dust, wind, aridity	Quaternary	Ca_2^+, Mg_2^+, NO_3^-, Cl^-	
	Glaciochemical-volcanic activity, sea salt	Ice cores	Quaternary	SO_4^{+2}—sulfate	353 et al. 1985; Zielinski 2000
	Electrical conductivity	Ice cores	Quaternary	Dust, circulation, volcanic events	Taylor et al. 1997
	Ammonium	Ice cores, biomass burning	Quaternary	NH_4^+	Mayewski et al. 1993; Taylor et al. 1997
	Loess (windblown sediment)	Terrestrial deposits	Pliocene-Quaternary	Monsoon influence	Porter 2001b
	$\delta^{18}O$ in speleothems	Speleothems	Quaternary		Wang et al. 2005; Fairchild et al. 2006
Atmospheric chemistry	CO_2 in air bubbles	Ice cores	Quaternary	CO_2 concentration	Petit et al. 1999; Siegenthaler et al. 2005
	CH_4 in air bubbles	Ice cores	Quaternary	Methane concentration and source	Chappellaz et al. 1997; Blunier and Brook 2001; Spahni et al. 2003

Climate Parameter*	Proxy Method	Primary Archives	Common Application	Comments	Reference
	Fossil leaf stomata	Nonmarine sediment	Paleozoic	Stomatal size, density	McElwain and Chaloner 1995; Kouwenberg et al. 2003
	$\delta^{11}B$ of forams	Marine sediment	Cenozoic	Ocean pH proxy	Pearson and Palmer 2000
	Pedogenic carbonates	Paleosols (soils)	Phanerozoic		Cerling 1992; Royer et al. 2001
	CO_2 concentrations	Marine sediment	Cenozoic		Pagani et al. 1999, 2005
	CO_2 and O_2 concentrations	Sediment $\delta^{13}C$	Mesozoic, Cenozoic	Indirect CO_2 and biogeochemistry	Katz et al. 2005
	N_2O nitrous oxide	Ice cores	Quaternary	Ocean and soil fluxes	Flückiger et al. 2004; Spahni et al. 2005
	Interhemisphere CH_4 gradient	Ice cores	Quaternary	Methane sources	Chappellaz et al. 1997; Dällenbach et al. 2000
	Atmospheric O_2 concentrations	Various materials	Paleozoic-Mesozoic	Modeling and empirical data	Belcher and McElwain 2008; Berner 2006; Falkowski et al. 2005; Bergman et al. 2004
SST[†]	Faunal/floral assemblages	Marine sediment	Cenozoic	Planktic forams, dinoflagellates	Various quantitative methods
	Sr/Ca in corals	Corals	Quaternary	Tropical SST and symbiont activity	Beck et al. 1992; Druffel 1997; Cohen et al. 2002; Sinclair 2005
	Alkenones (Uk37)	Marine sediment	Cenozoic	Long-chain molecules (biomarkers)	Bard 2001; Muller et al. 1998; Herbert 2003
	Mg/Ca in forams	Marine sediment	Cenozoic	Various species	Lea et al. 1999; Elderfield and Ganssen 2000; Barker et al. 2005
	Mg/Ca in forams	Polar, subpolar oceans	Late Cenozoic	*N. pachyderma*	Nurnberg et al. 1996
	Mg/Ca in corals	Tropical oceans	Quaternary		Mitsuguchi et al. 1996; Sinclair 2005
	$\delta^{18}O$ in corals	Tropical oceans	Late Quaternary		Grottoli and Eakin 2007
	$\delta^{44}Ca$	Forams	Cenozoic		Nagler et al. 2000
	$\delta^{18}O$ in mollusks	Mainly *Astarte islandica*	Mainly Holocene		Weidman et al. 1994; Marchitto et al. 2000; Schöne et al. 2005; Wanamaker et al. 2008
	Mollusk growth lines	*Astarte* floating chronology	Mainly Holocene	Fossil mollusks	Scourse et al. 2006
	Mollusk shells	Various aquatic organisms			Schone and Surge 2005

(continued)

Climate Parameter*	Proxy Method	Primary Archives	Common Application	Comments	Reference
Sea-surface hydrography	$\delta^{18}O$ in forams		Cenozoic	Precipitation	Rohling 2006, see precipitation
	$\delta^{18}O$ in corals		Quaternary	Precipitation	Grottoli and Eakin 2007
Ocean salinity	$\delta^{18}O$ and δD	Marine sediment	Various	"Net" deuterium excess	Rohling 2007
	$\delta^{18}O$ in forams	Enclosed marine basins	Quaternary	Precipitation, evaporation, circulation	Rohling 1999
	δD in algal biomarkers	Calcareous nannoplankton	Quaternary	Temperature also	Schouten et al. 2006; van der Meer et al. 2007
	Mg/Ca ratios in foraminifers	Planktic forams	Cenozoic		Ferguson et al. 2008
Ocean circulation	$\Delta^{14}C$ in corals		Holocene		Druffel 1997; Guilderson et al. 2004
	$\Delta^{14}C$ in mollusks		Holocene		Weidman and Jones 1993
	$\delta^{18}O$ and $\delta^{13}C$ in benthic forams	Marine sediment	Quaternary		Marchal and Curry 2008
Ocean pH	$\delta^{11}B$	Corals	Holocene		Sanyal et al. 1997; Pelejero et al. 2005; Gaillardet and Allègre 1995
	$\delta^{11}B$	Forams	Cenozoic	Also reflects boron isotope budget	Palmer et al. 1998; Pearson and Palmer 2000; Pagani et al. 2005a
Ocean alkalinity	Ba/Ca in forams		Quaternary		Lea 1995
Ocean carbonate saturation	Li/Ca lithium	Deep-sea forams	Cenozoic		Lear and Rosenthal 2006
Ocean productivity	$\delta^{15}N$ of nitrate	Deep-sea sediment	Mainly Quaternary	Phytoplankton nitrate utilization	Sigman et al. 1999
	Cd/P ratios	Deep-sea sediment	Mainly Quaternary		Elderfield and Rickaby 2000
	Pa/Th—radionuclides	Marine sediment	Mainly Quaternary	Windblown dust	Kumar et al. 1995
	Uranium concentration	Deep-sea, ocean-margin sediment	Quaternary	Complex chemical pathways	Rosenthal et al. 1995; Zheng et al. 2002
	^{10}Be, ^{230}Th, ^{231}Pa	Marine sediments	Quaternary	Also hydrothermal activity	Frank et al. 1994
	Foraminiferal assemblages	Marine sediment	Mainly Cenozoic		Fischer and Wefer 1999; Jorissen and Rohling 2000
	Dinoflagellate assemblages	Marine sediment	Mainly Quaternary	Also ocean hydrography	Rochon et al. 2008
	Radiolarian assemblages	Marine sediment	Cenozoic		Lazarus et al. 2008

Climate Parameter*	Proxy Method	Primary Archives	Common Application	Comments	Reference
Deep-sea bottom temperature	Mg/Ca in forams	Deep-sea forams	Cenozoic		Martin et al. 2002; Lear 2007
	Mg/Ca in ostracodes	Ostracodes	Cenozoic		Dwyer et al. 1995, 2002
Thermocline temperatures	Mg/Ca, other elements	Benthic forams	Cenozoic		Rosenthal et al. 1997
Deep-ocean circulation	Radiogenic isotopes of Nd, Pb, Hf, Os, Sr	Forams, ferromanganese crusts	Cenozoic	Weathering, circulation, climate	Ling et al. 2005; Frank 2002
	Sr—strontium	Forams, brachiopods	Phanerozoic	Other biogenic carbonates	Elderfield 1986; Veizer et al. 1999
	$^{143}Nd/^{144}Nd$ neodymium isotopes		Cenozoic, longer	$^{143}Nd/^{144}Nd$	Vance and Burton 1999
	$\delta^{13}C$ in forams		Cenozoic	Paleonutrient proxy	Curry and Oppo 2005
	Cd/Ca ratios in forams	Deep-sea sediment	Mainly Quaternary	Paleonutrient proxy	Boyle 1988; Rickaby and Elderfield 1999
	$^{231}Pa/^{230}Th$ ratios	Deep-sea sediment	Mainly Quaternary		McManus et al. 2004
	$\Delta^{14}C$ activity	Forams	Late Quaternary	Circulation and carbon cycling	Adkins and Boyle 1997; Broecker et al. 2007
	$\Delta^{14}C$ activity	Deep-sea corals	Late Quaternary		Adkins et al. 1998; Robinson et al. 2005
	$^{10}Be/^{9}Be$	Deep-sea sediment	Cenozoic	^{10}Be = cosmogenic origin, ^{9}Be = crustal erosion	
	Uranium series	Deep-sea sediment	Various	Various nuclides for productivity, sedimentation, circulation	Henderson and Anderson 2003
	Sortable silt	Deep-sea sediment	Late Quaternary	Bottom-current strength	McCave et al. 1995
$\delta^{18}O$ of seawater	$\delta^{18}O$ of porewater, chloride	Deep-sea porewater	Late Quaternary (LGM)	Ocean salinity	Adkins and Schrag 2001
Ocean anoxia	Molybdenum	Deep-sea sediment	Various timescales	Climate-driven anoxia	Crusius et al. 1996; Dean et al. 1999
Continental erosion, weathering	$^{87}Sr/^{86}Sr$ and Sr/Ca in forams	Deep-sea sediment, ferromanganese nodules	Neoproterozoic, Phanerozoic	Also hydrothermal origin	Veizer et al. 1999; Lear et al. 2003
	$^{187}Os/^{188}Os$		Various	Osmium isotope stratigraphy	Peucker-Ehrenbrink and Ravizza 2000; Klemm et al. 2005
Ice sheets	Glacial deposits	Lithological	Neoproterozoic to Quaternary		Various
	Glacial landscapes	Geomorphological	Mainly Quaternary		Various

(continued)

Climate Parameter*	Proxy Method	Primary Archives	Common Application	Comments	Reference
	Submarine scour marks	Geophysical, sidescan sonar	Quaternary		Vogt et al. 1994; Polyak et al. 2001; Kristofferson et al. 2004
	$\delta^{18}O$ of forams	Marine sediment	Cenozoic	Ice-volume proxy	Shackleton 1967; Lisiecki and Raymo 2005
	$\delta^{18}O$ of forams	Marine sediment	Quaternary	Ice volume, Red Sea	Siddall et al. 2003; Arz et al. 2007
	Tidal marsh forams	Coastal sediment	Quaternary	Forams in tidal peats	Scott 1978; Thomas and Varekamp 1991; van de Plassche 2003
Sea ice	Methanesulphonic acid	Ice cores	Last 800,000 years	From biogenic dimethylsulphide	Curran et al. 2003
	Dinoflagellates	Deep-sea sediment	Quaternary		de Vernal et al. 2001
	Ostracodes	Deep-sea sediment	Late Quaternary		Cronin et al. 1995
	Sediment mineralogy	Deep-sea sediment	Quaternary		Andrews and Eberl 2007
	Diatoms	Deep-sea sediment	Quaternary		McMinn et al. 2001
Icebergs (ice-rafted debris)	Hematite grains		Quaternary		Bond et al. 1992
	Vaious geochemical source proxies	Deep-sea sediment	Quaternary	Sm/Nd, Rb/Sr, Pb K/Ar isotopes	Hemming 2004
	Iron oxide sediment grains	Deep-sea sediment	Quaternary	Chemical finger-printing, distin-guish sea-ice from iceberg	Darby 2003; Darby and Bischof 2004; Darby and Zimmerman 2008
	Ice-rafted debris	Various sediment types	Quaternary	Lithological	Andrews 2000; Hemming 2004
	Detrital carbonate	Derived mainly from N. American bedrock	Quaternary		Bischof and Darby 1999; Andrews and Tedesco 1992; Andrews et al. 1998; Nørgaard-Pedersen et al. 2007
Solar activity	Cosmogenic isotopes, ^{10}Be, ^{36}Cl	Ice cores, sediment	Mainly last 100,000 years	Complicatd by solar modulation, geomagnetic intensity	Beer et al. 1988; Muscheler et al. 2005a; Snowball and Muscheler 2007
	Cosmogenic isotopes, ^{14}C	Trees, ice cores, sediment	Quaternary	Complicated by carbon cycling	
Glacial lake discharge	Mg/Ca, U/Ca, $^{87}Sr/^{86}Sr$	Planktic forams	Late Quaternary	Freshwater source tracer	Carlson et al. 2007a
	Glacial lake geology-geomorphology		Quaternary		Thorleifson 1996; Franzi et al. 2007
	$\delta^{18}O$ in forams		Quaternary		Boden et al. 1997

Climate Parameter*	Proxy Method	Primary Archives	Common Application	Comments	Reference
Paleo-elevation	$\delta^{18}O$ in lake sediment, fossil soil	Tibetan lakes	Cenozoic	Alternative mantle-thickening tectonic models	Rowley and Garzione 2007; Rowley et al. 2001; Rowley and Currie 2006; Molnar et al. 2006
	Fossil plants		Cenozoic		Spicer et al. 2003
	Pedogenic carbonates	Soils	Cenozoic		Quade et al. 2007

Note: Forams=benthic or planktic foraminifera. Subscripts indicate archive in which isotopic and trace element ratios are measured. LGM=Last Glacial Maximum; SST=sea-surface temperature.

*Most proxies are influenced by several climate-related factors (i.e., temperature, salinity). Here we categorize each on the basis of primary signal, recognizing complexity.

[†]Some species live at various depths in the upper ocean or on the bottom in shallow ocean.

References

Aagaard, K., F. Fahrbach, J. Meincke, and J. H. Swift. 1991. Saline outflow from the Arctic Ocean: Its contribution to the deep waters of the Greenland, Norwegian, and Iceland seas. Journal of Geophysical Research 96:20433–20441.

AchutaRao, K. M., B. D. Santer, P. J. Gleckler, et al. 2006. Variability of ocean heat uptake: Reconciling observations and models. Journal of Geophysical Research 111, C5:C05019, JC003136.

Ackert, R. P., Jr., R. A. Becker, B. S. Singer, et al. 2008. Patagonian glacier response during the late Glacial–Holocene transition. Science 321:392–395.

Adams, J. B., M. E. Mann, and C. M. Ammann. 2003. Proxy evidence for an El Niño–like response to volcanic forcing. Nature 426:274–278.

Adkins, J. F. and E. A. Boyle. 1997. Changing atmospheric $\Delta^{14}C$ and the record of deep water paleoventilation ages. Paleoceanography 12:337–344.

Adkins, J. F., E. A. Boyle, L. Keigwin, and E. Cortijo. 1997. Variability of the North Atlantic thermohaline circulation during the last interglacial period. Nature 390:154–156.

Adkins, J. F., H. Cheng, E. A. Boyle, E. R. M. Druffel, and R. L. Edwards. 1998. Deep-sea coral evidence for rapid changes in ventilation of the deep North Atlantic at 15.4 ka. Science 280:725–728.

Adkins, J. F. and D. P. Schrag. 2001. Pore fluid constraints on deep ocean temperature and salinity during the Last Glacial Maximum. Geophysical Research Letters 28:771–774.

Aerts, J. C. J. H., H. Renssen, P. J. Ward, et al. 2006. Sensitivity of global river discharges under Holocene and future climate conditions. Geophysical Research Letters 33:L19401, GL027493.

Agassiz, L. 1840. Etudes sur les Glaciers. Neuchatel.

Aharon, P. 2003. Meltwater flooding events in the Gulf of Mexico revisited: Implications for rapid climate change. Paleoceanography 18:PA000840.

Ahn, J. and E. J. Brook. 2007. Atmospheric CO_2 and climate from 65 to 30 ka B.P. Geophysical Research Letters 34:L10703, GL029551.

Ahn, J. and E. J. Brook. 2008. Atmospheric CO_2 and climate on millennial time scales during the last glacial period. Science 322:83–85.

Alley, R. B. 2007. Wally was right: Predictive ability of the North Atlantic "Conveyor Belt" hypothesis for abrupt climate change. Annual Review of Earth and Planetary Sciences 35:241–275.

Alley, R. B. and A. M. Agustsdottir. 2005. The 8k event: Cause and consequences of a major Holocene abrupt climate change. Quaternary Science Reviews 24:1123–1149.

Alley, R. B. and S. Anandakrishnan. 1995. Variations in melt-layer frequency in the GISP2 ice core: Implications for Holocene summer temperatures in central Greenland. Annals of Glaciology 40:341–349.

Alley, R. B., S. Anandakrishnan, and P. Jung. 2001. Stochastic resonance in the North Atlantic. Paleoceanography 16:190–198.

Alley, R. B., E. J. Brook, and S. Anandakrishnan. 2002. A northern lead in the orbital band: North-south phasing of ice age events. Quaternary Science Reviews 21:431–441.

Alley, R. B., P. U. Clark, P. Huybrechts, and I. Joughin. 2005. Ice-sheet and sea-level changes. Science 310:456–460.

Alley, R. B., K. Dupont, B. R. Parizek, et al. 2006. Outburst flooding and the initiation of ice-stream surges in response to climatic cooling: A hypothesis. Geomorphology 75:76–89.

Alley, R. B. and D. R. MacAyeal. 1994. Ice-rafted debris associated with binge/purge oscillations of the Laurentide Ice Sheet. Paleoceanography 9:503–511.

Alley, R. B., J. Marotzke, W. D. Nordhaus, et al. 2003. Abrupt climate change. Science 299:2005–2010.

Alley, R. B., P. A. Mayewski, T. Sowers, M. Starrer, K. C. Taylor, and P. U. Clark. 1997. Holocene climatic instability: A prominent widespread event at 8200 yr ago. Geology 25:483–486.

Alley, R. B., D. A. Meese, C. A. Shuman, et al. 1993. Abrupt increase in Greenland snow accumulation at the end of the Younger Dryas event. Nature 362:527–529.

Altabet, M. A., M. J. Higginson, and D. M. Murray. 2002. The effect of millennial-scale changes in Arabian Sea denitrification on atmospheric CO_2. Nature 415:159–162.

Ammann, C. M., F. Joos, D. S. Schimel, B. L. Otto-Bliesner, and R. A. Tomas. 2007. Solar influence on climate during the past millennium: Results from transient simulations with the NCAR Climate System Model. Proceedings of the National Academy of Sciences 104:3713–3718.

Ammann, C. M., G. A. Meehl, W. M. Washington, and C. S. Zender. 2003. A monthly and latitudinally varying volcanic forcing dataset in simulations of 20th century climate. Geophysical Research Letters 30:GL016875.

Ammann, C. M. and P. Naveau. 2003. Statistical analysis of tropical explosive volcanism occurrences over the last 6 centuries. Geophysical Research Letters 30:1210, doi:10.1029/2002GL016388.

Amundsen, K., S. Abella, E. Leopold, M. Stuiver, and S. Turner. 1994. Late-glacial and early sea-level fluctuations in the Central Puget Lowland, Washington, inferred from lake sediments. Quaternary Research 42:149–161.

An, Z. S., J. E. Kutzbach, W. L. Prell, and S. C. Porter. 2001. Evolution of Asian monsoons and phased uplift of the Himalaya-Tibetan plateau since late Miocene times. Nature 411:62–66.

An, Z. S. and S. C. Porter. 1997. Millennial-scale climatic oscillations during the last interglaciation in central China. Geology 25:603–606.

An, Z. S., S. C. Porter, J. E. Kutzbach, et al. 2000. Asynchronous Holocene optimum of the East Asian monsoon. Quaternary Science Reviews 19:743–762.

Anand, P., H. Elderfield, and M. H. Conte. 2003. Calibration of Mg/Ca thermometry in planktonic foraminifera from sediment trap time series. Paleoceanography 18:1050, 10.1029/2002PA000846.

Anbar, A. D. and A. H. Knoll. 2002. Proterozoic ocean chemistry and evolution: A bioinorganic bridge? Science 297:1137–1142.

Anbar, A. D. and O. Rouxel. 2007. Metal stable isotopes in paleoceanography. Annual Review of Earth and Planetary Sciences 35:717–746.

Andersen, K. K., A. Svensson, S. J. Johnsen, et al. 2006. The Greenland Ice Core Chronology 2005, 15–42 ka. Part 1: Constructing the time scale. Quaternary Science Reviews 25:3246–3257.

Anderson, D. M., J. T. Overpeck, and A. K. Gupta. 2002. Increase in the Asian southwest monsoon during the past four centuries. Science 297:596–599.

Anderson, J. B., S. S. Shipp, A. L. Lowe, J. Smith Wellner, and A. B. Mosola. 2002. The Antarctic Ice Sheet during the Last Glacial Maximum and its subsequent retreat history: A review. Quaternary Science Reviews 21:49–70.

Anderson, R. F., S. Ali, L. I. Bradtmiller, et al. 2009. Wind-driven upwelling in the Southern Ocean and the deglacial rise in atmospheric CO_2. Science 323:1443–1448.

Anderson, R. Y., A. Soutar, and T. C. Johnson. 1992. Long-term changes in El Niño/Southern Oscillation: Evidence from marine and lacustrine sediments. In H. F. Diaz and V. Markgraf, eds., El Niño: Historical and Paleoclimatic Aspects of the Southern Oscillation, pp. 419–433. Cambridge: Cambridge University Press.

Andersson, A. J., F. T. Mackenzie, and A. Lerman. 2005. Coastal ocean and carbonate systems in the high CO_2 world of the Anthropocene. American Journal of Science 305:875–918.

Andersson, A. J., F. T. Mackenzie, and A. Lerman. 2006. Coastal ocean CO_2–carbonic acid–carbonate sediment system of the Anthropocene. Global Biogeochemical Cycles 20:GB002506.

Andersson, C., B. Risebrobakken, E. Jansen, and S. O. Dahl. 2003. Late Holocene surface ocean conditions of the Norwegian Sea (Vøring Plateau). Paleoceanography 18:1044–1044.

Andrén, T., S. Bjorck, and S. Johnsen. 1999. Correlation of Swedish glacial varves with the Greenland (GRIP) oxygen isotope record. Journal of Quaternary Science 14:361–375.

Andrén, T., G. Lindeberg, and E. Andrén. 2002. Evidence of the final drainage of the Baltic Ice Lake in glacial varves from the Baltic Sea. Boreas 31:226–238.

Andrews, J. T. 1998. Abrupt changes (Heinrich events) in late Quaternary North Atlantic marine environments: A history and review of data and concepts. Invited Review. Journal of Quaternary Science 13:3–16.

Andrews, J. T. 2000. Icebergs and iceberg rafted detritus (IRD) in the North Atlantic: Facts and assumptions. Oceanography 13:100–108.

Andrews, J. T. 2007. The role of the Iceland Ice Sheet in the North Atlantic during the late Quaternary: A review and evidence from Denmark Strait. Journal of Quaternary Science doi:10:1002.

Andrews, J. T. and G. Dunhill. 2004. Early to mid-Holocene Atlantic Water influx and deglacial meltwater events, Beaufort Sea Slope, Arctic Ocean. Quaternary Research 61:14–21.

Andrews, J. T. and D. D. Eberl. 2007. Quantitative mineralogy of surface sediments on the Iceland shelf, and application to down-core studies of Holocene ice-rafted sediments. Journal of Sedimentary Research 77:469–479.

Andrews, J. T., J. Harðardóttir, J. S. Stoner, M. E. Mann, G. B. Kristjánsdóttir, and N. Koç. 2003. Decadal to millennial-scale periodicies in North Iceland Shelf sediments over the last 12,000 cal. yrs: Long-term North Atlantic oceanographic variability and solar forcing. Earth and Planetary Science Letters 210:453–465.

Andrews, J. T., L. Keigwin, F. Hall, and A. E. Jennings. 1999. Abrupt deglaciation events and Holocene palaeoceanography from high-resolution cores, Cartwright Saddle, Labrador Shelf, Canada. Journal of Quaternary Science 14:383–397.

Andrews, J. T., M. E. Kirby, A. Aksu, D. Barber, and D. Meese. 1998. Late Quaternary detrital carbonate (DC) layers in Baffin Bay marine sediments (67°–74°): Correlation with Heinrich events in the North Atlantic? Quaternary Science Reviews 17:1125–1137.

Andrews, J. T., B. Maclean, M. Kerwin, et al. 1995. Final stages in the collapse of the Laurentide Ice Sheet, Hudson Strait, Canada, NWT, ^{14}C AMS dates, seismic stratigraphy, and magnetic susceptibility logs. Quaternary Science Reviews 14:983–1004.

Andrews, J. T. and K. Tedesco. 1992. Detrital carbonate-rich sediments, northwest Labrador Sea: Implications for ice-sheet dynamics and iceberg rafting Heinrich events in the North Atlantic. Geology 20:1987–1990.

Andrews, J. T., K. Tedesco, and A. Jennings. 1993. Heinrich events: Chronology and processes, east-central Laurentide Ice Sheet and NW Labrador Sea. In W. Peltier, ed., *Ice in the Climate System*, pp. 167–186. Berlin: Springer-Verlag.

Andronova, N. G. and M. E. Schlesinger. 2000. Causes of global temperature changes during the 19th and 20th centuries. Geophysical Research Letters 27:2137–2140.

Annan, J. D. and J. C. Hargreaves. 2006. Using multiple observationally-based constraints to estimate climate sensitivity. Geophysical Research Letters 33:L06704.

Antevs, E. 1922. *The Recession of the Last Ice Sheet in New England*. New York: American Geographical Society.

Antevs, E. 1928. *The Last Glaciation with Special Reference to the Ice Sheet in North America*. New York: American Geographical Society.

Antevs, E. 1955. Varve and radiocarbon chronologies appraised by pollen data. Journal of Geology 63:495–499.

Antonov, J. I., S. Levitus, and T. P. Boyer. 2005. Thermosteric sea level rise, 1955–2003. Geophysical Research Letters 32:L12602, GL023112.

Appenzeller, C., T. F. Stocker, and M. Anklin. 1998. North Atlantic Oscillation dynamics recorded in Greenland ice cores. Science 282:446–448.

Archer, D. and A. Ganopolski. 2005. A movable trigger: Fossil fuel CO_2 and the onset of the next glaciation. Geochemistry, Geophysics, Geosystems 6:Q05003, doi:10.1029/2004GC000891.

Archer, D. and E. Maier-Reimer. 1994. Effect of deep-sea sedimentary calcite preservation on atmospheric CO_2 concentration. Nature 367:260–264.

Archer, D., A. Winguth, D. Lea, and N. Mahowald. 2000. What caused the glacial/interglacial atmospheric pCO_2 cycles? Reviews of Geophysics 38:159–189.

Archer, D. E. and B. Buffett. 2005. Time-dependent response of the global ocean clathrate reservoir to climatic and anthropogenic forcing. Geochemistry, Geophysics, Geosystems 6:Q03002, doi:10.1029/2004GC000854.

Arendt, A. A., K. A. Echelmeyer, W. D. Harrison, C. S. Lingle, and V. B. Valentine. 2002. Rapid wastage of Alaska glaciers and their contribution to rising sea level. Science 297:382–386.

Armand, L. and A. Leventer. 2003. Palaeo sea ice distribution: Its reconstruction and significance. In D. N. Thomas and G. S. Dieckmann, eds., *Sea Ice: An Introduction to Its Physics, Biology, Chemistry, and Geology*, pp. 333–372. Oxford: Blackwell Science.

Arrhenius, G. 1952. Sediment cores from the east Pacific: Properties of the sediment. Report of the Swedish Deep Sea Expedition, 1947–1948, 1:1–228. Goteborg: Blander.

Arrhenius, S. 1896. On the influence of carbonic acid in the air upon the temperature of the ground. Philosophical Magazine 41:237–245.

Arthur, M. A. and S. O. Schlanger. 1979. Cretaceous Oceanic Anoxic Events as causal factors in development of reef-reservoired giant oil fields. The American Association of Petroleum Geologists Bulletin 63:870–885.

Arthur, M. A., S. O. Schlanger, and H. C. Jenkyns. 1987. The Cenomanian–Turonian Oceanic Anoxic Event, II. Palaeo-oceanographic controls on organic-matter production and preservation. In J. Brooks and A. Fleet, eds., *Marine Petroleum Source Rocks*. pp. 401–420. London: Geological Society of London.

Arz, H., F. Lamy, J. Pätzold, P. J. Müller, and M. Prins. 2003. Mediterranean moisture source for an Early-Holocene humid period in the northern Red Sea. Science 300:118–121.

Arz, H. W., F. Lamy, A. Ganopolski, N. Nowaczyk, and J. Pätzold. 2007. Dominant northern hemisphere climate control over millennial-scale glacial sea-level variability. Quaternary Science Reviews 26:312–321.

Arz, H. W., J. Pätzold, and G. Wefer. 1999. The deglacial history of the western Tropical Atlantic as inferred from high resolution stable isotope records off northeastern Brazil. Earth and Planetary Science Letters 167:105–117.

Ashkenazy, Y. and E. Tziperman. 2006. Scenarios regarding the lead of equatorial sea surface temperature over global ice volume. Paleoceanography 21:PA2006, doi:10.1029/2005PA001232.

Asmeron, Y., V. Polyak, S. Burns, and J. Rassmussen. 2007. Solar forcing of Holocene climate: New insights from a speleothem record, southwestern United States. Geology 35:1–4.

Backman, J., M. Jakobsson, M. Frank, et al. 2008. Age model and core-seismic integration for the Cenozoic Arctic Coring Expedition sediments from the Lomonosov Ridge. Paleoceanography 23:PA1S03.

Backman, J., M. Jakobsson, R. Løvlie, L. Polyak, and L. Febo. 2004. Is the Central Arctic Ocean a sediment starved basin? Quaternary Science Reviews 23:1435–1454.

Backman, J., K. Moran, D. B. McInroy, L. A. Mayer, and the Expedition 302 Scientists. 2006. Proc. IODP, 302: College Station, Texas (Integrated Ocean Drilling Program Management International, Inc.). doi:10.2204/iodp.proc.302.2006.

Bagnato, S., B. K. Linsley, S. S. Howe, G. M. Wellington, and S. J. Salinger. 2005. Coral oxygen isotope records of interdecadal climate variations in the South Pacific Convergence Zone region. Geochemistry, Geophysics, Geosysystems 6:Q06001.

Baker, P. A., S. C. Fritz, J. Garland, and E. Ekdahl. 2005. Holocene hydrologic variation at Lake Titicaca, Bolivia/Peru, and its relationship to North Atlantic climate variation. Journal of Quaternary Science 20:655–662.

Baker, P. A., C. A. Rigsby, G. O. Seltzer, et al. 2001b. Tropical climate changes at millennial and orbital timescales on the Bolivian Altiplano. Nature 409:698–701.

Baker, P. A., G. O. Seltzer, S. C. Fritz, et al. 2001a. The history of South American tropical precipitation for the past 25,000 years. Science 291:640–643.

Baker, R. G. V. and R. J. Haworth. 2000a. Smooth or oscillating late Holocene sea-level curve? Evidence from cross-regional statistical regressions of fixed biological indicators. Marine Geology 163:353–365.

Baker, R. G. V. and R. J. Haworth. 2000b. Smooth or oscillating late Holocene sea-level curve? Evidence from the palaeozoology of fixed biological indicators in east Australia and beyond. Marine Geology 163:367–386.

Baker, R. G. V., R. J. Haworth, and P. G. Flood. 2005. An oscillating Holocene sea-level? Revisiting Rottnest Island, Western Australia, and the Fairbridge eustatic hypothesis. Journal of Coastal Research, Special Issue 42:3–14.

Baksi, A. K., V. Hsu, M. O. McWilliams, and E. Farrar. 1992. ^{40}Ar/^{39}Ar dating of the Brunhes-Matuyama geomagnetic field reversal. Science 256:356–357.

Balco, G., J. O. H. Stone, S. C. Porter, and M. Caffee. 2002. Cosmogenic-nuclide ages for New England coastal moraines, Martha's Vineyard and Cape Cod, Massachusetts, USA. Quaternary Science Reviews 21:2127–2135.

Baldwin, M. P. and T. J. Dunkerton. 1999. Propagation of the Arctic Oscillation from the stratosphere to the troposphere. Journal of Geophysical Research 104(D24):30937–30946.

Banks, M. R. 1981. Late Palaeozoic tillites of Tasmania. In M. J. Hambrey and W. B. Harland, eds., *Earth's Pre-Pleistocene Glacial Records*, pp. 495–501. Cambridge: Cambridge University Press.

Barber, D. C., A. Dyke, C. Hillaire-Marcel, et al. 1999. Forcing of the cold event of 8,200 years ago by catastrophic drainage of Laurentide lakes. Nature 400:344–348.

Bard, E. 2001. Comparison of alkenone estimates with other paleotemperature proxies. Geochemistry, Geophysics, Geosystems 2, doi:10.1029/2000GC000050.

Bard, E. and M. Frank. 2006. Climate change and solar variability: What's new under the sun? Earth and Planetary Science Letters 248:480–493.

Bard, E., M.-C. Guillemette, and R. Frauke. 2004. Present status of radiocarbon calibration and comparison records based on Polynesian corals and Iberian margin sediments. Radiocarbon 46:1189–1201.

Bard, E., B. Hamelin, M. Arnold, et al. 1996. Deglacial sea-level record from Tahiti corals and the timing of global meltwater discharge. Nature 382:241–244.

Bard, E., B. Hamelin, and R. G. Fairbanks. 1990a. U/Th ages obtained by mass spectrometry in corals from Barbados: Sea level during the past 130,000 years. Nature 346:456–458.

Bard, E., B. Hamelin, R. G. Fairbanks, and A. Zindler. 1990b. Calibration of the ^{14}C timescale over the past 30,000 years using mass spectrometric U-Th ages from Barbados corals. Nature 345:405–410.

Bard, E., G. Raisbeck, F. Yiou, and J. Jouzel. 2000b. Solar irradiance during the last 1200 years based on cosmogenic nuclides. Tellus 52B:985–992.

Bard, E. F. Rostek, J.-L. Turon, and S. Gendreau. 2000a. Hydrological impact of Heinrich events in the subtropical northeast Atlantic. Science 289:1321–1324.

Barker, P., M. K. Talbot, F. A. Street-Perrot, F. Marret, J. Scourse, and E. O. Odada. 2004. Late Quaternary climate variability in intertropical Africa. In R. W. Battarbee, F. Gasse, and C. E. Stickley, eds., *Past Climate Variability Through Europe and Africa*, vol. 6, *Developments in Palaeoenvironmental Research*, pp. 117–138. Dordrecht: Springer-Verlag.

Barker, P. A., F. A. Street-Perrott, M. J. Leng, et al. 2001. A 14,000 year oxygen isotope record from diatom silica in two alpine lakes on Mt. Kenya. Science 292:2307–2310.

Barker, P. F. 2001. Scotia Sea regional tectonic evolution: Implications for mantle flow and palaeocirculation. Earth-Science Reviews 55:1–39.

Barker, P. F. and E. Thomas. 2004. Origin, signature and paleoclimatic influence of the Antarctic Circumpolar Current. Earth-Science Reviews 66:143–166.

Barker, S., D. Archer, L. Booth, H. Elderfield, J. Henderiks, and R. E. M. Rickaby. 2006. Globally increased pelagic carbonate production during the Mid-Brunhes dissolution interval and the CO_2 paradox of MIS 11. Quaternary Science Reviews 25:3278–3293.

Barker, S., I. Cacho, H. Benway, and K. Tachikawa. 2005. Planktonic foraminiferal Mg/Ca as a proxy for past ocean temperatures: A methodological overview and data compilation for the Last Glacial Maximum. Quaternary Science Reviews 24:821–834.

Bar-Matthews, M., A. Ayalon, M. Gilmour, A. Matthews, and C. Hawkesworth. 2003. Sea-land oxygen isotopic relationships from planktonic foraminifera and speleothems in the Eastern Mediterranean region and their implications for paleorainfall during interglacial intervals. Geochimica et Cosmochimica Acta 67:3181–3199.

Bar-Matthews, M., A. Ayalon, A. Kaufman, and G. J. Wasserburg. 1999. The Eastern Mediterranean paleoclimate as a reflection of regional events: Soreq Cave, Israel. Earth and Planetary Science Letters 166:85–95.

Barnett, T. P., K. Hasselmann, M. Chelliah, et al. 1999. Detection and attribution of recent climate change: A status report. Bulletin of the American Meteorological Society 80:2631–2660.

Barnett, T. P., D. W. Pierce, K. M. AchutaRao, et al. 2005. Penetration of human-induced warming into the world's oceans. Science 309:284–287.

Barnett, T. P., D. W. Pierce, H. G. Hidalgo, et al. 2008. Human-induced changes in the hydrology of the western United States. Science 319:1080–1083.

Barnola, J.-M., P. Pimienta, D. Raynaud, and Y. S. Korotkevich. 1991. CO_2-climate relationship as deduced from the Vostok ice core: A re-examination based on new measurements and on a re-evaluation of the air dating. Tellus 43B:83–90.

Barnola, J.-M., D. Raynaud, Y. S. Korotkevich, and C. Lorius. 1987. Vostok ice core provides 160,000-year record of atmospheric CO_2. Nature 329:408–414.

Barnston, A. G., M. H. Glantz, and Y. He. 1999. Predictive skill of statistical and dynamical climate models in SST forecasts during the 1997–98 El Niño episode and the 1998 La Niña onset. Bulletin of the American Meteorological Society 80:217–243.

Barreiro, M., S. G. H. Philander, R. C. Pacanowski, and A. Federov. 2006. Simulations of warm tropical conditions with application to middle Pliocene atmospheres. Climate Dynamics 26:349–365.

Barrera, E., and C. C. Johnson, eds. 1999. *The Evolution of the Cretaceous Ocean-Climate System*. Geological Society of America Special Paper 332. Boulder, CO: Geological Society of America.

Barrett, P. F., F. Florindo, A. K. Cooper. 2006. Introduction to "Antarctic climate evolution: View from the margin." Palaeogeography, Palaeoclimatology, Palaeoecology 231:1–8.

Barrett, P. J., C. J. Adams, W. C. Macintosh, C. C. Swisher, and G. S. Wilson. 1992. Geochronologic evidence for Antarctic deglaciation around 3 million years ago. Nature 359:816–818.

Barrett, P. J., M. Sarti, and S. W. Wise. 2000. Studies from the Cape Roberts Project, Ross Sea, Antarctica, initial reports on CRP-3. Terra Antarctica 7(1/2):1–209.

Barron, E. J. 1986. Modeling in paleoceanography. Surveys in Geophysics 8:1–23.

Barron, E. J., P. J. Fawcett, W. H. Peterson, D. Pollard, and S. L. Thompson. 1995. A "simulation" of mid-Cretaceous climate. Paleoceanography 10:953–962.

Barron, E. J. and G. T. Moore. 1994. *Climate Model Application in Paleoenvironmental Analysis, Short Course #33.* Tulsa, OK: Society for Sedimentary Geology.

Barron, E. J., S. L. Thompson, and S. H. Schneider. 1981. An ice-free Cretaceous? Results from climate model simulations. Science 212:501–508.

Barron, E. J. and W. M. Washington. 1985. Warm Cretaceous climates—High atmospheric CO_2 as a plausible mechanism. In E. T. Sundquist and W. S. Broecker, eds., *The Carbon Cycle and Atmospheric CO_2: Natural Variations, Archean to Present,* pp. 546–553. Washington, DC: American Geophysical Union.

Barron, J. A. and D. Bukry. 2007. Solar forcing of Gulf of California climate during the past 2000 yr suggested by diatoms and silicoflagellates. Marine Micropaleontology 62:115–139.

Barron, J. A., D. Bukry, and J. L. Bischoff. 2004. High resolution paleoceanography of the Guaymas Basin, Gulf of California, during the past 15,000 years. Marine Micropaleontology 50:185–207.

Barron, J. A., L. Heusser, T. Herbert, and M. Lyle. 2003. High-resolution climatic evolution of coastal northern California during the past 16,000 years. Paleoceanography 18:PA000768.

Barrows, T. T., S. J. Lehman, L. K. Fifield, and P. De Deckker. 2007. Absence of cooling in New Zealand and the adjacent ocean during the Younger Dryas Chronozone. Science 318:86–89.

Bartdorff, O., K. Wallman, M. Latif, and V. Semenov. 2008. Phanerozoic evolution of atmospheric methane. Global Biogeochemical Cycles 22:GB1008,002985.

Bartlein, P. J., K. H. Anderson, P. M. Anderson, et al. 1998. Paleoclimate simulations for North America over the past 21,000 years: Features of the simulated climate and comparisons with paleoenvironmental data. Quaternary Science Reviews 17:549–585.

Bartoli, G., M. Sarnthein, M. Weinelt, H. Erlenkeuser, D. Garbe-Schönberg, and D. W. Lea. 2005. Final closure of Panama and the onset of northern hemisphere glaciation. Earth and Planetary Science Letters 237:33–44.

Bassinot, F. C., L. D. Labeyrie, E. Vincent, X. Quidelleur, N. J. Shackleton, and Y. Lancelot. 1994. The astronomical theory of climate and the age of the Brunhes-Matuyama magnetic reversal. Earth and Planetary Science Letters 126:91–108.

Battarbee, R. W., F. Gasse, and C. E. Stickley, eds. 2004. *Past Climate Variability Through Europe and Africa.* Dordrecht: Kluwer.

Bauch, H. A., H. Erlenkeuser, J. P. Helmke, and U. Struck. 2000. A paleoclimatic evaluation of marine oxygen isotope stage 11 in the high Northern Atlantic (Nordic seas). Global and Planetary Change 24:27–39.

Bauch, H. A. and Y. I. Polyakova. 2000. Late Holocene variations in Arctic shelf hydrology and sea-ice regime: Evidence from north of the Lena Delta. International Journal of Earth Science 89:569–577.

Bauch, H. A. and Y. I. Polyakova. 2003. Diatom-inferred salinity records from the Arctic Siberian Margin: Implications for fluvial runoff patterns during the Holocene. Paleoceanography 18:PA000847.

Bauch, H. A. and M. S. Weinelt. 1997. Surface water changes in the Norwegian Sea during last deglacial and Holocene times. Quaternary Science Reviews 16:1113–1124.

Bauer, E., M. Claussen, and V. Brovkin. 2003. Assessing climate forcings of the earth system for the past millennium. Geophysical Research Letters 30:1276, GL016639.

Baumgartner, S., J. Beer, J. Masarik, G. Wagner, L. Meynadier, and H.-A. Synal. 1998. Geomagnetic modulation of the ^{36}Cl flux in the GRIP ice core, Greenland. Science 279:1330–1332.

Beck, J. W., D. A. Richards, R. L. Edwards, et al. 2001. Extremely large variations of atmospheric ^{14}C concentration during the last glacial period. Science 292:2453–2458.

Beck, W., R. L. Edwards, E. Ito, et al. 1992. Sea-surface temperature from coral skeletal strontium/calcium ratios. Science 257:644–647.

Becker, B. 1993. An 11,000-year German oak and pine dendrochronology for radiocarbon calibration. Radiocarbon 35:201–213.

Beer, J., S. J. Johnsen, G. Bonani, et al. 1990. ^{10}Be peaks as time markers in polar ice cores. In E. Bard and W. S. Broecker, eds., *The Last Deglaciation: Absolute and Radiocarbon Chronologies,* pp. 141–153. Berlin: Springer-Verlag.

Beer, J., F. Joos, C. Lukaszyk, et al. 1988. ^{10}Be as an indicator of solar variability. In G. C. Castagnoli, ed., *Solar-Terrestrial Relationships and the Earth Environment in the Last Millennia,* pp. 221–233. New York: Elsevier/North Holland Physics Publishing.

Beer, J., F. Joos, C. Lukaszyk, et al. 1994. ^{10}Be as an indicator of solar variability and climate. In E. Nesme-Ribes, ed., *The Solar Engine and Its Influence on Terrestrial Atmospheres and Climate,* pp. 221–233. Berlin: Springer-Verlag.

Beer, J., W. Mende, and R. Stellmacher. 2000. The role of the sun in climate forcing. Quaternary Science Reviews 19:403–415.

Beer, J., R. Muscheler, G. Wagner, et al. 2002. Cosmogenic nuclides during isotope stages 2 and 3. Quaternary Science Reviews 21:1129–1139.

Beerling, D. J. 2002. Low atmospheric CO_2 levels during the Permo-Carboniferous glaciation inferred from fossil lycopsids. Proceedings of the National Academy of Sciences 99:12567–12571.

Beerling, D. J. and S. J. Brentnall. 2007. Numerical evaluation of mechanisms driving early Jurassic changes in global carbon cycling. Geology 35:247–250.

Beerling, D. J. and D. L. Royer. 2002. Fossil plants as indicators of the Phanerozoic global carbon cycle. Annual Review of Earth and Planetary Sciences 30:527–556.

Behl, R. J. and J. P. Kennett. 1996. Brief interstadial events in the Santa Barbara basin, NE Pacific, during the past 60 kyr. Nature 379:243–246.

Behrenfeld, M. J., E. Boss, D. A. Siegel, and D. M. Shea. 2005. Carbon-based ocean productivity and phytoplankton physiology from space. Global Biogeochemical Cycles 19:GB1006.

Behrenfeld, M. J., R. T. O'Malley, D. A. Siegel, et al. 2006. Climate-driven trends in contemporary ocean productivity. Nature 444:752–755.

Belcher, C. M. and J. C. McElwain. 2008. Limits for combustion in low O_2 redefine paleoatmospheric predictions for the Mesozoic. Science 321:1197–1200.

Belkin, M. 2004. Propagation of the "Great Salinity Anomaly" of the 1990s around the northern North Atlantic. Geophysical Research Letters 31:L08306.

Bell, G. D. and M. Chelliah. 2006. Leading tropical modes associated with interannual and multi-decadal fluctuations in North Atlantic hurricane activity. Journal of Climate 19:590–612.

Beltrami, H. 2002. Climate from borehole data: Energy fluxes and temperatures since 1500. Geophysical Research Letters 29:GL015702.

Bemis, B. E., H. J. Spero, J. Bijma, and D. W. Lea. 1998. Reevaluation of the oxygen isotopic composition of planktonic foraminifera: Experimental results and revised paleotemperature equations. Paleoceanography 13:150–160.

Bender, M., T. Sowers, M.-L. Dickson, et al. 1994. Climate correlations between Greenland and Antarctica during the past 100,000 years. Nature 372:663–666.

Bender, M. L., R. G. Fairbanks, F. W. Taylor, et al. 1979. Uranium-series dating of the Pleistocene reef tracts of Barbados, West Indies. Geological Society of America Bulletin 90:577–594.

Benestad, R. E. 2005. A review of the solar cycle length estimates. Geophysical Research Letters 32:GL023621.

Benson, L. V., M. S. Berry, E. A. Jolie, et al. 2007. Possible impacts of early-11th-, middle-12th-, and late-13th-century droughts on western Native Americans and the Mississippian Cahokians. Quaternary Science Reviews 26:336–350.

Benson, L. V., J. W. Burdett, M. Kashgarian, S. P. Lund, F. M. Phillips, and R. O. Rye. 1996. Climatic and hydrologic oscillations in the Owens Lake Basin and adjacent Sierra Nevada, California. Science 274:746–749.

Berelson, W. M., R. F. Anderson, J. Dymond, et al. 1997. Biogenic budgets of particle rain, benthic remineralization and sediment accumulation in the equatorial Pacific. Deep Sea Research Part II 44:2251–2282.

Berger, A. 1978. Long-term variation of daily insolation and Quaternary climatic changes. Journal of Atmospheric Sciences 35:2362–2367.

Berger, A., J. Imbrie, J. Hays, G. Kukla, and B. Saltzman, eds. 1984. Milankovitch and Climate: Understanding the Response to Astronomical Forcing. Dordrecht: D. Reidel.

Berger, A. and M. Loutre. 1992. Astronomical solutions for paleoclimate studies over the last 3 million years. Earth and Planetary Science Letters 111:369–382.

Berger, A. and M. F. Loutre. 1991. Insolation values for the climate of the last 10 million years. Quaternary Science Reviews 10:297–317.

Berger, A. and M. F. Loutre. 2003. Climate 400,000 years ago, a key to the future? In A. W. Droxler, R. Z. Poore, and L. H. Burckle, eds., Earth's Climate and Orbital Eccentricity: The Marine Isotope Stage 11 Question, pp. 17–26. Washington, DC: American Geophysical Union.

Berger, A., J. L. Melice, and M. F. Loutre. 2005. On the origin of the 100-kyr cycles in the astronomical forcing. Paleoceanography 20:PA4019.

Berger, W. and R. S. Kier. 1985. Glacial-Holocene changes in atmospheric CO_2 and the deep sea record. In J. E. Hansen and T. Takahashi, eds., Climate Processes and Climate Sensitivity, pp. 337–351. Washington, DC: American Geophysical Union.

Berger, W. H. 1979. Stable isotopes in foraminifera. In Cushman Foundation Short Course in Foraminifera, pp. 156–198. Washington, DC: Cushman Foundation for Foraminiferal Research.

Berger, W. H. 1982. Increase of carbon dioxide in the atmosphere during deglaciation: The coral reef hypothesis. Naturwissenshaften 69:87–88.

Berger, W. H. and E. Jansen. 1994. Mid-Pleistocene climate shift: The Nansen connection. In O. M. Johannessen, R. D. Muench, and J. E. Overland, eds., The Polar Oceans and Their Role in Shaping the Global Environment, pp. 295–311. Washington, DC: American Geophysical Union.

Berger, W. H. and L. D. Labeyrie. 1987. Abrupt climatic change—An introduction. In W. H. Berger and L. D. Labeyrie, eds., Abrupt Climatic Change: Evidence and Implications, pp. 3–22. Dordrecht: D. Reidel.

Berggren, W. A. 1982. Role of ocean gateways in climatic change. In Climate in Earth History, pp. 118–125. Washington, DC: National Academy Press.

Berggren, W. A. and C. D. Hollister. 1977. Plate tectonics and paleocirculation—commotion in the ocean. Tectonophysics 11:11–48.

Bergman, N. M., T. M. Lenton, and A. J. Watson. 2004. COPSE: A new model of biogeochemical cycling over Phanerozoic time. American Journal of Science 304:397–437.

Berner, R. A. 1990. Atmospheric carbon dioxide levels over Phanerozoic time. Science 249:1382–1386.

Berner, R. A. 1991. A model for atmospheric CO_2 over Phanerozoic time. American Journal of Science 291:339–376.

Berner, R. A. 1994. GEOCARB II: A revised model of atmospheric CO_2 over Phanerozoic time. American Journal of Science 294:56–91.

Berner, R. A. 2004. The Phanerozoic Carbon Cycle: CO_2 and O_2. New York: Oxford University Press.

Berner, R. A. 2006. GEOCARBSULF: A combined model for Phanerozoic atmospheric O_2 and CO_2. Geochimica et Cosmochimica Acta 70:5653–5664.

Berner, R. A. and Z. Kothavala. 2001. GEOCARB III: A revised model of atmospheric CO_2 over Phanerozoic time. American Journal of Science 301:182–204.

Bertrand, C., M. F. Loutre, M. Crucifix, and A. Berger. 2002. Climate of the last millennium: A sensitivity study. Tellus 54A:221–244.

Bianchi, G. G. and I. N. McCave. 1999. Holocene periodicity in North Atlantic climate and deep-ocean flow south of Iceland. Nature 397:515–517.

Bice, K. L., D. Birgel, P. A. Meyers, K. A. Dahl, K. Hinrichs, and R. D. Norris. 2006. A multiple proxy and model study of Cretaceous upper ocean temperatures and atmospheric CO_2 concentrations. Paleoceanography 21:PA2002.

Bice, K. L., B. T. Huber, and R. D. Norris. 2003. Extreme polar warmth during the Cretaceous greenhouse? Paradox of the late Turonian $\delta^{18}O$ record at DSDP Site 511. Paleoceanography 18:PA000848.

Bice, K. L. and R. D. Norris. 2002. Possible atmospheric CO_2 extremes of the middle Cretaceous (late Albian-Turonian). Paleoceanography 17:PA000778.

Bickle, M. J., H. J. Chapman, J. Bunbury, et al. 2005. The relative contributions of silicate and carbonate rocks to riverine Sr fluxes in the headwaters of the Ganges. Geochimica et Cosmochimica Acta 69:2221–2240.

Billups, K. 2002. Late Miocene through early Pliocene deep water circulation and climate change viewed from the subantarctic Southern Ocean. Palaeogeography, Palaeoclimatology, Palaeoecology 185:287–307.

Billups, K., H. Pälike, J. Channell, J. Zachos, and N. Shackleton. 2004. Astronomic calibration of the late Oligocene through early Miocene geomagnetic polarity time scale. Earth and Planetary Science Letters 224:33–44.

Billups, K. and D. F. Schrag. 2003. Application of benthic foraminiferal Mg/Ca ratios to questions of Cenozoic climate change. Earth and Planetary Science Letters 209:181–195.

Bindoff, N., J. Willebrand, et al. 2007. Chapter 5: Observations: Oceanic climate change and sea level. In S. Solomon, et al., eds., *Climate Change 2007: The Physical Science Basis*. Contribution of Working Group I to the Fourth Assessment Report of the Intergovernmental Panel on Climate Change, pp. 385–432. Cambridge: Cambridge University Press.

Bindschadler, R. A., M. A. King, R. B. Alley, S. Anandakrishnan, and L. Padman. 2003. Tidally controlled stick-slip discharge of a West Antarctic ice stream. Science 301:1087–1089.

Biondi, F., A. Gershunov, and D. R. Cayan. 2001. North Pacific decadal climate variability since 1661. Journal of Climate 14:5–10.

Birgel, D. and H. C. Hass. 2004. Oceanic and atmospheric variations during the last deglaciation in the Fram Strait (Arctic Ocean): A coupled high-resolution organic-geochemical and sedimentological study. Quaternary Science Reviews 23:29–47.

Birks, S. J., T. W. D. Edwards, and V. H. Remenda. 2007. Isotopic evolution of Glacial Lake Agassiz: New insights from cellulose and porewater isotopic archives. Palaeogeography, Palaeoclimatology, Palaeoecology 246:8–22.

Bischof, J. 2000. *Ice Drift, Ocean Circulation and Climate Change*. Chichester, UK: Springer Praxis.

Bischof, J. and D. Darby. 1999. Quaternary ice transport in the Canadian Arctic and extent of late Wisconsinan glaciation in the Queen Elizabeth Islands. Canadian Journal of Earth Science 36:2007–2022.

Bjerknes, J. 1966. A possible response of the atmospheric Hadley circulation to equatorial anomalies of ocean temperature. Tellus 18:820–829.

Bjerknes, J. 1969. Atmospheric teleconnections from the equatorial Pacific. Monthly Weather Review 97:163–172.

Björck, S. 1995. A review of the history of the Baltic Sea 12.0–8.0 ka BP. Quaternary International 27:19–40.

Björck, S., T. Andrén, S. Wastegard, G. Possnert, and K. Schoning. 2002. An event stratigraphy for the Last Glacial-Holocene transition in eastern middle Sweden: Results from investigations of varved clay and terrestrial sequences. Quaternary Science Reviews 21:1489–1501.

Björck, S. and G. Digerfeldt. 1989. Lake Mullsjon—A key site for understanding the final stage of the Baltic Lake drainage east of Mt. Billigen. Boreas 18:209–219.

Björck, S., B. Kromer, S. Johnsen, et al. 1996. Synchronized terrestrial-atmospheric deglacial records around the North Atlantic. Science 274:1155–1160.

Björck, S. and P. Möller. 1987. Late Weichselian environmental history in southeastern Sweden during the deglaciation of the Scandinavian Ice-Sheet. Quaternary Research 28:1–37.

Björck, S., M. Rundgren, Ó. Ingólfsson, and S. Funder. 1997. The Preboreal Oscillation around the Nordic Seas: Terrestrial and lacustrine responses. Journal of Quaternary Science 12:455–465.

Björck, S., M. J. C. Walker, L. C. Cwynar, et al., and INTIMATE Members. 1998. An event stratigraphy for the Last Termination in the North Atlantic region based on the Greenland ice-core record: A proposal by the INTIMATE group. Journal of Quaternary Science 13:283–292.

Black, D. E., M. A. Abahazi, R. C. Thunell, et al. 2007. An eight-century tropical Atlantic SST record from the Cariaco Basin: Baseline variability, twentieth-century warming, and Atlantic hurricane frequency. Paleoceanography 22:PA4204, PA001427.

Black, D. E., L. C. Peterson, J. T. Overpeck, et al. 1999. Eight centuries of Atlantic Ocean atmosphere variability. Science 286:1709–1713.

Blackmon, M., B. Boville, F. Bryan, et al. 2001. The Community Climate System Model. Bulletin of the American Meteorological Society 82:2357–2376.

Blanchon, P. and A. Eisenhauer. 2000. Multi-stage reef development on Barbados during the Last Interglaciation. Quaternary Science Reviews 20:1093–1112.

Blanchon, P. and J. Shaw. 1995. Reef drowning during the last deglaciation: Evidence for catastrophic sea-level rise and ice-sheet collapse. Geology 23:4–8.

Bloom, A. L., W. S. Broecker, J. M. A. Chappell, R. K. Matthews, and K. J. Mesolella. 1974. Quaternary sea level fluctuations on a tectonic coast: New ^{230}Th/^{234}U dates from the Huon Peninsula, New Guinea. Quaternary Research 4:185–205.

Bloom, A. L. and M. Stuiver. 1963. Submergence of the Connecticut coast. Science 139:332–334.

Blum, M. D., T. J. Misner, E. S. Collins, et al. 2001. Middle Holocene sea-level rise and highstand at +2 m, central Texas. Journal of Sedimentary Research 71:581–588.

Blunier, T. and E. J. Brook. 2001. Timing of millennial-scale climate change in Antarctica and Greenland during the last glacial period. Science 291:109–112.

Blunier, T., J. Chappellaz, J. Schwander, et al. 1998. Asynchrony of Antarctic and Greenland climate change during the last glacial period. Nature 394:739–743.

Blunier, T., J. Chappellaz, J. Schwander, B. Stauffer, and D. Raynaud. 1995. Variations in atmospheric methane concentrations during the Holocene Epoch. Nature 374:46–49.

Blunier, T., R. Spahni, J.-M. Barnola, et al. 2007. Synchronization of ice core records via atmospheric gases. Climate of the Past 3:325–330.

Bodèn, P., R. G. Fairbanks, J. D. Wright, and L. H. Burckle. 1997. High-resolution stable isotope records from southwest Sweden: The drainage of the Baltic Ice Lake and Younger Dryas ice marginal oscillation. Paleoceanography 12:39–49.

Bodiselitsch, B., C. Koeberl, S. Master, and W. U. Reimold. 2005. Estimating duration and intensity of Neoproterozoic snowball glaciations from Ir anomalies. Science 308:239–242.

Bohaty, S. M. and J. C. Zachos. 2003. Significant Southern Ocean warming event in the late middle Eocene. Geology 31:1017–1020.

Bonan, G. B. 2008. Forests and climate change: Forcings, feedbacks, and the climate benefits of forests. Science 320:1444–1448.

Bond, G., W. Broecker, S. Johnsen, et al. 1993. Correlations between climate records from North Atlantic sediments and Greenland ice. Nature 365:143–147.

Bond, G., H. Heinrich, W. Broecker, et al. 1992. Evidence for massive discharges of icebergs into the North Atlantic ocean during the last glacial period. Nature 360:245–249.

Bond, G., B. Kromer, J. Beer, et al. 2001. Persistent solar influence on North Atlantic climate during the Holocene. Science 294:2130–2136.

Bond, G. C. and R. Lotti. 1995. Iceberg discharges into the North Atlantic on millennial time scales during the last glaciation. Science 267:1005–1010.

Bond, G. C., W. Showers, M. Cheseby, et al. 1997. A pervasive millennial-scale cycle in North Atlantic Holocene and glacial climates. Science 278:1257–1266.

Bond, G. C., W. Showers, M. Elliot, et al. 1999. The North Atlantic's 1–2 kyr climate rhythm: Relation to Heinrich events, Dansgaard/Oeschger cycles and the Little Ice Age. In P. U. Clark, R. S. Webb, and L. D. Keigwin, eds., Mechanisms of Global Climate Change at Millennial Time Scales, pp. 35–58, Washington, DC: American Geophysical Union.

Bondevik, S., J. Mangerud, H. H. Birks, S. Gulliksen, and P. Reimer. 2006. Changes in North Atlantic radiocarbon reservoir ages during the Allerød and Younger Dryas. Science 312:1514–1517.

Bonfils, C., N. de Noblet-Ducoudré, J. Guiot, and P. Bartlein. 2004. Some mechanisms of mid-Holocene climate change in Europe, inferred from comparing PMIP models to data. Climate Dynamics 23:79–98.

Bony, S., J.-L. Dufresne, R. Colman, et al. 2006. How well do we understand and evaluate climate change feedback processes? Journal of Climate 19:3445–3482.

Bornemann, A., R. D. Norris, O. Friedrich, et al. 2008. Isotopic evidence for glaciation during the Cretaceous supergreenhouse. Science 319:189–192.

Bowen, D. Q. 1978. Quaternary Geology: A Stratigraphic Framework for Multidisciplinary Work. Oxford: Pergamon.

Bowen, D. Q., F. M. Phillips, A. M. McCabe, P. C. Knutz, and G. A. Sykes. 2002. New data for the Last Glacial Maximum in Great Britain and Ireland. Quaternary Science Reviews 21:89–101.

Bowen, G. J., D. J. Beerling, P. L. Koch, J. C. Zachos, and T. Quattlebaum. 2004. A humid climate state during the Palaeocene/Eocene thermal maximum. Nature 432: 495–499.

Boyle, E. A. 1988. Cadmium: Chemical tracer of deepwater paleoceanography. Paleoceanography 3:471–489.

Boyle, E. A. 1990. Quaternary deep water paleoceanography. Science 249:863–870.

Boyle, E. A. 1992. Cadmium and $\delta^{13}C$ paleochemical ocean distributions during stage 2 glacial maximum. Annual Review of Earth and Planetary Sciences 20:243–287.

Boyle, E. A. and L. D. Keigwin. 1982. Deep circulation of the North Atlantic over the last 200,000 years: Geochemical evidence. Science 218:784–787.

Boyle, E. A. and L. D. Keigwin. 1987. North Atlantic thermohaline circulation during the past 20,000 years linked to high latitude surface temperature. Nature 330:35–40.

Braconnot, P., B. Otto-Bliesner, S. Harrison, et al. 2007. Results of PMIP2 coupled simulations of the Mid-Holocene and Last Glacial Maximum—Parts 1 and 2: Experiments and large-scale features; feedbacks with emphasis on the location of the ITCZ and mid- and high latitude heat budget. Climate of the Past 3:261–277, 279–296.

Bradley, R. S. 1999. Paleoclimatology, Reconstructing Climates of the Quaternary. 2nd ed. San Diego, CA: Academic Press.

Bradley, R. S., K. R. Briffa, J. Cole, M. K. Hughes, and T. J. Osborn. 2003. The climate of the last millennium. In K. Alverson, R. S. Bradley, and T. F. Pedersen, eds., Paleoclimate, Global Change and the Future, pp. 105–141. Berlin: Springer-Verlag.

Bradley, R. S. and P. D. Jones, eds. 1992. Climate Since A.D. 1500. New York: Routledge.

Bradley, R. S. and P. D. Jones. 1993. "Little Ice Age" summer temperature variations: Their nature and relevance to recent global warming trends. The Holocene 3:367–376.

Bralower, T. J. 1988. Calcareous nannofossil biostratigraphy and assemblages of the Cenomanian/Turonian boundary interval: Implications for the origin and timing of oceanic anoxia. Paleoceanography 3:275–316.

Bralower, T. J., M. A. Arthur, R. M. Leckie, W. Sliter, D. J. Allard, and S. O. Schlanger. 1994. Timing and paleoceanography of oceanic dysoxia/anoxia in the late Barremian to early Aptian (early Cretaceous). Palaios 9:335–369.

Bralower, T. J., P. D. Fullagar, C. K. Paull, G. S. Dwyer, and R. M. Leckie. 1997. Mid Cretaceous strontium-isotope stratigraphy of deep-sea sections. Geological Society of America Bulletin 109:1421–1442.

Braun, H., M. Christl, Stefan Rahmstorf, et al. 2005. Possible solar origin of the 1,470-year glacial climate cycle demonstrated in a coupled model. Nature 438:208–211.

Bräuning, A. and B. Mantwill. 2004. Summer temperature and summer monsoon history on the Tibetan plateau during the last 400 years recorded by tree rings. Geophysical Research Letters 31:L24205, GL020793.

Breckenridge, A., T. Johnson, S. Beske-Diehl, and J. Mothersill. 2006. The timing of regional lateglacial events and postglacial sedimentation rates from Lake Superior. Quaternary Science Reviews 23:2355–2367.

Brewer, S., J. Guiot, and F. Torre. 2007. Mid-Holocene climate change in Europe: A data-model comparison. Climate of the Past 3:499–512.

Briffa, K., T. J. Osborn, F. H. Schweingruber, et al. 2001. Low frequency temperature variations from a northern tree-ring density network. Journal of Geophysical Research 106: 2929–2941.

Briffa, K., F. Schweingruber, P. Jones, and T. Osborn. 1998a. Reduced sensitivity of recent tree growth to temperature at high northern latitudes. Nature 391:678–682.

Briffa, K., F. Schweingruber, P. Jones, et al. 1998b. Trees tell of past climates: But are they speaking less clearly today? Philosophical Transactions of the Royal Society of London, Series B 353:65–73.

Briffa, K. R. 1994. Grasping at shadows? A selective review of the search for sunspot related variability in tree rings. In E. Nesme-Ribes, ed., The Solar Engine and Its Influence on Terrestrial Atmospheres and Climate, pp. 417–435. Berlin: Springer-Verlag.

Briffa, K. R. 2000. Annual climate variability in the Holocene: Interpreting the message from ancient trees. Quaternary Science Reviews 19:87–105.

Briffa, K. R., P. D. Jones, T. S. Bartholin, et al. 1992. Fennoscandian summers from A.D. 500: Temperature changes on short and long timescales. Climate Dynamics 7:111–119.

Briffa, K. R., T. J. Osborn, and F. H. Schweingruber. 2004. Large-scale temperature inferences from tree rings: A review. Global and Planetary Change 40:11–26.

Brinkhuis, H., S. Schouten, and Expedition Scientists. 2006. Episodic fresh surface waters in the Eocene Arctic Ocean. Nature 441:606–609.

Broecker, W. 2006. Was the Younger Dryas triggered by a flood? Science 312:1146–1148.

Broecker, W. S. 1982. Glacial to interglacial changes in ocean chemistry. Progress in Oceanography 11:151–197.

Broecker, W. S. 1994. Massive iceberg discharge triggers for global climate change. Nature 372:421–424.

Broecker, W. S. 1997. Thermohaline circulation, the Achilles heel of our climate system: Will man-made CO_2 upset the current balance? Science 278:1582–1588.

Broecker, W. S. 1998. Paleocean circulation during the last deglaciation: A bipolar seesaw. Paleoceanography 13:119–121.

Broecker, W. S. 2000. Was a change in thermohaline circulation responsible for the Little Ice Age? Proceedings of the National Academy of Sciences 97:1339–1342.

Broecker, W. S. 2001. Was the Medieval Warm Period global? Science 291:1497–1499.

Broecker, W. S. 2003. Does the trigger for abrupt climate change reside in the ocean or in the atmosphere? Science 300: 1519–1522.

Broecker, W. S., M. Andree, G. Bonani, et al. 1988b. Can the Greenland climatic jumps be identified in records from ocean and land? Quaternary Research 30:1–6.

Broecker, W. S., M. Andree, W. Wolfli, et al. 1988a. The chronology of the last deglaciation: Implications for the cause of the Younger Dryas Event. Paleoceanography 3:1–19.

Broecker, W. S., S. Barker, E. Clark, et al. 2004. Ventilation of the glacial deep Pacific Ocean. Science 306:1169–1172.

Broecker, W. S., G. Bond, M. Klas, G. Bonani, and W. Wolfli. 1990b. A salt oscillator in the glacial Atlantic? I. The concept. Paleoceanography 4:469–477.

Broecker, W. S., G. Bond, M. Klas, E. Clark, and J. McManus. 1992. Origin of the northern Atlantic's Heinrich events. Climate Dynamics 6:265–273.

Broecker, W. S. and E. Clark. 2003. Holocene atmospheric CO_2 increase as viewed from the seafloor. Global Biogeochemical Cycles 17:1052, GB001985.

Broecker, W. S., E. Clark, S. Barker, et al. 2007. Radiocarbon age of the late glacial deep water from the equatorial Pacific. Paleoceanography 22:PA2206.

Broecker, W. S., E. Clark, D. C. McCorckle, et al. 1999a. Evidence for a reduction in the carbonate ion content of the deep sea during the course of the Holocene. Paleoceanography 3:317.

Broecker, W. S. and G. H. Denton. 1989. The role of ocean-atmosphere reorganizations in glacial cycles. Geochimica et Cosmochimica Acta 53:2465–2501.

Broecker, W. S. and G. M. Henderson. 1998. The sequence of events surrounding Termination II and their implications for the cause of glacial-interglacial CO_2 changes. Paleoceanography 13:352–364.

Broecker, W. S., J. P. Kennett, B. P. Flower, et al. 1989. Routing of meltwater from the Laurentide Ice Sheet during the Younger Dryas cold episode. Nature 341:318–321.

Broecker, W. S., M. Klas, E. Clark, et al. 1990a. Accelerator mass-spectrometric radiocarbon measurements on foraminifera shells from deep-sea cores. Radiocarbon 32:119–133.

Broecker, W. S., J. Lynch-Stieglitz, E. Clark, I. Hajdas, and G. Bonani. 2001. What caused the atmosphere's CO_2 content to rise during the last 8000 years? Geochemistry, Geophysics, Geosystems 2:GC000177.

Broecker, W. S., K. Matsumoto, E. Clark, I. Hajdas, and G. Bonani. 1999b. Radiocarbon age differences between coexisting foraminiferal species. Paleoceanography 14:431–436.

Broecker, W. S. and T.-H. Peng. 1982. Tracers in the Sea. New York: Eldigio Press.

Broecker, W. S. and T.-H. Peng. 1989. The cause of glacial to interglacial atmospheric CO_2 change: A polar alkalinity hypothesis. Biogeochemical Cycles 3:215–239.

Broecker, W. S., D. M. Peteet, and D. Rind. 1985. Does the ocean-atmosphere system have more than one stable mode of operation? Nature 315:21–25.

Broecker, W. S. and T. F. Stocker. 2006. The Holocene CO_2 rise: Anthropogenic or natural? EOS, Transactions, American Geophysical Union 87:27–29.

Broecker, W. S. and D. L. Thurber. 1965. Uranium series dating of corals and oolites from Bahaman and Florida Key limestones. Science 149:58–60.

Broecker, W. S., D. L. Thurber, J. Goddard, et al. 1968. Milankovitch supported by precise dating of coral reefs and deep-sea sediments. Science 159:297–300.

Broecker, W. S. and J. van Donk. 1970. Insolation changes, ice volumes, and the O-18 record in deep-sea sediments. Reviews in Geophysics and Space Physics 8:169–198.

Brohan, P., J. J. Kennedy, I. Harris, S. F. B. Tett, and P. D. Jones. 2006. Uncertainty estimates in regional and global observed temperature changes: A new dataset from 1850. Journal of Geophysical Research 111:D12106, JD006548.

Brook, E. J., S. Harder, J.P. Severinghaus, E. Steig, and C. Sucher. 2000. On the origin and timing of rapid changes in atmospheric methane during the last glacial period. Global Biogeochemical Cycles 14:559–572.

Brook, E. J., J. Severinghaus, S. Harder, and M. Bender. 1999. Atmospheric methane and millennial-scale climate change. In P. U. Clark, R. S. Webb, and L. Keigwin, eds., Mechanisms of Global Climate Change at Millennial Timescales, pp. 165–176. Washington, DC: American Geophysical Union.

Brook, E. J., T. Sowers, and J. Orchado. 1996. Rapid variations in atmospheric methane concentration during the past 110,000 years. Science 273:1087–1091.

Brook, E. J., J. W. C. White, A. Schilla, et al. 2005. Timing of millennial-scale climate change at Siple Dome, West Antarctica, during the last glacial period. Quaternary Science Reviews 24:1333–1343.

Brook, E. J. and E. W. Wolff. 2006. The future of ice core science. EOS, Transactions, American Geophysical Union 87:39.

Brovkin, V., M. Claussen, E. Driesschaert, et al. 2006. Biogeophysical effects of historical land cover changes simulated by six earth system models of intermediate complexity. Climate Dynamics 26:587–600.

Brown, B., C. Gaina, and R. D. Müller. 2006. Circum-Antarctic paleobathymetry: Illustrated examples from Cenozoic to recent times. Palaeogeography, Palaeoclimatology, Palaeoecology 231:158–168.

Brown, E. T. and T. C. Johnson. 2005. Coherence between tropical East African and South American records of the Little Ice Age. Geochemistry, Geophysics, Geosystems 6:GC000959.

Brown, E. T., T. C. Johnson, C. A. Scholz, A. S. Cohen, and J. W. King. 2007. Abrupt change in tropical African climate linked to the bipolar seesaw over the past 55,000 years. Geophysical Research Letters 34:L20702, GL031240.

Brown, J., M. Collins, A. W. Tudhope, and T. Toniazzo. 2008. Modelling mid-Holocene tropical climate and ENSO variability: Towards constraining predictions of future change with palaeo-data. Climate Dynamics 30:19–36.

Brown, K. J., J. S. Clark, E. C. Grimm, et al. 2005. Fire cycles in North American interior grasslands and their relation to prairie drought. Proceedings of the National Academy of Sciences 102:8865–8870.

Brown, P. A. and J. P. Kennett. 1998. Megaflood erosion and meltwater plumbing changes during last North American deglaciation recorded in Gulf of Mexico sediments. Geology 26:599–602.

Brown, P. A., J. P. Kennett, and B. L. Ingram. 1999. Marine evidence for episodic Holocene megafloods in North America and the northern Gulf of Mexico. Paleoceanography 14:498–510.

Browning, J. V., K. G. Miller, P. P. McLaughlin, et al. 2006. Quantification of the effects of eustasy, subsidence, and sediment supply on Miocene sequences, Mid-Atlantic margin of the United States. Geological Society of America Bulletin 118:567–588.

Browning, J. V., K. G. Miller, and D. K. Pak. 1996. Global implications of lower to middle Eocene sequence boundaries on the New Jersey coastal plain: The icehouse cometh. Geology 24:639–642.

Bryan, S. E., T. R. Riley, D. A. Jerram, P. T. Leat, and C. J. Stephens. 2002. Silicic volcanism: An under-valued component of large igneous provinces and volcanic rifted margins. In M. A. Menzies, et al., eds., *Magmatic Rifted Margins*, pp. 99–118. Boulder, CO: Geological Society of America.

Bryden, H. L., H. R. Longworth, and S. A. Cunningham. 2005. Slowing of the Atlantic meridional overturning circulation at 25°N. Nature 438:655–657.

Buesseler, K. O., J. E. Andrews, S. M. Pike, and M. A. Charette. 2004. The effects of iron fertilization on carbon sequestration in the Southern Ocean. Science 304:414–417.

Buesseler, K. O. and R. S. Lampitt. 2008. Preface, Introduction to "Understanding the ocean's biological pump: Results from VERTIGO." Deep-Sea Research II 55:1519–1521.

Buffett, B. and D. E. Archer. 2004. Global inventory of methane clathrate: Sensitivity to changes in environmental conditions. Earth and Planetary Science Letters 227:185–199.

Buhay, W. M. and R. N. Betcher. 1998. Paleohydrologic implications of ^{18}O-enriched Lake Agassiz water. Journal of Paleolimnology 19:285–296.

Bull, D., A. E. S. Kemp, and G. P. Weedon. 2000. A 160-k.y.-old record of El Niño–Southern Oscillation in marine production and coastal runoff from Santa Barbara Basin, California, USA. Geology 28:1007–1010.

Burdett, J. W., J. P. Grotzinger, and M. A. Arthur. 1990. Did major changes in the stable-isotope composition of Proterozoic seawater occur? Geology 18:227–230.

Bürger, G. 2007. Comment on "The spatial extent of 20th-century warmth in the context of the past 1200 years." Science 316:1844a.

Burns, S. J., D. Fleitmann, A. Matter, J. Kramers, and A. A. Al-Subbary. 2003. Indian Ocean climate and an absolute chronology over Dansgaard/Oeschger events 9 to 13. Science 301:1365–1367.

Burton, K. W. 2006. Global weathering variations inferred from marine radiogenic isotope records. Journal of Geochemical Exploration 88:262–265.

Burton, K. W., H.-F. Lang, and R. K. O'Nions. 1997. Closure of the Central American Isthmus and its effect on deep-water formation in the North Atlantic. Nature 386:382–386.

Busalacchi, A. and T. N. Palmer. 2006. Preface. Special Section: 1st WCRP Climate Variability and Predictability (CLIVAR) Science Conference. Journal of Climate 19:4975–4976.

Butzin, M., M. Prange, and G. Lohmann. 2005. Radiocarbon simulations for the glacial ocean: The effects of wind stress, Southern Ocean sea ice and Heinrich events. Earth and Planetary Science Letters 235:45–61.

Cabanes, C., A. Cazenave, and C. Le Provost. 2001. Sea level rise during past 40 years determined from satellite and in situ observations. Science 294:840–842.

Cai, Y., Z. An, H. Cheng, et al. 2006. High-resolution absolute-dated Indian Monsoon record between 53 and 36 ka from Xiaobailong Cave, southwestern China. Geology 34:621–624.

Caillon, N., J. P. Severinghaus, J. Jouzel, et al. 2003. Timing of atmospheric CO_2 and Antarctic temperature changes across Termination III. Science 299:1728–1731.

Caldeira, K., R. A. Berner, and E. T. Sundquist. 1999. Seawater pH and atmospheric carbon dioxide. Science 286:2043.

Caldeira, K. and M. E. Wicket. 2003. Anthropogenic carbon and ocean pH. Nature 425:365.

Calvo, E., J. F. Marshall, C. Pelejero, et al. 2007. Interdecadal climate variability in the Coral Sea since 1710 A.D. Palaeogeography, Palaeoclimatology, Palaeoecology 248:190–201.

Came, R. E., J. M. Eiler, J. Veizer, et al. 2007. Coupling of surface temperature and atmospheric CO_2 concentrations during the Palaeozoic era. Nature 449:198–201.

Came, R. E., D. W. Oppo, and W. B. Curry. 2003. Atlantic Ocean circulation during the Younger Dryas: Insights from a new Cd/Ca record from the western subtropical South Atlantic. Paleoceanography 18:PA000888.

Camoin, G. F., M. Colonna, L. F. Montaggioni, et al. 1997. Holocene sea level changes and reef development in the southwestern Indian Ocean. Coral Reefs 16:247–259.

Camoin, C. G., Y. Iryu, and D. McInroy and expedition scientists. 2007. Expedition 310 Summary. Proceedings of the Integrated Ocean Drilling Program, Washington, D.C., vol. 310, doi:10.2204/iodp.proc.310.101.2007.

Cane, M. A. 1986. El Niño. Annual Review of Earth and Planetary Sciences 14:43–70.

Cane, M. A. 2005. The evolution of El Niño, past and future. Earth and Planetary Science Letters 230:227–240.

Cane, M. A. and A. C. Clement. 1999. A role for the tropical Pacific coupled ocean-atmosphere system on Milankovich and millenial timescales. Part II: Global impacts. In P. U. Clark, R. S. Webb, and L. D. Keigwin, eds., *Mechanisms of Millennial Scale Global Climate Change*, pp. 373–383. Washington, DC: American Geophysical Union.

Cane, M. A., M. Khodri, P. Braconnot, et al. 2006. Progress in Paleoclimate Modeling. Journal of Climate 19:5031–5057.

Cane, M. A. and P. Molnar. 2001. Closing of the Indonesian seaway as a precursor to east African aridification around 3–4 million years ago. Nature 411:157–162.

CAPE–Last Interglacial Project Members. 2006. Last interglacial Arctic warmth confirms polar amplification of climate change. Quaternary Science Reviews 25:1383–1400.

CAPE Project Members. 2006. Simulating Arctic climate warmth and icefield retreat in the last interglaciation. Science 311:1751–1753.

Capo, R. C. and D. J. DePaolo. 1990. Seawater strontium isotopic variations from 2.5 million years ago to present. Science 249:51–55.

Carlson, A., A. N. LeGrande, D. W. Oppo, et al. 2008. Rapid early Holocene deglaciation of the Laurentide Ice Sheet. Nature Geoscience 1:620–624.

Carlson, A. E., P. U. Clark, B. A. Haley, et al. 2007a. Geochemical proxies of North American freshwater routing during the Younger Dryas. Proceedings of the National Academy of Sciences 104:6556–6561.

Carlson, A. E., P. U. Clark, G. M. Raisbeck, and E. J. Brook. 2007b. Rapid Holocene deglaciation of the Labrador Sector of the Laurentide Ice Sheet. Journal of Climate 20:5126–5133.

Carriquiry, J. D., M. J. Risk, and H. P. Schwarcz. 1994. Stable isotope geochemistry of corals from Costa Rica as a proxy indicator of the El Niño/Southern Oscillation (ENSO). Geochimica et Cosmochimica Acta 58:335–351.

Castagnoli, G. C., ed. 1988. Solar-Terrestrial Relationships and the Earth Environment in the Last Millennia. New York: Elsevier/North-Holland Physics Publishing.

Castaneda, I. S., J. Werne, and T. C. Johnson. 2007. Wet and arid phases in the southeast African tropics since the Last Glacial Maximum. Geology 35:823 826.

Castellano, E., S. Becaglia, J. Jouzelb, et al. 2004. Volcanic eruption frequency over the last 45 ky as recorded in Epica–Dome C ice core (East Antarctica) and its relationship with climatic changes. Global and Planetary Change 42:195–205.

Catlin, D. C. and R. Buick. 2006. Oxygen and life in the Precambrian. Geobiology 4:225–226.

Cayan, D. R., S. A. Kammerdiener, M. D. Dettinger, J. M. Caprio, and D. H. Peterson. 2001. Changes in the onset of spring in the western United States. Bulletin of the American Meteorological Society 82:399–415.

Cazenave, A. and R. S. Nerem. 2004. Present-day sea level change: Observations and causes. Reviews of Geophysics 42:RG3001, RG000139.

Cecil, C. B. and N. T. Edgar, eds. 2003. Climate Controls on Stratigraphy: SEPM Special Publication No. 77. Tulsa, OK: Society for Economic Paleontologists and Mineralogists.

Cerling, T. 1991. Carbon dioxide in the atmosphere: Evidence from Cenozoic and Mesozoic paleosols. American Journal of Science 291:377–400.

Cerling, T. 1992. Use of carbon isotopes as an indicator of the $P(CO_2)$ of the paleoatmosphere. Global Biogeochemical Cycles 6:307–314.

Cerling, T. E. 1999. Paleorecords of C4 plants and ecosystems. In R. F. Sage and R. K. Monson, eds., C4 Plant Biology, pp. 445–469. San Diego, CA: Academic Press.

Cerling, T. E., J. Quade, Y. Wang, and J. R. Bowman. 1989. Carbon isotopes in soils and paleosols as ecology and paleoecology indicators. Nature 361:344–345.

Chagnon, S. A. 1999. Impacts of 1997–98 El Niño generated weather in the United States. Bulletin of the American Meteorological Society 80:1819–1827.

Chaisson, W. and A. C. Ravelo. 2000. Pliocene development of the East-West hydrographic gradient in the Equatorial Pacific. Paleoceanography 15:497–505.

Chamberlin, T. C. 1899. An attempt to frame a working hypothesis of the cause of glacial periods on an atmospheric basis. Journal of Geology 7:545–584, 667–685, 751–787.

Chandler, M., H. Dowsett, and A. Haywood. 2008. The PRISM Model/Data Cooperative: Mid-Pliocene data-model comparisons. PAGES Newsletter 16:24–25.

Chandler, M. A., S. J. Richards, and M. J. Shopsin. 2005. EdGCM: Enhancing climate science education through climate modeling research projects. In Proceedings of the 85th Annual Meeting of the American Meteorological Society, 14th Symposium on Education, San Diego, CA, January 8–14, 2005, p. P1.5.

Chandler, M. A., D. Rind, and R. S. Thompson, 1994. Joint investigations of the middle Pliocene climate II: GISS GCM Northern Hemisphere results. Global and Planetary Change 9:197–219.

Chandler, M. A. and L. E. Sohl. 2000. Climate forcings and the initiation of low-latitude ice sheets during the Neoproterozoic Varanger glacial interval. Journal of Geophysical Research 105:20737–20756.

Chao, B. F., Y. H. Wu, and Y. S. Li. 2008. Impact of artificial reservoir water impoundment on global sea level. Science 320:212–214.

Chao, W. C. and B. Chen. 2001. The origin of monsoons. Journal of the Atmospheric Sciences 58:3497–3507.

Chapman, D. H. 1937. Late-glacial and postglacial history of the Champlain Valley. American Journal of Science, 5th series, 34(200):89–124.

Chapman, M. R. and M. A. Maslin. 1999. Low-latitude forcing of meridional temperature and salinity gradients in the subpolar North Atlantic and the growth of glacial ice sheets. Geology 27:875–878.

Chapman, M. R. and N. J. Shackleton. 2000. Evidence of 550-year and 1000-year cyclicity in North Atlantic circulation patterns during the Holocene. The Holocene 10:287–291.

Chappell, J. 1983. Evidence for smoothly falling sea level relative to north Queensland, Australia, during the past 6,000 yr. Nature 302:406–408.

Chappell, J. 2002. Sea level changes forced ice breakouts in the last glacial cycle: New results from coral terraces. Quaternary Science Reviews 21:1229–1240.

Chappellaz, J., J.-M. Barnola, D. Raynaud, Y. S. Korotkevich, and C. Lorius. 1990. Ice-core record of atmospheric methane over the past 160,000 years. Nature 345:127–131.

Chappellaz, J., T. Blunier, S. Kints, et al. 1997. Changes in the atmospheric CH_4 gradient between Greenland and Antarctica during the Holocene. Journal of Geophysical Research 102(D13):15987–15997.

Chappellaz, J., T. Blunier, D. Raynaud, et al. 1993. Synchronous changes in atmospheric CH_4 and Greenland climate between 40 and 8 kyr BP. Nature 366:443–445.

Charman, D. J., A. Blundell, R. C. Chiverrell, D. Hendon, and P. G. Langdon. 2006. Compilation of non-annually resolved Holocene proxy climate records: Stacked Holocene peatland palaeo-water table reconstructions from northern Britain. Quaternary Science Reviews 25:336–350.

Charney, J. et al. 1979. *Carbon Dioxide and Climate: A Scientific Assessment*. Washington, DC: U.S. National Academy of Sciences.

Cheddadi, R., G. Yu, J. Guiot, S. P. Harrison, and I. C. Prentice. 1997. The climate of Europe 6000 years ago. Climate Dynamics 13:1–9.

Chen, J. L., C. R. Wilson, D. D. Blankenship, and B. D. Tapley. 2006. Antarctic mass rates from GRACE. Geophysical Research Letters 33:L11502, GL026369.

Cheng, H., R. L. Edwards, M. T. Murrell, and T. M. Benjamin. 1998. Uranium-thorium-protactinium dating systematics. Geochimica et Cosmochimica Acta 62:3437–3452.

Chiang, J. C. H. and C. M. Bitz. 2005. Influence of high latitude ice cover on the marine intertropical convergence zone. Climate Dynamics 25:477–496.

Chiu, T.-C., R. G. Fairbanks, L. Cao, and R. A. Mortlock. 2007. Analysis of the atmospheric ^{14}C record spanning the past 50,000 years derived from high-precision ^{230}Th/^{234}U/^{238}U and ^{231}Pa/^{235}U and ^{14}C dates on fossil corals. Quaternary Science Reviews 26:18–36.

Christensen, J. N., A. N. Halliday, L. V. Godfrey, J. R. Hein, and D. K. Rea. 1997. Climate and ocean dynamics and the lead isotope records in Pacific ferromanganese crusts. Science 277:913–918.

Chung, S.-L., C.-H. Lo, T.-Y. Lee, et al. 1998. Diachronous uplift of the Tibetan Plateau starting 40 Myr ago. Nature 394:769–773.

Church, J. A., and N. J. White. 2006. A 20th century acceleration in global sea-level rise. Geophysical Research Letters 33:L01602, GL024826.

Church, J. A., N. J. White, R. Coleman, K. Lambeck, and J. X. Mitrovica. 2004. Estimates of the regional distribution of sea-level rise over the 1950 to 2000 period. Journal of Climate 17:2609–2625.

Church, J. A., N. J. White, and J. R. Hunter. 2006. Sea level rise at tropical Pacific and Indian Ocean islands. Global and Planetary Change 53:155–168.

Clague, J. 2005. Status of the Quaternary. Quaternary Science Reviews 24:2424–2425.

Clapperton, C. 1993. *Quaternary Geology and Geomorphology of South America*. Amsterdam: Elsevier.

Clark, P. U. 2002. Early deglaciation in the tropical Andes. Science 298:7.

Clark, P. U., R. B. Alley, L. D. Keigwin, J. M. Licciardi, S. J. Johnsen, and H. Wang. 1996. Origin of the first global meltwater pulse following the Last Glacial Maximum. Paleoceanography 11:563–577.

Clark, P. U., R. B. Alley, and D. Pollard. 1999. Northern hemisphere ice-sheet influences on global climate change. Science 286:1104–1111.

Clark, P. U., A. Archer, D. Pollard, et al. 2006. The middle Pleistocene transition: Characteristics, mechanisms, and implications for long-term changes in atmosphere pCO$_2$. Quaternary Science Reviews 25:3150–3184.

Clark, P. U. and P. J. Bartlein. 1995. Correlation of late Pleistocene glaciation in the western United States with North Atlantic Heinrich events. Geology 23:483–486.

Clark, P. U. and P. F. Karrow. 1984. Late Pleistocene water bodies in the St. Lawrence Lowland, New York, and regional correlations. Geological Society of America Bulletin 95:805–813.

Clark, P. U., S. J. Marshall, G. K. C. Clarke, et al. 2001. Freshwater forcing of abrupt climate change during the last glaciation. Science 239:283–287.

Clark, P. U., A. M. McCabe, A. C. Mix, and A. J. Weaver. 2004. Rapid rise of sea level 19,000 years ago and its global implications. Science 304:1141–1144.

Clark, P. U., J. X. Mitrovica, G. A. Milne, and M. E. Tamisiea. 2002. Sea-level fingerprinting as a direct test for the source of global meltwater pulse 1A. Science 295:2438–2441.

Clark. P. U. and D. Pollard. 1998. Origin of the middle Pleistocene transition by ice sheet erosion of regolith. Paleoceanography 13:1–9.

Clarke, G. K., D. Leverington, J. Teller, and A. Dyke. 2003. Superlakes, megafloods, and abrupt climate change. Science 301:922–923.

Clarke, G. K. C., D. W. Leverington, J. T. Teller, and A. S. Dyke. 2004. Paleohydraulics of the last outburst flood from glacial Lake Agassiz and the 8200 BP cold event. Quaternary Science Reviews 23:389–407.

Clausen, H. B. and C. U. Hammer. 1988. The Laki and Tambora eruptions as revealed in Greenland ice cores from 11 locations. Annals of Glaciology 10:16–22.

Clausen, H. B., C. Hammer, C. S. Huidberg, et al. 1997. A comparison of the volcanic records over the past 4000 years from the Greenland ice core project and Dye 3 Greenland ice cores. Journal of Geophysical Research 102(C12):26707–26723.

Clemens, S. C. 2005. Millennial-band climate spectrum resolved and linked to centennial-scale solar cycles. Quaternary Science Reviews 24:521–531.

Clemens, S. C., W. Prell, D. Murray, G. Shimmield, and G. Weedon. 1991. Forcing mechanisms of the Indian Ocean monsoon. Nature 353:720–725.

Clemens, S. C. and R. Tiedemann. 1997. Eccentricity forcing of Pliocene–early Pleistocene climate revealed in a marine oxygen-isotope record. Nature 385:801–804.

Clement, A. C. and M. A. Cane. 1999. A role for the tropical Pacific coupled ocean-atmosphere system on Milankovich and millenial timescales. Part I: A modeling study of tropical Pacific variability. In P. U. Clark, R. S. Webb, and L. D. Keigwin, eds., *Mechanisms of Millenial Scale Global Climate Change*. Washington, DC: American Geophysical Union.

Clement, A. C., M. A. Cane, and R. Seager. 2001. An orbitally driven tropical source for abrupt climate change. Journal of Climate 14:2369–2375.

Clement, A. C., A. Hall, and A. J. Broccoli. 2004. The importance of precessional signals in the tropical climate. Climate Dynamics 22:327–341.

Clement, A. C., R. Seager, and M. A. Cane. 1999. Orbital controls on ENSO and the tropical climate. Paleoceanography 14:441–456.

Clement, A. C., R. Seager, and M. A. Cane. 2000. Suppression of El Niño during the mid-Holocene by changes in the earth's orbit. Paleoceanography 15:731–737.

Clement, A. C., R. Seager, M. A. Cane, and S. E. Zebiak. 1996. An ocean dynamical thermostat. Journal of Climate 9:2190–2196.

CLIMAP Project Members. 1976. The surface of the ice-age earth. Science 191:1131–1137.

CLIMAP Project Members. 1981. *Seasonal Reconstruction of the Earth's Surface During the Last Glacial Maximum,* Boulder, CO: Geological Society of America.

CLIMAP Project Members. 1984. The last interglacial ocean. Quaternary Research 21:123–224.

Cline, R. M. and J. D. Hays, eds. 1976. *Investigation of Late Quaternary Pale-oceanography and Paleoclimatology.* Boulder, CO: Geological Society of America.

Coale, K. H., K. S. Johnson, F. P. Chavez, et al. 2004. Southern Ocean iron enrichment experiment: Carbon cycling in high- and low-Si waters. Science 304:408–414.

Coates, A. G., M.-P. Aubry, W. A. Berggren, L. C. Collins, and M. Kunk. 2003. Early Neogene history of the Central American arc from Bocas del Toro, western Panama. Geological Society of America Bulletin 115:271–287.

Coates, A. G., L. C. Collins, M.-P. Aubry, and W. A. Berggren. 2004. The geology of the Darien, Panama, and the late Miocene-Pliocene collision of the Panama arc with north-western South America. Geological Society of America Bulletin 116:1327–1344.

Coates, A. G., J. B. Jeremy, L. S. Collins, et al. 1992. Closure of the Isthmus of Panama: The near-shore marine record of Costa Rica and western Panama. Geological Society of America Bulletin 104:814–828.

Coates, A. G. and J. A. Obando. 1996. Geological evolution of the Central American Isthmus. In J. B. C. Jackson, A. F. Budd, and A. G. Coates, eds., *Evolution and Environment in Tropical America,* pp. 21–56. Chicago: University of Chicago Press.

Cobb, K. M., C. D. Charles, H. Cheng, and R. L. Edwards. 2003. The El Niño–Southern Oscillation and tropical Pacific climate during the last millennium. Nature 424:271–276.

Cobb, K. M., C. D. Charles, and D. E. Hunter. 2001. A central tropical Pacific coral demonstrates Pacific, Indian and Atlantic decadal climate connections. Geophysical Research Letters 28:2209–2212.

Coffin, M. F. and O. Eldholm. 1994. Large igneous provinces: Crustal structure, dimensions, and external consequences. Reviews of Geophysics 32:1–36.

Coffin, M. F. and O. Eldholm. 2005. Large igneous provinces. In R. C. Selley, R. Cocks, and I. R. Plimer, eds., *Encyclopedia of Geology,* pp. 315–323. Oxford: Elsevier.

Cogley, J. G. 2005. Mass and energy balances of glaciers and ice sheets. In *Encyclopedia of Hydrological Sciences, Part 14: Snow and Glacier Hydrology.* New York: John Wiley & Sons, doi: 10.1002/0470848944.hsa171.

Cohen, A. L. 2003. *Paleolimnology: The History and Evolution of Lake Systems.* New York: Oxford University Press.

Cohen, A. L., K. E. Owens, G. D. Layne, and N. Shimizu. 2002. The effect of algal symbionts on the accuracy of Sr/Ca paleotemperatures from coral. Science 296:331–333.

Cohen, A. S., A. L. Coe, and D. B. Kemp. 2007. The late Palaeocene, early Eocene and Toarcian (early Jurassic) carbon isotope excursions: A comparison of their time scales, associated environmental changes, causes and consequences. Journal of the Geological Society (London) 164:1093–1108.

COHMAP Members. 1988. Climatic changes of the last 18,000 years: Observations and model simulations. Science 241:1043–1052.

Cole, J. E., R. B. Dunbar, T. R. McClanahan, and N. A. Muthiga. 2000. Tropical Pacific forcing of decadal SST variability in the western Indian Ocean over the past two centuries. Science 287:617–619.

Cole, J. E., D. Rind, and R. G. Fairbanks. 1993. Isotopic response to interannual climatic variability simulated by an atmospheric general circulation model. Quaternary Science Reviews 12:387–406.

Cole-Dai, J., E. Mosley-Thompson, and L. Thompson. 1997. Annually resolved southern hemisphere volcanic history from two Antarctic ice cores. Journal of Geophysical Research 102:16761–16771.

Collins, L. S., A. G. Coates, W. A. Berggren, M.-P. Aubry, and J. Zhang. 1996. The late Miocene Panama Isthmian Strait. Geology 24:687–690.

Colman, S. M., K. A. Clark, L. Clayton, A. K. Hansel, and C. E. Larsen. 1994. Deglaciation, lake levels, and meltwater discharge in the Lake Michigan basin. Quaternary Science Reviews 13:879–890.

Colman, S. M., J. A. Peck, E. Karabanov, et al. 1995. Continental climate response to orbital forcing from biogenic silica records in Lake Baikal. Nature 378:769–771.

Comiso, J. C. 2006. Abrupt decline in the Arctic winter sea ice cover. Geophysical Research Letters 33:L18504, GL027341.

Condie, K. C. 2001. *Mantle Plumes and Their Record in Earth History.* Cambridge: Cambridge University Press.

Condie, K. C., D. Abbott, and D. J. Des Marais. 2002. Preface. Journal of Geodynamics 34:159–162.

Condon, D., M. Zhu, S. Bowring, et al. 2005. U–Pb ages from the Neoproterozoic Doushantuo Formation, China. Science 308:95–98.

Conte, M. H., M.-A. Sicre, C. Rühlemann, et al. 2006. Global temperature calibration of the alkenone unsaturation index (UK'37) in surface waters and comparison with surface sediments. Geochemistry Geophysics Geosystems 7, doi:10.1029/2005GC001054.

Cook, E. R. 2003. Multi-proxy reconstructions of the North Atlantic Oscillation (NAO) Index: A critical review and a new well-verified winter NAO Index reconstructed back to AD 1400. In J. W. Hurrell, Y. Kushnir, G. Ottersen, and M. Visbeck, eds., *The North Atlantic Oscillation: Climatic Significance and Environmental Impact,* pp. 63–79. Washington, DC: American Geophysical Union.

Cook, E. R., K. R. Briffa, D. K. Meko, D. S. Graybill, and G. Funkhouser. 1995. The "segment length curse" in long tree-ring chronology development for paleoclimatic studies. The Holocene 5:229–237.

Cook, E. R., R. D. D'Arrigo, and K. R. Briffa. 1998. A reconstruction of the North Atlantic Oscillation using tree-ring chronologies from North America and Europe. The Holocene 8:9–17.

Cook, E. R., R. D. D'Arrigo, and M. E., Mann. 2002. A well-verified, proxy reconstruction of the winter North Atlantic Oscillation index since A.D. 1400. Journal of Climate 15:1754–1764.

Cook, E. R., J. Esper, and R. D. D'Arrigo. 2004a. Extra-tropical northern hemisphere land temperature variability over the past 1000 years. Quaternary Science Reviews 23:2063–2074.

Cook, E. R., D. K. Meko, D. W. Stahle, and M. K. Cleaveland. 1996. Tree-ring reconstructions of past drought across the coterminous United States: Tests of a regression method and calibration/verification results. In Tree Rings, Environment, and Humanity. Special issue of Radiocarbon: 155–169. (J. S. Dean, D. M. Meko, and T. W. Swetnam, eds.)

Cook, E. R., D. K. Meko, D. W. Stahle, and M. K. Cleaveland. 1999. Drought reconstructions for the continental United States. Journal of Climate 12:1145–1162.

Cook, E. R., R. Seager, M. A. Cane, and D. H. Stahle. 2007. North American drought: Reconstructions, causes, and consequences. Earth-Science Reviews 81:93–134.

Cook, E. R., C. A. Woodhouse, C. M. Eakin, D. M. Meko, and D. W. Stahle. 2004b. Long-term aridity changes in the western United States. Science 306:1015–1018.

Coope, G. R. 1977. Fossil coleopteran assemblages as sensitive indicators of climatic changes during the Devensian (last) cold stage. Philosophical Transactions of the Royal Society of London, Series B, Biological Sciences 280:313–340.

Coope, G. R. 1994. The response of insect faunas to glacial-interglacial climatic fluctuation. Philosophical Transactions of the Royal Society of London, Series B, 344:19–26.

Corrège, T. 2006. Sea surface temperature and salinity reconstruction from coral geochemical tracers. Palaeogeography, Palaeoclimatology, Palaeoecology 232:408–428.

Corliss, B. H., A. S. Hunt, and L. D. Keigwin. 1982. Benthonic foraminiferal faunal and isotopic data for the Holocene evolution of the Champlain Sea. Quaternary Research 17:325–338.

Cortijo, E., L. Labeyrie, L. Vidal, et al. 1997. Changes in sea surface hydrology associated with Heinrich event 4 in the North Atlantic Ocean between 40 and 60°N. Earth and Planetary Science Letters 146:29–45.

Cortijo, E., S. J. Lehman, L. Keigwin, M. Chapman, and D. Paillard. 1999. Changes in meridional temperature and salinity gradients in the North Atlantic Ocean (30°N–72°N) during the last interglacial period. Paleoceanography 14:23–33.

Courtillot, V. E. and P. R. Renne. 2003. On the ages of flood basalt events. Comptes Rendus Geoscience 335:113–140.

Covey, C., L. C. Sloan, and M. I. Hoffert. 1996. Paleoclimate data constraints on climate sensitivity: The paleocalibration method. Climatic Change 32:165–184.

Cox, P. and C. Jones. 2008. Illuminating the modern dance of climate and CO_2. Science 321:1642–1644.

Coxall, H. K., P. A. Wilson, H. Pälike, C. H. Lear, and J. Backman. 2005. Rapid stepwise onset of Antarctic glaciation and deeper calcite compensation in the Pacific Ocean. Nature 433:53–57.

Craig, H. 1965. The measurement of oxygen isotope paleotemperature. In E. Tongiori, ed., Second Conference on Oceanographic Studies and Paleotemperatures, pp. 161–182. Spoleto, Italy: Consiglio Naz della Richerche.

Craig, H. and L. I. Gordon. 1965. Deuterium and oxygen-18 variations in the ocean and the marine atmosphere. Proceedings of the Spoleto Conference on Stable Isotopes in Oceanographic Studies and Palaeotemperatures 2:1–87.

Cramer, B. S., J. D. Wright, D. V. Kent, and M.-P. Aubry. 2003. Orbital climate forcing of $\delta^{13}C$ excursions in the late Paleocene–early Eocene (chrons C24n–C25n). Paleoceanography 18:1097, doi:10.1029/2003PA000909.

Croll, J. 1864. On the physical cause of the change of climate during geological epochs. Philosophical Magazine 28:121–137.

Croll, J. 1875. Climate and Time. New York: Appleton & Co.

Cronin, T. M. 1977. Late Wisconsin marine environments of the Champlain Valley (New York, Quebec). Quaternary Research 7:238–253.

Cronin, T. M. 1989. Paleozoogeography of post-glacial Ostracoda from northeastern North America. In N. R. Gadd, ed., Late Quaternary Development of the Champlain Sea Basin, pp. 125–144. Ottawa: Canadian Geological Association.

Cronin, T. M. 1991. Pliocene paleoceanography of the North Atlantic Ocean based on marine Ostracoda. Quaternary Science Reviews 10:175–188.

Cronin, T. M. 1999. Principles of Paleoclimatology. New York: Columbia University Press.

Cronin, T. M. and H. J. Dowsett. 1990. A quantitative micropaleontologic method for shallow marine paleoclimatology: Application to Pliocene deposits of the western North Atlantic. Marine Micropaleontology 16:117–147.

Cronin, T. M. and H. J. Dowsett, eds. 1991. Pliocene climates. Quaternary Science Reviews 10(2–3).

Cronin, T. M., G. S. Dwyer, P. A. Baker, J. Rodriguez-Lazaro, and D. M. DeMartino. 2000. Orbital and suborbital variability in deep North Atlantic bottom water temperature obtained from Mg/Ca ratios in the ostracode Krithe. Palaeogeography, Palaeoclimatology, Palaeoecology 162:45–57.

Cronin, T. M., G. S. Dwyer, T. Kamiya, S. Schwede, and D. A. Willard. 2003. Medieval Warm Period, Little Ice Age and 20th century temperature variability from Chesapeake Bay. Global and Planetary Change 36:17–29.

Cronin, T. M., T. Edgar, G. Brooks, et al. 2007a. Sea level rise in Tampa Bay. EOS, Transactions, American Geophysical Union 88:117–118.

Cronin, T. M., T. R. Holtz Jr., R. Stein, et al. 1995. Late Quaternary paleoceanography of the Eurasian Basin, Arctic Ocean. Paleoceanography 10:259–281.

Cronin, T. M., P. L. Manley, S. Brachfeld, et al. 2008. Impacts of post-glacial lake drainage events and revised chronology of the Champlain Sea episode 13–9 ka. Palaeogeography, Palaeoclimatology, Palaeoecology 262:46–60.

Cronin, T. M., B. J. Szabo, T. A. Ager, J. E. Hazel, and J. P. Owens. 1981. Quaternary climates and sea levels of the U.S. Atlantic coastal plain. Science 211:233–240.

Cronin, T. M., R. Thunell, G. S. Dwyer, et al. 2005. Multiproxy evidence of Holocene climate variability from estuarine sediments, eastern North America. Paleoceanography 20:PA4006.

Cronin, T. M., P. R. Vogt, D. A. Willard, et al. 2007b. Rapid sea level rise and ice sheet response to 8,200-year climate event. Geophysical Research Letters 34:L20603, GL031318.

Cronin, T. M., R. Whatley, A. Wood, et al. 1993. Evidence for elevated mid-Pliocene temperatures in the Arctic Ocean based on marine ostracoda. Paleoceanography 8:161–173.

Crowell, J. C. 1999. Pre-Mesozoic Ice Ages: Their Bearing on Understanding the Climate System. Boulder, CO: Geological Society of America.

Crowley, T. J. 1996. Pliocene climates: The nature of the problem. Marine Micropaleontology 27:3–12.

Crowley, T. J. 2000. Causes of climate change over the past 1000 years. Science 289:270–277.

Crowley, T. J. 2002. Cycles, cycles everywhere. Science 295:1473–1474.

Crowley, T. J. and S. K. Baum. 1992. Modeling late Paleozoic glaciation. Geology 20:507–510.

Crowley, T. J., S. K. Baum, K.-Y. Kim, G. C. Hegerl, and W. T. Hyde. 2003. Modeling ocean heat content changes during the last millennium. Geophysical Research Letters 30:1932, doi:10.1029/2003GL017801.

Crowley, T. J. and R. A. Berner. 2001. CO_2 and climate change. Science 292:870–872.

Crowley, T. J., W. T. Hyde, and W. R. Peltier. 2001. CO_2 levels required for deglaciation of the "near-snowball" earth. Geophysical Research Letters 28:283–286.

Crowley, T. J. and K.-Y. Kim. 1996. Comparison of proxy records of climate change and solar forcing. Geophysical Research Letters 23:359–362.

Crowley, T. J. and T. S. Lowery. 2000. How warm was the Medieval Warm period? Ambio 29:51–54.

Crowley, T. J. and G. R. North. 1991. *Paleoclimatology*. New York: Oxford University Press.

Crowley, T. J., G. Zielinski, B. Vinther, et al. 2008. Volcanism and the Little Ice Age. PAGES Newsletter 16:22–23.

Crucifix, M., M. F. Loutre, and A. Berger. 2005. Commentary on "The Anthropogenic greenhouse era began thousands of years ago." Climatic Change 69:419–426.

Crucifix, M., M. F. Loutre, and A. Berger. 2006. The climate response to the astronomical forcing. Space Science Reviews 125:213–226.

Crusius, J., S. Calvert, T. Pedersen, and D. Sage. 1996. Rhenium and molybdenum enrichments in sediments as indicators of oxic, suboxic, and sulfidic conditions of deposition. Earth and Planetary Science Letters 145:65–78.

Crutzen, P. J. and E. F. Stoermer. 2000. The "Anthropocene." International Geosphere Biosphere Programme Newsletter 41:17–18.

Cubasch, U., G. A. Meehl, G. J. Boer, et al. 2001. Projections of future climate change. In J. T. Houghton, et al., eds., *Climate Change 2001: The Scientific Basis. Contribution of Working Group I to the Third Assessment Report of the Intergovernmental Panel on Climate Change*, pp. 527–582. Cambridge: Cambridge University Press.

Cuffey, K. M., G. D. Clow, R. B. Alley, et al. 1995. Large arctic temperature change at the Wisconsin-Holocene transition. Science 270:455–458.

Cullen, H. M. and P. B. deMenocal. 2000. North Atlantic influence on Tigris-Euphrates streamflow. International Journal of Climatology 20:853–863.

Culver, S. J., D. V. Ames, D. R. Corbett, et al. 2006. Foraminiferal and sedimentary record of late Holocene barrier island evolution, Pea Island, North Carolina. Journal of Coastal Research 22:406–416.

Culver, S. J. and B. P. Horton. 2005. Infaunal marsh foraminifera from the Outer Banks, North Carolina, U.S.A. Journal of Foraminiferal Research 35:148–170.

Cumming, B. F., K. R. Laird, J. R. Bennett, J. P. Smol, and A. K. Salomon. 2002. Persistent millennial-scale shifts in moisture regimes in western Canada during the past six millennia. Proceedings of the National Academy of Sciences 99:16117–16121.

Cunningham, S. A., T. Kanzow, D. Rayner, et al. 2007. Temporal variability of the Atlantic Meridional Overturning Circulation at 26.5°N. Science 317:935–938.

Curran, M. A. J., T. D. van Ommen, V. I. Morgan, K. L. Phillips, and A. S. Palmer. 2003. Ice core evidence for Antarctic sea ice decline since the 1950s. Science 302:1203–1206.

Curry, R., B. Dickson, and I. Yashayaev. 2003. A change in the freshwater balance of the Atlantic Ocean over the past four decades. Nature 426:826–829.

Curry, R. and C. Mauritzen. 2005. Dilution of the northern North Atlantic Ocean in recent decades. Science 308:1772–1774.

Curry, W. B. and D. W. Oppo. 1997. Synchronous high-frequency oscillations in tropical sea surface temperatures and North Atlantic deep water production during the last glacial cycle. Paleoceanography 12:1–14.

Curry, W. B. and D. W. Oppo. 2005. Glacial water mass geometry and the distribution of $\delta^{13}C$ of ΣCO_2 in the Western Atlantic Ocean. Paleoceanography 20:PA1017, PA001021.

Cutler, K. B., S. C. Gray, G. S. Burr, et al. 2004. Radiocarbon calibration and comparison to 50 kyr BP with paired ^{14}C and ^{230}Th dating of corals from Vanuatu and Papua New Guinea. Radiocarbon 46:1127–1160.

Czaja, A., A. W. Robinson, and T. Huck. 2003. The role of Atlantic ocean–atmosphere coupling in affecting North Atlantic Oscillation variability. In J. W. Hurrell, Y. Kushnir, G. Ottersen, and M. Visbeck, eds., *The North Atlantic Oscillation: Climatic Significance and Environmental Impact*, pp. 147–172. Washington, DC: American Geophysical Union.

Daigle, T. A. and D. S. Kaufman. 2008. Holocene climate inferred from glacier extent, lake sediment and tree rings at Goat Lake, Kenai Mountains, Alaska USA. Journal of Quaternary Science doi:10.1002/jqs.1166.

Dällenbach, A., T. Blunier, J. Flückiger, et al. 2000. Changes in the atmospheric CH_4 gradient between Greenland and Antarctica during the last glacial and the transition to the Holocene. Geophysical Research Letters 27:1005–1008.

Damon, P. E. and A. N. Peristykh. 1999. Solar cycle length and 20th century northern hemisphere warming: Revisited. Geophysical Research Letters 26:2469–2472.

Damon, P. E. and A. N. Peristykh. 2005. Solar forcing of global temperature since AD 1400. Climatic Change 68:101–111.

Damon, P. E. and C. P. Sonett. 1991. Solar and terrestrial components of the atmospheric C-14 variation spectrum. In C. P. Sonett, M. S. Giampapa, and M. S. Matthews, eds., *The Sun in Time*, pp. 360–388. Tuscon: University of Arizona Press.

Dansgaard, W. 1964. Stable isotopes in precipitaiton. Tellus 16:436–468.

Dansgaard, W., H. B. Clausen, N. Gundestrup, et al. 1982. A new Greenland deep ice core. Science 218:1273–1277.

Dansgaard, W., S. J. Johnsen, H. B. Clausen, D. Dahl-Jensen, N. Gundestrup, and C. U. Hammer. 1984. North Atlantic climatic oscillations revealed by deep Greenland ice cores. In J. E. Hansen and T. Takahashi, eds., *Climate Processes and Climate Sensitivity*, pp. 288–298. Washington, DC: American Geophysical Union.

Dansgaard, W., S. J. Johnsen, H. B. Clausen, et al. 1993. Evidence for general instability of climate from a 250-kyr ice-core record. Nature 364:218–220.

Dansgaard, W., S. J. Johnsen, I. Møller, and C. C. Langway Jr. 1969. One thousand centuries of climatic record from Camp Century on the Greenland Ice Sheet. Science 166:377–381.

Dansgaard, W., S. J. Johnsen, N. Reeh, et al. 1975. Climate changes, Norseman, and modern man. Nature 255:24–28.

Dansgaard, W. and H. Oeschger. 1989. Past environmental long-term records from the Arctic. In H. Oeschger and C. C. Langway Jr., eds., *The Environmental Record in Glaciers and Ice Sheets*, pp. 287–318. Chichester, UK: John Wiley & Sons.

Dansgaard, W., J. W. C. White, and S. J. Johnsen. 1989. The abrupt termination of the Younger Dryas climatic event. Nature 339:532–534.

Darby, D., J. Bischof, R. Spielhagen, S. A. Marshall, and S. W. Herman. 2002. Arctic ice export events and their potential impact on global climate during the late Pleistocene. Paleoceanography 17:1–17.

Darby, D., L. Polyak, and H. A. Bauch. 2006. Past glacial and interglacial conditions in the Arctic Ocean and marginal seas—A review. Progress in Oceanography 71:129–144.

Darby, D. A. 2003. Sources of sediment found in sea ice from the western Arctic Ocean, new insights into processes of entrainment and drift patterns. Journal of Geophysical Research 108(C8):13-1–13-10.

Darby, D. A. 2008. Arctic perennial ice cover over the last 14 million years. Paleoceanography 23:PA1S07.

Darby, D. A. and P. Zimmerman. 2008. Ice-rafted detritus events in the Arctic during the last glacial interval, and the timing of the Innuitian and Laurentide Ice Sheet calving events. Polar Research 27:114–127.

D'Arrigo, R. D., E. R. Cook, M. E. Mann, and G. C. Jacoby. 2003. Tree-ring reconstructions of temperature and sea-level pressure variability associated with the warm-season Arctic Oscillation since AD 1650. Geophysical Research Letters 30:1549, GL017250.

D'Arrigo, R. D., E. R. Cook, R. J. Wilson, R. Allan, and M. E. Mann. 2005a. On the variability of ENSO over the past six centuries. Geophysical Research Letters 32:GL03711.

D'Arrigo, R. D., R. Villalba, and G. Wiles. 2001. Tree-ring estimates of Pacific decadal climate variability. Climate Dynamics 18:219–224.

D'Arrigo, R. D., and R. Wilson. 2006. On the Asian expression of the PDO. International Journal of Climatology 26:1607–1617.

D'Arrigo, R. D., R. Wilson, C. Deser, et al. 2005b. Tropical–north Pacific climate linkages over the past four centuries. Journal of Climate 18:5253–5265.

D'Arrigo, R. D., R. Wilson, and G. Jacoby. 2006. On the long-term context for late twentieth century warming. Journal of Geophysical Research 111(D3):D03103.

D'Arrigo, R. D., R. Wilson, B. Liepert, and P. Cherubini. 2008. On the "divergence problem" in northern forests: A review of the tree-ring evidence and possible causes. Global and Planetary Change 60:289–305.

Davis, B. A. S., S. Brewer, A. C. Silverman, et al. 2003. The temperature of Europe during the Holocene reconstructed from pollen data. Quaternary Science Reviews 22:1701–1716.

Davis, C. H., Y. Li, J. R. McConnell, M. M. Frey, and E. Hanna. 2005. Snowfall-driven growth in East Antarctic Ice Sheet mitigates recent sea-level rise. Science 308:1898–1901.

Davis, J. and B. Laird. 1977. The effects of Tropical Storm Agnes on the Chesapeake Bay estuarine ecosystem. In E. P. Ruzecki, et al., eds., The Effects of Tropical Storm Agnes on the Chesapeake Bay Estuarine System, pp. 1–29. Baltimore, MD: The Johns Hopkins University Press.

Davis, O. 1984. Multiple thermal maxima during the Holocene. Science 225:617–619.

de Abreu, L., F. Abrantes, N. Shackleton, et al. 2005. Ocean climate variability in the eastern North Atlantic during interglacial Marine Isotope Stage 11: A partial analogue for the Holocene? Paleoceanography 20:PA3009.

Dean, W. E. and M. A. Arthur, eds. 1998. Stratigraphy and Paleoenvironments of the Cretaceous Western Interior Seaway. Tulsa, OK: Society of Economic Paleontologists and Mineralogists.

Dean, W. E. and J. W. Gardner. 1982. Origin and geochemistry of redox cycles of Jurassic to Eocene age, Cape Verde Basin (DSDP Site 367), Continental Margin of North-West Africa. In S. O. Schlanger and M. B. Cita, eds., Nature and Origin of Carbon-Rich Facies, pp. 55–78. London: Academic Press.

Dean, W. E., D. Z. Piper, and L. C. Peterson. 1999. Molybdenum accumulation in Cariaco basin sediment over the past 24 k.y.: A record of water-column anoxia and climate. Geology 27:507–510.

DeAngelis, H. and P. Skvarca. 2003. Glacier surge after ice shelf collapse. Science 299:1560–1562.

De Boer, P. L. and D. G. Smith, eds. 1994. Orbital Forcing and Cyclic Sequences. Oxford: Blackwell.

DeConto, R. M. and D. Pollard. 2003a. A coupled climate-ice sheet modeling approach to the early Cenozoic history of the Antarctic Ice Sheet. Palaeogeography, Palaeoclimatology, Palaeoecology 198:39–52.

DeConto, R. M. and D. Pollard. 2003b. Rapid Cenozoic glaciation of Antarctica induced by declining atmospheric CO_2. Nature 421:245–249.

DeConto, R. M., D. Pollard, P. A. Wilson, H. Pälike, C. H. Lear, and M. Pagani. 2008. Thresholds for Cenozoic bipolar glaciation. Nature 455:652–657.

Deevey, E. S. and R. F. Flint. 1957. Postglacial hypsithermal interval. Science 125:182–184.

DeGeer, G. 1912. A geochronology of the last 12,000 years. 11th International Geological Congress Stockholm 1910, Compte Rendu 1:241–258.

Dekens, P. S., D. W. Lea, D. K. Pak, and H. J. Spero. 2002. Core top calibration of Mg/Ca in tropical foraminifera: Refining paleo-temperature estimation. Geochemistry, Geophysics, Geosystems 3:1022, doi:10.1029/2001GC000200.

Dekens, P. S., A. C. Ravelo, and M. D. McCarthy. 2007. Warm upwelling regions in the Pliocene warm period. Paleoceanography 22:PA3211.

Delaney, M. L., E. R. M. Druffel, and L. Linn. 1993. Seasonal cycles of manganese and cadmium in Galapagos coral. Geochimica et Cosmochimica Acta 57:347–354.

De La Rocha, C. L. 2006. Opal-based isotopic proxies of paleoenvironmental conditions. Global Biogeochemical Cycles 20:GB4S09.

De La Rocha, C. L. and D. J. DePaolo. 2000. Isotopic evidence for variations in the marine calcium cycle over the Cenozoic. Science 289:1176–1178.

Delmas, R. J. 1992. Environmental records from ice cores. Reviews of Geophysics 30:1–21.

Delmas, R. J., J.-M. Ascencio, and M. Legrand. 1980. Polar ice evidence that atmospheric CO_2 20,000 yr B.P. was 50% of present. Nature 284:155–157.

Delmas, R. J., M. Legrand, A. J. Aristarain, and F. Zanolini. 1985. Volcanic deposits in Antarctic snow and ice. Journal of Geophysical Research 90:901–920.

Delworth, T. L. and K. W. Dixon. 2000. Implications of the recent trend in the Arctic/North Atlantic Oscillation for the North Atlantic thermohaline circulation. Journal of Climate 13:3721–3727.

Delworth, T. L. and M. E. Mann. 2000. Observed and simulated multi-decadal variability in the northern hemisphere. Climate Dynamics 16:661–676.

Delworth, T. L., A. Rosati, R. J. Stouffer, et al. 2006. GFDL's CM2 Global Coupled Climate Models. Part I: Formulation and simulation characteristics. Journal of Climate 19:643–674.

deMenocal, P. B. 1995. Plio-Pleistocene African climate. Science 270:53–59.

deMenocal, P. B. 2001. Cultural responses to climate change during the late Holocene. Science 292:667–673.

deMenocal, P. B., J. Ortiz, T. Guilderson, and M. Sarnthein. 2000. Coherent high- and low-latitude climate variability during the Holocene warm period. Science 288:2198–2202.

Demicco, R. V., T. K. Lowenstein, and L. A. Hardie. 2003. Atmospheric pCO_2 since 60 Ma from records of seawater pH, calcium, and primary carbonate mineralogy. Geology 31:793–796.

Denis, D., X. Crosta, S. Zaragosi, O. Romero, B. Martin, and V. Mas. 2006. Seasonal and subseasonal climate changes recorded in laminated diatom ooze sediments, Adélie Land, East Antarctica. The Holocene 16:1137–1147.

Denman, K. L., G. Brasseur, et al. 2007. Chapter 7: Couplings between changes in the climate system and biogeochemistry. In S. Solomon, et al., eds., *Climate Change 2007: The Physical Science Basis*. Contribution of Working Group I to the Fourth Assessment Report of the Intergovernmental Panel on Climate Change, pp. 499–587. Cambridge: Cambridge University Press.

Denny, C. S. 1974. Pleistocene geology of the northeastern Adirondack region. Professional Paper 786. New York. U.S. Geological Survey.

Denton, G. H., R. B. Alley, G. C. Cromer, and W. S. Broecker. 2005. The role of seasonality in abrupt climate change. Quaternary Science Reviews 24:1159–1182.

Denton, G. H. and C. H. Hendy. 1994. Younger Dryas–age advance of Franz Josef Glacier in the southern Alps of New Zealand. Science 264:1434–1437.

Denton, G. H. and T. J. Hughes, eds. 1981. *The Last Great Ice Sheets*. New York: John Wiley & Sons.

Denton, G. H. and T. J. Hughes. 2002. Reconstructing the Antarctic Ice Sheet at the Last Glacial Maximum. Quaternary Science Reviews 21:193–202.

Denton, G. H. and W. Karlén. 1973. Holocene climatic variations—Their pattern and possible causes. Quaternary Research 3:155–205.

Denton, G. H., T. V. Lowell, C. J. Heusser, et al. 1999. Geomorphology, stratigraphy, and radiocarbon chronology of Llanquihue drift in the area of the Southern Lake District, Seno Reloncaví, and Isla Grande de Chiloé, Chile. Geografiska Annaler Series A Physical Geography 81A:167–229.

Denton, G. H., M. L. Prentice, and L. H. Burckle. 1991. Cainozoic history of the Antarctic Ice Sheet. In R. J. Tingey, ed., *The Geology of Antarctica*, pp. 365–433. Oxford: Clarendon Press.

Denton, G. H., D. E. Sugden, D. R. Marchant, B. L. Hall, and T. I. Wilch. 1993. East Antarctic Ice Sheet sensitivity to Pliocene climatic change from a Dry Valleys perspective. Geografiska Annaler 75A:155–204.

Derry, L. A. and C. France-Lanord. 1996. Neogene Himalayan weathering history and river $^{87}Sr/^{86}Sr$: Impact on the marine Sr record. Earth and Planetary Science Letters 142:59–74.

Deser, C. and M. L. Blackmon. 1993. Surface climate variations over the North Atlantic Ocean during winter: 1900–89. Journal of Climate 6:1743–1753.

Deser, C., A. S. Phillips, and J. W. Hurrell. 2004. Pacific interdecadal climate variability: Linkages between the Tropics and the North Pacific during Boreal Winter since 1900. Journal of Climate 17:3109–3124.

Dethier, D. P., F. Pessl Jr., R. F. Keuler, M. A. Balzarini, and D. R. Pevear. 1995. Late Wisconsinan glaciomarine deposition and isostatic rebound, northern Puget Lowland, Washington. Geological Society of America Bulletin 107:1288–1303.

Dettinger, M. D. and D. R. Cayan. 1995. Large-scale atmospheric forcing of recent trends toward early snowmelt in California. Journal of Climate 8:606–623.

de Vernal, A., M. Henry, J. Matthiessen, et al. 2001. Dinoflagellate cyst assemblages as tracers of sea-surface conditions in the northern North Atlantic, Arctic and sub-Arctic seas: The new "n = 677" database and application for quantitative paleoceanographical reconstruction. Journal of Quaternary Science 16:681–699.

de Vernal, A. and C. Hillaire-Marcel. 2006. Provincialism in trends and high frequency changes in the northwest North Atlantic during the Holocene. Global and Planetary Change 54:263–290.

de Vernal, A. and C. Hillaire-Marcel. 2008. Natural variability of Greenland climate, vegetation, and ice volume during the past million years. Science 320:1622–1625.

de Vernal, A., C. Hillaire-Marcel, and G. Bilodeau. 1996. Reduced meltwater outflow from the Laurentide ice margin during the Younger Dryas. Nature 381:774–777.

de Villiers, S., B. K. Nelson, and A. R. Chivas. 1995. Biological controls on coral Sr/Ca and $\delta^{18}O$ reconstructions of sea surface temperatures. Science 269:1247–1249.

de Vries, H. 1958. Variation in the concentration of radiocarbon with time and location on Earth. Proceedings Koninklijk Nederlands Akademie van Weterschappen, Series B 61: 94–102.

D'Hondt, S. and M. A. Arthur. 1996. Late Cretaceous oceans and the cool tropic paradox. Science 271:1838–1841.

Diaz, H. F. and G. N. Kiladis. 1992. Atmospheric teleconnections associated with the extreme phases on the Southern Oscillation. In H. F. Diaz and V. Markgraf, eds., *El Niño: Historical and Paleoclimatic Aspects of the Southern Oscillation*, pp. 7–28. Cambridge: Cambridge University Press.

Diaz, H. F. and V. Markgraf, eds. 1992. *El Niño: Historical and Paleoclimatic Aspects of the Southern Oscillation*. Cambridge: Cambridge University Press.

Diaz, H. F. and Y. M. Tourre. 2008. Variability in El Niño–Southern Oscillation patterns and potential climate effects. EOS, Transactions, American Geophysical Union 89:220.

Dickens, G. R., M. M. Castillo, and J. G. C. Walker. 1997. A blast of gas in the latest Paleocene: Simulating first-order effects of massive dissociation of oceanic methane hydrate. Geology 25:259–262.

Dickens, G. R., J. R. O'Neil, D. K. Rea, and R. M. Owen. 1995. Dissociation of oceanic methane hydrate as a cause of the carbon isotope excursion at the end of the Paleocene. Paleoceanography 10:965–971.

Dickinson, W. R. 2001. Paleoshoreline record of relative Holocene sea levels on Pacific Islands. Earth-Science Reviews 5:191–234.

Edmond, J. M. and Y. Huh. 2003. Non–steady state carbonate recycling and implications for the evolution of atmospheric PCO_2. Earth and Planetary Science Letters 216:125–139.

Edwards, R. L., J. H. Chen, T.-L. Ku, and G. J. Wasserberg. 1987. Precise timing of the last interglacial period from mass spectrometric determination of Thorium-230 in corals. Science 236:1547–1553.

Ehleringer, J. R., M. D. Dearing, and T. E. Cerling, eds. 2005. *A History of Atmospheric CO₂ and Its Impacts on Plants, Animals, and Ecosystems.* New York: Springer Verlag.

Ehlers, J. and P. L. Gibbard. 2003. Extent and chronology of glaciations. Quaternary Science Reviews 22:1561–1568.

Ehrmann, W. 1998. Implications of late Eocene to early Miocene clay mineral assemblages in McMurdo Sound (Ross Sea, Antarctica) on paleoclimate and ice dynamics. Palaeogeography, Palaeoclimatology, Palaeoecology 139: 213–231.

Ehrmann, W., M. Setti, and L. Marinoni. 2005. Clay minerals in Cenozoic sediments off Cape Roberts (McMurdo Sound, Antarctica) reveal palaeoclimatic history. Palaeogeography, Palaeoclimatology, Palaeoecology 229:187–211.

Eiríksson, J., H. Bára Bartels-Jónsdóttir, A. G. Cage, et al. 2006. Variability of the North Atlantic Current during the last 2000 years based on shelf bottom water and sea surface temperatures along an open ocean/shallow marine transect in western Europe. The Holocene 16:1017–1029.

Eiríksson, J., K. L. Knudsen, H. Haflidason, and P. Henriksen. 2000. Late-glacial and Holocene palaeoceanography of the North Icelandic shelf. Journal of Quaternary Science 15:23–42.

Ekart, D. D., T. E. Cerling, I. P. Montañez, and N. J. Tabor. 1999. A 400 million year carbon isotope record of pedogenic carbonate: Implications for paleoatmospheric carbon dioxide. American Journal of Science 299:805–827.

Elderfield, H. 1986. Strontium isotope stratigraphy. Palaeogeography, Palaeoclimatology, Palaeoecology 57:71–90.

Elderfield, H. and G. Ganssen. 2000. Past temperature and $\delta^{18}O$ of surface ocean waters inferred from foraminiferal Mg/Ca ratios. Nature 405:442–444.

Elderfield, H. and R. E. M. Rickaby. 2000. Oceanic Cd/P ratio and nutrient utilisation in the glacial Southern Ocean. Nature 405:305–310.

Elderfield, H., J. Yuam, P. Anand, T. Kiefer, and B. Nyland. 2006. Calibrations for benthic foraminiferal Mg/Ca paleothermometry and the carbonate ion hypothesis. Earth and Planetary Science Letters 250:633–649.

Eldrett, J. S., I. C. Harding, P. A. Wilson, E. Butler, and A. P. Roberts. 2007. Continental ice in Greenland during the Eocene and Oligocene. Nature 446:176–179.

Eleson, J. W. and T. J. Bralower. 2005. Evidence of changes in surface water temperature and productivity at the Cenomanian/Turonian Boundary. Micropaleontology 51:319–332.

Elias, S. A. 1994. *Quaternary Insects and Their Environments.* Washington, DC: Smithsonian Institution Press.

Elias, S. A., ed. 2006. *Encyclopedia of Quaternary Science.* 4 vols. Amsterdam: Elsevier.

Elkibbi, M. and J. A. Rial. 2001. An outsider's review of the astronomical theory of the climate: Is the eccentricity-driven insolation the main driver of the ice ages? Earth-Science Reviews 56:161–177.

Elliot, M., L. Labeyrie, G. Bond, E. Cortijo, J. L. Turon, N. Tisnerat, and J. C. Duplessy. 1998. Millennial-scale iceberg discharges in the Irminger Basin during the last glacial period: Relationship with the Heinrich events and environmental settings. Paleoceanography 13:433–446.

Elliot, M., L. Labeyrie, and J. C. Duplessy. 2002. Changes in North Atlantic deep-water formation associated with the Dansgaard-Oeschger temperature oscillations (10–60ka). Quaternary Science Reviews 21:1153–1165.

Ellison, C. R. W., M. R. Chapman, and I. R. Hall. 2006. Surface and deep ocean interactions during the cold climate event 8200 years ago. Science 312:1929–1932.

Emanuel, K. 2005. Increasing destructiveness of tropical cyclones over the past 30 years. Nature 436:686–688.

Emile-Geay, J., M. Cane, R. Seager, A. Kaplan, and P. Almasi. 2007. El Niño as a mediator of the solar influence on climate. Paleoceanography 22:PA3210, PA001304.

Emiliani, C. 1955. Pleistocene temperatures. Journal of Geology 63:538–578.

Emiliani, C., S. Gartner, B. Lidz, et al. 1975. Paleoclimatological analysis of late Quaternary cores from the northeastern Gulf of Mexico. Science 189:1083–1088.

Enfield, D. B., A. M. Mestas-Nuñez, and P. J. Trimble. 2001. The Atlantic multi-decadal oscillation and its relation to rainfall and river flows in the continental U.S. Geophysical Research Letters 28:2077–2080.

England, P. and P. Molnar. 1990. Surface uplift, uplift of rocks, and exhumation of rocks. Geology 18:1173–1177.

Enzel, Y., L. L. Ely, S. Mishra, et al. 1999. High-resolution Holocene environmental changes in the Thar Desert, northwestern India. Science 284:125–128.

EPICA Community Members. 2004. Eight glacial cycles from an Antarctic ice core. Nature 429:623–628.

EPICA Community Members. 2006. One-to-one coupling of glacial climate variability in Greenland and Antarctica. Nature 444:195–198.

Epstein, S., R. Buchsbaum, H. A. Lowenstam, and H. C. Urey. 1953. Revised carbonate-water isotopic temperature scale. Geological Society of America Bulletin 64:1315–1326.

Erba, E. 2004. Calcareous nannofossils and Mesozoic oceanic anoxic events. Marine Micropaleontology 52:85–106.

Erbacher, J., B. T. Huber, R. D. Norris, and M. Markey. 2001. Increased thermohaline stratification as a possible cause for an ocean anoxic event in the Cretaceous period. Nature 409:325–327.

Erbacher, J., J. Thurow, and R. Littke. 1996. Evolution patterns of radiolaria and organic matter variations: A new approach to identify sea-level changes in mid-Cretaceous pelagic environments. Geology 24:499–502.

Ericson, D. B. and G. Wollin. 1968. Pleistocene climates and chronology in deep-sea sediments. Science 162:1227–1229.

Esper, J., E. R. Cook, and F. H. Schweingruber. 2002a. Low-frequency signals in long tree-ring chronologies for reconstructing past temperature variability. Science 295:2250–2253.

Esper, J., D. C. Frank, R. J. S. Wilson, and K. R. Briffa. 2005a. Effect of scaling and regression on reconstructed temperature amplitude for the past millennium. Geophysical Research Letters 32:L07711.

Esper, J., D. C. Frank, R. J. S. Wilson, U. Buntgen, and K. Treydte. 2007. Uniform growth trends among central Asian low-and high-elevation juniper tree sites. Trees 21:141–150.

Esper, J., F. H. Schweingruber, and M. Winiger. 2002b. 1300 years of climatic history for Western Central Asia inferred from tree-rings. The Holocene 12:267–277.

Esper, J., R. J. S. Wilson, D. C. Frank, et al. 2005b. Climate: Past ranges and future changes. Quaternary Science Reviews 24:2164–2166.

Etheridge, D. M., L. P. Steele, R. J. Francey, and R. L. Langenfelds. 1998. Atmospheric methane between 1000 AD and present: Evidence of anthropogenic emissions and climatic variability. Journal of Geophysical Research. 103:15979–15993.

Etheridge, D. M., L. P. Steele, R. L. Langenfields, R. J. Francey, J.-M. Barnola, and V. I. Morgan. 1996. Natural and anthropogenic changes in atmospheric CO_2 over the last 1000 years from air in Antarctic ice and firn. Journal of Geophysical Research 101:4115–4128.

Etienne, J. L., P. A. Allen, R. Rieu, and E. Le Guerroué. 2007. Neoproterozoic glaciated basins: A critical review of the "Snowball Earth" hypothesis by comparison with Phanerozoic glaciations. In M. J. Hambrey, et al., eds., Glacial Processes and Products. International Association of Sedimentologists, Special Publications. New York: John Wiley.

Evans, D. J. A., C. D. Clark, and W. A. Mitchell. 2005. The last British Ice Sheet: A review of the evidence utilised in the compilation of the Glacial Map of Britain. Earth-Science Reviews 70:253–312.

Evans, M. N., M. A. Cane, D. P. Schrag, et al. 2001b. Support for tropically-driven Pacific decadal variability based on paleo-proxy evidence. Geophysical Research Letters 28: 3689–3692.

Evans, M. N., A. Kaplan, and M. A. Cane. 2002. Pacific sea surface temperature field reconstruction from coral delta ^{18}O data using reduced space objective analysis. Paleoceanography 17:PA000590.

Evans, M. N., A. Kaplan, M. A. Cane, and R. Villalba. 2001a. Globality and optimality in climate field reconstructions from proxy data. In V. Markgraf, ed., Inter-hemispheric Climate Linkages, pp. 53–72. Cambridge: Cambridge University Press.

Exon, N. F., J. P. Kennett, M. J. Malone, et al., eds. 2001. Proceedings Ocean Drilling Program Initial Reports, vol. 189. College Station, TX: Ocean Drilling Program.

Exon, N. F., J. P. Kennett, M. J. Malone, et al. 2002. Drilling reveals climatic consequences of Tasmanian Gateway opening. EOS, Transactions, American Geophysical Union 83:253–259.

Eyles, N. and N. Januszczak. 2004. "Zipper-rift": A tectonic model for Neoproterozoic glaciations during the breakup of Rodinia after 750 Ma. Earth-Science Reviews 65:1–73.

Eynaud, F., S. Zaragosi, J. Scourse, et al. 2007. Deglacial laminated facies on the NW European continental margin: The hydrographic significance of British-Irish Ice Sheet deglaciation and Fleuve Manche paleoriver discharges. Geochemistry, Geophysics, Geosystems 8:GC001496.

Fairbanks, R. and R. K. Matthews. 1978. The marine oxygen isotope record in Pleistocene coral, Barbados, West Indies. Quaternary Research 10:181–196.

Fairbanks, R. G. 1989. A 17,000-year glacio-eustatic sea level record: Influence of glacial melting rates on the Younger Dryas event and deep-ocean circulation. Nature 342:637–642.

Fairbanks, R. G. and R. E. Dodge. 1979. Annual periodicity of the $^{18}O/^{16}O$ and $^{13}C/^{12}C$ ratios in the coral Montastrea annularis. Geochimica et Cosmochimica Acta 43:1009–1020.

Fairbanks, R. G., R. A. Morlock, T.-C. Chiu, et al. 2005. Radiocarbon calibration curve spanning 0 to 50,000 years BP based on paired $^{230}Th/^{234}U/^{238}U$ and ^{14}C dates on pristine corals. Quaternary Science Reviews 24:1781–1796.

Fairbridge, R. W. 1961. Eustatic changes in sea-level. Physics and Chemistry of the Earth 4:99–185.

Fairchild, I. J. and M. J. Kennedy. 2007. Neoproterozoic glaciation in the earth system. Journal of the Geological Society 164:895–921.

Fairchild, I. J., C. L. Smith, A. Baker, et al. 2006. Modification and preservation of environmental signals in speleothems. Earth-Science Reviews 75:105–153.

Falkowski, P. G., M. E. Katz, A. J. Milligan, et al. 2005. The rise of oxygen over the past 205 million years and the evolution of large placental mammals. Science 309:2202–2204.

Fanning, C. M. and P. K. Link. 2004. U-Pb SHRIMP ages of Neoproterozoic (Sturtian) glaciogenic Pocatello Formation, southeastern Idaho. Geology 32:881–884.

Farley, K. A. and D. B. Patterson. 1995. A 100-kyr periodicity in the flux of extraterrestrial $_3He$ to the sea floor. Nature 378:600–603.

Farmer, G. L., D. Barber, and J. Andrews. 2003. Provenance of late Quaternary ice-proximal sediments in the North Atlantic: Nd, Sr and Pb isotopic evidence. Earth and Planetary Science Letters 209:227–243.

Farrell, J. W., T. F. Pedersen, S. E. Calvert, and B. Nielsen. 1995. Glacial-interglacial changes in nutrient utilization in the equatorial Pacific Ocean. Nature 377:514–516.

Fassell, M. L., and T. J. Bralower. 1999. Warm, equable mid-Cretaceous: Stable isotope evidence. In E. Barrera and C. C. Johnson, eds., The Evolution of the Cretaceous Ocean Climate System, pp. 121–142. Boulder, CO: Geological Society of America.

Federov, A. V., P. S. Dekens, M. McCarthy, et al. 2006. The Pliocene paradox (mechanisms for a permanent El Niño). Science 312:1485–1489.

Federov, A. V. and S. G. Philander. 2000. Is El Niño changing? Science 288:1997–2002.

Feely, R. A., C. L. Sabine, J. M. Hernandez-Ayon, D. Ianson, and B. Hales. 2008. Evidence for upwelling of corrosive "acidified" water onto the continental shelf. Science 320:1490–1492.

Feely, R. A., C. L. Sabine, K. Lee, et al. 2004. Impact of anthropogenic CO_2 on the $CaCO_3$ system in the oceans. Science 305:362–366.

Feng, Z.-D., C. B. An, and H. B. Wang. 2006. Holocene climatic and environmental changes in the arid and semi-arid areas of China: A review. The Holocene 16:119–130.

Ferguson, J. E., G. M. Henderson, M. Kucera, and R. E. M. Rickaby. 2008. Systematic change of foraminiferal Mg/Ca ratios across a strong salinity gradient. Earth and Planetary Science Letters 265:153–166.

Fetterer, F., K. Knowles, W. Meier, and M. Savoie. 2002. Sea Ice Index. Boulder, CO: National Snow and Ice Data Center. Digital media (updated 2008).

Field, D. B., T. R. Baumgartner, C. D. Charles, V. Ferreira-Bartrina, and M. D. Ohman. 2006. Planktonic foraminifera of the California current reflect 20th-century warming. Science 311:63–66.

Finney, B. P., C. A. Schlotz, T. C. Johnson, S. Trumbore, and J. Southon. 1996. Late Quaternary lake level change of Lake Malawi. In T. C. Johnson and E. O. Odada, eds., The Limnology,

Chronology, and Paleoclimatology of East African Lakes, pp. 495–508. Toronto: Gordon and Breach.

Firestone, R. B., A. West, J. P. Kennett, et al. 2007. Evidence for an extraterrestrial impact 12,900 years ago that contributed to the megafaunal extinctions and the Younger Dryas cooling. Proceedings of the National Academy of Sciences 104: 16016–16021.

Fischer, A. and D. J. Bottjer. 1991. Orbital forcing and sedimentary sequences. Journal of Sedimentary Petrology 61: 1063–1069.

Fischer, A. F. 1981. Climatic oscillations in the biosphere. In M. H. Nitecki, ed., *Biotic Crises in Ecological and Evolutionary Time,* pp. 103–131. New York: Academic Press.

Fischer, A. G. and M. A. Arthur. 1977. Secular variations in the pelagic realm, in Deepwater Carbonate Environments. Society of Economic Paleontologists and Mineralogists Special Publication 25:19–50 (H. E. Cook and P. Enos, eds.).

Fischer, G. and G. Wefer, eds. 1999. *Uses of Proxies in Paleoceanography: Examples from the South Atlantic.* Berlin: Springer-Verlag.

Fischer, H., T. Kumke, G. Lohmann, et al., eds. 2004. *The Climate in Historical Times: Towards a Synthesis of Holocene Proxy Data and Climate Models.* Berlin: Springer-Verlag.

Fischer, H. and B. Mieding. 2005. A 1,000-year ice core record of interannual to multidecadal variations in atmospheric circulation over the North Atlantic. Climate Dynamics 25:65–74.

Fischer, H., M.-L. Siggaard-Andersen, U. Ruth, R. Rothlisberger, and E. Wolff. 2007. Glacial/interglacial changes in mineral dust and sea-salt records in polar ice cores: Sources, transport, and deposition. Reviews of Geophysics 45:RG1002.

Fischer, H., M. Wahlen, J. Smith, D. Mastroianni, and B. Deck. 1999. Ice core records of atmospheric CO_2 around the last three glacial terminations. Science 283:1712–1714.

Fisher, D. A., R. M. Koerner, J. C. Bourgeois, et al. 1998. Penny Ice Cap cores, Baffin Island, Canada, and the Wisconsinan Foxe Dome connection: Two states of Hudson Bay ice cover. Science 279:692–695.

Fisher, D. A., R. M. Koerner, and N. Reeh. 1995. Holocene climatic records from Agassiz Ice Cap, Ellesmere Island, NWT, Canada. The Holocene 5:19–24.

Fisher, T. G. 2003. Chronology of glacial Lake Agassiz meltwater routed to the Gulf of Mexico. Quaternary Research 59: 271–276.

Fisher, T. G., T. V. Lowell, and H. M. Loope. 2006. Comment on "Alternative routing of Lake Agassiz overflow during the Younger Dryas: New dates, paleotopography, and a re-evaluation" by Teller et al. (2005). Quaternary Science Reviews 25:1137–1141.

Fisher, T. G., D. G. Smith, and J. T. Andrews. 2002. Preboreal oscillation caused by a glacial Lake Agassiz flood. Quaternary Science Reviews 21:873–878.

Fleitmann, D., S. J. Burns, A. Mangini, et al. 2007. Holocene ITCZ and Indian monsoon dynamics recorded in stalagmites from Oman and Yemen (Socotra). Quaternary Science Reviews 26:170–188.

Fleitmann, D., S. J. Burns, M. Mudelsee, et al. 2003. Holocene forcing of the Indian Monsoon recorded in a stalagmite from southern Oman. Science 300:1737–1739.

Fleitmann, D., S. J. Burns, U. Neff, et al. 2004. Palaeoclimatic interpretation of high-resolution oxygen isotope profiles derived from annually laminated speleothems from Southern Oman. Quaternary Science Reviews 23:935–945.

Fleming, K., P. Johnston, D. Zwartz, et al. 1998. Refining the eustatic sea-level curve since the Last Glacial Maximum using far- and intermediate-field sites. Earth and Planetary Science Letters 163:327–342.

Florindo, F., A. K. Cooper, and P. E. O'Brien. 2003. Introduction to "Antarctic Cenozoic paleoenvironments: Geologic record and models." Palaeogeography, Palaeoclimatology, Palaeoecology 198:1–9.

Flower, B. P., D. W. Hastings, H. W. Hill, and T. M. Quinn. 2004. Phasing of deglacial warming and Laurentide Ice Sheet meltwater in the Gulf of Mexico. Geology 32:597–600.

Flower, B. P. and J. P. Kennett. 1990. The Younger Dryas cool episode in the Gulf of Mexico. Paleoceanography 5:949–961.

Flower, B. F. and J. P. Kennett. 1993a. Middle Miocene ocean–climate transition: High resolution oxygen and carbon isotopic records from DSDP Site 588A, southwest Pacific. Paleoceanography 8:811–843.

Flower, B. F. and J. P. Kennett. 1993b. Relations between Monterey Formation deposition and middle Miocene global cooling: Naples Beach section, California. Geology 21:877–880.

Flower, B. F. and J. P. Kennett. 1994. The middle Miocene climatic transition: East Antarctic Ice Sheet development, deep ocean circulation and global carbon cycling. Palaeogeography, Palaeoclimatology, Palaeoecology 108:537–555.

Flower, B. F. and J. P. Kennett. 1995. Middle Miocene Deepwater Paleoceanography in the Southwest Pacific: Relations with East Antarctic Ice Sheet development. Paleoceanography 10:1095–1112.

Flower, B. P., J. C. Zachos, and H. Paul. 1997. Milankovitch-scale variability recorded near the Oligocene/Miocene boundary: Hole 929A. Proceedings Ocean Drilling Program Leg 154: 433–439.

Flückiger, J., T. Blunier, B. Stauffer, et al. 2004. N_2O and CH_4 variations during the last glacial epoch: Insight into global processes. Global Biogeochemical Cycles 18:GB1020.

Flückiger, J., A. Dällenbach, T. Blunier, et al. 1999. Variations in atmospheric N_2O concentration during abrupt climatic changes. Science 285:227–230.

Flückiger, J., E. Monnin, B. Stouffer, et al. 2002. High-resolution Holocene N_2O ice core record and its relationship with CH_4 and CO_2. Global Biogeochemical Cycles 16:1010, GB001417.

Fogt, R. J. and D. H. Bromwich. 2006. Decadal variability of the ENSO teleconnection to the high-latitude South Pacific governed by coupling with the Southern Annular Mode. Journal of Climate 19:979–997.

Foley, J. A., R. DeFries, G. P. Asner, et al. 2005. Global consequences of land use. Science 309:570–574.

Foley, J. A., I. C. Prentice, N. Ramankutty, et al. 1996. An integrated biosphere model of land surface processes, terrestrial carbon balance and vegetation dynamics. Global Biogeochemical Cycles 10:603–628.

Folland, C. K., D. E. Parker, A. W. Colman, and R. Washington. 1999. Large scale modes of ocean surface temperature since the late nineteenth century. In A. Navarra, ed., *Beyond El Niño: Decadal and Interdecadal Climate Variability,* pp. 73–102. New York: Springer-Verlag.

Folland, C. K., N. A. Rayner, S. J. Brown, et al. 2001. Global temperature change and its uncertainties since 1861. Geophysical Research Letters 28:2621–2624.

Folland, C. K., J. A. Renwick, M. J. Salinger, and A. B. Mullan. 2002. Relative influences of the IPO and ENSO on the South Pacific Convergence Zone. Geophysical Research Letters 29:GL014201.

Fontanier, C., A. Mackensen, F. J. Jorissen, et al. 2006. Stable oxygen and carbon isotopes of live benthic foraminifera from the Bay of Biscay: Microhabitat impact and seasonal variability. Marine Micropaleontology 58:159–183.

Forest, C. E., P. H. Stone, A. P. Sokolov, M. R. Allen, and M. D. Webster. 2002. Quantifying uncertainties in climate system properties with the use of recent climate observations. Science 295:113–117.

Forman, S. L., D. J. Lubinski, O. Ingólfsson, et al. 2004. A review of postglacial emergence on Svalbard, Franz Josef Land and Novaya Zemlya, northern Eurasia. Quaternary Science Reviews 23:1391–1434.

Forman, S. L., R. Oglesby, and R. S. Webb. 2001. Temporal and spatial patterns of Holocene dune activity on the Great Plains of North America: Megadroughts and climate links. Global and Planetary Change 29:1–29.

Forster, A., S. Schouten, M. Baas, and J. S. Sinninghe Damste. 2007a. Mid-Cretaceous (Albian Santonian) sea surface temperature record of the tropical Atlantic Ocean. Geology 35:919–922.

Forster, P., V. Ramaswamy, et al. 2007b. Chapter 2: Changes in atmospheric constituents and inradiative forcing. In S. Solomon, et al., eds., *Climate Change 2007: The Physical Science Basis*. Contribution of Working Group I to the Fourth Assessment Report of the Intergovernmental Panel on Climate Change, pp. 129–234. Cambridge: Cambridge University Press.

Foukal, P., C. Fröhlich, H. Spruit, and T. M. L. Wigley Jr. 2006. Variations in solar luminosity and their effect on the earth's climate. Nature 443:161–166.

Fowler, A. M. 2008. ENSO history recorded in *Agathis australis* (kauri) tree rings. Part B: 423 years of ENSO robustness. International Journal of Climatology 28:21–35.

Frakes, L. A., J. E. Francis, and J. I. Syktus. 1992. *Climate Modes of the Phanerozoic*. New York: Cambridge University Press.

Francis, J. A., E. Hunter, J. R. Key, and X. Wang. 2005. Clues to variability in Arctic minimum sea ice extent. Geophysical Research Letters 32:L21501, GL024376.

François, R., M. A. Altabet, E.-F. Yu, et al. 1997. Contribution of Southern Ocean surface-water stratification to low atmospheric CO_2 concentrations during the last glacial period, Nature 389:929–935.

Frank, D. and J. Esper. 2005. Temperature reconstructions and comparisons with instrumental data from a tree-ring network for the European Alps. International Journal of Climatology 25:1437–1454.

Frank, D., J. Esper, and E. R. Cook. 2007. Adjustment for proxy number and coherence in a large-scale temperature reconstruction. Geophysical Research Letters 34:L16709.

Frank, M. 2002. Radiogenic isotopes: Tracers of past ocean circulation and erosional input. Reviews of Geophysics 40:1–38.

Frank, M., J. Backman, M. Jakobsson, et al. 2008. Beryllium isotopes in central Arctic Ocean sediments over the past 12.3 million years: Stratigraphic and paleoclimatic implications. Paleoceanography 23:PA1S02.

Frank, M., J.-D. Eckhardt, A. Eisenhauer, et al. 1994. Beryllium 10, thorium 230, and protactinium 231 in Galapagos microplate sediments: Implications of hydrothermal activity and paleoproductivity changes during the last 100,000 years. Paleoceanography 9:559–578.

Frank, M., R. Gersonde, and M. R. van der Loeff. 2000. Similar glacial and interglacial export bioproductivity in the Atlantic sector of the Southern Ocean: Multiproxy evidence and implications for glacial atmospheric CO_2. Paleoceanography 15:642–658.

Frank, M. and R. K. O'Nions. 1998. Souces of Pb for Indian Ocean ferromanganese crusts: A record of Himalayan erosion? Earth and Planetary Science Letters 158:121–130.

Frank, M., R. K. O'Nions, J. R. Hein, and V. K. Banakar. 1999a. 60 Ma records of major elements and Pb-Nd isotopes from hydrogenous ferromanganese crusts: Reconstruction of seawater paleochemistry. Geochimica et Cosmochimica Acta 63:1689–1708.

Frank, M., B. C. Reynolds, and R. K. O'Nions. 1999b. Nd and Pb isotopes in the Atlantic and Pacific water masses before and after closure of the Panama gateway. Geology 27:1147–1150.

Frank, M., T. van de Flierdt, A. N. Halliday, et al. 2003. Evolution of deepwater mixing and weathering inputs in the central Atlantic Ocean over the past 33 Myr. Paleoceanography 18:1091.

Frank, M., N. Whiteley, T. van de Flierdt, B. C. Reynolds, and K. O'Nions. 2006. Nd and Pb isotope evolution of deep water masses in the eastern Indian Ocean during the past 33 Myr. Chemical Geology 226:264–279.

Franzi, D. A., J. A. Rayburn, P. K. L. Knuepfer, and T. M. Cronin. 2007. *Late Quaternary History of Northeastern New York and Adjacent Parts of Vermont and Quebec. 70th Annual Northeast Friends of the Pleistocene Guidebook*. Plattsburgh, NY.

Free, M. and A. Robock. 1999. Global warming in the context of the Little Ice Age. Journal of Geophysical Research 104:19057–19070.

Freeman, K. H. and J. M. Hayes. 1992. Fractionation of carbon isotopes by phytoplankton and estimates of ancient CO_2 levels. Global Biogeochemical Cycles 6:185–198.

Fricker, H. A., T. Scambos, R. Bindschadler, and L. Padman. 2007. An active subglacial water system in West Antarctica mapped from space. Science 315:1544–1548.

Friis-Christensen, E. and K. Lassen. 1991. Length of the solar cycle: An indicator of solar activity closely associated with climate. Science 254:698–700.

Fritts, H. C. 1991. *Reconstructing Large-Scale Climatic Patterns from Tree Ring Data*. Tucson: University of Arizona Press.

Fritz, S. C. 2008. Deciphering climatic history from lake sediments. Journal of Paleolimnology 39:5–16.

Fröhlich, C. and J. Lean. 2004. Solar radiative output and its variability: Evidence and mechanisms. The Astronomy and Astrophysics Reviews 12:273–320.

Fronval, T. and E. Jansen. 1996. Rapid changes in ocean circulation and heat flux in the Nordic seas during the last interglacial period. Nature 383:806–810.

Gagan, M. K., L. K. Ayliffe, J. W. Beck, et al. 2000. New views of tropical paleoclimates from corals. Quaternary Science Reviews 19:45–64.

Gagan, M. K., L. K. Ayliffe, D. Hopley, et al. 1998. Temperature and surface-ocean water balance of the mid-Holocene tropical western Pacific. Science 279:1014–1018.

Gagan, M. K., A. R. Chivas, and P. J. Isdale. 1994. High resolution isotopic records from corals using ocean temperatures and mass-spawning chronometers. Earth and Planetary Science Letters 121:549–558.

Gagan, M. K., E. J. Hendy, S. G. Haberle, and W.S. Hantoro. 2004. Post-glacial evolution of the Indo-Pacific Warm Pool and El Niño–Southern Oscillation. Quaternary International 118/119:127–143.

Gaillardet, J. and J. C. Allègre. 1995. Boron isotopic composition of corals: Seawater or diagenesis record? Earth and Planetary Science Letters 136:665–676.

Gallup, C., R. L. Edwards, and R. G. Johnson. 1994. The timing of high sea levels over the past 200,000 years. Science 263: 796–800.

Ganachaud, A. and C. Wunsch. 2000. Improved estimates of global ocean circulation, heat transport and mixing from hydrographic data. Nature 408:453–457.

Ganachaud, A. and C. Wunsch. 2003. Large-scale ocean heat and freshwater transports during the World Ocean Circulation Experiment. Journal of Climate 16:696–705.

Ganopolski, A. and S. Rahmstorf. 2002. Abrupt glacial climate changes due to stochastic resonance. Physical Review Letters 88:038501.

Garabato, A. C. N., K. L. Polzin, B. A. King, K. J. Heywood, and M. Visbeck. 2004. Widespread intense turbulent mixing in the Southern Ocean. Science 303:210–213.

Gard, G. 1993. Late Quaternary coccoliths at the North Pole: Evidence of ice-free conditions and rapid sedimentation in the central Arctic Ocean. Geology 21:227–230.

Garzione, C. N., G. D. Hoke, J. C. Libarkin, et al. 2008. Rise of the Andes. Science 320:1304–1307.

Gasse, F. 2000. Hydrological changes in the African tropics since the Last Glacial Maximum. Quaternary Science Reviews 19:189–211.

Gasse, F. and E. Van Campo. 1994. Abrupt post-glacial climate events in West Asia and North Africa monsoon domains. Earth and Planetary Science Letters 126:435–456.

Gat, J. R. 1996. Oxygen and hydrogen isotopes in the hydrologic cycle. Annual Review of Earth and Planetary Sciences 24: 225–262.

Gedalof, Z. and N. Mantua. 2002. A multi-century perspective of variability in the Pacific Decadal Oscillation: New insights from tree rings and corals. Geophysical Research Letters 29:2204, GL015824.

Gedalof, Z. and D. J. Smith. 2001. Interdecadal climate variability and regime-scale shifts in Pacific North America. Geophysical Research Letters 28:1515–1518.

Gehrels, W. R., D. F. Belnap, S. Black, and R. M. Newnham. 2002. Rapid sea-level rise in the Gulf of Maine, USA, since AD 1800. The Holocene 12:383–389.

Gehrels, W. R., J. R. Kirby, A. Prokoph, et al. 2005. Onset of recent rapid sea-level rise in the western Atlantic Ocean. Quaternary Science Reviews 24:2083–2100.

Gehrels, W. R., W. A. Marshall, M. J. Gehrels, et al. 2006. Rapid sea-level rise in the North Atlantic Ocean since the first half of the 19th century. The Holocene 16:948–964.

Gehrels, W. R., G. A. Milne, J. R. Kirby, R. T. Patterson, and D. F. Belknap. 2004. Late Holocene sea-level changes and isostatic crustal movements in Atlantic Canada. Quaternary International 120:79–89.

Geirsdóttir, A., J. Hardardóttir, and J. T. Andrews. 2000. Late-Holocene terrestrial glacial history of Miki and I. C. Jacobsen Fjords, East Greenland. The Holocene 10:123–134.

Genthon, C., J.-M. Barnola, D. Raynaud, et al. 1987. Vostok ice core: Climatic response to CO_2 and orbital forcing changes over the last climatic cycle. Nature 329:414–418.

Genty, D., D. Blamart, R. Ouahdi, et al. 2003. Precise dating of Dansgaard-Oeschger climate oscillations in western Europe from stalagmite data. Nature 421:833–837.

Gergis, J. L. and A. M. Fowler. 2006. How unusual was late 20th century El Niño–Southern Oscillation (ENSO)? Assessing evidence from tree-ring, coral, ice-core and documentary palaeoarchives, A.D. 1525–2002. Advances in Geoscience 6:173–179.

Gersonde, R., A. Abelmann, U. Brathauer, et al. 2003. Last glacial sea surface temperatures and sea-ice extent in the Southern Ocean (Atlantic-Indian sector): A multiproxy approach. Paleoceanography 18:1061, PA000809.

Gersonde, R., X. Crosta, A. Abelman, and L. Armand. 2005. Sea-surface temperature and sea ice distribution of the Southern Ocean at the EPILOG Last Glacial Maximum—A circum-Antarctic view based on siliceous microfossil records. Quaternary Science Reviews 24:869–896.

Ghil, M. and S. Childress. 1987. *Topics in Geophysical Fluid Dynamics: Atmospheric Dynamics, Dynamo Theory and Climate Dynamics*. New York: Springer-Verlag.

Ghosh, P., C. N. Garzione, and J. M. Eiler. 2006. Rapid uplift of the Altiplano revealed through ^{13}C-^{18}O bonds in paleosol carbonates. Science 311:511–515.

Gibbs, M. T., E. J. Barron, and L. R. Kump. 1997. An atmospheric pCO_2 threshold for glaciation in the late Ordovician. Geology 25:447–450.

Gibbs, S. J., T. J. Bralower, P. R. Bown, J. C. Zachos, and L. M. Bybell. 2006. Shelf and open-ocean calcareous phytoplankton assemblages across the Paleocene-Eocene Thermal Maximum: Implications for global productivity gradients. Geology 34:233–236.

Gibson, J. A. E., T. Trull, P. D. Nichols, R. E. Summons, and A. McMinn. 2003. Sedimentation of ^{13}C-rich organic matter from Antarctic sea-ice algae: A potential indicator of past sea-ice extent. Geology 27:331–334.

Gilbert, G. K. 1890. *Lake Bonneville*. U.S. Geological Survey Memoir 1.

Gilbert, G. K. 1895. Sedimentary measurement of geologic time. Journal of Geology 3:121–127.

Gildor, H. and E. Tziperman. 2000. Sea ice as the glacial cycles' climate switch: Role of seasonal and orbital forcing. Paleoceanography 15:605–615.

Gildor, H. and E. Tziperman. 2001. A sea-ice switch mechanism for the 100 kyr glacial cycles. Journal of Geophysical Research (Oceans) 106(C5):9117–9133.

Gillespie, A. and P. Molnar. 1995. Asynchronous maximum advances of mountain and continental glaciers. Reviews of Geophysics 33:311–364.

Gillett, N. P., H. F. Graf, and T. J. Osborn. 2003. Climate change and the North Atlantic Oscillation. In J. W. Hurrell, Y. Kushnir, G. Ottersen, and M. Visbeck, eds., *The North Atlantic Oscillation: Climatic Significance and Environmental Impact*, pp. 193–209. Washington, DC: American Geophysical Union.

Giraudeau, J., A. E. Jennings, and J. T. Andrews. 2004. Timing and mechanisms of surface and intermediate water circulation changes in the Nordic Seas over the last 10000 cal. years: A view from the North Iceland shelf. Quaternary Science Reviews 23:2127–2139.

Gleissberg, W. 1966. Ascent and descent in the eighty-year cycles of solar activity. Journal of the British Astronomical Association 76:265–270.

Glueck, M. F. and C. W. Stockton. 2001. Reconstruction of the North Atlantic Oscillation, 1429–1983. International Journal of Climatology 21:1453–1465.

Goddéris, Y., Y. Donnadieu, C. Dessert, et al. 2007. Coupled modeling of global carbon cycle and climate in the Neoproterozoic: Links between Rodinia breakup and major glaciations. Compte Rendue Geoscience 339:212–222.

Goddéris, Y. and L. M. François. 1995. The Cenozoic evolution of the strontium and carbon cycles: Relative importance of continental erosion and mantle exchange. Chemical Geology 126:169–190.

Goddéris, Y. and L. M. François. 1995. The Cenozoic evolution of the strontium and carbon cycles: Relative importance of continental erosion and mantle exchanges. Chemical Geology 126:169–190.

Goldenberg, S. B., C. W. Landsea, A. M. Mestas Nuñez, and W. M. Gray. 2001. The recent increase in Atlantic hurricane activity: Causes and implications. Science 293:474–479.

Goldstein, S. L., and S. R. Hemming. 2003. Long-lived isotopic tracers in oceanography, paleoceanography, and ice-sheet dynamics. In H. D. Holland, et al., eds., *Treatise on Geochemistry: The Oceans and Marine Geochemistry*, pp. 453–489. Amsterdam: Elsevier.

Golonka, J. and D. W. Ford. 2000. Pangean (late Carboniferous–middle Jurassic) paleoenvironment and lithofacies. Palaeogeography, Palaeoclimatology, Palaeoecology 161:1–34.

Gomez, B., L. Carter, N. A. Trustrum, A. S. Palmer, and A. P. Roberts. 2004. El Niño–Southern Oscillation signal associated with middle Holocene climate change in intercorrelated terrestrial and marine sediment cores, North Island, New Zealand. Geology 32:653–656.

Gonzalez, J. L. and T. E. Törnqvist. 2006. Coastal Louisiana in crisis: Subsidence or sea level rise. EOS, Transactions, American Geophysical Union 87:493–508.

Goodkin, N. F., K. A. Hughen, A. L. Cohen, and S. R. Smith. 2005. The Little Ice Age at Bermuda from a growth-dependent calibration of Coral Sr/Ca. Paleoceanography 20 PA4016, PA001140.

Goodman, J. C. 2006. Through thick and thin: Marine and meteoric ice in a "Snowball Earth" climate. Geophysical Research Letters 33:L16701.

Goosse, H., O. Arzel, J. Luterbacher, et al. 2006. The origin of the European "Medieval Warm Period." Climate of the Past 2:99–113.

Goosse, H. and T. Fichefet. 1999. Importance of ice–ocean interactions for the global ocean circulation: A model study. Journal of Geophysical Research 104:23337–23355.

Goosse, H., V. Masson-Delmotte, H. Renssen, et al. 2004. A late medieval warm period in the Southern Ocean as a delayed response to external forcing? Geophysical Research Letters 31:L06203.

Goosse, H. and H. Renssen. 2005. Simulating the evolution of the ice cover in the Southern Ocean during the last 150 years: Implications of the long memory of the ocean. International Journal of Climatology 25:569–579.

Goosse, H., H. Renssen, A. Timmermann, and R. S. Bradley. 2005. Internal and forced climate variability during the last millennium: A model-data comparison using ensemble simulations. Quaternary Science Reviews 24:1345–1360.

Gornitz, V., ed. 2009. *Encyclopedia of Paleoclimatology and Ancient Environments*. New York: Springer.

Gouretski, V. and K. P. Koltermann. 2007. How much is the ocean really warming? Geophysical Research Letters 34:L01610.

Gradstein, F., J. Ogg, and A. Smith, eds. 2004. *A Geologic Time Scale 2004*. Cambridge: Cambridge University Press.

Gray, S. T., L. J. Graumlich, J. L. Betancourt, and G. T. Pederson. 2004. A tree-ring based reconstruction of the Atlantic Multidecadal Oscillation since 1567 A.D. Geophysical Research Letters 31:L12205.

Grebmeier, J. M., J. E. Overland, S. E. Moore, et al. 2006. A major ecosystem shift in the Northern Bering Sea. Science 311: 1461–1464.

Gregory, J. M., R. J. Stouffer, S. C. B. Raper, P. A. Stott, and N. A. Raynor. 2002. An observationally based estimate of the climate sensitivity. Journal of Climate 15:3117–3121.

Grimm, E. C., G. L. Jacobson Jr., W. A. Watts, B. C. S. Hansen, and K. A. Maasch. 1993. A 50,000-year record of climate oscillations from Florida and its temporal correlation with the Heinrich Events. Science 261:198–200.

Grootes, P. M., M. Stuiver, J. W. C. White, S. Johnsen, and J. Jouzel. 1993. Comparison of oxygen isotopic records from the GISP2 and GRIP Greenland ice cores. Nature 366:552–554.

Grottoli, A. G. and C. M. Eakin. 2007. A review of modern coral $\delta^{18}O$ and ^{14}C proxy records. Earth-Science Reviews 81:67–91.

Grotzinger, J. P. 1990. Geochemical model for Proterozoic stromatolite decline. American Journal of Science 290A:80–103.

Grousset, F. E., L. Labeyrie, J. A. Sinko, et al. 1993. Patterns of ice-rafted detritus in the glacial north Atlantic (40–55°N). Paleoceanography 8:175–192.

Grousset, F. E., M. Parra, A. Bory, et al. 1998. Saharian wind regimes traced by the Sr-Nd isotopic composition of the subtropical Atlantic sediments: Last glacial maximum vs. today. Quaternary Science Reviews 17:395–409.

Grousset, F. E., C. Pujol, L. Labeyrie, G. Auffret, and A. Boelaert. 2000. Were the North Atlantic Heinrich events triggered by the behaviour of the European ice sheets? Geology 28:123–126.

Grove, J. M. 1988. *The Little Ice Age*. London: Methuen.

Guilderson, T. P., R. G. Fairbanks, and J. L. Rubenstone. 1994. Tropical temperature variations since 20,000 years ago: Modulating interhemispheric climate change. Science 263:663–665.

Guilderson, T. P., D. P. Schrag, and M. A. Cane. 2004. Surface water mixing in the Solomon Sea as documented by a high-resolution coral ^{14}C record. Journal of Climate 17:1147–1156.

Guiot, J., J. J. Boreux, P. Braconnot, F. Torre, and PMIP participating groups. 1999. Data-model comparisons using fuzzy logic in palaeoclimatology. Climate Dynamics 15:569–581.

Guo, Z., S. Peng, Q. Hao, P. E. Biscaye, Z. An, and T. Liu. 2004. Late-Miocene-Pliocene development of the Asian aridification as recorded in the Red-Earth Formation in northern China. Global and Planetary Change 44:135–145.

Guo, Z., W. F. Ruddiman, Q. Z. Hao, et al. 2002. Onset of Asian desertification by 22 Myr ago inferred from loess deposits in China. Nature 416:159–163.

Gwiazda, R. H., S. R. Hemming, and W. S. Broecker. 1996a. Tracking the sources of icebergs with lead isotopes: The provenance of ice-rafted debris in Heinrich layer 2. Paleoceanography 11:77–93.

Gwiazda, R. H., S. R. Hemming, and W. S. Broecker. 1996b. Provenance of icebergs during Heinrich event 3 and the contrast to their sources during other Heinrich episodes. Paleoceanography 11:371–378.

Haake, F. W. and U. Pflaumann. 1989. Late Pleistocene foraminiferal stratigraphy on the Vøring Plateau, Norwegian Sea. Boreas 18:343–356.

Haeberli, W. and H. Holzhauser. 2003. Alpine glacier mass changes during the past two millennia. PAGES Newsletter 11:13–15.

Hagelberg, T. K., G. Bond, and P. deMenocal. 1994. Milankovitch band forcing of sub-Milankovitch climate variability during the Pleistocene. Paleoceanography 9:545–558.

Häkkinen, S. and P. B. Rhines. 2004. Decline of subpolar North Atlantic circulation during the 1990s. Science 304:555–559.

Hald, M., C. Andersson, H. Ebbesen, et al. 2007. Variations in temperature and extent of Atlantic water in the northern North Atlantic during the Holocene. Quaternary Science Reviews 26:3423–3440.

Hald, M., H. Ebbesen, M. Forwick, et al. 2004. Holocene paleoceanography and glacial history of the west Spitsbergen area, Euro-Arctic margin. In E. Jansen, P. deMenocal, and F. Grousset, eds., Holocene Climate Variability: A Marine Perspective, pp. 2075–2088. Oxford: Pergamon.

Haley, B. A., M. Frank, R. F. Spielhagen, and A. Eisenhauer. 2008. Influence of brine formation on Arctic Ocean circulation over the past 15 million years. Nature Geoscience 1:68–72.

Halfman, J. D., T. C. Johnson, and B. P. Finney. 1994. New AMS dates, stratigraphic correlation and decadal climatic cycles for the past 4 ka at Lake Turkana, Kenya. Palaeogeography, Palaeoclimatology, Palaeoecology 111:83–98.

Hall, A. and M. Visbeck. 2002. Synchronous variability in the southern hemisphere atmosphere, sea ice and ocean resulting from the annular mode. Journal of Climate 15:3043–3057.

Hall, B. L. 2007. Late-Holocene advance of the Collins Ice Cap, King George Island, South Shetland Islands. The Holocene 17:1253–1258.

Hall, I. R., G. G. Bianchi, and J. R. Evans. 2004. Centennial to millennial Holocene climate-deepwater linkage in the North Atlantic. Quaternary Science Reviews 23:1529–1536.

Hall, I. R., S. B. Moran, R. Zahn, et al. 2006. Accelerated drawdown of meridional overturning in the late-glacial Atlantic triggered by transient pre-H event freshwater perturbation. Geophysical Research Letters 33:L16616.

Hall, J. M. and L.-H. Chan. 2004a. Li/Ca in multiple species of benthic and planktonic foraminifera: Thermocline, latitudinal, and glacial-interglacial variation. Geochimica et Cosmochimica Acta 68:529–545.

Hall, J. M. and L.-H. Chan. 2004b. Ba/Ca in benthic foraminifera: Reconstruction of thermocline and mid-depth circulation in the North Atlantic during the last glaciation. Paleoceanography 19:PA4018.

Hall, N. M. J. and P. J. Valdes. 1997. A GCM simulation of the climate 6000 years ago. Journal of Climate 10:3–17.

Hallam, A. 1992. Phanerozoic Sea-Level Changes. New York: Columbia University Press.

Halpert, M. S. and C. F. Ropelewski. 1992. Surface temperature patterns associated with the southern oscillation. Journal of Climate 5:577–593.

Halpert, M. S. and T. M. Smith. 1994. The global climate for March–May 1993: Mature ENSO conditions persist and a blizzard blankets the Eastern United States. Journal of Climate 7:1772–1793.

Halverson, G. P., P. F. Hoffman, D. P. Schrag, A. C. Maloof, and A. H. N. Rice. 2005. Toward a Neoproterozoic composite carbon-isotope record. Geological Society of America Bulletin 117:1181–1207.

Hambrey, M. J. and W. B. Harland. 1981. Earth's Pre-Pleistocene Glacial Record. Cambridge: Cambridge University Press.

Hambrey, M. J. and W. B. Harland. 1985. The late Proterozoic glacial era. Palaeogeography, Palaeoclimatology, Palaeoecology 51:255–272.

Hambrey, M. J., P.-N. Webb, D. M. Harwood, and L. A. Krissel. 2003. Neogene glacial record from the Sirius Group of the Shackleton Glacier region, central Tranantarctic Mountains, Antarctica. Geological Society of America Bulletin 115:994–1015.

Hammer, C. U. 1977. Past volcanism revealed by Greenland Ice Sheet impurity. Nature 270:482–486.

Hammer, C. U. 1984. Traces of Icelandic eruptions in the Greenland Ice Sheet. Jökull 34:51–65.

Hammer, C. U. 1989. Dating by physical and chemical seasonal variation and reference horizons. In H. Oeschger and C. C. Langway Jr., eds., The Environmental Record in Glaciers and Ice Sheets, pp. 99–121. Chichester, UK: John Wiley & Sons.

Hammer, C. U., H. B. Clausen, and W. Dansgaard. 1980. Greenland Ice Sheet evidence of post-glacial volcanism and its climatic impact. Nature 288:230–235.

Hanebuth, T., K. Stattegger, and P. M. Grootes. 2000. Rapid flooding of the Sunda Shelf: A late-glacial sea-level record. Science 288:1033–1035.

Hanna, E., P. Huybrechts, I. Janssens, et al. 2005. Runoff and mass balance of the Greenland Ice Sheet: 1958–2003. Journal of Geophysical Research-Atmospheres 110:D13108.

Hanna, E., R. McConnell, S. Das, J. Cappelen, and A. Stevens. 2006. Observed and modelled Greenland Ice Sheet snow accumulation, 1958–2003, and links with regional climate forcing. Journal of Climate 19:344–358.

Hannachi, A., I. T. Jolliffe, and D. B. Stephenson. 2007. Empirical orthogonal functions and related techniques in atmospheric science: A review. International Journal of Climatology 27:1119–1152.

Hannachi, A., I. T. Jolliffe, D. B. Stephenson, and N. Trendafilov. 2005. International Journal of Climatology 26:7–28.

Hansen, J., A. Lacis, D. Rind, et al. 1984. Climate sensitivity: Analysis of feedback mechanisms. Climate Processes and Climate Sensitivity, pp. 130–163. Washington, DC: American Geophysical Union.

Hansen, J., L. Nazarenko, R. Ruedy, et al. 2005. Earth's energy imbalance: Confirmation and implications. Science 308: 1431–1435.

Hansen, J. and M. Sato. 2004. Greenhouse gas growth rates. Proceedings of the National Academy of Sciences 101:16109–16114.

Hansen, J., M. Sato, P. Kharecha, et al. 2007. Climate change and trace gases. Philosophical Transactions of the Royal Society of London, Series A 365:1925–1954.

Haq, B. U., J. Hardenbol, and P. R. Vail. 1987. The chronology of fluctuating sea level since the Triassic. Science 235:1156–1167.

Haq, B. U., J. Hardenbol, and P. R. Vail. 1988. Mesozoic and Cenozoic chronostratigraphy and cycles of sea-level change. Society of Economic Paleontologists and Mineralogists (Special Publication) 42:71–108.

Haq, B. U. and S. R. Schutter. 2008. A chronology of Paleozoic sea-level changes. Science 322:64–68.

Hardas, P. and J. Mutterlose. 2007. Calcareous nannofossil assemblages of Oceanic Anoxic Event 2 in the equatorial Atlantic: Evidence of an eutrophication event. Marine Micropaleontology 66:52–69.

Hare, S. R. and N. J. Mantua. 2000. Empirical evidence for North Pacific regime shifts in 1977 and 1989. Progress in Oceanography 47:103–145.

Hare, S. R., N. J. Mantua, and R. C. Francis. 1999. Inverse production regimes: Alaskan and West Coast Pacific salmon. Fisheries 21:6–14.

Hargreaves, J., S. L. Weber, P. Braconnot, and J. Guiot, eds. 2006. Modelling Late Quaternary Climate. Climate of the Past 2 (entire volume).

Harland, W. B. 1964. Evidence of late Precambrian glaciation and its significance. In A. E. M. Nairn, ed., *Problems in Palaeoclimatology*, pp. 119–149. London: Interscience.

Harms, U., C. Koeberl, and M. D. Zoback, eds. 2007. *Continental Scientific Drilling: A Decade of Progress, and Challenges for the Future*. Berlin: Springer-Verlag.

Harris, R. N. and D. S. Chapman. 2001. Mid-latitude (30°–60°N) climatic warming inferred by combining borehole temperatures with surface air temperatures. Geophysical Research Letters 28:747–750.

Harrison, D. E. and M. Carson. 2007. Is the world ocean warming? Upper-ocean temperature trends: 1950–2000. Journal of Physical Oceanography 37:174–187.

Harrison, K. G. 2000. Role of increased marine silica input on paleo-pCO_2 levels. Paleoceanography 15:292–298.

Harrison, S. P., D. Jolly, F. Laarif, et al. 1998. Intercomparison of simulated global vegetation distributions in response to 6 kyr BP orbital forcing. Journal of Climate 11:2721–2742.

Harrison, S. P., J. E. Kutzbach, Z. Liu, et al. 2003. Mid-Holocene climate of the Americas: A dynamical response to changed seasonality. Climate Dynamics 20:663–688.

Harrison, S. P., G. Yu, and P. E. Tarasov. 1996. The late Quaternary lake-level record from northern Eurasia. Quaternary Research 45:138–159.

Harrison, T. M., P. Copeland, W. S. F. Kidd, and A. Yin. 1992. Raising Tibet. Science 255:1668–1670.

Hassan, F. A. 1981. Historical Nile floods and their implications for climatic change. Science 212:1142–1145.

Hastings, D. W., A. Russell, and S. Emerson. 1998. Foraminiferal magnesium in G. sacculifer as a paleotemperature proxy. Paleoceanography 13:161–169.

Hathorne, E. C. and R. H. James. 2006. Temporal record of lithium in seawater: A tracer for silicate weathering. Earth and Planetary Science Letters 246:393–406.

Haug, G. H., D. Gunther, L. C. Peterson, et al. 2003. Climate and the collapse of Maya civilization. Science 299:1731–1735.

Haug, G. H., K. A. Hughen, L. C. Peterson, D. M. Sigman, and U. Röhl. 2001b. Southward migration of the Intertropical Convergence Zone through the Holocene. Science 293: 1304–1308.

Haug, G. H. and R. Tiedemann. 1998. Effect of the formation of the Isthmus of Panama on Atlantic Ocean thermohaline circulation. Nature 393:673–676.

Haug, G. H., R. Tiedemann, R. Zahn, and A. C. Ravelo. 2001a. Role of Panama uplift on oceanic freshwater balance. Geology 29:207–210.

Hay, W. W., E. Soeding, R. M. DeConto, and C. N. Wold. 2002. The late Cenozoic uplift-climate change paradox. International Journal of Earth Science 91:746–774.

Hayes, J. M., B. M. Popp, R. Takigiku, and M. W. Johnson. 1989. An isotopic study of biogeochemical relationships between carbonates and organic carbon in the Greenhorn Formation. Geochimica et Cosmochimica Acta 53:2961–2972.

Hays, J. D., J. Imbrie, and N. J. Shackleton. 1976. Variations in the earth's orbit: Pacemaker of the ice ages. Science 194:1121–1132.

Haywood, A. M., P. Dekens, A. C. Ravelo, and M. Williams. 2005. Warmer tropics during the mid-Pliocene? Evidence from alkenone paleothermometry and a fully coupled ocean-atmosphere GCM. Geochemistry, Geophysics, Geosystems 6:Q03010.

Haywood, A. M. and P. J. Valdes. 2004. Modelling Pliocene warmth: Contribution of atmosphere, oceans and cryosphere. Earth and Planetary Science Letters 218:363–377.

Haywood, A. M., P. J. Valdes, and V. L. Peck. 2007b. A permanent El Niño like state during the Pliocene? Paleoceanography 22:PA1213.

Haywood, A. M., P. J. Valdes, and B. W. Sellwood. 2002. Magnitude of climate variability during middle Pliocene warmth: A palaeoclimate modelling study. Palaeogeography, Palaeoclimatology, Palaeoecology 188:1–24.

Haywood, A. M., P. J. Valdes, and M. Williams. 2007a. The mid Pliocene Warm Period: A test-bed for integrating data and models. In M. Williams, et al., eds., *Deep-Time Perspectives on Climate Change: Marrying the Signal from Computer Models and Biological Proxies*. The Micropalaeontological Society, Geological Society Special Publication, pp. 443–458. London: Geological Society of London.

Head, M. J., and P. L. Gibbard, eds. 2005. *Early–Middle Pleistocene Transitions: The Land–Ocean Evidence*. London: Geological Society of London.

Healey, S. and R. Thunell. 2004. Millennial-scale variability in western subtropical North Atlantic surface and deep water circulation during Marine Isotope Stages 11 and 12. Paleoceanography 19:PA1013.

Hearty, P. J., J. T. Hollin, A. C. Neumann, M. J. O'Leary, and M. McCulloch. 2007. Global sea-level fluctuations during the Last Interglaciation (MIS 5e). Quaternary Science Reviews 26:2090–2112.

Hearty, P. J., P. Kindler, H. Cheng, and R. L. Edwards. 1999. A +20 m middle Pleistocene sea-level highstand (Bermuda and the Bahamas) due to partial collapse of Antarctic ice. Geology 27:375–378.

Hebbeln, D., K.-L. Knudsen, R. Gyllencreutz, et al. 2006. Late Holocene coastal hydrographic and climate changes in the eastern North Sea. The Holocene 16:987–1001.

Hegerl, G. C., T. J. Crowley, M. Allen, et al. 2007. Detection of human influence on a new, validated 1500-year temperature reconstruction. Journal of Climate 20:650–666.

Hegerl, G. C., T. J. Crowley, S. K. Baum, K.-Y. Kim, and W. T. Hyde. 2003. Detection of volcanic, solar and greenhouse gas signals in paleo-reconstructions of northern hemisphere temperature. Geophysical Research Letters 30:1242.

Hegerl, G. C., T. J. Crowley, W. T. Hyde, and D. J. Warme. 2006. Climate sensitivity constrained by temperature reconstructions over the past seven centuries. Nature 440:1029–1032.

Heinrich, H. 1988. Origin and consequence of cyclic ice rafting in the northeast Atlantic Ocean during the past 130,000 years. Quaternary Research 29:142–152.

Held, I. M. and B. J. Soden. 2000. Water vapor feedback and global warming. Annual Review of Energy and the Environment 25:441–475.

Held, I. M. and B. J. Soden. 2006. Robust response of the hydrological cycle to global warming. Journal of Climate 19:5686–5699.

Heller, F. and T. S. Liu. 1982. Magnetostratigraphic dating of loess deposits in China. Nature 300:431–433.

Helmke, J. P. and H. A. Bauch. 2003. Comparison of glacial and interglacial conditions between the polar and the subpolar North Atlantic Region over the past five climate cycles. Paleoceanography 18:PA000794.

Hemming, S. R. 2004. Heinrich events: Massive late Pleistocene detritus layers of the North Atlantic and their global climate imprint. Review of Geophysics 42:1–43.

Hemming, S. R., W. S. Broecker, W. D. Sharp, et al. 1998. Provenance of Heinrich layers in core V28-82, northeastern Atlantic: ^{40}Ar/^{39}Ar ages of ice-rafted hornblende, Pb isotopes in feldspar grains, and Nd-Sr-Pb isotopes in the fine sediment fraction. Earth and Planetary Science Letters 164:317–333.

Hemming, S. R., R. H. Gwiazda, J. T. Andrews, W. S. Broecker, A. E. Jennings, and T. Onstott. 2000. ^{40}Ar/^{39}Ar and Pb-Pb study of individual hornblende and feldspar grains from southeastern Baffin Island glacial sediments: Implications for the provenance of the Heinrich layers. Canadian Journal of Earth Sciences 37:879–890.

Hemming, S. R. and I. Hajdas. 2003. Ice rafted detritus evidence from ^{40}Ar/^{39}Ar ages of individual hornblende grains for evolution of the southeastern Laurentide Ice Sheet since 43 ^{14}Cky. Quaternary International 99–100:29–43.

Henderiks, J. and M. Pagani. 2007. Refining ancient carbon dioxide estimates: Significance of coccolithophore cell size for alkenone-based pCO$_2$ records. Paleoceanography 22:PA3202.

Henderiks, J. and R. E. M. Rickaby. 2007. A coccolithophore concept for constraining the Cenozoic carbon cycle. Biogeosciences 4:323–329.

Henderson, G. M. 2002. New oceanic proxies for paleoclimate. Earth and Planetary Science Letters 203:1–13.

Henderson, G. M. and R. F. Anderson. 2003. The U-series toolbox for paleoceanography. Reviews in Mineralogy and Geochemistry 52:493–531.

Henderson, G. M., L. F. Robinson, K. Cox, and A. L. Thomas. 2006. Recognition of non-Milankovitch sea-level highstands at 185 and 343 thousand years ago from U-Th dating of

Bahamas sediment. Quaternary Science Reviews 25:3346–3358.

Henderson, G. M. and N. C. Slowey. 2000. Evidence from U-Th dating against northern-hemisphere forcing of the penultimate deglaciation. Nature 404:61–66.

Hendy, E. J., M. Gagan, C. A. Alibert, et al. 2002. Abrupt decrease in tropical Pacific sea surface salinity at end of Little Ice Age. Science 295:1511–1514.

Hendy, I. L. and J. P. Kennett. 2000. Dansgaard-Oeschger cycles and the California current system: Planktonic foraminiferal response to rapid climate change in Santa Barbara Basin, Ocean Drilling Program Hole 893A. Paleoceanography 15:30–42.

Hendy, I. L., J. P. Kennett, E. B. Roark, and B. L. Ingram. 2002. Apparent synchroneity of submillennial scale climate events between Greenland and Santa Barbara Basin, California from 30–10 ka. Quaternary Science Reviews 21:1167–1184.

Hendy, I. L., T. F. Pedersen, J. P. Kennett, and R. Tada, 2004. Intermittent existence of a southern Californian upwelling cell during submillennial climate change of the last 60 kyr. Paleoceanography 19:PA3007.

Herbert, T. D. 1997. A long history of marine carbon cycle modulation by orbital climatic change. Proceedings of the National Academy of Sciences 94:8362–8369.

Herbert, T. D. 2003. Alkenone Paleotemperature Determinations. In H. Elderfield and K. K. Turekian, eds., Treatise in Marine Geochemistry, pp. 391–432. Amsterdam: Elsevier.

Heroy, D. C., C. Sjunneskog, and J. B. Anderson. 2008. Holocene climate change in the Bransfield Basin, Antarctic Peninsula: Evidence from sediment and diatom analysis. Antarctic Science 20:69–87.

Herrmann, A. D., M. E. Patzkowsky, and D. Pollard. 2003. Obliquity forcing with 8–12 times preindustrial levels of atmospheric pCO$_2$ during the late Ordovician glaciation. Geology 31:485–488.

Herweijer, C., R. Seager, and E. R. Cook. 2006. North American droughts of the mid to late nineteenth century: A history, simulation and implication for Medieval drought. The Holocene 16:159–171.

Herweijer, C., R. Seager, E. R. Cook, and J. Emile-Geay. 2007. North American droughts of the last millennium from a gridded network of tree-ring data. Journal of Climate 20:1353–1376.

Hesse, P. P. and G. H. McTainsh. 1999. Last Glacial Maximum to early Holocene wind strength in the mid-latitudes of the southern hemisphere from aeolian dust in the Tasman Sea. Quaternary Research 52:343–349.

Hesselbo, S. P., D. R. Gröcke, H. C. Jenkyns, et al. 2000. Massive dissociation of gas hydrate during a Jurassic oceanic event. Nature 406:392–395.

Hesselbo, S. P., H. C. Jenkyns, L. V. Duarte, and L. C. V. Oliveira. 2007. Carbon-isotope record of the early Jurassic (Toarcian) Oceanic Anoxic Event from fossil wood and marine carbonate (Lusitanian Basin, Portugal). Earth and Planetary Science Letters 253:455–470.

Hewitt, C. D. and J. B. F. Mitchell. 1998. A fully coupled GCM simulation of the climate of the mid-Holocene. Geophysical Research Letters 25:361–364.

Higgens, S. M., R. F. Anderson, F. Marcantonio, P. Schlosser, and M. Stute. 2002. Sediment focusing creates 100-ka cycles in interplanetary dust accumulation on the Ontong Java Plateau. Earth and Planetary Science Letters 203:383–397.

Higgins, J. A. and D. P. Schrag. 2003. Aftermath of a Snowball Earth. Geochemistry, Geophysics, Geosystems 4:GC000403.

Higgins, J. A. and D. P. Schrag. 2006. Beyond methane: Towards a theory for the Paleocene–Eocene Thermal Maximum. Earth and Planetary Science Letters 245:523–537.

Hilgen, F., H. Brinkhuis, and W.-J. Zachariasse. 2006. Unit stratotypes for global stages: The Neogene perspective. Earth-Science Reviews 74:113–125.

Hilgen, F. J. 1991a. Astronomical calibration of Gauss to Matuyama sapropels in the Mediterranean and implication for the Geomagnetic Polarity Time Scale. Earth and Planetary Science Letters 104:226–244.

Hilgen, F. J. 1991b. Extension of the astronomically calibrated (polarity) time scale to the Miocene/Pliocene boundary. Earth and Planetary Science Letters 107:349–368.

Hilgen, F. J., W. Krijgsman, C. G. Langereis, F. J. Lourens, A. Santarelli, and W.-J. Zachariasse. 1995. Extending the astronomical (polarity) time scale into the Miocene. Earth and Planetary Science Letters 136:495–510.

Hilgen, F. J. and C. G. Langereis. 1989. Periodicities of $CaCO_3$ cycles in the Pliocene of Sicily: discrepancies with the quasi-periods of the earth's orbital cycles? Terra Nova 1:409–415.

Hill, H., B. Flower, T. Quinn, D. Hollander, and T. Guilderson. 2006. Laurentide Ice Sheet meltwater and abrupt climate change during the last glaciation. Paleoceanography 21:PA1006, PA001186.

Hillaire-Marcel, C. and G. Bilodeau. 2000. Instabilities in the Labrador Sea water mass structure during the last climatic cycle. Canadian Journal of Earth Science 37:795–809.

Hillaire-Marcel, C. and A. de Vernal. 2008. Stable isotope clues to episodic sea ice formation in the glacial North Atlantic. Earth and Planetary Science Letters 268:143–150.

Hillaire-Marcel, C., G. Gariepy, B. Ghaleb, et al. 1996. U-series measurements in Tyrrhenian deposits from Mallorca— Further evidence for two last-interglacial high sea levels in the Balearic Islands. Quaternary Science Reviews 15:63–75.

Hillaire-Marcel, C. and S. Occhietti. 1980. Chronology, paleogeography, and paleoclimatic significance of the late and post-glacial events in eastern Canada. Zeitschrift fur Geomorphologie 24:373–392.

Hillebrand, C.-D. and G. Cortese. 2006. Polar stratification: A critical view from the Southern Ocean. Palaeogeography, Palaeoclimatology, Palaeoecology 242:240–252.

Hinnov, L. A. 2000. New perspectives on orbitally forced stratigraphy. Annual Review of Earth and Planetary Sciences 28:419–475.

Hinnov, L. A. and J. G. Ogg. 2007. Cyclostratigraphy and the astronomical timescale. Stratigraphy 4:239–251.

Hodell, D., C. D. Charles, and U. S. Ninnemann. 2000. Comparison of interglacial stages in the South Atlantic sector of the southern ocean for the past 450 kyr: Implications for Marine Isotope Stage (MIS) 11. Global and Planetary Change 24:7–26.

Hodell, D. and F. Woodruff. 1994. Variations in the strontium isotopic ratio of seawater during the Miocene: Stratigraphic and geochemical implications. Paleoceanography 9:405–426.

Hodell, D. A., M. Brenner, J. H. Curtis, and T. Guilderson. 2001. Solar forcing of drought frequency in the Maya lowlands. Science 292:1367–1370.

Hodell, D. A., J. H. Curtis, and M. Brenner. 1995. Possible role of climate in the collapse of classic Maya civilization. Nature 375:391–394.

Hodell, D. A., P. A. Mueller, J. A. McKenzie, and G. A. Mead. 1989. Strontium isotope stratigraphy and geochemistry of the late Neogene ocean. Earth and Planetary Science Letters 92:165–178.

Hoelzle, M., T. Chinn, D. Stumm, et al. 2007. The application of glacier inventory data for estimating past climate change effects on mountain glaciers: A comparison between the European Alps and the Southern Alps of New Zealand. Global and Planetary Change 56:69–82.

Hoffert, M. I. and C. Covey. 1992. Deriving global climate sensitivity from palaeoclimate reconstructions. Nature 360:573–576.

Hoffman, P. F. 2005. 28th DeBeers Alexander Du Toit Memorial Lecture: On Cryogenian (Neoproterozoic) ice-sheet dynamics and the limitations of the glacial sedimentary record. South African Journal of Geology 108:557–576.

Hoffman, P. F., G. P. Halverson, E. W. Domack, J. M. Husson, J. A. Higgins, and D. P. Schrag. 2007. Are basal Ediacaran (635 Ma) post-glacial "cap dolostones" diachronous? Earth and Planetary Science Letters 254:114–131.

Hoffman, P. F., A. J. Kaufman, G. P. Halverson, and D. P. Schrag. 1998. A Neoproterozoic Snowball Earth. Science 281:1342–1346.

Hoffman, P. F. and D. P. Schrag. 2002. The Snowball Earth hypothesis: Testing the limits of global change. Terra Nova 14:129–155.

Hoffmann, G., E. Ramirez, J. D. Taupin, et al. 2003. Coherent isotope history of Andean ice cores over the last century. Geophysical Research Letters 30:1179, GL014870.

Hoffmann, K.-H., D. J. Condon, S. A. Bowring, and J. L. Crowley. 2004. U-Pb zircon date from the Neoproterozoic Ghaub Formation, Namibia: Constraints on Marinoan glaciation. Geology 32:817–820.

Holbourn, A. and W. Kuhnt. 2001. No extinctions during Oceanic Anoxic Event 1B: The Aptian-Albian benthic foraminifera record of ODP Let 171. In D. Kroon, R. D. Norris, and A. Klaus, eds., Western North Atlantic Palaeogene and Cretaceous Palaeoceanography, pp. 73–92. London: Geological Society of London.

Holbourn, A., W. Kuhnt, H. Kawamura, et al. 2005a. Orbitally-paced paleoproductivity variations in the Timor Sea and Indonesian throughflow variability during the last 460 kyr. Paleoceanography 20:PA3002, doi:2004pa00109.

Holbourn, A., W. Kuhnt, M. Schulz, and H. Erlenkeuser. 2005b. Impacts of orbital forcing and atmospheric carbon dioxide on Miocene ice-sheet expansion. Nature 438:483–487.

Holbourn, A., W. Kuhnt, M. Schulz, J.-A. Flores, and N. Andersen. 2007. Orbitally-paced climate evolution during the middle Miocene "Monterey" carbon isotope excursion. Earth and Planetary Science Letters 261:534–550.

Holgate, S. J. 2007. On the decadal rates of sea level change during the twentieth century. Geophysical Research Letters 34:L01602.

Holgate, S. J. and P. L. Woodworth. 2004. Evidence for enhanced coastal sea level rise during the 1990s. Geophysical Research Letters 31:L07305.

Holland, D. M., R. H. Thomas, B. De Young, M. H. Ribergaard, and B. Lyberth. 2008. Acceleration of Jakobshavn Isbrae triggered by warm subsurface ocean waters. Nature Geoscience 1:659–664.

Holland, G. J. and P. J. Webster. 2007. Heightened tropical cyclone activity in the North Atlantic: Natural variability or climate trend? Philosophical Transactions of the Royal Society of London, Series A:doi:1098.

Holland, H. D. 1984. *The Chemical Evolution of the Atmosphere and Oceans*. Princeton, NJ: Princeton University Press.

Holland, H. D. 2003. The geologic history of seawater. In H. Elderfield, ed., *Treatise on Geochemistry*, vol. 6, *The Oceans and Marine Geochemistry*, pp. 583–625. Amsterdam: Elsevier.

Holland, M. M. and C. M. Bitz. 2003. Polar amplification of climate change in coupled models. Climate Dynamics 21:221–232.

Holmgren, K., J. A. Lee-Thorp, G. R. J. Cooper, et al. 2003. Persistent millennial-scale climatic variability over the past 25,000 years in Southern Africa. Quaternary Science Reviews 22:2311–2326.

Holton, J. R. and R. S. Lindzen. 1972. An updated theory for the quasibiennial cycle of the tropical stratosphere. Journal of Atmospheric Science 29:1076–1080.

Hong, Y. T., B. Hong, Q. H. Lin, et al. 2003. Correlation between Indian Ocean summer monsoon and North Atlantic climate during the Holocene. Earth and Planetary Science Letters 211:371–380.

Hong, Y. T., Z. G. Wang, H. B. Jang, et al. 2001. A 6000-year record of changes in drought and precipitation in northeastern China based on a ^{13}C time series from peat cellulose. Earth and Planetary Science Letters 185:111–119.

Hopley, D., S. G. Smithers, and K. E. Parnell. 2007. *The Geomorphology of the Great Barrier Reef: Development, Diversity and Change*. Cambridge: Cambridge University Press.

Horton, T. W., D. J. Sjostrom, M. J. Abruzzese, et al. 2004. Spatial and temporal variation of Cenozoic surface elevation in the Great Basin and Sierra Nevada. American Journal of Science 304:862–888.

Hou, S., J. Chappellaz, J. Jouzel, et al. 2007. Summer temperature trend over the past two millennia using air content in Himalayan ice. Climate of the Past 3:89–95.

Houghton, R. A. 2003. Revised estimates of the annual net flux of carbon to the atmosphere from changes in land use and land management 1850–2000. Tellus 55B:378–390.

Houghton, R. A. 2007. Balancing the carbon budget. Annual Review of Earth and Planetary Sciences 35:313–347.

Hovan, S. A., D. K. Rea, and N. G. Pisias. 1991. Late Pleistocene continental climate and oceanic variability recorded in northwest Pacific sediments. Paleoceanography 6:349–370.

Howard, W. R. 1997. A warm future in the past. Nature 388: 418–419.

Howat, I. M., I. Joughin, and T. A. Scambos. 2007. Rapid changes in ice discharge from Greenland outlet glaciers. Science 315:1559–1561.

Hoyt, D. V. and K. H. Schatten. 1997. *The Role of the Sun in Climate Change*. New York: Oxford University Press.

Hu, F. S., E. Ito, T. A. Brown, B. B. Curry, and D. R. Engstrom. 2001. Pronounced climatic variations during the last two millennia in the Alaska Range. Proceedings of the National Academy of Sciences 98:10552–10556.

Hu, F. S., D. Kaufman, S. Yoneji, et al. 2003. Cyclic variation and solar forcing of Holocene climate in the Alaskan subarctic. Science 301:1890–1893.

Huang, Y., B. Shuman, Y. Wang, and T. Webb III. 2004. Hydrogen isotope ratios of individual lipids in lake sediments as novel tracers of climatic and environmental change: A surface sediment test. Journal of Paleolimnology 31:363–375.

Huber, B. T., D. A. Hodell, and C. P. Hamilton. 1995. Mid- to late Cretaceous climate of the southern high latitudes: Stable isotopic evidence for minimal equator-to-pole thermal gradients. Geological Society of America Bulletin 107: 1164–1191.

Huber, B. T., R. M. Leckie, R. D. Norris, T. J. Bralower, and E. CoBabe. 1999. Foraminiferal assemblage and stable isotopic change across the Cenomanian-Turonian boundary in the subtropical North Atlantic. Journal of Foraminiferal Research 29:392–417.

Huber, B. T., R. D. Norris, and K. G. MacLeod. 2002. Deep sea paleotemperature record of extreme warmth during the Cretaceous. Geology 30:123–126.

Huber, C., M. Leuenberger, R. Spahni, et al. 2006. Isotope calibrated Greenland temperature record over Marine Isotope Stage 3 and its relation to CH_4. Earth and Planetary Science Letters 243:504–519.

Huber, M. 2008. A hotter greenhouse? Science 321:353–354.

Huber, M., H. Brinkhuis, C. E. Stickley, et al. 2004. Eocene circulation of the Southern Ocean: Was Antarctica kept warm by subtropical waters? Paleoceanography 19:PA4026.

Huber, M. and R. Caballero. 2003. Eocene El Niño: Evidence for robust tropical dynamics in the "hothouse." Science 299: 877–881.

Hughen, K., J. Southon, S. Lehman, C. Bertrand, and J. Turnbull. 2006. Marine-derived ^{14}C calibration and activity record for the past 50,000 years updated from the Cariaco Basin. Quaternary Science Reviews 25:3216–3227.

Hughen, K. A., M. G. L. Baillie, E. Bard, et al. 2004b. Marine04-marine radiocarbon age calibration, 0–26 cal kyr BP. Radiocarbon 46:1059–1086.

Hughen, K. A., T. I. Eglinton, L. Xu, and M. Makou. 2004c. Abrupt tropical vegetation response to rapid climate changes. Science 304:1955–1959.

Hughen, K. A., S. Lehman, J. Southon, et al. 2004a. ^{14}C activity and global carbon cycle changes over the past 50,000 Years. Science 303:202–207.

Hughen, K. A., J. T. Overpeck, S. J. Lehman, et al. 1998. Deglacial changes in ocean circulation from an extended radiocarbon calibration. Nature 391:65–68.

Hughen, K. A., J. T. Overpeck, L. C. Peterson, and S. Trumbore. 1996. Rapid climate changes in the tropical Atlantic regions during the last deglaciation. Nature 380:51–54.

Hughen, K. A., D. P. Schrag, S. B. Jacobsen, and W. Hantoro. 1999. El Niño during the last interglacial period recorded by a fossil coral from Indonesia. Geophysical Research Letters 26: 3129–3132.

Hughen, K. A., J. R. Southon, S. J. Lehman, and J. T. Overpeck. 2000. Synchronous radiocarbon and climate shifts during the last deglaciation. Science 290:1951–1954.

Hughes, M. K. and H. F. Diaz, eds. 1994a. *The Medieval Warm Period*. Dordrecht: Kluwer.

Hughes, M. K. and H. F. Diaz. 1994b. Was there a "Medieval Warm Period," and if so where and when? Climatic Change 26:109–142.

Hulbe, C. L. 1997. An ice shelf mechanism for Heinrich layer production. Paleoceanography 12:711–717.

Hulbe, C. L. 2001. How ice sheets flow. Science 294:2300–2301.

Hulbe, C. L. and M. A. Fahnestock. 2007. Century-scale discharge stagnation and reactivation of the Ross ice streams, West Antarctica. Journal of Geophysical Research, Earth Surface 112:JF000603.

Hulbe, C. L. and D. R. MacAyeal. 1999. A new numerical model of coupled inland ice sheet, ice stream and ice shelf flow and its application to the West Antarctic Ice Sheet. Journal of Geophysical Research 104(B11):25349–25366.

Hulbe, C. L., D. R. MacAyeal, G. H. Denton, J. Kleman, and T. V. Lowell. 2004. Catastrophic ice shelf breakup as the source of Heinrich event icebergs. Paleoceanography 19:PA1004.

Hüls, M. and R. Zahn. 2000. Millennial-scale sea surface temperature variability in the western tropical North Atlantic from planktonic foraminifera. Paleoceanography 15:659–678.

Hulton, N. R. J., R. S. Purves, R. D. McCulloch, D. E. Sugden, and M. J. Bentley. 2002. The Last Glacial Maximum and deglaciation in southern South America. Quaternary Science Reviews 21:233–241.

Hunt, A. S. and A. E. Rathburn. 1988. Microfaunal assemblages of southern Champlain Sea piston cores. In N. R. Gadd, ed., The Late Quaternary Development of the Champlain Sea Basin, pp. 145–154. Ottawa: Geological Association of Canada.

Hunt, B. G. 2006. The Medieval Warm Period, the Little Ice Age and simulated climatic variability. Climate Dynamics 27:677–694.

Hunt, B. G. and T. I. Elliott 2006. Climatic trends. Climate Dynamics 26:567–585.

Hunter, S. E., D. Wilkinson, D. A. V. Stow, et al. 2007. The Eirik Drift: A longterm barometer of North Atlantic Deep water flux south of Cape Farewell, Greenland. In A. Viana and M. Rebesco, eds., Economic and Paleoceanographic Significance of Contourite Deposits, pp. 245–264. London: Geological Society of London.

Huntington, T. G. 2006. Evidence for intensification of the global water cycle: Review and synthesis. Journal of Hydrology 319:83–95.

Huon, S., F. E. Grousset, D. Burdloff, G. Barboux, and A. Mariotti. 2002. Sources of fine-sized organic matter in North Atlantic Heinrich layers: $\delta^{13}C$ and $\delta^{15}N$ tracers. Geochimica et Cosmochimica Acta 66:223–239.

Hurrell, J. W. 1995. Decadal trends in the North Atlantic Oscillation: Regional temperatures and precipitation. Science 269:676–679.

Hurrell, J. W. and R. R. Dickson. 2004. Climate variability over the North Atlantic. In N. C. Stenseth, et al., eds., Marine Ecosystems and Climate Variation—the North Atlantic, pp. 15–31. Oxford: Oxford University Press.

Hurrell, J. W., Y. Kushnir, G. Ottersen, and M. Visbeck, eds. 2003a. The North Atlantic Oscillation: Climatic Significance and Environmental Impact. Washington, DC: American Geophysical Union.

Hurrell, J. W., Y. Kushnir, and M. Visbeck. 2001. The North Atlantic Oscillation. Science 291:603–605.

Hurrell, J. W., Y. Kushnir, M. Visbeck, and G. Ottersen. 2003b. An Overview of the North Atlantic Oscillation. In J. W. Hurrell, Y. Kushnir, G. Ottersen, and M. Visbeck, eds., The North Atlantic Oscillation: Climatic Significance and Environmental

Impact, pp. 1–35. Washington, DC: American Geophysical Union.

Hurrell, J. W. and H. Van Loon. 1997. Decadal variations in climate associated with the North Atlantic Oscillation. Climate Change 36:301–326.

Hurrell J. W., M. Visbeck, A. Busalacchi, et al. 2006. Atlantic climate variability and predictability: A CLIVAR perspective. Journal of Climate 19:5100–5121.

Huybers, P. 2005. Comment on "Hockey sticks, principal components, and spurious significance" by S. McIntyre and R. McKitrick. Geophysical Research Letters 32:L20705.

Huybers, P. 2006. Early Pleistocene glacial cycles and the integrated summer insolation forcing. Science 313: 508–511.

Huybers, P. and W. Curry. 2006. Links between annual, Milankovitch, and continuum temperature variability. Nature 441:329–332.

Huybers, P. and C. Wunsch. 2005. Obliquity pacing of the late Pleistocene glacial terminations. Nature 434:491–494.

Huybrechts, P. 2002. Sea level changes at the LGM from ice-dynamic reconstructions of the Greenland and Antarctic ice sheets during the glacial cycles. Quaternary Science Reviews 21:203–231.

Huybrechts, P. 2006. Numerical modeling of ice sheets through time. In P. G. Knight, ed., Glacier Science and Environmental Change, pp. 406–412. Oxford: Blackwell Publishing.

Huybrechts, P., T. Payne, and EISMINT Intercomparison Group. 1996. The EISMINT benchmarks for testing ice sheet models. Annals of Glaciology 23:1–12.

Hyde, W. T., T. J. Crowley, S. K. Baum, and W. R. Peltier. 2000. Neoproterozoic "Snowball Earth" simulations with a coupled climate/ice-sheet model. Nature 405:425–429.

Imbrie, J., A. Berger, E. Boyle, et al. 1993. On the structure and origin of major glaciation cycles. 2. The 100,000-year cycle. Paleoceanography 8:699–735.

Imbrie, J., E. Boyle, S. Clemens, et al. 1992. On the structure and origin of major glaciation cycles. 1. Linear responses to Milankovitch forcing. Paleoceanography 7:701–738.

Imbrie, J., J. D. Hays, D. G. Martinson, et al. 1984. The orbital theory of Pleistocene climate: Support from a revised chronology of the marine ^{18}O record. In A. Berger, J. Imbrie, J. Hays, G. Kukla, and B. Saltzman, eds., Milankovitch and Climate: Understanding the Response to Astronomical Forcing, pp. 269–305. Dordrecht: D. Reidel.

Imbrie, J. and K. P. Imbrie. 1979. Ice Ages: Solving the Mystery. Cambridge, MA: Harvard University Press.

Imbrie, J. and J. Z. Imbrie. 1980. Modeling the climatic response to orbital variations. Science 207:943–953.

Imbrie, J. and N. G. Kipp. 1971. A new micropaleontological method for quantitative paleoclimatology: Application to a late Pleistocene Caribbean core. In K. K. Turekian, ed., Late Cenozoic Glacial Ages, pp. 71–181. New Haven, CT: Yale University Press.

Imbrie, J., A. McIntyre, and A. Mix. 1989. Oceanic response to orbital forcing in the late Quaternary: Observational and experimental strategies. In A. Berger, ed., Climate and Geosciences, pp. 121–164. Norwell, MA: Kluwer.

Indermüle, A., E. Monnin, B. Stauffer, T. F. Stocker, and M. Wahlen. 2000. Atmospheric CO_2 concentration from 60 to 20 kyr BP from the Taylor Dome Ice Core, Antarctica. Geophysical Research Letters 27:735–738.

Indermüle, A., T. F. Stocker, F. Joos, et al. 1999. Holocene carbon-cycle dynamics based on CO_2 trapped in ice at Taylor Dome, Antarctia. Nature 398:121–126.

Ingólfsson, O. 2004. Quaternary glacial and climate history of Antarctica. In J. Ehlers and P. L. Gibbard, eds., *Quaternary Glaciations of the World, Part III*, pp. 3–43. Dordrecht: Kluwer.

Ingólfsson, O. and C. Hordt. 1999. The Antarctic contribution to Holocene global sea level rise. Polar Research 18:323–330.

International Ad Hoc Detection and Attribution Group. 2005. Detecting and attributing external influences on the climate system: A review of recent advances. Journal of Climate 18:1291–1314.

IPCC WG I, Intergovernmental Panel on Climate Change. 2007. *Climate Change 2007: The Physical Science Basis.* Contribution of Working Group I to the Fourth Assessment, Report of the Intergovernmental Panel on Climate Change. S. Solomon et al., eds. Cambridge: Cambridge University Press.

IPCC WG II, Intergovernmental Panel on Climate Change. 2007. *Climate Change 2007: Impacts, Adaptation and Vulnerability.* Contribution of Working Group II to the Fourth Assessment, Report of the Intergovernmental Panel on Climate Change. M. L. Parry et al., eds. Cambridge: Cambridge University Press.

Irizarry-Ortiz, M. M., G. Wang, and E. A. B. Eltahir. 2003. Role of the biosphere in the mid-Holocene climate of West Africa. Journal of Geophysical Research 108:4042, JD000989.

Isbell, J. L., P. A. Lenaker, R. A. Askin, M. F. Miller, and L. E. Babcock. 2003. Reevaluation of the timing and extent of late Paleozoic glaciation in Gondwana: Role of the Transantarctic Mountains. Geology 31:977–980.

Isbell, J. L., M. F. Miller, L. E. Babcock, and S. T. Hasiotis. 2001. Ice-marginal environment and ecosystem prior to initial advance of the late Palaeozoic ice sheet in the Mount Butters area of the central Transantarctic Mountains, Antarctica. Sedimentology 48:953–970.

Ishii, M., M. Kimoto, K. Sakamoto, and S.-I. Iwasaki. 2006. Steric sea level changes estimated from historical ocean subsurface temperature and salinity analyses. Journal of Oceanography 62:155–170.

Ito, E. 2002. Mg/Ca, Sr/Ca, $\delta^{18}O$, $\delta^{13}C$ chemistry of Quaternary lacustrine ostracode shells from the North American continental interior. In J. A. Holmes and A. R. Chivas, eds., *Applications of the Ostracoda to Quaternary Research*, pp. 267–278. Washington, DC: American Geophysical Union.

Ivany, L. C., S. Van Simaeys, E. W. Domack, and S. D. Samson. 2006. Evidence for an earliest Oligocene ice sheet on the Antarctic Peninsula. Geology 34:377–380.

Iversen, J. 1954. The late-glacial flora of Denmark and its relationship to climate and soil. Danmazks Geologische Undersogelse Series II 75:1–175.

Jackson, M. G., N. Oskaarsson, R. G. Tronnes, et al. 2005. Holocene loess deposition in Iceland: Evidence for millennial-scale atmosphere-ocean coupling in the North Atlantic. Geology 33:509–512.

Jaffrés, J. B. D., G. A. Shields, and K. Wallmann. 2007. The oxygen isotope evolution of seawater: A critical review of a long-standing controversy and an improved geological water cycle model for the past 3.4 billion years. Earth-Science Reviews 83:83–122.

Jain, S. and L. S. Collins. 2007. Trends in Caribbean paleoproductivity related to the Neogene closure of the Central American seaway. Marine Micropaleontology 63:57–74.

Jakobsson, M., J. Backman, B. Rudels, et al. 2007a. The early Miocene onset of a ventilated circulation regime in the Arctic Ocean. Nature 447:986–990.

Jakobsson, M., S. Björck, G. Alm, et al. 2007b. Reconstructing the Baltic Ice Dammed Lake: Bathymetry, area and volume. Global and Planetary Change 57:355–370.

Jakobsson, M., J. V. Gardner, P. Vogt, et al. 2005. Multibeam bathymetric and sediment profiler evidence for ice grounding on the Chukchi Borderland, Arctic Ocean. Quaternary Research 63:150–160.

Jakobsson M., R. Macnab, L. Mayer, et al. 2008. An improved bathymetric portrayal of the Arctic Ocean: Implications for ocean modeling and geological, geophysical and oceanographic analyses. Geophysical Research Letters 35:L07602, GL033520.

Jansen, E. 1987. Rapid changes in the inflow of Atlantic water into the Norwegian Sea at the end of the last glaciation. In W. H. Berger and L. D. Labeyrie, eds., *Abrupt Climatic Change*, pp. 299–310. Dordrecht: D. Reidel.

Jansen, E. and K. R. Bjorklund. 1985. Surface ocean circulation in the Norwegian Sea 15,000 B.P. to present. Boreas 14:243–257.

Jansen, E., J. A. Overpeck, et al. 2007. Chapter 6: Paleoclimate. In S. Solomon, et al., eds., *Climate Change 2007: The Physical Science Basis.* Contribution of Working Group I to the Fourth Assessment Report of the Intergovernmental Panel on Climate Change, pp. 433–498. Cambridge: Cambridge University Press.

Jansen, E. and J. Sjoholm. 1991. Reconstruction of glaciation over the past 6 Myr from ice-borne deposits in the Norwegian Sea. Nature 349:600–603.

Jansen, E. and T. Veum. 1990. Evidence for two-step deglaciation and its impact on North Atlantic deep-water circulation. Nature 343:612–616.

Jansen, J. H. F., A. Kuijpers, and S. R. Troelstra. 1986. A mid-Brunhes climatic event: Long-term changes in global atmosphere and ocean circulation. Science 32:619–622.

Jansson, K. N. and J. Kleman. 2004. Early Holocene glacial lake meltwater injections into Labrador Sea and Ungava Bay. Paleoceanography 19:PA1001, PA000943.

Jasper, J. P. and J. M. Hayes. 1990. A carbon isotope record of CO_2 levels during the late Quaternary. Nature 347:462–464.

Jenkins, G. S. 2004a. High obliquity as an alternative hypothesis to early and late Proterozoic extreme climate conditions. In G. S. Jenkins, et al., eds., *The Extreme Proterozoic: Geology, Geochemistry, and Climate*, pp. 183–192. Washington, DC: American Geophysical Union.

Jenkins, G. S. 2004b. A review of Neoproterozoic climate modeling studies. In G. S. Jenkins, et al., eds., *The Extreme Proterozoic: Geology, Geochemistry, and Climate*, pp. 73–78. Washington, DC: American Geophysical Union.

Jenkyns, H. C. 1980. Cretaceous anoxic events: From continents to oceans. Journal of the Geological Society of London 137:171–188.

Jenkyns, H. C. and C. J. Clayton. 1997. Lower Jurassic epicontinental carbonates and mudstones from England and Wales:

Chemostratigraphic signals and the early Toarcian anoxic event. Sedimentology 44:687–706.

Jennerjahn, T. C., V. Ittekkot, H. W. Arz, et al. 2004. Asynchronous terrestrial and marine signals of climate change during Heinrich events. Science 306:2236–2239.

Jennings, A., K. L. Knudsen, M. Hald, C. V. Hansen, and J. T. Andrews. 2002. A mid-Holocene shift in Arctic sea-ice variability on the East Greenland Shelf. The Holocene 12:49–58.

Jensen, K. 1935. Archaeological dating in the history of North Jutland's vegetation. Acta Archaeologica 5:185–214.

Jensen, K. G., A. Kuijpers, N. Koç, and J. Heinemeier. 2004. Diatom evidence of hydrographic changes and ice conditions in Igaliku Fjord, South Greenland, during the past 1500 years. The Holocene 14:152–164.

Jevrejeva, S., J. C. Moore, A. Grinsted, and P. L. Woodworth. 2008. Recent global sea level acceleration started over 200 years ago? Geophysical Research Letters 35:L08715, GL033611.

Ji, J., J. Shen, W. Balsam, J. Chen, L. Liu, and X. Liu. 2005. Asian monsoon oscillations in the northeastern Qinghai-Tibet Plateau since the late glacial as interpreted from visible reflectance of Qinghai Lake sediments. Earth and Planetary Science Letters 233:61–70.

Jiang, D., H. Wang, Z. Ding, X. Lang, and H. Drange. 2005, Modeling the middle Pliocene climate with a global atmospheric general circulation model. Journal of Geophysical Research 110:D14107.

Jirikowic, J. L. and P. E. Damon. 1994. The Medieval solar activity maximum. Climatic Change 26:309–316.

Johnsen, S. J., H. B. Clausen, W. Dansgaard, et al. 1992. Irregular glacial interstadials recorded in a new Greenland ice core. Nature 359:311–313.

Johnsen, S. J., D. Dahl-Jensen, W. Dansgaard, and N. Gundstrup. 1995. Greenland palaeotemperature derived from GRIP borehole temperatures and ice core isotope profiles. Tellus 45B:624–630.

Johnsen, S. J., D. Dahl-Jensen, N. Gundestrup, et al. 2001. Oxygen isotope and palaeotemperature records from six Greenland ice-core stations: Camp Century, Dye-3, GRIP, GISP2, Renland and NorthGRIP. Journal of Quaternary Sciences 16:299–307.

Johnsen, S. J., W. Dansgaard, and J. W. C. White. 1989. The origin of Arctic precipitation under present and glacial conditions. Tellus 418:452–468.

Johnson, T. C., E. T. Brown, J. McManus, et al. 2002. A high-resolution paleoclimate record spanning the past 25,000 years in southern East Africa. Science 296:113–115.

Johnson, T. C., C. A. Scholtz, M. R. Talbot, et al. 1996. Late Pleistocene dessication of Lake Victoria and rapid expansion of cichlid fishes. Science 313:803–807.

Jones, C. E. and H. C. Jenkyns. 2001. Seawater strontium isotopes, oceanic anoxic events, and seafloor hydrothermal activity in the Jurassic and Cretaceous. American Journal of Science 301:112–149.

Jones, P. D., R. S. Bradley, and J. Jouzel, eds. 1996. *Climate Variations and Forcing Mechanisms of the Last 2000 Years.* Berlin: Springer-Verlag.

Jones, P. D., K. R. Briffa, T. P. Barnett, and S. F. B. Tett. 1998. High-resolution palaeoclimatic records for the past millennium—Interpretation, integration and comparison with general circulation model control-run temperatures. The Holocene 8:455–471.

Jones, P. D., T. Jonsson, and D. Wheeler. 1997. Extensions to the North Atlantic Oscillation using early instrumental pressure observations from Gibraltar and south-west Iceland. International Journal of Climatology 17:1433–1450.

Jones, P. D. and M. E. Mann. 2004. Climate over millennia. Reviews of Geophysics 42:RG2002.

Jones, P. D. and A. Moberg. 2003. Hemispheric and large-scale surface air temperature variations: An extensive revision and an update to 2001. Journal of Climate 16:206–223.

Jones, P. D., M. New, D. E. Parker, S. Martin, and I. G. Rigor. 1999. Surface air temperature and its changes over the past 150 years. Reviews of Geophysics 37:173–199.

Jones, P. D., T. J. Osborn, and K. R. Briffa. 2001. The evolution of climate over the last millennium: Calibrating the isotopic paleothermometer. Science 292:662–667.

Joos, F., S. Gerber, I. C. Prentice, B. L. Otto-Bliesner, and P. J. Valdes. 2004. Transient simulation of Holocene atmospheric carbon dioxide and terrestrial carbon since the Last Glacial Maximum. Global Biogeochemical Cycles 18:GB2002, GB002156.

Joos, F. and R. Spahni. 2008. Rates of change in natural and anthropogenic radiative forcing over the past 20,000 years. Proceedings of the National Academy of Sciences 105: 1425–1430.

Jorissen, F. J. and E. J. Rohling. 2000. Faunal perspectives on paleoproductivity. Marine Micropaleontology 40:131–134.

Joussaume, S., K. E. Taylor, P. Braconnot, et al. 1999. Monsoon changes for 6000 years ago: Results of 18 simulations from the Paleoclimate Modeling Intercomparison Project (PMIP). Geophysical Research Letters 26:859–862.

Jouzel, J. 1999. Calibrating the isotopic paleothermometer. Science 286:910–911.

Jouzel, J., C. Lorius, J. R. Petit, et al. 1987. Vostok ice core: A continuous iso-topic temperature record over the last climatic cycle (160,000 years). Nature 329:403–408.

Jouzel, J., V. Masson-Delmotte, O. Cattani, et al. 2007. Orbital and millennial Antarctic climate variability over the past 800,000 years. Science 317:793–796.

Jouzel, J., L. Merlivat, and C. Lorius. 1982. Deuterium excess in an east Antarctic ice core suggests higher relative humidity at the oceanic surface during the Last Glacial Maximum. Nature 299:688–691.

Jouzel, J., J. R. Petit, J.-M. Barnola, et al. 1992. The last deglaciation in Antarctica: Further evidence for a "Younger Dryas" type climatic event. In E. Bard and W. S. Broecker, eds., *The Last Deglaciation: Absolute and Radiocarbon Chronologies,* pp. 229–266. Berlin: Springer-Verlag.

Jouzel, J., M. Stievenard, S. J. Johnsen, et al. 2007. The GRIP deuterium excess record. Quaternary Science Reviews 26:1–17.

Jouzel, J., R. Vaikmae, J. R. Petit, et al. 1995. The two-step shape and timing of the last deglaciation in Antarctica. Climate Dynamics 11:151–161.

Juckes, M. N., M. R. Allen, K. R. Briffa, et al. 2007. Millennium temperature reconstruction intercomparison and evaluation. Climate of the Past 3:591–609.

Jung, S. J. A., G. R. Davies, G. M. Ganssen, and D. Kroon. 2004. Synchronous Holocene sea surface temperature and rainfall variations in the Asian monsoon system. Quaternary Science Reviews 23:2207–2218.

Jungclaus, J. H., M. Botzet, H. Haak, et al. 2006. Ocean circulation and tropical variability in the Coupled Model ECHAM5/MPI-OM2006. Journal of Climate 19:3952–3972.

Kaiser, J., F. Lamy, and D. Hebbeln. 2005. A 70-kyr sea surface temperature record off southern Chile (Ocean Drilling Program Site 1233). Paleoceanography 20:PA4009.

Kandiano, E. S. and H. A. Bauch. 2007. Phase relationship and surface water mass change in the Northeast Atlantic during Marine Isotope Stage 11 (MIS 11). Quaternary Research 68:445–455.

Kanfoush, S. L., D. A. Hodell, C. D. Charles, et al. 2000. Millennial-scale instability of the Antarctic Ice Sheet during the last glaciation. Science 288:1815–1818.

Kaplan, A., M. A. Cane, Y. Kushnir, and A. C. Clement. 1998. Analysis of global sea surface temperatures 1856–1991. Journal of Geophysical Research 103:18567–18589.

Karlstrom, K. E., S. A. Bowring, C. M. Dehler, et al. 2000. The Chuar Group of the Grand Canyon: Record of breakup of Rodinia, associated change in the global carbon cycle, and ecosystem expansion by 740 Ma. Geology 28:619–622.

Karrow, P. F., A. Dreimanis, and P. J. Barnett. 2000. A proposed diachronic revision of late Quaternary time-stratigraphic classification in the eastern and northern Great Lakes area. Quaternary Research 54:1–12.

Karrow, P. F. and S. Occhietti. 1989. Chapter 4: Quaternary geology of the St. Lawrence Lowlands. In R. J. Fulton, ed., *Quaternary Geology of Canada and Greenland*, vol. K-1, *The Geology of North America*, pp. 321–389. Ottawa: Geological Survey of Canada.

Kaser, G., J. G. Cogley, M. B. Dyurgerov, M. F. Meier, and A. Ohmura. 2006. Mass balance of glaciers and ice caps: Consensus estimates for 1961–2004. Geophysical Research Letters 33:L19501, GL027511.

Kaspi, Y., R. Sayag, and E. Tziperman. 2004. A "triple sea-ice state" mechanism for the abrupt warming and synchronous ice sheet collapses during Heinrich events. Paleoceanography 19:PA3004.

Kasting, J. F. 1987. Theoretical constraints on oxygen and carbon dioxide concentrations in the Precambrian atmosphere. Precambrian Research 34:205–228.

Kasting, J. F. 1993. Earth's early atmosphere. Science 259:920–926.

Kasting, J. F. 2005. Methane and climate during the Precambrian Era. Precambrian Research 137:119–129.

Kasting, J. F. and T. P. Ackerman. 1986. Climatic consequences of very high CO_2 levels in Earth's early atmosphere. Science 234:1383–1385.

Kasting, J. F. and M. Howard. 2006. Atmospheric composition and climate on the early earth. Philosophical Transactions of the Royal Society of London, Series B:1733–1742.

Kasting, J. F., M. T. Howard, K. Wallman, et al. 2006. Paleoclimates, ocean depth, and the oxygen isotopic composition of seawater. Earth and Planetary Science Letters 252:82–93.

Katz, M. E., J. D. Wright, K. G. Miller, et al. 2005. Biological overprint of the geological carbon cycle. Marine Geology 217:323–338.

Kawamura, K., T. Nakazawa, S. Aoki, et al. 2003. Atmospheric CO_2 variations over the last three glacial-interglacial climatic cycles deduced from the Dome Fuji deep ice core, Antarctica using a wet extraction technique. Tellus 55B:126–137.

Kawamura, K., F. Parrenin, L. Lisiecki, et al. 2007. Northern hemisphere forcing of climatic cycles in Antarctica over the past 360,000 years. Nature 448:912–917.

Kaye, C. A. and E. S. Barghoorn. 1964. Late Quaternary sea-level change and crustal rise at Boston, Massachusetts, with notes on the autocompaction of peat. Geological Society of America Bulletin 75:63–80.

Kearney, M. S. 1996. Sea-level change during last thousand years in Chesapeake Bay. Journal of Coastal Research 12:977–983.

Keeling, C. D. 1973. Industrial production of carbon dioxide from fossil fuels and limestone. Tellus 5:174–198.

Keeling, C. D., R. B. Bacastow, A. E. Bainbridge, et al. 1976. Atmospheric carbon dioxide variations at Mauna Loa Observatory, Hawaii. Tellus 28:538–551.

Keeling, R. F. and B. B. Stephens. 2001. Antarctic sea ice and the control of Pleistocene climate instability. Paleoceanography 16:112–131, 330–334.

Keigwin, L. D. 1982. Isotopic paleoceanography of the Caribbean and East Pacific: Role of Panama Uplift in late Neogene time. Science 217:350–353.

Keigwin, L. D. 1996. The Little Ice Age and Medieval Warm Period in the Sargasso Sea. Science 274:1504–1508.

Keigwin, L. D. 2004. Radiocarbon and stable isotope constraints on Last Glacial Maximum and Younger Dryas ventilation in the western North Atlantic. Paleoceanography 19:PA001029.

Keigwin, L. D. and E. A. Boyle. 1999. Surface and deep ocean variability in the northern Sargasso Sea during Marine Isotope Stage 3. Paleoceanography 14:164–170.

Keigwin, L. D. and E. A. Boyle. 2000. Detecting Holocene changes in thermohaline circulation. Proceedings of the National Academy of Sciences 97:1343–1346.

Keigwin, L. D. and E. A. Boyle. 2008. Did North Atlantic overturning halt 17,000 years ago? Paleoceanography 23:PA1101, 001500.

Keigwin, L. D. and G. A. Jones. 1989. Glacial-Holocene stratigraphy, chronology, and paleoceanographic observations on some North Atlantic sediment drifts. Deep-Sea Research 36:845–867.

Keigwin, L. D. and G. A. Jones. 1994. Western North Atlantic evidence for millennial-scale changes in ocean circulation and climate. Journal of Geophysical Research 99(C6):12397–12410.

Keigwin, L. D., G. A. Jones, S. J. Lehman, and E. A. Boyle. 1991. Deglacial meltwater discharge, North Atlantic deep circulation, and abrupt climate change. Journal of Geophysical Research, Oceans 96:16811–16826.

Keigwin, L. D. and R. S. Pickart. 1999. Slope water current over the Laurentian Fan on interannual to millennial time scales. Science 286:520–523.

Keigwin, L. D., J. Sachs, Y. Rosenthal, and E. A. Boyle. 2005. The 8200 year B.P. event in the slope water system, western subpolar North Atlantic. Paleoceanography 20:PA001074.

Kelley, J. T., D. F. Belknap, and J. F. Daly. 2001. Comment on "North Atlantic climate-ocean variations and sea level in Long Island Sound, Connecticut, since 500 cal yr A.D." Quaternary Research 55:105–107.

Kemp, A. E. S. and R. C. Dugdale. 2006. The role of diatom production and Si flux and burial in the regulation of global cycles. Global Biogeochemical Cycles 20:GB4S01.

Kemp, D. B., A. L. Coe, A. S. Cohen, and L. Schwark. 2005. Astronomical pacing of methane release in the early Jurassic period. Nature 437:396–399.

Kemp, D. B., A. L. Coe, A. S. Cohen, and L. Schwark. 2006. Palaeoceanography: Methane release in the early Jurassic period (reply). Nature 441:E5–E6.

Kendall, B. S., R. A. Creaser, G. M. Ross, and D. Selby. 2004. Constraints on the timing of Marinoan "Snowball Earth" glaciation by ^{187}Re–^{187}Os dating of a Neoproterozoic post-glacial black shale in Western Canada. Earth and Planetary Science Letters 222:729–740.

Kennedy, M. J., N. Christie-Blick, and A. R. Prave. 2001a. Carbon isotopic composition of Neoproterozoic glacial carbonates as a test of paleoceanographic models for Snowball Earth phenomena. Geology 29:1135–1138.

Kennedy, M. J., N. Christie-Blick, and L. E. Sohl. 2001b. Are Proterozoic cap carbonates and isotopic excursions a record of gas hydrate destabilization following Earth's coldest intervals? Geology 29:443–446.

Kennedy, M. J., B. Runnegar, A. R. Prave, K. H. Hoffman, and M. Arthur. 1998. Two or four Neoproterozoic glaciations? Geology 26:1059–1063.

Kennett, J. P. 1977. Cenozoic evolution of Antarctic glaciation, the circum-Antarctic oceans and their impact on global paleoceanography. Journal of Geophysical Research 82:3843–3859.

Kennett, J. P. 1990. The Younger Dryas cooling event: An introduction. Paleoceanography 5:891–895.

Kennett, J. P., K. G. Cannariato, I. L. Hendy, and R. J. Behl. 2000a. Carbon isotopic evidence for methane hydrate instability during Quaternary interstadials. Science 288:128–133.

Kennett, J. P., K. G. Cannariato, I. L. Hendy, and R. J. Behl. 2000b. Methane Hydrates in Quaternary Climate Change: The Clathrate Gun Hypothesis. Washington, DC: American Geophysical Union.

Kennett, J. P., K. Elmstrom, and N. L. Penrose. 1985. The deglaciation in the Orca Basin Gulf of Mexico: High-resolution planktonic foraminifera. Palaeogeography, Palaeoclimatology, Palaeoecology 50:189–216.

Kennett, J. P. and N. P. Exon. 2004. Paleoceanographic evolution of the Tasmanian Seaway and its climatic implications. In N. F. Exon, J. P. Kennett, and M. J. Malone, eds., The Cenozoic Southern Ocean: Tectonics, Sedimentation and Climate Change Between Australia and Antarctica, pp. 345–367. Washington, DC: American Geophysical Union.

Kennett, J. P. and N. J. Shackleton. 1975. Laurentide Ice Sheet meltwater recorded in the Gulf of Mexico. Science 188:147–150.

Kennett, J. P. and N. J. Shackleton. 1976. Oxygen isotopic evidence for the development of the psychrosphere 38 Myr ago. Nature 260:513–515.

Kennett, J. P. and L. D. Stott. 1991. Abrupt deep sea warming, paleoceanographic changes and benthic extinctions at the end of the Paleocene. Nature 353:225–229.

Kerr, R. A. 2000. A North Atlantic climate pacemaker for the centuries. Science 288:1984–1985.

Kerwin, M. W. 1996. A regional stratigraphic isochron (ca. 8000 ^{14}C yr B.P.) from final deglaciation of Hudson Strait. Quaternary Research 46:89–98.

Kiefer, T., S. Lorenz, M. Schulz, et al. 2002. Response of precipitation over Greenland and the adjacent ocean to North-Pacific warm spells during the last glacial period: Implications for Dansgaard-Oeschger cycles. Terra Nova 14:295–300.

Kiefer, T., M. Sarnthein, H. Erlenkeuser, P. Grootes, and A. Roberts. 2001. North Pacific response to millennial-scale changes in ocean circulation over the last 65 ky. Paleoceanography 16:179–189.

Kiehl, J. and K. Trenberth. 1997. Earth's annual global mean energy budget. Bulletin of the American Meteorological Society 78:197–206.

Kienast, M., S. Steinke, K. Stattegger, and S. E. Calvert. 2001. Synchronous tropical South China Sea SST change and Greenland warming during deglaciation. Science 291: 2132–2134.

Kim, J.-H., H. Meggers, N. Rimbu, et al. 2007. Impacts of the North Atlantic gyre circulation on Holocene climate off northwest Africa. Geology 35:387–390.

Kim, J.-H., N. Rimbu, S. J. Lorenz, et al. 2004. North Pacific and North Atlantic sea-surface temperature variability during the Holocene. Quaternary Science Reviews 23: 2141–2154.

Kindler, P. and P. J. Hearty. 2000. Elevated marine terraces from Eleuthera (Bahamas) and Bermuda: Sedimentological, petrographic, and geochronological evidence for important deglaciation events during the middle Pleistocene. Global and Planetary Change 24:41–58.

Kirschvink, J. L. 1992. Late Proterozoic low-latitude global glaciation: The Snowball Earth. In J. W. Schopf and C. C. Klein, eds., The Proterozoic Biosphere: A Multidisciplinary Study, pp. 51–52. Cambridge: Cambridge University Press.

Kitagawa, H. and J. van der Plicht. 1998. Atmospheric radiocarbon calibration to 45,000 yrs BP: Late glacial fluctuations and cosmogenic isotope production. Science 279:1187–1100.

Kitzberger, T., P. M. Brown, E. K. Heyerdahl, T. W. Swetnam, and T. T. Veblen. 2007. Contingent Pacific–Atlantic Ocean influence on multicentury wildfire synchrony over western North America. Proceedings of the National Academy of Sciences 104:543–548.

Kleiven, H. F., C. Kissel, C. Laj, U. S. Ninnemann, T. O. Richter, and E. Cortijo. 2007. Reduced North Atlantic deep water coeval with the glacial Lake Agassiz freshwater outburst. Science 319:60–64.

Klemm, V., S. Levasseur, M. Frank, J. R. Hein, and A. N. Halliday. 2005. Osmium isotope stratigraphy of a marine ferromanganese crust. Earth and Planetary Science Letters 238:42–48.

Kleypas, J. A., R. A. Feely, V. J. Fabry, et al. 2006. Impacts of Ocean Acidification on Coral Reefs and Other Marine Calcifiers: A Guide for Future Research. Report of a workshop sponsored by the National Science Foundation, the National Oceanographic and Atmospheric Administration, and the U.S. Geological Survey.

Kleypas, J. A. and C. Langdon. 2006. Coral reefs and changing seawater chemistry. In J. T. Phinney, et al., eds., Coral Reefs and Climate Change: Science and Management, pp. 73–110. Washington, DC: American Geophysical Union.

Kluge, T., T. Marx, D. Scholz, et al. 2008. A new tool for palaeoclimate reconstruction: Noble gas temperatures from fluid inclusions in speleothems. Earth and Planetary Science Letters 269:407–414.

Knauth, L. P. 2005. Temperature and salinity history of the Precambrian ocean: Implications for the course of microbial evolution. Palaeogeography, Palaeoclimatology, Palaeoecology 219:53–69.

Knauth, L. P. and D. R. Lowe. 2003. High Archean climatic temperature inferred from oxygen isotope geochemistry of cherts in the 3.5 Ga Swaziland Supergroup, South Africa. Geological Society of America Bulletin 115:566–580.

Knies, J., H.-P. Kleiber, J. Matthiessen, C. Müller, and N. Nowaczyk. 2001. Marine ice-rafted debris records constrain maximum extent of Saalian and Weichselian ice sheets along the northern Eurasian margin. Global and Planetary Change 31:45–64.

Knies, J., J. Matthiessen, C. Vogt, and R. Stein. 2002. Evidence of "mid-Pliocene (~3 Ma) global warmth" in the eastern Arctic Ocean and implications for the Svalbard/Barents Sea Ice Sheet during the late Pliocene and early Pleistocene (~3–1.7 Ma). Boreas 31:82–93.

Knight, J. R., R. J. Allan, C. K. Folland, M. Vellinga, and M. E. Mann. 2005. A signature of persistent natural thermohaline circulation cycles in observed climate. Geophysical Research Letters 32:L20708.

Knoll, A. H. 2003. *Life on a Young Planet: The First Three Billion Years of Evolution on Earth*. Princeton, NJ: Princeton University Press.

Knoll, A. H. and S. B. Carroll. 1999. The early evolution of animals: Emerging views from comparative biology and geology. Science 284:2129–2137.

Knoll, A. H., J. M. Hayes, A. J. Kaufman, K. Swett, and I. B. Lambert. 1986. Secular variation in carbon isotope ratios from Upper Proterozoic successions of Svalbard and East Greenland. Nature 321:831–838.

Knowles, N., M. Dettinger, and D. Cayan. 2006. Trends in snowfall versus rainfall for the Western United States. Journal of Climate 19:4545–4559.

Knox, F. and M. B. McElroy. 1984. Changes in atmospheric CO_2: Influence of the marine biota at high latitudes. Journal of Geophysical Research 89:4629–4637.

Knudsen, K. L., H. Jiang, J. Hansen, et al. 2004. Environmental changes off North Iceland during the deglaciation and the Holocene: Foraminifera, diatoms, and stable isotopes. Marine Micropaleontology 50:273–305.

Knudsen, K. L., M. K. B. Søndergaard, J. Eiríksson, and H. Jiang. 2008. Holocene thermal maximum off North Iceland: Evidence from benthic and planktonic foraminifera in the 8600–5200 cal year BP time slice. Marine Micropaleontology 67:120–142.

Knutson, T. R., J. J. Sirutis, S. T. Garner, I. M. Held, and R. E. Tuleya. 2007. Simulation of the recent multidecadal increase of Atlantic hurricane activity using an 18-km-grid regional model. Bulletin of the American Meteorological Society 88:1549–1565.

Koç, N. and E. Jansen. 1994. Response of the high-latitude northern hemisphere to orbital climatic forcing: Evidence from the Nordic Seas. Geology 22:523–526.

Koç, N., E. Jansen, M. Hald, and L. Labeyrie. 1996. Lateglacial-Holocene sea-surface temperatures and gradients between the North Atlantic and the Norwegian Sea: Implications for the Nordic heat pump. In J. T. Andrews, W. E. N. Austin, and H. E. Bergsten, eds., *The Late Glacial Paleoceanography of the North Atlantic Margins*, pp. 177–185. London: Geological Society of London.

Koch, J., and J. J. Clague. 2006. Are insolation and sunspot activity the primary drivers of global Holocene glacier fluctuations? PAGES Newsletter 14:20–21.

Koch, P. L., W. C. Clyde, R. P. Hepple, et al. 2003. Isotopic records of carbon cycle and climate change across the Paleocene-Eocene boundary from the Bighorn Basin, Wyoming. In S. L. Wing, et al., eds., *Causes and Consequences of Globally Warm Climates in the Early Paleogene*, pp. 49–64. Boulder, CO: Geological Society of America.

Koch, P. L., J. C. Zachos, and P. D. Gingerich. 1992. Correlation between isotope records in marine and continental carbon reservoirs near the Palaeocene/Eocene boundary. Nature 358:319–322.

Koç Karpuz, N. and E. Jansen. 1992. A high resolution diatom record of the last deglaciation from the southeastern Norwegian Sea: Documentation of rapid climatic changes. Paleoceanography 5:557–580.

Kohfeld, K. E. and S. P. Harrison. 2000. How well can we simulate past climates? Evaluating the models using global environmental data sets. Quaternary Science Reviews 19:321–346.

Köhler, P., H. Fischer, G. Munhoven, and R. E. Zeebe. 2005. Quantitative interpretation of atmospheric carbon records over the last glacial termination. Global Biogeochemical Cycles 19:GB4020.

Kominz, M. 1984. Oceanic ridge volumes and sea-level change—An error analysis. In J. S. Schlee, ed., *Interregional Unconformities and Hydrocarbon Accumulation*, pp. 109–127. Tulsa, OK: American Association of Petroleum Geologists.

Kominz, M. A., K. G. Miller, and J. V. Browning. 1998. Long-term and short-term global Cenozoic sea-level estimates. Geology 26:311–314.

Kominz, M. A. and S. F. Pekar. 2001. Oligocene eustasy from two-dimensional sequence stratigraphic backstripping. Geological Society of America Bulletin 113:291–304.

Korte, C., T. Jasper, H. W. Kozur, and J. Veizer. 2005. $\delta^{18}O$ and $\delta^{13}C$ of Permian brachiopods: A record of seawater evolution and continental glaciation. Palaeogeography, Palaeoclimatology, Palaeoecology 224:333–351.

Kortekaas, M., A. S. Murray, P. Sandgren, and S. Björck. 2006. OSL chronology for a sediment core from the southern Baltic Sea: A continuous sedimentation record since deglaciation. Quaternary Geochronology 2:95–101.

Kortenkamp, S. J. and S. F. Dermott. 1998. A 100,000-year periodicity in the accretion rate of interplanetary dust. Science 280:874–876.

Koutavas, A., P. B. deMenocal, G. C. Olive, and J. Lynch-Stieglitz. 2006. Mid-Holocene El Niño–Southern Oscillation (ENSO) attenuation revealed by individual foraminifera in eastern tropical Pacific sediments. Geology 34:993–996.

Koutavas, A. and J. Lynch-Stieglitz. 2003. Glacial-interglacial dynamics of the eastern equatorial Pacific cold tongue-Intertropical Convergence Zone system reconstructed from oxygen isotope records. Paleoceanography 18:PA000894.

Koutavas, A. and J. Lynch-Stieglitz. 2004. Variability of the marine ITCZ over the eastern Pacific during the past 30,000 years: Regional perspective and global context. In R. Bradley and H. Diaz, eds., *The Hadley Circulation: Present Past and Future*, pp. 347–369. New York: Springer.

Koutavas, A., J. Lynch-Stieglitz, T. M. Marchitto Jr., and P. S. Julian. 2002. El Niño–like pattern in ice age tropical Pacific sea surface temperature. Science 297:226–230.

Kouwenberg, L. L. R., J. C. McElwain, W. M. Kürschner, et al. 2003. Stomatal frequency adjustment of four conifer species to historical changes in atmospheric CO_2. American Journal of Botany 90:610–619.

Kristjánsdóttir, G. B., J. S. Stoner, A. E. Jennings, J. T. Andrews, and K. Gronveld. 2007. Geochemistry of Holocene cryptotephras from the North Iceland Shelf (MD99-2269): Intercalibration with radiocarbon and palaeomagnetic chronostratigraphies. The Holocene 17:155–176.

Kristoffersen, Y., B. Coakley, W. Jokat, et al. 2004. Seabed erosion on the Lomonosov Ridge, central Arctic Ocean: A tale of deep draft icebergs in Eurasia Basin and the influence of Atlantic water inflow on iceberg motion? Paleoceanography 19:1–14.

Krom, M. D., J. D. Stanley, R. A. Cliff, and J. C. Woodward. 2002. Nile River sediment fluctuations over the past 7000 yr and their key role in sapropel development. Geology 30:71–74.

Kromer, B., M. Spurk, S. Remmele, M. Barbetti, and V. Toniello. 1998. Segments of atmospheric ^{14}C change as derived from late Glacial and early Holocene floating tree-ring series. Radiocarbon 40:351–358.

Kröpelin, S., D. Verschuren, A.-M. Lézine, et al. 2008. Climate-driven ecosystem succession in the Sahara: The past 6000 years. Science 320:765–768.

Kucera, M., A. Rosell-Melé, R. Schneider, C. Waelbroeck, and M. Weinelt. 2005. Multiproxy approach for the reconstruction of the glacial ocean surface (MARGO). Quaternary Science Reviews 24:813–819.

Kucharik, C. J., J. A. Foley, C. Delire, et al. 2000. Testing the performance of a dynamic global ecosystem model: Water balance, carbon balance, and vegetation structure. Global Biogeochemical Cycles 14:795–825.

Kudrass, H. R., A. Hofmann, H. Doose, K. Emeis, and H. Erlenkeuser. 2001. Modulation and amplification of climatic changes in the northern hemisphere by the Indian summer monsoon during the last 80 k.y. Geology 29:63–66.

Kuhlemann, J., E. J. Rohling, I. Krumrei, et al. 2008. Regional synthesis of Mediterranean atmospheric circulation during the Last Glacial Maximum. Science 321:1338–1340.

Kuhnt, W., A. Holbourn, R. Hall, M. Zuvela, and R. Käse. 2004. Neogene history of the Indonesian throughflow. In P. Clift, et al., eds., Continent-Ocean Interactions Within the East Asian Marginal Seas, pp. 299–320. Washington, DC: American Geophysical Union.

Kuiper, K. F., A. Deino, F. J. Hilgen, et al. 2008. Synchronizing rock clocks of earth history. Science 320:500–504.

Kukla, G. 1987. Loess stratigraphy in central China. Quaternary Science Reviews 6:191–219.

Kukla, G. and Z. An. 1989. Loess stratigraphy in Central China. Palaeogeography, Palaeoclimatology, Palaeoecology 72:203–225.

Kumar, N., R. F. Anderson, R. A. Mortlock, et al. 1995. Increased biological productivity and export production in the glacial Southern Ocean. Nature 378:675–680.

Kump, L. R. 2001. What drives climate? Nature 408:651–652.

Kump, L. R. 2002. Reducing uncertainty about carbon dioxide as a climate driver. Nature 419:188–190.

Kump, L. R., A. Pavlov, A., and M. A. Arthur. 2005. Massive release of hydrogen sulfide to the surface ocean and atmosphere during intervals of oceanic anoxia. Geology 33:397–400.

Kump, L. R. and D. Pollard. 2008. Amplification of Cretaceous warmth by biological cloud feedbacks. Science 320:195.

Kuper, R. and S. Kröpelin. 2006. Climate-controlled Holocene occupation in the Sahara: Motor of Africa's evolution. Science 313:803–807.

Kürschner, W. M., J. van der Burgh, H. Visscher, and D. L. Dilcher. 1996. Oak leaves as biosensors of late Neogene and early Pleistocene paleoatmospheric CO_2 concentrations. Marine Micropaleontology 27:299–312.

Kurtz, A. C., L. R. Kump, M. A. Arthur, J. C. Zachos, and A. Paytan. 2003. Early Cenozoic decoupling of the global carbon and sulfur cycles. Paleoceanography 18:1090, doi:10.1029/2003PA000908.

Kushnir, Y., W. A. Robinson, I. Blade, et al. 2002. Atmospheric GCM response to extratropical SST anomalies: Synthesis and evaluation. Journal of Climate 15:2233–2256.

Kushnir, Y. and J. M. Wallace. 1989. Interaction of low- and high-frequency transients in forecast experiment with a general circulation model. Journal of Atmospheric Sciences 46:1411–1418.

Kutzbach, J. E., P. J. Bartlein, J. A. Foley, et al. 1996. Potential role of vegetation feedback in the climate sensitivity of high-latitude regions. Global Biogeochemical Cycles 10:727–736.

Kutzbach, J. E. and P. Behling. 2004. Comparison of simulated changes of climate in Asia for two scenarios: Early Miocene to present, and present to future enhanced greenhouse. Global and Planetary Change 41:157–165.

Kutzbach, J. E. and P. J. Guetter. 1986. The influence of changing orbital parameters and surface boundary conditions on climate simulations for the past 18000 years. Journal of the Atmospheric Sciences 43:1726–1759.

Kutzbach, J. E. and Z. Liu. 1997. Response of the African monsoon to orbital forcing and ocean feedbacks in the middle Holocene. Science 278:440–443.

Kutzbach, J. E., W. F. Ruddiman, and W. L. Prell. 1997. Possible effects of Cenozoic uplift and CO_2 lowering on global and regional hydrology. In W. F. Ruddiman, ed., Tectonic Uplift and Climate Change, pp. 149–170. New York: Plenum.

Kutzbach, J. E. and A. F. Street-Perrott. 1985. Milankovitch forcing of fluctuations in the level of tropical lakes from 18 to 0 kyr BP. Nature 317:130–134.

Labeyrie, L., J. Cole, K. Alverson, and T. Stocker. 2003. The history of climate dynamics in the late Quaternary. In K. D. Alverson, R. S. Bradley, and T. F. Pedersen, eds., Paleoclimate, Global Change and the Future, pp. 33–61. New York: Springer.

Labeyrie, L. D., J.-C. Duplessy, and P. L. Blanc. 1987. Variations in mode of formation and temperature of oceanic deep waters over the past 125,000 years. Nature 327:477–482.

Labeyrie, L., H. Leclaire, C. Waelbroeck, et al. 1999. Temporal variability of the surface and deep waters of the N-W Atlantic Ocean at orbital & millennial scales. In P. U. Clark, R. S. Webb, and L. D. Keigwin, eds., Mechanisms of Millennial-Scale Climate Change, pp.77–98. Washington, DC: American Geophysical Union.

Labeyrie, L., C. Waelbroeck, E. Cortijo, E. Michel, and J. C. Duplessy. 2005. Changes in deep water hydrology during the last deglaciation. Compte Rendu Geoscience 337:919–927.

Ladurie, E. L. 1971. Times of Feast, Times of Famine. New York: Doubleday.

Laird, K. R., S. C. Fritz, E. C. Grimm, and P. C. Mueller. 1996. Century-scale paleoclimatic reconstruction from Moon Lake, a closed basin lake in the Northern Great Plains. Limnology and Oceanography 41:890–902.

Laj, C., C. Kissel, A. Mazaud, J. E. T. Channell, and J. Beer. 2000. North Atlantic paleointensity stack since 75 kal (NAPIS-75) and the duration of the Laschamp event. Philosophical Transactions of the Royal Societyof London, Series A 358:1009–1025.

Laj, C., C. Kissel, A. Mazaud, E. Michel, R. Muscheler, and J. Beer. 2002. Geomagnetic field intensity, North Atlantic Deep Water circulation and atmospheric ^{14}C during the last 50 kyr. Earth and Planetary Science Letters 200:177–190.

Lamb, H. H. 1965. The early Medieval Warm Epoch and its sequel. Palaeogeography, Palaeoclimatology, Palaeoecology 1:13–27.

Lamb, H. H. 1970. Volcanic dust in the atmosphere; with a chronology and assessment of its meteorological significance. Philosophical Transactions of the Royal Society of London Series A 266:425–533.

Lamb, H. H. 1977. *Climate History and the Future*, vol. 2, *Climate Present, Past, and Future*. London: Methuen &. Co.

Lamb, H. H. 1983. Update of the chronology of assessments of the volcanic dust veil index. Climate Monitoring 12:79–90.

Lamb, H. H. 1995. *Climate History and the Modern World*, 2nd ed. London: Routledge.

Lambeck, K. 1993. Glacial rebound and sea-level change: An example of a relationship between mantle and surface processes. Tectonophysics 223:15–37.

Lambeck, K. 1997. Sea-level change along the French Atlantic and Channel Coasts since the time of the Last Glacial Maximum. Palaeogeography, Palaeoclimatology, Palaeoecology 129:1–22.

Lambeck, K. 2002. Sea level change from mid Holocene to recent time: An Australian example with global implications. In J. X. Mitrovica and B. Vermeersen, eds., *Ice Sheets, Sea Level and the Dynamic Earth*, pp. 33–50. Washington, DC: American Geophysical Union.

Lambeck, K., F. Antonioli, A. Purcell, and S. Silenzi. 2004b. Sea-level change along the Italian coast for the past 10,000 years. Quaternary Science Reviews 23:1567–1598.

Lambeck, K., M. Anzidei, F. Antonioli, A. Benini, and A. Esposito. 2004a. Sea level in Roman time in the central Mediterranean and implications for recent change. Earth and Planetary Science Letters 224:563–575.

Lambeck, K. and E. Bard. 2000. Sea-level change along the French Mediterranean coast for the past 30,000 years. Earth and Planetary Science Letters 175:203–222.

Lambeck, K. and J. Chappell. 2001. Sea level change through the last glacial cycle. Science 292:679–686.

Lambert, F., B. Delmonte, J. R. Petit, et al. 2008. Dust-climate couplings over the past 800,000 years from the EPICA Dome C ice core. Nature 452:616–619.

Lamy, F., J. Kaiser, U. Ninnemann, et al. 2004. Antarctic timing of surface water changes off Chile and Patagonian Ice Sheet response. Science 304:1959–1962.

Landais, A., J.-M. Barnola, V. Masson-Delmotte, et al. 2004a. A continuous record of temperature evolution over a sequence of Dansgaard-Oeschger events during Marine Isotopic Stage 4 (76 to 62 kyr BP). Geophysical Research Letters 31:L22211.

Landais, A., V. Masson-Delmotte, J. Jouzel, et al. 2006. The glacial inception as recorded in the NorthGRIP Greenland ice core: Timing, structure and associated abrupt temperature changes. Climate Dynamics 26:273–284.

Landais, A., J. P. Steffensen, N. Caillon, et al. 2004b. Evidence for stratigraphic distortion in the Greenland Ice Core Project (GRIP) ice core during Event 5e1 (120 kyr BP) from gas isotopes. Journal of Geophysical Research 109:D06103, JD004193.

Landsea, C. W. 2005. Hurricanes and global warming. Nature 438:E11–13.

Landsea, C. W. 2007. Counting Atlantic tropical cyclones back to 1900. EOS, Transactions, American Geophysical Union 88:197.

Landsea, C. W., B. A. Harper, K. Hoarau, and J. A. Knaff. 2006. Can we detect trends in extreme tropical cyclones? Science 313:452–454.

Landsea, C. W. and J. A. Knaff. 2000. How much skill was there in forecasting the very strong 1997–98 El Niño? Bulletin of the American Meteorological Society 81:2107–2119.

Landvik, J. Y., S. Bondevik, A. Elverhøi, et al. 1998. The Last Glacial Maximum of the Barents Sea and Svalbard area: Ice sheet extent and configuration. Quaternary Science Reviews 17:43–75.

Lang, C., M. Leuenberger, J. Schwander, and S. Johnsen. 1999. 16°C rapid temperature variation in central Greenland 70,000 years ago. Science 286:934–937.

Langdon, P. G., K. E. Barber, and P. D. M. Hughes. 2003. A 7500 year peat-based palaeoclimatic reconstruction and evidence for an 1100 year cyclicity of surface wetness from Temple Hill Moss, Pentland Hills, Southeast Scotland. Quaternary Science Reviews 22:259–274.

Langway, C. C., Jr., H. O. Oeschger, and W. Dansgaard, eds. 1985. *Greenland Ice Cores: Geophysics, Geochemistry, and the Environment*. Washington, DC: American Geophysical Union.

Langway, C. C., Jr., K. Osada, H. B. Clausen, C. U. Hammer, and H. Shoji. 1995. A 10-century comparison of prominent bipolar volcanic events in ice cores. Journal of Geophysical Research 100:16241–16247.

Larsen, C. E. and I. Clark. 2006. A search for scale in sea-level studies. Journal of Coastal Research 22:788–800.

Larsen, H. C., A. D. Saunders, P. D. Clift, J. Beget, W. Wei, S. Spezzaferri, and ODP Leg 152 Scientific Party. 1994. Seven million years of glaciation in Greenland. Science 264:952–955.

Larson, R. L. 1991a. Latest pulse of Earth: Evidence for a mid-Cretaceous superplume. Geology 19:547–550.

Larson, R. L. 1991b. Geological consequences of superplumes. Geology 19:963–966.

Lasalle, P. and J. Elson. 1975. Emplacement of the St. Narcisse moraine as a climatic event in eastern Canada. Quaternary Research 5:621–625.

Laskar, J. 1999. The limits of Earth orbital calculations for geological time-scale use. Philosophical Transactions of the Royal Society of London, Series A 357:1735–1759.

Laskar, J., F. Joutel, and F. Boudin. 1993. Orbital, precessional, and insolation quantities for the earth from –20 Myr to +10 Myr. Astronomy and Astrophysics 270:522–533.

Laskar, J., P. Robutel, F. Joutel, et al. 2004. A long-term numerical solution for the insolation quantities of the earth. Astronomy and Astrophysics 428:261–285.

Lassen, S. J., A. Kuijpers, H. Kunzendorf, et al. 2004a. Late-Holocene Atlantic bottom-water variability in Igaliku Fjord, South Greenland, reconstructed from foraminifera faunas. The Holocene 14:165–171.

Last, W. M., J. T. Teller, and R. M. Forester. 1994. Paleohydrology and paleochemistry of Lake Manitoba, Canada: The isotope and ostracode records. Journal of Paleolimnology 12:269–282.

Latif, M., C. Böning, J. Willebrand, et al. 2006. Is the thermohaline circulation changing? Journal of Climate 19:4631–4637.

Latif, M., C. Böning, J. Willebrand, et al. 2007. Decadal to multidecadal variability of the Atlantic MOC: Mechanisms and predictability. In A. Schmittner, J. C. H. Chiang, and S. R. Hemming, eds., *Ocean Circulation: Mechanisms and Impacts—Past and Future Changes of Meridional Overturning*, pp. 149–166. Washington, DC: American Geophysical Union.

Latif, M., E. Roeckner, M. Botzet, et al. 2004. Reconstructing, monitoring, and predicting multidecadal-scale changes in the North Atlantic thermohaline circulation with sea surface temperature. Journal of Climate 17:1605–1614.

Latimer, J. C. and G. Filippelli. 2002. Eocene to Miocene terrigenous inputs, paleoproductivity, and the onset of the Antarctic Circumpolar Current: Geochemical evidence from ODP Leg 177, Site 1090. Paleogeography, Paleoclimatology, and Paleoecology 182:151–164.

Lawrence, K. T., Z.-H. Liu, and T. D. Herbert. 2006. Evolution of the eastern tropical Pacific through Plio-Pleistocene glaciation. Science 312:79–83.

Lawver, L. A. and L. M. Gahagan. 2003. Evolution of Cenozoic seaways in the circum-Antarctic region. Palaeogeography, Palaeoclimatology, Palaeoecology 198:11–37.

Lazarus, D., B. Bittniok, L. Diester-Haass, et al. 2008. Radiolarian and sedimentologic paleoproductivity proxies in late Pleistocene sediments of the Benguela Upwelling System, ODP Site 1084. Marine Micropaleontology 68:223–235.

Lea, D. 1995. A trace metal perspective on the evolution of the Antarctic circumpolar deep water chemistry. Paleoceanography 10:733–745.

Lea, D. W. 2004. The 100,000 year cycle in tropical SST, greenhouse forcing, and climate sensitivity. Journal of Climate 17:2170–2179.

Lea, D. W., T. A. Mashiotta, and H. J. Spero. 1999. Controls on magnesium and strontium uptake in planktonic foraminifera determined by live culturing. Geochimica et Cosmochimica Acta 63:2369–2380.

Lea, D. W., D. K. Pak, C. L. Belanger, et al. 2006. Paleoclimate history of Galápagos surface waters over the last 135,000 yr. Quaternary Science Reviews 25:1152–1167.

Lea, D. W., D. K. Pak, L. C. Peterson, and K. A. Hughen. 2003. Synchroneity of tropical and high-latitude Atlantic temperatures over the last glacial termination. Science 301:1361–1364.

Lea, D. W., D. K. Pak, and H. J. Spero. 2000. Climate impact of late Quaternary equatorial Pacific sea surface temperature variations. Science 289:1719–1724.

Lea, D. W., G. T. Shen, and E. A. Boyle. 1989. Coralline barium records temporal variability in equatorial Pacific upwelling. Nature 340:373–576.

Lean, J. 2000. Evolution of the sun's spectral irradiance since the Maunder Minimum. Geophysical Research Letters 27:2425–2428.

Lean, J., J. Beer, and R. Bradley. 1995. Reconstruction of solar irradiance since 1610: Implications for climate change. Geophysical Research Letters 22:3195–3198.

Lean, J., G. Rottman, J. Harder, and G. Kopp, 2005. SORCE contributions to new understanding of global change and solar variability. Solar Physics 230:27–53.

Lean, J., A. Skumanich, and O. White. 1992. Estimating the sun's radiative output during the Maunder Minimum. Geophysical Research Letters 19:1591–1594.

Lear, C. H. 2007. Mg/Ca paleothermometry: A new window into Cenozoic climate change. In M. Williams, et al., eds., *Deep-Time Perspectives on Climate Change: Marrying the Signal from Computer Models and Biological Proxies*, pp. 313–332. London: Geological Society of London.

Lear, C. H., T. R. Bailey, P. N. Pearson, H. K. Coxall, and Y. Rosenthal. 2008. Cooling and ice growth across the Eocene-Oligocene transition. Geology 36:251–254.

Lear, C. H., H. Elderfield, and P. A. Wilson. 2000. Cenozoic deep-sea temperatures and global ice volumes from Mg/Ca in benthic foraminiferal calcite. Science 287:269–272.

Lear, C. H., H. Elderfield, and P. A. Wilson. 2003. A Cenozoic seawater Sr/Ca record from benthic foraminiferal calcite and its application in determining global weathering fluxes. Earth and Planetary Science Letters 208:69–84.

Lear, C. H. and Y. Rosenthal. 2006. Benthic foraminiferal Li/Ca: Insights into Cenozoic seawater carbonate saturation state. Geology 34:985–988.

Leather, J., P. A. Allen, M. D. Brasier, and A. Cozzi. 2002. Neoproterozoic Snowball Earth under scrutiny: Evidence from the Fiq glaciation of Oman. Geology 30:891–894.

Leckie, R. M., T. Bralower, and R. Cashman. 2002. Oceanic anoxic events and plankton evolution: Biotic response to tectonic forcing during the mid-Cretaceous. Paleoceanography 17:PA000623.

Leder, J. J., P. K. Swart, A. M. Szmant, and R. E. Dodge. 1996. The origin of variations in the isotopic record of scleractinian corals—1. Oxygen. Geochimica et Cosmochimica Acta 60:2857–2870.

Ledwell, J. L., E. T. Montgomery, K. L. Polzin, L. C. St. Laurent, R. W. Schmitt, and J. M. Toole. 2000. Evidence for enhanced mixing over rough topography in the abyssal ocean. Nature 403:179–182.

LeGrande, A. N., and G. A. Schmidt. 2006. Global gridded data set of the oxygen isotopic composition in seawater. Geophysical Research Letters 33:L12604.

LeGrande, A. N., and G. A. Schmidt. 2008. Ensemble, water-isotope enabled, coupled general circulation modeling insights into the 8.2-kyr event. Paleoceanography 23:PA3207, 001610.

LeGrande, A. N., G. A. Schmidt, D. T. Shindell, et al. 2006. Consistent simulations of multiple proxy responses to an abrupt climate change event. Proceedings of the National Academy of Sciences 103:837–842.

Lehman, S. 1993. Ice sheets, wayward winds and sea change. Nature 365:108–109.

Lehman, S. J. and L. D. Keigwin. 1992. Sudden changes in North Atlantic circulation during the last deglaciation. Nature 356:757–762.

Lekens, W. A. H., H. P. Sejrup, H. Haflidason, J. Knies, and T. Richter. 2006. Meltwater and ice rafting in the southern Norwegian Sea between 20 and 40 calendar kyr B.P.: Implications for Fennoscandian Heinrich events. Paleoceanography 21:PA3013.

Lekens, W. A. H., H. P. Sejrup, H. Haflidason, et al. 2005. Laminated sediments preceding Heinrich event 1 in the northern North Sea and southern Norwegian Sea: Origin, processes and regional linkage. Marine Geology 216:27–50.

Lemke, P., J. Ren, et al. 2007. Chapter 4: Observations: Changes in snow, ice and frozen ground. In S. Solomon, et al., eds., *Climate Change 2007: The Physical Science Basis*. Contribution of Working Group I to the Fourth Assessment Report of the Intergovernmental Panel on Climate Change, pp. 337–383. Cambridge: Cambridge University Press.

Le Treut, H. and M. Ghil. 1983. Orbital forcing, climatic interactions, and glaciation cycles. Journal of Geophysical Research 88:5167–5190.

Le Treut, H., R. Somerville, et al. 2007. Chapter 1: Historical overview of climate change science. In S. Solomon, et al., eds., *Climate Change 2007: The Physical Science Basis*. Contribution of Working Group I to the Fourth Assessment Report of the Intergovernmental Panel on Climate Change, pp. 93–127. Cambridge: Cambridge University Press.

Leuenberger, M., C. Lang, and J. Schwander. 1999. $\delta^{15}N$ measurements as a calibration tool for the paleothermometer and gas-ice age differences. A case study for the 8200 B.P. event on GRIP ice. Journal of Geophysical Research 104(D18):22163–22170.

Leuschner, D. C. and F. Sirocko. 2000. The low-latitude monsoon climate during Dansgaard-Oeschger cycles and Heinrich Events. Quaternary Science Reviews 19:243–254.

Leventer, A., E. Domack, A. Barkoukis, B. McAndrews, and J. Murray. 2002. Laminations from the Palmer Deep: A diatom-based interpretation. Paleoceanography 17:1–15.

Leventer, A., E. W. Domack, S. E. Ishman, et al. 1996. Productivity cycles of 200–300 years in the Antarctic Peninsula region: Understanding linkages among the sun, atmosphere, oceans, sea ice, and biota. Geological Society of America Bulletin 108:1626–1644.

Leventer, A., D. F. Williams, and J. P. Kennett. 1982. Dynamics of the Laurentide Ice Sheet during the last deglaciation: Evidence from the Gulf of Mexico. Earth and Planetary Science Letters 59:11–17.

Leverett, F. 1932. *Quaternary Geology of Minnesota and Parts of Adjacent States*. Washington, DC: U.S. Government Printing Office.

Leverett, F. and F. B. Taylor. 1915. *The Pleistocene of Indiana and Michigan and History of the Great Lakes*. Washington, DC: U.S. Government Printing Office.

Leverington, D. W., J. D. Mann, and J. T. Teller. 2000. Changes in the bathymetry and volume of glacial Lake Agassiz between 11,000 and 9300 ^{14}C yr BP. Quaternary Research 54:174–181.

Leverington, D. W., J. D. Mann, and J. T. Teller. 2002a. Changes in the bathymetry and volume of glacial Lake Agassiz between 9200 and 7700 ^{14}C yr BP. Quaternary Research 57:244–252.

Leverington, D. W. and J. T. Teller. 2003. Paleotopographic reconstructions of the eastern outlets of glacial Lake Agassiz. Canadian Journal of Earth Sciences 40:1259–1278.

Leverington, D. W., J. T. Teller, and J. D. Mann. 2002b. A GIS model for reconstruction of late Quaternary landscapes from isobase data and modern topography. Computers and Geosciences 28:631–639.

Levesque, A., F. E. Mayle, I. R. Walker, and L. C. Cwynar. 1993. A previously unrecognized late-glacial cold event in eastern North America. Nature 361:623–626.

Levi, C., L. Labeyrie, F. Bassinot, et al. 2007. Low-latitude hydrological cycle and rapid climate changes during the last deglaciation. Geochemistry, Geophysics, Geosystems 8:Q05N12, GC001514.

Levitus, S., J. Antonov, and T. Boyer. 2005. Warming of the world ocean, 1955–2003. Geophysical Research Letters 32:L02604.

Levitus, S., J. I. Antonov, T. F. Boyer, and C. Stephens. 2000. Warming of the world ocean. Science 287:2225–2229.

Levitus, S., J. I. Antonov, J. Wang, et al. 2001. Anthropogenic warming of Earth's climate system. Science 292:267–270.

Lewis, C. F. M. and T. W. Anderson. 1989. Oscillations of levels and cool phases of the Laurentian Great Lakes caused by inflows from glacial Lakes Agassiz and Barlow-Ojibway. Journal of Paleolimnology 2:99–146.

Lewis, C. F. M., T. C. Moore Jr., D. K. Rea, et al. 1994. Lakes of the Huron basin: Their record of runoff from the Laurentide Ice Sheet. Quaternary Science Reviews 13:891–922.

Lewis, J. P., A. J. Weaver, and M. Eby. 2006. Deglaciating the Snowball Earth: Sensitivity to surface albedo. Geophysical Research Letters 33:L23604.

Lewis, J. P., A. J. Weaver, and M. Eby. 2007. Snowball versus Slushball Earth: Dynamic versus nondynamic sea ice? Journal of Geophysical Research 112:C11014.

Li, Z. X., S. V. Bogdanova, A. S. Collins, et al. 2008. Assembly, configuration, and break-up history of Rodinia: A synthesis. Precambrian Research 160:179–210.

Licciardi, J. M., J. T. Teller, and P. U. Clark. 1999. Freshwater routing by the Laurentide Ice Sheet during the last deglaciation. In P. U. Clark, L. Keigwin, and P. Webb., eds., *Mechanisms of Millennial-Scale Global Climate Change*, pp. 177–201. Washington, DC: American Geophysical Union.

Lindsay, R. W. and J. Zhang, 2005. The thinning of arctic sea ice, 1988–2003: Have we passed a tipping point? Journal of Climate 18:4879–4894.

Lindzen, R. S. 1997. Can increasing carbon dioxide cause climate change? Proceedings of the National Academy of Sciences 94:8335–8342.

Ling, H.-F., S-Y. Jiang, M. Frank, et al. 2005. Differing controls over the Cenozoic Pb and Nd isotope evolution of deepwater in the central North Pacific Ocean. Earth and Planetary Science Letters 232:345–361.

Linsley, B. K., R. B. Dunbar, G. M. Wellington, and D. A. Mucciarone. 1994. A coral-based reconstruction of intertropical convergence zone variability over Central America since 1707. Journal of Geophysical Research 99(C5):9977–9994.

Linsley, B. K., A. Kaplan, Y. Gouriou, et al. 2006. Tracking the extent of the South Pacific Convergence Zone since the early 1600s. Geochemistry, Geophysics, Geosystems 7:Q05003, GC001115.

Linsley, B. K., L. Ren, R. B. Dunbar, and S. S. Howe. 2000b. ENSO and decadal-scale climate variability at 10°N in the Eastern Pacific from 1893 to 1994: A coral-based reconstruction from Clipperton Atoll. Paleoceanography 15:322–335.

Linsley, B. K., G. M. Wellington, and D. P. Schrag. 2000a. Decadal sea surface temperature variability in the subtropical South Pacific from 1726 to 1997 AD. Science 290:1145–1148.

Linsley, B. K., G. M. Wellington, D. P. Schrag, et al. 2004. Coral evidence for changes in the amplitude and spatial pattern of South Pacific interdecadal climate variability over the last 300 years. Climate Dynamics 22:1–11.

Lisiecki, L. E. and M. E. Raymo. 2005. A Plio-Pleistocene stack of 57 globally distributed benthic $\delta^{18}O$ records. Paleoceanography 20:PA1003.

Litzow, M. A. 2006. Climate regime shifts and community reorganization in the Gulf of Alaska: How do recent shifts compare with 1976/1977? ICES Journal of Marine Science 63:1386–1396.

Liu, K. B. and M. L. Fearn. 2000. Reconstruction of prehistoric landfall frequencies of catastrophic hurricanes in northwestern Florida from lake sediment records. Quaternary Research 54:238–245.

Liu, T. S. and Z. L. Ding. 1998. Chinese loess and the paleomonsoon. Annual Review of Earth and Planetary Sciences 26:111–145.

Liu, Z., M. A. Altabet, and T. D. Herbert. 2005. Glacial-Interglacial Modulation of Eastern Tropical North Pacific Denitrification over the last 1.8-Myr. Geophysical Research Letters 32:L23607.

Liu, Z. and T. D. Herbert. 2004. High latitude signature in eastern equatorial Pacific climate during the early Pleistocene epoch. Nature 427:720–723.

Liu, Z., M. Notaro, J. Kutzbach, and N. Liu. 2006. Assessing global vegetation-climate feedbacks from the observation. Journal of Climate 19:787 814.

Liu, Z., Y. Wang, R. Gallimore, et al. 2007. Simulating the transient evolution and abrupt change of Northern Africa atmosphere–ocean–terrestrial ecosystem in the Holocene. Quaternary Science Reviews 26:1816–1837.

Livermore, R., C.-D. Hillenbrand, M. Meredith, and G. Eagles. 2007. Drake Passage and Cenozoic climate: An open and shut case? Geochemistry, Geophysics, Geosystems 8:Q01005.

Livermore, R., A. Nankivell, G. Eagles, and P. Morris. 2005. Paleogene opening of Drake Passage. Earth and Planetary Science Letters 236:459–470.

LoDico, J. M., B. F. Flower, and T. M. Quinn. 2006. Sub-millennial scale climatic and hydrologic variability in the Gulf of Mexico during the early Holocene. Paleoceanography 21:10.1029, PA001243.

Lorius, C., J. Jouzel, D. Raynaud, J. Hansen, and H. Le Treut. 1990. The ice-core record: Climate sensitivity and future greenhouse warming. Nature 347:139–145.

Lorius, C., J. Jouzel, C. Ritz, et al. 1985. A 150,000-year climatic record from Antarctic ice. Nature 316:591–596.

Lough, J. M. and D. J. Barnes. 1990. Intra-annual timing of density band formation of *Porites* coral from the central Great Barrier Reef. Journal of Experimental Marine Biology and Ecology 135:35–47.

Lough, J. M. and D. J. Barnes. 1992. Comparisons of skeletal density variations in *Porites* from the Great Barrier Reef. Journal of Experimental Marine Biology and Ecology 155:1–25.

Loulergue, L., A. Schilt, R. Spahni, et al. 2008. Orbital and millennial-scale features of atmospheric CH_4 over the past 800,000 years. Nature 453:383–386.

Lourens, L. J., A. Antonarakou, F. J. Hilgen, et al. 1996. Evaluation of the Plio-Pleistocene astronomical timescale. Paleoceanography 11:391–413.

Lourens, L. J., F. Hilgen, N. J. Shackleton, J. Laskar, and D. Wilson. 2004. Chapter 21: The Neogene Period. In F. Gradstein, J. Ogg, and A. Smith, eds., *A Geologic Time Scale 2004*, pp. 409–440. Cambridge: Cambridge University Press.

Lourens, L. J., A. Sluijs, D. Kroon, et al. 2005. Astronomical pacing of late Palaeocene to early Eocene global warming events. Nature 435:1083–1087.

Loutre, M. F. and A. Berger. 2000. Future climatic changes: Are we entering an exceptionally long interglacial? Climatic Change 46:61–90.

Loutre, M. F. and A. Berger. 2003. Marine Isotope Stage 11 as an analogue for the present interglacial. Global and Planetary Change 36:209–217.

Lowell, T. V. 2000. As climate changes so do glaciers. Proceedings of the National Academy of Science 97:1351–1354.

Lowell, T. V., T. Fisher, G. Comer, et al. 2005. Testing the Lake Agassiz meltwater trigger for the Younger Dryas. EOS, Transactions, American Geophysical Union 86:365–373.

Lowell, T. V., C. J. Heusser, B. G. Andersen, et al. 1995. Interhemispheric correlation of late Pleistocene glacial events. Science 269:1541–1549.

Lowell, T. V. and M. A. Kelly. 2008. Was the Younger Dryas global? Science 321:348–349.

Lowell, T. V., G. J. Larson, J. D. Hughes, and G. H. Denton. 1999. Age verification of the Lake Gribben forest bed and the Younger Dryas advance of the Laurentide Ice Sheet. Canadian Journal of Earth Science 36:383–393.

Lowenstein, T. K. and R. V. Demicco. 2006. Elevated Eocene atmospheric CO_2 and its subsequent decline. Science 313:1928.

Luckman, B. H. 2000. The Little Ice Age in the Canadian Rockies. Geomorphology 32:357–384.

Luckman, B. H. and R. J. S. Wilson. 2005. Summer temperatures in the Canadian Rockies during the last millennium: A revised record. Climate Dynamics 24:131–143.

Lund, D. C. and W. B. Curry. 2004. Late Holocene variability in Florida Current surface density: Patterns and possible causes. Paleoceanography 19:PA4001.

Lund, D. C. and W. B. Curry. 2006. Florida Current surface temperature and salinity variability during the last millennium. Paleoceanography 21:PA2009.

Lund, D. C., J. Lynch-Stieglitz, and W. B. Curry. 2006. Gulf Stream density structure and transport during the past millennium. Nature 444:601–604.

Luterbacher, J., D. Dietrich, E. Xoplaki, M. Grosjean, and H. Wanner. 2004. European seasonal and annual temperature variability, trends, and extremes since 1500. Science 303: 1499–1503.

Luterbacher, J., C. Schmutz, D. Gyalistras, E. Xoplaki, and H. Wanner. 1999. Reconstruction of monthly NAO and EU indices back to AD 1675. Geophysical Research Letters 26:2745–2748.

Luterbacher, J., E. Xoplaki, D. Dietrich, et al. 2002. Extending North Atlantic Oscillation reconstruction back to 1500. Atmospheric Science Letters 2:114–124.

Lüthi, D., et al. 2008. High-resolution carbon dioxide concentration record 650,000–800,000 years before present. Nature 453:379–382.

Lyle, M., S. Gibbs, T. C. Moore Jr., and D. K. Rea. 2007. Late Oligocene initiation of the Antarctic Circumpolar Current: Evidence from the South Pacific. Geology 35:691–694.

Lyle, M., A. Olivarez Lyle, J. Backman, and A. Tripati. 2005. Biogenic sedimentation in the Eocene equatorial Pacific: The stuttering greenhouse and Eocene carbonate compensation depth. In M. Lyle, et al., eds., *Proceedings of the Ocean Drilling Program Scientific Results*, vol. 199, pp. 1–35. College Station, TX: Ocean Drilling Program.

Lynch-Stieglitz, J. 2004. Hemispheric asynchrony of abrupt climate change. Science 304:1919–1920.

Lynch-Stieglitz, J., J. F. Adkins, W. B. Curry, et al. 2007. Atlantic meridional overturning circulation during the Last Glacial Maximum. Science 316:66–69.

MacAyeal, D. R. 1993a. Binge/purge oscillations of the Laurentide Ice Sheet as a cause of the North Atlantic's Heinrich events. Paleoceanography 8:775–784.

MacAyeal, D. R. 1993b. A low-order model of the Heinrich event cycle. Paleoceanography 8:767–773.

MacAyeal, D. R., V. Rommelaere, P. Huybrechts, et al. 1996. An ice-shelf model test based on the Ross Ice Shelf. Annals of Glaciology 23:46–51.

MacDonald, G. M., D. W. Beilman, K. V. Kremenetski, Y. Sheng, L. C. Smith, and A. A. Velichko. 2006. Rapid early development of circumarctic peatlands and atmospheric CH_4 and CO_2 variations. Science 314:285–288.

MacDonald, G. M. and R. A. Case. 2005. Variations in the Pacific Decadal Oscillation over the past millennium. Geophysical Research Letters 32:L08703.

Mackensen, A., H.-W. Hubberten, T. Bickert, G. Fischer, and D. K. Futterer. 1993. $\delta^{13}C$ in benthic foraminiferal tests of *Fontbotia wuellerstorfi* (Schwager) relative to $\delta^{13}C$ of dissolved inorganic carbon in Southern Ocean deep water: Implications for glacial ocean circulation models. Paleoceanography 6:587–610.

Maclaren, C. 1842. The glacial theory of Professor Agassiz of Neuchatel. American Journal of Science 42:346–365.

Magny, M. 2004. Holocene climatic variability as reflected by mid-European lake-level fluctuations, and its probable impact on prehistoric human settlements. Quaternary International 113:65–79.

Magny, M., C. Begeot, J. Guiot, A. Marguet, and Y. Billaud. 2003. Reconstruction and palaeoclimatic interpretation of mid-Holocene vegetation and lake-level changes at Saint-Jorioz, Lake Annecy, French Pre-Alps. The Holocene 13:265–275.

Maier-Reimer, E., U. Mikolajewicz, and T. J. Crowley. 1990. Ocean General Circulation model sensitivity experiment with an open Central American isthmus. Paleoceanography 5:49–366.

Manabe, S. and R. J. Stouffer. 1988. Two stable equilibria of a coupled ocean-atmosphere model. Journal of Climate 1:841–866.

Manabe, S. and R. J. Stouffer. 1995. Simulation of abrupt climate change induced by freshwater input to the North Atlantic Ocean. Nature 378:165–167.

Manabe, S. and R. J. Stouffer. 1997. Coupled ocean-atmosphere model response to freshwater input: Comparison to Younger Dryas event. Paleoceanography 12:321–336.

Manabe, S. and R. J. Stouffer. 1999. The role of thermohaline circulation in climate. Tellus 51A:91–109.

Manabe, S. and R. J. Stouffer. 2000. Study of abrupt climate change by a coupled ocean-atmosphere model. Quaternary Science Reviews 19:285–299.

Mangerud, J. 2004. Ice sheet limits on Norway and the Norwegian continental shelf. In J. Ehlers and P. Gibbard, eds., *Quaternary Glaciation-Extent and Chronology: Europe*, vol. 1. Amsterdam: Elsevier.

Mangerud, J., S. T. Anderson, B. E. Birklund, and J. J. Donner. 1974. Quaternary stratigraphy of Norden, a proposal for terminology and classification. Boreas 3:109–128.

Mangerud, J., V. I. Astakhov, M. Jacobsson, and J. I. Svendsen. 2001a. Huge Ice-Age lakes in Russia. Journal of Quaternary Science 16:773–777.

Mangerud, J., V. I. Astakhov, A. Murray, and J. I. Svendsen. 2001b. The chronology of a large ice-dammed lake and the Barents–Kara Ice Sheet advances, Northern Russia. Global and Planetary Change 31:319–334.

Mangerud, J., M. Jakobsson, H. Alexanderson, et al. 2004. Ice-dammed lakes and rerouting of the drainage of northern Eurasia during the last glaciation. Quaternary Science Reviews 23:1313–1332.

Mann, M. E. 2007. Climate over the past two millennia. Annual Review of Earth and Planetary Sciences 35:111–136.

Mann, M. E., R. S. Bradley, and M. K. Hughes. 1998. Global scale temperature patterns and climate forcing over the past six centuries. Nature 392:779–788.

Mann, M. E., R. S. Bradley, and M. K. Hughes. 1999. Northern hemisphere temperatures during the past millennium: Inferences, uncertainties, and limitations. Geophysical Research Letters 26:759–762.

Mann, M. E., R. S. Bradley, and M. K. Hughes. 2000. Long-term variability in the El Niño Southern Oscillation and associated teleconnections. In H. F. Diaz and V. Markgraf, eds., *El Niño and the Southern Oscillation: Multiscale Variability and Global and Regional Impacts*, pp. 357–412. Cambridge: Cambridge University Press.

Mann, M. E., M. A. Cane, S. E. Zebiak, and A. Clement. 2005a. Volcanic and solar forcing of the tropical Pacific over the past 1000 years. Journal of Climate 18:447–456.

Mann, M. E. and K. Emanuel. 2006. Atlantic hurricane trends linked to climate change. EOS, Transactions, American Geophysical Union 87:233–244.

Mann, M. E., K. A. Emanuel, G. J. Holland, and P. J. Webster. 2007. Atlantic tropical cyclones revisited. EOS, Transactions, American Geophysical Union 88:349–350.

Mann, M. E. and P. D. Jones. 2003. Global surface temperatures over the past two millennia. Geophysical Research Letters 30:1820, GL017814.

Mann, M. E., J. Park, and R. S. Bradley. 1995. Global inter-decadal and century-scale climate oscillations during the past five centuries. Nature 378:266–270.

Mann, M. E. and S. Rutherford. 2002. Climate reconstruction using "pseudoproxies." Geophysical Research Letters 29:1501, doi:10.1029/2001GL014554.

Mann, M. E., S. Rutherford, E. Wahl, and C. Ammann. 2005b. Testing the fidelity of methods used in proxy-based reconstructions of past climate. Journal of Climate 18:4097–4107.

Mann, M. E., Z. Zhang, M. K. Hughes, et al. 2008. Proxy-based reconstructions of hemispheric and global surface temperature variations over the past two millennia. Proceedings of the National Academy of Sciences 105:13252–13257.

Mantua, N. J., S. R. Hare, Y. Zhang, et al. 1997. A Pacific interdecadal climate oscillation with impacts on salmon production. Bulletin of the American Meteorological Society 78:1069–1079.

Marcantonio, F., R. F. Anderson, M. Stute, et al. 1996. Extraterrestrial ^3He as a tracer of marine sediment transport and accumulation. Nature 383:705–707.

Marchal, O., I. Cacho, T. F. Stocker, et al. 2002. Apparent long-term cooling of the sea surface in the Northeast Atlantic and Mediterranean during the Holocene. Quaternary Science Reviews 21:455–483.

Marchal, O. and W. B. Curry. 2008. On the abyssal circulation in the glacial Atlantic. Journal of Physical Oceanography doi:10.1175/2008JPO3895.1.

Marchal, O., R. François, T. F. Stocker, and F. Joos. 2000. Ocean thermohaline circulation and sedimentary ^{231}Pa/^{230}Th ratio. Paleoceanography 15:625–641.

Marchant, D. R. and G. H. Denton. 1996. Miocene and Pliocene paleoclimate of the Dry Valleys region, Southern Victoria land: A geomorphological approach. Marine Micropaleontology 27:253–271.

Marchitto, T. M., W. B. Curry, and D. W. Oppo. 1998. North Atlantic gyre ventilation and intermediate water formation during the last glaciation and Younger Dryas. Nature 393:557–561.

Marchitto, T. M. and P. B. deMenocal. 2003. Late Holocene variability of upper North Atlantic deep water temperature and salinity. Geochemistry, Geophysics, Geosystems 4:GC000598.

Marchitto, T. M., G. A. Jones, G. A. Goodfriend, and C. R. Weidman. 2000. Precise temporal correlation of Holocene mollusk shells using sclerochronology. Quaternary Research 53:236–246.

Marchitto, T. M., S. J. Lehman, J. D. Ortiz, J. Flückiger, and A. van Geen. 2007. Marine radiocarbon evidence for the mechanism of deglacial atmospheric CO_2 rise. Science 316:1456–1459.

Marchitto, T. M. and K.-Y. Wei. 1995. History of Laurentide meltwater flow to the Gulf of Mexico during the last deglaciation, as revealed by reworked calcareous nannofossils. Geology 23:779–782.

Marinov, I., A. Gnanadesikan, J. R. Toggweiler, and J. L. Sarmiento. 2006. The Southern Ocean biogeochemical divide. Nature 441:964–967.

Markgraf, V. 1991. Younger Dryas in southern South America. Boreas 20:63–69.

Markgraf, V. 1993. Younger Dryas in southernmost South America—An update. Quaternary Science Reviews 12:351–355.

Marko, P. B. 2002. Fossil calibration of molecular clocks and the divergence times of geminate species pairs separated by the Isthmus of Panama. Molecular Biology and Evolution 19:2005–2031.

Marlow, J. R., C. B. Lange, G. Wefer, and A. Rosell-Melé. 2000. Upwelling intensification as part of the Pliocene-Pleistocene climate transition. Science 290:2288–2291.

Marotzke, J. 2000. Abrupt climate change and thermohaline circulation: Mechanisms and predictability. Proceedings of the National Academy of Sciences 97:1347–1350.

Marshall, C. H., R. A. Pielke, L. T. Steyaert, and D. A. Willard. 2004. The impact of anthropogenic land-cover change on the Florida peninsula sea breezes and warm season sensible weather. Monthly Weather Review 132:28–52.

Marshall, G. J. 2003. Trends in the Southern Annular Mode from observations and reanalysis. Journal of Climate 16:4134–4143.

Marshall, G. J. and W. M. Connolley. 2006. Effect of changing Southern Hemisphere winter sea surface temperatures on Southern Annular Mode strength. Geophysical Research Letters 33:L17717.

Marshall, G. J., P. A. Stott, J. Turner, et al. 2004. Causes of exceptional atmospheric circulation changes in the southern hemisphere. Geophysical Research Letters 31:L14205.

Marshall, L. G., S. D. Webb, J. J. Sepkoski Jr., and D. M. Raup. 1982. Mammalian evolution and the Great American Interchange. Science 215:1351–1357.

Marshall, S. J. and G. K. C. Clarke. 1997. A continuum mixture model of ice stream thermomechanics in the Laurentide Ice Sheet: 2. Application to the Hudson Strait Ice Stream. Journal of Geophysical Research [Solid Earth] 102:20615–20637.

Marshall, S. J. and G. K. C. Clarke. 1999. Modeling North American freshwater runoff through the last glacial cycle. Quaternary Research 52:300–315.

Martin, J. H. 1990. Glacial-interglacial CO_2 change: The iron hypothesis. Palaeoceanography 5:1–13.

Martin, P., D. Archer, and D. Lea. 2005. Role of deep sea temperatures in the carbon cycle during the Last Glacial. Paleoceanography 20:PA2015.

Martin, P. A., D. W. Lea, Y. Rosenthal, et al. 2002. Quaternary deep sea temperature histories derived from benthic foraminiferal Mg/Ca. Earth and Planetary Science Letters 198:193–209.

Martinson, D. G., N. G. Pisias, J. D. Hays, et al. 1987. Age dating and the orbital theory of the Ice Ages: Development of a high-resolution 0 to 300,000-year chronostratigraphy. Quaternary Research 27:1–29.

Martrat, B., J. O. Grimalt, C. Lopez-Martinez, et al. 2004. Abrupt temperature changes in the western Mediterranean over the past 250,000 years. Science 306:1762–1765.

Martrat, B., J. O. Grimalt, N. J. Shackleton, et al. 2007. Four climate cycles of recurring deep and surface water destabilizations on the Iberian Margin. Science 317:502–507.

Maslin, M. A. and S. J. Burns. 2000. Reconstruction of the Amazon basin effective moisture availability over the last 14,000 years. Science 290:2285–2287.

Maslin, M. A. and A. J. Ridgwell. 2005. Mid-Pleistocene revolution and the "eccentricity myth." Geological Society of London, Special Publications 247:19–34.

Masson, V., R. Cheddadi, P. Braconnot, S. Joussaume, D. Texier, and PMIP participants. 1999. Mid-Holocene climate in Europe: What can we infer from PMIP mode-data comparisons? Climate Dynamics 15:163–182.

Masson, V., F. Vimeux, J. Jouzel, et al. 2000. Holocene climate variability in Antarctica based on 11 ice-core isotopic records. Quaternary Research 54:348–358.

Masson-Delmotte, V., J. Jouzel, A. Landais, et al. 2005. GRIP deuterium excess reveals rapid and orbital-scale changes in Greenland moisture origin. Science 309:118–121.

Masson-Delmotte, V., M. Kageyama, P. Braconnot, et al. 2006. Past and future polar amplification of climate change: Climate model intercomparisons and ice-core constraints. Climate Dynamics 27:437–440.

Masson-Delmotte, V., B. Stenni, and J. Jouzel. 2004. Common millennial-scale variability of Antarctic and Southern Ocean temperatures during the past 5000 years reconstructed from the EPICA Dome C ice core. The Holocene 14:145–151.

Matthes, F. E. 1939. Report of Committee on Glaciers, April 1939. Transactions of the American Geophysical Union 20:518–523.

Matthews, R. K. and R. Z. Poore. 1980. Tertiary δ^{18}O and glacio-eustatic sea level fluctuations. Geology 8:501–504.

Maunder, E. W. 1922. The prolonged sunspot minimum, 1645–1715. The British Astronomical Association Journal 32:140–145.

Mauritzen, C. 1996. Production of dense overflow waters feeding the North Atlantic across the Greenland-Scotland Ridge. Part 1: Evidence for a revised circulation scheme. Deep-Sea Research 43:769–806.

Mayewski, P. A., L. D. Meeker, M. S. Twickler, et al. 1997. Major features and forcing of high latitude northern hemisphere atmospheric circulation over the last 110,000 years. Journal of Geophysical Research 102(C12):26345–26366.

Mayewski, P. A., L. D. Meeker, S. Whitlow, et al. 1993. The atmosphere during the Younger Dryas. Science 261:195–197.

Mayewski, P. A., E. E. Rohling, J. C. Stager, et al. 2004. Holocene climate variability. Quaternary Research 62:243–255.

Mayewski, P. A., M. S. Twickler, S. I. Whitlow, et al. 1996. Climate change during the last deglaciation in Antarctica. Science 272:1636–1638.

McArthur, J. M. and R. J. Howarth. 2004. Strontium isotope stratigraphy. In F. M. Gradstein, J. G. Ogg, and A. Smith, eds., A Geological Timescale 2004, pp. 96–105. Cambridge: Cambridge University Press.

McCabe, G. J. and M. D. Dettinger. 1999. Decadal variations in the strength of ENSO teleconnections with precipitation in the western United States. International Journal of Climatology 19:1399–1410.

McCabe, G. J. and M. A. Palecki. 2006. Multidecadal climate variability of global lands and oceans. International Journal of Climatology 26:849–865.

McCabe, G. J., M. A. Palecki, and J. L. Betancourt. 2004. Pacific and Atlantic Ocean influences on multidecadal drought frequency in the United States. Proceedings of the National Academy of Sciences 101:4136–4141.

McCabe, M., J. Knight, and S. McCarron. 1998. Evidence for Heinrich event 1 in the British Isles. Journal of Quaternary Science 13:549–568.

McCave, I. N. and I. R. Hall. 2006. Size sorting in marine muds: Processes, pitfalls and prospects for paleoflow-speed proxies. Geochemistry, Geophysics, Geosystems 7:Q10N05, GC001284.

McCave, I. N., B. Manighetti, and N. A. S. Beveridge. 2002. Circulation in the glacial North Atlantic inferred from grain-size measurements. Nature 374:149–152.

McCave, I. N., B. Manighetti, and S. O. Robinson. 1995. Sortable silt and sediment size/composition slicing: Parameters for paleocurrent speed and paleoceanography. Paleoceanography 10:593–610.

McCay, G. A., A. R. Prave, G. I. Alsop, and A. E. Fallick. 2006. Glacial trinity: Neoproterozoic Earth history within the British-Irish Caledonides. Geology 34:909–912.

McClymont, E. L. and A. Rosell-Melé. 2005. Links between the onset of modern Walker circulation and the mid-Pleistocene climate transition. Geology 33:389–392.

McConnaughey, T. A. 1989. C-13 and O-18 isotopic disequilibria in biological carbonates: I. Patterns. Geochimica et Cosmochimica Acta 53:151–163.

McCorkle, D., J. M. Bernhard, C. J. Hintz, et al. 2008. The carbon and oxygen stable isotopic composition of cultured benthic foraminifera. Geological Society, London, Special Publications 303:135–154.

McCorkle, D. C., L. D. Keigwin, B. C. Corliss, and S. R. Emerson. 1990. The influence of microhabitats on the carbon isotopic composition of deep-sea sediments. Paleoceanography 5:161–186.

McCormac, B. M., ed. 1983. Weather and Climate Responses to Solar Variations. Boulder: Colorado Associated University Press.

McCracken, K. G., G. A. M. Dreschhoff, D. F. Smart, and M. A. Shea. 2001. Solar cosmic ray events for the period 1561–1994: 2. The Gleissberg periodicity. Journal of Geophysical Research 106(A10):21599–21609.

McCullough, M., S. Fallon, T. Wyndham, et al. 2003. Coral record of increased sediment flux to the inner Great Barrier Reef since European settlement. Nature 421:727–730.

McDermott, F. 2004. Palaeoclimate reconstruction from stable isotope variations in speleothems: A review. Quaternary Science Reviews 23:901–918.

McDermott, F., H. P. Schwarz, and P. J. Rowe. 2006. Isotopes in speleothems. In M. J. Leng, ed., Isotopes in Palaeoenvironmental Research, pp. 185–226. Dordrecht: Springer.

McDougall, I. and T. M. Harrison. 1999. Geochronology and Thermochronology by the 40Ar/39Ar Method. Oxford: Oxford University Press.

McElwain, J. C. 1998. Do fossil plants signal palaeoatmospheric carbon dioxide concentration in the geological past? Philosophical Transactions of the Royal Society of London, Series B 353:83–96.

McElwain, J. C., D. J. Beerling, and F. I. Woodward. 1999. Fossil plants and global warming at the Triassic-Jurassic boundary. Science 285:1386–1390.

McElwain, J. C. and W. G. Chaloner. 1995. Stomatal density and index of fossil plants track atmospheric carbon dioxide in the Palaeozoic. Annals of Botany 76:385–395.

McElwain, J. C. and W. G. Chaloner. 1996. The fossil cuticle as a skeletal record of environmental change. Palaios 11:376–388.

McElwain, J. C., J. Wade-Murphy, and S. P. Hesselbo. 2005. Changes in carbon dioxide during an oceanic anoxic event linked to intrusion into Gondwana coals. Nature 435:479–482.

McGarry, S. F. and A. Baker. 2000. Organic acid fluorescence: Applications to speleothem palaeoenvironmental reconstruction. Quaternary Science Reviews 19:1087–1101.

McGregor, H. V., M. Dima, H. W. Fischer, and S. Mulitza. 2007. Rapid 20th-century increase in coastal upwelling off northwest Africa. Science 315:637–639.

McGregor, H. V. and M. K. Gagan. 2004. Western Pacific coral δ18O records of anomalous Holocene variability in El Niño–Southern Oscillation. Geophysical Research Letters 31:L11204, GL019972.

McIntyre, A. and B. Molfino. 1996. Forcing of Atlantic equatorial and subpolar millennial cycles by precession. Nature 274:1867–1870.

McIntyre, S. and R. McKitrick. 2003. Corrections to the Mann et al. (1998) proxy data base and Northern Hemispheric average temperature series. Energy Environment 14:751–771.

McKay, C. P. 2000. Thickness of tropical ice and photosynthesis on a Snowball Earth. Geophysical Research Letters 27:2153–2156.

McManus, J. F., R. F. Anderson, W. S. Broecker, M. Q. Fleisher, and S. M. Higgins. 1998. Radiometrically determined sedimentary fluxes in the sub-polar North Atlantic during the last 140,000 years. Earth and Planetary Science Letters 155:29–43.

McManus, J. F., G. C. Bond, W. S. Broecker, S. Johnsen, L. Labeyrie, and S. Higgins. 1994. High-resolution climate records from the North Atlantic during the last interglacial. Nature 371:326–329.

McManus, J. F., R. François, J.-M. Gherardi, L. D. Keigwin, and S. Brown-Leger. 2004. Collapse and rapid resumption of Atlantic meridional circulation linked to deglacial climate changes. Nature 428:834–837.

McManus, J. F., D. Oppo, J. Cullen, and S. Healey. 2003. Marine Isotope Stage 11 (MIS 11): Analog for Holocene and future climate? In Andre W. Droxler, Richard Z. Poore, and Lloyd H. Burckle, *Earth's Climate and Orbital Eccentricity: The Marine Isotope Stage 11 Question*, pp. 69–85. Washington, DC: American Geophysical Union.

McManus, J. F., D. W. Oppo, and J. L. Cullen. 1999. A 0.5 million-year record of millennial-scale climate variability in the North Atlantic. Science 283:971–975.

McMillan, M. E., P. L. Heller, and S. L. Wing. 2006. History and causes of post-Laramide relief in the Rocky Mountain orogenic plateau. Geological Society of America Bulletin 118:393–405.

McMinn, A., W. Howard, and D. Roberts. 2001. Late Pliocene dinoflagellate cyst and diatom analysis from a high resolution sequence in DSDP Site 594, Chatham Rise, southwest Pacific. Marine Micropaleontology 43:207–221.

McPhaden, M. J. 2004. Evolution of the 2002-03 El Niño. Bulletin of the American Meteorological Society 85:677–695.

McPhaden, M. J., S. E. Zebiak, and M. H. Glantz. 2006. ENSO as an integrating concept in earth science. Science 314:1740–1745.

Medina-Elizalde, M. and D. W. Lea. 2005. The mid-Pleistocene transition in the tropical Pacific. Science 310:1009–1012.

Meehl, G. A., T. F. Stocker, et al. 2007. Chapter 10: Global climate projections. In S. Solomon, et al., eds., *Climate Change 2007: The Physical Science Basis*. Contribution of Working Group I to the Fourth Assessment Report of the Intergovernmental Panel on Climate Change, pp. 749–845. Cambridge: Cambridge University Press.

Meese, D. A., A. J. Gow, R. B. Alley, et al. 1997. The Greenland Ice Sheet Project 2 depth-age scale: Methods and results. Journal of Geophysical Research 102:26411–26423.

Meese, D. A., A. J. Gow, P. Grootes, et al. 1994. The accumulation record from the GISP2 core as an indicator of climate change throughout the Holocene. Science 266:1680–1682.

Meier, M. F., M. B. Dyurgerov, U. K. Rick, et al. 2007. Glaciers dominate eustatic sea-level rise in the 21st century. Science 317:1064–1067.

Meier, M. F. and J. M. Wahr. 2002. Sea level is rising: Do we know why? Proceedings of the National Academy of Sciences 99:6524–6526.

Meinen, C. S., M. O. Baringer, and S. L. Garzoli. 2006. Variability in deep Western Boundary Current transports: Preliminary results from 26.5°N in the Atlantic. Geophysical Research Letters 33:L17610, GL026965.

Meissner, K. J. and P. U. Clark. 2006. Impact of floods versus routing events on the thermohaline circulation. Geophysical Research Letters 33:L15704, GL026705.

Mekik, F., R. François, and M. Soon. 2007. A novel approach to dissolution correction of Mg/Ca-based paleothemometry in the tropical Pacific. Paleoceanography 22:PA001504.

Meltzer, D. J. 1999. Human responses to middle Holocene (Altithermal) climates on the North American Great Plains. Quaternary Research 52:404–416.

Ménot, G., E. Bard, F. Rostek, et al. 2006. Early reactivation of European rivers during the last deglaciation. Science 313:1623–1625.

Mercer, J. H. 1969. The Allerød oscillation: A European climatic anomaly. Arctic and Alpine Research 1:227–234.

Mesolella, K. J., R. K. Matthews, W. S. Broecker, and D. L. Thurber. 1968. The astronomical theory of climatic change: Barbados data. Journal of Geology 77:250–274.

Mestas-Nuñez, A. M. and A. Miller. 2006. Interdecadal variability and climate change in the eastern tropical Pacific: A review. Progress in Oceanography 69:267–284.

Meyers, P. A. 2006. Paleoceanographic and paleoclimatic similarities between Mediterranean sapropels and Cretaceous black shales. Palaeogeography, Palaeoclimatology, Palaeo ecology 235:305–320.

Meyers, S. R. and B. B. Sageman. 2007. Quantification of deep-time orbital forcing by average spectral misfit. American Journal of Science 307:773–792.

Miall, A. D. 1997. *The Geology of Stratigraphic Sequences*. Berlin: Springer.

Mickler, P. J., L. A. Stern, and J. L. Banner. 2006. Large kinetic isotope effects in modern speleothems. Geological Society of America Bulletin 118:65–81.

Mii, H.-S., E. L. Grossman, and T. E. Yancey. 1999. Carboniferous isotope stratigraphies of North America: Implications for Carboniferous paleoceanography and Mississippian glaciation. Geological Society of America Bulletin 111:960–973.

Mikaloff Fletcher, S. E., N. Gruber, A. R. Jacobson, et al. 2006. Inverse estimates of anthropogenic CO_2 uptake, transport, and storage by the ocean. Global Biogeochemical Cycles 20, doi:10.1029/2005GB002530.

Mikolajewicz, U., E. Maier-Reimer, T.J. Crowley, and K.-Y. Kim. 1993. Effect of Drake and Panamanian gateways on the circulation of an ocean model. Paleoceanography 8:409–426.

Milankovitch, M. 1941. Kanon der Erdbestrahlung und seine Andwendung auf das Eiszeitenproblem. Royal Serbian Academy Special Publication 133:1–633. Trans. 1969. Israel Program for Scientific Translation, U.S. Department of Commerce.

Miller, G. H., A. P. Wolfe, E. J. Steig, et al. 2001. The Goldilocks dilemma: Big ice, little ice, or "just-right" ice in the Eastern Canadian Arctic. Quaternary Science Reviews 21:33–48.

Miller, K. G., R. G. Fairbanks, and G. S. Mountain. 1987. Tertiary oxygen isotope synthesis, sea level history, and continental margin erosion. Paleoceanography 2:1–19.

Miller, K. G., M. A. Kominz, J. V. Browning, et al. 2005. The Phanerozoic record of global sea-level change. Science 312:1293–1298.

Miller, K. G., G. S. Mountain, the Leg 150 Shipboard Party, and Members of the New Jersey Coastal Plain Drilling Project. 1996. Drilling and dating New Jersey Oligocene-Miocene sequences: Ice volume, global sea level, and Exxon records. Science 271:1092–1094.

Miller, K. G., P. J. Sugarman, J. V. Browning, et al. 2003. A chronology of late Cretaceous sequences and sea-level history: Glacioeustasy during the Greenhouse World. Geology 31: 585–588.

Miller, K. G., P. J. Sugarman, J. V. Browning, et al. 2004. Upper Cretaceous sequences and sea-level history, New Jersey Coastal Plain. Geological Society of America Bulletin 116:368–393.

Miller, K. G., J. D. Wright, and R. G. Fairbanks. 1991. Unlocking the icehouse: Oligocene-Miocene oxygen isotopes, eustasy, and margin erosion. Journal of Geophysical Research 96:6829–6848.

Miller, L. and B. C. Douglas. 2004. Mass and volume contributions to twentieth-century global sea level rise. Nature 428:406–409.

Milliman, J. D. and K. O. Emery. 1968. Sea levels during the past 35,000 years. Science 162:1121–1123.

Milly, P. C. D., J. Betancourt, M. Falkenmark, et al. 2008. Stationarity is dead: Whither water management? Science 319:573–574.

Minnis, P., E. F. Harrison, L. L. Stowe, et al. 1993. Radiative climate forcing by the Mount Pinatubo eruption. Science 259:1411–1415.

Mitrovica, J. X., N. Gomez, and P. U. Clark. 2009. The sea-level fingerprint of West Antarctic collapse. Science 323:753.

Mitrovica, J. X. and W. R. Peltier. 1991. On postglacial geoid subsidence over the equatorial oceans. Journal of Geophysical Research 96:1419–1422.

Mitsuguchi, T., E. Matsumoto, O. Abe, T. Uchida, and P. J. Isdale. 1996. Mg/Ca thermometry in coral skeletons. Science 274:961–963.

Mix, A. 2006. Running hot and cold in the eastern equatorial Pacific. Quaternary Science Reviews 25:1147–1149.

Mix, A. C. 1992. The marine oxygen isotope record: Constraints on timing and extent of ice-growth events (120–65 ka). In P. U. Clark and P. D. Lea, eds., *The Late Interglacial-Glacial Transition in North America*, pp. 19–30. Boulder, CO: Geological Society of America.

Mix, A. C., E. Bard, and R. Schneider. 2001. Environmental processes of the ice age: Land, oceans, glaciers (EPILOG). Quaternary Science Reviews 20:627–657.

Miyahara, H., K. Masuda, Y. Muraki, et al. 2004. Cyclicity of solar activity during the Maunder Minimum deduced from radiocarbon content. Solar Physics 224:317–322.

Moberg, A., D. M. Sonechkin, K. Holmgren, N. M. Datsenko, and W. Karlèn. 2005. Highly variable northern hemisphere temperatures from low- and high-resolution proxy data. Nature 433:613–617.

Molnar, P. 2004. Late Cenozoic increase in accumulation rates of terrestrial sediment: How might climate change have affected erosion rates? Annual Review of Earth and Planetary Sciences 32:67–89.

Molnar, P. 2007. An examination of evidence used to infer late Cenozoic "uplift" of mountain belts and other high terrain: What scientific question does such evidence pose? Journal of the Geological Society of India 70:395–410.

Molnar, P. 2008. Closing of the Central American Seaway and the ice age: A critical review. Paleoceanography 23:PA2201, PA001574.

Molnar, P. and M. A. Cane. 2002. El Niño's tropical climate and teleconnections as a blueprint for pre–ice age climates. Paleoceanography 17:1548, PA000663.

Molnar, P. and M. A Cane. 2007. Pre–ice age El Niño–like global climate: Which El Niño? Geosphere 3:337–365.

Molnar, P. and P. England. 1990. Late Cenozoic uplift of mountain ranges and global climate change: Chicken or egg? Nature 346:29–34.

Molnar, P., G. A. Houseman, and P. C. England. 2006. Palaeo-altimetry of Tibet. Nature 444:E4.

Monaghan, A. J. and D. H. Bromwich. 2008. Advances in describing recent Antarctic climate variability. Bulletin of the American Meteorological Society 89:1295–1306.

Monnin, E., A. Indermühle, A. Dällenbach, et al. 2001. Atmospheric CO_2 concentrations over the last glacial termination. Science 291:112–114.

Monnin, E., E. J. Steig, U. Siegenthaler, et al. 2004. Evidence for substantial accumulation rate variability in Antarctica during the Holocene through synchronization of CO_2 in the Taylor Dome, Dome C and DML ice cores. Earth and Planetary Science Letters 224:45–54.

Montaggioni, F. L. 2005. History of Indo-Pacific coral reef systems since the last glaciation: Development patterns and controlling factors. Earth-Science Reviews 71:1–75.

Montañez, I. P., N. J. Tabor, D. Niemeier, et al. 2007. CO_2-forced climate and vegetation instability during late Paleozoic deglaciation. Science 315:87–91.

Moore, J. K., M. R. Abbott, J. G. Richman, and D. M. Nelson. 2000. The Southern Ocean at the Last Glacial Maximum: A strong sink for atmospheric carbon dioxide. Global Biogeochemical Cycles 14:455–475.

Moore, T. C., Jr., J. G. C. Walker, D. K. Rea, et al. 2000. The Younger Dryas interval and outflow from the Laurentide Ice Sheet. Paleoceanography 15:9–18.

Mora, C. I., S. G. Driese, and L. A. Colarusso. 1996. Middle to late Paleozoic atmospheric CO_2 levels from soil carbonate and organic matter. Science 271:1105–1107.

Moran, K., J. Backman, H. Brinkhuis, et al. 2006. The Cenozoic palaeoenvironment of the Arctic Ocean. Nature 44:601–605.

Moreno, P. I., G. L. Jacobson Jr., T. V. Lowell, and G. H. Denton. 2001. Interhemispheric climate links revealed by a late-glacial cooling episode in southern Chile. Nature 409:804–808.

Morgan, V., M. Delmotte, T. van Ommen, et al. 2002. Relative timing of deglacial climate events in Antarctica and Greenland. Science 297:1862–1864.

Morgan, V. and T. D. van Ommen. 1997. Seasonality in late-Holocene climate from ice-core records. The Holocene 7:351–354.

Moriya, K., P. A. Wilson, O. Friedrich, J. Erbacher, and H. Kawahata. 2007. Testing for ice sheets during the mid-Cretaceous greenhouse using glassy foraminiferal calcite from the mid-Cenomanian tropics on Demerara Rise. Geology 35:615–618.

Moros, M., K. Emeis, B. Risebrobakken, et al. 2004. Sea surface temperatures and ice rafting in the Holocene North Atlantic: Climate influences on northern Europe and Greenland. Quaternary Science Reviews 23:2113–2126.

Morrill, C., J. T. Overpeck, and J. E. Cole. 2003. A synthesis of abrupt changes in the Asian summer monsoon since the last deglaciation. The Holocene 13:465–476.

Morton, R. A., J. G. Paine, and M. D. Blum. 2000. Responses of stable bay margins and barrier island systems to Holocene sea level changes, western Gulf of Mexico. Journal of Sedimentary Research 70:478–490.

Moses, C. S., P. K. Swart, and B. E. Rosenheim. 2006. Evidence of multidecadal salinity variability in the eastern tropical North Atlantic. Paleoceanography 21:PA3010, PA001257.

Mosley-Thompson, E. 1996. Holocene climate changes recorded in an East Antarctica ice core. In P. D. Jones, R. S. Bradley, and J. Jouzel, eds., Climate Variations and Forcing Mechanisms of the Last 2000 Years, pp. 263–279. Berlin: Springer-Verlag.

Mosley-Thompson, E., J. R. McConnell, R. C. Bales, et al. 2001. Local to regional-scale variability of Greenland accumulation from PARCA cores. Journal of Geophysical Research (Atmospheres) 106(D24):33839–33852.

Mosley-Thompson, E., C. R. Readinger, P. Craigmile, L. G. Thompson, and C. A. Calder. 2005. Regional sensitivity of Greenland precipitation to NAO variability. Geophysical Research Letters 32:L24707, GL024776.

Mosley-Thompson, E. and L. G. Thompson. 2003. Ice core paleoclimate histories from the Antarctic Peninsula: Where do we go from here? In E. Domack, et al., eds., Antarctic Peninsula Climate Variability: Historical and Paleoenvironmental Perspectives, pp. 115–127. Washington, DC: American Geophyiscal Union.

Mosley-Thompson, E., L. G. Thompson, J. Dai, M. E. Davis, and P. N. Lin. 1993. Climate of the last 500 years: High resolution ice core records. Quaternary Science Reviews 12:419–430.

Mosley-Thompson, E., L. G. Thompson, and P.-N. Lin. 2006. A multi-century perspective on 20th century climate change with new contributions from high Arctic and Greenland (PARCA) cores. Annals of Glaciology 43:42–48.

Moy, C. M., G. O. Seltzer, D. T. Rodbell, and D. M. Anderson. 2002. Variability of El Niño/Southern Oscillation activity at millennial timescales during the Holocene epoch. Nature 420:162–165.

Mudelsee, M. and M. E. Raymo. 2005. Slow dynamics of the northern hemisphere glaciation. Paleoceanography 20:PA4022.

Mudelsee, M. and M. Schulz. 1997. The mid-Pleistocene climate transition: Onset of 100 ka cycle lags ice volume build-up by 280 ka. Earth and Planetary Science Letters 151:117–123.

Muhs, D. R. 2002. Evidence for the timing and duration of the last interglacial period from high-precision uranium-series ages of corals on tectonically stable coastlines. Quaternary Research 58:36–40.

Muhs, D. R., K. R. Simmons, and B. Steinke. 2002. Timing and warmth of the last interglacial period: New U-series evidence from Hawaii and Bermuda and a new fossil compilation for North America. Quaternary Science Reviews 21:1355–1383.

Muhs, D. R. and B. J. Szabo. 1994. New uranium-series ages of the Waimanalo Limestone, Oahu, Hawaii: Implications for sea level during the last interglacial period. Marine Geology 118:315–326.

Muller, P. J., G. Kirst, G. Ruhland, I. von Storch, and A. Rosell-Melé. 1998. Calibration of the alkenone paleotemperature index Uk37 based on core-tops from the eastern South Atlantic and the global ocean (60°N–60°S). Geochimica et Cosmochimica Acta 62:1757–1772.

Muller, R. A. and G. F. MacDonald. 1995. Glacial cycles and orbital inclination. Nature 377:107–108.

Muller, R. A. and G. J. MacDonald. 1997a. Spectrum of the 100 kyr glacial cycle: Orbital inclination, not eccentricity. Proceedings of the National Academy Sciences 94: 8329–8334.

Muller, R. A. and G. J. MacDonald. 1997b. Glacial cycles and astronomical forcing. Science 277:215–218.

Muller, R. A. and G. J. MacDonald. 2000. Ice Ages and Astronomical Causes: Data, Spectral Analysis, and Mechanisms. London: Springer-Praxis.

Müller, R. D., M. Sdrolias, C. Gaina, B. Steinberger, and C. Heine. 2008. Long-term sea-level fluctuations driven by ocean basin dynamics. Science 319:1357–1362.

Munk, W. 1997. Once again: One again—Tidal friction. Progress in Oceanography 40:7–35.

Munk, W. 2002. Twentieth century sea level: An enigma. Proceedings of the National Academy of Sciences 99: 6550–6555.

Munk, W. and C. Wunsch 1998. Abyssal recipes II: Energetics of tidal and wind mixing. Deep-Sea Research I 45:1977–2010.

Murray, R., M. Leinen, D. Murray, A. Mix, and C. Knowlton. 1995. Terrigenous Fe input and biogenic sedimentation in the glacial and interglacial equatorial Pacific Ocean. Global Biogeochemical Cycles 9:667–684.

Muscheler, R. and J. Beer. 2006. Solar forced Dansgaard/Oeschger events? Geophysical Research Letters 33:L2070.

Muscheler, R., J. Beer, P. W. Kubik, and H.-A. Synal. 2005a. Geomagnetic field intensity during the last 60,000 years based on ^{10}Be & ^{36}Cl from the Summit ice cores and ^{14}C. Quaternary Science Reviews 24:1849–1860.

Muscheler, R., J. Beer, G. Wagner, and R. C. Finkel. 2000. Changes in deep-water formation during the Younger Dryas cold period inferred from a comparison of ^{10}Be and ^{14}C records. Nature 408:567–570.

Muscheler, R., J. Beer, G. Wagner, et al. 2004. Changes in the carbon cycle during the last deglaciation as indicated by the comparison of ^{10}Be and ^{14}C records. Earth and Planetary Science Letters 219:325–340.

Muscheler, R., F. Joos, J. Beer, et al. 2007. Solar activity during the last 1000 yr inferred from radionuclide records. Quaternary Science Reviews 26:82–97.

Muscheler, R., F. Joos, S. A. Müller, and I. Snowball. 2005b. How unusual is today's solar activity? Nature 436:E1–E4.

Nagler, T. F., A. Eisenhauer, A. Muller, C. Hemleben, and J. Kramers. 2000. The δ^{44}Ca-temperature calibration on fossil and cultured Globigerinoides sacculifer: New tool for reconstruction of past sea surface temperatures. Geochemistry, Geophysics, Geosystems 1: doi:10.1029/2000GC000091.

Naish, T. R., R. Powell, R. Levy, et al. 2007. A record of Antarctic climate and ice sheet history recovered. EOS, Transactions, American Geophysical Union 88:557–568.

Naish, T. R., K. J. Woolfe, P. J. Barrett, et al. 2001. Orbitally induced oscillations in the East Antarctic Ice Sheet at the Oligocene/Miocene boundary. Nature 413:719–723.

Nakagawa, T., H. Kitagawa, Y. Yasuda, et al. 2005. Pollen/event stratigraphy of the varved sediment of Lake Suigetsu, central Japan from 15,701 to 10,217 SG vyr BP (Suigetsu varve years before present): Description, interpretation, and correlation with other regions 2005. Quaternary Science Reviews 24:1691–1701.

Nakagawa, T., H. Kitagawa, Y. Yasuda, et al. and Yangtze River Civilization Program Members. 2003. Asynchronous climate changes in the North Atlantic and Japan during the last termination. Science 299:688–691.

National Academy of Sciences. 2006. *Surface Temperature Reconstructions for the Last 2,000 Years.* Washington, DC: National Academy Press.

National Research Council. 2002. *Abrupt Climate Change: Inevitable Surprises.* Washington, DC: National Academy Press.

Nees, S., A. V. Altenbach, H. Kassens, and J. Thiede. 1997. High-resolution record of foraminiferal response to late Quaternary sea-ice retreat in the Norwegian-Greenland Sea. Geology 25:659–662.

Neff, U., S. J. Burns, A. Mangini, et al. 2001. Strong coherence between solar variability and the monsoon in Oman between 9 and 6 kyr ago. Nature 411:290–293.

Neftel, A., H. Oeschger, T. Staffelbach, and B. Stauffer. 1988. CO_2 record in the Byrd ice core 50,000–5,000 years BP. Nature 331:609–611.

Neftel, A. E., H. Oeschger, J. Schwander, B. Stauffer, and R. Zumbrunn. 1982. Ice core measurements give atmospheric CO_2 content during the past 40,000 years. Nature 295: 220–223.

Nerem, R. S. and G. T. Mitchum. 2002. Estimates of vertical crustal motion derived from differences of TOPEX/POSEI-DON and tide gauge sea level measurements. Geophysical Research Letters 29:1934, GL015037.

Nesme-Ribes, E., ed. 1994. *The Solar Engine and Its Influence on Terrestrial Atmospheres and Climate.* Berlin: Springer-Verlag.

Neumann, A. C. and P. J. Hearty. 1996. Rapid sea-level changes at the close of the last interglacial (substage 5e) recorded in Bahamian island geology. Geology 24:775–778.

Newhall, C. and S. Self. 1982. The Volcanic Explosivity Index (VEI): An estimate of explosive magnitude for historical volcanism. Journal of Geophysical Research 87:1231–1238.

Newton, A., R. Thunell, and L. Stott. 2006. Climate and hydro-graphic variability in the Indo-Pacific Warm Pool during the last millennium. Geophysical Research Letters 33:L19710.

Nicoll, K. 2004. Recent environmental change and prehistoric human activity in Egypt and northern Sudan. Quaternary Science Reviews 23:561–580.

Nikitima, D. L., J. E. Pizzuto, R. A. Schwimmer, and K. W. Ramsey. 2000. An updated Holocene sea-level curve for the Delaware coast. Marine Geology 171:7–20.

Nordt, L., S. Atchley, and S. Dworkin. 2002. Paleosol barometer indicates extreme fluctuations in atmospheric CO_2 across the Cretaceous-Tertiary boundary. Geology 30:703–706.

Nordt, L., S. Athchley, and S. Dworkin. 2003. Terrestrial evidence for two intense greenhouse events in the latest Cretaceous. GSA Today 13:4–9.

Nørgaard-Pedersen, N., N. Mikkelsen, and Y. Kristoffersen. 2007. Arctic Ocean record of last two glacial-interglacial cycles off north Greenland/Ellesmere Island—Implications for glacial history. Marine Geology 244:93–108.

Nørgaard-Pedersen, N., R. F. Spielhagen, H. Elenkeuser, et al. 2003. The Arctic Ocean during the Last Glacial Maximum: Atlantic and polar domains of surface water mass distribution and ice cover. Paleoceanography 18:1–19.

Nørgaard-Pedersen, N., R. F. Spielhagen, J. Thiede, and H. Kassens. 1998. Central Arctic surface ocean environment during the past 80,000 years. Paleoceanography 13:193–204.

Norris, R. D., K. L. Bice, E. A. Magno, and P. A. Wilson. 2002. Jiggling the tropical thermostat in the Cretaceous hothouse. Geology 30:299–302.

Norris, R. D. and P. A. Wilson. 1998. Low-latitude sea-surface temperatures for the mid-Cretaceous and the evolution of planktonic foraminifera. Geology 26:823–826.

North Greenland Ice Core Project Members (NGRIP). 2004. High-resolution record of northern hemisphere climate extending into the last interglacial period. Nature 431: 147–151.

Notaro, M., Z. Liu, R. Gallimore, et al. 2005. Simulated and observed pre-industrial to modern vegetation and climate changes. Journal of Climate 18:3650–3671.

Notaro, M., Z. Liu, and J. W. Williams. 2006. Observed vegetation-climate feedbacks in the United States. Journal of Climate 19:763–786.

Notaro, M., S. Vavrus, and Z. Liu. 2007. Global vegetation and climate change due to future increases in CO_2 as projected by a fully coupled model with dynamic vegetation. Journal of Climate 20:70–90.

Nott, J. and M. Hayne. 2001. High frequency of "super-cyclones" along the Great Barrier Reef over the past 5,000 years. Nature 413:508–512.

Nurnberg, D., J. Bijma, and C. Hemleben. 1996. Assessing the reliability of magnesium in foraminiferal calcite as a proxy for water mass temperatures. Geochimica et Cosmochimica Acta 60:803–814.

Nyberg, J., B. A. Malmgren, A. Winter, et al. 2007. Low Atlantic hurricane activity in the 1970s and 1980s compared to the past 270 years. Nature 447:698–701.

Nydick, K. R., A. B. Bidwell, E. Thomas, and J. C. Varekamp. 1995. A sea level rise curve from Guilford, Connecticut, USA. Marine Geology 124:137–159.

O'Brien, S. R., P. A. Mayewski, L. D. Meeker, et al. 1995. Complex-ity of Holocene climate as reconstructed from a Greenland ice core. Science 270:1962–1964.

Occhietti, S. and P. J. H. Richard. 2003. Effet réservoir sur les Ages ^{14}C de Champlain à la transition Pléistocène-Holocene révision de la chronologie de la déglaciation au Québec méridional. Géographie Physique et Quaternaire 57:115–138.

Oerlemans, J. 1982. A model of the Antarctic Ice Sheet. Nature 297:550–553.

Oerlemans, J. 2005. Exacting a climate signal from 169 glacier records. Science 308:675–677.

Oeschger, H., J. Beer, U. Siegenthaler, B. Stauffer, W. Dansgaard, and C. C. Langway. 1984. Late glacial climate history from ice cores. In J. E. Hansen and T. Takahashi, eds., *Climate Processes and Climate Sensitivity*, pp. 299–306. Washington, DC: American Geophysical Union.

Oeschger, H. and C. C. Langway Jr., eds. 1989. *The Environmental Record in Glaciers and Ice Sheets.* Chichester, UK: John Wiley & Sons.

Ohkouchi, N., T. I. Eglinton, L. D. Keigwin, and J. M. Hayes. 2002. Spatial and temporal offsets between proxy records in a sediment drift. Science 298:1224–1227.

Ohmura, A. 2004. *Cryosphere During the Twentieth Century.* Washington, DC: American Geophysical Union.

Olausson, E. 1965. Evidence of climatic changes in North Atlantic deep-sea cores with remarks on isotopic palaeotemperature analysis. Progress in Oceanography 3:221–252.

Olsen, P. E. 1984. Periodicity of lake-level cycles in the late Triassic Lockatong Formation of the Newark Basin (Newark Supergroup, New Jersey and Pennsylvania). In A. Berger, J. Imbrie, J. Hays, G. Kukla, and B. Saltzman, eds., *Milanko-*

vitch and Climate, NATO Symposium, Part 1, pp. 129–146. Dordrecht: D. Reidel.

Olsen, P. E. and D. V. Kent. 1996. Milankovitch climate forcing in the tropics of Pangea during the late Triassic. Palaeogeography, Palaeoclimatology, Palaeoecology 122:1–26.

Olson, S. L. and P. J. Hearty. 2009. A sustained +21 m sea-level highstand during MIS 11 (400 ka): Direct fossil and sedimentary evidence from Bermuda. Quaternary Science Reviews 28:271–285.

Oman, L., A. Robock, G. Stenchikov, G. A. Schmidt, and R. Ruedy. 2005. Climatic response to high-latitude volcanic eruptions. Journal of Geophysical Research 110:D13103.

O'Nions, R. K., M. Frank, F. von Blanckenburg, and H.-F. Ling. 1998. Secular variation of Nd and Pb isotopes in ferromanganese deposits from the Atlantic, Indian, and Pacific oceans. Earth and Planetary Science Letters 159:183–191.

Opdyke, B. N. and J. C. G. Walker. 1992. Return of the coral reef hypothesis: Basin to shelf partitioning of $CaCO_3$ and its effect on atmospheric CO_2. Geology 20:733–736.

Oppo, D. W. and M. Horowitz. 2000. Glacial deepwater hydrography: South Atlantic benthic Cd/Ca and $\delta^{13}C$ evidence. Paleoceanography 15:147–160.

Oppo, D. W., M. Horowitz, and S. J. Lehman. 1997. Marine core evidence for reduced deep water production during Termination II followed by a relatively stable substage 5e (Eemian). Paleoceanography 12:51–63.

Oppo, D. W., L. D. Keigwin, J. F. McManus, and J. L. Cullen, 2001. Evidence for millennial scale variability during Marine Isotope Stage 5 and Termination II. Paleoceanography 16:280–292.

Oppo, D. W. and S. J. Lehman. 1993. Mid-depth circulation of the subpolar North Atlantic during the Last Glacial Maximum. Science 259:1148–1152.

Oppo, D. W. and S. J. Lehman. 1995. Suborbital timescale variability of North Atlantic deep water during the past 200,000 years. Paleoceanography 10:901–910.

Oppo, D. W., B. K. Linsley, Y. Rosenthal, S. Dannenmann, and L. Beaufort. 2003a. Orbital and suborbital climate variability in the Sulu Sea, western tropical Pacific. Geochemistry, Geophysics, Geosystems 4:GC000260.

Oppo, D. W., J. F. McManus, and J. L. Cullen. 2003b. Palaeoceanography: Deepwater variability in the Holocene epoch. Nature 422:277–278.

Oppo, D. W., G. A. Schmidt, and A. N. LeGrande. 2007. Seawater isotope constraints on tropical hydrology during the Holocene. Geophysical Research Letters 34:L13701, GL030017.

Ortiz, J. D., S. O'Connell, J. DelViscio, et al. 2004. Enhanced marine productivity off western North America during warm climate intervals of the past 52 kyr. Geology 32:521–524.

Osborn, T. J. and K. R. Briffa, 2004. The real color of climate change? Science 306:621–622.

Osborn, T. J. and K. R. Briffa. 2006. The spatial extent of 20th-century warmth in the context of the past 1200 years. Science 311:841–844.

Osborn, T. J. and K. R. Briffa. 2007. Response to comment on "The spatial extent of 20th-century warmth in the context of the past 1200 years." Science 316:1844b.

Otto-Bliesner, B. L., C. D. Hewitt, T. M. Marchitto, et al. 2007. Last Glacial Maximum ocean thermohaline circulation: PMIP2 model intercomparisons and data constraints. Geophysical Research Letters 34:L12706, GLO29475.

Overland, J. E. and M. Wang. 2005. The Arctic climate paradox: The recent decrease of the Arctic Oscillation. Geophysical Research Letters 32:L06701, GL021752.

Overpeck, J. E., D. Anderson, S. Trumbore, and W. Prell. 1996. The southwest Indian monsoon over the last 18,000 years. Climate Dynamics 12:213–225.

Overpeck, J. E., K. Hughen, D. Hardy, et al. 1997. Arctic environmental change of the last four centuries. Science 278: 1251–1256.

Overpeck, J. E. and R. Webb. 2000. Nonglacial rapid climate events: Past and future. Proceedings of the National Academy of Sciences 97:1335–1338.

Pagani, M., M. A. Arthur, and K. H. Freeman. 1999. Miocene evolution of atmospheric carbon dioxide. Paleoceanography 14:273–292.

Pagani, M., K. Caldeira, D. Archer, and J. C. Zachos. 2006a. An ancient carbon mystery. Science 314:1556–1557.

Pagani, M., K. H. Freeman, K. Ohkouchi, and K. Caldeira. 2002. Comparison of water column [CO_2aq] with sedimentary alkenone-based estimates: A test of the alkenone-CO_2 proxy. Paleoceanography 17:1069, doi:10.1029/2002PA000756.

Pagani, M., N. Pedentchouk, M. Huber, and Expedition Scientists. 2006b. Arctic hydrology during global warming at the Palaeocene-Eocene thermal maximum. Nature 442: 671–675.

Pagani, M., J. C. Zachos, K. H. Freeman, B. Tripple, and S. Bohaty. 2005. Marked decline in atmospheric carbon dioxide concentrations during the Paleogene. Science 309:600–603.

Pahnke, K. and R. Zahn. 2005. Southern hemisphere water mass conversion linked with North Atlantic climate variability. Science 307:1741–1746.

Pahnke, K., R. Zahn, H. Elderfield, and M. Schulz. 2003. 340,000-year centennial-scale marine record of southern hemisphere climatic oscillation. Science 301:948–952.

Paillard, D. 1998. The timing of Pleistocene glaciations from a simple multiple-scale climate model. Nature 391:378–381.

Paillard, D. 2001. Glacial cycles: Toward a new paradigm. Reviews of Geophysics 39:325–346.

Paillard, D. and F. Parrenin. 2004. The Antarctic Ice Sheet and the triggering of deglaciations. Earth and Planetary Science Letters 227:263–271.

Pair, D. and C. G. Rodrigues. 1993. Late Quaternary deglaciation of the southwestern St. Lawrence Lowland, New York and Ontario. Geological Society of America Bulletin 105: 1151–1164.

Pälike, H., J. Frazier, and J. C. Zachos. 2006a. Extended orbitally forced palaeoclimatic records from the equatorial Atlantic Ceara Rise. Quaternary Science Reviews 5:3138–3149.

Pälike, H., R. D. Norris, J. O. Herrle, et al. 2006b. The heartbeat of the Oligocene climate system. Science 314:1894–1898.

Palmer, M. R., P. N. Pearson, and S. J. Cobb. 1998. Reconstructing past ocean pH-depth profiles. Science 282:1468–1471.

Palmer, T. N., F. J. Doblas-Reyes, A. Weisheimer, and M. J. Rodwell. 2008. Toward seamless prediction. Bulletin of the American Meteorological Society 89:459–470.

Pancost, R. D. and C. S. Boot. 2004. The palaeoclimatic utility of terrestrial biomarkers in marine sediments. Marine Chemistry 92:239–261.

Pap, J. M., P. Fox, and C. Frohlich, eds. 2004. *Solar Variability and Its Effect on Climate*. Washington, DC: American Geophysical Union.

Parent, M. and S. Occhietti. 1999. Late Wisconsin deglaciation and Champlain Sea invasion in the St. Lawrence Valley, Quebec. Géographie Physique et Quaternaire 53:117–135.

Parkinson, C. L., D. J. Cavalieri, P. Gloersen, H. J. Zwally, and J. C. Comiso. 1999. Arctic sea ice extents, areas and trends, 1978–1996. Journal of Geophysical Research 104:20837–20856.

Parrenin, F. and D. Paillard. 2003. Amplitude and phase of glacial cycles from a conceptual model. Earth and Planetary Science Letters 214:243–250.

Parrish, J. T. 1998. *Interpreting Pre-Quaternary Climate from the Geological Record*. New York: Columbia University Press.

Partin, J. W., K. M. Cobb, J. F. Adkins, B. Clark, and D. P. Fernandez. 2007. Millennial-scale trends in west Pacific warm pool hydrology since the Last Glacial Maximum. Nature 449:452–455.

Paul, F., A. Kääb, M. Maisch, T. Kellenberger, and W. Haeberli. 2004. Rapid disintegration of Alpine glaciers observed with satellite data. Geophysical Research Letters 31:L21402, GL020816.

Pauling, A., J. Luterbacher, C. Casty, and H. Wanner. 2005. 500 years of gridded high resolution precipitation reconstructions over Europe and the connection to large-scale circulation. Climate Dynamics 26:387–405.

Pauling, A., J. Luterbacher, and H. Wanner. 2003. Evaluation of proxies for European and North Atlantic temperature field reconstructions. Geophysical Research Letters 30:1787, GL017589.

Pavlov, A. A., M. T. Hurtgen, J. F. Kasting, and M. A. Arthur. 2003. Methane-rich Proterozoic atmosphere. Geology 31:87–90.

Payton, C. E., ed. 1977. *Stratigraphic Interpretation of Seismic Data*. Tulsa, OK: American Association of Petroleum Geologists.

Peacock, S., E. Lane, and J. M. Restrepo. 2006. A possible sequence of events for the generalized glacial-interglacial cycle. Global Biogeochemical Cycles 20:GB2010.

Pearce, C. R., A. S. Cohen, A. L. Coe, and K. W. Burton. 2008. Molybdenum isotope evidence for global ocean anoxia coupled with perturbations to the carbon cycle during the Early Jurassic. Geology 36:231–234.

Pearson, P. N. and M. R. Palmer. 2000. Atmospheric carbon dioxide concentrations over the past 60 million years. Nature 406:695–699.

Pearson, P. N., B. E. van Dongen, C. J. Nicholas, et al. 2007. Stable warm tropical climate through the Eocene Epoch. Geology 35:211–214.

Peck, V. L., I. R. Hall, R. Zahn, and J. D. Scourse. 2007. Progressive reduction in NE Atlantic intermediate water ventilation prior to Heinrich events: Response to NW European ice sheet instabilities? Geochemistry, Geophysics, Geosystems 8:Q01N10.

Peck, V. L., I. R. Hall, R. Zahn, et al. 2006. High resolution evidence for linkages between NW European ice sheet instability and Atlantic meridional overturning circulation. Earth and Planetary Science Letters 243:476–488.

Pecker, J.-C. and K. Runcorn, eds. 1990. The earth's climate and variability of the sun over recent millennia: Geophysical, astronomical, and archaeological aspects. Philosophical Transactions of the Royal Society of London 330:395–697.

Pedersen, T. F. and S. E. Calvert. 1990. Anoxia versus productivity: What controls the formation of organic carbon-rich sediments and sedimentary rocks? American Association of Petroleum Geologists Bulletin 74:456–466.

Pegram, W. J. and K. K. Turekian. 1999. The osmium isotopic composition change of Cenozoic sea water as inferred from a deep-sea core corrected for meteoric contributions. Geochimica et Cosmochimica Acta 63:4053–4058.

Pekar, S. F., N. Christie-Blick, M. A. Kominz, and K. G. Miller. 2001. Evaluating the stratigraphic response to eustasy from Oligocene strata in New Jersey. Geology 29:55–58.

Pekar, S. F., N. Christie-Blick, M. A. Kominz, and K. G. Miller. 2002. Calibrating eustasy to oxygen isotopes for the early icehouse world of the Oligocene. Geology 30:903–906.

Pekar, S. F. and R. M. DeConto. 2006. High-resolution ice-volume estimates for the early Miocene: Evidence for a dynamic ice sheet in Antarctica. Palaeogeography, Palaeoclimatology, Palaeoecology 231:101–109.

Pekar, S. F., A. M. Fuller, and S. Li. 2005, Glacioeustatic changes in the early and middle Eocene (51–42 Ma) greenhouse world based on shallow-water stratigraphy from ODP Leg 189 Site 1171 and oxygen isotope records. Geological Society of America Bulletin 117:1081–1093.

Pekar, S. F. and K. G. Miller. 1996. New Jersey Oligocene "icehouse" sequences (ODP Leg 150) correlated with global $\delta^{18}O$ and Exxon eustatic records. Geology 24:567–570.

Pelejero, C., E. Calvo, M. T. McCulloch, J. F. Marshall, M. K. Gagan, J. M. Lough, and B. N. Opdyke. 2005. Preindustrial to modern interdecadal variability in coral reef pH. Science 309:2204–2207.

Pelejero, C., J. O. Grimalt, S. Heilig, M. Kienast, and L. Wang. 1999. High-resolution U_{37}^K temperature reconstructions in the South China Sea over the past 220 kyr. Paleoceanography 14:224–231.

Peltier, W. R. 2002. On eustatic sea level history: Last Glacial Maximum to Holocene. Quaternary Science Reviews 21:377–396.

Peltier, W. R. 2004. Global glacial isostasy and the surface of the ice-age earth: The ICE-5G (VM2) model and GRACE. Annual Review of Earth and Planetary Sciences 32:111–149.

Peltier, W. R. 2005. On the hemispheric origins of meltwater pulse 1a. Quaternary Science Reviews 24:1655–1671.

Peltier, W. R. and R. G. Fairbanks. 2006. Global glacial ice volume and Last Glacial Maximum duration from an extended Barbados sea level record. Quaternary Science Reviews 25:3322–3337.

Peltier, W. R., I. Shennan, R. Drummond, and B. Horton. 2002. On the postglacial isostatic adjustment of the British Isles and the shallow viscoelastic structure of the earth. Geophysical Journal International 148:443–475.

Peristykh, A. N. and P. E. Damon. 2003. Persistence of the Gleissberg 88-year solar cycle over the last 12,000 years: Evidence from cosmogenic isotopes. Journal of Geophysical Research 108(A1):1003, JA009390.

Peteet, D. M., ed. 1993. Global Younger Dryas? Quaternary Science Reviews 12:277–355.

Peteet, D. M. 1995. Global Younger Dryas? Quaternary International 28:93–104.

Peterson, B. J., R. M. Holmes, J. W. McClelland, et al. 2002. Increasing river discharge to the Arctic Ocean. Science 298:2171–2173.

Peterson, B. J., J. McClelland, R. Curry, et al. 2006. Trajectory shifts in the Arctic and subarctic freshwater cycle. Science 313:1061–1066.

Peterson, L. C., G. H. Haug, K. A. Hughen, and U. Röhl. 2000. Rapid changes in the hydrologic cycle of the tropical Atlantic during the last glacial. Science 290:1947–1951.

Peterson, L. C., J. T. Overpeck, N. G. Kipp, and J. Imbrie. 1991. A high-resolution late Quaternary upwelling record from the anoxic Cariaco Basin, Venezuela. Paleoceanography 6:99–119.

Petit, J. R., I. Basile, A. Leruyuet, et al. 1997. Four climate cycles in Vostok ice core. Nature 387:359–360.

Petit, J. R., J. Jouzel, D. Raynaud, et al. 1999. Climate and atmospheric history of the past 420,000 years from the Vostok ice core, Antarctica. Nature 399:429–436.

Peucker-Ehrenbrink, B. and M. W. Miller. 2006. Marine $^{87}Sr/^{86}Sr$ record mirrors the evolving upper continental crust. Geochimica et Cosmochimica Acta 70:A487–A487.

Peucker-Ehrenbrink, B. and G. Ravizza 2000. The marine osmium isotope record. Terra Nova 12:205–219.

Peyron, O., D. Jolly, R. Bonnefille, A. Vincens, and J. Guiot. 2000. Climate of East Africa 6000 ^{14}C Yr B.P. as inferred from Pollen Data. Quaternary Research 54:90–101.

Pflaumann, U., M. Sarnthein, M. Chapman, et al. 2003. Glacial North Atlantic: Sea-surface conditions reconstructed by GLAMAP 2000. Paleoceanography 18:1065, PA000774.

Pfuhl, H. A. and I. N. McCave. 2005. Evidence for late Oligocene establishment of the Antarctic Circumpolar Current. Earth and Planetary Science Letters 235:715–728.

Philander, S. G. and A. Federov. 2003. Role of tropics in changing the response to Milankovich forcing some three million years ago. Paleoceanography 18:1045, doi:10.1029/2002PA000837.

Philander, S. G. H. 1983. El Niño Southern Oscillation phenomena. Nature 302:295–301.

Philander, S. G. H. 1990. *El Niño, La Niña, and the Southern Oscillation.* San Diego, CA: Academic Press.

Phleger, F. B. 1976. Interpretations of late Quaternary foraminifera in deep-sea cores. Progress in Micropaleontology: 263–276.

Pielke, R. A., Jr. 2005. Are there trends in hurricane destruction? Nature 438:E11.

Pielke, R. A., Jr., C. Landsea, M. Mayfield, J. Laver, and R. Pasch. 2005. Hurricanes and global warming. Bulletin of the American Meteorology Society 86:1571–1575.

Pierrehumbert, R. T. 2000. Climate change and the tropical Pacific: The sleeping dragon wakes. Proceedings of the National Academy of Sciences 97:1355–1358.

Pierrehumbert, R. T. 2002. The hydrologic cycle in deep-time climate problems. Nature 419:191–198.

Pierrehumbert, R. T. 2004. High levels of atmospheric carbon dioxide necessary for the termination of global glaciation. Nature 429:646–649.

Pilgrim, L. 1904. Versuch einer rechnerischen Behandlung des Eiszeitenproblems. Jahreschefte fur Vaterlandische Naturkunde in Wurttemberg 60.

Piotrowski, A. M., S. L. Goldstein, S. R. Hemming, and R. G. Fairbanks. 2004. Intensification and variability of ocean thermohaline circulation through the last deglaciation. Earth and Planetary Science Letters 225:205–220.

Piotrowski, A. M., S. L. Goldstein, S. R. Hemming, and R. G. Fairbanks. 2005. Temporal relationships of carbon cycling and ocean circulation at glacial boundaries. Science 307: 1933–1938.

Pirazzoli, P. A. 1991. *World Atlas of Holocene Sea-Level Changes.* Amsterdam: Elsevier.

Pirazzoli, P. A. 1996. *Sea-Level Changes, the Last 20,000 Years.* Chichester, UK: John Wiley & Sons.

Pirazzoli, P. 2005. A review of possible eustatic, isostatic and tectonic contributions in eight late-Holocene relative sea-level histories from the Mediterranean area. Quaternary Science Reviews 24:1989–2001.

Pisias, N. G. and J. Imbrie. 1986. Orbital geometry, CO_2 and Pleistocene climate. Oceanus 29:43–49.

Pisias, N. G., D. G. Martinson, T. C. Moore Jr., et al. 1984. High resolution stratigraphic correlation of benthic oxygen isotopic records spanning the last 300,000 years. Marine Geology 56:119–136.

Pisias, N. G. and A. C. Mix. 1997. Spatial and temporal oceanographic variability of the eastern equatorial Pacific during the late Pleistocene: Evidence from radiolarian microfossils. Paleoceanography 12:381–393.

Pisias, N. G. and T. C. Moore Jr. 1981. The evolution of Pleistocene climate: A time series approach. Earth and Planetary Science Letters 52:450–458.

Pisias, N. G. and N. J. Shackleton. 1984. Modeling the global climate response to orbital forcing and atmospheric carbon dioxide changes. Nature 310:757–759.

Pitman, W. C. 1978. Relationship between eustasy and stratigraphic sequences of passive margins. Geological Society of America Bulletin 89:1389–1403.

Plag, H.-P. 2006. Recent relative sea level trends: An attempt to quantify the forcing factors. Philosophical Transactions of the Royal Society of London, Series A 364:821–844.

Plumb, K. A. 1991. New Precambrian time scale. Episodes 14:139–140.

Pollard, D. and R. M. DeConto. 2005. Hysteresis in Cenozoic Antarctic ice sheet variations. Global and Planetary Change 45:9–21.

Pollard, D. and J. F. Kasting. 2005. Snowball Earth: A thin-ice solution with flowing sea glaciers. Journal of Geophysical Research 110:C07010.

Pollard, R. T., I. Salter, R. J. Sanders, et al. 2009. Southern Ocean deep-water carbon export enhanced by natural iron fertilization. Nature 457:577–580.

Polyak, L., D. A. Darby, J. Bischof, M. and Jakobsson. 2007. Stratigraphic constraints on late Pleistocene glacial erosion and deglaciation of the Chukchi margin, Arctic Ocean. Quaternary Research 67:235–245.

Polyak, L., M. H. Edwards, B. J. Coakley, and M. Jakobsson. 2001. Ice shelves in the Pleistocene Arctic Ocean inferred from glaciogenic deep-sea bedforms. Nature 410: 453–457.

Polyak, V. J. and Y. Asmerom. 2001. Late Holocene climate and cultural changes in the southwestern United States. Science 294:148–151.

Polzin, K. L., J. M. Toole, J. R. Ledwell, and R. W. Schmitt. 1997. Spatial variability of turbulent mixing in the abyssal ocean. Science 276:93–96.

Poore, R. Z., H. J. Dowsett, S. Verardo, and T. M. Quinn. 2003. Millennial- to century-scale variability in Gulf of Mexico Holocene climate records. Paleoceanography 18:1048, PA000868.

Poore, R. Z., L. Osterman, W. B. Curry, and R. L. Phillips. 1999. Late Pleistocene and Holocene meltwater events in the western Arctic Ocean. Geology 27:759–762.

Poore, R. Z., M. J. Pavich, and H. D. Grissino-Mayer. 2005. Record of the North American southwest monsoon from Gulf of Mexico sediment cores. Geology 33:209–212.

Poore, R. Z., R. L. Phillips, and H. J. Rieck. 1993. Paleoclimate record for Northwind Ridge, Western Arctic Ocean. Paleoceanography 8:149–159.

Poore, R. Z. and L. C. Sloan. 1996. Introduction: Climates and climate variability of the Pliocene. Marine Micropaleontology 27:1–2.

Popp, B. N., E. A. Laws, R. R. Bidigare, et al. 1998. Effect of phytoplankton cell geometry on carbon isotopic fractionation. Geochimica et Cosmochimica Acta 62:69–77.

Porter, S. C. 2000a. Onset of neoglaciation in the southern hemisphere. Journal of Quaternary Science 15:395–408.

Porter, S. C. 2000b. High-resolution paleoclimatic information from Chinese eolian sediments based on grayscale intensity profiles. Quaternary Research 53:70–77.

Porter, S. C. 2001a. Snowline depression in the tropics during the last glaciation. Quaternary Science Reviews 20:1067–1091.

Porter, S. C. 2001b. Chinese loess record of monsoon climate during the last glacial-interglacial cycle. Earth-Science Reviews 54:115–128.

Porter, S. C. and G. H. Denton. 1967. Chronology of neoglaciation in the North American cordillera. American Journal of Science 265:177–210.

Porter, S. C. and Z. Weijian. 2006. Synchronism of Holocene East Asian monsoon variations and North Atlantic drift-ice tracers. Quaternary Research 65:443–449.

Poulsen, C. J., A. S. Gendaszek, and R. L. Jacob. 2003. Did the rifting of the Atlantic Ocean cause the Cretaceous thermal maximum? Geology 31:115–118.

Poulsen, C. J., R. L. Jacob, R. T. Pierrehumbert, and T. T. Huynh. 2002. Testing paleogeographic controls on a Neoproterozoic Snowball Earth. Geophysical Research Letters 29:10.1029/2001GL014352.

Power, S., T. Casey, C. Folland, A. Colman, and V. Mehta. 1999. Interdecadal modulation of the impact of ENSO on Australia. Climate Dynamics 15:319–324.

Powers, L. A., T. C. Johnson, J. P. Werne, et al. 2005. Large temperature variability in the southern African tropics since the Last Glacial Maximum. Geophysical Research Letters 32:L08706, GL022014.

Prahl, F. G., L. A. Muelhausen, and D. L. Zahnle. 1988. Further evaluation of long-chain alkenones as indicators of paleoceanography conditions. Geochimica et Cosmochimica Acta 53:2303–2310.

Prahl, F. G. and S. G. Wakeham. 1987. Calibration of unsaturation patterns in long chain keytone compositions for paleotemperature assessment. Nature 330:367–369.

Prange, M. and M. Schultz. 2004. A coastal upwelling seesaw in the Atlantic Ocean as a result of the closure of the Central American seaway. Geophysical Research Letters 31:L17207.

Prell, W. L. and J. E. Kutzbach. 1987. Monsoon variability over the past 150,000 years. Journal of Geophysical Research 92:8411–8425.

Prell, W. L. and J. E. Kutzbach. 1992. Sensitivity of the Indian monsoon to forcing parameters and implications for its evolution. Nature 360:647–652.

Prell, W. L. and J. E. Kutzbach. 1997. The impact of Tibet-Himalayan elevation on the sensitivity of the monsoon climate system to changes in solar radiation. In W. F. Ruddiman, ed., *Tectonic Uplift and Climate Change*, pp. 171–201. New York: Plenum.

Premoli Silva, I., E. Erba, G. Salvini, C. Locatelli, and D. Verga. 1999. Biotic changes in Cretaceous oceanic anoxic events of the Tethys. Journal of Foraminiferal Research 29:352–370.

Prentice, I. C., D. Jolly, and BIOME 6000 Participants. 2000. Mid-Holocene and glacial-maximum vegetation geography of the northern continents and Africa. Journal of Biogeography 27:507–519.

Prentice, I. C. and T. Webb III. 1998. BIOME 6000: Reconstructing global mid-Holocene vegetation patterns from palaeoecological records. Journal of Biogeography 25:997–1005.

Proctor, C. J., A. Baker, and W. L. Barnes. 2002. A three thousand year record of North Atlantic climate. Climate Dynamics 19:449–454.

Proctor, C. J., A. Baker, W. L. Barnes, and M. A. Gilmour. 2000. A thousand year speleothem proxy record of North Atlantic climate from Scotland. Climate Dynamics 16:815–820.

Prokopenko, A. A., L. A. Hinnov, D. F. Williams, and M. I. Kuzmin. 2006. Orbital forcing of continental climate during the Pleistocene: A complete astronomically tuned climatic record from Lake Baikal, SE Siberia. Quaternary Science Reviews 25:3431–3457.

Prokopenko, A. A., E. B. Karabanov, D.F. Williams, et al. 2001. Biogenic silica record of the Lake Baikal response to climate forcing during the Brunhes chron. Quaternary Research 55:123–132.

Prothero, D., L. C. Ivany, E. Nesbitt, eds. 2003. *From Greenhouse to Icehouse: The Marine Eocene-Oligocene Transition.* New York: Columbia University Press.

Pucéat, E., C. Lécuyer, Y. Donnadieu, et al. 2007. Fish tooth $\delta^{18}O$ revising late Cretaceous meridional upper ocean water temperature gradients. Geology 35:107–110.

Pucéat, E., C. Lécuyer, S. M. F. Sheppard, et al. 2003. Thermal evolution of Cretaceous Tethyan marine waters inferred from oxygen isotope composition of fish tooth enamels. Paleoceanography 18:7.1–7.12.

Quade, J., T. E. Cerling, and J. R. Bowman. 1989. Systematic variations in the carbon and oxygen isotopic composition of pedogenic carbonate along elevation transects in the southern Great Basin, United States. Geological Society of America Bulletin 101:464–475.

Quade, J., C. Garzione, and J. Eiler. 2007. Paleoelevation reconstruction using pedogenic carbonates. Reviews in Mineralogy and Geochemistry 66:53–87.

Quade, J., L. Roe, P. G. DeCelles, and T. P. Ojha. 1997. The late Neogene $^{87}Sr/^{86}Sr$ record of lowland Himalayan rivers. Science 276:1828–1831.

Quadrelli, R. and J. M. Wallace. 2004. A simplified linear framework for interpreting patterns of northern hemisphere wintertime climate variability. Journal of Climate 17:3728–3744.

Quinn, T. M. and D. E. Sampson. 2002. A multiproxy approach to reconstructing sea surface conditions using coral skeleton geochemistry. Paleoceanography 17:1062, PA000528.

Quinn, T. M., F. W. Taylor, T. J. Crowley, et al. 1998. A multicentury coral stable isotope record from a New Caledonia coral: Interannual and decadal sea surface temperature variability in the southwest Pacific since 1657 A.D. Paleoceanography 13:412–426.

Quinn, W. H. 1992. A study of Southern Oscillation–related climatic activity for A.D. 622–1990 incorporating Nile River flood data. In H. F. Diaz and V. Markgraf, eds., *El Niño: Historical and Paleoclimatic Aspects of the Southern Oscillation*, pp. 119–149. Cambridge: Cambridge University Press.

Quinn, W. H., V. T. Neal, and S. E. Antunez de Mayolo. 1987. El Niño occurrences over the past four and a half centuries. Journal of Geophysical Research 92:14449–14461.

Quinn, W. H., D. Q. Zopf, K. S. Short, and R. T. Kuoyang. 1978. Historical trends and statistics of the Southern Oscillation, El Niño, Indonesian droughts. Fishery Bulletin 76:663–678.

Rahmstorf, S. 1994. Rapid climate transitions in a coupled ocean-atmosphere model. Nature 372:82–85.

Rahmstorf, S. 1995. Bifurcations of the Atlantic thermohaline circulation in response to changes in the hydrological cycle. Nature 378:145–149.

Rahmstorf, S. 1996. On the freshwater forcing and transport of the Atlantic thermohaline circulation. Climate Dynamics 12:799–811.

Rahmstorf, S. 2002. Ocean circulation and climate during the past 120,000 years. Nature 419:207–214.

Rahmstorf, S. 2003. Timing of abrupt climate change: A precise clock. Geophysical Research Letters 30:1510–1514.

Rahmstorf, S. 2007. A semi-empirical approach to projecting future sea-level rise. Science 315:368–370.

Rahmstorf, S., M. Crucifix, A. Ganopolski, et al. 2005. Thermohaline circulation hysteresis: A model intercomparison. Geophysical Research Letters 32:L23605.

Raisbeck, G. M., F. Yiou, J. Jouzel, and T. F. Stocker. 2007. Direct north-south synchronization of abrupt climate change record in ice cores using beryllium 10. Climate of the Past Discussions 3:755–769.

Raisbeck, G. M., F. Yiou, J. Jouzel, et al. 1992. [10]Be deposition at Vostok, Antarctica during the last 50,000 years and its relationship to possible cosmogenic production variations during this period. In E. Bard and W. S. Broecker, eds., *The Last Deglaciation: Absolute and Radiocarbon Chronologies*, pp. 125–139. Berlin: Springer-Verlag.

Ram, M. and M. R. Stoltz. 1999. Possible solar influences on the dust profile of the GISP2 ice core from central Greenland. Geophysical Research Letters 26:1763–1764.

Ramanathan, V. and W. Collins. 1991. Thermodynamic regulation of ocean warming by cirrus clouds deduced from observations of the 1987 El Niño. Nature 351:27–32.

Ramaswamy, V., O. Boucher, J. Haigh, et al. 2001. Radiative forcing of climate change. In J. T Houghton, et al., eds., *Climate Change 2001: The Scientific Basis. Contribution of Working Group I to the Third Assessment Report of the Intergovernmental Panel on Climate Change*, pp. 351–416. Cambridge: Cambridge University Press.

Rampino, M. R. and S. Self. 1992. Volcanic winter and accelerated glaciation following the Toba super-eruption. Nature 359:50–52.

Rampino, M. R., S. Self, and R. B. Stothers. 1988. Volcanic winters. Annual Review of Earth and Planetary Sciences 16:73–99.

Rampino, M. R., R. B. Stothers, and S. Self. 1985. Climatic effects of volcanic eruptions. Nature 313:272.

Ramsey, C. B. 1998. Probability and dating. Radiocarbon 40:461–474.

Ramsey, C. B., J. van deer Plicht, and B. Weniger. 2001. "Wiggle-matching" radiocarbon dates. Radiocarbon 43:381–389.

Randall, D. A., R. A. Wood, S. Bony, et al. 2007. Chapter 8: Climate models and their evaluation. In S. Solomon, et al., eds., *Climate Change 2007: The Physical Basis,* Contribution of Working Group I to the Fourth Assessment Report of the Intergovernmental Panel on Climate Change. Cambridge: Cambridge University Press.

Rashid, H. and E. Boyle. 2007. Mixed-layer deepening during Heinrich events: A multi-planktonic foraminiferal $\delta^{18}O$ approach. Science 318:439–441.

Rashid, H., B. P. Flower, R. Z. Poore, and T. M. Quinn. 2007. A ~25 ka monsoon variability record from the Andaman Sea. Quaternary Science Reviews 26:2586–2597.

Rashid, H., R. Hesse, and D. J. W. Piper. 2003. Evidence for an additional Heinrich event between H5 and H6 in the Labrador Sea, Paleoceanography 18:1077, doi:10.1029/2003PA000913.

Rasmussen, E. M., X. Wang, and C. F. Ropelewski. 1990. The biennial component of ENSO variability. Journal of Marine Systems 1:71–96.

Rasmussen, S. O., K. K. Andersen, A. M. Svensson, et al. 2006. A new Greenland ice core chronology for the last glacial termination. Journal of Geophysical Research 111:D06102, JD006079.

Rasmussen, T. L., E. Balboa, E. Thomsen, L. Labeyrie, and T. C. E. van Weering. 1999. Climate records and changes in deep outflow from the Norwegian Sea. Terra Nova 11:60–61.

Rasmussen, T. L., D. W. Oppo, E. Thomsen, and S. J. Lehman. 2003. Deep sea records from the southeast Labrador Sea: Ocean circulation changes and ice-rafting events during the last 160,000 years. Paleoceanography 18:1018, PA000736.

Rasmussen, T. L., E. Thomsen, L. Labeyrie, and T. C. E. van Weering. 1996. Circulation changes in the Faeroe-Shetland Channel correlating with cold events during the last glacial period (58–10 ka). Geology 24:937–940.

Rasmussen, T. L., T. C. E. van Weering, and L. Labeyrie. 1997. Climatic instability, ice sheets and ocean dynamics at high northern latitudes during the last glacial period (58–10 ka BP). Quaternary Science Reviews 16:71–80.

Rasmusson, E. M. and T. H. Carpenter. 1982. Variations in tropical sea surface temperature and surface wind fields associated with Southern Oscillation/El Niño. Monthly Weather Review 110:354–383.

Rasmusson, E. M., X. Wang, and C. F. Ropelewski. 1990. The biennial component of ENSO variability. Journal of Marine Systems 1:71–96.

Ravelo, A. C., D. H. Andreasen, M. Lyle, A. O. Lyle, and M. W. Wara. 2004. Regional climate shifts caused by gradual global cooling in the Pliocene epoch. Nature 429:262–267.

Ravelo, A. C., P. S. Dekens, and M. McCarthy. 2006. Evidence for El Niño–like conditions during the Pliocene. GSA Today 16:4–11.

Rayburn, J. A., D. A. Franzi, and P. L. K. Knuepfer. 2007. Evidence from the Lake Champlain Valley for a later onset of the Champlain Sea and implications for late glacial meltwater routing to the North Atlantic. Palaeogeography, Palaeoclimatology, Palaeoecology 246:62–74.

Rayburn, J. A., P. L. K. Knuepfer, and D. A. Franzi. 2005. A series of large, late Wisconsinan meltwater floods through the Champlain and Hudson valleys, New York State, USA. Quaternary Science Reviews 24:2410–2419.

Raymo, M. E. 1991. Geochemical evidence supporting T. C. Chamberlin's theory of glaciation. Geology 19:344–347.

Raymo, M. E. 1994a. The initiation of northern hemisphere glaciation. Annual Review of Earth and Planetary Sciences 22:353–383.

Raymo, M. E. 1994b. The Himalayas, organic carbon burial, and climate in the Miocene. Paleoceanography 9:399–404.

Raymo, M. E. 1997. The timing of major climate terminations. Paleoceanography 12:577–585.

Raymo, M. E., B. Grant, M. Horowitz, and G. H. Rau. 1996. Mid Pliocene warmth: Stronger greenhouse and stronger conveyor. Marine Micropaleontology 27:313–326.

Raymo, M. E., D. Hodell, and E. Jansen. 1992. Response of deep ocean circulation to initiation of northern hemisphere glaciation. Paleoceanography 7:645–672.

Raymo, M. E., L. Lisiecki, and K. Nisancioglu. 2006. Plio-Pleistocene ice volume, Antarctic climate, and the global $\delta^{18}O$ record. Science 313:492–495.

Raymo, M. E., D. W. Oppo, and W. Curry. 1997. The mid-Pleistocene climate transition: A deep sea carbon isotopic perspective. Paleoceanography 12:546–559.

Raymo, M. E., W. F. Ruddiman, J. Backman, S. M. Clemens, and D. G. Martinson. 1989. Late Pliocene variation in northern hemisphere ice sheets and North Atlantic deep water circulation. Paleoceanography 4:413–446.

Raymo, M. E., W. F. Ruddiman, and P. N. Froelich. 1988. Influence of late Cenozoic mountain building on ocean geochemical cycles. Geology 16:649–653.

Raymo, M. E., W. F. Ruddiman, N. J. Shackleton, and D. W. Oppo. 1990. Evolution of Atlantic-Pacific $\delta^{13}C$ gradients over the last 2.5 m.y. Earth and Planetary Science Letters 97:353–368.

Raynaud, D., J.-M. Barnola, J. Chappellaz, et al. 2000. The ice record of greenhouse gases: A view in the context of future changes. Quaternary Science Reviews 19:9–17.

Raynaud, D., J.-M. Barnola, R. Souchez, et al. 2005. The record for marine isotopic stage 11. Nature 436:39–40.

Raynaud, D., J. Chappellaz, J.-M. Barnola, Y. S. Korotkevich, and C. Lorius. 1988. Climatic and CH_4 cycle implications of glacial-interglacial CH_4 change in the Vostok ice core. Nature 333:655–657.

Raynaud, D., J. Jouzel, J.-M. Barnola, et al. 1993. The ice record of greenhouse gases. Science 259:926–934.

Raynaud, D., V. Lipenkov, B. Lemieux, et al. 2007. The local insolation signature of air content in Antarctic ice: A new step toward an absolute dating of ice records. Earth and Planetary Science Letters 261:337–349.

Rayner, N. A., P. Brohan, D. E. Parker, et al. 2006. Improved analyses of changes and uncertainties in sea surface temperature measured in situ since the mid-nineteenth century: The HadSST2 dataset. Journal of Climate 19:446–469.

Rayner, N. A., D. E. Parker, E. B. Horton, et al. 2003. Global analyses of SST, sea ice and night marine air temperature since the late nineteenth century. Journal of Geophysical Research 108:4407, JD002670.

Rea, D. K. 1994. The paleoclimatic record provided by eolian deposition in the deep sea: The geologic history of wind. Reviews of Geophysics 32:159–195.

Rea, D. K. and M. W. Lyle. 2005. Paleogene calcite compensation depth in the eastern subtropical Pacific: Answers and questions. Paleoceanography 20:PA1012.

Reimer, P. and K. Hughen. 2008. Palaeoclimate: Tree rings floating on ice cores. Nature Geoscience 1:218–219.

Reimer, P. J., M. G. L. Baillie, E. Bard, et al. 2004. IntCal04 terrestrial radiocarbon age calibration, 0–26 cal kyr BP. Radiocarbon 46:1029–1058.

Remenda, V. H., J. A. Cherry, and T. W. D. Edwards. 1994. Isotopic composition of old ground water from Lake Agassiz: Implications for late Pleistocene climate. Science 226:1975–1978.

Renssen, H., H. Goosse, and T. Fichefet. 2007. Simulation of Holocene cooling events in a coupled climate model. Quaternary Science Reviews 26:2019–2029.

Renssen, H., H. Goosse, T. Fichefet, and J.-M. Campin. 2001. The 8.2 kyr BP event stimulated by a global atmosphere-sea-ice-ocean model. Geophysical Research Letters 28:1567–1570.

Renssen, H., H. Goosse, T. Fichefet, V. Masson-Delmotte, and N. Koç. 2005a. The Holocene climate evolution in the high-latitude southern hemisphere simulated by a coupled atmosphere-sea ice-ocean-vegetation model. The Holocene 15:951–964.

Renssen, H., H. Goosse, T. Fichefet, et al. 2005b. Simulating the Holocene climate evolution at northern high latitudes using a coupled atmosphere-sea ice-ocean-vegetation model. Climate Dynamics 24:23–43.

Renssen, H., H. Goosse, and R. Muscheler. 2006. Coupled climate model simulation of Holocene cooling events: Oceanic feedback amplifies solar forcing. Climate of the Past 2:79–90.

Retallack, G. J. 1990. *Soils of the Past: An Introduction to Paleopedology.* London: Harper Collins Academic.

Retallack, G. J. 2005. Pedogenic carbonate proxies for amount and seasonality of precipitation in paleosols. Geology 33:333–336.

Reuer, M. K., E. A. Boyle, and J. E. Cole. 2003. A mid-century reduction in tropical upwelling inferred from coralline trace element proxies. Earth and Planetary Science Letters 210:437–452.

Revelle, R. and H. E. Suess. 1957. Carbon dioxide exchange between atmosphere and ocean and the question of an increase of atmospheric CO_2 during past decades. Tellus 9:18–27.

Reynolds, B. C., M. Frank, and R. K. O'Nions. 1999. Nd- and Pb-isotope time series from Atlantic ferromanganese crust: Implications for changes in provenance and paleocirculation over the last 8 Myr. Earth and Planetary Science Letters 109:11–23.

Rial, J. A. 1999. Pacemaking the ice ages by frequency modulation of Earth's orbital eccentricity. Science 285:564–568.

Rial, J. A. 2004. Abrupt climate change: Chaos and order at orbital and millennial scales. Global and Planetary Change 41:95–109.

Richard, P. J. H. and S. Occhietti. 2005. ^{14}C chronology for ice retreat and inception of Champlain Sea in the St. Lawrence Lowlands, Canada. Quaternary Research 63:353–358.

Richey, J. N., R. Z. Poore, B. P. Flower, and T. M. Quinn. 2007. 1400 yr multiproxy record of climate variability from the northern Gulf of Mexico. Geology 35:423–426.

Rickaby, R. E. M. and H. Elderfield. 1999. Planktonic foraminiferal Cd/Ca: Paleonutrients or paleotemperature? Paleoceanography 14:293–323.

Rickaby, R. E. M., M. J. Greaves, and H. Elderfield. 2000. Cd in planktonic and benthic foraminiferal shells determined by thermal ionisation mass spectrometry. Geochimica et Cosmochimica Acta 64:1229–1236.

Rickaby, R. E. M. and P. Halloran. 2005. Cool La Niña during the warmth of the Pliocene? Science 307:1948–1952.

Ridge, J. C., M. R. Besonon, M. Brochu, et al. 1999. Varve, paleomagnetic, and ^{14}C chronologies for late Pleistocene events in New Hampshire and Vermont (USA). Géographie Physique et Quaternaire 53:79–105.

Ridgwell, A. J., M. J. Kennedy, and K. Caldeira. 2003. Carbonate deposition, climate stability, and Neoproterozoic ice ages. Science 302:859–862.

Ridgwell, A., A. Watson, and M. Raymo. 1999. Is the spectral signature of the 100 kyr glacial cycle consistent with a Milankovitch origin? Paleoceanography 14: 437–440.

Rieu, R., P. A. Allen, M. Plötze, and T. Pettke. 2007. Climatic cycles during a Neoproterozoic "snowball" glacial epoch. Geology 35:299–302.

Rignot, E., J. Bamber, M. van den Broeke, et al. 2008. Recent mass loss of the Antarctic Ice Sheet from dynamic thinning. Nature Geoscience doi:10.1038/ngeo102.

Rignot, E., G. Casassa, P. Gogineni, et al. 2004. Accelerated ice discharge from the Antarctic Peninsula following the collapse of Larsen B ice shelf. Geophysical Research Letters 31:L18401, GL020697.

Rignot, E. and P. Kanagaratnam. 2006. Changes in the velocity structure of the Greenland Ice Sheet. Science 311: 986–990.

Rignot, E. and R. H. Thomas. 2002. Mass balance of polar ice sheets. Science 297:1502–1506.

Rigor, I. and J. M. Wallace. 2004. Variations in the age of Arctic sea-ice and summer sea-ice extent. Geophysical Research Letters 31:L09401, GL019492.

Rigor, I., J. M. Wallace, and R. L. Colony. 2002. Response of sea ice to Arctic Oscillation. Journal of Climate 15:2648–2663.

Rind, D. 2002. The sun's role in climate variability. Science 296:673–677.

Rind, D. 2008. The consequences of not knowing low- and high-latitude climate sensitivity. Bulletin of the American Meteorological Society 2008:855–864.

Rind, D. and M. Chandler. 1991. Increased ocean heat transports and warmer climate. Journal of Geophysical Research 96:7437–7461.

Rind, D., J. Perlwitz, and P. Lonergan. 2005. AO/NAO response to climate change: 1. Respective influences of stratospheric and tropospheric climate changes. Journal of Geophysical Research 110:D12107.

Rind, D., D. Peteet, W. Broecker, A. McIntyre, and W. Ruddiman. 1986. The impact of cold North Atlantic sea surface temperatures on climate: Implications for the Younger Dryas cooling (11–10 k). Climate Dynamics 1:3–53.

Rind, D., G. Russell, and W. F. Ruddiman. 1997. The effects of uplift on ocean-atmosphere circulation. In W. F. Ruddiman, ed., Tectonic Uplift and Climate. New York: Plenum Press, pp. 123–147.

Rind, D., D. Shindell, J. Perlwitz, et al. 2004. The relative importance of solar and anthropogenic forcing of climate change between the Maunder Minimum and the present. Journal of Climate 17:906–929.

Rinterknecht, V. R., P. U. Clark, G. M. Raisbeck, et al. 2006. The last deglaciation of the southeastern sector of the Scandinavian Ice Sheet. Science 311:1449–1452.

Risebrobakken, B., E. Jansen, C. Andersson, E. Mjelde, and K. Hevrøy. 2003. A high-resolution study of Holocene paleoclimatic and paleoceanographic changes in the Nordic Seas. Paleoceanography 18:1017, PA000764.

Rittenour, T. M., J. Brigham-Grette, and M. E. Mann. 2000. El Niño–like climate teleconnections in New England during the late Pleistocene. Science 288:1039–1042.

Roberts, N. 1998. The Holocene: An Environmental History, 2nd ed. Oxford: Blackwell.

Robinson, L. F., J. F. Adkins, L. D. Keigwin, et al. 2005. Radiocarbon variability in the western North Atlantic during the last deglaciation. Science 310:1469–1473.

Robinson, R. S., B. G. Brunelle, and D. M. Sigman. 2004. Revisiting nutrient utilization in the glacial Antarctic: Evidence from a new method for diatom-bound N isotopic analysis. Paleoceanography 19:PA3001.

Robinson, R. S., A. Mix, and P. Martinez. 2007. Southern Ocean control on the extent of denitrification in the southeast Pacific over the last 70 ka. Quaternary Science Reviews 26:201–212.

Robinson, S., M. Maslin, and I. N. McCave. 1995. Magnetic susceptibility variations in upper Pleistocene deep-sea sediments of the NE Atlantic: Implications for ice rafting and paleocirculation at the Last Glacial Maximum. Paleoceanography 10:221–250.

Robock, A. 2000. Volcanic eruptions and climate. Reviews of Geophysics 38:191–219.

Robock, A. and M. Free. 1995. Ice cores as an index of global volcanism from 1850–present. Journal of Geophysical Research 100:11549–11567.

Robock, A. and M. Free. 1996. The volcanic record in ice cores for the past 2000 years. In P. D. Jones, R. S. Bradley, and J. Jouzel, eds., Climate Variations and Forcing Mechanism of the Last 2000 Years, pp. 533–546. New York: Springer-Verlag.

Roche, D., D. Paillard, and E. Cortijo. 2004. Constraints on the duration and freshwater release of Heinrich event 4 through isotope modeling. Nature 432:379–382.

Roche, D. M., T. M. Dokken, H. Goosse, H. Renssen, and S. L. Weber. 2007. Climate of the Last Glacial Maximum: Sensitivity studies and model-data comparison with the LOVECLIM coupled model. Climate of the Past 3:205–224.

Rochon, A., F. Eynaud, and A. de Vernal, eds. 2008. Dinocysts as tracers of hydrological conditions and productivity along ocean margins. Marine Micropaleontology 68.

Rodbell, D. T., G. O. Seltzer, D. M. Anderson, et al. 1999. An ~15,000-year record of El Niño–driven alluviation in southwestern Ecuador. Science 283:516–520.

Rodrigues, C. G. and G. Vilks. 1994. The impact of glacial lake runoff on the Goldthwait and Champlain Seas: The relationship between Glacial Lake Agassiz runoff and the Younger Dryas. Quaternary Science Reviews 13:923–944.

Roe, G. H. 2006. In defense of Milankovitch. Geophysical Research Letters 33:L24703, GL027817.

Roe, G. H. and M. B. Baker. 2007. Why is climate sensitivity so unpredictable? Science 318:629–632.

Rohling, E. J. 1999. Environmental controls on salinity and $\delta^{18}O$ in the Mediterranean. Paleoceanography 14:706–715.

Rohling, E. J. 2006. Oxygen isotope composition of seawater. In S. A. Elias, ed., *Encyclopedia of Quaternary Science,* vol. 3, pp. 1748–1756. Amsterdam: Elsevier.

Rohling, E. J. 2007. Progress in palaeosalinity: Overview and presentation of a new approach. Paleoceanography 22:PA3215, PA001437.

Rohling, E. J. and S. Cooke. 1999. Stable oxygen and carbon isotope ratios in foraminiferal carbonate shells. In B. K. Sen Gupta, ed., *Modern Foraminifera*, pp. 239–258. Dordrecht: Kluwer.

Rohling, E. J., M. Fenton, F. J. Jorissen, et al. 1998. Magnitudes of sea-level lowstands of the past 500,000 years. Nature 394: 162–165.

Rohling, E. J., K. Grant, C. Hemleben, et al. 2008. High rates of sea-level rise during the last interglacial period. Nature Geoscience 1:38–42.

Rohling, E. J., R. Marsh, N. C. Wells, M. Siddall, and N. R. Edwards. 2004. Similar meltwater contributions to glacial sea level changes from Antarctic and northern ice sheets. Nature 430:1016–1021.

Rohling, E. J. and H. Pälike. 2005. Centennial-scale climate cooling with a sudden cold event around 8,200 years ago. Nature 434:975–979.

Rooth, C. 1982. Hydrology and ocean circulation. Progress in Oceanography 11:131–149.

Ropelewski, C. F. and M. S. Halpert. 1986. North American precipitation and temperature patterns associated with the El Niño/Southern Oscillation (ENSO). Monthly Weather Review 114:2352–2362.

Ropelewski, C. F. and M. S. Halpert. 1987. Global and regional scale precipitation patterns associated with the El Niño/Southern Oscillation. Monthly Weather Review 115: 1606–1626.

Rose, J. 2007. The use of time units in Quaternary Science, Quaternary Science Reviews 26:1193.

Rosell-Melé, A., E. Bard, K.-C. Emeis, et al. 2004. Sea surface temperature anomalies in the oceans at the LGM estimated from the alkenone-UK37 index: Comparison with GCMs. Geophysical Research Letters 31:L03208.

Rosell-Melé, A., M. A. Maslin, J. R. Maxwell, and P. Schaeffer. 1997. Biomarker evidence for "Heinrich" events. Geochimica et Cosmochimica Acta 61:1671–1678.

Rosenheim, B. E., P. K. Swart, S. R. Thorrold, and P. Willenz. 2005. Salinity changes in the subtropical Atlantic: Secular increase and teleconnections to the North Atlantic Oscillation. Geophysical Research Letters 32:L02603, GL021499.

Rosenthal, Y., E. A. Boyle, L. Labeyrie, and D. Oppo. 1995. Glacial enrichments of authigenic Cd and U in Subantarctic sediments: A climatic control on the elements' oceanic budget? Paleoceanography 10:395–413.

Rosenthal, Y., E. A. Boyle, and N. Slowey. 1997. Temperature control on the incorporation of magnesium, strontium, fluorine, and cadmium into benthic foraminiferal shells from Little Bahama Bank: Prospects for thermocline paleoceanography. Geochimica et Cosmochimica Acta 61:3633–3643.

Rosenthal, Y., D. W. Oppo, and B. K. Linsley. 2003. The amplitude and phasing of climate change during the last deglaciation in the Sulu Sea, western equatorial Pacific. Geophysical Research Letters 30, doi:10.1029/2002GL016612.

Rosenthal, Y., S. Perron-Cashman, C. H. Lear et al. 2004. Interlaboratory comparison study of Mg/Ca and Sr/Ca

measurements in planktonic foraminifera for paleoceanographic research. Geochemistry, Geophysics, Geosystems 5:Q04D09, GC000650.

Rothman, D. H. 2002. Atmospheric carbon dioxide levels for the past 500 million years. Proceedings of the National Academy of Sciences 99:4167–4171.

Rothrock, D. A., Y. Yu, and G. A. Maykut. 1999. Thinning of the Arctic sea-ice cover. Geophysical Research Letters 26: 3469–3472.

Rothrock, D. A. and J. Zhang. 2005. Arctic Ocean sea ice volume: What explains its recent depletion? Journal of Geophysical Research 110:C01002, doi:10.1029/2004JC002282.

Rothrock, D. A., J. Zhang, and Y. Yu. 2003. The arctic ice thickness anomaly of the 1990s: A consistent view from observations and models. Journal of Geophysical Research 108(C3):3083, doi:10.1029/2001JC001208.

Rousseau, D.-D., J. Guiot, and R. Bonnefille, eds. 2008. Data/model interactions: The biological perspective of understanding past global changes. Climate of the Past (Special Issue).

Rousseau, D.-D., G. Kukla, and J. McManus. 2006. What is what in the ice and the ocean? Quaternary Science Reviews 25:2025–2030.

Rowley, D. B. and B. S. Currie. 2006. Palaeo-altimetry of the late Eocene to Miocene Lunpola basin, central Tibet. Nature 439:677–681.

Rowley, D. B. and C. N. Garzione. 2007. Stable isotope-based paleoaltimetry. Annual Review of Earth and Planetary Sciences 35:463–508.

Rowley, D. B., R. T. Pierrehumbert, and B. S. Currie. 2001. A new approach to stable isotope-based paleoaltimetry: Implications for paleoaltimetry and paleohypsometry of the High Himalaya since the late Miocene. Earth and Planetary Science Letters 188:253–268.

Royal Society of London Working Group. 2005. Ocean acidification due to increasing atmospheric carbon dioxide. http://royalsociety.org/displaypagedoc.asp?id=13539.

Royden, B., C. Burchfiel, and R. D. van der Hilst. 2008. The geological evolution of the Tibetan Plateau. Science 321: 1054–1058.

Royer, D. L. 2006. CO_2-forced climate thresholds during the Phanerozoic. Geochimica et Cosmochimica Acta 70: 5665–5675.

Royer, D. L., R. A. Berner, and D. J. Beerling. 2001b. Phanerozoic atmospheric CO_2 change: Evaluating geochemical and paleobiological approaches. Earth-Science Reviews 54: 349–392.

Royer, D. L., R. A. Berner, I. P. Montañez, N. J. Tabor, and D. J. Beerling. 2004. CO_2 as a primary driver of Phanerozoic climate change. GSA Today 14:4–10.

Royer, D. L., R. A. Berner, and J. Park. 2007. Climate sensitivity constrained by CO_2 concentrations over the past 420 million years. Nature 446:530–532.

Royer, D. L., S. L. Wing, D. J. Beerling, et al. 2001a. Paleobotanical evidence for near present day levels of atmospheric CO_2 during part of the Tertiary. Science 292:2310–2313.

Rozanski, K., L. Araguas-Araguas, and R. Gonfiantini. 1993. Isotopic patterns in modern global precipitation. In P. K. Swart, et al., eds., *Climate Change in Continental Isotopic Records*, pp. 1–36. Washington, DC: American Geophysical Union.

Ruddiman, W. F. 1977. Late Quaternary deposition of ice-rafted sand in the subpolar North Atlantic (lat 40° to 65°N). Geological Society of America Bulletin 88:1813–1827.

Ruddiman, W. F. 2003. The Anthropogenic greenhouse era began thousands of years ago. Climatic Change 61:261–293.

Ruddiman, W. F. 2006a. Ice-driven CO_2 feedback on ice volume. Climate of the Past Discussions 2:43–78.

Ruddiman, W. F. 2006b. Orbital changes and climate. Quaternary Science Reviews 25:3092–3112.

Ruddiman, W. F. 2007. The early anthropogenic hypothesis: Challenges and responses. Reviews of Geophysics 45:RG000207R.

Ruddiman, W. F. and J. E. Kutzbach. 1989. Forcing of late Cenozoic northern hemisphere climate by plateau uplift in southeast Asia and the American southwest. Journal of Geophysical Research 94(D15):18409–18427.

Ruddiman, W. F. and A. McIntyre. 1973. Time-transgressive deglacial retreat of polar waters from the North Atlantic. Quaternary Research 3:117–130.

Ruddiman, W. F. and A. McIntyre. 1981. Oceanic mechanisms for amplification of the 23,000-year ice-volume cycle. Science 212:617–627.

Ruddiman, W. F. and W. L. Prell. 1997. Introduction to the uplift–climate connection. In W. F. Ruddiman, ed., *Tectonic Uplift and Climate Change*, pp. 3–15. New York: Plenum Press.

Ruddiman, W. F. and M. E. Raymo. 1988. Northern hemisphere climate regimes during the past 3 Ma: Possible tectonic connections. Philosophical Transactions of the Royal Society of London, Series B 318:411–430.

Ruddiman, W. F., M. E. Raymo, D. G. Martinson, B. M. Clement, and J. Backman. 1989. Pleistocene evolution: Northern hemisphere ice sheets and North Atlantic Ocean. Paleoceanography 4:353–412.

Ruddiman, W. F., C. D. Sancetta, and A. McIntyre. 1977. Glacial/interglacial response rate of subpolar North Atlantic waters to climatic change: The record in oceanic sediments. Philosophical Transactions of the Royal Society of London, Series B 280:119–142.

Rudels, B., E. Fahrbach, J. Meincke, G. Budéus, and P. Eriksson. 2002. The East Greenland Current and its contribution to the Denmark Strait overflow. ICES Journal of Marine Science 59:1133–1154.

Rühlemann, C., S. Mulitza, G. Lohmann, et al. 2004. Intermediate depth warming in the tropical Atlantic related to weakened thermohaline circulation: Combining paleoclimate and modeling data for the last deglaciation. Paleoceanography 19:PA1025.

Rühlemann, C., S. Mulitza, P. J. Müller, G. Wefer, and R. Zahn. 1999. Warming of the tropical Atlantic Ocean and slowdown of thermohaline circulation during the last deglaciation. Nature 402:511–514.

Russell, J., M. R. Talbot, and B. J. Haskell. 2003a. Mid-Holocene climate change in Lake Bosumtwi, Ghana. Quaternary Research 60:133–141.

Russell, J. M. and T. C. Johnson. 2005a. Late Holocene climate change in the North Atlantic and equatorial Africa: Millennial-scale ITCZ migration. Geophysical Research Letters 32:GL023295.

Russell, J. M. and T. C. Johnson. 2005b. A high-resolution geochemical record from Lake Edward, Uganda Congo and the timing and causes of tropical African drought during the late Holocene. Quaternary Science Reviews 24:1375–1389.

Russell, J. M., T. C. Johnson, K. R. Kelts, T. Lærdald, and M. R. Talbot. 2003b. An 11000-year lithostratigraphic and paleohydrologic record from equatorial Africa: Lake Edward, Uganda-Congo. Palaeogeography, Palaeoclimatology, Palaeoecology 193:25–49.

Russell, J. M., D. Verschuren, and H. Eggermont. 2007. Spatial complexity of "Little Ice Age" climate in East Africa: Sedimentary records from two crater lake basins in western Uganda. The Holocene 17:183–193.

Russell, P. B., J. M. Livingston, R. F. Pueschel, et al. 1996. Global to microscale evolution of the Pinatubo volcanic aerosol, derived from diverse measurements and analyses. Journal of Geophysical Research 101:18745–18763.

Rutberg, R., S. R. Hemming, and S. L. Goldstein. 2000. Reduced North Atlantic Deep Water flux to the glacial Southern Ocean inferred from neodymium isotope ratios. Nature 405:935–938.

Rutherford, S. and S. D'Hondt. 2000. Early onset and tropical forcing of 100,000-year Pleistocene glacial cycles. Nature 408:72–75.

Rutherford, S., M. E. Mann, T. J. Osborn, et al. 2005. Proxy based northern hemisphere surface temperature reconstructions: Sensitivity to methodology, predictor network, target season, and target domain. Journal of Climate 18:2308–2329.

Rutter, N. W. and Z. Ding. 1993. Palaeoclimates and monsoon variations interpreted from micromorphogenic features of the Baoji palaeosols, China. Quaternary Science Reviews 12:853–862.

Ryan, W. B. F. and M. B. Cita. 1977. Ignorance concerning episodes of ocean-wide stagnation. Marine Geology 23:197–215.

Rye, R., P. H. Kuo, and H. D. Holland. 1995. Atmospheric carbon dioxide concentrations before 2.2 billion years ago. Nature 378:603–605.

Sabine, C. L., R. A. Feely, N. Gruber, et al. 2004. The oceanic sink for anthropogenic CO_2. Science 305:367–371.

Sachs, J. P. 2007. Cooling of Northwest Atlantic slope waters during the Holocene. Geophysical Research Letters 34:L03609, GL028495.

Sachs, J. P., R. F. Anderson, and S. J. Lehman. 2001. Glacial surface temperatures of the southeast Atlantic Ocean. Science 293:2077–2079.

Sachs, J. P. and S. J. Lehman. 1999. Subtropical North Atlantic temperatures 60,000 to 30,000 years ago. Science 286:756–759.

Sachse, D., J. Radke, and G. Gleixner. 2005. δD values of individual n-alkanes from terrestrial plants along a climatic gradient—Implications for the sedimentary biomarker record. Organic Geochemistry 37:469–483.

Saenger, C., T. M. Cronin, R. Thunell, and C. Vann. 2006. Modeling river discharge and precipitation from estuarine salinity in the northern Chesapeake Bay: Application to Holocene paleoclimte. The Holocene 16:1–11.

Sagan, C. and C. Chyba. 1997. The early faint sun paradox: Organic shielding of ultraviolet-labile greenhouse gases. Science 276:1217–1221.

Sahagian, D. L. 2005. Paleoelevation measurement: Combining proxies and approaches. EOS, Transactions, American Geophysical Union 86:500.

Sahagian, D. L. and J. E. Maus. 1994. Basalt vesicularity as a measure of atmospheric pressure and paleoelevation. Nature 372:449–451.

Sahagian, D. L., O. Pinous, A. Olferiev, V. Zakaharov, and A. Beisel. 1996. Eustatic curve for the Middle Jurassic–Cretaceous based on Russian platform and Siberian stratigraphy; zonal resolution. American Association of Petroleum Geologists Bulletin 80:1433–1458.

Sahagian, D. L., A. A. Proussevitch, and W. D. Carlson. 2002. Timing of Colorado Plateau uplift: Initial constraints from vesicular basalt-derived paleoelevations. Geology 30:807–810.

Saint-Laurent, D. 2004. Palaeoflood hydrology: An emerging science. Progress in Physical Geography 28:531–543.

Saltzman, B. 2002. *Dynamical Paleoclimatology*. San Diego, CA: Academic Press.

Saltzman, B. and K. A. Maasch. 1991. A first-order global model of late Cenozoic climatic change. Climate Dynamics 5: 201–210.

Saltzman, B. and A. Sutera. 1987. The mid-Quaternary climatic transition as the free response of a three-variable dynamical model. Journal of Atmospheric Science 44:236–241.

Saltzman, B. and M. Verbitsky. 1993. Multiple instabilities and modes of glacial rhythmicity in the Plio-Pleistocene: A general theory of late Cenozoic climatic change. Climate Dynamics 9:1–15.

Sandweiss, D. H., J. B. Richardson III, E. J. Reitz, H. B. Rollins, and K. A. Maasch. 1996. Geoarchaeological evidence from Peru for a 5000 years B.P. onset of El Niño. Science 273: 1531–1533.

Santer, B. D., T. M. L. Wigley, P. J. Gleckler, et al. 2006. Forced and unforced ocean temperature changes in Atlantic and Pacific tropical cyclogenesis regions. Proceedings of the National Academy of Sciences 103:13905–13910.

Sanyal, A., N. G. Hemming, W. S. Broecker, and G. N. Hanson. 1997. Changes in pH in the eastern equatorial Pacific across stage 5–6 boundary based on boron isotopes in foraminifera. Global and Planetary Change 11:125–133.

Sarmiento, J. and J. R. Toggweiler. 1984. A new model for the role of the oceans in determining atmospheric PCO_2. Nature 308:621–624.

Sarmiento, J. L., T. M. C. Hughes, R. J. Stouffer, and S. Manabe. 1998. Simulated response of the ocean carbon cycle to anthropogenic climate warming. Nature 393:245–249.

Sarmiento, J. L., R. D. Slater, R. T. Barber, et al. 2004. Response of ocean ecosystems to climate warming. Global Biogeochemical Cycles 18:GB3003.

Sarnthein, M. and A. V. Altenbach. 1995. Late Quaternary changes in surface water and deep water masses of the Nordic Seas and north-eastern North Atlantic: A review. Geological Rundschau 84:89–107.

Sarnthein, M., R. Gersonde, S. Niebler, et al. 2003a. Overview of glacial Atlantic Ocean mapping (GLAMAP 2000). Paleoceanography 18:1030, PA000769.

Sarnthein, M., J. P. Kennett, J. Chappell, et al. 2000. Exploring late Pleistocene climate variations. EOS, Transactions, American Geophysical Union 81:625, 629–630.

Sarnthein, M., K. Stattegger, D. Dreger, et al. 2001. Fundamental modes and abrupt changes in North Atlantic circulation and climate over the last 60 ky—Numerical modelling and reconstruction. In P. Schäfer, W. Ritzrau, M. Schlüter, and J. Thiede, eds., *The Northern North Atlantic: A Changing Environment*, pp. 365–410. Heidelberg: Springer-Verlag.

Sarnthein, M. and R. Tiedemann. 1990. Younger Dryas–style cooling events at glacial terminations I–IV at ODP Site 658: Associated benthic $\delta^{13}C$ anomalies constrain meltwater hypothesis. Paleoceanography 5:1041–1055.

Sarnthein, M., S. Van Kreveld, H. Erlenkeuser, et al. 2003b. Centennial-to-millennial-scale periodicities of Holocene climate and sediment injections off the western Barents shelf, 75° N. Boreas 32:447–461.

Sarnthein, M., K. Winn, S. J. A. Jung, et al. 1994. Changes in east Atlantic deepwater circulation over the last 30,000 years: Eight time slice reconstructions. Paleoceanography 9: 209–267.

Sato, M., J. E. Hansen, M. P. McCormick, and J. B. Pollack. 1993. Stratospheric aerosol optical depths 1850–1990. Journal of Geophysical Research 98:22987–22994.

Savarino, J., A. Romero, J. Cole-Dai, S. Bekki, and M. H. Thiemens. 2003. UV induced mass-independent sulfur isotope fractionation in stratospheric volcanic sulfate. Geophysical Research Letters 30:2131, doi:10.1029/2003GL018134.

Sayag, R., E. Tziperman, and M. Ghil. 2004. Rapid switch-like sea ice growth and land ice–sea ice hysteresis. Paleoceanography 19:PA1021, PA000946.

Scaife, A. A., C. K. Folland, L. V. Alexander, A. Moberg, and J. R. Knight. 2008. European climate extremes and the North Atlantic Oscillation. Journal of Climate 21:72–83.

Scambos, T. A., J. Bohlander, C. Shuman, and P. Skvarca. 2004. Glacier acceleration and thinning after ice shelf collapse in the Larsen B embayment, Antarctica. Geophysical Research Letters 31:GL020670.

Scambos, T. A., C. Hulbe, M. A. Fahnestock, and J. Bohlander. 2000. The link between climate warming and breakup of ice shelves in the Antarctic Peninsula. Journal of Glaciology 46:516–530.

Schaefer, H. and M. J. Whiticar. 2008. Potential glacial-interglacial changes in stable carbon isotopic ratios of methane sources and sink fractionation. Global Biogeochemical Cycles 22:GB1000, 002889.

Schaefer, H., M. J. Whiticar, E. J. Brook, et al. 2006. Ice record of $\delta^{13}C$ for atmospheric CH_4 across the Younger Dryas–Preboreal transition. Science 313:1109–1112.

Schaefer, J., G. H. Denton, D. J. A. Barrell, et al. 2006. Near-synchronous interhemispheric termination of the Last Glacial Maximum in mid-latitudes. Science 312:1510–1513.

Scheffer, M., V. Brovkin, and P. Cox. 2006. Positive feedback between global warming and atmospheric CO_2 concentration inferred from past climate change. Geophysical Research Letters 33:L10702.

Schefuß, E., S. Schouten, and R. R. Schneider. 2005. Climatic controls on central African hydrology during the past 20,000 years. Nature 437:1003–1006.

Schefuß, E., J. S. Sinninghe Damsté, and J. H. F. Jansen. 2004. Forcing of tropical Atlantic sea surface temperatures during the mid-Pleistocene transition. Paleoceanography 19:PA4029, PA000892.

Scher, H. D. and E. E. Martin. 2004. Circulation in the Southern Ocean during the Paleogene inferred from neodymium isotopes. Earth and Planterary Science Letters 228:391–405.

Scher, H. D. and E. E. Martin. 2006. Timing and climatic consequences of the opening of Drake Passage. Science 312:428–430.

Scherer, R. P., A. Aldahar, S. Tulaczyk, et al. 1998. Pleistocene collapse of the West Antarctic Ice Sheet. Science 281:82–85.

Schimel, D., J. Melillo, H. Tian, et al. 2000. Contribution of increasing CO_2 and climate to carbon storage by ecosystems in the United States. Science 287:1922–1925.

Schimmelmann, A., A. L. Sessions, and M. Mastalerz. 2006. Hydrogen isotopic (D/H) composition of organic matter during diagenesis and thermal maturation. Annual Review of Earth and Planetary Sciences 34:501–533.

Schlanger, S. O., M. A. Arthur, H. C. Jenkyns, and P. A. Scholle. 1987. The Cenomanian-Turonian oceanic anoxic event, I. Stratigraphy and distribution of organic carbon-rich beds and the marine $\delta^{13}C$ excursion. In J. Brooks and A. Fleet, eds., *Marine Petroleum Source Rocks*, pp. 347–375. London: Geological Society of London.

Schlanger, S. O. and H. C. Jenkyns. 1976. Cretaceous oceanic anoxic events: Causes and consequences. Geologie en Mijnbouw 55:179–184.

Schlesinger, M. E. and N. Ramankutty. 1994. An oscillation in the global climate system of period 65–70 years. Nature 367:723–726.

Schlesinger, M. E., N. Ramankutty, N. Andronova, and M. Margolis. 2000. Temperature oscillations in the North Atlantic. Science 289:547b.

Schmidt, G. A. 2007. The physics of climate modeling. Physics Today (January):72–73.

Schmidt, G. A., R. Ruedy, J. E. Hansen, et al. 2006. Present day atmospheric simulations using GISS ModelE: Comparison to in-situ, satellite and reanalysis data. Journal of Climate 19:153–192.

Schmidt, M. W., H. J. Spero, and D. W. Lea. 2004. Links between salinity variation in the Caribbean and North Atlantic thermohaline circulation. Nature 428:160–163.

Schmidtz, B. and V. Pujalte. 2007. Abrupt increase in seasonal extreme precipitation at the Paleocene-Eocene boundary. Geology 35:215–218.

Schmieder, F., T. von Dobeneck, and U. Bleil. 2000. The Mid-Pleistocene climate transition as documented in the deep South Atlantic Ocean: Initiation, interim state and terminal event. Earth and Planetary Science Letters 179:539–550.

Schmitt, R. W., J. R. Ledwell, E. T. Montgomery, K. L. Polzin, and J. M. Toole. 2005. Enhanced diapycnal mixing by salt fingers in the thermocline of the tropical Atlantic. Science 308:685–688.

Schmittner, A., E. J. Brook, and J. Ahn. 2007a. Impact of the ocean's overturning circulation on atmospheric CO_2. In A. Schmittner, J. Chiang, and S. Hemming, eds., *Ocean Circulation: Mechanisms and Impacts*, pp. 315–334. Washington, DC: American Geophysical Union.

Schmittner, A., J. Chiang, and S. Hemming, eds. 2007b. *Ocean Circulation: Mechanisms and Impacts*. Washington, DC: American Geophysical Union.

Schmittner, A., E. D. Galbraith, S. W. Hostetler, T. F. Pedersen, and R. Zhang. 2007a. Large fluctuations of dissolved oxygen in the Indian and Pacific oceans during Dansgaard-Oeschger oscillations caused by variations of North Atlantic Deep Water subduction. Paleoceanography 22:PA3207.

Schmittner, A., A. Oschlies, H. D. Matthews, and E. D. Galbraith. 2008. Future changes in climate, ocean circulation, ecosystems, and biogeochemical cycling simulated for a business-as-usual CO_2 emission scenario until year 4000 AD. Global Biogeochemical Cycling 22:GB1013, 002953.

Schmittner, A., M. Sarnthein, H. Kinkel, et al. 2004. Global impact of the Panamanian seaway closure. EOS, Transactions, American Geophysical Union 85(49):526.

Schneider, B. and A. Schmittner. 2006. Simulating the impact of the Panamanian seaway closure on ocean circulation, marine productivity and nutrient cycling. Earth and Planetary Science Letters 246:367–380.

Schneider, N. and B. D. Cornuelle. 2005. The forcing of the Pacific decadal index. Journal of Climate 18:4355–4373.

Scholl, D. W., F. C. Craighead, and M. Stuiver. 1969. Florida submergence curve revisited: Its relation to coastal sedimentation rates. Science 163:562–564.

Scholl, D. W., and M. Stuiver. 1967. Recent submergence of southern Florida: A comparison with adjacent coasts and eustatic data. Geological Society of America Bulletin 78:437–454.

Schöne, B. R., J. Fiebig, M. Pfeiffer, et al. 2005. Climate records from a bivalve Methuselah (*Arctica islandica*, Mollusca; Iceland). Palaeogeography, Palaeoclimatology, Palaeoecology 228:130–148.

Schöne, B. R. and D. Surge. 2005. Looking back over skeletal diaries—High-resolution environmental reconstructions from accretionary hard parts of aquatic organisms. Palaeogeography, Palaeoclimatology, Palaeoecology 228:1–3.

Schoof, C. 2007. Ice sheet grounding line dynamics: Steady states, stability and hysteresis. Journal of Geophysical Research 112:F03S28, 000664.

Schott, W. 1935. Die Foraminiferen in dem äquatorialen Teil des Atlantischen Ozeans. Wissenschaftliche Ergebnisse der deutschen Atlantik Expedition auf dem Forschungs- und Vermessungsschiff Meteor 1925–1927:43–134.

Schouten, S., J. Eldrett, D. R. Greenwood, et al. 2008. Onset of long-term cooling of Greenland near the Eocene-Oligocene boundary as revealed by branched tetraether lipids. Geology 36:147–150.

Schouten, S., E. C. Hopmans, A. Forster, et al. 2003. Extremely high sea-surface temperatures at low latitudes during the middle Cretaceous as revealed by archaeal membrane lipids. Geology 31:1069–1072.

Schouten, S., E. C. Hopmans, E. Schefuß, and J. S. Sinninghe Damsté. 2002. Distributional variations in marine crenarchaeotal membrane lipids: A new tool for reconstructing ancient sea water temperatures? Earth and Planetary Science Letters 204:265–274.

Schouten, S., J. Ossebaar, K. Schreiber, et al. 2006. The effect of temperature, salinity and growth rate on the stable hydrogen isotopic composition of long chain alkenones produced by *Emiliania huxleyi* and *Gephyrocapsa oceanica*. Biogeosciences 3:113–119.

Schrag, D. P., R. A. Berner, P. F. Hoffman, and G. P. Halverson. 2002. On the initiation of a Snowball Earth. Geochemistry, Geophysics, Geosystems 3:GC000266.

Schulz, K. G. and R. E. Zeebe. 2006. Pleistocene glacial terminations triggered by synchronous changes in Southern and Northern Hemisphere insolation: The insolation canon hypothesis. Earth and Planetary Science Letters 249: 326–336.

Schulz, M. 2002. On the 1470-year pacing of Dansgaard-Oeschger warm events. Paleoceanography 17:1014, PA000571.

Schulz, M., W. H. B. Berger, M. Sarnthein, and P. M. Grootes. 1999. Amplitude variations of 1470-year climate oscillations during the last 100,000 years linked to fluctuations of continental ice mass. Geophysical Research Letters 26:3385–3388.

Schulz, M. and A. Paul. 2002. Holocene climate variability on centennial-to-millennial time scales: 1. Climate records from the North-Atlantic realm. In G. Wefer, W. H. Berger, K. E. Behre, and E. Jansen, eds., *Climate Development and History of the North Atlantic Realm*, pp. 41–54. Heidelberg: Springer.

Schwabe, A. N. 1844. Sonnen-Beobachtungen in Jahr 1843. Astionomische Nachrichten 21:233. (Cited in Hoyt and Schatten 1997.)

Schwander, J., T. Sowers, J.-M. Barnola, et al. 1997. Age scale of the air in the summit ice: Implication for glacial-interglacial temperature change. Journal of Geophysical Research 102:19483–19493.

Schwander, J. and B. Stauffer. 1984. Age differences between polar ice and air trapped in its bubbles. Nature 311:45–47.

Schwarcz, H. P. 1989. Uranium series dating of Quaternary deposits. Quaternary International 1:7–17.

Schweingruber, F. H. 1996. *Tree Rings and Environment: Dendroecology.* Berne, Switzerland: Paul Haupt AG Bern.

Scileppi, E. and J. P. Donnelly. 2007. Sedimentary evidence of hurricane strikes in western Long Island, New York. Geochemistry, Geophysics, Geosystems 8:Q06011.

Scotese, C. R. 1997. *Continental Drift,* 7th ed. Arlington, TX: PALEOMAP Project.

Scotese, C. R., A. J. Boucot, and W. S. McKerrow. 1999. Gondwanan palaeogeography and palaeoclimatology. Journal of African Earth Sciences 28:99–114.

Scotese, C. R. and W. W. Sager. 1988. Mesozoic and Cenozoic plate tectonic reconstructions. Tectonophysics 155:27–48.

Scott, D. B. 1978. Vertical zonations of marsh foraminifera as accurate indicators of former sea-levels. Nature 272:528–531.

Scourse, J., C. Richardson, G. Forsythe, et al. 2006. First cross-matched floating chronology from the marine fossil record: Data from growth lines of the long-lived bivalve mollusc *Arctica islandica.* The Holocene 16:967–974.

Scourse, J. D., I. R. Hall, I. N. McCave, J. R. Young, and C. Sugdon. 2000. The origin of Heinrich layers: Evidence from H2 for European precursor events. Earth and Planetary Science Letters 182:187–195.

Seidenkrantz, M.-S., L. Bornmalm, S. J. Johnsen, et al. 1996. Two-step deglaciation at the oxygen isotope stage 6/5E transition: The Zeifen-Kattegat climate oscillation. Quaternary Science Reviews 15:77–90.

Sejrup, H. P., H. Haflidason, I. Aarseth, et al. 1994. Late Weichselian glaciation history of the northern North Sea. Boreas 23:1–13.

Selby, D. and R. A. Creaser. 2005. Direct radiometric dating of the Devonian-Carboniferous timescale boundary using the Re-Os black shale geochronometer. Geology 33:545–548.

Self, S. 2006. The effects and consequences of very large explosive volcanic eruptions. Philosophical Transactions of the Royal Society of London Series A 364:2073–2097.

Seltzer, G., D. Rodbell, and S. Burns. 2000. Isotopic evidence for late Quaternary climatic change in tropical South America. Geology 28:35–38.

Seltzer, G. O., P. Baker, S. Cross, R. Dunbar, and S. C. Fritz. 1998. High-resolution seismic reflection profiles from Lake Titicaca, Peru-Bolivia: Evidence for Holocene aridity in the tropical Andes. Geology 26:167–170.

Seltzer, G. O., D. T. Rodbell, P. A. Baker, et al. 2002. Early warming of tropical South America at the last glacial-interglacial transition. Science 296:1685–1686.

Selvaraj, K., C. T. A. Chen, and J.-Y. Lou. 2007. Holocene East Asian monsoon variability: Links to solar and tropical Pacific forcing. Geophysical Research Letters 34:L01703, GL028155.

Senum, G. I. and J. S. Gaffney. 1985. A re-examination of the tropospheric methane cycle: Geophysical implications, In E. T. Sundquist and W. S. Broecker, eds., *The Carbon Cycle and Atmospheric CO_2: Natural Variations, Archean to Present,* pp. 61–69. Washington, DC: American Geophysical Union.

Sepulchre, P., G. Ramstein, F. Fluteau, et al. 2006. Tectonic uplift and eastern Africa aridification. Science 313:1419–1423.

Serreze, M. C., and J. A. Francis. 2006. The arctic amplification debate. Climatic Change 76:241–264.

Serreze, M. C., M. M. Holland, and J. Stroeve. 2007. Perspectives on the Arctic's shrinking sea-ice cover. Science 315:1533–1536.

Severinghaus, J. P. and E. J. Brook. 1999. Abrupt climate change at the end of the last glacial period inferred from trapped air in polar ice. Science 286:930–934.

Severinghaus, J. P., T. Sowers, E. J. Brook, R. B. Alley, and M. L. Bender. 1998. Timing of abrupt climate change at the end of the Younger Dryas interval from the thermally fractioned gases in polar ice. Nature 391:141–146.

Shackleton, N. J. 1967. Oxygen isotope analyses and Pleistocene temperatures re-assessed. Nature 215:15–17.

Shackleton, N. J. 2000. The 100,000-year ice-age cycle identified and found to lag temperature, carbon dioxide, and orbital eccentricity. Science 289:1897–1902.

Shackleton, N. J., A. Berger, and W. R. Peltier. 1990. An alternative astronomical calibration of the lower Pleistocene timescale based on ODP Site 677. Transactions of the Royal Society of Edinburgh 81:251–261.

Shackleton, N. J., S. Crowhurst, T. Hagelberg, N. G. Pisias, and D. A. Schneider. 1995a. A new late Neogene time scale: Application to Leg 138 sites. In N. G. Pisias, et al., eds., *A New Late Neogene Timescale: Application to Leg 138 Sites,* pp. 73–101. College Station, TX: Ocean Drilling Program.

Shackleton, N. J., S. J. Crowhurst, G. P. Weedon, and J. Laskar. 1999. Astronomical calibration of Oligocene-Miocene time. Philosophical Transactions of the Royal Society of London, Series A 357:1907–1929.

Shackleton, N. J., T. K. Hagelberg, and S. J. Crowhurst. 1995b. Evaluating the success of astronomical tuning: Pitfalls of using coherence as a criterion for assessing pre-Pleistocene timescales. Paleoceanography 10:693–698.

Shackleton, N. J. and J. P. Kennett. 1975. Paleotemperature history of the Cenozoic and the initiation of Antarctic glaciation: Oxygen and carbon isotope analyses in DSDP Sites 277, 279, and 281. Initial Report of the Deep Sea Drilling Project 29:743–755.

Shackleton, N. J. and N. D. Opdyke. 1973. Oxygen isotope and paleomagnetic stratigraphy of equatorial Pacific core

V28-238: Oxygen isotope temperatures and ice volumes on a 105 and 106 year scale. Quaternary Research 3:39–55.

Shackleton, N. J. and N. D. Opdyke. 1976. Oxygen isotopes and paleomagnetic stratigraphy of Pacific core V28-239: Late Pliocene to latest Pleistocene: An investigation of late Quaternary paleoceanography and paleoclimatology. Geological Society of America Memoir 145:449–464.

Sheldon, N. D. 2006, Precambrian paleosols and atmospheric CO_2 levels. Precambrian Research 147:148–155.

Shemesh, A., D. Hodell, X. Crosta, et al. 2002. Sequence of events during the last deglaciation in Southern Ocean sediments and Antarctic ice cores. Paleoceanography 17:PA000599.

Shen, C., W.-C. Wang, W. Gong, and Z. Hao. 2006. A Pacific Decadal Oscillation record since 1470 AD reconstructed from proxy data of summer rainfall over eastern China. Geophysical Research Letters 33:L03702, GL024804.

Shen, G. T. and E. A. Boyle. 1988. Determination of lead, cadmium, and other trace metals in annually-banded corals. Chemical Geology 67:47–62.

Shen, G. T., J. E. Cole, D. W. Lea, et al. 1992. Surface ocean variability at Galapagos from 1936–1982: Calibration of geochemical tracers in corals. Paleoceanography 7:563–583.

Shen, G. T. and R. B. Dunbar. 1995. Environmental controls on uranium in reef corals. Geochimica et Cosmochimica Acta 59:2009–2024.

Shen, G. T., R. B. Dunbar, G. M. Wellington, M. W. Colgan, and P. W. Glynn. 1991. Paleochemistry of manganese in corals from the Galapagos Islands. Coral Reefs 10:91–101.

Shennan, I. and S. Hamilton. 2006. Relative sea-level observations in western Scotland since the Last Glacial Maximum for testing models of glacial isostatic land movements and ice-sheet reconstructions. Journal of Quaternary Science 21:601–613.

Shennan, I., S. L. Hamilton, C. Hillier, and S. A. Woodroffe. 2005. 16,000-year record of near-field relative sea-level changes, northwest Scotland, United Kingdom. Quaternary International 133–134:95–106.

Shennan, I. and B. P. Horton. 2002. Relative sea-level changes and crustal movements of the UK. Journal of Quaternary Science 16:511–526.

Shennan, I., W. R. Peltier, R. Drummond, and B. P. Horton. 2002. Global to local scale parameters determining relative sea-level changes and the post-glacial isostatic adjustment of Great Britain. Quaternary Science Reviews 21:397–408.

Shepard, F. P. 1963. Thirty-five thousand years of sea level. In T. Clements, ed., *Essays in Marine Geology in Honor of K. O. Emery*, pp. 1–10. Los Angeles: University of Southern California Press.

Shepard, F. P. 1964. Sea level changes in the past 6000 years: Possible archeological significance. Science 143:574–576.

Shepherd, A., and D. Wingham. 2007. Recent sea-level contributions of the Antarctic and Greenland ice sheets. Science 315:1529–1532.

Shepherd, A., D. Wingham, and E. Rignot. 2004. Warm ocean is eroding West Antarctic Ice Sheet. Geophysical Research Letters 31:L23402, GL021106.

Shepherd, A., D. J. Wingham, A. J. Payne, and P. Skvarca. 2003. Larsen Ice Shelf has progressively thinned. Science 302:856–859.

Sheppard, P. R., P. E. Tarasov, L. J. Graumlich, et al. 2004. Annual precipitation since 515 BC reconstructed from living and fossil juniper growth of northeastern Qinghai Province, China. Climate Dynamics 23:869–881.

Sher, A. V., S. A. Kuzmina, T. V. Kuznetsova, and L. D. Sulerzhitsky. 2005. New insights into the Weichselian environment and climate of the East Siberian Arctic, derived from fossil insects, plants and mammals. Quaternary Science Reviews 24:533–569.

Sheth, H. C. 1999. Flood basalts and large igneous provinces from deep mantle plumes: Fact, fiction, and fallacy. Tectonophysics 311:1–29.

Sheth, H. C. 2005. Were the Deccan flood basalts derived in part from ancient oceanic crust within the Indian continental lithosphere? Gondwana Research 8:109–127.

Sheth, H. C. 2007. "Large igneous provinces (LIPs)": Definition, recommended terminology and a hierarchical classification. Earth-Science Reviews 85:117–124.

Shevenell, A. E., J. P. Kennett, and D. W. Lea. 2004. Middle Miocene Southern Ocean cooling and Antarctic cryosphere expansion. Science 305:1766–1770.

Shields, G. A. 2007. A normalized seawater strontium isotope curve and the Neoproterozoic-Cambrian chemical weathering event. Earth Discussions 2:69–84.

Shin, S.-I., P. D. Sardeshmukh, and R. S. Webb. 2006. Understanding the Mid-Holocene climate. Journal of Climate 19:2801–2817.

Shindell, D., R. L. Miller, G. A. Schmidt, and L. Pandolfo. 1999. Simulation of recent northern winter climate trends by greenhouse-gas forcing. Nature 399:452–456.

Shindell, D., G. A. Schmidt, M. E. Mann, and G. Faluvegi. 2004. Dynamic winter climate response to large tropical volcanic eruptions since 1600. Journal of Geophysical Research 109:D05104.

Shindell, D., G. A. Schmidt, M. E. Mann, D. Rind, and A. Waple. 2001. Solar forcing of regional climate change during the Maunder Minimum. Science 294:2149–2152.

Shindell, D., G. A. Schmidt, R. L. Miller, and M. E. Mann. 2003. Volcanic and solar forcing of climate change during the preindustrial era. Journal of Climate 16:4094–4107.

Shindell, D. T. and G. A. Schmidt. 2004. Southern hemisphere climate response to ozone changes and greenhouse gas increases. Geophysical Research Letters 31:L18209.

Shuman, B., P. Bartlein, N. Logar, P. Newby, and T. W. Webb III. 2002. Parallel climate and vegetation responses to the early Holocene collapse of the Laurentide Ice Sheet. Quaternary Science Reviews 21:1793–1805.

Sicre, M.-A., J. Jacob, U. Ezat, et al. 2008. Decadal variability of sea surface temperatures off North Iceland over the last 2000 years. Earth and Planetary Science Letters 268:137–142.

Siddall, M., E. Bard, E. J. Rohling, and C. Hemleben. 2006. Sea-level reversal during Termination II. Geology 34:817–820.

Siddall, M., E. J. Rohling, A. Almogi-Labin, et al. 2003. Sea-level fluctuations during the last glacial cycle. Nature 423:853–858.

Siddall, M., T. F. Stocker, T. Blunier, et al. 2007. Marine Isotope Stage (MIS) 8 millennial variability stratigraphically identical to MIS 3. Paleoceanography 22:PA1208.

Siegenthaler, U. and J. L. Sarmiento. 1993. Atmospheric carbon dioxide and the ocean. Nature 365:119–125.

Siegenthaler, U., T. F. Stocker, E. Monnin, et al. 2005. Stable carbon cycle–climate relationship during the late Pleistocene. Science 310:1313–1317.

Siegenthaler, U. and T. Wenk. 1984. Rapid atmospheric CO_2 variations and ocean circulation. Nature 308:624–626.

Sigman, D. M., M. A. Altabet, D. C. McCorkle, R. François, and G. Fisher. 1999. The delta N-15 of nitrate in the Southern Ocean: Consumption of nitrate in surface waters. Global Biogeochemical Cycles 13:1149–1166.

Sigman, D. M. and E. A. Boyle. 2000. Glacial/interglacial variations in atmospheric carbon dioxide. Nature 407:859–869.

Sigman, D. M. and G. H. Haug. 2003. Biological pump in the past. In H. Elderfield, ed., *Treatise on Geochemistry*, vol. 6, *The Oceans and Marine Geochemistry*, pp. 491–528. New York: Elsevier.

Sigman, D. M., S. L. Jaccard, and G. H. Haug. 2004. Polar ocean stratification in a cold climate. Nature 428:59–63.

Sigman, D. M., S. J. Lehman, and D. W. Oppo. 2003. Evaluating mechanisms of nutrient depletion and ^{13}C enrichment in the intermediate-depth Atlantic during the last ice age. Paleoceanography 18:1072, doi:10.1029/2002PA000818.

Sikes, E. L. and L. D. Keigwin. 1994. Equatorial Atlantic sea surface temperature for the last 30 kyr: A comparison of UA37, $\delta^{18}O$ and foraminiferal assemblage temperature estimates. Paleoceanography 9:31–45.

Simpkin, T. and L. Siebert. 1994. *Volcanoes of the World*. 2nd ed. Tucson, AZ: Geoscience Press.

Sinclair, D. J. 2005. Correlated trace element "vital effects" in tropical corals: A new geochemical tool for probing biomineralization. Geochimica et Cosmochimica Acta 69:3265–3284.

Sinclair, D. J., B. Williams, and M. Risk. 2006. A biological origin for climate signals in corals–trace element "vital effects" are ubiquitous in Scleractinian coral skeletons. Geophysical Research Letters 33:L17707.

Singer, C., J. Shulmeister, and B. McLea. 1998. Evidence against a significant Younger Dryas cooling event in New Zealand. Science 281:812–814.

Skinner, L. C. and N. J. Shackleton. 2005. Rapid transient changes in northeast Atlantic deep water ventilation age across Termination I. Paleoceanography 19:PA2005.

Sloan, L. C. and E. J. Barron. 1992. A comparison of Eocene climate model results to quantified paleoclimatic interpretations. Palaeogeography, Palaeoclimatology, Palaeoecology 93:183–202.

Sloan, L. C., T. J. Crowley, and D. Pollard. 1996. Modeling of middle Pliocene climate with the NCAR GENESIS general circulation model. Marine Micropaleontology 27:51–61.

Slowey, N. C., G. M. Henderson, and W. B. Curry. 1996. Direct U-Th dating of marine sediments from the two most recent interglacial periods. Nature 383:242–244.

Sluijs, A., S. Schouten, M. Pagani, and Expedition Scientists. 2006. Subtropical Arctic Ocean temperatures during the Palaeocene/Eocene thermal maximum. Nature 441:610–613.

Smith, F. A. and K. H. Freeman. 2006. Influence of physiology and climate on δD of leaf wax n-alkanes from C3 and C4 grasses. Geochimica et Cosmochimica Acta 70:1172–1187.

Smith, J. A., G. O. Seltzer, D. L. Farber, D. T. Rodbell, and R. C. Finkel. 2005. Early local Last Glacial Maximum in the tropical Andes. Science 308:678–681.

Smith, T. M. and R. W. Reynolds. 2004. Improved extended reconstruction of SST (1854–1997). Journal of Climate 17:2466–2477.

Snowball, I. and R. Muscheler. 2007. Palaeomagnetic intensity data: An Achilles heel of solar activity reconstructions. The Holocene 17:851–859.

Soden, B. J. and I. M. Held. 2006. An assessment of climate feedbacks in coupled ocean-atmosphere models. Journal of Climate 19:3354–3360.

Soden, B. J., D. L. Jackson, V. Ramaswamy, M. D. Schwarzkopf, and X. Huang. 2005. The radiative signature of upper tropospheric moistening. Science 310:841–844.

Sohl, L. E., and M. A. Chandler. 2007. Reconstructing Neoproterozoic palaeoclimates using a combined data/modelling approach. In M. Williams, et al., eds., *Deep-Time Perspectives on Climate Change: Marrying the Signal from Computer Models and Biological Proxies*, pp. 61–80. London: Geological Society of London.

Solanki, S. K. and M. Fligge. 1999. A reconstruction of total solar irradiance since 1700. Geophysical Research Letters 26:2465–2468.

Solanki, S. K., M. Schüssler, and M. Fligge. 2000. Secular evolution of the sun's large-scale magnetic field since the Maunder Minimum. Nature 408:445–447.

Solanki, S. K., I. G. Usoskin, B. Kromer, M. Schüssler, and J. Beer. 2004. Unusual activity of the sun during recent decades compared to the previous 11,000 years. Nature 431:1084–1087.

Sonett, C. P., M. S. Giampapa, and M. S. Matthews, eds. 1992. *The Sun in Time*. Tucson: University of Arizona Press.

Soon, W. and S. Baliunas. 2003. Proxy climatic and environmental changes of the past 1000 years. Climate Research 23:89–110.

Sowers, T., R. B. Alley, and J. Jubenville. 2003. Ice core records of atmospheric N_2O covering the last 106,000 years. Science 301:945–948.

Sowers, T. and M. Bender. 1995. Climate records covering the last deglaciation. Science 269:210–214.

Sowers, T., M. Bender, L. Labeyrie, et al. 1993. A 135,000-year Vostok-SPECMAP common temporal framework. Paleoceanography 8:737–766.

Sowers, T., M. Bender, D. Raynaud, Y. S. Korotkevich, and J. Orchado. 1991. The $\delta^{18}O$ of atmospheric G2 from air inclusions in the Vostok ice core: Timing of CO_2 and ice volume changes during the penultimate deglaciation. Paleoceanography 6:679–696.

Sowers, T., E. Brook, D. Etheridge, et al. 1997. An inter-laboratory comparison of techniques for extracting and analyzing gases in ice cores. Journal of Geophysical Research 102:26527–26539.

Spahni, R., J. Chappellaz, T. F. Stocker, et al. 2005. Atmospheric methane and nitrous oxide of the late Pleistocene from Antarctic ice cores. Science 310:1317–1321.

Spahni, R., J. Schwander, J. Flückiger, et al. 2003. The attenuation of fast atmospheric CH_4 variations recorded in polar ice cores. Geophysical Research Letters 30:GL017093.

Speijer, R. P. and A.-M. M. Morsi. 2002. Ostracode turnover and sea-level changes associated with the Paleocene-Eocene thermal maximum. Geology 30:23–26.

Spell, T. L. and I. McDougall. 1992. Revisions to the age of the Brunhes-Matuyama boundary and the Pleistocene geomagnetic timescale. Geophysical Research letters 19:1181–1192.

Spero, H. 1992. Do planktic foraminifera accurately record changes in the carbon isotopic composition of seawater CO_2? Marine Micropaleontology 19:275–285.

Spero, H. and D. Williams. 1988. Extracting environmental information from planktonic foraminiferal δ¹³C data. Nature 335:717–719.

Spero, H. J., J. Bijma, D. W. Lea, and B. E. Bemis. 1997. Effect of seawater carbonate concentration on planktonic foraminiferal carbon and oxygen isotopes. Nature 390:497–500.

Spero, H. J., K. M. Mielke, E. M. Kalve, D. W. Lea, and D. K. Pak. 2003. Multispecies approach to reconstructing eastern equatorial Pacific thermocline hydrography during the past 360 kyr. Paleoceanography 18:1022.

Spicer, R. A., N. B. W. Harris, M. Widdowson, et al. 2003. Constant elevation of southern Tibet over the past 15 million years. Nature 421:622–624.

Spielhagen, R. F., K.-H. Baumann, H. Erlenkeuser, et al. 2004. Arctic Ocean deep-sea record of northern Eurasian ice sheet history. Quaternary Science Reviews 23:1455–1483.

Spielhagen, R. F., H. Erlenkeuser, and C. Siegert. 2005. History of freshwater runoff across the Laptev Sea (Arctic) during the last deglaciation. Global and Planetary Change 48:187–207.

Spörer, G. 1889. Uber die Periodicitat de Sonnenflecken seit dem Jahr 1618. Nova Acta der Ksl. Leop.-Carol Deutschen Akademie der Natuifoischei 53:283–324. (Cited in Hoyt and Schatten 1997.)

Spötl, C. and A. Mangini. 2002. Stalagmite from the Austrian Alps reveals Dansgaard-Oeschger events during Isotope Stage 3: Implications for the absolute chronology of Greenland ice cores. Earth and Planetary Science Letters 203:507–518.

Sprovieri, R., E. Di Stefano, A. Incarbona, and D. W. Oppo. 2006. Suborbital climate variability during Marine Isotope Stage 5 in the Mediterranean basin: Evidence from planktonic foraminifera and calcareous nannofossil relative abundance fluctuations. Quaternary Science Reviews 25:2332–2342.

Stahle, D. W., R. D. D'Arrigo, P. J. Krusic, et al. 1998b. Experimental dendroclimatic reconstruction of the Southern Oscillation. Bulletin of the American Meteorological Society 79:2137–2152.

Stahle, D. W., M. K. Cleaveland, D. B. Blanton, M. D. Therrell, and D. A. Gay. 1998a. The Lost Colony and Jamestown Droughts. Science 280:564–567.

Stanford, J. D., E. J. Rohling, S. E. Hunter, et al. 2006. Timing of meltwater pulse 1a and climate responses to meltwater injections. Paleoceanography 21:PA4103.

Stanley, S. and L. A. Hardie. 1998. Secular oscillations in the carbonate mineralogy of reef-building and sediment-producing organisms driven by tectonically forced shifts in seawater chemistry. Palaeogeography, Palaeoclimatology, Palaeoecology 144:3–19.

Stapor, F. W. and G. W. Stone. 2005. Reply to comment "Validity of sea-level indicators" by E. G. Otvos. Marine Geology 217:189–201.

Stauffer, B., T. Blunier, A. Dällenbach, et al. 1998. Atmospheric CO_2 concentration and millennial-scale climate change during the last glacial period. Nature 392:59–65.

Stauffer, B., H. Hofer, H. Oeschger, J. Schwander, and U. Siegenthaler. 1984. Atmospheric CO_2 concentration during the last glaciation. Annals of Glaciology 5:160–164.

Stauffer, B., E. Lochbronner, H. Oeschger, and J. Schwander. 1988. Methane concentration in the glacial atmosphere was only half that of preindustrial Holocene. Nature 332:812–814.

Stea, R. R. and R. J. Mott. 1998. Deglaciation of Nova Scotia: Stratigraphy and chronology of lake sediment cores and buried organic sections. Géographie Physique et Quaternaire 52:3–21.

Steffensen, J. P., K. A. Andersen, M. Bigler, et al. 2008. High-resolution Greenland ice core data show abrupt climate change happens in few years. Science 321:680–684.

Steig, E., P. M. Grootes, and M. Stuiver. 1994. Seasonal precipitation timing and ice core records. Science 266:1885–1886.

Steig, E. J., E. J. Brook, J. W. C. White, et al. 1998. Synchronous climate changes in Antarctica and the North Atlantic. Science 282:92–95.

Steig, E. J., D. L. Morse, E. D. Waddington, et al. 2000. Wisconsinan and Holocene climate history from an ice core at Taylor Dome, Western Ross Embayment, Antarctica. Geografiska Annaler: Series A, Physical Geography 82a(2–3):213–235.

Stein, R., C. Schubert, C. Vogt, and D. Fütterer. 1994. Stable isotope stratigraphy, sedimentation rates, and salinity changes in the latest Pleistocene to Holocene eastern central Arctic Ocean. Marine Geology 119:333–355.

Steineck, P. L. and E. Thomas. 1996. The latest Paleocene crisis in the deep-sea: Ostracode succession at Maud Rise, Southern Ocean. Geology 24:583–586.

Steinke, S., M. Kienast, and T. Hanebuth. 2003. On the significance of sea-level variations and shelf paleo-morphology in governing sedimentation in the southern South China Sea during the last deglaciation. Marine Geology 201:179–206.

Stenchikov, G., A. Robock, V. Ramaswamy, et al. 2002. Arctic oscillation response to the 1991 Mount Pinatubo eruption: Effects of volcanic aerosols and ozone depletion. Journal of Geophysical Research 107(D24):4803, doi:10.1029/2002JD002090.

Stenni, B., J. Jouzel, V. Masson-Delmotte, et al. 2003. A late-glacial high-resolution site and source temperature record derived from the EPICA Dome C isotope records (East Antarctica). Earth and Planetary Science Letters 217:183–195.

Stenni, B., V. Masson-Delmotte, S. Johnsen, et al. 2001. An oceanic cold reversal during the last deglaciation. Science 293:2074–2077.

Steph, S., R. Tiedemann, M. Prange, et al. 2006. Changes in Caribbean surface hydrography during the Pliocene shoaling of the Central American seaway. Paleoceanography 21:PA4221.

Stephens, B. B. and R. F. Keeling. 2000. The influence of Antarctic sea ice on glacial-interglacial CO_2 variations. Nature 404:171–174.

Stephenson, D. B., H. Wanner, S. Brönnimann, and J. Luterbacher. 2003. The history of scientific research on the North Atlantic Oscillation. In J. W. Hurrell, Y. Kushnir, G. Ottersen, and M. Visbeck, eds., *The North Atlantic Oscillation: Climatic Significance and Environmental Impact*, pp. 37–50. Washington, DC: American Geophysical Union.

Stewart, I., D. R. Cayan, and M. Dettinger. 2005. Changes towards earlier streamflow timing across western North America. Journal of Climate 18:1136–1155.

Stickley, C. E., H. Brinkhuis, S. A. Schellenberg, et al. 2004. Timing and nature of the deepening of the Tasmanian Gateway. Paleoceanography 19:PA4027.

Stine, A. R., P. Huybers, and I. Y. Fung. 2009. Changes in the phase of the annual cycle of surface temperature. Nature 457:435–440.

Stirling, C. H., T. M. Esat, M. T. McCulloch, and K. Lambeck. 1995. High-precision U-series dating of corals from Western Australia and implications for the timing and duration of the

Last Interglacial. Earth and Planetary Science Letters 135: 115–130.

St. John, K. E. K. 2008. Cenozoic ice-rafting history of the central Arctic Ocean: Terrigenous sands on the Lomonosov Ridge. Paleoceanography 23:PA1505.

St. John, K. E. K. and L. A. Krissek. 2002. The late Miocene to Pleistocene ice-rafting history of southeast Greenland. Boreas 31:28–35.

Stocker, T. F. 2000. Past and future reorganisations in the climate system. Quaternary Science Reviews 19:301–319.

Stocker, T. F. and S. J. Johnsen. 2003. A minimum thermodynamic model for the bipolar seesaw. Paleoceanography 18:1087, doi:10.1029/2003PA000920.

Stocker, T. F. and O. Marchal. 2000. Abrupt climate change in the computer: Is it real? Proceedings of the U.S. National Academy of Sciences 97:1362–1365.

Stommel, H. 1961. Thermohaline convection with two stable regimes of flow. Tellus 13:224–230.

St-Onge, G., J. S. Stoner, and C. Hillaire-Marcel. 2003. Holocene paleomagnetic records from the St. Lawrence Estuary, eastern Canada: Centennial- to millennial-scale geomagnetic modulation of cosmogenic isotopes. Earth and Planetary Science Letters 209:113–130.

Storey, M., R. A. Duncan, and C. C. Swisher III. 2007. Paleocene-Eocene Thermal Maximum and the opening of the Northeast Atlantic. Science 316:587–589.

Stott, L., K. Cannariato, R. Thunell, G. Haug, A. Koutavas, and S. Lund. 2004. Decline of surface temperature and salinity in the western tropical Pacific Ocean in the Holocene epoch. Nature 431:56–59.

Stott, L., C. Poulsen, S. Lund, and R. Thunell. 2002. Super ENSO and global climate oscillations at millennial time scales. Science 297:222–226.

Stott, L., A. Timmermann, and R. Thunell. 2007. Southern hemisphere and deep-sea warming led deglacial atmospheric CO_2 rise and tropical warming. Science 318:435–438.

Stouffer, R. J. and S. Manabe. 2003. Equilibrium response of thermohaline circulation to large changes in atmospheric CO_2 concentration. Climate Dynamics 20:759–773.

Stouffer, R. J., J. Yin, J. M. Gregory, et al. 2006. Investigating the causes of the response of the thermohaline circulation to past and future climate changes. Journal of Climate 19:1365–1387.

Street-Perrott, F. A. and R. A. Perrott. 1990. Abrupt climate fluctuations in the tropics: The influence of Atlantic Ocean circulation. Nature 343:607–612.

Street-Perrott, F. A. and R. A. Perrott. 1993. Holocene vegetation, lake levels and climate in Africa. In H. E. Wright Jr., et al., eds., Global Climates Since the Last Glacial Maximum, pp. 318–356. Minneapolis: University of Minnesota Press.

Stroeve, J. C., M. M. Holland, W. Meier, T. Scambos, and M. Serreze. 2007. Arctic sea ice decline: Faster than forecast. Geophysical Research Letters 34:L09501, GL029703.

Stroeve, J. C., M. C. Serreze, F. Fetterer, et al. 2005. Tracking the Arctic's shrinking ice cover: Another extreme minimum in 2004. Geophysical Research Letters 32:L04501.

Stroeven, A. P. and J. Kleman. 1999. Age of Sirius Group on Mount Feather, McMurdo Dry Valleys, Antarctica, based on glaciological inferences from the overridden mountain range of Scandinavia. Global and Planetary Change 23:231–247.

Stromberg, B. 1992. The final stage of the Baltic Ice Lake. In A.-M. Robertsson, et al., eds., Quaternary Stratigraphy, Glacial Morphology, and Environmental Changes, Sveriges Geologiska Undersokning 81:347–354.

Stuiver, M. 1965. Carbon-14 content of the 18th and 19th century wood: Variations correlated with sunspot activity. Science 149:533–537.

Stuiver, M. 1993. A note on single-year calibration of the AD radiocarbon timescale. Radiocarbon 35:67–72.

Stuiver, M. and T. F. Braziunas. 1987. Tree cellulose $^{13}C/^{12}C$ isotope ratios and climatic change. Nature 328:58–60.

Stuiver, M. and T. F. Braziunas. 1989. Atmospheric ^{14}C and century-scale solar oscillations. Nature 338:405–408.

Stuiver, M. and T. F. Braziunas. 1993. Sun, ocean, climate and atmospheric $^{14}CO_2$: An evaluation of causal and spectral relationships. The Holocene 3:289–305.

Stuiver, M., T. F. Braziunas, and P. M. Grootes. 1997. Is there evidence for solar forcing of climate in the GISP2 oxygen isotope record? Quaternary Research 48:259–266.

Stuiver, M., P. M. Grootes, and T. F. Braziunas. 1995. The GISP2 $\delta^{18}O$ climate record of the past 16,500 years and the role of the sun, ocean, and volcanoes. Quaternary Research 44:341–354.

Stuiver, M. and P. D. Quay. 1980. Changes in atmospheric carbon-14 attributed to a variable sun. Science 207:11–19.

Stuiver, M. and P. J. Reimer. 1993. Extended ^{14}C data base and revised CALIB 3.0 ^{14}C age calibration program. Radiocarbon 35:215–230.

Stuiver, M., P. J. Reimer, E. Bard, et al. 1998. INTCAL98 radiocarbon age calibration, 24,000–0 cal BP. Radiocarbon 40:1041–1083.

Suess, H. 1965. Secular variations of the cosmic ray produced carbon-14 in the atmosphere and their interpretations. Journal of Geophysical Research 70:5935–5952.

Suess, H. 1968. Climatic changes, solar activity and the cosmic-ray production rate of natural radiocarbon. Meteorological Monographs 8:146–150.

Sugden, D. E. 1996. The East Antarctic Ice Sheet: Unstable ice or unstable ideas? Transactions of the Institute of British Geographers, new series 21:443–454.

Sugden, D. E. and G. H. Denton. 2004. Cenozoic landscape evolution of the Convoy Range to Mackay Glacier area, Transantarctic Mountains: Onshore to offshore synthesis. Geological Society of America Bulletin 116:840–857.

Sugden, D. E., G. H. Denton, and D. R. Marchant. 1995. Landscape evolution of the Dry Valleys, Transantarctic Mountains. Journal of Geophysical Research 100:9949–9967.

Sugden, D. E., D. R. Marchant, and G. H. Denton, eds. 1993. The case for a stable East Antarctic ice sheet. Geografiska Annaler 75A:151–351.

Suggate, R. P. 1990. Late Pliocene and Quaternary glaciations of New Zealand. Quaternary Science Reviews 9:175–197.

Sun, Y. and Z. An. 2005. Late Pliocene-Pleistocene changes in mass accumulation rates of eolian deposits on the central Chinese Loess Plateau. Journal of Geophysical Research 110:D23101.

Sun, Y., S. C. Clemens, Z. An, and Z. Yu. 2006. Astronomical timescale and palaeoclimatic implication of stacked 3.6-Myr monsoon records from the Chinese Loess Plateau. Quaternary Science Reviews 25:33–48.

Sundquist, E. T. 1985. Geological perspectives on carbon dioxide and the carbon cycle. In E. T. Sundquist and W. S. Broecker, eds., The Carbon Cycle and Atmospheric CO_2: Natural Variations, Archean to Present, pp. 5–60. Washington, DC: American Geophysical Union.

Sundquist, E. T. and W. S. Broecker, eds. 1985. *The Carbon Cycle and Atmospheric CO₂: Natural Variations, Archaen to Present.* Washington, DC: American Geophysical Union.

Sutcliffe, O. E., J. A. Dowdeswell, R. J. Whittenton, J. N. Theron, and J. Craig. 2000. Calibrating the late Ordovician glaciation and mass extinction by the eccentricity cycles of Earth's orbit. Geology 28:967–970.

Sutton, R. T. and D. L. R. Hodson. 2005. Atlantic Ocean forcing of North American and European summer climate. Science 309:115–118.

Svendsen, J. I., H. Alexanderson, V. I. Astakhov, et al. 2004. Late Quaternary ice sheet history of northern Eurasia. Quaternary Science Reviews 23:11–13.

Svensen, H., S. Planke, A. Malthe-Sorenssen, et al. 2004. Release of methane from a volcanic basin as a mechanism for initial Eocene global warming. Nature 429:542–545.

Svensson, A., K. K. Andersen, M. Bigler, et al. 2006. The Greenland ice core chronology 2005, 15–42 ka. Part 2: Comparison to other records. Quaternary Science Reviews 25:3258–3267.

Swart, P. K. 1983. Carbon and oxygen isotope fractionation in scleractinian corals: A review. Earth-Science Reviews 19:51–80.

Swart, P. K., R. E. Dodge, and H. J. Hudson. 1996b. A 240-year stable oxygen and carbon isotopic record in a coral from South Florida: Implications for the prediction of precipitation in southern Florida. Palaios 11:362–375.

Swart, P. K., G. F. Healy, R. E. Dodge, et al. 1996a. The stable oxygen and carbon isotopic record from a coral growing in Florida Bay: A 160 year record of climatic and anthropogenic influence. Palaeogeography, Palaeoclimatology, Palaeoecology 123:219–237.

Swart, P. K., K. C. Lohman, J. McKenzie, and S. Savin, eds. 1993. *Climate Change in Continental Isotope Records.* Washington, DC: American Geophysical Union.

Swetnam, T. W. 1995. Fire history and climate change in giant *Sequoia* groves. Science 262:885–889.

Szabo, B. J. 1985. Uranium-series dating of fossil corals from marine sediments of southeastern United States Atlantic Coastal Plain. Geological Society of America Bulletin 96:398–406.

Szabo, B. J., K. R. Ludwig, D. R. Muhs, and K. R. Simmons. 1994. Thorium-230 age of corals and duration of the last interglacial sea-level high stand on Oahu, Hawaii. Science 266: 93–96.

Takashima, R., H. Nishi, B. T. Huber, and R. M. Leckie. 2006. Greenhouse world and the Mesozoic Ocean. Oceanography 19:82–92.

Talley, L. D. 2002. Ocean circulation. In M. C. MacCracken and J. S. Perry, eds., *Encyclopedia of Global Environmental Change,* pp. 1–23. Chichester, UK: John Wiley & Sons.

Tarasov, L. and W. R. Peltier. 2005. Arctic freshwater forcing of the Younger Dryas cold reversal. Nature 435:662–665.

Taylor, F. and A. McMinn. 2001. Evidence from diatoms for Holocene climate fluctuation along the east Antarctic margin. The Holocene 11:455–466.

Taylor, F. and A. McMinn. 2002. Late Quaternary diatom assemblages from Prydz Bay, eastern Antarctica. Quaternary Research 57:151–161.

Taylor, K. C., P. A. Mayewski, R. B. Alley, et al. 1997. The Holocene–Younger Dryas transition recorded at Summit, Greenland. Science 278:825–827.

Taylor, K. C., J. W. C. White, J. P. Severinghaus, et al. 2004. Abrupt climate change around 22 ka on the Siple Coast of Antarctica. Quaternary Science Reviews 23:7–15.

Tedesco, K. and R. Thunell. 2003. High resolution tropical climate record for the last 6000 years. Geophysical Research Letters 30:1891, GL017959.

Teller, J. T. 1990. Volume and routing of late-glacial runoff from the southern Laurentide Ice Sheet. Quaternary Research 34:12–23.

Teller, J. T., M. Boyd, Z. Yang, P. S. G. Kor, and A. M. Fard. 2005. Alternative routing of Lake Agassiz during the Younger Dryas: New dates, paleotopography and a re-evaluation. Quaternary Science Reviews 24:1890–1905.

Teller, J. T. and D. W. Leverington. 2004. Glacial Lake Agassiz: A 5000-year history of change and its relationship to the δ¹⁸O record of Greenland. Geological Society of America Bulletin 116:729–742.

Teller, J. T., D. Leverington, and J. Mann. 2002. Freshwater outbursts to the oceans from glacial Lake Agassiz and their role in climate change during the last deglaciation. Quaternary Science Reviews 21:879–887.

Teller, J. T. and L. H. Thorleifson. 1983. The Lake Agassiz–Lake Superior connection. In J. T. Teller and L. Clayton, eds., *Glacial Lake Agassiz,* pp. 261–290. Ottawa: Geological Association of Canada.

Thieler, E. R., B. Butman, W. C. Schwab, et al. 2007. A catastrophic meltwater flood event and the formation of the Hudson Shelf Valley. Palaeogeography, Palaeoclimatology, Palaeoecology 246:120–136.

Thierstein, H. R. and W. H. Berger. 1978. Injection events in ocean history. Nature 276:461–466.

Thomas, D., T. J. Bralower, and C. E. Jones. 2003. Neodymium isotopic reconstruction of late Paleocene–early Eocene thermohaline circulation. Earth and Planetary Science Letters 209:309–322.

Thomas, E. 1998. The biogeography of the late Paleocene benthic foraminiferal extinction. In M.-P. Aubry, S. Lucas, and W. A. Berggren, eds., *Late Paleocene–Early Eocene Biotic and Climatic Events in the Marine and Terrestrial Records,* pp. 214–243. New York: Columbia University Press.

Thomas, E. 2003. Extinction and food at the sea floor: A high-resolution benthic foraminiferal record across the initial Eocene Thermal Maximum, Southern Ocean Site 690. In S. Wing, P. Gingerich, B. Schmitz, and E. Thomas, eds., *Causes and Consequences of Globally Warm Climates of the Paleogene,* pp. 319–332. Boulder, CO: Geological Society of America.

Thomas, E. 2007. Cenozoic mass extinctions in the deep sea: What disturbs the largest habitat on Earth? In S. Monechi, R. Coccioni, and M. Rampino, eds., *Large Ecosystem Perturbations: Causes and Consequences,* pp. 1–23. Boulder, CO: Geological Society of America.

Thomas, E. and N. J. Shackleton. 1996. The Palaeocene-Eocene benthic foraminiferal extinction and stable isotope anomalies. Geological Society of London Special Publication 101:401–441.

Thomas, E., and J. C. Varekamp. 1991. Paleo-environmental analyses of marsh sequences (Clinton, Connecticut): Evidence for punctuated rise in relative sea level during the latest Holocene. Journal of Coastal Research, Special Issue 11: 125–158.

Thomas, E. K. and J. P. Briner. 2009. Climate of the past millennium inferred from varved proglacial lake sediments on northeast Baffin Island, Arctic Canada. Journal of Paleolimnology 41:209–224.

Thomas, E. R., E. W. Wolff, R. Mulvaney, et al. 2007. The 8.2 ka event from Greenland ice cores. Quaternary Science Reviews 26:70–81.

Thomas, R., E. Rignot, G. Casassa, et al. 2004. Accelerated sea-level rise from West Antarctica. Science 306:255–258.

Thompson, D. W. J., S. Lee, and M. P. Baldwin. 2003. Atmospheric processes governing the northern hemisphere annular mode/ North Atlantic Oscillation. In J. W. Hurrell, Y. Kushnir, G. Ottersen, and M. Visbeck, eds., The North Atlantic Oscillation: Climatic Significance and Environmental Impact, pp. 85–89. Washington, DC: American Geophysical Union.

Thompson, D. W. J. and J. M. Wallace. 1998. The Arctic Oscillation signature in the wintertime geopotential height and temperature fields. Geophysical Research Letters 25:1297–1300.

Thompson, D. W. J. and J. M. Wallace. 2000. Annular modes in the extratropical circulation, Part I: Month to month variability. Journal of Climate 13:1000–1016.

Thompson, D. W. J. and J. M. Wallace. 2001. Regional climate impacts of the northern hemisphere annular mode. Science 293:85–88.

Thompson, L. G., S. Hastenrath, and B. Morales Arnao. 1979. Climatic ice core records from the tropical Quelccaya ice cap. Science 203:1240–1243.

Thompson, L. G., E. Mosley-Thompson, J. F. Bolzan, and B. R. Koci. 1985. A 1500-year record of tropical precipitation in ice cores from Quelccaya ice cap, Peru. Science 229:971–973.

Thompson, L. G., E. Mosley-Thompson, H. Brecher, et al. 2006a. Abrupt tropical climate change: Past and present. Proceedings of the National Academy of Sciences 103:10536–10543.

Thompson, L. G., E. Mosley-Thompson, W. Dansgaard, and P. M. Grootes. 1986. The Little Ice Age as recorded in the stratigraphy of the tropical Quelccaya Ice Cap. Science 234:361–364.

Thompson, L. G., E. Mosley-Thompson, M. E. Davis, et al. 1989. Holocene–Late Pleistocene climatic ice core records from Qinghai-Tibetan Plateau. Science 246:474–477.

Thompson, L. G., E. Mosley-Thompson, M. E. Davis, et al. 1995a. A 1000 year climatic ice-core record from the Guliya Ice Cap, China: Its relationship to global climate variability. Annals of Glaciology 21:175–181.

Thompson, L. G., E. Mosley-Thompson, M. E. Davis, et al. 1995b. Late-glacial stage and Holocene tropical ice core records from Huascarán, Peru. Science 269:46–48.

Thompson, L. G., E. Mosley-Thompson, M. E. Davis, et al. 2002. Kilimanjaro ice core records: Evidence of Holocene climate change in tropical Africa. Science 298:589–593.

Thompson, L. G., E. Mosley-Thompson, and B. Morales Arnao. 1984. El Niño–Southern Oscillation events recorded in the stratigraphy of the tropical Quelccaya Ice Cap, Peru. Science 226:50–53.

Thompson, L. G., T. Yao, M. E. Davis, et al. 1997. Tropical climate instability: The last glacial cycle from a Qinghai-Tibetan ice core. Science 276:1821–1825.

Thompson, L. G., T. Yao, M. E. Davis, et al. 2006b. Holocene climate variability archived in the Puruogangri ice cap on the central Tibetan Plateau. Annals of Glaciology 43:61–69.

Thompson, R. S. and R. F. Fleming. 1996. Middle Pliocene vegetation: Reconstructions, paleoclimate inferences and boundary conditions for climate modeling. Marine Micropaleontology 27:27–50.

Thompson, W. G. and S. L. Goldstein. 2005. Open-system coral ages reveal persistent suborbital sea-level cycles. Science 308:401–404.

Thompson, W. G. and S. L. Goldstein. 2006. A radiometric calibration of the SPECMAP timescale. Quaternary Science Reviews 25:3207–3215.

Thompson, W. G., M. W. Spiegelman, S. L. Goldstein, and R. C. Speed. 2003. An open-system model for U-series age determinations of fossil corals. Earth and Planetary Science Letters 210:365–381.

Thomson, D. 1995. The seasons, global temperature, and precession. Science 268:59–68.

Thorleifson, L. H. 1996. Review of Lake Agassiz history. In J. T. Teller, et al., eds., Sedimentology, Geomorphology and History of the Central Lake Agassiz Basin, pp. 55–84. Winnipeg, Manitoba Field Trip Guidebook B2, Geological Association of Canada/Mineralogical Association of Canada Annual Meeting May 27–29.

Thunell, R. C. and A. B. Kepple. 2004. Glacial-holocene $\delta^{15}N$ record from the Gulf of Tehuantepec, Mexico: Implications for denitrification in the eastern equatorial Pacific and changes in atmospheric N_2O. Global Biogeochemical Cycles 18:1011, GB002028.

Tian, J., D. M. Nelson, and F. S. Hu. 2006. Possible linkages of late-Holocene drought in the North American mid-continent to Pacific Decadal Oscillation and solar activity. Geophysical Research Letters 33:L23702.

Tibert, N. E. and R. M. Leckie. 2004. High-resolution estuarine sea level cycles from the late Cretaceous: Amplitude constraints using agglutinated foraminifera. Journal of Foraminiferal Research 34:130–143.

Tiedemann, R., M. Sarnthein, and N. J. Shackleton. 1994. Astronomic timescale for the Pliocene Atlantic $\delta^{18}O$ and dust flux records of Ocean Drilling Program site 659. Paleoceanography 9:619–638.

Tiedemann, R., A. Sturm, S. Steph, S. P. Lund, and J. S. Stoner. 2006. Astronomically calibrated timescales from 6 to 2.5 Ma, and benthic isotope stratigraphies of sites 1236, 1237, 1239, and 1241, Ocean Drill. In R. Tiedemann, et al., eds., Proceedings of the Ocean Drilling Program Scientific Results, pp. 1–69. College Station, TX: Ocean Drilling Program.

Tinsley, B. A. 2003. Solar activity and the earth's climate. EOS, American Geophysical Union Transactions 84:77–83.

Toggweiler, J. R., J. L. Russell, and S. Carson. 2006. Mid-latitude westerlies, atmospheric CO_2 and climate change during the Ice Ages. Paleoceanography 21:PA2005.

Törnqvist, T. E., S. J. Bick, J. L. González, K. Van der Borg, and A. F. M. De Jong. 2004. Tracking the sea-level signature of the 8.2 ka cooling event: New constraints from the Mississippi Delta. Geophysical Research Letters 31:L23309, GL021429.

Törnqvist, T. E., M. H. M. Van Ree, R. Van't Veer, and B. Van Geel. 1998. Improving methodology for high-resolution reconstruction of sea-level rise and neotectonics by paleoecological analysis and AMS ^{14}C dating of basal peats. Quaternary Research 49:72–85.

Törnqvist, T. E., D. J. Wallace, J. E. A. Storms, et al. 2008. Mississippi Delta subsidence primarily caused by compaction of Holocene strata. Nature Geoscience 1:173–176.

Torsvik, T. H. 2003. The Rodinia jigsaw puzzle. Science 300:1379–1381.

Torsvik, T. H. and L. R. M. Cocks. 2004. Earth geography from 400 to 250 million years: A palaeomagnetic, faunal and facies review. Journal of the Geological Society of London 161:555–572.

Torsvik, T. H., C. Gaina, and T. F. Redfield. 2008. Antarctica and global paleogeography: From Rodinia, through Gondwanaland and Pangea, to the birth of the Southern Ocean and the opening of gateways. In A. Cooper, et al., eds., *Antarctica: A Keystone in a Changing World*, Proceedings of the 10th International Symposium on Antarctic Earth Sciences, pp. 125–140. Washington, DC: National Academies Press.

Toscano, M. A. and I. G. Macintyre. 2003. Corrected western Atlantic sea-level curve for the last 11,000 years based on calibrated ^{14}C dates from *Acropora palmata* framework and intertidal mangrove peat. Coral Reefs 22:257–270.

Trenberth, K. 1993. *Climate System Modeling*. Cambridge: Cambridge University Press.

Trenberth, K. E. 2005. Uncertainty in hurricanes and global warming. Science 308:1753–1754.

Trenberth, K. E., G. W. Branstator, D. Karoly, et al. 1998. Progress during TOGA in understanding and modeling global teleconnections associated with tropical sea surface temperatures. Journal of Geophysical Research 103:14291–14324.

Trenberth, K. E., J. M. Caron, D. P. Stepaniak, and S. Worley. 2002. The evolution of ENSO and global atmospheric surface temperatures. Journal of Geophysical Research 107:D8, JD000298.

Trenberth, K. E. and J. W. Hurrell. 1994. Decadal atmosphere-ocean variations in the Pacific. Climate Dynamics 9:303–319.

Trenberth, K. E., P. D. Jones, et al. 2007. Observations: Surface and atmospheric climate Change. In S. Solomon, et al., eds., *Climate Change 2007: The Physical Science Basis*. Contribution of Working Group I to the Fourth Assessment Report of the Intergovernmental Panel on Climate Change, pp. 235–336. Cambridge: Cambridge University Press.

Trenberth, K. E. and B. L. Otto-Bliesner. 2003. Toward integrated reconstructions of past climates. Science 300:589–591.

Trenberth, K. E. and D. J. Shea. 1987. On the evolution of the Southern Oscillation. Monthly Weather Review 115:3078–3096.

Trenberth, K. E. and D. J. Shea. 2006. Atlantic hurricanes and natural variability in 2005. Geophysical Research Letters 33:L12704.

Trenberth, K. E. and D. P. Stepaniak. 2001. Indices of El Niño evolution. Journal of Climate 14:1697–1701.

Trenberth, K. E., D. P. Stepaniak, and J. M. Caron. 2000. The global monsoon as seen through the divergent atmospheric circulation. Journal of Climate 13:3969–3993.

Treydte, K. S., G. H. Schleser, G. Helle, et al. 2006. The twentieth century was the wettest period in northern Pakistan over the past millennium. Nature 440:1179–1182.

Tripati, A., J. Backman, H. Elderfield, and P. Ferretti. 2005. Eocene bipolar glaciation associated with global carbon cycle changes. Nature 436:341–346.

Tripati, A. and H. Elderfield. 2005. Deep-sea temperature and circulation changes at the Paleocene-Eocene Thermal Maximum. Science 308:1894–1898.

Trouet, V., J. Esper, N. E. Graham, et al. 2009. Persistent positive North Atlantic Oscillation mode dominated the Medieval Climate Anomaly. Science 324:78–80.

Truffer, M. and M. Fahnestock. 2007. Rethinking ice sheet time scales. Science 315:1508–1510.

Tudhope, A. W., C. P. Chilcott, M. T. McCulloch, et al. 2001. Variability in the El Niño–Southern Oscillation through a glacial-interglacial cycle. Science 291:1511–1517.

Tuenter, E., S. L. Weber, F. J. Hilgen, and L. J. Lourens. 2005. Sea-ice feedbacks on the climatic response to precession and obliquity forcing. Geophysical Research Letters 32:L24704, GL024122.

Turney, C., M. Baillie, S. Clemens, et al. 2005. Testing solar forcing of pervasive Holocene climate cycles. Journal of Quaternary Science 20:511–518.

Tyson, P., R. Fuchs, C. Fu, et al., eds. 2002. *Global-Regional Linkages in the Earth System*. New York: Springer-Verlag.

Tziperman, E. 1997. Inherently unstable climate behaviour due to weak thermohaline ocean circulation. Nature 386:592–595.

Tziperman, E. and H. Gildor. 2003. On the mid-Pleistocene transition to 100-kyr glacial cycles and the asymmetry between glaciation and deglaciation times. Paleoceanography 18:1001.

Tziperman, E., M. Raymo, P. Huybers, and C. Wunsch. 2006. Consequences of pacing the Pleistocene 100 kyr ice ages by nonlinear phase locking to Milankovitch forcing. Paleoceanography 21:PA4206.

Ulmishek, G. F. and H. D. Klemme. 1990. Depositional controls, distribution, and effectiveness of world's petroleum source rocks. U.S. Geological Survey Bulletin 1931.

Upham, W. 1895. *The Glacial Lake Agassiz*. U.S. Geological Survey Monograph 45, Washington, DC.

Urban, F. E., J. E. Cole, and J. T. Overpeck. 2000. Influence of mean climate change on climate variability from a 155-year tropical Pacific coral record. Nature 407:989–993.

Urey, H. C. 1947. The thermodynamic properties of isotopic substances. Journal of the Chemical Society (London) (April):562–581.

Urey, H. C., H. A. Lowenstam, S. Epstein, and C. R. McKinney. 1951. Measurement of paleotemperatures and temperatures of the upper Cretaceous of England, Denmark, and the southeastern United States. Geological Society of America Bulletin 62:399–416.

Vaganov, E. A., K. J. Anchukaitis, and M. N. Evans. 2009. How well understood are the processes that create dendroclimatic records? A mechanistic model of the climatic control on conifer tree-ring growth dynamics. In M. K. Hughes, T. W. Swetnam, and H. F. Diaz, eds., *Dendroclimatology: Progress and Prospects*, chap. 3. New York: Springer.

Vail, P. R., R. M. Mitchum Jr., R. G. Todd, et al. 1977. Seismic stratigraphy and global changes of sea level. American Association of Petroleum Geologists Memoir 26:49–212.

Vance, D. and K. W. Burton. 1999. Neodymium isotopes in planktonic foraminifera: A record of the response of continental weathering and ocean circulation rates to climate change. Earth and Planetary Science Letters 173:365–379.

van de Plassche, O. 2000. North Atlantic climate-ocean variations and sea level in Long Island Sound, Connecticut, since 500 cal yr A.D. Quaternary Research 53:89–97.

van de Plassche O., K. van de Borg, and A. F. M. de Jong 1998. Sea-level climate correlation during the past 1400 yr. Geology 26:319–322.

van de Plassche, O., G. van der Schrier, S. L. Weber, W. R. Gehrels, and A. J. Wright. 2003. Sea-level variability in the northwest Atlantic during the past 1500 years: A delayed response to solar forcing? Geophysical Research Letters 30:1921, GL017558.

Van der Burgh, J., H. Visscher, D. Dilcher, and W. M. Kurschner. 1993. Paleoatmospheric signatures in Neogene fossil leaves. Science 260:1788–1790.

Van der Meer, M. T. J., M. Baas, W. I. C. Rijpstra, et al. 2007. Hydrogen isotopic compositions of long-chain alkenones record freshwater flooding of the Eastern Mediterranean at the onset of sapropel deposition. Earth and Planetary Science Letters 262:594–600.

van de Schootbrugge, B., J. M. McArthur, T. R. Bailey, et al. 2005. Toarcian oceanic anoxic event: An assessment of global causes using belemnite carbon isotope records. Paleoceanography 20:PA3008.

van de Wal, R. S. W., W. Boot, M. R. van den Broeke, et al. 2008. Large and rapid melt-induced velocity changes in the ablation zone of the Greenland Ice Sheet. Science 321:111–113.

van Kreveld, S. M. Sarnthein, H. Erlenkeuser, et al. 2000. Potential links between surging ice sheets, circulation changes, and the Dansgaard-Oeschger cycles in the Irminger Sea, 60–18 kyr. Paleoceanography 15:425–442.

Van Sickel, W. A., M. A. Kominz, K. G. Miller, and J. V. Browning. 2004. Late Cretaceous and Cenozoic sea-level estimates: Backstripping analysis of borehole data, onshore New Jersey. Basin Research 16:451–465.

Varekamp, J. C. and E. Thomas. 1998. Climate change and the rise and fall of sea level over the millennium. EOS, Transactions, American Geophysical Union 79:69, 74–75.

Varekamp, J. C., E. Thomas, and O. van de Plassche. 1992. Relative sea-level rise and climate change over the last 1500 years. Terra Nova 4:293–304.

Vaughn, D. G. and C. S. M. Doake. 1996. Recent atmospheric warming and retreat of ice shelves on the Antarctic Peninsula. Nature 379:328–330.

Vecchi, G. A., B. J. Soden, A. T. Wittenberg, et al. 2006. Weakening of tropical Pacific atmospheric circulation due to anthropogenic forcing. Nature 441:73–76.

Veeh, H. H. and J. Chappell. 1970. Astronomical theory of climate change: Support from New Guinea. Science 167: 862–865.

Veevers, J. J. 2004. Gondwanaland from 650–500 Ma assembly through 320 Ma merger in Pangea to 185–100 Ma breakup: Supercontinental tectonics via stratigraphy and radiometric dating. Earth-Science Reviews 68:1–132.

Veizer, J. 1989. Strontium isotopes in seawater through time. Annual Review of Earth and Planetary Sciences 17:141–167.

Veizer, J., D. Ala, K. Azmy, et al. 1999. $^{87}Sr/^{86}Sr$, $\delta^{13}C$ and $\delta^{18}O$ evolution of Phanerozoic seawater. Chemical Geology 161:59–88.

Veizer, J., Y. Goddéris, and L. M. François. 2000. Evidence for decoupling of atmospheric CO_2 and global climate during the Phanerozoic eon. Nature 408:698–701.

Verdon, D. C. and S. W. Franks. 2006. Long-term behaviour of ENSO: Interactions with the PDO over the past 400 years inferred from paleoclimate records. Geophysical Research Letters 33:L06712, GL025052.

Verschuren, D., K. R. Briffa, P. Hoelzmann, et al. 2004. Holocene climate variability through Europe and Africa: A Time Stream 1 synthesis. In R. W. Battarbee, F. Gasse, and C. E. Stickley, eds., Past Climate Variability through Europe and Africa, pp. 567–582. Dordrecht: Kluwer.

Verschuren, D., K. R. Laird, and B. Cumming. 2000. Rainfall and drought in equatorial east Africa during the past 1,100 years. Nature 403:410–414.

Veum, T. 1992. Water mass exchange between the North Atlantic and the Norwegian Sea during the past 28,000 years. Nature 356:783.

Via, R. K. and D. J. Thomas. 2006. Evolution of Atlantic thermohaline: Early Oligocene circulation onset of deep-water production in the North Atlantic. Geology 34:441–444.

Viau, A. E., K. Gajewski, P. Fines, D. E. Atkinson, and M. C. Sawada. 2002. Widespread evidence of 1500 yr climate variability in North America during the past 14000 yr. Geology 30:455–458.

Viau, A. E., K. Gajewski, M. C. Sawada, and P. Fines. 2006. Millennial-scale temperature variations in North America during the Holocene. Journal of Geophysical Research-Atmosphere 111:D09102, JD006031.

Vidal, L. and H. W. Arz. 2004. Oceanic climate variability at millennial time scales: Modes of climate connections. In R. W. Battarbee, F. Gasse, and C. E. Stickley, eds., Past Climate Variability through Europe and Africa, pp. 31–40. Dordrecht: Kluwer.

Vidal, L., L. Labeyrie, E. Cortijo, et al. 1997. Evidence for changes in the North Atlantic Deep Water linked to meltwater surges during the Heinrich events. Earth and Planetary Science Letters 146:13–26.

Vidal, L., L. Labeyrie, and T. C. E. van Weering. 1998. Benthic $\delta^{18}O$ records in the North Atlantic over the last glacial period (60–10ky): Evidence for brine formation. Paleoceanography 13:245–251.

Vidal, L., R. R. Schneider, O. Marchal, et al. 1999. Link between the North and South Atlantic during the Heinrich events of the last glacial period. Climate Dynamics 15:909–919.

Vincent, E. and W. H. Berger. 1985. Carbon dioxide and polar cooling in the Miocene: The Monterey hypothesis. In E. T. Sundquist and W. S Broecker, eds., The Carbon Cycle and Atmospheric CO_2: Natural Variations, Archean to Present, pp. 455–468. Washington, DC: American Geophysical Union.

Visbeck, M. E., P. Chassignet, R. G. Curry, et al. 2003. The ocean's response to North Atlantic Oscillation variability. In J. W. Hurrell, Y. Kushnir, G. Ottersen, and M. Visbeck, eds., The North Atlantic Oscillation: Climatic Significance and Environmental Impact, pp. 113–146. Washington, DC: American Geophysical Union.

Visser, K., R. Thunell, and L. Stott. 2003. Magnitude and timing of temperature change in the Indo-Pacific warm pool during deglaciation. Nature 421:152–154.

Voelker, A. H. L., M. Sarnthein, P. M. Grootes, et al. 1998. Correlation of marine ^{14}C ages from the Nordic Seas with the GISP2 isotope record: Implications for radiocarbon calibration beyond 25 ka BP. Radiocarbon 40:517–534.

Voelker, A. H. L. and Workshop Participants. 2002. Global distribution of centennial-scale records for Marine Isotope Stage (MIS) 3: A database. Quaternary Science Reviews 21:1185–1212.

Vogt, P. R., K. Crane, and E. Sundvor. 1994. Deep Pleistocene iceberg plow marks on the Yermak Plateau: Sidescan and 3.5 kHz evidence for thick calving ice fronts and a possible marine ice sheet in the Arctic Ocean. Geology 22:403–406.

Vollweiler, N., D. Scholz, C. Mühlinghaus, A. Mangini, and C. Spötl. 2006. A precisely dated climate record for the last 9 kyr from three high alpine stalagmites, Spannagel Cave, Austria. Geophysical Research Letters 33:L20703, GL027662.

von Grafenstein, U. 2002. Oxygen-isotope studies of ostracods from deep lakes. In J. A. Holmes and A. R. Chivas, eds., Applications of the Ostracoda to Quaternary Research, pp. 249–266. Washington, DC: American Geophysical Union.

von Grafenstein, U., H. Erlenkeuser, A. Brauer, J. Jouzel, and S. Johnsen. 1999. A mid-European decadal isotope-climate record from 15,500 to 5,000 years B.P. Science 284:1654–1657.

von Grafenstein, U., H. Erlenkeuser, J. Muller, J. Jouzel, and S. Johnsen. 1998. The short cold period 8,200 years ago documented in oxygen isotope records of precipitation in Europe and Greenland. Climate Dynamics 14:73–81.

von Storch, H. and E. Zorita. 2005. Comment on "Hockey sticks, principal components, and spurious significance" by S. McIntyre and R. McKitrick. Geophysical Research Letters 32:L20701.

von Storch, H., E. Zorita, J. M. Jones, et al. 2004. Reconstructing past climate from noisy data. Science 306:679–682.

Vuille, M., R. S. Bradley, M. Werner, R. Healy, and F. Keimig. 2003. Modeling $\delta^{18}O$ in precipitation over the tropical Americas: 1. Interannual variability and climatic controls. Journal of Geophysical Research 108:4174.

Vuille, M., M. Werner, R. Bradley, and F. Keimig, 2005. Stable isotopes in precipitation in the Asian monsoon region. Journal of Geophysical Research 110:D23108, JD006022.

Wade, B. S. and H. Pälike. 2004. Oligocene climate dynamics. Paleoceanography 19:PA4019, doi:10.1029/2004PA001042.

Wadley, M. R. and G. R. Bigg. 2006. Are "Great Salinity Anomalies" advective? Journal of Climate 19:1080–1087.

Waelbroeck, C., J.-C. Duplessy, E. Michel, et al. 2001. The timing of the last deglaciation in North Atlantic climate records. Nature 412:724–727.

Wagner, F., B. Aaby, and H. Visscher. 2002. Rapid atmospheric CO_2 changes associated with the 8,200-years-B.P. cooling event. Proceedings of the National Academy of Sciences 99:12011–12014.

Wagner, F., S. J. P. Bohncke, D. L. Dilcher, et al. 1999. Century-scale shifts in early Holocene CO_2 concentration. Science 284:1971–1973.

Wagner, G., J. Beer, C. Laj, et al. 2000. Chlorine-36 evidence for the Mono Lake event in the Summit GRIP ice core. Earth and Planetary Science Letters 181:1–6.

Wagner, G., C. Laj, J. Beer, et al. 2001. Reconstruction of the paleoaccumulation rate of central Greenland during the last 75 kyr using the cosmogenic radionuclides ^{36}Cl and ^{10}Be and geomagnetic field intensity data. Earth and Planetary Science Letters 193:515–521.

Wagner, T., J. O. Herrle, J. S. Sinninghe Damsté, et al. 2008. Rapid warming and salinity changes of Cretaceous surface waters in the subtropical North Atlantic. Geology 36:203–206.

Wahl, E. R., D. M. Ritson, and C. M. Ammann. 2006. Comment on "Reconstructing past climate from noisy data." Science 312:529.

Wakelin, S. L., P. L. Woodworth, R. A. Flather, and J. A. Williams. 2003. Sea-level dependence on the NAO over the NW European Continental Shelf. Geophysical Research Letters 30:1403, GL017041.

Walker, G. T. 1924. Correlation in seasonal variations of weather. IX. A further study of world weather. Memoirs of the Indian Meteorological Department 24:275–332. (Cited in Philander 1990.)

Walker, G. T. and E. W. Bliss. 1932. World weather V. Memoirs of the Royal Meteorological Society 4:53–84. (Cited in Philander 1990.)

Walker, J. C. G., P. B. Hays, and J. F. Kasting. 1981. A negative feedback mechanism for the long-term stabilization of Earth's surface temperature. Journal of Geophysical Research (Oceans) 86:9776–9782.

Walker, M. 2005. Quaternary Dating Methods: An Introduction. New York: John Wiley & Sons.

Wallace, J. M. and D. S. Gutzler. 1981. Teleconnections in the geopotential height field during the northern hemisphere winter. Monthly Weather Review 109:784–811.

Wanamaker, A. D., Jr., K. J. Kreutz, B. R. Schöne, et al. 2008. Coupled North Atlantic slope water forcing on Gulf of Maine temperatures over the past millennium. Climate Dynamics 31:183–194, doi:10.1007/s00382-007-0344-8.

Wang, C., X. Zhao, Z. Liu, et al. 2008. Constraints on the early uplift history of the Tibetan Plateau. Proceedings of the National Academy of Sciences 105:4987–4992.

Wang, H., L. R. Follmer, and J. C. Liu. 2000. Isotopic evidence of paleo–El Niño–Southern Oscillation cycles in loess-paleosol record in the central United States. Geology 28: 771–774.

Wang, L., M. Sarnthein, P. M. Grootes, and H. Erlenkeuser. 1999. Millennial reoccurrence of century-scale abrupt events of East Asian monsoon: A possible heat conveyor for the global deglaciation. Paleoceanography 14:725–731.

Wang, P., S. Clemens, L. Beaufort, P. Braconnot, et al. 2005. Evolution and variability of the Asian monsoon system: State of the art and outstanding issues. Quaternary Science Reviews 24:595–629.

Wang, X., A. S. Auler, R. L. Edwards, et al. 2006. Interhemispheric anti-phasing of rainfall during the last glacial period. Quaternary Science Reviews 25:3391–3403.

Wang, Y., L. A. Mysak, and N. T. Roulet. 2005. Holocene climate and carbon cycle dynamics: Experiments with the "green" McGill Paleoclimate Model. Global Biogeochemical Cycles 19:GB3022, GB002484.

Wang, Y. J., H. Cheng, R. L. Edwards, et al. 2001. A high-resolution absolute-dated late Pleistocene monsoon record from Hulu Cave, China. Science 294:2345–2348.

Wang, Y. J., H. Cheng, R. L. Edwards, et al. 2005. The Holocene Asian monsoon: Links to solar changes and North Atlantic climate. Science 308:854–857.

Wanless, H. R. and F. P. Shepard. 1936. Sea level and climate changes related to late Paleozoic cycles. Geological Society of America Bulletin 47:1177–1206.

Wanner, H., J. Beer, J. Bütikofer, et al. 2008. Mid- to Late Holocene climate change: An overview. Quaternary Science Reviews 27:1791–1828.

Wara, M. W., A. C. Ravelo, and M. L. Delaney. 2005. Permanent El Niño–like conditions during the Pliocene warm period. Science 309:758–761.

Warren, S. G., R. E. Brandt, T. C. Grenfell, and C. P. McKay. 2002. Snowball Earth: Ice thickness on the tropical ocean. Journal of Geophysical Research 107(C10):3167.

Watanabe, O., J. Jouzel, S. Johnsen, et al. 2003. Homogeneous climate variability across east Antarctica over the past three glacial cycles. Nature 422:509–512.

Watanabe, T., A. Winter, and T. Oba. 2001. Seasonal changes in sea surface temperature and salinity during the Little Ice Age in the Caribbean Sea deduced from Mg/Ca and O-18/O-16 ratios in corals. Marine Geology 173:21–35.

Weart, S. 2003. *The Discovery of Global Warming*. Cambridge, MA: Harvard University Press.

Weaver, A. J. 1995. Driving the ocean conveyor. Nature 378:135–136.

Weaver, A. J., M. Eby, M. Kienast, and O. A. Saenko. 2007. Response of the Atlantic meridional overturning circulation to increasing atmospheric CO_2: Sensitivity to mean climate state. Geophysical Research Letters 34:L05708, doi:10.1029/2006GL028756.

Weaver, A. J., M. Eby, E. C. Wiebe, et al. 2001. The UVic Earth System Climate Model: Model description, climatology and application to past, present and future climates. Atmosphere-Ocean 39:361–428.

Weaver, A. J., O. A. Saenko, P. U. Clark, and J. X. Mitrovica. 2003. Meltwater Pulse 1A from Antarctica as a trigger of the Bølling-Allerød warm interval. Science 299:1709–1713.

Webb, P.-N. and D. M. Harwood. 1991. Late Cenozoic glacial history of the Ross Embayment, Antarctica. Quaternary Science Reviews 10:215–223.

Webb, P.-N., D. M. Harwood, B. C. McKelvey, J. H. Mercer, and L. D. Stott. 1984. Cenozoic marine sedimentation and ice-volume variation on the east Antarctic craton. Geology 12:287–291.

Webb, T., III and J. E. Kutzbach. 1998. An introduction to late Quaternary climates: Data syntheses and model experiments. Quaternary Science Reviews 17:465–471.

Weddle, T. K. and M. J. Retelle, eds. 2001. *Deglacial History and Relative Sea-Level Changes, Northern New England and Adjacent Canada*. Boulder, CO: Geological Society of America.

Wefer, G., W. H. Berger, G. Siedler, and D. Webb, eds. 1996. *The South Atlantic: Present and Past Circulation*. Berlin: Springer.

Wefer, G., S. Mulitza, and V. Ratmeyer, eds. 2004. *The South Atlantic in the Late Quaternary: Reconstructions of Material Budgets and Current Systems*. Berlin: Springer.

Wehmiller, J. F., K. R. Simmons, H. Cheng, et al. 2004. Uranium-series coral ages from the US Atlantic Coastal Plain—the "80ka problem" revisited. Quaternary International 120:3–14.

Weidick, A. 1995. Greenland. In R. S. Williams Jr. and J. G. Ferrigno, eds., *Satellite Image Atlas of Glaciers of the World*. U.S. Geological Survey Professional Paper 1386-C. Washington, DC: U.S. Government Printing Office.

Weidman, C. R. and G. A. Jones. 1993. Development of the mollusc Arctica islandica as a palaeoceanographic tool for reconstructing annual and seasonal records of $\Delta^{14}C$ and $\delta^{18}O$ in the mid-to-high latitude North Atlantic Ocean. In *Isotope Techniques in the Study of Past and Current Environmental Changes in the Hydrosphere and the Atmosphere*, pp. 461–470. Vienna: International Atomic Energy Agency.

Weidman, C. R., G. A. Jones, and K. C. Lohmann. 1994. The long-lived mollusc Arctica islandica: A new paleoceanographic tool for the reconstruction of bottom temperatures for the continental shelves of the northern North Atlantic Ocean. Journal of Geophysical Research 99:18305–18314.

Weijers, J. W. H., E. Schefuß, S. Schouten, and J. S. Sinninghe Damste. 2007. Coupled thermal and hydrological evolution of tropical Africa over the last deglaciation. Science 315:1701–1704.

Weldeab, S., R. R. Schneider, and M. Kölling. 2006. Deglacial sea surface temperature and salinity increase in the western tropical Atlantic in synchrony with high latitude climate instabilities. Earth and Planetary Science Letters 241:699–706.

Weldeab, S., R. R. Schneider, M. Kölling, and G. Wefer. 2005. Holocene African droughts relate to eastern equatorial Atlantic cooling. Geology 33:981–984.

Wellington, G. M. and R. B. Dunbar. 1995. Stable isotopic signature of El Niño–Southern Oscillation events in eastern tropical Pacific reef corals. Coral Reefs 14:5–25.

Wellington, G. M., R. B. Dunbar, and G. Merlen. 1996. Calibration of stable oxygen isotope signatures in Galapagos coral. Paleoceanography 11:467–480.

Wenzel, M. and J. Schröter. 2007. The global ocean mass budget in 1993–2003 estimated from sea level change, Journal of Physical Oceanography 37:203–213.

Westerhold, T., U. Röhl, J. Laskar, et al. 2007. On the duration of magnetochrons C24r and C25n and the timing of early Eocene global warming events: Implications from the Ocean Drilling Program Leg 208 Walvis Ridge depth transect. Paleoceanography 22:PA2201.

Wentz, F. J., L. Ricciardulli, K. Hilburn, and C. Mears. 2007. How much more rain will global warming bring? Science 317:233–235.

Whetton, P., R. Allan, and I. Rutherford. 1996. Historical ENSO teleconnections in the eastern hemisphere: A comparison with latest El Niño series of Quinn. Climatic Change 32:103–109.

Whetton, P. and I. Rutherford. 1994. Historical ENSO teleconnections in the eastern hemisphere. Climatic Change 28:221–253.

Whittlesey, C. 1868. Depression of the ocean during the ice period. American Association for the Advancement of Science Proceedings 16:92–97.

Wiersma, A. P. and H. Renssen. 2006. Model-data comparison for the 8.2 ka BP event: Confirmation of a forcing mechanism by catastrophic drainage of Laurentide lakes. Quaternary Science Reviews 25:63–88.

Wignall, P. B., J. M. McArthur, C. T. S. Little, and A. Hallam. 2006. Palaeoceanography: Methane release in the early Jurassic period. Nature 441:E5, doi:10.1038.

Wiles, G. C., D. J. Barclay, P. E. Calkin, and T. V. Lowell. 2008. Century to millennial-scale temperature variations for the last two thousand years inferred from glacial geologic records of southern Alaska. Global and Planetary Change 60:115–125.

Wiles, G. C., G. G. Jacoby, K. K. Davi, and R. P. McAllister. 2002, Late Holocene glacial fluctuations in the Wrangell Mountains, Alaska. Geological Society of America Bulletin 114:896–908.

Wilkinson, B. H. and T. Algeo. Sedimentary carbonate record of Ca-Mg cycling at the earth's surface. American Journal of Science 289:1158–1194.

Willard, D. A., C. E. Bernhardt, D. A. Korejwo, and S. R. Meyers. 2005. Impact of millennial scale climate variability on eastern North American terrestrial ecosystems: Pollen-based reconstructions. Global and Planetary Change 47:17–35.

Willemse, N. W. and T. E. Tørnqvist. 1999. Holocene century-scale temperature variability from west Greenland lake records. Geology 27:580–584.

Williams, D. F., J. Peck, E. B. Karabanov, et al. 1997. Lake Baikal record of continental climate response to orbital insolation during the past 5 million years. Science 278:1114–1117.

Williams, G. E. 2000. Geological constraints on the Precambrian history of Earth's rotation and the Moon's orbit. Reviews of Geophysics 38:37–59.

Williams, R. S., Jr. and J. G. Ferrigno, editors and contributing authors. 1988–. Satellite image atlas of glaciers of the world: U.S. Geological Survey Professional Paper 1386-A-K.

Willson, R. C. 1997. Total solar irradiance trend during solar cycles 21 and 22. Science 277:1963–1965.

Willson, R. C. and A. V. Mordvinov. 2003. Secular total solar irradiance trend during solar cycles 21–23. Geophysical Research Letters 30:1199, GL016038.

Wilson, G. S., G. A. Barron, A. C. Ashworth, et al. 2002. The Mount Feather Diamicton of the Sirius Group: An accumulation of indicators of Neogene Antarctic glacial and climatic history. Palaeogeography, Palaeoclimatology, Palaeoecology 182:117–131.

Wilson, P. A., R. D. Norris, and M. J. Cooper. 2002. Testing the Cretaceous greenhouse hypothesis using glassy foraminiferal calcite from the core of the Turonian tropics on Demerara Rise. Geology 30:607–610.

Wilson, R., D. Frank, J. Topham, K. Nicolussi, and J. Esper. 2005. Spatial reconstruction of summer temperatures in Central Europe for the last 500 years using annually resolved proxy records: Problems and opportunities. Boreas 34:490–494.

Wilson, R., A. Tudhope, P. Brohan, et al. 2006. 250 years of reconstructed and modeled tropical temperatures. Journal of Geophysical Research 111:C10007.

Wilson, R. J. S., R. D'Arrigo, B. Buckley, et al. 2007. A matter of divergence: Tracking recent warming at hemispheric scales using tree-ring data. Journal of Geophysical Research 112:D17103.

Wilson, R. J. S., B. H. Luckman, and J. Esper. 2005. A 500-year dendroclimatic reconstruction of spring/summer precipitation from the lower Bavarian forest region, Germany. International Journal of Climatology 25:611–630.

Winckler, G., R. F. Anderson, M. O. Fleischer, D. McGee, and N. Mahowald. 2008. Covariant glacial-interglacial dust fluxes in the equatorial Pacific and Antarctica. Science 320:93–96.

Wing, S. L., G. J. Harrington, F. A. Smith, et al. 2005. Transient floral change and rapid global warming at the Paleocene-Eocene boundary. Science 310:993–996.

Winograd, I. J., T. B. Coplen, J. M. Landwehr, et al. 1992. Continuous 500,000-year climate record from vein calcite in Devils Hole, Nevada. Science 258:255–260.

Winograd, I. J., T. B. Coplen, B. J. Szabo, and A. C. Riggs. 1988. A 250,000-year climatic record from Great Basin vein calcite: Implications for Milankovitch theory. Science 242:1275–1280.

Winograd, I. J., J. M. Landwehr, K. R. Ludwig, T. B. Coplen, and A. C. Riggs. 1997. Duration and structure of the last four interglaciations. Quaternary Research 48:141–154.

Winter, A., T. Oba, H. Ishioroshi, T. Watanabe, and J. R. Christy. 2000. Tropical sea surface temperatures: Two-to-three degrees cooler than present during the Little Ice Age. Geophysical Research Letters 27:3365–3368.

Wohlfarth, B. 1996. The chronology of the Last Termination: A review of radiocarbon-dated, high-resolution terrestrial stratigraphies. Quaternary Science Reviews 15:267–284.

Wohlfarth, B., S. Björck, and G. Possnert. 1995. The Swedish timescale: A potential calibration tool for the radiocarbon time scale during the late Weichselian. Radiocarbon 37:347–359.

Wohlfarth, B. and G. Possnert. 2000. AMS ^{14}C measurements from the Swedish varved clays. Radiocarbon 42:323–333.

Wold, C. N. 1994. Cenozoic sediment accumulation on drifts in the northern North Atlantic. Palaeoceanography 9:917–941.

Wolf, R. 1868. Astronomische Mittheilungen. N.p. (Cited in Hoyt and Schatten 1997.)

Wolff, E. W. 2007. When is the "present."? Quaternary Science Reviews 26:3023–3024.

Wolff, E. W., H. Fischer, F. Fundel, et al. 2006. Southern ocean sea ice extent, productivity and iron flux over the last eight glacial cycles. Nature 440:491–496.

Wolff, E. W., H. Fischer, and R. Rothlisberger. 2009. Glacial terminations as southern warmings without northern control. Nature Geoscience 2:206–209, doi:10.1038/NGE0442.

Wollenburg, J. E., J. Knies, and A. Mackensen. 2004. High-resolution paleoproductivity fluctuations during the past 24 kyr as indicated by benthic foraminifera in the marginal Arctic Ocean. Palaeogeography, Palaeoclimatology, Palaeoecology 204:209–238.

Woodhouse, C. A. and J. T. Overpeck. 1998. 2000 years of drought variability in the central United States. Bulletin of the American Meteorological Society 79:2693–2714.

Woodruff, F. and S. Savin. 1989. Miocene deepwater oceanography. Paleoceanography 4:87–140.

Woodruff, F. and S. Savin. 1991. Mid-Miocene isotope stratigraphy in the deep sea: High-resolution correlations, paleoclimate cycles, and sediment preservation. Paleoceanography 6:755–806.

Woodworth, J. B. 1905. Ancient Water Levels of the Champlain and Hudson Valleys. Albany: New York State Education Department.

Woodworth, P. L. 2006. Some important issues to do with long-term sea level change. Philosophical Transactions of the Royal Society of London, Series A 364:787–803.

Woodworth, P. L. and R. Player. 2003. The permanent service for mean sea level: An update to the 21st century. Journal of Coastal Research 19:287–295.

Woods, T. N. and J. Lean. 2007. Anticipating the next decade of sun-earth system variations. EOS, American Geophysical Union Transactions 88:457–458.

Wright, H. E., Jr., J. E. Kutzbach, T. Webb III, et al. 1993. Global Climates Since the Last Glacial Maximum. Minneapolis: University of Minnesota Press.

Wright, J. D. and K. G. Miller. 1992. Miocene stable isotope stratigraphy, Site 747, Kerguelen Plateau. In S. W. Wise Jr., et al., eds., Proceedings of the Ocean Drilling Program Scientific Results, vol. 120, Part B, pp. 855–866. College Station, TX: Ocean Drilling Program.

Wunsch, C. 1998. The work done by the wind on the oceanic general circulation. Journal of Physical Oceanography 28:2332–2340.

Wunsch, C. 1999. The interpretation of short climate records, with comments on the North Atlantic and Southern Oscillation. Bulletin of the American Meteorological Society 80:245–255.

Wunsch, C. 2000. On the sharp spectral lines in the climate record and the millennial peak. Paleoceanography 15: 417–424.

Wunsch, C. 2002. What is the thermohaline circulation? Science 298:1179–1181.

Wunsch, C. 2003a. Greenland-Antarctic phase relations and millennial time-scale climate fluctuation in the Greenland ice cores. Quaternary Science Reviews 22:1631–1646.

Wunsch, C. 2003b. The spectral description of climate change including the 100 Kyr energy. Climate Dynamics 20:353–363.

Wunsch, C. 2004. Quantitative estimate of the Milankovitch-forced contribution to observed Quaternary climate change. Quaternary Science Reviews 23:1001–1012.

Wunsch, C. 2006. Abrupt climate change: An alternative view. Quaternary Research 65:191–203.

Wunsch, C. and R. Ferrari. 2004. Vertical mixing, energy, and the general circulation of the oceans. Annual Review Fluid Mechanics 36:281–314.

Wunsch, C. and P. Heimbach. 2006. Decadal changes in the North Atlantic meridional overturning and heat flux. Journal of Physical Oceanography 36:2012–2024.

Wunsch, C. and P. Heimbach. 2007. Practical global oceanic state estimation. Physica D 230:197–208.

Wunsch, C. and P. Heimbach. 2008. How long to oceanic tracer and proxy equilibrium? Quaternary Science Reviews 27:637–651.

Wunsch, C., R. Ponte, and P. Heimbach. 2007. Decadal trends in sea level patterns: 1993–2004. Journal of Climate 20:5889–5991.

Wyrtki, K. 1973. Teleconnections in the equatorial Pacific. Science 180:66–68.

Wyrtki, K. 1975. El Niño—The dynamic response of the equatorial Pacific Ocean to atmospheric forcing. Journal of Physical Oceanography 5:572–584.

Yancheva, G., N. R. Nowaczyk, J. Mingram, et al. 2007. Influence of the intertropical convergence zone on the east Asian monsoon. Nature 445:74–77.

Yang, S., H. Odah, and J. Shaw. 2000. Variations in the geomagnetic dipole moment over the last 12000 years. Geophysical Journal International 140:158–162.

Yasuhara, M., T. M. Cronin, P. B. deMenocal, H. Okahashi, and B. K. Linsley. 2008. Abrupt climate change and collapse of deep-sea ecosystems. Proceedings of the National Academy of Sciences 105:1556–1560.

Yin, A. 2006. Cenozoic tectonic evolution of the Himalayan orogen as constrained by along-strike variation of structural geometry, exhumation history, and foreland sedimentation. Earth-Science Reviews 76:1–131.

Yin, A. and T. M. Harrison. 2000. Geologic evolution of the Himalayan-Tibetan orogen. Annual Review of Earth and Planetary Sciences 28:211–280.

Yokoyama, Y., K. Lambeck, P. P. J. De Deckker, and L. K. Fifield. 2000. Timing of the Last Glacial Maximum from observed sea-level minima. Nature 406:713–716.

Yoshimori, M., T. F. Stocker, C. C. Raible, and M. Renold. 2005. Externally forced and internal variability in ensemble climate simulation of the Maunder Minimum. Journal of Climate 18:4253–4270.

Yu, E.-F., R. François, and M. Bacon. 1996. Similar rates of modern and last-glacial ocean thermohaline circulation inferred from radiochemical data. Nature 379:689–694.

Yu, S.-Y., B. E. Berglund, P. Sandgren, and K. Lambeck. 2007. Evidence for a rapid sea-level rise 7600 yr ago. Geology 35:891–894.

Yu, Y., G. A. Maykut, and D. A. Rothrock. 2004. Changes in the thickness distribution of Arctic sea ice between 1958–1970 and 1993–1997. Journal of Geophysical Research 109:C08004, JC001982.

Yu, Y. T., T. Yang, J. Li, et al. 2006. Millennial-scale Holocene climate variability in the NW China drylands and links to the tropical Pacific and the North Atlantic. Palaeogeography, Palaeoclimatology. Palaeoecology 233:149–162.

Yu, Z. and E. Ito. 1999. Possible solar forcing of century-scale drought frequency in the northern great plains. Geology 27:263–266.

Yuan, D. X., H. Cheng, R. L. Edwards, et al. 2004. Timing, duration, and transitions of the last interglacial Asian monsoon. Science 304:575–578.

Zachos, J. C., M. A. Arthur, T. J. Bralower, and H. J. Spero. 2002. Palaeoclimatology (communication arising): Tropical temperatures in greenhouse episodes. Nature 419:897–898.

Zachos, J. C., G. R. Dickens, and R. E. Zeebe. 2008. An early Cenozoic perspective on greenhouse warming and carbon-cycle dynamics. Nature 451:279–283.

Zachos, J. C., B. F. Flower, and H. Paul. 1997. Orbitally paced climate oscillations across the Oligocene/Miocene boundary. Nature 388:567–570.

Zachos, J. C. and L. R. Kump. 2005. Carbon cycle feedbacks and the initiation of Antarctic glaciation in the earliest Oligocene. Global and Planetary Change 47:51–66.

Zachos, J. C., M. Pagani, L. Sloan, E. Thomas, and K. Billups. 2001. Trends, rhythms, and aberrations in the global climate 65 Ma to present. Science 292:686–691.

Zachos, J. C., T. M. Quinn, and K. A. Salamy. 1996. High-resolution deep-sea foraminiferal stable isotope records of the Eocene-Oligocene climate transition. Paleoceanography 11:251–266.

Zachos, J. C., U. Röhl, S. A. Schellenberg, et al. 2005. Rapid acidification of the ocean during the Paleocene-Eocene Thermal Maximum. Science 308:1611–1615.

Zachos, J. C., S. Schouten, S. Bohaty, et al. 2006. Extreme warming of mid-latitude coastal ocean during the Paleocene-Eocene Thermal Maximum: Inferences from TEX86 and isotope data. Geology 34:737–740.

Zachos, J. C., M. W. Wara, S. Bohaty, et al. 2003. A transient rise in tropical sea surface temperature during the Paleocene-Eocene thermal maximum. Science 302:1551–1554.

Zagwijn, W. H. 1961. Vegetation, climate and radiocarbon datings in the late Pleistocene of the Netherlands. I. Eemian and Early Weischelian. Mededelingen Geologische Stichttmg, new series 14:15–45.

Zahn, R., J. Schönfeld, H. Kudrass, M. Park, H. Erlenkeuser, and P. Grootes. 1997. Thermohaline instability in the North Atlantic during meltwater events: Stable isotope and ice-rafted detritus records from core SO75-26KL, Portuguese margin. Paleoceanography 12:696–710.

Zaragosi, S., F. Eynaud, C. Pujol, et al. 2001. Initiation of the European deglaciation as recorded in the northwestern Bay of Biscay slope environments (Meriadzek Terrace and Trevelyan

Escarpment): A multi-proxy approach. Earth and Planetary Science Letters 188:493–507.

Zebiak, S. E. and M. A. Cane. 1987. A model El Niño/Southern Oscillation. Monthly Weather Review 115:2262–2278.

Zemp, M., W. Haeberli, M. Hoelzle, and F. Paul. 2006. Alpine glaciers to disappear within decades? Geophysical Research Letters 33:L13504, GL026319.

Zeng, N. 2003. Glacial-interglacial atmospheric CO_2 change—The glacial burial hypothesis. Advances in Atmospheric Science 20:677–693.

Zeng, N. 2007. Quasi-100ky glacial-interglacial cycles triggered by subglacial burial carbon release. Climate of the Past 3:135–153.

Zhang, R., T. L. Delworth, and I. M. Held. 2007. Can the Atlantic Ocean drive the observed multidecadal variability in northern hemisphere mean temperature? Geophysical Research Letters 34:L02709.

Zhang, X. B., F. W. Zwiers, G. C. Hegerl, et al. 2007. Detection of human influence on twentieth-century precipitation trends. Nature 448:461–465.

Zheng, Y., R. F. Anderson, A. van Geen, and M. Q. Fleisher. 2002. Remobilization of authigenic uranium in marine sediments by bioturbation. Geochimica et Cosmochimica Acta 66:1759–1772.

Zhou, S., A. J. Miller, J. Wang, and J. K. Angell. 2001. Trends of NAO and AO and their associations with stratospheric processes. Geophysical Research Letters 28:4107–4110.

Ziegler, A. M., M. L. Hulver, and D. B. Rowley. 1996. Permian world topography and climate. In I. P. Martini, ed., *Late Glacial and Postglacial Environmental Changes—Quaternary,* *Carboniferous-Permian, and Proterozoic.* New York: Oxford University Press.

Ziegler, A. M., C. R. Scotese, and S. F. Barrett. 1983. Mesozoic and Cenozoic paleogeographic maps. In P. Broche and J. Sundermann, eds., *Tidal Friction and the Earth's Rotation II.* Berlin: Springer-Verlag.

Zielinski, G. A. 2000. Use of paleo-records in determining variability within the volcanism-climate system. Quaternary Science Reviews 19:417–438.

Zielinski, G. A., P. A. Mayewski, L. D. Meeker, S. Whitlow, and M. Twickler. 1996. A 110,000-year record of explosive volcanism from the GISP2 (Greenland) ice core. Quaternary Research 45:109–118.

Zielinski, G. A., P. A. Mayewski, L. D. Meeker, et al. 1994. Record of volcanism since 7000 B.C. from the GISP2 Greenland ice core and implications for the volcano-climate system. Science 264:948–952.

Zielinski, G. A., P. A. Mayewski, L. D. Meeker, et al. 1997. Volcanic aerosol records and tephrachronology of the summit Greenland ice cores. Journal of Geophysical Research 102:26625–26640.

Zimov, S. A., E. A. G. Schuur, and F. S. Chapin III. 2006. Permafrost and the global carbon budget. Science 312:1612–1613.

Zwally, H. J., W. Abdalati, T. Herring, et al. 2002. Surface melt-induced acceleration of Greenland Ice-Sheet flow. Science 297:218–222.

Zweck, C. and P. Huybrechts. 2003. Modeling the marine extent of northern hemisphere ice sheets during the last glacial cycle. Annals of Glaciology 37:173–180.

Index